PEROVSKITE MATERIALS - SYNTHESIS, CHARACTERISATION, PROPERTIES, AND APPLICATIONS

Edited by **Likun Pan** and **Guang Zhu**

Perovskite Materials - Synthesis, Characterisation, Properties, and Applications
http://dx.doi.org/10.5772/60469
Edited by Likun Pan and Guang Zhu

Contributors

Xinhua Zhu, Heng Wu, Hitoshi Ohsato, Carlito Jr. Ponseca, Yoon Hee Jeong, Abhijit Biswas, Ki-Seok Kim, Rajender Singh, Ramesh Ade, Michele Cadelano, Giovanni Bongiovanni, Valerio Sarritzu, Nicola Sestu, Daniela Marongiu, Feipeng Chen, Roberto Piras, Francesco Quochi, Andrea Mura, Michele Saba, Jinguang Cheng, Gina Pecchi, Nestor Escalona, I Tyrone Ghampson, Ruddy Morales, Qing Shen, Yuhei Ogomi, Taro Toyoda, Kenji Yoshino, Shuzi Hayase, Diego Pereira Tarragó, Berta Moreno, Eva Chinarro Martín, Vânia De Sousa, Sheng Hsiung Chang, Hsin-Ming Cheng, Sheng-Hui Chen, Kuen-Feng Lin, Feng Li, Piotr Dulian, Bernabé Marí, Krishna R. Adhikari, Nada F. Atta, Ahmed Galal, Ekram Ekram H. El-Ads, Ekram El-Ads, Idoia Ruiz De Larramendi, Nagore Ortiz-Vitoriano, Teofilo Rojo, Isaen B. Dzul-Bautista, Juan Carlos Rendon-Angeles, Zully Matamoros-Veloza, Jorge López-Cuevas, Kazumichi Yanagisawa, Montoya-Cisneros Karla Lorena, Martin Schmal, Fabio Souza Toniolo, Yiguo Su, Xiaojing Wang, Chunfang Du, Junyu Lang, Jinbo Yang, M.S. Kim, W. Grover Coors

Published by InTech

Janeza Trdine 9, 51000 Rijeka, Croatia

Publishing Process Manager Danijela Duric

Technical Editor InTech DTP team

Cover InTech Design team

First published February, 2016

Printed in Croatia

Additional hard copies can be obtained from orders@intechopen.com

Perovskite Materials - Synthesis, Characterisation, Properties, and Applications, Edited by Likun Pan and Guang Zhu

p. cm.

ISBN 978-953-51-2245-6

For EU product safety concerns:
IN TECH d.o.o., Prolaz Marije Krucifikse Kozulić 3, 51000 Rijeka, Croatia,
info@intechopen.com or visit our website at intechopen.com.

Meet the editors

Dr. Likun Pan received his PhD in 2005 at Nanyang Technological University, Singapore, and currently works at the Department of Physics, East China Normal University as a professor. His research interest includes the synthesis and properties of nanomaterials and their applications in energy and environmental fields. He has published more than 180 journal articles with over 4000 citations and his current H-index is 36. Dr. Pan has severed as editorial board member of several international journals and as technical chair or general secretary in several international conferences.

Prof. Guang Zhu received his PhD in 2012 at East China Normal University, China. He was a postdoctoral fellow at Northwestern University, USA, and later joined Suzhou University as a associate professor, China. His current research interests include materials synthesis, characterization, and applications in photocatalysis, electrosorption, and solar cells.

Contents

Preface

Perovskite materials have attracted much attention because of their promising physical properties in colossal magnetoresistance, superconductivity, multi-ferroelectricity, metal-insulator transition, charge/orbital ordering, photoelectricity, etc. Over the past two decades, remarkable efforts have been conducted worldwide to explore various techniques to prepare perovskite materials for many applications. Difference in properties of perovskite structured materials has provided new insight and direction for researchers to meet the different applications. Herein, this book was written in part as a survey of the state of the art research activities that focus on the rational synthesis, structural characterization, properties, and applications of perovskite materials. In part, it is also an effort to bridge this knowledge gap and to better understand perovskite materials characterization and their applications.

The book with 21 chapters has been organized to accommodate the vision outlined above via summarizing the current state of the perovskite materials: synthesis, characterization, properties, and applications. Most chapters have been structured to include (1) a review on the actual knowledge and (2) cutting-edge research results. Thus this book is an essential source of reference for scientists with research fields in energy, physics, chemistry and materials. It is also suitable reading matter for graduate students.

The Book Editors would like to thank all who kindly contributed their high quality chapters for this book and to InTech staff for their kind help and cooperation. We are also indebted to the InTech editorial and production team.

Likun Pan
Engineering Research Center for Nanophotonics and Advanced Instrument,
Ministry of Education, Department of Physics East China Normal University,
P. R. China

Guang Zhu
Anhui Key Laboratory of Spin Electron and Nanomaterials
Suzhou University, Suzhou,
P. R. China

Section 1

Perovskite Materials: Synthesis and Characterization

Chapter 1

Solid-State Mechanochemical Syntheses of Perovskites

Piotr Dulian

Additional information is available at the end of the chapter

http://dx.doi.org/10.5772/61521

Abstract

The chapter presents the possibility of applying high-energy ball milling techniques to carry out the synthesis of ceramics with perovskite structure, thereby eliminating prolonged use of high temperatures in their preparation.

On the examples of alkaline-earth metal perovskites, an influence of the most significant parameters of mechanochemical treatment on their forming process and product quality was illustrated. For the first time, it was done that the contamination of the product derived from the attrition of the grinding media and internal parts of the vial can constitute modifiers of the functional properties of produced electroceramics.

Dielectric characteristics of mechanochemically produced materials as well as using high-temperature solid synthesis were compared.

Keywords: Mechanochemistry, High-energy Ball Milling, Perovskites, Complex Oxides, Electroceramics

1. Introduction

The family of chemical compounds with perovskite-type structure due to the unique electrical properties comprises a broad range of electrotechnical materials – dielectrics by semiconductors, superionic conductors, conductors with combined ionic and electron conductivity to high-temperature superconductors [1, 2].

Moreover, as is known, these compounds in certain temperature ranges have piezoelectric, pyroelectric, ferroelectric, antiferroelectric, paraelectric, ferromagnetic, or paramagnetic properties [3–6]. Also important is the simplicity of their crystalline structure, chemical composition, and the synthesis of these compounds in monocrystalline or polycrystalline form. It is easy to modify the structure and thus the properties of perovskites. Even a slight change of their ideal crystal structure and chemical composition may result in the appearance of new, not only electrical but also other, properties such as catalytic or mechanical. Therefore, it is very important to select the method of their production. The synthesis of polycrystalline titanates with perovskite structure, due to the fact that these compounds are hardly fusible materials, is carried out at high temperatures by solid-phase reaction. However, such a temperature causes the appearance of sinters and agglomerates what hinder fine crystalline product formation. This is extremely important because the morphology and grain size in electroceramics directly affects on their properties [7–12]. This problem can be solved by different ways, e.g., using a sol–gel method [13]. This technique owing to the thorough mixing of the precursor materials in solution and the relatively lower temperature of crystallization (1,200 K) allows to obtain a homogenous material with small, fine crystals and excellent chemical stoichiometry. Unfortunately, the sol–gel method is complex and requires advanced, very clean equipment, and organometallic reagents that are not only expensive but also environmentally hazardous. There are also known other synthesis techniques such as co-precipitation or hydrothermal methods. However, many of these methods enable to synthesize perovskite-type ceramics with fine crystals; they are unpopular because their complexities and costs preclude their use in a large-scale industrial fabrication [14–20]. Alternatively, this kind of ceramics can be produced by high-energy ball milling at room temperature [21–28]. This technique leads to the activation and/or synthesis of new compounds. Activation of solid powders in this case is based on the high degree of fragmentation and a large number of structural defects. However, in order to obtain the crystalline milling products, sometimes the subsequent heat treatment is needed. Then, much lower temperature may be used than in conventional methods. Moreover, using the mechanochemical method, it is possible to control the grain morphology of ceramics by selection of appropriate process parameters. Such a simplicity and large control of process parameters make it an excellent alternative to expensive and complex manufacturing techniques of advanced ceramic materials.

2. General aspects of mechanochemistry

The essence of mechanical treatment is the impact of moving grinding media with grains of ground material and the interaction between the grains of powder. During these processes depending on the type of the mill and the applied milling parameters, the energy supplied to the material is in the range of 0.1 to 100 MJ/kg. Transfer of this energy is precisely localized in the collision zone at the moment of the collision between the grinding media. During the collision, the kinetic energy of the grinding media is absorbed by a small volume (approx. 1 mm^3 of ground powder) and is immediately converted into elastic energy. Resulting stresses cause the destruction of the ground material. Depending on the physical nature of the ground powder, mainly hardness and thermodynamic conditions, the cracks of crystals occur, resulting in the reduction of the grain size, and/or a mutual merging of the particles.

All these phenomena intensify the diffusion processes in solids accelerating the chemical reactions. It is caused by the forces of collisions, strike/hit, compression, and friction occurring between the grinding media and the ground material and between the grinding media and the walls of the reactor [29–35]. In consequence, the reactions take place without the need of diffusion of substrates through the product layer because interfacial contact of ground materials is periodically renewed.

2.1. Process parameters of mechanochemical treatment

Fabrication of different materials by a mechanochemical treatment is a complex process because it is influenced by many factors. Generally, they are connected with energetics and/or the environment of milling. The amount of energy supplied to the material during mechanical treatment mainly depends on the type of mill. There are various kinds of construction solutions of the mills [31, 36–37]. The type of mill should be chosen taking into account the advantages and disadvantages of each device. The decisive parameter for the kinetics of the mechanochemical processes is the rotation speed of the reactor [38]. The rotation speed or impact, in the case of vibratory mills, transfer directly into interaction frequency of grinding media with powder particles and their speed inside the reactor. Milling energy highly depends on the BPR factor (*ball to powder ratio*), which expresses the ratio (e.g., mass) of grinding media to the ground material. Although this relationship is not linear, this is related to the degree of the reactor filling with balls (a large number of them makes the movement of the balls more complex) [29, 39]. Grinding media lose their energy due to frequent collisions among themselves. Too small or too large volume of ball in the vial reduces the efficiency of the milling process. It should be pointed that both high speed and the large number of balls cause increase in the reactor temperature. Higher temperature can be beneficial to the phenomenon, stimulates the diffusion of the atoms in the solids, and also increases the degree of ductility of the steel, which leads to a faster wear. All parameters relating to energetics of the grinding unit should be taken into account when planning the mechanical treatment process.

The selection of the process parameters directly affects the properties of the obtained phases. Depending on the desired final effect of high-energy, ball milling can/must be used different time, the atmosphere, and the medium of mechanochemical treatment. Determination of mechanochemical synthesis duration is rather simple – depends on the time required to form the desired phases. The processes associated with mechanical activation of solids, such as structure modification, deposition of active catalytic phase on carrier or simple communition need own individual milling time, therefore it must be selected experimentally.

Mechanochemical processes are often carried out under a protective atmosphere or in vacuum. This prevents milled material from the additional reactions with air components such as oxygen or nitrogen. Negative phenomenon of agglomeration of grains as a result of high-energy milling process can be reduced by using water or alcohol as a medium [21, 40–52].

The kinetics of mechanochemical reactions can be also controlled through various types of precursors. This is particularly important in the case of exothermic reactions, which very often have an explosive nature (SHS reaction) as in the case of metallothermic processes [53–55]. The use one of the substrates not in the form of oxide (i.e., as a ready reactant) but as hydroxosalt or carbonate needs heat to its stepwise decomposition into the oxide (endothermic process), increasing the time of availability of the reagent for the synthesis. This type of process is called by Avvakumow [56] as "soft mechanochemistry".

3. Direct mechanochemical syntheses of titanates

The continuous desire to limit great inconvenience to the natural environment and the cost of preparation of a variety of functional materials and the need for materials with more "sophisticated" properties makes the mechanochemical treatment an interesting alternative technique to the conventional ones. The possibility of applying this method for the preparation of high-quality ceramic materials with perovskite structure is presented below. In Figures 1 and 2 are shown the influence of various process parameters of high-energy ball milling and the nature of the reactants on the dynamics and properties of the obtained products, respectively.

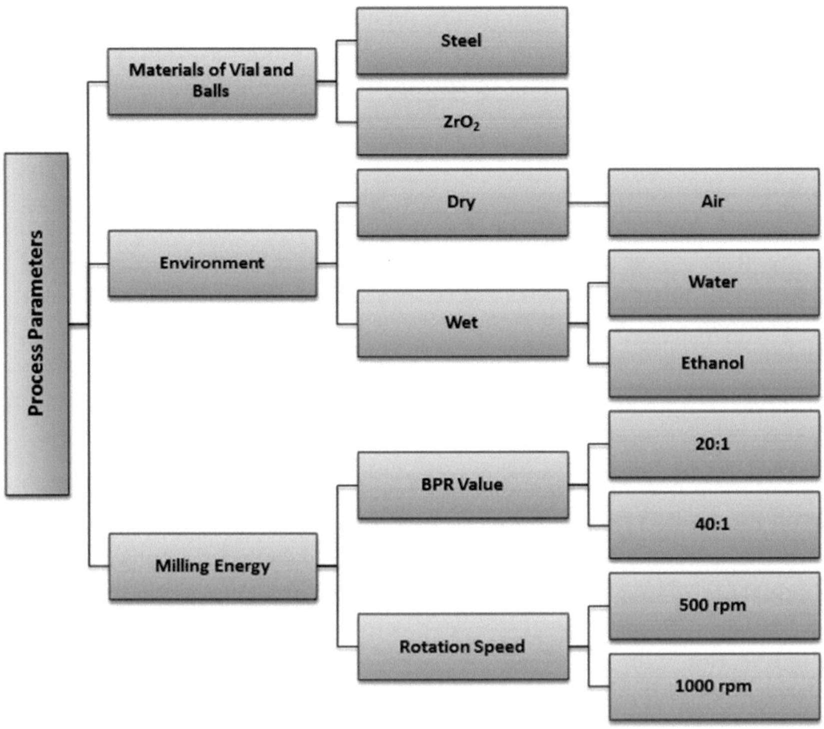

Figure 1. Various milling conditions in mechanochemical syntheses of perovskites

4. Syntheses of perovskites of alkaline earth metals – $MTiO_3$ (M = Ca, Sr, Ba)

Mechanochemical synthesis of calcium titanate

Synthesis of compounds with perovskite structure in the system CaO-TiO_2 is carried out in the solid phase without the need of high-temperature processing (for processing details – see:

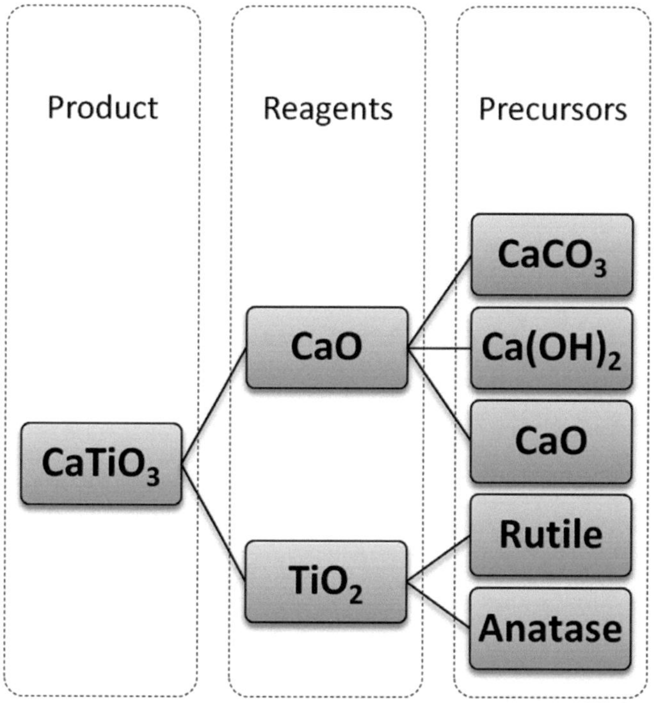

Figure 2. Different oxides' precursors to syntheses of complex oxides ($CaTiO_3$)

Appendix A1). However, the formation of a crystalline product is highly conditioned by several milling process parameters (see: Section 2.1). The influence of parameters relating to milling energy, such as the BPR value and rotation speed on the product formation time, is illustrated in Figure 3. It is worth to pay attention to the synthesis time, which using the appropriate conditions can be only about 1 h, i.e., substantially shorter than the other methods of synthesis.

Decrease in the values of rotation speed or BPR parameter increases the synthesis time. The type of TiO_2 (rutile or anatase) is also important for the kinetics of process. Mechanochemically obtained calcium titanate is characterized by grain size less than 100 nm and good homogeneity in terms of particle size. In this case, it is not necessary to make the subsequent high-temperature treatment process, which eliminates the problem of agglomerates formation and an excessive non-uniform grain growth (Figure 4). This has a meaning for properties (e.g., a ferroelectric or catalytic) of materials.

Commonly used wet environment, during the high-energy ball milling (ethanol or water), in order to reduce the negative phenomenon of agglomeration acts negatively on the course of the synthesis reaction because of hindering the reactants' phase contact. However, high-energy milling activates the reactants by particles size reduction and creation of the crystal defects which facilitates the diffusion of atoms, lowering the temperature of subsequent calcination process (see Figure 5).

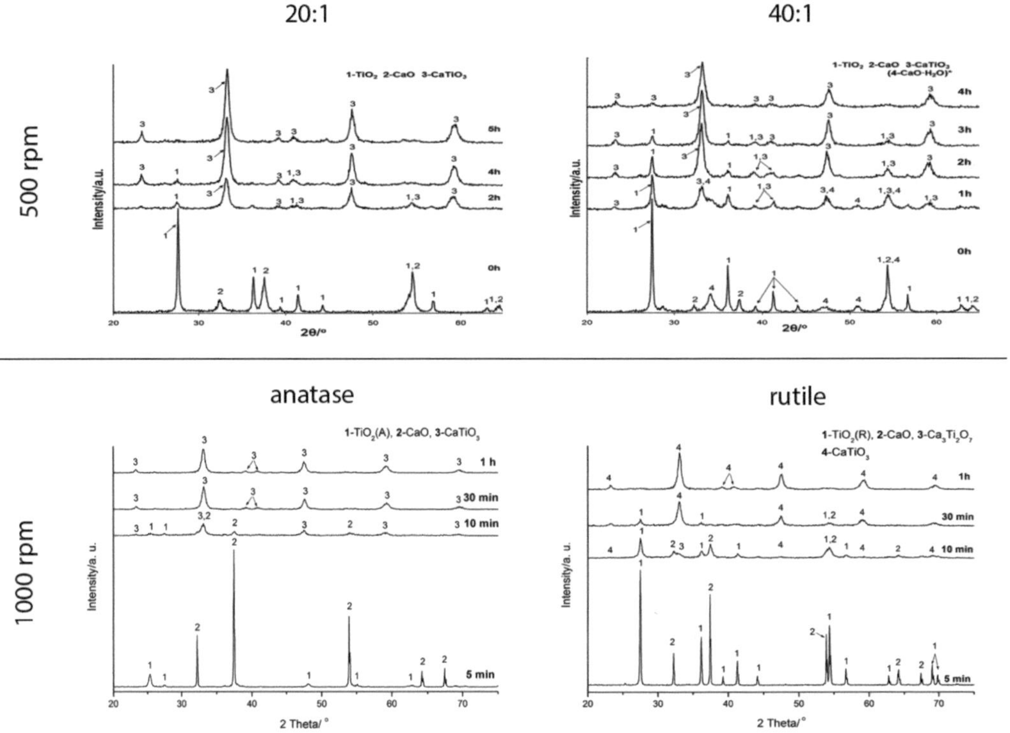

Figure 3. XRD patterns of calcium titanate – illustration of milling conditions (BPR = 20:1 and 40:1; rpm = 500 and 1000) and two forms of TiO_2 for $CaTiO_3$ synthesis

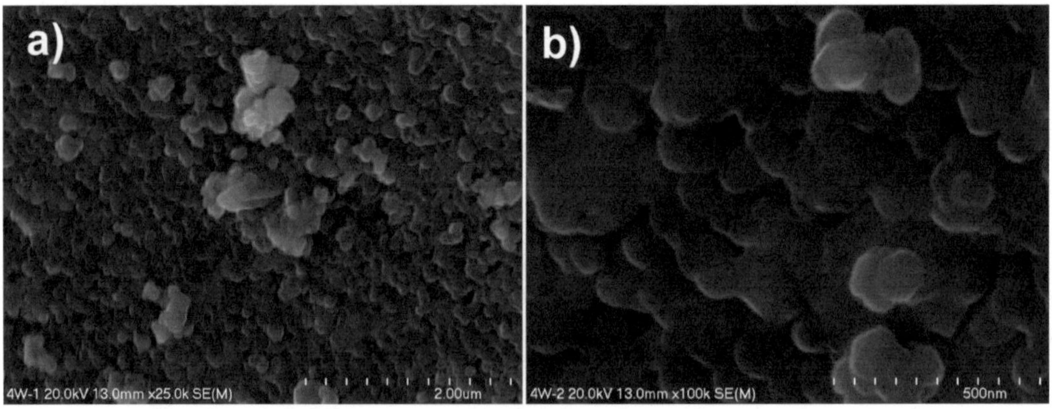

Figure 4. SEM images of mechanochemically synthesized $CaTiO_3$; (a) and (b) – different magnification

In the systems of $CaCO_3$-TiO_2 and $Ca(OH)_2$-TiO_2, synthesis is much more difficult. It is a consequence of two reactions: (1) and (2) or (1*) and (2). Synthesis is limited by the decomposition rate of carbonate or calcium hydroxide to the oxide.

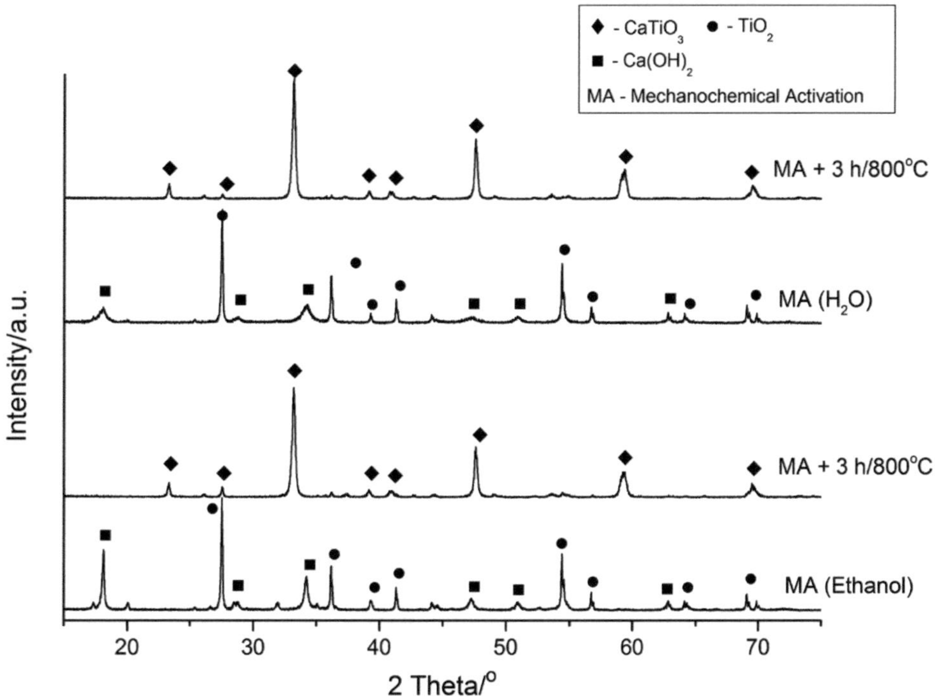

Figure 5. XRD patterns of CaO-TiO$_2$ system after mechanochemical activation in different media (water and alcohol) and subsequent calcination

$$\begin{aligned} &CaCO_{3(s)} \leftrightarrow CaO_{(s)} + CO_{2(g)} && \Delta H > 0 \\ &Ca(OH)_{2(s)} \leftrightarrow CaO_{(s)} + H_2O_{(g)} && \Delta H > 0^* \end{aligned} \quad (1)$$

$$CaO_{(s)} + TiO_{2(s)} \rightarrow CaTiO_{3(s)} \qquad \Delta H < 0 \quad (2)$$

The example of mechanochemical synthesis of calcium titanate in $CaCO_3$-TiO_2 system is presented in Figure 6.

As can be seen, in order to obtain a monophase product after 4 h of milling, a subsequent heat treatment at 800°C was necessary to use [40].

The preparation of other alkaline-earth metal titanates, e.g., barium and strontium, by a mechanochemical synthesis proceeds in analogy to the above-mentioned $CaTiO_3$ example.

The abovementioned experimental results indicate that the method of mechanochemical synthesis may be used to produce high-quality perovskite ceramics. Thus, short time of the synthesis, the use of substrates in the oxide form, and a lack of using the high-temperature treatment significantly reduce both manufacturing costs and negative influence on environment.

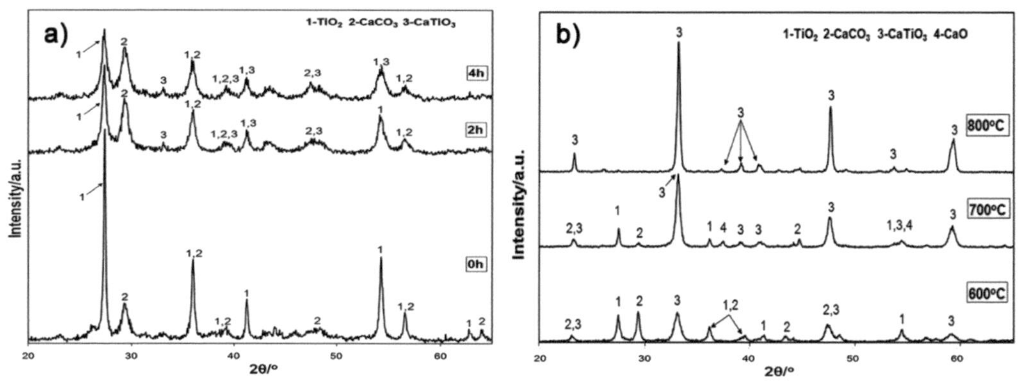

Figure 6. XRD patterns of calcium titanate formation: (a) up to 4 h of milling, (b) subsequent calcination at different temperatures

5. Modification of perovskite-type electroceramics properties

Apart from the rapid and relatively simple synthesis of perovskite compounds, mechanochemical method also allows to modify their chemical and physical properties. Such modifications can be made by changing the chemical composition of perovskites, production of ceramics of an adequate morphology (grain size, defected, etc.), and also using as additives the contaminations originating from wear down of the grinding media (Examples 1–3).

5.1. Formation of solid solution by doping various cations

Example 1. $Ba_{1-x}Sr_xTiO_3$ ($0.0 \leq x \leq 0.4$)

XRD patterns of the synthesized compounds are shown in Figure 7. They indicate the influence of the presence of Sr^{2+} on the crystallinity of products. Any phase of strontium oxide was found even in the case of strontium concentration equal to $x = 0.4$ in $Ba_{1-x}Sr_xTiO_3$.

Different ionic radii of Ba^{2+} (1.36 Å) and Sr^{2+} (1.16 Å) induce distortion of lattice. A close look at a slow scan of reflection (e.g., from 31.0° to 32.5°) shown as an inset which indicates that there is a slight shift of this reflection to higher 2θ angles. This confirms the substitution of Sr^{2+} ions in the $BaTiO_3$ lattice. It might be concluded that single phases of $BaTiO_3$, $Ba_{0.8}Sr_{0.2}TiO_3$, and $Ba_{0.6}Sr_{0.4}TiO_3$ can be successfully prepared by the high-energy ball milling process. (Synthesis conditions – see: Appendix A2)

Below it is shown an additional excellent example that, in the case of mechanochemical synthesis, should not be based on results from only one technique. The negative phenomenon of agglomeration, often occurring during milling, may affect the interpretation of the particle size and specific surface area of the material. Performed for all solid solutions, the analysis of the of particle size distribution by laser diffraction method showed that in each case about 10 microns is the dominant fraction. The specific surface area measurements (BET method) of each solid solutions clearly indicated that this area is growing with the increasing concentra-

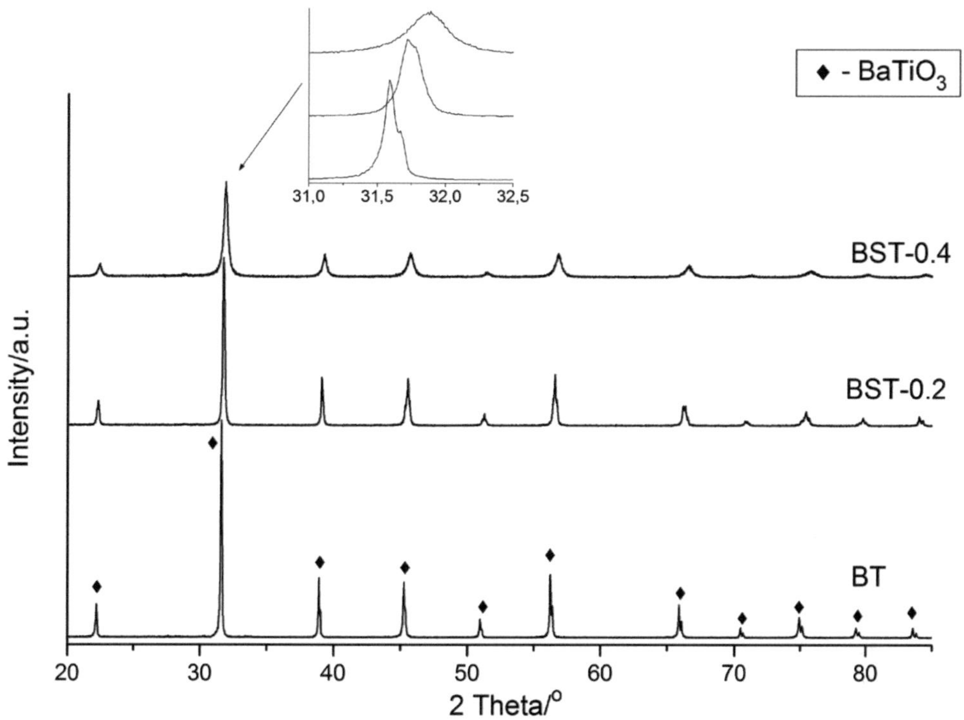

Figure 7. XRD patterns of mechanochemically synthesized $BaTiO_3$ (BT) and $Ba_{0.8}Sr_{0.2}TiO_3$ (BST-0.2), $Ba_{0.6}Sr_{0.4}TiO_3$ (BST-0.4) [59]

tion of modifier – from 7.16 m²/g for pure $BaTiO_3$ to 34.99 m²/g for $Ba_{0,6}Sr_{0,4}TiO_3$ indicating on diminishing grains size. This was also confirmed by microscopic observation (SEM). The size of grains varies from 500 to approx. 100 nm.

Dielectric properties of $Ba_{1-x}Sr_xTiO_3$

The dielectric properties of $BaTiO_3$ (BT), $Ba_{0.8}Sr_{0.2}TiO_3$ (BST-0.2), and $Ba_{0.6}Sr_{0.4}TiO_3$ (BST-0.4) ceramics are described by the temperature dependence of the real (ε') and imaginary (ε") parts of electric permittivity. These properties were determined at selected frequencies of the electric field (1 MHz).

For the ε'/T dependence (Figure 8a), with increase in the strontium concentration, the *Curie* temperature T_C gradually shifts toward lower value, and the peak of this transition becomes broader.

For $BaTiO_3$ (BT) sample a classic paraelectric–ferroelectric (PE–FE) phase transition at 368 K occurs simultaneously with the change from a cubic to tetragonal structure. At lower temperature, the peak has diffused character which can be explained by the presence of small amount of impurities (ZrO_2) from the reaction vial and grinding media. For the samples of solid solution BST-0.2 and BST-0.4, the ε'(T) plots show a diffusion nature of PE–FE phase transitions. The value of ε' maximum for BST-0.2 ceramics is about three times smaller and for BST-0.4

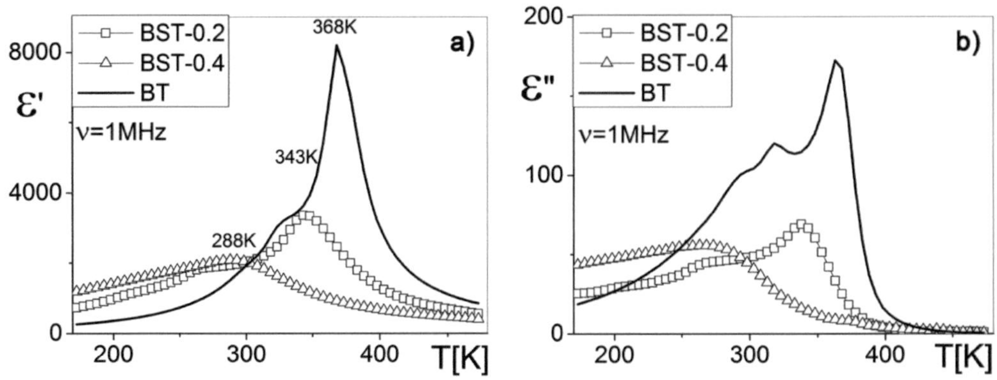

Figure 8. The temperature dependence of real part of dielectric permittivity (ε′) a) and imaginary part of dielectric permittivity (ε") b) for $BaTiO_3$ (BT), $Ba_{0.8}Sr_{0.2}TiO_3$ (BST-0.2), and $Ba_{0.6}Sr_{0.4}TiO_3$ (BST-0.4) samples

four times than in the case of BT sample. The phase transitions for $Ba_{0.8}Sr_{0.2}TiO_3$ (BST-0.2) and $Ba_{0.6}Sr_{0.4}TiO_3$ (BST-0.4) samples occur at temperature 343 and 288 K, respectively.

The energy loss of the electric field represented by imaginary part of electrical permittivity (ε") is tied to a structural phase change (Figure 8b). The temperature of maximum ε"(T) correlates with the temperature of maximum ε'(T).

Example 2. $(Ba_{1-x}Na_x)(Ti_{1-x}Nb_x)O_3$ ***(0.0 ≤ x ≤ 0.15)***

Figure 9 shows a comparison of X-ray powder diffraction patterns of $BaTiO_3$ ceramics and $(Ba_{1-x}Na_x)(Ti_{1-x}Nb_x)O_3$ for: $x = 0.01$; $x = 0.04$; $x = 0.15$ obtained by mechanochemical method. Visible shifts of the main diffraction reflections confirm the formation of appropriate solid solutions. In addition, these materials are characterized by a uniform grain size, approx. 500 nm. All are characterized by clearly defined grain boundaries and the lack of sinters. The example of morphology images of these materials in comparison with ceramics synthesized by conventional high-temperature method is presented in Figure 10.

Preparation of $(Ba_{1-x}Na_x)(Ti_{1-x}Nb_x)O_3$ solid solution by conventional high-temperature solid-phase synthesis requires a long-term heating of the mixture starting materials at high temperature. Using mechanochemical method significantly reduces the synthesis time. The monophase product was obtained after 1.5 h of high-energy milling and as in the previous example without the need for subsequent calcination.

Dielectric properties of $(Ba_{1-x}Na_x)(Ti_{1-x}Nb_x)O_3$

Comparing the dielectric properties of the same materials produced by two methods (Figure 11), i.e., mechanochemical and high-temperature syntheses, can draw the following conclusions:

- $BaTiO_3$ and the ceramic solid solution of BNTNx for the composition of $x = 0.01$ are classical ferroelectrics with a sharp PE–FE phase transition.

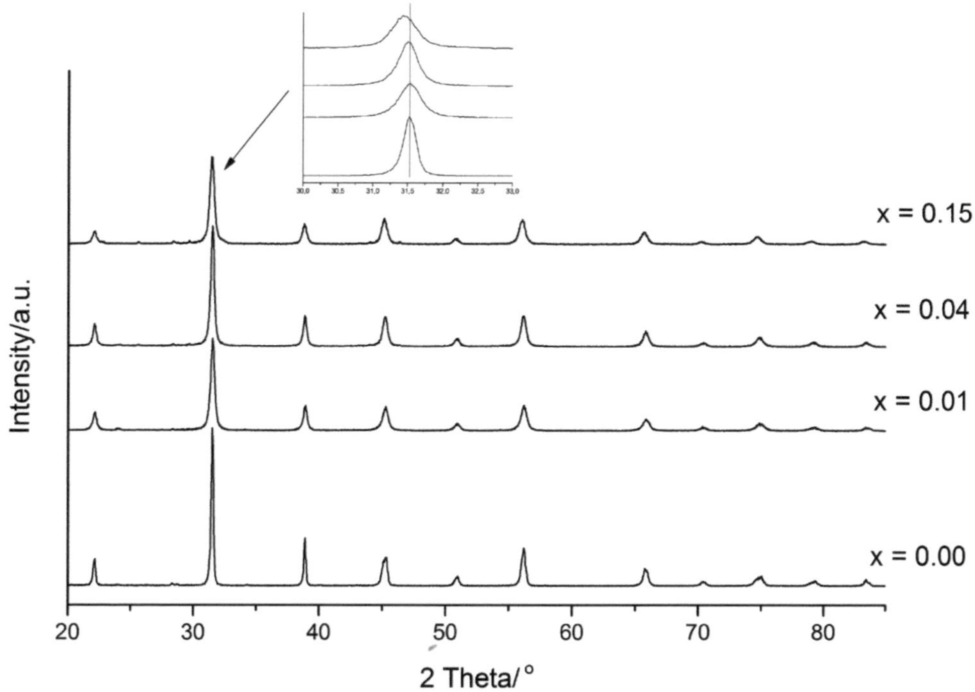

Figure 9. XRD patterns of mechanochemically synthesized $(Ba_{1-x}Na_x)(Ti_{1-x}Nb_x)O_3$ for: $x = 0.01$; $x = 0.04$; $x = 0.15$ solid solutions

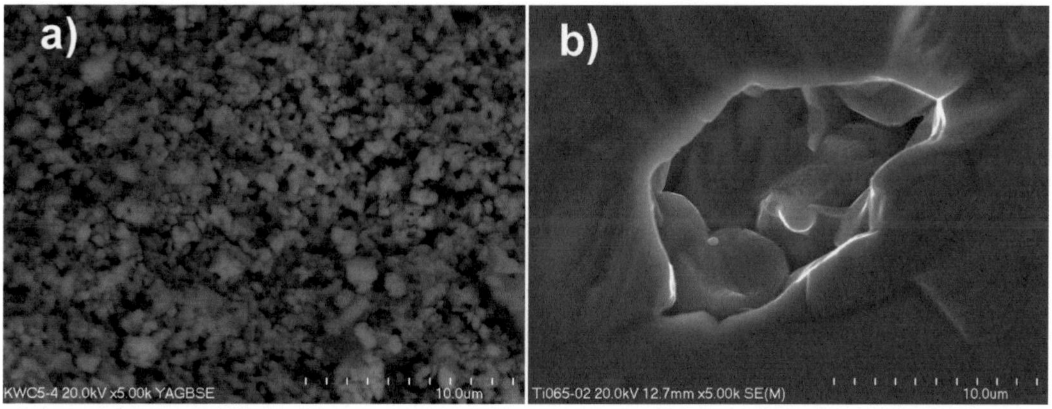

Figure 10. The SEM micrographs of microstructure of BNTNx (for $x = 0.04$) sample surface, (a) mechanochemical, (b) high temperature

- The increase in the value of x in the BNTNx samples causes a diffuseness of the phase transition. Such behavior is a result of the different valency of substituted ions in both cationic sublattices.

- For composition of $x = 0.15$, a strong dispersion of the dielectric permittivity maximum is observed. The obtained results clearly indicate that BNTNx ($x = 0.15$) sample is relaxor ferroelectric.
- Materials produced by mechanochemical synthesis method is characterized by better parameters, e.g., higher value of dielectric permittivity, less diffused character of para–ferroelectric phase transition.

The quick and simple synthesis without any thermal operation and better functional properties of products show the advantages of mechanochemistry. More information can be found in the works [57, 58, 61].

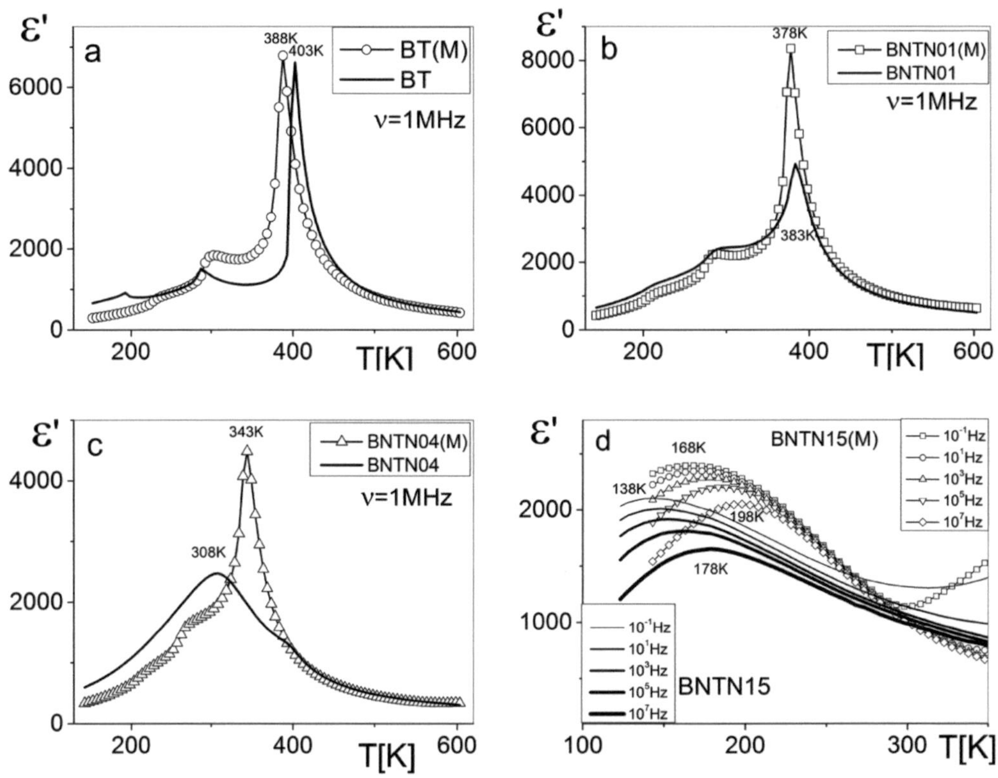

Figure 11. The temperature dependence of the real part of the complex dielectric permittivity for (a) BT, (b) BNTNx; x = 0.01, (c) x = 0.04, (d) x = 0.15 samples obtained by conventional (solid line) and mechanochemical (open symbol) methods [61]

Example 3. Influence of impurities from the milling process on the properties of $BaTiO_3$ (A) and $CaCu_3Ti_4O_{12}$ (B)

As mentioned in the chapter introduction, negative phenomenon of all processes associated with the high-energy milling is the attrition of some construction elements such as grinding media and/or inner coating of vial. This is a significant problem to technologists because such impurities are difficult to remove from milling products.

However, this effect can be used to modify the properties of materials. By selecting a suitable material of grinding media and grinding vial, it is possible to modify for example catalytic or electrical properties of ceramics.

Possibilities of the use of such impurities from the milling process are illustrated on perovskite compounds with high technological importance, i.e., $BaTiO_3$ and $CaCu_3Ti_4O_{12}$. The results are presented for comparison with analogous materials produced by high-temperature solid phase synthesis.

a. *$BaTiO_3$*

Barium titanate was prepared mechanochemically using grinding media and grinding vial made of steel and zirconium oxide. Conditions of synthesis – see: appendix A3.

In Figure 12 is presented room temperature X-ray powder diffraction for mechanochemically and high-temperature produced $BaTiO_3$. There are considerable differences in the intensities and in half-width of the diffraction reflexes of ceramics and indicating on the various morphologies of the grains. This is confirmed by SEM microscopic observation (Figure 13). The samples after mechanical treatment have much smaller particles (about 500 nm) in comparison to conventionally produced $BaTiO_3$, which is characterized by large, uneven in size and shape of the grains.

Materials prepared by these two methods also differ in terms of size BET surface area. The mechanochemically obtained powders were characterized by surface area of approx. 7 m^2/g, $BaTiO_3$ prepared by high-temperature synthesis due to the presence, among others, of sinters have a much smaller specific surface area of 0.25 m^2/g.

$BaTiO_3$ synthesized using the high-temperature method had tetragonal symmetry at room temperature, in contrast to that obtained mechanochemically, which was cubic. The high degree of communition and the large number of defects caused by high-energy ball milling limits long-range order in the crystallographic structure, and this prevents phase transitions. Calcination of the powder for 1 h at 1,373 K eliminates this problem [21].

The amount of impurities – in both cases the iron or zirconium oxide – in samples prepared mechanochemically was approx. 1% wt. Because the barium titanate is a model ferroelectrics, below is illustrated the effect of the presence of these impurities on these properties. Figure 14 presents the temperature dependence of the dielectric permittivity for all ceramics. Samples are marked on the system as: $BaTiO_3$/T – high-temperature synthesis, $BaTiO_3$/Zr and $BaTiO_3$/Fe – mechanochemical synthesis.

In the case of $BaTiO_3$/T, the temperature at which ε' is the highest represents the paraelectric–ferroelectric (PE–FE) phase transition. At all frequencies of the electric field, a classical, sharp transition can be seen at 403 K, which corresponds to a structural shift between cubic and tetragonal phases. At 288 K there is another maximum, however its value is around fourtimes smaller than the transition at 403 K. The observed temperature of this phase transition for

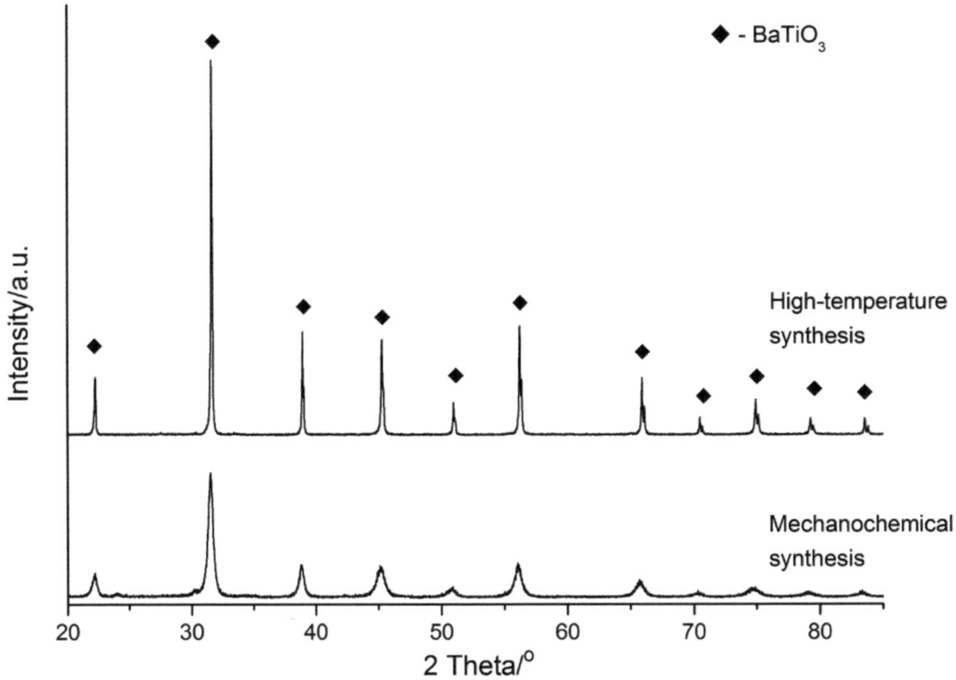

Figure 12. Comparison of X-ray powder patterns of $BaTiO_3$ obtained by two different synthesis methods [21]

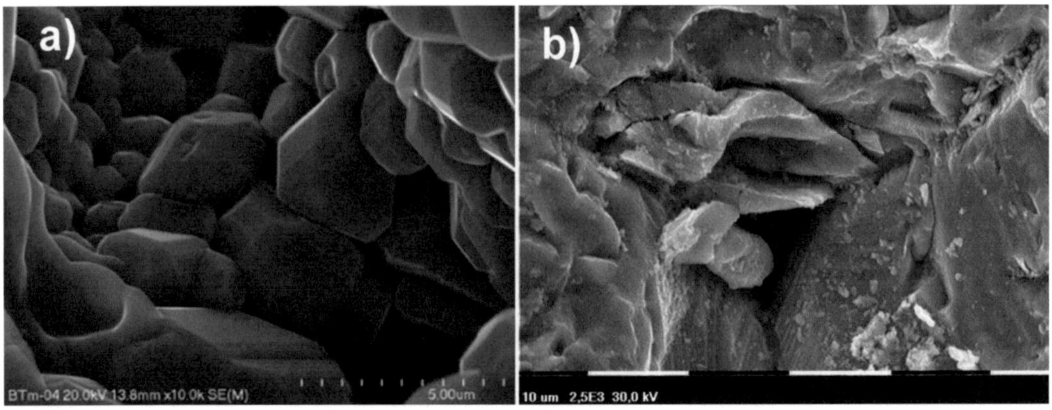

Figure 13. SEM photomicrographs of $BaTiO_3$ powders obtained by (a) mechanochemical and (b) high-temperature solid-state syntheses [21]

$BaTiO_3$ varies from the literature by 5–20 K [8, 16, 17, 20]. It is possible to surmise that transition at 288 K which corresponds to a change from a tetragonal structure to an orthorhombic one.

For $BaTiO_3$/Zr ceramics, the PE–FE transition occurs at 368 K. This transition is somewhat diffused, and the frequency of the electrical field a little bit changes the behavior of the material.

http://dx.doi.org/10.5772/61521

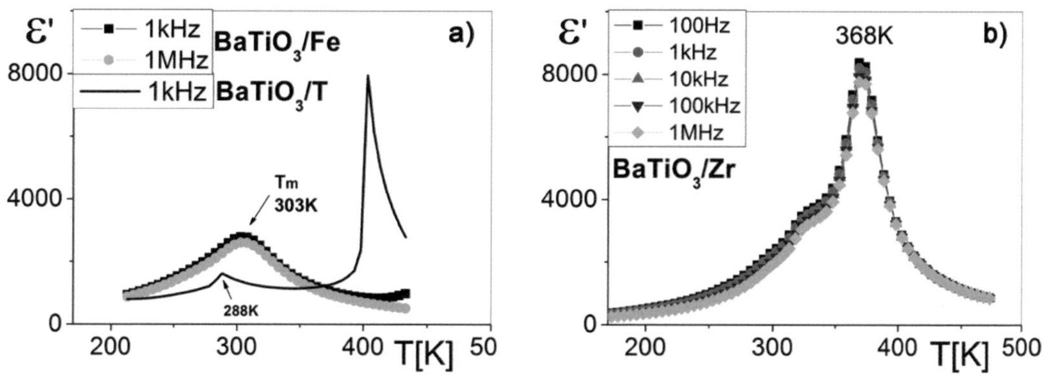

Figure 14. The temperature dependence of real part of dielectric permittivity (ε') for $BaTiO_3$/T, $BaTiO_3$/Fe, and $BaTiO_3$/Zr samples [21]

A further lowering of the temperature causes greater peak diffusing compared to that observed in "pure" $BaTiO_3$. There is no maximum observed at 288 K.

The $BaTiO_3$/Fe product has a characteristically diffused peak at the PE–FE transition, the most diffused of any of the samples tested. The observed maximum of ε' approaches a value of 3,000 at a temperature of 303 K, nearly 100 K lower than the same transition in $BaTiO_3$/T. Measurements at different frequencies of electrical field show that this material does not have the properties of a relaxor.

It is possible to postulate that the diffused character of the phase transition in $BaTiO_3$/Zr and $BaTiO_3$/Fe ceramics is caused by the presence of the impurities from the milling process.

b. $CaCu_3Ti_4O_{12}$

Figure 15 shows the XRD patterns for the mechanochemically synthesized $CaCu_3Ti_4O_{12}$ (CCTO/Fe and CCTO/Zr, reactor and balls of steel, and ZrO_2, respectively) and by high-temperature treatment (CCTO/T). As can be seen, all samples are monophase – CCTO. Syntheses details are given in Appendix A3.

Significant differences in the intensities and widths of diffraction reflexes of materials produced by high-temperature and mechanochemical route indicate different degrees of crystallinity and grain morphology. These observations were confirmed by SEM studies (Figure 16a, b).

High-temperature synthesis method results in grain growth (1.5–2.5 μm) and sintering of grains giving uneven distribution of particle size (Figure 16a). In addition, in material at the grain boundaries are visible places with high concentration of copper caused by the presence of sinters and a change in the oxidation state of copper ions during calcination at high temperature. This phenomenon is well known and this is the main problem during production of this material by methods where high-temperature processing is used. Such a ceramic is chemically inhomogeneous in volume of the particles and at their boundaries. $CaCu_3Ti_4O_{12}$ prepared mechanochemically is characterized by a uniform size distribution in the range of

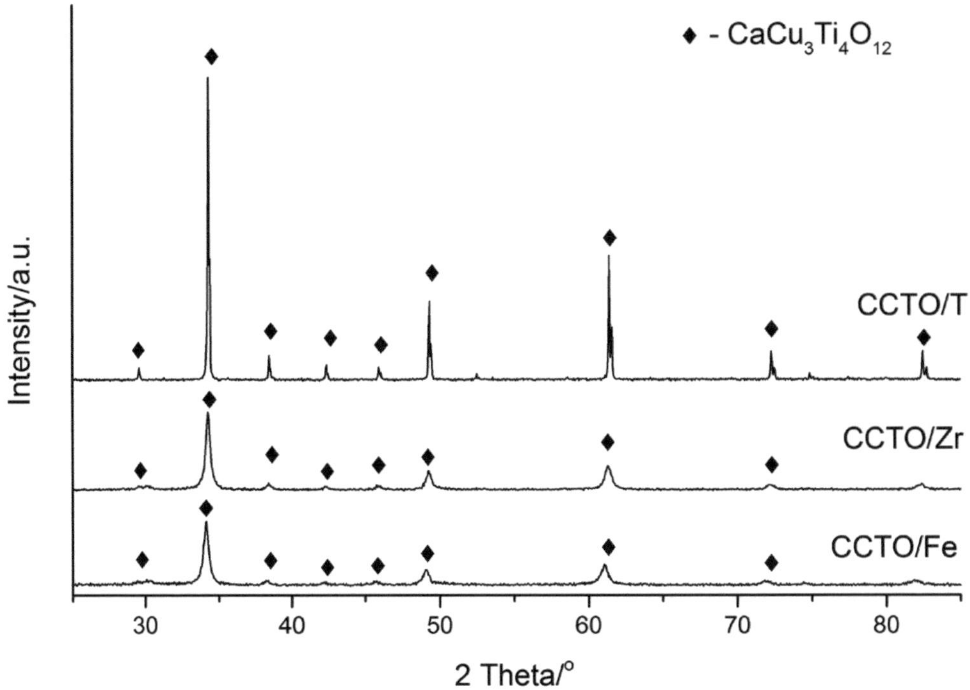

Figure 15. XRD patterns of mechanochemically synthesized $CaCu_3Ti_4O_{12}$ (CCTO/Fe and CCTO/Zr, reactor and balls of steel and ZrO_2, respectively) and by high-temperature treatment (CCTO/T) [60]

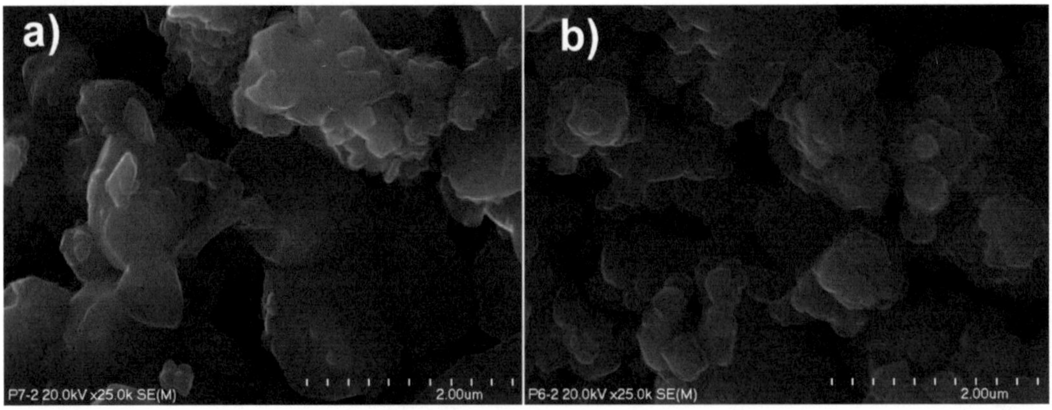

Figure 16. SEM images of $CaCu_3Ti_4O_{12}$: (a) high-temperature treatment (CCTO/T), (b) mechanochemical synthesis (CCTO/Zr) [60]

100–500nm and the lack of the high-temperature processing during synthesis eliminates the problem of chemical inhomogeneity of the material (Figure 16b). Grain boundary problem in ceramics $CaCu_3Ti_4O_{12}$ and their influence on its dielectric properties are widely discussed in the papers [62–65]. Dielectric properties of studied perovskite-related $CaCu_3Ti_4O_{12}$ compound synthesized under different conditions as a temperature relationship of the real component of dielectric permittivity (ε') and dielectric loss (tan δ) at the field frequency of 1 kHz are presented in Figure 17a and b, respectively.

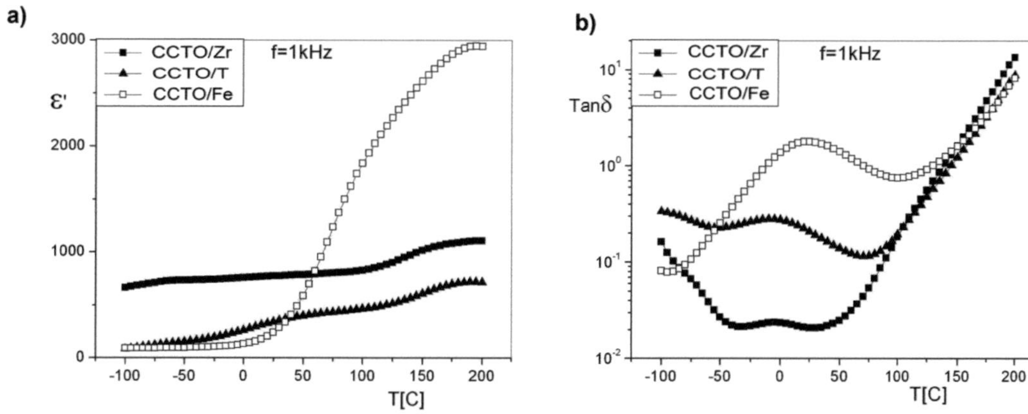

Figure 17. Temperature relationship of the real component of (a) dielectric permittivity (ε') and (b) dielectric loss (tan δ) at the field frequency of 1 kHz for $CaCu_3Ti_4O_{12}$ synthesized under different conditions [60]

The presence of small amount of zirconia (CCTO/Zr) causes the smallest changes in the electric permittivity (ε') versus temperature, similar to CCTO/T nature. From the practical point of view, such effect as well as very low and stable value of dielectric loss (tan δ) in the temperature range of –50–50°C is very convenient. Material behaves differently with the presence of metallic iron (CCTO/Fe). From the ambient temperature, a significant increase in the value of ε' is observed. At 200°C, ε' reaches a value of about 3,000. However, high values of dielectric loss (tan δ), indicating the conversion of electrical energy into heat, disqualify this ceramics for practical applications, e.g., as capacitor material.

6. Summary

Mechanochemistry being one of the easiest and cheapest methods of producing nanomaterials also enables the synthesis of advanced ceramics with perovskite structure. It is an interesting alternative to other methods to produce such compounds. As is shown in the presented examples, the reactions between the substrates in the form of oxide are going exclusively by high-energy milling without requiring long-term calcination at high temperature. This helps to avoid many technological difficulties and problems related to powder morphology and properties. Using mechanochemical synthesis of $CaCu_3Ti_4O_{12}$ eliminates problems associated with inhomogeneity of grains and intergranular boundaries. Consequently, the product prepared mechanochemically has much better functional properties than the same obtained by conventional high-temperature solid phase synthesis. The same is the case of the popular electroceramics PZT ($PbZr\text{-}TiO_3$), due to the high volatility of lead at high temperature. Using other synthesis techniques, it is difficult to maintain the stoichiometry of the compound. Applying for this purpose, the mechanochemical treatment not only omitted this problem but it is also able to produce nanometric ceramics.

In spite of the synthesis of perovskite compounds, mechanochemical technique can be applied also to modification of their properties. Selection of appropriate conditions for high-energy ball milling process is very important.

Thus, mechanochemistry realized by high-energy ball milling becomes recently the more environmentally acceptable for perovskite processing mainly for the following reasons:

- Simplify the synthesis process to one step
- Reduce the cost of chemicals and/or heat used in a traditional way
- Give the micro/nanosized products.

7. Appendix

A1.

Mechanochemical treatment was provided using two different high-energy laboratory planetary mills with vial (250 ml) and balls (10 mm diameter) made of Ni-Cr steel:

1. *Pulverisette-6* (by Fritsch GmbH) with a rotation speed 500 rpm.
2. *Activator-2S* (by Activator Corp. Novosibirsk) a rotation speed 1,000 rpm.

The subsequent calcination processes of powders were carried out in Nabertherm HTC 03/15 laboratory furnace in air atmosphere.

A2.

Mechanochemical treatment was provided using Fritsch *Pulverisette-6* planetary ball mill with vial (250 ml) and balls (10 mm diameter) made of ZrO_2. Milling parameters: rpm = 500; ball to powder mass ratio: BPR = 20:1; milling time: 1.5 h; atmosphere: air.

Preparation of reference samples by high-temperature solid phase reaction. The reactants powders were ground together in a stoichiometric ratio in an agate mortar, pressed into pellet discs 4 mm thick and of 12 mm in diameter and sintered in air with the use of a Nabertherm HTC 03/15 laboratory furnace for 12 h at the temperature of 1,350°C.

A3.

Mechanochemical treatment was provided using Fritsch *Pulverisette-6* planetary ball mill with vial (250 ml) and balls (10 mm diameter) made of ZrO_2 and steel. Milling parameters: rpm = 500; ball to powder mass ratio: BPR = 20:1; milling time: 1.5 h; atmosphere: air.

Preparation of reference samples by high-temperature solid phase reaction – as in the A2.

The phase identification and physicochemical characteristics of milling products were determined using the following methods:

- X-ray powder diffraction (XRD) patterns were recorded on a Philips
- X'Pert or Philips PW 1830 diffractometers (CuKα).
- Morphology of the surface were observed using a the scanning electron microscope (SEM) HITACHI S-4700 instrument with X-ray EDS analysis and the Philips SEM 525.
- Specific surface area (BET) of samples was determined using Accelerated Surface Area and Porosimetry Analyzer (ASAP) 2020 from Micrometrics.
- The elemental compositions of the samples were measured by an energy-dispersive X-ray fluorescent spectroscopy (XRF) using a XRF Bruker S4 Explorer instrument.
- The dielectric spectroscopy measurements in the frequency domain were performed at frequency ranging from 20 Hz to 1 MHz with the use of an LCR Agilent 4284A meter and a Quatro Krio 4.0 temperature control system. The dielectric property measurements were carried out at temperature between 473 and 173 K with a 5 K step. To perform the dielectric measurements, the samples of 7 mm in diameter and 2.5 mm thick were covered with silver electrodes.

Acknowledgements

The author would like to thank Professor Krystyna Wieczorek-Ciurowa of the Cracow University of Technology, Poland, for introduction into mechanochemical science and many hours of valuable discussion helpful during the preparation of this manuscript. For scientific collaboration, thanks also to Professor Czesław Kajtoch and Dr Wojciech Bąk of the Pedagogical University of Cracow.

This study was supported by the National Science Centre Poland, Project DEC-2012/05/N/ST8/03764.

Author details

Piotr Dulian*

Address all correspondence to: piotrdulian@chemia.pk.edu.pl

Faculty of Chemical Engineering and Technology, Cracow University of Technology, Poland

References

[1] Setter N, Wasser R. Electroceramic materials. Acta Mater 2000;48(1):151–78. DOI: 10.1016/S1359-6454(99)00293-1

[2] Waser R. Modeling of electroceramics – applications and prospects. J Eur Ceramic Soc 1999;19(6):655–64. DOI: 10.1016/S0955-2219(98)00293-3

[3] Scott JF. Applications of modern ferroelectrics. Science 2007;315(5814):954–9. DOI: 10.1126/science.1129564

[4] Eerestein W, Mathur ND, Scott JF. Multiferroic and magnetoelectric materials. Nature 2006;442(7104):759–65. DOI: 10.1038/nature05023

[5] Nuraje N, Su K. Perovskite ferroelectric nanomaterials. Nanoscale 2013;5(19):8752–80. DOI: 10.1039/c3nr02543h

[6] Cohen RE. Origin of ferroelectricity in perovskite oxides. Nature 1992;358(6382):136–8. DOI: 10.1038/358136a0

[7] Li S, Eastman JA, Li Z, Foster CM, Newnham RE, Cross LE. Size effects in nanostructured ferroelectrics. Phys Lett A: Gen Atom Solid State Phys 1996;212(6):341–6. DOI: 10.1016/0375-9601(96)00077-1

[8] Kretschmer R, Binder K. Surface effects on phase transitions in ferroelectrics and dipolar magnets. Phys Rev B 1979;20(3):1065–76. DOI: 10.1103/PhysRevB.20.1065

[9] Cottam MG, Tilley DR, Zeks B. Theory of surface modes in ferroelectrics. J Phys C: Solid State Phys 1984;17(10):1793–823. DOI: 10.1088/0022-3719/17/10/022

[10] Binder K. Surface effects on phase transitions in ferroelectrics and antiferroelectrics. Ferroelectrics 1981;35(1):99–104. DOI: 10.1080/00150198108017670

[11] Liu J, Chen W, Wang B, Zheng Y. Theoretical methods of domain structures in ultrathin ferroelectric films: a review. Materials 2014;6(9):6502–68. DOI: 10.3390/ma7096502

[12] Haertling GH. Ferroelectric ceramics: history and technology. J Am Ceramic Soc 1999;82(4):797–818.

[13] Somani V, Kalita SJ. Synthesis and characterization of nanocrystalline barium strontium titanate powder via sol-gel processing. J Electroceramics 2007;18(1):57–65. DOI: 10.1007/s10832-007-9008-7

[14] Senna M. A straight way toward phase pure complex oxides. J Eur Ceramic Soc 2005;25(12):1977–84. DOI: 10.1016/j.jeurceramsoc.2005.03.002

[15] Barboux P, Griesmar P, Ribot F, Mazerolles L. Homogeneity-related problems in solution derived powders. J Solid State Chem 1995;117(2):343–50. DOI: 10.1006/jssc.1995.1283

[16] Bocquet JF, Chhor K, Pommier C. Barium titanate powders synthesis from solvothermal reaction and supercritical treatment. Mater Chem Phys 1999;57(3):272–80. DOI: 10.1016/S0254-0584(98)00233-8

[17] Dutta PK, Asiaie R, Akbar SA, Zhu W. Hydrothermal synthesis and dielectric properties of tetragonal BaTiO3. Chem Mater 1994;6(9):1542–8. DOI: 10.1021/cm00045a011

[18] Sengodan R, Chandar Shekar B. Characterization of barium strontium titanate (BST) single crystal nanorods prepared by wet chemical method. Optoelectronics Adv Mater Rapid Commun 2014;8(7):617–21.

[19] Somani V, Kalita SJ. Synthesis and characterization of nanocrystalline barium strontium titanate powder via sol-gel processing. J Electroceramics 2007;18(1):57–65. DOI: 10.1007/s10832-007-9008-7

[20] Kaur J, Kotnala RK, Verma KC. Surfactant free hydrothermal synthesis, electrical, optical and ferroelectric properties of BaTiO3 nanoparticles. J Optoelectronics Adv Mater 2012;14(3):219–23.

[21] Piotr D, Wojciech B, Wieczorek-Ciurowa K, Czeslaw K. Controlled mechanochemical synthesis and properties of a selected perovskite-type electroceramics. Mater Sci Poland 2013;31(3):462–70. DOI: 10.2478/s13536-013-0126-4

[22] Zhang Q, Saito F. A review on mechanochemical syntheses of functional materials. Adv Powder Technol 2012;23(5):523–31. DOI: 10.1016/j.apt.2012.05.002

[23] Wieczorek-Ciurowa K, Gamrat K. Mechanochemical syntheses as an example of green processes. J Thermal Analysis Calorimetry 2007;88(1):312–7. DOI: 10.1007/s10973-006-8098-9

[24] Senna M. Solid state mechanochemical processes for better electroceramics. Acta Chim Sloven 2014;61(3):425–31.

[25] Wongmaneerung R, Chaisan W, Khamman O, Yimnirun R, Ananta S. Potential of vibro-milling technique for preparation of electroceramic nanopowders. Ceramics Int. 2008;34(4):813–7. DOI: 10.1016/j.ceramint.2007.09.090

[26] Murty BS, Ranganathan S. Novel materials synthesis by mechanical alloying/milling. Int Mater Rev 1998;43(3):101–41. DOI: 10.1179/095066098790105654

[27] Baláž P, Baláž M, Bujňáková Z. Mechanochemistry in technology: from minerals to nanomaterials and drugs. Chem Eng Technol 2014;37(5):747–56. DOI: 10.1002/ceat.201300669

[28] Wieczorek-Ciurowa K, Dulian P, Bak W, Kajtoch C. High-energy ball milling as a green chemistry method for modification of CaTiO3 electrical properties. Przemysl Chem 2011;90(7):1400–3.

[29] Venkataraman KS, Narayanan KS. Energetics of collision between grinding media in ball mills and mechanochemical effects. Powder Technol 1998;96(3):190–201. DOI: 10.1016/S0032-5910(97)03368-8

[30] Menyhart M, Miskiewicz L. Communition and structural changes in a jet mill. Powder Technol 1976;15(2):261–6. DOI: 10.1016/0032-5910(76)80055-1

[31] Baláž P, Achimovičová M, Baláž M, Billik P, Cherkezova-Zheleva Z, Criado JM, Delogu F, Dutková E, Gaffet E, Gotor FJ, Kumar R, Mitov I, Rojac T, Senna M, Streletskii A, Wieczorek-Ciurowa K. Hallmarks of mechanochemistry: from nanoparticles to technology. Chem Soc Rev 2013;42(18):7571–637. DOI: 10.1039/c3cs35468g

[32] Erdemolu M, Baláž P. An overview of surface analysis techniques for characterization of mechanically activated minerals. Mineral Process Extract Metallurgy Rev 2012;33(1):65–88. DOI: 10.1080/08827508.2010.542582

[33] Beyer MK, Clausen-Schaumann H. Mechanochemistry: The mechanical activation of covalent bonds. Chem Rev 2005;105(8):2921–48. DOI: 10.1021/cr030697h

[34] James SL, Adams CJ, Bolm C, Braga D, Collier P, Friić T, Grepioni F, Harris KDM, Hyett G, Jones W, Krebs A, MacK J, Maini L, Orpen AG, Parkin IP, Shearouse WC, Steed JW, Waddell DC. Mechanochemistry: opportunities for new and cleaner synthesis. Chem Soc Rev 2012;41(1):413–47. DOI: 10.1039/c1cs15171a

[35] Boldyreva E. Mechanochemistry of inorganic and organic systems: what is similar, what is different? Chem Soc Rev 2013;42(18):7719–38. DOI: 10.1039/c3cs60052a

[36] Suryanarayana C. Mechanical alloying and milling. Prog Mater Sci 2001;46(1):1–184. DOI: 10.1016/S0079-6425(99)00010-9

[37] Ivanov E, Suryanarayana C. Materials and process design through mechanochemical routes. J Mater Syn Process 2000;8(3):235–44. DOI: 10.1023/A:1011372312172

[38] Sopicka-Lizer M, editor. High-energy ball milling: mechanochemical processing of nanopowders. 1st Editionst ed. Woodhead Publishing Limited; 2010. 440 p. DOI: 10.1533/9781845699444

[39] Rumpf H. Physical aspects of communition and new formulation of a law of communition. Powder Technol 1973;7(3):145–59. DOI: 10.1016/0032-5910(73)80021-X

[40] Wieczorek-Ciurowa K, Dulian P, Nosal A, Domagała J. Effects of reagents' nature on mechanochemical synthesis of calcium titanate. J Thermal Analysis Calorimetry 2010;101(2):471–7. DOI: 10.1007/s10973-010-0802-0

[41] Boldyrev VV. Mechanochemistry and mechanical activation of solids. Solid State Ionics 1993;63–65(C):537–43. DOI: 10.1016/0167-2738(93)90157-X

[42] Hainovsky NG, Pavlyukhin YuT, Boldyrev VV. Application of mechanical activation to the synthesis of high temperature superconductors. Mater Sci Eng B 1991;8(4):283–6. DOI: 10.1016/0921-5107(91)90048-Z

[43] Koch CC. The synthesis and structure of nanocrystalline materials produced by mechanical attrition: a review. Nanostructured Mater 1993;2(2):109–29. DOI: 10.1016/0965-9773(93)90016-5

[44] Wieczorek-Ciurowa K, Rakoczy J, Błońska-Tabero A, Filipek E, Nizioł J, Dulian P. Mechanochemical synthesis of double vanadate in Cu-Fe-V-O system and its physicochemical and catalytic properties. Catalys Today 2011;176(1):314–7. DOI: 10.1016/j.cattod.2010.12.007

[45] Wieczorek-Ciurowa K. Illustration of selected functional inorganic materials fabricated by mechanochemical approaches. Acta Phys Polon A 2014;126(4):1049–52. DOI: 10.12693/APhysPolA.126.1049

[46] Wieczorek-Ciurowa K, Gamrat K. Some aspects of mechanochemical reactions. Mater Sci Poland 2007;25(1):219–32.

[47] Rougier A, Soiron S, Haihal I, Aymard L, Taouk B, Tarascon J-M. Influence of grinding on the catalytic properties of oxides. Powder Technol 2002;128(2):139–47. DOI: 10.1016/S0032-5910(02)00191-2

[48] Ullah M, Ali ME, Hamid SBA. Surfactant-assisted ball milling: a novel route to novel materials with controlled nanostructure: a review. Rev Adv Mater Sci 2014;37(1):1–14.

[49] Janot R, Guérard D. Ball-milling in liquid media: applications to the preparation of anodic materials for lithium-ion batteries. Prog Mater Sci 2005;50(1):1–92. DOI: 10.1016/S0079-6425(03)00050-1

[50] Balogh J, Bujdosó L, Kaptás D, Kemény T, Vincze. The role of impurities in the crystallization of ball-milled amorphous alloys. Mater Sci Forum 1998;269(1):63–8.

[51] Courtney TH, Maurice D. Process modeling of the mechanics of mechanical alloying. Scripta Materialia 1996;34(1):5–11. DOI: 10.1016/1359-6462(95)00464-5

[52] Gotor FJ, Achimovicova M, Real, C, Balaz P. Influence of the milling parameters on the mechanical work intensity in planetary mills. Powder Technol 2013;233:1–7. DOI: 10.1016/j.powtec.2012.08.031

[53] Wieczorek-Ciurowa K, Gamrat K, Shirokov Ju G. Mechanism of mechanochemical reactions in malachite-active metal systems. Thermochim Acta 2003;400(1):221–5. DOI: 10.1016/S0040-6031(02)00495-1

[54] Takacs L. Self-sustaining reactions induced by ball milling. Prog Mater Sci 2002;47(4): 355–414. DOI: 10.1016/S0079-6425(01)00002-0

[55] Wieczorek-Ciurowa K, Gamrat K, Paryło M, Shirokov Ju G. The influence of aluminium and aluminium oxide on the effects of mechanical activation of nickel hydroxocarbonate. J Thermal Analys Calorimetry 2002;69(1):237–45. DOI: 10.1023/A: 1019958328002

[56] Avvakumov EG, Karakchiev LG, Gusev AA, Vinokurova OB. Preparation of disperse ceramic materials by soft mechanochemical synthesis. Mater Sci Forum 2002;386:245–50.

[57] Bak, W, Dulian P, Sitko D, Garbarz-Glos B, Kajtoch C, Wieczorek-Ciurowa K, Smeltere I. Effect of variable valence ion doping on the dielectric properties of BaTiO3-based materials. Ferroelectrics 2014;464(1):35–41. DOI: 10.1080/00150193.2014.892810

[58] Dulian P, Bak W, Wieczorek-Ciurowa K, Kajtoch C. Dielectric properties of vanadium doped barium titanate synthesized via high-energy ball milling. Mater Sci Poland 2014;32(2):257–63. DOI: 10.2478/S13536-013-0185-6

[59] Dulian P, Bak WB, Wieczoek-Ciurowa K, Kajtoch C. Dielectric behaviour of BaTiO3-SrTiO3 solid solutions fabricated by high-energy ball milling. Key Eng Mater 2014;605:63–66. DOI: 10.4028/www.scientific.net/KEM.605.63

[60] Dulian P, Bak W, Wieczorek-Ciurowa K, Kajtoch C. Comparative studies of dielectric properties of Ca0.25Cu 0.75TiO3 ceramics produced by mechanochemical synthesis and in the way of high-temperature treatment. Key Eng Mater 2013;543:326–9. DOI: 10.4028/www.scientific.net/KEM.543.326

[61] Bak W, Dulian P, Grabarz-Glos B, Kajtoch Cz, Wieczorek-Ciurowa K. Dielectric behaviour of (Ba1-xNax)(Ti1-xNbx)O3 ceramics obtained by conventional and mechanochemical synthesis. Ferroelectrics. Forthcoming.

[62] Dulian P, Bak W, Kajtoch Cz, Wieczorek-Ciurowa K. A high-energy ball milling as a useful technique for the synthesis of CaCu3Ti4O12 electroceramics. Acta Phys Polon A 2014;126(4):931–7. DOI: 10.12693/APhysPolA.126.931

[63] Lunkenheimer P, Fichtl R, Ebbinghaus SG, Loidl A. Nonintrinsic origin of the colossal dielectric constants in $CaCu_3Ti_4O_{12}$. Phys Rev B Condens Matter Mater Phys 2004;70(17):1–4. DOI: 10.1103/PhysRevB.70.172102

[64] Adams TB, Sinclair DC, West AR. Giant barrier layer capacitance effects in $CaCu_3Ti_4O_{12}$ ceramics. Adv Mater 2002;14(18):1321–3. DOI: 10.1002/1521-4095(20020916)14:18<1321::AID-ADMA1321>3.0.CO;2-P

[65] Bender BA, Pan M-J. The effect of processing on the giant dielectric properties of $CaCu_3Ti_4O_{12}$. Mater Sci Eng B: Solid-State Mater Adv Technol 2005;117(3):339–47. DOI: 10.1016/j.mseb.2004.11.019

Synthesis of Perovskite Oxides by Hydrothermal Processing – From Thermodynamic Modelling to Practical Processing Approaches

Juan Carlos Rendón-Angeles, Zully Matamoros-Veloza, Karla Lorena Montoya-Cisneros, Jorge López Cuevas and Kazumichi Yanagisawa

Additional information is available at the end of the chapter

http://dx.doi.org/10.5772/61568

Abstract

The present chapter aims to provide a conscious review of the principles associated with the chemical reactions and the control of the parameters related to the hydrothermal processing of perovskite-structured compounds. Highlights on fundamental principles of the thermodynamic modelling coupled with the relevant technical expertise gained during the past two decades are discussed. Achievements conducted in the early 1990s on thermodynamic modelling of hydrothermal reactions, leading to the estimation of the chemical reaction equilibrium occurring under specific conditions, i.e. above 100°C and 0.1 MPa, are discussed. Additional efforts resulted in different thermodynamic models that predict crystal growth kinetics and the stability for particle nucleation; the models based on chemical population balance approaches are also considered. However, these models do not apply for perovskite compounds containing rare earth elements that crystallize under hydrothermal conditions above 250°C, i.e. orthorhombic lanthanum chromite perovskite. Hence, the final part comprises a literature survey for the experimental research work conducted on various perovskite species produced via hydrothermal treatments, emphasizing the relevant conditions that led to the stoichiometric single-phase crystallization.

Keywords: Perovskite materials, Hydrothermal synthesis, Crystallization, Thermodynamic modelling, Solid-Liquid Equilibrium

1. Introduction

Over the past two decades, remarkable efforts have been conducted worldwide to explore various techniques to prepare perovskite-structured ceramic oxide materials. Perovskite oxides exhibit a broad variety of functional properties, namely ferroelectricity, piezoelectricity, pyroelectricity and non-linear dielectric behaviour, among others. The functionality of perovskite-structured compounds depends on the relationship established between the specific crystalline structure and the composition of its major constituents. This particular family of inorganic compounds has a large range of interesting physical properties applicable to the design and preparation of different electronic devices with applications in charge storage, non-volatile memories, transducers, actuators and infrared detection. Particular efforts focused on optimizing the physical properties of perovskite-structured compounds have recently taken place by numerous research groups worldwide. Recently, most studies are based on establishing a correlation between the crystalline structure and the chemical stoichiometry of the major constituents. These have led to improvements in the functional properties of the ideal ABX_3 compound with cubic structure (space group).

Another relevant subject that attracted the interest of various research groups concerns with the development of environmentally friendly chemical processes for producing perovskite compounds at laboratory and large scales. The hydrothermal technique has become one of the most suitable chemical processing routes in terms of energy consumption and environmental friendliness for the preparation of a vast number of perovskite compounds with specific crystalline structures and chemical compositions. In the present review, the state-of–the-art associated with this technique is discussed in terms of processing aspects involved in the crystallization of perovskite compounds. In particular for the case of alkaline earth titanate perovskite oxides, these compounds had been used as a standard for developing thermodynamic modelling approaches. These models proposed consider the particular chemical reaction equilibriums that occur under hydrothermal conditions and lead the control of perovskite particle crystallization. Various stability crystallization diagrams were calculated by simulating a complex reaction system containing solid and ionic species in the aqueous phase employing the thermodynamic models. Likewise, a comprehensive analysis is based on the expertise gained during the past two decades by several research groups. With regards to the correlation of the fundamentals principles and the technical aspects involved in the hydrothermal synthesis, a vast number of perovskite compounds are addressed. The present review intends to give guidance for the researchers, as well as to encourage the newcomers from research fields such as Material Science, Chemistry and Physics, by directing their attention towards the key points for conducting hydrothermal reactions for preparing perovskite compounds.

2. Hydrothermal synthesis of perovskite materials

2.1. Structure of perovskite materials, applications and synthesis processes

2.1.1. The crystalline structure of mineral perovskite

The perovskite oxide family has been studied extensively because of the diverse properties exhibited by a material when a considerable number of different atoms produce the typical atomic arrangement associated with this crystalline structure. Gustav Rose discovered the $CaTiO_3$ mineral in 1839 at the Ural Mountains in Russia, and the mineral was named by the Russian mineralogist Lev Aleksevich Perovski [1]. The "perovskite" term applied to a particular group of inorganic compounds having similar crystalline structure, and the ABX_3 chemical stoichiometry, the major constituent of the compound is $CaTiO_3$. This compound is constituted by divalent (A^{2+}) and tetravalent (B^{4+}) cations. The small B^{4+} cation is corner-sharing bonded with six X^{2-} anions, resulting in an octahedral BX_6 unit forms the ideal face-centered cubic close packed unit cell. Likewise, the large A cation located at the 12-fold coordination sites produces eight octahedral points in the corners of the cube (Figure 1a). In contrast, the simplest atomic arrangement of the ideal cubic cell unit involves a different cation distribution. In this structure the A cations are in the corners of the cube, while the B cation located in the center of the cube is coordinated with six anions in the face-centred positions of the cubic unit cell (Figure 1b). This ideal cubic perovskite is not very commonly found in the mineral reservoirs, because even the mineral species has slight distortions in the atomic ordering in the cubic structure. The study of the perovskite crystalline structure firstly conducted by Goldsmith in 1920 led to propose various fundamental aspects that correlate the crystalline structure and chemical composition parameters [1,2]. One of the most important principles proposed was the tolerance factor. The tolerance factor is one tool normally used for predicting the structural arrangement and stability of a particular perovskite composition, either from the chemical and physical points of view. This factor is evaluated before selecting the adequate processing route for perovskite preparation.

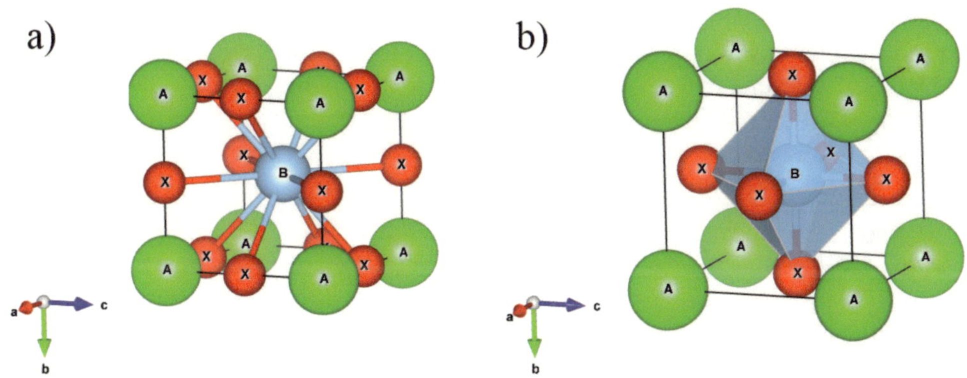

Figure 1. Structural representation of the ideal cubic perovskite showing the (a) cubic A unit cell and (b) cubic B unit cell.

The compounds that belong to the idealized cubic (fcc) structure is $SrTiO_3$ because the strontium titanate oxide exhibits the atomic packing shown in Figure 1b. The ideal cubic structure has a space group. In agreement with the Wyckoff positions, the atomic distribution is as follows: the A atoms are in x, y, z coordinates ½, ½, ½ (Figure 1b) while the B atom is located at 0, 0, 0 (Figure 1b), and the X atoms in 3d are located at spatial positions ½, 0, 0; 0, ½, 0, and 0, 0, ½. The lattice parameter "a_0" of this perovskite structure is 3.905 Å. This crystalline structure undergoes a series of distortions; one of this is caused by a deficiency in the Sr^{2+} ions (A) in the framework, producing a ReO_3-type structure [2]. Another distortion occurs when the ReO_3 structure transforms into dense packing due to the octahedral unit rotation, producing a structural geometry resembling the hexagonal close packing type, i.e. RhF_3. The void at the centre is in an octahedral surrounding coordination; when this octahedral hole is occupied, the ilmenite structure ($FeTiO_3$) is obtained. The slight distortions attainable in the cubic structure are due to the displacement of ions from the ideal positions, producing a variation of a few tenths of an Å. Therefore, the final symmetry varies considerably between different materials. A mechanism that likely promotes these distortions is related to the capability of perovskite to accommodate a great variety of atoms, due to the flexibility of its crystalline structure [1–3]. An additional mechanism is the Jahn–Teller effect.

2.1.2. Tolerance factor, orthorhombic and hexagonal crystalline unit cells

The ionic radii differences produce the structural distortions on perovskite structure, as was determined in the pioneering research work conducted by Goldschimidt [1–3]. The equation that geometrically correlates the ionic radii (r_A, r_B and r_O) for the ideal cubic cell with the lattice parameter a_0 is given as:

$$a = \sqrt{2}\left(r_A + r_O\right) = 2\left(r_A + r_O\right) \tag{1}$$

The expression that involves the unit cell length ratio is known as the Goldschmidt´s **tolerance factor t**, which is employed to estimate the distortion level attained by a particular perovskite-structured compound. Fundamentally, it considers the ionic radii of the constituent atoms that fit the chemical stoichiometry and enhances a pure ionic bonding between them; the mathematical expression for the tolerance factor is given in Eq. (2).

$$t = \frac{(r_A + r_O)}{\sqrt{2}(r_A + r_O)} \tag{2}$$

According to Eq. (2), the ideal perovskite cubic structure has a "t" value equal to 1, which can be calculated for $SrTiO_3$ with ionic radii of r_A = 1.44 Å, r_B = 0.605 Å and r_O = 1.40 Å. The value of t is less than 1 when the A ionic radius is slightly small, and structurally the octahedral unit $[BO_6]$ tilts forward filling the additional space. From Eq. (2), the grade of tolerance where the ideal cubic structure is attainable in perovskite compounds is $0.89 < t < 1$ [2–4].

Orthorhombic ABO_3 perovskites are among the most important constituents of the Earth´s crust. These compounds have been under exhaustive study due to the wide variety of their functional properties. The distortion of the cubic cell results in the formation of the orthorhombic structure. This process occurs by the tilting of the BO_6 octahedra, but this distortion is not detectable as the temperature increases because the tilt angle decreases [5]. A typical orthorhombic structure that doubles in dimensions the cubic one is shown in Figure 2.

http://dx.doi.org/10.5772/61568

Orthorhombic RMO_3 perovskites (where R = rare earth element or Y, M = 3d-block transition metal) manifests a high intrinsic orthorhombic distortion when the R^{3+} ionic radius is approximately 1.11 Å, and this distortion decreases when the radii of R^{3+} is greater than 1.11 Å. Among the major constituents of the orthorhombic-structured perovskites that have been under exhaustive crystallography studies are $RFeO_3$, $RTiO_3$, RVO_3, $RMnO_3$ and $RNiO_3$. Complementary studies were conducted to determine the correlation between the Jahn–Teller cooperative orbital orderings of M cations with the usual site distortions [6].

The hexagonal-structured perovskite compounds have tolerance factor values above 1. In particular, this occurs when either the A-cation or the B-cation are either too large or too small. The differences on the atomic radii distort the perovskite cubic structure forming the hexagonal perovskites, as it is shown in Figure 2b. In this structure, closely packed layers constituted by octahedra units bonded by face sharing promote the hexagonal structure. The stability of the hexagonal crystalline structure constituted by face-sharing octahedra is lower than that for the one formed by corner-sharing octahedra. However, some hexagonal perovskites belonging to this group overcome this restriction due to the metal–metal bonding between the B ions corresponding to the BX_6 octahedra; these bonds are strong enough so that the metal–metal repulsion is overcome [1]. As a consequence of these stability restrictions, hexagonal perovskites are less commonly found compared with the cubic perovskites. $BaNiO_3$ is one of the hexagonal perovskites that adopt the space group $\boldsymbol{P6_3/mm}$. The representation of the unit cell structure is shown in Figure 2b. This shows the chains of face-sharing BX_6 octahedra orientated along the **c**-axis. Just like the cubic perovskites, hexagonal perovskites can also undergo distortions leading to a variety of structures.

On the other hand, the cubic cell is also susceptible to undergo a small structural deformation producing rhombohedral symmetry, but the deformation does not produce a marked enlargement of the unit cell. Hence, the unit cell likely contains at least one or two rhombohedral polyhedral units with angles of $\alpha \sim 90°$ or $\alpha \sim 60°$. However, the anions are displaced provoking the formation of a large unit cell with $\alpha \sim 60°$.

The tetragonal-structured perovskite, $BaTiO_3$, is probably one known example of a ferroelectric perovskite that is a stable phase at room temperature. In this structure, the TiO_6 octahedra are slightly distorted (one Ti-O bond at 1.86 Å, four at 2.00 Å and a longer one at 2.17 Å). The barium atom is in coordination with four oxygen atoms at 2.80 Å, four at 2.83 Å and four more at 2.88 Å. Another tetragonal perovskites ($PbHfO_3$, $SrPbO_3$, $SrZrO_3$, $AgTaO_3$, etc.) are isotypic with $BaTiO_3$ and possess single-molecular cells. However, a vast number of these materials exhibit the tetragonal structure at elevated temperatures, which makes crystallographic analyses difficult to conduct. In general, a significant number of perovskite-like materials have several polymorphic transformations. Some of these are important regarding their physical properties and applications. The compounds that exhibit this behaviour are $BaTiO_3$ and $KNbO_3$; the following transformation sequence was determined to occur by increasing temperature: rhombohedral → orthorhombic → tetragonal → cubic.

The lowest temperature crystalline compounds (orthorhombic, tetragonal and cubic) have ferroelectric properties. These particular phase transitions are reversible in nature, and all the polymorphic forms exhibit a pseudo-cubic unit cell with $a_0 \sim 4$ Å. Consequently, the polymorphic variation on the perovskite structure further affects the crystal chemistry of this group of materials [4–6].

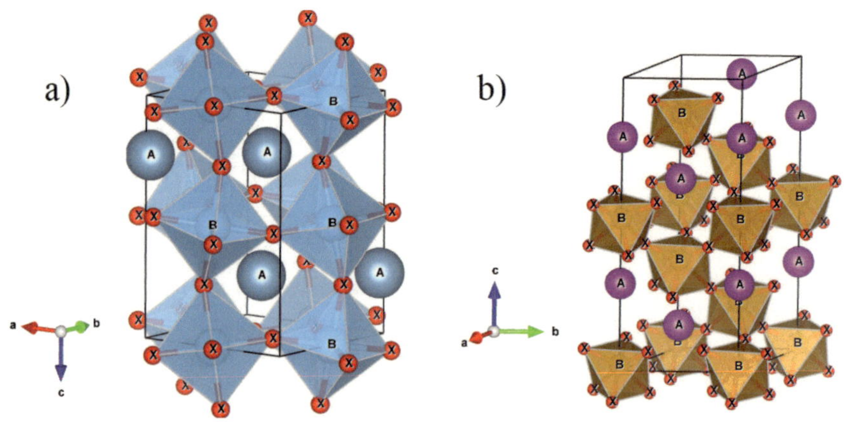

Figure 2. Typical perovskite (a) orthorhombic and (b) hexagonal structural unit cells.

2.1.3. Applications of perovskite materials

Although the physical properties of perovskites-structured materials are not the primary concern of the present review, some relevant ones are discussed below. Since the functionality of perovskites materials was discovered over five decades ago, hundreds of research works have directed forward elucidating the physical and chemical properties of perovskites. These studies have provided pertinent information regarding the fundamentals of the chemical and physical aspects that enhance the structural distortions in ABO_3 materials. The former literature also suggests that there are still surprises to discover for this particular group of compounds, in particular, the perovskites with low tolerance factors; for instance, those comprised in the phase stability diagram $FeTiO_3$-$LiNbO_3$-*Pnma* [7]. A vast number of elements in the periodic table are likely located at either A or B unit cell sites. This fact provides an enormous range of compounds with structural similarity and a variety of properties. Among the most important properties are ferroelectricity ($BaTiO_3$), ferromagnetism (Sr_2FeRuO_6), weak ferromagnetism ($LaFeO_3$), colossal magnetoresistance [8–12], superconductivity ($Ba_{0.6}K_{0.4}BiO_3$) [13] and large thermal conductivity ($LaCoO_3$). Insulating to metallic transitions have a particular interest in the design of devices for thermistor applications ($LaCoO_3$), the fluorescence is applicable for laser devices ($LaAlO_3$:Nd), and transport properties have attracted the attention of research for the development of high-temperature thermoelectric power devices (La_2CuO_4) [1,5,7–14].

Perovskite materials have been investigated for applications involving the preparation of solid electrolytes. The compounds that have been used in various electrochemical devices due to their high performance are barium cerate ($BaCeO_3$) and barium zirconate ($BaZrO_3$). Solid electrolyte performs three essential functions: (1) separates the anode from the cathode in the electrochemical cell (oxidizing and reducing sides), (2) it can operate as electronic insulator enhancing the flow of electric current through an external circuit and (3) high ionic conduction coefficient, required to provide the control of the electric current flow in the external circuit [15]. Likewise, the proton-conducting ceramics applications are classified according to two basic functions: (1) the material is capable to generate an electromotive force when undergoing a chemical potential gradient and (2) capability for electrochemical ion transport (hydrogen or oxygen) enhanced by an external power source. The employment of an ionic proton conductor as an electrolyte in devices operating under chemical potential gradient provides a device capable of producing electric energy [18,19]. The development of the solid oxide fuel cell (SOFC) electrical power sources was derived from this electrochemical principle. The SOFC is a conversion energy device that generates electricity via an electrochemical reaction occurring at temperatures above 800°C. The chemical reaction that takes place between the fuel (methane, hydrogen, natural gas) and the oxidizing agent (oxygen from air) is the motion to produce the electricity [15,18]. In this particular electrochemical converter device, the principal components responsible for the redox reaction are the electrodes; the ion transport takes place in all the cell constituents (solid electrolyte and electrodes). In contrast with the rechargeable battery devices, the SOFC does not need to be recharged, and it only requires to be continuously fed with a particular fuel for electricity generation [15–20].

Among the miscellaneous applications, such as photocatalytic reactions, some perovskites have been employed. Photocatalytic oxidation was extensively investigated by using various titanates and cobaltites. Titanate perovskite materials ($SrTiO_3$ and $BaTiO_3$) were found to exhibit a strong photocatalytic effect, in contrast with that determined for the easily reducible $LaCoO_3$. Perovskite materials have demonstrated a notable performance as hosts in laser devices because in the A site of the structure some rare earth elements (Nd^{3+} or Sm^{3+}) are incorporated. These elements are some of the major constituents employed to produce laser ion devices. Moreover, various perovskite oxides possess high electrical resistivity, which makes them useful for the preparation of dielectric materials that are poor conductors of electricity, which are widely used for electrical insulation in electrical power systems. Additionally, these materials can be used for energy storage purposes as in capacitors. Typical conductor and semiconductor oxides are characterized to contain either B-ions with a net charge lower than the stable B^{4+}, or B-ions having two different net charges. Typical materials that have been considered to be good conductors or semiconductors are $LaVO_3$, $LaTiO_3$ and $SrMoO_3$. Miscellaneous applications of dielectric perovskites include mechanical actuation, sound generation, in materials subjected to dimensional changes as a result of an applied voltage (piezoelectrics), for transducers as in condensers, for piezoelectric microphones, for detecting changes in temperature as in pyroelectrics, and as liquid crystals employed for alphanumeric displays.

Perovskite oxides materials are also used as oxygen sensors. The mechanism involved in this type of sensors is electrical conductivity in nature, which enhances the oxygen adsorption in the crystalline structure. The electrical conductivity is proportional to oxygen partial pressure and to the concentration of vacant sites in the X site of ABX_3. One compound that exhibits a high sensitivity to oxygen motion is $SrTiO_3$. Furthermore, $SrSnO_3$ is a promising material for combustion monitoring-sensors. The design of functional materials requires an understanding of the relationship between chemical composition and crystalline structure [15–21].

2.1.4. Synthesis of perovskite materials

The preparation of perovskite oxides has been the subject of considerable research because the design and development of new technological properties require the preparation of materials with special physical morphology. Some examples of the new developments include thin films, porous solids, monodispersed powders with nanometric size, among others. In addition to this, the interest for specific structural ordering exhibiting chains that enhances unidirectional properties has prompted various research groups to pursue the preparation of new perovskite oxides with specific compositions [22,23]. The conventional method of solid-state reaction (ceramic method) was broadly used to produce perovskite oxides powders during the early 1940s. An analogous method, the flux method, which is also based on solid-state reaction, has been employed for the preparation of single crystals of various perovskite oxides [23]. Both methods are adequate to synthesize oxide compounds because they allow achieving the proper conditions in an easy way, due to the reaction proceeding under air atmosphere and ambient pressure [24]. Common steps involved in the solid-state reaction methods are homogeneous grinding and mixing. Oxide and carbonate solid reactants are among the most employed. These are mixed in stoichiometric amounts prior to the conduction of a heat treatment at elevated temperatures (< 1000°C) for long periods of time (several hours or days). The grinding and heating cycles are repeated until the desired pure phase is successfully produced. Any doubt exists in regards of the usefulness and easiness of the universal method, but the achievement of the conditions for atomic diffusion could be a problem. Hence, to accelerate the atomic diffusion for preventing the formation of metastable phases, the process must proceed at high temperatures. However, this has the disadvantage of promoting the volatility of reactants such as lead oxide (in the case of compounds containing Pb^{2+} as in the $PbTiO_3$). In addition, the need of regrinding the materials several times involves high energy and time consumption, as these parameters have a strong influence on achieving the desired product functionality [22–24].

Recently, the solid-state metathesis (SSM) reaction method has emerged as an efficient processing route for synthesizing a broad range of non-oxide compounds. The process occurring under thermodynamic equilibrium conditions is more effective than the conventional solid-state reaction method [22,25]. The SSM reaction method is a variant of the SHS process in which the chemical reactions are conducted very rapidly. A typical SSM reaction involves an ionic exchange taking place between the reactants to produce thermodynamically stable perovskites. Along with the reaction, a remarkable enthalpy change occurs, and high adiabatic reaction temperatures are reached. The solid-state metathesis reaction is conducted using alkaline precursors, chalcogenide, silicide or boride; the salt is reacted with a metal halide according to the general [reaction 3]

$$A_pB_q + CX_p \rightarrow BX_q + pAX \tag{3}$$

where A = Li, Na, K, Mg, Ca, Sr, or Ba; B = B, Si, N, P, As, Sb, Bi, O, S, Se, or Te; C = transition group, principal group or actinide metal, and X = halogen.

The effectiveness of transferring the heat required achieving the chemical reactions under solid-state conditions for the methods mentioned above constitutes one of their main disadvantages from the point of view of the energy cost. One of the new processing techniques that have overcome this problem is the one known as "microwave irradiation". During the microwave processing conditions, the heating is directly supplied to the solid reactants, because it proceeds from the interaction at a molecular level between the electromagnetic field and some of the elements constituting the solid. This interaction varies depending on factors such as the dielectric and magnetic properties of the solid reactants, frequency and microwave power generation, microwave permeability and size and density of the materials [24]. The microwave heating improves the reaction kinetics in a range between 10 and 1,000 folds in comparison with the conventional and the SSM routes, resulting in significant differences associated with the crystalline structure and properties of the reaction products produced. Recently, the synthesis of various perovskite-structured compounds has been under extensive study by microwave processing. Table 1 summarizes the processing details regarding the synthesis of some single-phase perovskite materials produced under microwave irradiation [24].

Chemists in material research are constantly seeking to develop new processing methods with lower costs of energy consumption and that are more environmentally friendly. The novel methods are remarkable functional, allowing us to synthesize perovskite oxides under milder conditions, in comparison with the conventional solid-state process [26–28]. The ceramic materials produced by non-conventional techniques, such as soft chemistry routes, have a more homogeneous distribution of constituents, are purer, can be produced with a wide range of partial dopant substitutions and with various shapes (i.e. coatings and fibers), which are not really obtainable via conventional methods [29]. Soft chemical processing routes also allow the preparation of bulk monoliths without the use of intermediate powders, which occurs by drying a precursor gel at supercritical conditions [29].

Coprecipitation and sol–gel routes emerged since the early 1960s as processing alternatives for ceramic powders, including some perovskite compounds. Certain novelty aspects are inherent in these techniques, which are associated with the control of physical aspects such as size, the state of aggregation and purity of the produced compounds. The functionality of these processes is based on the principles related to the mechanism of formation of colloids in aqueous media at very low temperatures. In the aqueous "sol–gel" processing, the starting chemical reagents, i.e. salt solutions containing specific cation species, undergo chemical treatments at mild temperatures (< 100°C) to form stable colloid dispersions, or sols, when mixed with an acid or basic solution. The particle size of the sols varies between 1 nm and 1 μm. The sols can be prepared from highly hydrolyzable salt reagents such as $ZrCl_4$ or $TiCl_4$, i.e. in the case of perovskite zirconate or titanate. In the "sol–gel" processing, the crystallization step occurs at calcination temperatures much lower than those associated with the preparation of the same crystalline phase via the solid-state process. Prior to this step, the separation of the gel from the sol is conducted by a preliminary dehydration stage [29,30]. Enormous efforts have been carried out over the past three decades aiming to develop new alternative soft chemistry routes that allow us to reduce the sol–gel crystallization temperature and to control the stoichiometry of the reaction products. One of the processing routes recently proposed involves the reaction of a hydroxide reagent precursor, either NaOH or KOH, with $Nb(OH)_5 \cdot xH_2O$. The dissolution of Nb_2O_5 in an HF solution was the preliminary step to prepare the precursor of niobium. The hydroxide powder mixture containing stoichiometric amounts of the precursor reagents (sources of K and Nb) was heat-treated at different temperatures ranging from 200 to 700°C for 6 h in air. The orthorhombic $KNbO_3$ perovskite phase was produced without the formation of any by-products at a temperature of 700°C while the $NaNbO_3$ powders were prepared at a temperature as low as 500°C [30].

Chemical reagents supplying A and B cations		Reaction product	Microwave		Conventional	
A cation supplier	B cation supplier		Power (W)	Time (min)	Temperature (°C)	Time (min)
Li_2CO_3	Nb_2O_5	$LiNbO_3$	800	15	500	720
Na_2CO_3		$NaNbO_3$	800	17	1,250	*
K_2CO_3		$KNbO_3$	800	12	1,000	1,800
$BaCO_3$	TiO_2	$BaTiO_3$	1,000	25	1,400	*
$PbNO_3$		$PbTiO_3$	600	9	360	480

*Note: Time not specified, the product was obtained after several periods of ball milling.

Table 1. Summary of the processing conditions employed for the preparation of niobate and titanate perovskites via microwave and conventional solid-state processing methods (data taken from reference [24]).

In approaches based on the preparation of polymer precursors using organic solvent solutions, namely the Pechini method and its alternative process polymerized complex method (PC). A polymer containing the cation precursors of the perovskite compound is produced via a chemical reaction occurring between suitable organometallic precursors and the liquid solvent. The polymer precursor is produced by a treatment conducted at low temperature (200–400°C), leading to a compositional homogenization of the elements allocated in the polymer structure [31–33]. The amorphous polymer precursors are usually calcinated to promote the crystallization of the stable $KNbO_3$ [31,32] and $BaTiO_3$ [33] ceramic phases. Furthermore, the polymer precursors can be designed, in terms of their constituents and composition, so as to improve their combustion capability. Based on this principle, a combustion technique has been proposed as a novel soft chemical route. This technique is of industrial interest because it allows processing large batch volumes of powder. Furthermore, the production of nanocrystalline oxide ceramics powders proceeds at lower calcination temperature in a very short time. The produced particles have the highest degree of purity and characteristics such as narrow particle size distribution, higher surface area that enhances its sinterability [32,33]. The combustion technique is based on the exothermic decomposition of a fuel-oxidant precursor. The reaction promotes the formation of a fine monodispersed powder with perovskite structure, or a partially decomposed precursor containing considerable amounts of carbon traces. The results after processing depend on the precursor and in the fuel-to-oxidant ratio used to conduct the combustion. During the decomposition, the decomposition of the organic compound facilities the rapid increase of the temperature coupled with gas production, resulting in the coalescence of particles inconsequence short diffusion pathways as well. Hence, a foamed porous aggregate formed by a pure-phase nanoparticles agglomerated can be obtained at low temperatures [32,33].

Miscellaneous processing techniques involving non-equilibrium reaction conditions are also of industrial interest to produce perovskite oxides. The glycothermal method is a novel technique involving crystallization of oxide particles, i.e., $KNbO_3$, and which is conducted at supercritical conditions, depending on the chosen organic solvent. At supercritical conditions, the organic solvent capability is similar to that of the normal polar liquids, but it exhibits better transport properties (viscosity, diffusivity and thermal conductivity, among others) [34]. This technique is reliable in terms of reproducibility and environmental aspects. However, the high cost of the organometallic reagent precursors and the restricted types of organic solvents are some disadvantages associated with this technique. An analogous technique that employs only water as the solvent, hydrothermal processing has been widely used for more than three decades to synthesize a numerous variety of perovskite materials. This technology was explored in the engineering fields of crystal growth and metal leaching in the middle of the 20th century. Hitherto, the hydrothermal technology broadly covers various interdisciplinary fields of materials science. In the solid-state chemistry field, the hydrothermal media have recently been exploited for preparing vast type of ceramic compounds. This technology provides an efficient reaction environment for synthesizing perovskite powders, due to the effect of a combination of parameters, such as solvent media, temperature and pressure, on the ionic reaction equilibrium. The conventional hydrothermal (CH) method is an efficient route that enhances the crystallization of micro/nanometric morphology controlled and crystal growth-oriented particles. In addition to that, this method depends on the inorganic salts solubility in water under variable temperature and pressure conditions. Another fundamental factor that has a marked influence enhancing the heterogeneous reactions in this process is the vapour pressure. Hence, a detailed state-of-the-art regarding this technique is discussed in the next sections. A particular emphasis that considers the theoretical principles associated with thermodynamic modelling of the heterogeneous chemical reaction equilibrium associated with a particular hydrothermal environment is addressed. Additionally, the link between the theoretical principles and the extensive practical expertise gained over the past decades regarding the hydrothermal synthesis of the most representative perovskite compounds is further discussed in the final section of the present review.

2.1.5. *Definition of hydrothermal synthesis*

The word "hydrothermal" comes from the etymological root of the Greek word "hydrous" that means water while "thermal" means heat. The "hydrothermal" term has a purely geological origin. In the middle of the 18th century, the Scottish Geologist Sir Roderick Murchinson introduced this term to the Scientific Society. In practical terms, he described that the formation of mineral species that occurred on Earth's crust is due to the reaction of water exposed to conditions of elevated temperature and pressures. Another technical definition accepted by the Scientific Society refers the "hydrothermal" term as any heterogeneous chemical reaction, occurring in the presence of a solvent media. The reactions occur inside a hermetical sealed vessel system at temperatures above 25°C and pressure of 0.1 MPa. Under hydrothermal conditions, it does not matter whether the solvent is aqueous or non-aqueous. The crystallization process of solid phases under hydrothermal conditions is carried out at autogenous pressure, achieved by the saturated vapour pressure of the fluid at the specified temperature and composition of the hydrothermal solution. In this concern, in terms of industrial and commercial processing, mild operating conditions are preferred, for example, temperatures of treatment below 350°C and pressures less than 50 MPa [35, 36]. The processing parameter that allows the transition from mild to severe reaction conditions during a hydrothermal treatment is the lining material strength of the autoclave vessel. At severe treatment conditions in highly concentrated acidic or basic solutions, the lining material might undergo a rapid corrosion process. The progress in the experimental work research in this area of investigation has allowed improving the understanding of how is the behaviour of the chemical reactions generated in the hydrothermal media. The crystallization of several materials of oxide and non-oxide has been made with adequate optimization of the experimental parameters during the hydrothermal treatment ($T < 200°C$ and $P < 1.5$ MPa). The recent scientific and technological achievements have made the hydrothermal synthesis more economical. One example is the synthesis of ceramic particles, which can be prepared in a single step, using advanced pressure reactor technology coupled with processing methodologies proposed for a wide number of inorganic compounds [37].

2.2. Thermodynamic modelling associated with the hydrothermal synthesis of perovskite oxides

In the early 1990s, a great interest to prepare monodispersed perovskite oxide fine particles with controlled morphology was the main concern of the chemical community. One chemical technique that provides an adequate environment to accomplish the preparation of a substantial number of ceramic materials is the hydrothermal processing. Because this method combines the dynamic interaction of processing parameters such as solvent type, temperature and pressure that governs the ionic mobility, this particular technique involves numerous simultaneous chemical reactions, which normally take place in an aqueous system comprising the interactions of dissolved and solid species. The reaction product obtained consists of an anhydrous single-phase crystalline oxide or multicomponent phases. The huge diversity of precursor chemical reactants (i.e. water, soluble salts, hydroxides and oxides, among others) allows preparing solvents that can be employed as hydrothermal media. This processing parameter promotes a broad range of reaction pathways to achieve the crystallization of a solid in a particular hydrothermal reaction system. In general, phase-pure oxides (perovskite-structured) with specific stoichiometry, particle size and morphology can be hydrothermally produced in one step processing, even using low-cost reagents at mild temperatures and pressures, five-fold lower than those required for conventional processing techniques.

Hitherto, the practical fundamentals derived from the expertise gained over the past five decades. Regarding the processing of inorganic compounds by the hydrothermal technology, to take an adequate advantage of the technology's novelty, one must bear in mind some critical key processing factors such as the selection of a suitable precursor system (highly reactive and cost-effective) required for the optimization of the chemical equilibrium that enhances the crystallization of the desired phase. The effectiveness of the selected experimental approach can be evaluated selecting the suitable chemical precursor concentration and mixing ratios, pH of the hydrothermal media, temperature and pressure level [38,39]. This procedure is relatively complex and time-consuming due to the numerous variables involved. Hence, to determine the effectiveness of the precursor hydrothermal system, an approach based on thermodynamic modelling to simulate the hydrothermal reactions has been developed since the early 1990s [40]. OLI Systems Inc. (USA) developed an algorithm that included the thermodynamic basis to simulate the chemical reactions. The model is capable to calculate thermochemical data with high consistency and accuracy for a broad number of perovskite compounds, including the single-phase $CaTiO_3$, $SrTiO_3$, $SrZrO_3$, $PbTiO_3$, $BaTiO_3$, as well as some selected solid solutions of $Ba_{1-x}Sr_xTiO_3$ and $PbZr_{1-x}Ti_xO_3$ [41–49]. In addition to that, these studies have produced relevant information regarding the behaviour of solutions under modifying pressure and temperature conditions. The most features investigated under hydrothermal conditions are solubility, stability and yield product amount, reaction mechanisms, among others.

2.2.1. *Thermodynamic model*

Today, our knowledge of the physical chemistry of the hydrothermal processes is enriched by the experience acquisition in the area of hydrothermal solution chemistry. The role of the solvent during the hydrothermal treatment needs to be understood because it depends on the experimental variables temperature and pressure. Also, the solvent reactivity is affected under hydrothermal conditions by some interrelated parameters such as the structure at critical, supercritical and subcritical conditions, dielectric constant, pH variation, viscosity, coefficient of expansion and density. In general, the thermodynamic modelling applied to a hydrothermal system constitutes a potential tool for predicting the concentration and activities of ionic or neutral substances used as solvents. These include a reaction system constituted for some solid, vapour and non-aqueous phases. A particular methodology involving a model based on the theory of electrolyte systems is established; the model enables a quantitative description of both the phase and the ionic equilibrium and provides values of the activity coefficients. Furthermore, the models provide data about the standard-state properties of all the substances involved in the reaction model, coupled with the Gibbs energy excess (referred to a non-ideal solution). Rafal and Zemaitis have given a complete interpretation of the experimental design fundamentals for thermodynamic modelling of the hydrothermal systems, as described in Ref. [39].

The hydrothermal medium is an environment for the inorganic particle synthesis. The chemical interactions between solid and fluid phases are affected mainly by variables such as solvent liquid, temperature, reaction interval, pH of the solvent fluid, the initial concentration of the precursor feedstock and occasionally pressure. The correct selection of these parameters is crucial to determinate the principal production conditions that allow the formation of single-phase stoichiometric compounds, namely ABX3. Pioneering hydrothermal research investigations were conducted empirically, establishing the experimental parameters in a trial-and-error fashion, resulting in an inaccurate approximation for a practical particle synthesis. The hydrothermal reaction system becomes complex when highly concentrated solutions rich in ions are employed, or if multiple heterogeneous chemical reactions alternatively proceed in the hydrothermal media. In this particular case, the ionic concentration related to the chemical equilibrium promoted in the hydrothermal fluid is strongly dependent on ionic species activity coefficients [38]. At the early 1990s, Lenka and Riman reported the first rigorous approach that is in essential a practical thermodynamic model of ionic species hydrolyzed in a hydrothermal reaction system [40]. The model is capable of predicting or optimizing the experimental conditions, namely feedstock composition, solution pH, temperature and pressure, to minimize the Edisonial trial and error design. Additionally, this model produces a series of phase stability diagrams based on the interaction of the main parameters, such as feedstock concentration, solvent pH and its concentration.

The parameters required to obtain a hydrothermal phase stability diagram are the equilibrium concentrations of the ionic species added in the system as a function of temperature, pressure and the initial content of the precursor feedstock. Initially, the number of independent chemical reaction equilibria produced under hydrothermal conditions is denoted by "k". The specific jth reaction ($j = 1,..., k$) imply n_j different chemical substances; these are represented by $A_i^{(j)}$ ($i = 1,..., k$). Thus, the number of possible reactions occurring in an infinitesimal point in the hydrothermal system can be given by the following expression 4.

$$\sum_{i=1}^{n_j} v_i^{(j)} A_i^{(j)} = 0, \qquad j = 1,......k. \tag{4}$$

While the equilibrium state of any jth reaction is determined using the expression for the variation in the standard Gibbs energy

$$\Delta G_j^0 = \sum_{i=1}^{n_j} v_i^{(j)} \Delta G_f^0\left(A_i^{(j)}\right) = -RTlnK_j\left(T,P\right), \qquad j = 1,...k. \tag{5}$$

Where $G^0(A_i^{(j)})$ corresponds to the standard Gibbs energy for the substances formation ($A_i^{(j)}$) and K_j represents the equilibrium constant of the jth reaction.

The equilibrium constant of any reaction is related to the molality "m", which is the used concentration unit. This can be expressed as

$$K_j(T,P)=\prod_{i=1}^{n_j}\left(m_{A_i^{(j)}}\gamma_{A_i^{(j)}}\right)^{v_i^{(j)}}, \qquad (j=1,\ldots,k) \tag{6}$$

where $\gamma_{A_i^{(j)}}$ represents the activity coefficient of the ionic species $A_i^{(j)}$. To solve Eqs. 4 and 5, the mass and electro neutrality equilibria must be established, coupled with the standard Gibbs energy values associated to the formation and activity coefficients. In parallel to the solution of these equations, the Helgeson-Kirkham-Flowers-Tanger (HKFT) equation of state must be taken into account. The HKFT equation provides the standard-state thermodynamic functions of aqueous, ionic and neutral species as a function of both temperature and pressure up to 1000°C and 500 MPa, respectively. In addition, the values of the standard Gibbs energy of formation are calculated from the standard Gibbs energy, ΔG_f^0 enthalpy ΔH_f^0 of formation and entropy S^o at a reference temperature (usually, 25°C, 298.15 K). Furthermore, parameters that must be considered are the partial volume V° and the heat capacity C_p^0 as functions of temperature. The algorithm OLI Systems Inc employs a Debye-Hükel term for net ionic species interactions, a modified Bromley expression (Bromley-Zemaitis) for a short-range ionic interactions and the Pitzer expression of ion-neutral molecule interactions (as described in detail in Ref. [38]). Some restrictions in the use of the Pitzer term are taken into account in a simplified version of the model algorithm, and also without considering three-body interactions. The OLI software data bank contains vast information associated with the standard thermodynamic values for several liquids, gaseous and solid substances; data from other sources should strictly be evaluated before using them. The information of solid substances is obtained from the JANAF/NBS databases, as well as from some other particular references such as Medvedev [43] and Robie [44] in Ref. [38]. For the case of ceramic materials, especially perovskite oxides, i.e. lead and alkaline-earth titanates, thermochemical data of pure substances have been reported by Brain [40].

Solubility databases of solid species in pure water and other solutions, namely alkaline and/or acidic solutions, as a function of temperature, are the best reference data to obtain the standard-state properties. These data can be taken from the books Thermochemical Data of Pure Substances and Solubilities of Inorganics and Metal Organic Compounds (Refs. 45 and 46 in Ref. [38]). However, precise and accordant thermochemical data regarding the solubility of solid compounds can be obtained by regression fitting performed using the OLI software [38,40].

The methodology proposed by Lenka and Riman [38] to determine the chemical reaction equilibria is associated with a set of feedstock precursors. For example, an organic salt (barium acetate) or an inorganic salt (barium nitrate) and an oxide containing a tetravalent cation are used for the preparation of a double cation perovskite oxide (ABX_3). The set of chemical reactions that are likely to occur in the hydrothermal process are summarized in Table 2. This table contains the equilibria equations of 37 aqueous and solid substances that might prevail in the Ba-Ti-H_2O system. One fact that deserves emphasis is related to the usage of HNO_3 and KOH because the addition of acidic or alkaline reagents is required to control the pH of the hydrothermal fluid. The presence of these solutions strongly affects the crystallization of the oxide species. The most common reagents employed to adjust the pH of the hydrothermal reaction media are alkaline solutions, such as hydroxide solutions and ammonia. In order to complete the set of modelling data, the gaseous species produced during the hydrothermal treatment needs to be calculated as well. In general, for the synthesis of perovskite compounds under specific T–P conditions, the gaseous species play a secondary role in their formation. It is necessary to remark that even a relatively simple hydrothermal system (Ba-Ti-H_2O) includes a vast variety of species, which lead to the formation of a lot of equations for resolving.

The estimated entropies corresponding to the possible reactions in the hydrothermal medium are given in Table 3. Once thermodynamic parameters are determined at room temperature condition, their values at different conditions must be determined using the approach described in Ref. [40]. The complete set of data referring to the ionic species standard-state properties are used to obtain the Ba-Ti system stability diagram. The data show the predominant phases at the conditions of T and pH as a function of the net content of Ba $\left(m_{Ba_T}\right)$ or Ti $\left(m_{Ti_T}\right)$. The standard-state values of all the species involved in the reaction system are calculated assuming that inner pressure in the hydrothermal vessels is autogenously controlled.

The iterative practice of the model using different A and B site feedstock concentrations, in conjunction with the addition of pH adjusting agents (mineralizer or solvent), allows to estimate diagrams that correlates product yield and crystalline phase stability.

$H_2O = H^+ + OH^-$	$Ba(CH_3COO)^+ = Ba^{2+} + (CH_3COO)^-$
$H_2O_{(g)} = H^+ + OH^-$	$Ba(CH_3COO)_{2(aq)} = Ba(CH_3COO)^+ + (CH_3COO)^-$
$TiO_{2(s)} + OH^- = TiOH^{3+} + H^+$	$Ba(CH_3COO)_3{}^{2-} = Ba^{2+} + 3(CH_3COO)^-$
$Ti^{4+} + H_2O = TiOH^{3+} + H^+$	$Ba(CH_3COO)_{2(s)} = Ba^{2+} + 2(CH_3COO)^-$
$TiOH^{3+} + H_2O = Ti(OH)_2{}^{2+} + H^+$	$BaNO_3{}^+ = Ba^{2+} + NO_3{}^-$
$Ti(OH)_2{}^{2+} + H_2O = Ti(OH)^{3+} + H^+$	$Ba(NO_3)_{2(aq)} = BaNO_3{}^+ + NO_3{}^-$
$Ti(OH)^{3+} + H_2O = Ti(OH)_{4(aq)} + H^+$	$Ba(NO_3)_{2(s)} = Ba^{2+} + 2NO_3{}^-$
$Ti(OH)_{4(aq)} = TiO_{2(s)} + 2H_2O$	$HCH_3COO_{(g)} = HCH_3COO_{(aq)}$
$TiO_{2(rutile)} + 2H_2O = Ti(OH)_{4(aq)}$	$HCH_3COO_{(aq)} = H^+ + (CH_3COO)^-$
$TiO_{2(anatase)} + 2H_2O = Ti(OH)_{4(aq)}$	$(CH_3COO)_{2(aq)} = 2HCH_3COO_{(aq)}$
$Ba(OH)_{2(s)} = Ba^{2+} + OH^-$	$(CH_3COO)_{2(vap)} = (CH_3COO)_{2(aq)}$
$BaO_{(s)} + 2H^+ = Ba^{2+} + H_2O$	$HNO_{3(g)} = HNO_{3(aq)}$
$CO_{2(g)} = CO_{2(aq)}$	$HNO_{3(aq)} = H^+ + NO_3{}^-$
$CO_{2(aq)} + H_2O = H^+ + HCO_{3-}$	$KOH_{(s)} = K^+ + OH^-$
$HCO_{3-} = H^+ + CO_3{}^{2-}$	$KOH_{(s)} \cdot 2H_2O = K^+ + OH^- + 2H_2O$
$BaCO_{3(s)} = Ba^{2+} + CO_3{}^{2-}$	$KCH_3COO_{(aq)} = K^+ + (CH_3COO)^-$
$BaHCO_{3+} = Ba^{2+} + HCO_3{}^-$	$KNO_{3(aq)} = K^+ + NO_3{}^-$
$BaTiO_{3(s)} + H_2O = Ba^{2+} + 2OH^- + TiO_{2(s)}$	$KTiO_{3(s)} + 3H_2O = 2K^+ + Ti(OH)_{4(aq)} + 2OH^-$
$Ba_2TiO_{4(s)} + 2H_2O = 2Ba^{2+} + 4OH^- + TiO_{2(s)}$	

Table 2. Reaction equilibria in the Ba-Ti-H_2O hydrothermal system using TiO_2, $Ba(CH_3COO)_2$ and/or $Ba(NO_3)_2$ as precursor feedstock [38,40].

	Ionic species						
	H^+	$BaOH^+$	$BaHCO_3^+$	Ba^{2+}	Ti^{4+}	$Ti(OH)^{3+}$	$Ti(OH)_2^{2+}$
ΔG_f^0 /(kJmol^{-1})	0	-716.72	-1153.5	-560.78	-354.18	-614.0	-869.56
ΔH_f^0 /(kJmol^{-1})	0	-749.35	-1207.2	-537.64	-	-	-
S^0 /(Jmol^{-1}K^{-1}	0	55.5	195.9	9.62	-456.5	-189.85	-40.8
C_p^0 /(Jmol^{-1}K^{-1})	0	44.0	-	-12.3	-	-	-
10^6 V^0/(m^{3}mol^{-1})	0	-	-	-12.6	-	-	-
lit.	0	14, 21	16	16, 21	14, 21	14, 39	14, 39

	Ionic species				
	$Ti(OH)_3^+$	OH^-	CO_3^{2-}	HCO_3^-	$HTiO_3^-$
ΔG_f^0 /(kJmol^{-1})	-1092.5	-157.30	-527.98	-586.94	-955.88
ΔH_f^0 /(kJmol^{-1})	-	-230.03	-675.23	-689.93	-
S^0 /(Jmol^{-1}K^{-1})	56.9	-10.7	-50.0	98.4	117.3
C_p^0 /(Jmol^{-1}K^{-1})	-	-	-290.8	-35.4	-
10^6 V^0/(m^{3}mol^{-1})	-	-	-21.0	102.9	-
lit.	14, 39	11	16, 21	16, 21	38

	Aqueous species			
	H_2O	$Ti(OH)_4$	CO_2	$BaCO_3$
ΔG_f^0 /(kJmol^{-1})	-237.25	-1118.38	-385.97	-1103.9
ΔH_f^0 /(kJmol^{-1})	-285.83	-1511.26	-413.8	-1196.0
S^0 /(Jmol^{-1}K^{-1})	70.0	-54.8	117.6	66.9
C_p^0 /(Jmol^{-1}K^{-1})	75.3	-50.2	243.1	-
lit.	11	14	16, 22	16

	Solid species			
	$Ba(OH)_2$	$BaCO_3$	$BaTiO_3$	BaO
ΔG_f^0 /(kJmol^{-1})	-855.17	-1164.8	-1572.4	-525.35
ΔH_f^0 /(kJmol^{-1})	-941.40	-1244.7	-1659.8	-553.54
S^0 /(Jmol^{-1}K^{-1})	108.8	112.1	107.9	70.4
C_p^0 /(Jmol^{-1}K^{-1})	97.9	-	102.5	-
a /(Jmol^{-1}K^{-1})	116.8	89.96	121.5	53.30
10^3 b/(Jmol^{-1}K^{-2})	18.44	46.28	8.535	4.351
10^{-5} c/(Jmol^{-1}K)	-18.41	-16.36	19.16	8.301
10^6 V^0/(m^{3}mol^{-1})	38.19	45.81	38.80	25.59
lit.	14, 13b, 43	16	15, 41, 37	15, 41, 44

	Solid species			
	$Ba(OH)_2 \bullet 8H_2O$	Ba_2TiO_4	$TiO_2(rutile)$	$TiO_2(anatase)$
ΔG_f^0 /(kJmol^{-1})	-2779.9	-2132.9	-890.70	-883.27
ΔH_f^0 /(kJmol^{-1})	-3328.4	-2243.0	-946.01	-938.92
S^0 /(Jmol^{-1}K^{-1})	422.6	196.6	50.3	49.9
C_p^0 /(Jmol^{-1}K^{-1})	-	-152.6	55.1	-55.3
a /(Jmol^{-1}K^{-1})	-	179.9	62.85	75.04
10^3 b/(Jmol^{-1}K^{-2})	-	6.694	11.38	0.0
10^{-5} c/(Jmol^{-1}K)	-	-29.12	-9.897	-17.63
10^6 V^0/(m^{3}mol^{-1})	144.7	-	18.82	20.52
lit.	14, 45	15, 41	16	15, 42, 44

Table 3. Relevant species in the Ba-Ti hydrothermal systems and their standard state properties at 25°C (298.15 K).

2.2.2. *Stability diagrams calculated from the thermodynamic model*

The stability diagrams have been developed for a vast number of hydrothermal systems including those research works related to the preparation of double component oxides. The stability diagram of perovskite-structured compounds, single-phase stoichiometry ABX_3 [41–47] and some relevant solid solutions [48, 49] are the most investigated applying thermodynamic modelling under hydrothermal conditions. Stability diagrams supply pertinent information regarding the ranges of equilibrium conditions at which several aqueous and/or solid species are stable in the hydrothermal reaction system. The molality of the aqueous ionic precursor includes the sum of equilibrium concentrations corresponding to dissolved ionic species in the hydrothermal media, but this term does not take into account the ionic species that precipitate from the solution. The stability diagrams also provide details related to the optimum conditions enabling incipient crystallization of the solid phases involved in the reaction. This process occurs when the ionic constituents that form the solid phase reach the supersaturation conditions in the hydrothermal fluid promoting its nucleation. However, the determination of the experimental reaction conditions required for an assumed yield of a perovskite reaction product from these diagrams is not straightforward.

The diagrams are constituted for two fields, as can be seen in Figure 3a: the solid lines indicate the boundary of incipient solid crystallization while the dashed lines denote the location where two ionic species in the liquid have equal concentrations. The stability phase diagrams are usually calculated solving the equilibrium and balance equations for the compositions of starting feedstock that accounts for the total Ba (m_{Ba_T}) as function of pH or Ti (m_{Ti_T}) as function of pH, respectively. Typical examples of these diagrams calculated for the Ba-Ti-H_2O system were produced at two different temperatures, 25 and 90°C. The latter temperature has been empirically determined to correspond with optimum for the hydrothermal processing of perovskite particles of $BaTiO_{3(s)}$ [40]. In this typical example, the stability diagrams were calculated at 25°C (lines 1) and 90°C (line 2) by using an ideal approximation, the approach

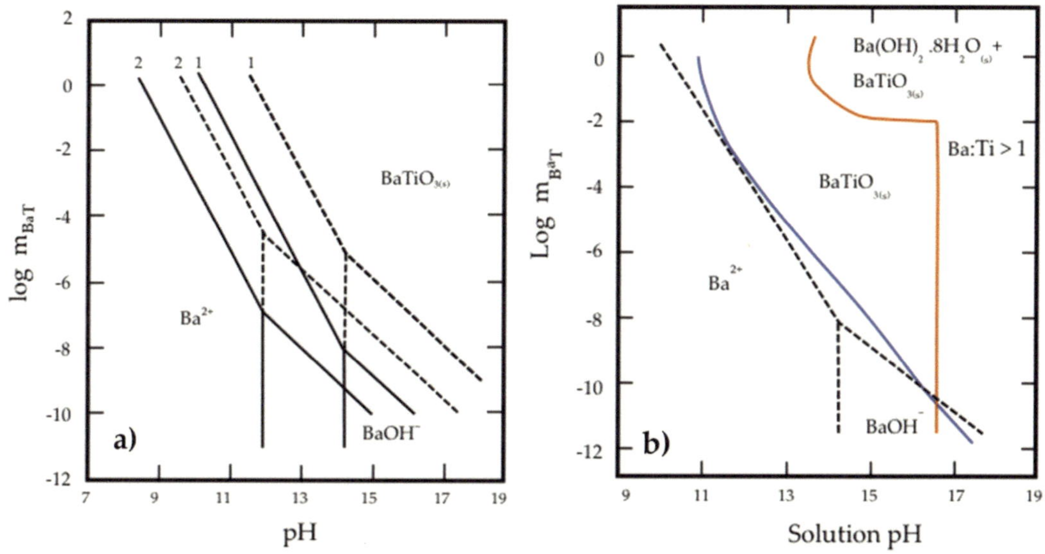

Figure 3. Ba-Ti hydrothermal system stability diagrams calculated (a) using an ideal-solution approximation at 25°C (1) and 90°C (2); the solid and dashed lines denote the results calculated using data from Barin and Naumove, Refs. 15 and 37 in Ref. [40], respectively. And (b) at 25°C using modelled activity coefficients; state standard data were taken from Barin; the dashed line corresponds to the ideal-solution results taken from Lenka and Riman [40].

assumes that all the activity coefficients of the reaction are equal to one (Figure 3a). The variation in the boundaries of stability for the phases contained in the diagram are due to the data set of standard state properties employed to solve the equilibrium of the related species. This particular diagram illustrates the effect of deviation with respect to the ideal-solution boundaries, which were calculated using data for standard state properties from two different sources. Differences on the standard data might markedly shift the boundaries between the stability regions, as is shown in Figure 3a.

Figure 3b gives the results of the complete thermodynamic modelling conducted at 25°C, in which solid lines represent the boundaries of ionic and solid species. For comparison purposes, this diagram includes the boundary (dashed line) calculated with the aid of the ideal-solution approximation. It deserves to be emphasized that significant differences existing between the phase limits calculated using both complete and simplified (ideal solution) models. However, for the ideal-solution model, the bound between ionic species of Ba^{2+} and solid $BaTiO_{3(s)}$, as well as between ionic Ba^{2+}–$BaOH^{+}$, and ionic $BaOH^{+}$–solid $BaTiO_{3(s)}$, are well defined. In contrast, the phase bound determined from the complete model do not follow straight-line functions. The curvature trend is significantly marked at higher concentrations of the aqueous species (Ba (m_{Ba_T}) >10^{-4}). Furthermore, when the solution non-ideality is considered to carry out the modelling, the bounds shifted towards higher pH values. In particular, the phase boundary between Ba^{2+}–$BaOH^{+}$ is shifted by approximately 2 pH units, while that for Ba^{2+}–$BaTiO_{3(s)}$ is changed approximately in 1 pH unit. This behaviour also occurs at higher temperatures between 100 and 200°C.

The stability diagrams do not supply information related to the experimental conditions required to promote the total reaction of the precursors. Assuming that under certain hydrothermal conditions, a total 100% yield of the crystallized reaction product must be produced. Analogous yield diagrams can establish a more practical insight of the hydrothermal processing. This tool has a great potential for establishing the optimum conditions required to achieve a total consumption of the precursors to produce high yields of the desired compound (e.g. ABX_3). Therefore, the yield analysis takes into account the initial input concentration of the precursors, because the equilibrium concentration of species in a saturated solution is of primary concern for the hydrothermal process, because the feedstock must be transformed into a phase-pure product. The yield diagrams consist of sections where a specific amount of the desired product crystallizes involving the total reaction of the precursors. In other words, where the yield product is at least equal to an assumed value, namely 99%, at the stability boundary, the product yield is very small, because only an incipient nucleation of the product occurs at this point. The yield increases as the hydrothermal reaction process is driven beyond the solubility curve into the solid–liquid region. Hence, the product yield is determined by dividing the number of moles of the product by the total number of moles corresponding to the input metal precursors [38, 41–43].

Complex phase diagrams derived by thermodynamic modelling can be developed following the theory and methodology described above. These diagrams include both stability and yield fields and are calculated using the ideal solution and complete models. However, single stability and yield diagrams are portrayed against the solution pH, because this is an independent variable associated with the hydrothermal processing. The concentration of a pH-adjusting solution required to achieve the crystallization of the desired product is not indicated in the diagram. One typical example was derived for the Sr–Zr system, and it is shown in Figure 4. The stability and yield diagrams were calculated for the preparation of $SrZrO_3$ employing strontium hydroxide and zirconium oxide as precursors at 200°C. The diagram was determined at a fixed mixing Sr/Zr ratio of 1 which is plotted as a variable in the y coordinate axis, while the molality of the alkaline (KOH) is portrayed in the x coordinate axis. The solid line in the diagram (Figure 4) indicates the starting point of the crystallization of the $SrZrO_3$ perovskite phase, although $SrZrO_3$ and ZrO_2 might coexist in the field above the solid line as the reaction tends to move closer to the shaded area. The remaining content of ZrO_2 is gradually consumed leading to an increased amount of $SrZrO_3$ produced. The shaded area is related to the optimum conditions, in terms of precursor feedstock: molality of $Sr(OH)_2$ and KOH required to obtain the pure-phase $SrZrO_3$ at a yield above 99% at the treatment temperature of 200°C [40–46].

The thermodynamic model of heterogeneous aqueous ionic liquid is a practical tool to prepare stability and yield diagrams. This powerful tool capable of estimating the effect of various processing conditions is likely towards to conduct the crystallization of smart perovskite ceramic materials in a more cost-effective way. In particular, it is possible to determine adequate conditions such as reaction temperature, pH, input precursor concentration and mixing ratios. The theoretical predictions allow researchers to formulate processing guidelines for finding optimum synthetic conditions to produce perovskite-type multicomponent compounds.

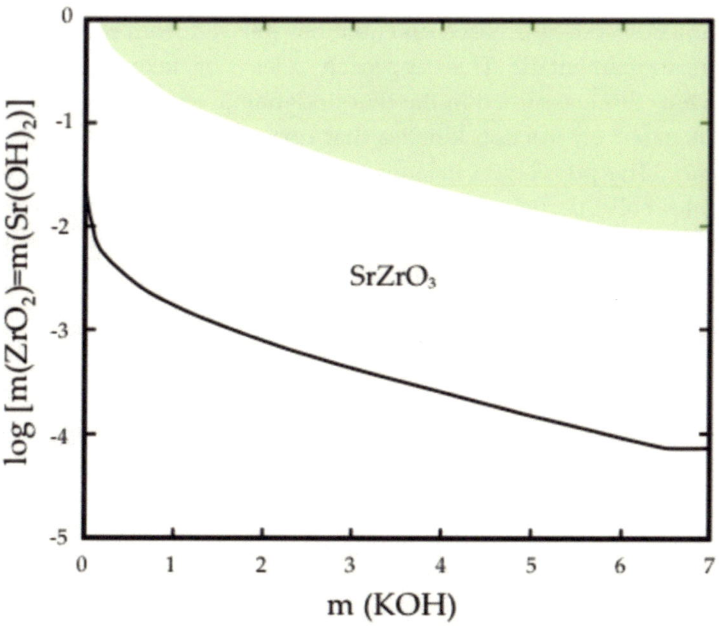

Figure 4. Sr-Zr system phase stability and yield diagram calculated as a function of the solvent content ([KOH]) at 200°C for a precursor mixing ratio Sr/Zr = 1; the precursor source of strontium selected was $Sr(OH)_2$ [40].

2.3. New model based on kinetic precipitation and balance population approaches

2.3.1. Hydrothermal crystallization of ABX_3 based on kinetic and population balance equation

New models based on kinetic and population balance equations have received a tremendous attention. In the past decade, academic and industrial communities have devoted efforts to propose the models and evaluate their functionality to estimate the factors affecting the particle formation in a wide variety of processes. Additionally, the control of particle size of the final particles prepared via hydrothermal crystallization constitutes a potential advantage in comparison with other synthesis processes. Under hydrothermal conditions, the particle size distribution that grows from a solution strongly depends on the nuclei formation rates and subsequent particle growth. At an accelerated nucleation rate, the total number of particles produced is large while their size is relatively small. In a particular hydrothermal reaction system, the rates of nucleation and growth depend on supersaturation. Processing parameters such as precursor feedstock concentrations, temperature, and mixing conditions typically influence the supersaturation. Therefore, the final particle size distribution can be tailored by adjusting the hydrothermal synthesis conditions.

In the past decade, various efforts were conducted in order to present different approaches accounting for the phenomena related with the particle formation of perovskite-structured compounds. Testino et al. have proposed the first meticulous kinetic approach devoted to the formation of $BaTiO_3$ particles, the particle formation is based on a solution precipitation reaction [50–51]. The new model derived for the crystallization of $BaTiO_3$ particles in an aqueous system is based on equations derived from basic nucleation, growth and aggregation fundamentals. This approach does not take into account empirical relationships such as those assumed in the thermodynamic equilibrium models. In contrast, the new model is based on reaction kinetics that correlates a mass balance coupled with a population balance. The population balance concerns the evaluation of some entities in a reaction system, i.e. solid particles or, events that might dictate the particle crystallization behaviour in the hydrothermal system under study. The performance of individual entities depends on variables associated with an appropriate reaction environment; consequently, the population balance equation might be correlated with balance equations that include the environmental variables [50, 52, 53].

The theoretical framework of the kinetic model was conceived by the exhaustive preliminary investigation of the aspects associated with the formation of $BaTiO_3$ particles in an aqueous system [51]. The concentration of barium in the hydrothermal system promotes a variation in the rate of formation of $BaTiO_3$ particles. When the content of Ba^{2+} increases, it decreases the particle size of the crystallized particles, and it also affects the chemical reaction when the temperature is varied; this effect has not been clarified. The model was developed taking into consideration the formation kinetics of perovskite $BaTiO_3$ particles from dilute solutions of $BaCl_2$ and $TiCl_4$ ($\leq$0.1 M) at a value of pH = 14 between 80–90°C. The overall chemical reaction experimentally studied is as follows:

$$BaCl_{2(aq)} + TiCl_{4(aq)} + 6NaOH_{(aq)} \rightarrow BaTiO_{3(s)} + 6NaCl_{(aq)} + 3H_2O_{(l)} \quad (7)$$

where (aq) denotes the hydrolyzed salt in water. The new kinetic model that is applicable to the preparation of different perovskite compounds was correlated with experimental data collected at various alkaline media concentration, temperature and Ba/Ti ratio. The preliminary experimental analysis allowed the researchers to establish the stages and details of the mechanism associated with the nucleation, growth and particle aggregation [50, 51]. In this reaction system, the influence of non-ideality of the aqueous solution, net particle superficial area and thermodynamic properties were not considered.

In agreement with the analysis conducted by Testino et al., chemical reaction 7 occurs in two steps. (i) In the first step, the formation of a titanium hydroxide gel (THG) phase proceeds rapidly. Consequently, (ii) a slower reaction takes place between the THG phase and the Ba^{2+} ions dissolved in solution. When the supersaturation state of solute has reached under the second step of the reaction, the $BaTiO_3$ nuclei spontaneously precipitate. The $BaTiO_3$ crystallization mechanism controls the overall kinetics involved in the hydrothermal synthesis process. The kinetic study of reaction 7 determined that the evolution resembles sigmoidal curve behaviour, and the particle growth progress observations strongly support a nucleation and growth mechanism occurring in a second step. The global reaction proceeds via a two-step reaction mechanism, which achieves the formation of $BaTiO_3$ via solution precipitation in mild alkaline conditions [50, 51]. Furthermore, the global mechanism also applies to the hydrothermal synthesis of barium titanate at temperatures above 100°C. When the THG suspension obtained by adding an alkaline media (NaOH or KOH) to the aqueous solution containing both $BaCl_2$ and $TiCl_4$ is used as the precursor; the same process is likely to occur when amorphous titanium hydroxide is produced by hydrolyzing a titanium (TiO_2) precursor in a $Ba(OH)_2$ solution. Hence, the global reaction involving both nucleation and growth of the perovskite is given by the following reaction

$$Ti(OH)_{4(aq)} + Ba^{2+}{}_{(aq)} + OH^{-}{}_{(aq)} \rightarrow BaTiO_{3(s)} + 3H_2O_{(l)} \tag{8}$$

It can be assumed that local equilibrium conditions are carried on at the surface of the growing crystal, involving the equilibrium between the Ba^{2+} and $BaOH^{-}$ ionic species, which are correlated by Eq. (9)

$$Ba^{2+}{}_{(aq)} + OH^{-}{}_{(aq)} \rightarrow BaOH^{-}{}_{(aq)} \tag{9}$$

The supersaturation state associated with reaction 8 is interpreted as

$$S = \frac{\left[Ti(OH)_4\right]\left[Ba^{2+}\right]\left[OH^{-}\right]^2}{K_S} \tag{10}$$

The supersaturation solution grade is independent of the solvent species involved in the precipitation reaction; therefore, the equilibrium conditions are kept at the solid/liquid interface for an extended period. The $Ti(OH)_4$ species are provided by the THG precursor phase dissolution. A global mechanism that proceeds in two steps conducts this process. At the first step, the disruption of the polymeric network of $Ti(OH)_4$ takes place by the nucleophilic attack of the OH^- ions occurring at Ti atom positions (denoted as Ti*, in Eq. 11). The second step of reaction comprises the reaction between oxygen ions (O*) produced during the decomposition of $Ti(OH)_4$ with water molecules, restoring the hydroxyl ions. The dissolution rate, which is reversible and rate controlled at the first step, can be expressed by

$$\frac{d\left[Ti_{gel}\right]}{dt} = k_1\left[Ti^{*}\right]\left[OH^{-}\right] - k_{-1}\left[O^{*}\right]\left[Ti(OH)_4\right] \tag{11}$$

where the term $[Ti_{gel}]$ corresponds to the number of moles of titanium contained in the gel precursor in the reacting suspension. Additionally, k_1 and k_{-1} are correlated to the kinetic constants of the direct and reverse process, respectively. When [Ti*] and [O*] are proportional to $[Ti_{gel}]$, Eq. (8) can be rewritten as

$$-\frac{d\left[Ti_{gel}\right]}{dt}=k_1\left[Ti_{gel}\right]\left[OH^-\right]\left\{1-\frac{K}{K_S}\left[Ti\left(OH\right)_4\right]\right\} \tag{12}$$

where k is the rate constant for THG dissolution, and K is the reciprocal of the equilibrium constant overall reaction

$$TiO\left(OH\right)_{2(gel)}+Ba^{2+}{}_{(aq)}+2OH^-{}_{(aq)}\rightarrow BaTiO_{3(s)}+2H_2O_{(l)} \tag{13}$$

This general approach likely describes the mechanism applied to the classical hydrothermal process, in which crystalline titanium oxide particles (TiO_2) are suspended in aqueous $Ba(OH)_2$. In this reaction system, the net surface area of the titanium oxide particles affects Ti^{4+} ionic species produced via the dissolution of TiO_2 powder and its dissolution rate.

2.3.2. Mass balance expression

The complete model expression is constituted by two additional terms; the first one takes into account a global mass balance of the ionic species involved in the formation of the ABX_3 single phase. The mathematical expression that involves the mass balance of one of the precursors, i.e. titanium ions, in the hydrothermal system can be written as [50]

$$-\frac{d\left[Ti_{aq}\right]}{dt}=\frac{d\left[Ti_{BT}\right]}{dt}+\frac{d\left[Ti_{gel}\right]}{dt} \tag{14}$$

In this differential equation, the term $[Ti_{aq}]$ indicates the concentration of the aqueous titanium species present in the reaction system, which typically is assigned to the $[Ti(OH)_4]$. Likewise, the content of the precipitated $BaTiO_3$ at a certain time of reaction is $[Ti_{BT}]$. Thus, the first term in Eq. (14) is correlated with the $BaTiO_3$ particle formation rate promoted by nucleation and growth events.

$$\frac{d\left[Ti_{BT}\right]}{dt}=\left\{\hat{B}+\hat{G}\right\} \tag{15}$$

Eq. (16) corresponds to the global equation that combines the nucleation and growth with the dissolution rate of the THG given by Eq. (12). This expression involves the crystallization rate of $BaTiO_3$ by nucleation and growth events coupled with the simultaneous consumption of the THG phase, which is analogous to the yield product term in the thermodynamic models. If the Ba/Ti molar ration in the THG is denoted by R_{gel}, the mass balance can be given by Eq. (17), in which $[Ba_{aq}]$ is the absolute concentration of Ba ions corresponding to the sum of $[Ba^{2+}]$ and $[BaOH^+]$ species.

http://dx.doi.org/10.5772/61568

$$\frac{d\left[Ti_{aq}\right]}{dt} = \mathrm{k}\left[Ti_{gel}\right]\left[OH^{-}\right]\left(1 - \frac{K}{K_S}\left[Ti(OH)_4\right]\right) - \left\{\hat{B} + \hat{G}\right\} \tag{16}$$

$$\frac{d\left[Ba_{aq}\right]}{dt} = R_{gel}\frac{d\left[Ti_{gel}\right]}{dt} - \left\{\hat{B} + \hat{G}\right\} \tag{17}$$

2.3.3. Population balance

The population balance equation (PBE) is related to net particle number continuity. It was the concept introduced to complete the kinetic model. The last term takes into account the evolution of the particle size distribution for a particular cluster of particles subjected to nucleation, growth, coalescence and disruption processes. According to Testino et al. [50], the expression can be given as

$$\frac{\partial n(L,t)}{\partial t} + \frac{\partial}{\partial L}\left[G(L,t)n(L,t)\right] = B(n) - D(n) \tag{18}$$

In the PBE equation, $n(L,t)$ corresponds to the density of particles at a particular time t in the particle size range L to $L+\mathrm{d}L$. While G represents the isotropic growth rate, B and D are the particle rates of genesis and extinction, respectively. The net particles number per unit volume is defined at an infinitesimal period as dN, and the density function is $n = \mathrm{d}N/\mathrm{d}L$. Equation 15 is related to the rate of particles accumulation in the range L to L+dL. This range is a function of the particle growth rate and specifically of the speed at which particles of that size are directly created or removed. In order to solve this equation, the kernel aggregation function coupling with the PBE and four particle size distribution moments were considered. Because these are the quantities used to solve PBE problems [50, 52–54].

The new kinetic model derived for $BaTiO_3$ particles is likely applicable to a broad number of perovskite systems, following the methodology described above. In either case, a raw dried titanium hydroxide or Ti-gel formed as an intermediate product are used as a precursor reactant. The nucleation and growth theory for supersaturated solutions correlates the rates of nucleation and growth with some physical parameters such as temperature, the surface tension of the solid phase and diffusion coefficient. In the model, nucleation is considered to occur in two steps, in the initial stage a homogeneous nucleation proceeds in the supersaturated solution. After the conclusion of the preliminary step, a secondary nucleation is then enhanced promoted by the variation of solute saturation in the hydrothermal media. The secondary nucleation controls the creation of new embryos on the surface of $BaTiO_3$ particles, which are already present in the suspension as a result of nucleation and growth events. The secondary nucleation promotes the rapid increase in the $BaTiO_3$ formation rate after the initial period dominated by primary nucleation. Moreover, secondary nucleation leads the poly-crystalline nature of the final particles. Diffusion-controlled growth is the process involved in the coarsening of the elementary crystals that constitute the polycrystalline particles. The new algorithm has a potential for calculating the net number of events (moments) under particular reaction conditions. These events are of particular concern because these define the particles' aspect such as total number, dimension, surface and volume of particles per liquid suspension unit volume. The new model based on a PBE analysis yields an uncertain approximation of the real process, but hinders the introduction of completely empirical relationships between growth rate and supersaturation state.

In general, one limitation of the kinetic model is the use of PBEs considering one internal variable, i.e. the size of the primary crystallites. Consequently, a more meticulous approach proposed takes into account the use of PBEs with at least two internal coordinates, for example, the size of the primary particles and the size of the polycrystalline particles. New approaches might provide a better description by taking into account the interactions between the elementary crystallites affecting the overall precipitation process. However, the application of PBEs with more than one internal coordinate to precipitation problems is still at an early stage of development. Recently, Marchisio has presented a detailed mathematical model involving the precipitation of $BaTiO_3$ in aqueous solution [54]. This particular model is based on a bivariate population balance equation. The new approach overcomes the limitations considered for a mono-variate population balance equation, namely the crystallite size and particle size. The new improved model (not reported here) shows qualitative results similar to those previously discussed for the mono-variate model. However, quantitatively the modelling results are very different from the predictions obtained with the mono-variate PBE. This complete model can be used to identify new model parameters associated with the mechanism involved in the nucleation, growth and aggregation stages of $BaTiO_3$ hydrothermal synthesis [54].

In the past decades, an increasing interest in the production of submicron powders with uniform particle size distribution (PSD) through solution precipitation methods, including the hydrothermal powder processing, has been the main concern in the chemical engineering field, particularly, the search for the adequate operating conditions for preparing perovskite nanoparticles, similar to those examples described in the following section. However, modelling works on thermodynamics coupled with the kinetic aspects of particle formation and its evolution have also been conducted by various researcher groups in different areas [38–42, 50–54]. The motivations for developing new optimized mathematical models are multiples. Indeed, the mathematical models described in the present review can be used to estimate the hydrothermal conditions for particle formation and the evolution of mechanisms associated with this process. Also, these can be used to set up the optimal operating conditions to prepare a powder with desired characteristics, following a product engineering philosophy. Finally, there is an enormous potential to employ both thermodynamic and kinetic models, by implementing them in computational fluid dynamics models, to scale up the precipitation processes from laboratory to an industrial scale. However, new researchers in the hydrothermal field must bear in mind that some caution must be exercised, when applying the proposed models to diverse perovskite systems, because modelling results might vary depending on the reaction temperature and concentration of precursors. These factors might lead to erroneous conclusions associated with the reaction mechanism operating for a particular reaction system.

Moreover, the models assume that during the nucleation and growth events, the hydrothermal system is perfectly mixed, leading to a precipitation rate being controlled by the chemical reaction kinetics. Although these inferences are acceptable in small laboratory vessels, these might not operate for larger industrial reactors, where mixing problems might frequently rise due to slow and less-efficient agitation systems. The precipitation rate is one of the problems that commonly meets in the scale-up of precipitation processes.

2.4. Hydrothermal processing of perovskite-structured ceramic oxides

According to the literature survey, the synthesis under hydrothermal conditions of different types of perovskite compounds was triggered during the 20th century. Nowadays, this synthesis method constitutes an important tool for materials processing. Because hydrothermal synthesis offers some processing advantages, which enhances the preparation of monodispersed nanoparticles with controlled size and morphology. In the present review, the experimental details for the synthesis of a broad number of perovskites are addressed based on their compositional aspects, namely single substitution in either A or B sites. Some double perovskites and their related solid solutions are also included.

The synthesis of single-phase stoichiometric perovskite ABX_3 has been possible under hydrothermal conditions by the employment of a broad number of atomic elements. Among the elements incorporated at the A site of the perovskite structure are the elements of IA (K, Ag, Na) and IIA (Ca, Sr, Ba) groups of the periodic table. Some bivalent elements (Bi, Pb, Rb) have been incorporated in this site as well. Experimental results demonstrated that trivalent rare earth elements, namely La, Pr, Nd, Sm, Eu, Gd, Tb, Dy, Ho, Er, Tm, Yb, Lu or Y, are able to be located at the A site, while the B site could be occupied by stable transition metals elements (Ti, Fe, Zr, Ta, Nb, Bi, Sn, Co, Ru, Cr and Mn). The next sections of the present review give an account of the main experimental aspects related to the hydrothermal processing of perovskite-structured ceramic oxides. Particular emphasis is placed on the practical aspects associated with the selection of mineralizers and their concentrations, solution pH, choice of precursors, and reaction intervals and temperatures. This review intends to present a more realistic and practical state-of-the-art on the optimization of the synthesis of the major perovskite-structured compounds that have been extensively produced via hydrothermal processing. It deserves emphasis that succeeding on the crystallization of pure single phase perovskite oxides under hydrothermal conditions depends on major functional parameters such as the nature of precursors and mineralizer. The main aspects related to the hydrothermal processing of perovskites are discussed in terms of the type of the site substitution, focusing only on single-phase perovskite oxides. Starting from the synthesis of major A-site-ordered perovskite-structured oxides ($ATiO_3$, $AFeO_3$, $AZrO_3$, etc.) and a few solid solutions, and subsequently proceeding to the synthesis of B-site-ordered perovskites and, finally, to the synthesis of A and/ B-site-ordered perovskite oxides.

2.4.1. Synthesis of A-site-ordered perovskite titanates ($ATiO_3$)

In recent years, much progress has been made in the synthesis of A-site-ordered perovskite titanates ($ATiO_3$; A= Ba, Pb, Ca and Sr; hereafter the compounds containing the latter cations are referred as BT, PT, CT and ST, respectively) under hydrothermal conditions. In the synthesis of BT particles, various authors have employed different methodologies to prepare powders with perovskite structure under hydrothermal conditions. These methods, in general, enable us to synthesize this kind of perovskite oxides in various aqueous alkaline solutions using various mineralizers. One experimental procedure developed for the synthesis of BT powders use $Ba(OH)_2$ as the precursor of barium and pH adjustment agent, instead of using a strong alkaline mineralizer (KOH or NaOH). The BT particles synthesized at a low temperature of 80°C in a solution with a pH of 8 exhibited a spherical morphology and cubic ABX_3 structure. The chemical reaction promoted under hydrothermal conditions involved the use of chemically modified Ti-peroxo-hydroxide and $Ba(OH)_2$ precursors. The Ti-peroxo-complex precursor was rapidly dissolved in the reaction media ($Ba(OH)_2$), subsequently enhancing the crystallization of BT particles, providing fast and mild powder synthesis conditions [55].

The crystallization of BT particles was found to be carried out with similar morphologic and structural aspects when $BaCl_2$ is employed as a precursor and using NaOH as mineralizer [56]. The optimal concentration of the NaOH mineralizer to conduct the synthesis was 6 M; it allowed to obtain the BT perovskite single phase between the range of 120 and 200°C for reaction periods of 1–4 h. Reaction kinetics influences the crystallization of BT at low temperatures and short reaction times. However, the thermal equilibrium plays a crucial role at elevated temperatures (200°C) and/or prolonged reaction times (4 h) that control the particle crystallization. In other study conducted under similar conditions of synthesis [57], different metal titanates ($ATiO_3$, A= Ba, Sr, Ca), among them BT particles, were also prepared. For this case, the reaction synthesis was developed at highly alkaline conditions with a value of pH > 12. The excess amount of NaOH provided an alkaline environment for the dissolution of TiO_2 and its subsequent reaction to produce BT. They found that the formed Ti-OH complexes reacted at 160°C for 72 h with the Ba^{2+} ions in the solution to form the single-phase BT perovskite oxide.

Other methodologies have been employed to prepare BT powders exhibiting morphologies and particle size, using KOH as a mineralizer. Recently, nanocrystalline particles with a cubic shape of BT were prepared using organic compounds; these promoted the control of the particle size during the crystallization step. In this case, the hydrothermal media consisted in a KOH solution prepared with 11 mmol of reagent (2.2 M) [58]. In contrast, different perovskite-structured BT oxides were prepared from altered titanium isopropoxide and barium acetate salts controlling the pH of the suspension adding a 2.0 M KOH solution [59]. Under these conditions, phase-pure perovskite BT powders were successfully synthetized at 150°C for 18 h in an extremely alkaline feedstock media. The pH of the suspension was 13; it was concluded from this study that a strong alkaline solvent is necessary for the production of pure-phase perovskites under hydrothermal conditions. In a subsequent study, the effect of varying the molar [KOH]/[Ti-isopropoxide] ratio during the reaction synthesis was investigated. The principal aim was to determine the impact of this processing factor in the crystallinity and particle size of the BT particles [60]. In this particular reaction system, the nucleation and growth processes are affected by the electric charge micelles produced on the surface of the BT particles during the hydrothermal process. The increment in OH^- concentration caused by dissolution of KOH in the reaction medium improves the tendency to BT particles separation, because the generation of electrostatic repulsion forces acting in-between negatively charged surfaces hinders the formation of BT agglomerates [60].

Later on, Xiao and co-workers [61] developed another methodology to synthesize BT powders with spherical morphology. In this approach was involved the usage of a surfactant agent such as ethylene glycol in a 1 M KOH solution. After 200°C for 12 h, the experimental results showed evidence of the formation reaction by-product, namely traces of $BaCO_3$, which was removed by acid washing. The synthesis of BT with dendritic shapes without any surfactants was conducted using KOH solutions with various concentrations of 0.1, 0.3, 0.7 and 1 M, keeping other reaction conditions unchanged (200°C, 12 h). It was found that the KOH concentration was a crucial parameter that enhances the formation of the BT dendrites [62]. When the concentration of KOH increased, the particle morphology changed from dendritic to a sphere-like shape. This phenomenon was explained based on the fact that dendritic growth tends to take place as the system is driven farther away from equilibrium. Therefore, when low KOH concentrations were used as solvent, the alkalinity of the media did not allow the system to reach the equilibrium conditions. This fact promoted the formation of the BT dendrites rather that the sphere-shaped particles [62].

Under microwave-hydrothermal conditions, the synthesis of BT powders was accelerated in comparison with conventional hydrothermal synthesis. Zhu et al. [63] satisfactorily synthesized perovskite BT nanowires in different reaction media using pure water and mixtures of water–ethylene glycol solutions. The pH value of the resultant mixture was adjusted to 14.0 adding 1.1217 M of KOH before the hydrothermal treatment. A substantial reduction of the reaction time was achieved using a highly concentrated alkaline solvent, allowing the synthesis of pure-phase BT particles to proceed for 50 min at 150°C. The alkalinity of the hydrothermal media had a significant influence on the crystal structure of the single-phase BT and affected the morphology of the BT powders produced by hydrothermal-microwave-assisted treatments [64]. The BT nanoparticles prepared under microwave-assisted sol-hydrothermal conditions exhibited a varied morphology accounting from elongated particles to a ring-like shape. The differences in particle morphology occurred by changing the KOH concentration from 0.25 to 5 M. The reaction involves the metal hydrous complex gel formation that is promoted by dissolution of the precursors, and the subsequent recrystallization when the solvent reaches the supersaturated state. The nuclei were rapidly transformed into BT nanoparticles because of the high pH and high reaction temperature. The nanoparticles that resulted from the high saturation of the reaction solution dissolved in the solution forming additional nuclei avoiding a marked particle growth [64].

On the other hand, the synthesis of PT particles has been extensively investigated under different hydrothermal conditions. Ohara et al. [65] conducted one of the pioneering research works for preparing PT perovskites. The optimum conditions for preparing pure PT single-phase fine fibers were 150°C for variable reaction times (24 to 72 h). The PT fibers were produced selecting a molar Pb/Ti ratio = 1, and Ti ions were supplied by adding potassium titanate $2K_2O \cdot 11TiO_2 \cdot 3H_2O$ as the precursor material. These authors also demonstrate that the hydrothermal technique provides the possibility to prepare PT fine powders with spherical morphology. The morphological particle transition from fibrous to spherical was obtained via an ion-exchange reaction between Pb^{2+} ions and K^+ ions. This phenomenon occurs under hydrothermal conditions at high pH, 150°C, and pressure (0.4–0.5 MPa). The formation of the PT phase requires a Pb^{2+} concentration higher than that of K^+ in the system. Likewise, by employing, additionally to the KOH concentration, other mineralizer salts, such as $NaNO_3$, KNO_3 and $LiNO_3$ in the reaction media, it was possible to obtain tetragonal perovskite PT nanosheets [66]. The synthesis of pure perovskite PT with tetragonal structure was carried out in a suspension with KOH and $NaNO_3$ concentrations of 1 and 4 M, respectively. According to these studies, it was found that the PT synthesis preferentially occurred under stoichiometric equilibrium conditions.

Miscellaneous research works were carried out to investigate the favourable conditions for synthesizing CT and single-phase RCT doped with rare earth elements (R = Eu) perovskites. These were compounds prepared using the soft chemical process microwave-hydrothermal method without the usage of organic surfactants [67, 68]. The synthesis of pure-phase orthorhombic CT and RCT was successfully carried out in a reaction media with a pH value of 14 using KOH as the mineralizer. The presence of the alkaline solution promoted the co-precipitation of a complex hydroxide constituted by the $TiO(OH)_2$-$Ca(OH)_2$-$Eu(OH)_3$ during the early stage of the reaction. During the formation of either CT or RCT compounds, the direct rotational water excitation that is achieved by microwave radiation produced the release of uncoupling OH^- groups from the complex hydroxide. Consequently, calcium and $Ti(OH)_4$ clusters rapidly interact themselves in the aqueous media as a result of the preliminary OH removal process. Therefore, at a pH $\geq$ 9, the heating promoted by microwave radiation achieved the CT and RCT nucleation, respectively. Under these conditions, the crystallization kinetics of both perovskite-structured compounds was accelerated one to two orders of magnitude. Thus, it was presumed that the diffusion of calcium and titanium clusters is higher under hydrothermal conditions than at ambient pressure and temperature. This phenomenon was explained to proceed by effective particle collisions producing irreversible oriented attachments that offer favourable thermodynamic and kinetics conditions for the crystallization of CT and RCT, in addition to the particle shape control.

Regarding the synthesis of ST perovskite, it was conducted in KOH solutions with concentrations as low as 0.1 M. The ST nuclei precipitated incorporating mild concentrated $Sr(OH)_2$ solutions in the hydrothermal system, and many Sr^{2+} ions remained in solution after the crystallization process. However, a large amount of $Sr(OH)_2$ remained when the KOH concentration was increased up to 1 M, and the mass transport rate of ions that feed the growing crystals was accelerated. Therefore, a variety of crystal surfaces have the opportunity to grow due to the relatively larger rate of ionic mass transport, which allowed the formation of cubic-like-shaped ST aggregated crystals [69]. Figure 5 gives the typical micrographs of ST perovskite powders hydrothermally prepared at 250°C for various reaction intervals [70]. Finally, the hydrothermal synthesis of some solid solutions was satisfactorily carried out under conditions similar to those mentioned above. Wei et al. [71] prepared pure-phase $Pb_{0.70}La_{0.30}TiO_3$ fine powders with cubic structure by employing co-precipitated Pb-La-Ti-OH_X hydroxide and KOH solutions with different concentrations of 2, 4 and 6 M at 220°C for 36 h. Additionally, in another set of experiments, the synthesis of $Bi_{0.5}Na_{0.5}TiO_3$ (BNT) was found to occur in highly concentrated alkaline hydrothermal media of 12 M NaOH. This concentration is the minimum required to achieve the crystallization of single-phase BNT particles at a lower temperature as 160°C for a short time of 3 h [72].

Figure 5. SEM Micrographs of ST perovskite obtained under hydrothermal conditions at 250°C in a KOH solution (5 M) after (a) 0.08, (b) 1 and (c) 24 h [70].

2.4.2. *Synthesis of ferrite type perovskite (AFeO$_3$)*

Recently, the synthesis of multiferroic materials such as perovskite BF has received intensive scientific attention because of its magnetic and ferroelectric properties. A particular interest has been paid for preparing submicron BF powders with the assistance of NaOH mineralizer at lower temperatures (150–190 °C). The reactant reagents used were $Bi(NO_3)_3$ and $FeCl_3$ [73]. In this particular case, the crystallization of the perovskite-structured BF powders with hexagonal structure occurred in low concentrated NaOH solutions, 0.03–0.12 M, at 170°C for 16 h. The BT perovskite phase crystallization and particle morphology were affected by processing parameters such as solvent solution concentration, reaction temperature and time. The parameter that strongly affects the crystallization step was the concentration of the solvent solution of NaOH. At a low concentration of NaOH media, below 0.03 M, a large number of water molecules and a few $[Na(H_2O)_n)]^+$ cations coexist in the aqueous media. The presence of these species reduces the rate of the crystallization of the amorphous precursor colloid phase that contains Bi_{3+} and Fe_{3+} ions. Therefore, the principal mechanism associated with the nucleation and growth stages involves an in-situ gel transformation process. The amorphous colloid gel transformation occurs under early and intermediate step of the crystallization process. This fact is likely to proceed when water molecules are removed from the structural gel network enabling the nucleation of small embryos. The gradual increase of temperature and solvent concentration promotes the presence of a few number molecules of water together with a large number of $[Na(H_2O)_n]_+$ cations. These species accelerate the dissolution rate of the amorphous precursor, enhancing a rapid dissolution–recrystallization process. Once the solution was supersaturated, nucleation and crystallization took place faster in the saturated solvent solution [73].

The control of the BF particle morphology can be achieved under hydrothermal conditions. BF pure-phase microplates with rhombohedral structure were preferentially synthesized at 200°C for 8 h, using $C_6H_{10}BiNO_8$ as a Bi precursor reactant and particle surface modifier [74]. The synthesis was carried out in a suspension obtained adding a 0.4 M KOH solution. The formation of the perovskite BF rhombohedral structure was achieved by a self-assembly process coupled with coarsening stage promoted by the Ostwald ripening growth mechanism. A process comprising different steps explained the production of the BF pure phase. The nuclei are generated in the supersaturated solution yielding the growth of $Bi_{25}FeO_{40}$ spherical-shaped nanoparticles during the first of the reaction. Then, these spherical nanoparticles underwent an oriented aggregation due to the preferential 2D plane growth that transformed the spheres into BF nanosheet particles. The adsorption of the $C_6H_{10}BiNO_8$ might modify the formation mechanism because this organic compound prevents the contact between the facets on which the adsorption selectively occurred. Finally, the gradual particle growth produced the formation of microplates that occurred via a repeated dissolution–crystallization mechanism.

The hydrothermal synthesis of BF has also been performed using highly concentrated alkaline solutions. A study directed towards to produce large-scale polyhedral BF particles was conducted at 200°C for 12 h using KOH solutions with concentrations varying between 1 to 9 M [75]. The total OH^- ions concentration affected the agglomeration rate and the particle morphology control. The variation of KOH solvent concentration, reaction time, heating and cooling rates were considered to determine the particle morphology changes of the BF submicron particles. The formation of the large-scale BF polyhedron occurred due to the selective particle aggregation that was promoted via the dissolution–crystallization mechanism. When the concentration of solute reached the critical supersaturation required to enhance the embryo nucleation, the hydrothermal conditions simultaneously preceded the nucleation and growth of small crystallites and subsequently these BT particles nanoparticles underwent a marked agglomeration. Likewise, at low KOH concentrations, the solvent is not capable of controlling the particle surface energy, in consequence the BF nanoparticles underwent a marked agglomeration forming isotropic sphere aggregates. However, when saturated KOH solutions were used, the KOH favoured the control of the particle surface energy, and the particle growth was dominated via an oriented attachment process, which conducted the formation of euhedral-shaped particles.

In a different set of hydrothermal experiments, the crystallization of BF particles was obtained 220°C for 6 h in a 4 M KOH solution [76]. In contrast, above a critical KOH concentration, the formation of secondary impurity phases was favoured. Likewise, the formation of pure BF was strongly dependent on the KOH solvent concentration. During the progress of the hydrothermal treatment, the hydroxides species of $Bi(OH)_3$ and $Fe(OH)_3$ were further dissolved in the solvent (KOH), and those reacted at high temperatures and pressures. The solvent saturation provoked the precipitation of the chemically stable ceramic oxide particles. The particle coarsening continues gradually as far as the hydrothermal system maintains a steady supersaturated state. Therefore, the dissolution and crystallization process progressed in the supersaturated fluid because the system tends to reach stability by itself. It was argued that the dissociation of bismuth and iron hydroxide coupled with the formation of complexes ionic could hinder the growth of BF crystallites and limit the size of BF particles to the submicron range. Other studies related to the synthesis of BF where performed for Jiang et al. [77]. These authors carried out the hydrothermal synthesis of BF single crystals using hydrate sodium bismuth oxide ($NaBiO_3\,nH_2O$), $Fe(NO_3)_3$ and KOH. Rhombohedral-structured perovskite was obtained with a K/Bi molar ratio of 180 at 180°C for 7 days. However, chemical analyses showed that the molar ratio of the reaction product Bi:K:Fe was equal to 0.96:0.03:1.00, also detecting a small amount of potassium. From these results, the chemical composition of BF large crystals could be deduced to be $Bi_{0.96}K_{0.03}FeO_3$. Taking into account the starting compound with Bi^{5+}, the charge balance of this compound may be obtained by the presence of a mixed valence Bi atoms, resulting in the formation of a compound of $Bi^{3+}{}_{0.915}Bi^{5+}{}_{0.045}K^{+}{}_{0.03}Fe^{3+}O_3$.

On the other hand, the synthesis of BF particles has been possible to carry out at low temperatures with the aid of polymers and surfactants such as PVA [78]. The nanoparticles with spherical morphology have an average diameter of about 10 nm; these were satisfactorily synthesized at 160°C for 9 h in a suspension containing 15 ml of PVA. In a second run of experiments a KOH solution without surfactant addition was used as solvent. The pH value of the residual solution after the hydrothermal treatment was 8; this value is high because a highly concentrated KOH (12 M) was added prior the treatment. The polymer decreased the growing speed of BF nuclei enhancing the formation of nanoparticles. In accordance with the former results, the authors concluded that the polymer addition limited the crystal growth, and the crystal growth rate must be slower than the rate associated with the BF embryo nucleation [78]. Furthermore, experimental works were conducted to study the particle morphology control of BF by using $BiCl_3$, $FeCl_3$ and KOH [79]. In the method investigated, the concentration of the KOH solutions varied in the range of 0.2 to 0.7 M, and the hydrothermal synthesis was carried out at standard conditions at 180°C for 24 h. The control of particle morphology was investigated employing organic surfactant with a concentration of 0.2 M (PEG, EDTA, CTAB and PVP). The surfactant was added to the KOH solution prior the mixing of the Bi and Fe precursors. The usage of concentrated KOH solutions produced a marked agglomeration of particles, which gradually dispersed at long reaction intervals. The morphology of the particles also gradually becomes more regular and homogeneous. A mechanism based on the presence of electrostatic charge micelles is likely to promote the dispersion and morphology control of the BF particles. The micelles formed when a large amount of OH^- cover the surface of the growing BF particles, generates a network of negatively charged nuclei. The repulsive forces between the nuclei are strong, and after the repulsive force reaches its maximum level, it conduces to particle isolation. Therefore, the decrease of KOH concentration in the hydrothermal system provokes the loss of the repulsive force that causes the marked particle aggregation [78].

Other studies focusing on the synthesis of BF particles via hydrothermal processing were performed in reaction media containing organic solvents, namely ethanol or acetone. Recently, Chen et al. [80–81] studied the crystallization of BF particles at low temperature using these organic solvents, which were mixed with the 4 M KOH solution. Well-crystallized pure rhombohedral structure BF nanoparticles were prepared at 120°C for 16 h with an ethanol/water ratio of 4:3 [80]. In this typical hydrothermal process, the ratio ethanol/water had a significant role in the formation of pure BF. Later on, BF powders were synthesized with the aid of acetone, using bismuth and iron nitrates as reactants and a KOH solution concentration of 7 M. The study was directed towards to investigate the effect of solvent (KOH/acetone) on the chemical stability of the reaction product. The BF powders were prepared by using acetone as a solvent to dissolve the nitrate. Then, the suspension was hydrothermally treated at 130°C for 12 h; these conditions yield to the formation of nanometric-sized cubic-shaped particles of BF [81]. The preferential formation of the BF powders occurring at very low temperature was explained as follows: (1) the metal elements have different electronegativities, 2.02 for Bi and 1.83 for Fe. This difference might cause a variation in the hydrolysis rates of the salts employed; in practice bismuth salts hydrolyze more easily in water in comparison with the iron salt. The hydrolysis of $Bi(NO_3)_3$ rapidly occurred resulting in the subsequent formation of a white insoluble precipitate as it is expressed by chemical reaction 19.

$$Bi\left(NO_3\right)_{3(aq)} + H_2O_{(aq)} \rightarrow BiONO_{3(s)} + 2HNO_{3(aq)} \tag{19}$$

The usage of acetone as solvent might limit the hydrolysis of bismuth salt, keeping the element as ionic species with their different electronegativities. Therefore, it might reduce the energy required for successful dissolution and precipitation to form the BF phase during the hydrothermal process. Because the presence of insoluble $BiONO_3$ does not proceed in the system, in contrast only $Bi(OH)_3$ and $Fe(OH)_3$ hydroxides (acting as precursors) precipitates. (2) The increase of the vapour pressure in the vessels is likely to proceed due to the evaporation temperature of acetone, which is 56.05°C. The excess of pressure derived from the evaporation of acetone might accelerate the processes of dissolution and crystallization, and, finally, it leads to the formation of pure BF powders at low temperature. (3) A surface tension phenomena produced by the liquid phase can control the reaction mechanism associated with the crystallization of BF particles. The mechanism is likely to proceed due to the surface tension differences existing between acetone, water and absolute ethanol (23.02, 77.82 and 22.39 m/Nm, respectively). Under hydrothermal conditions, if the surface tension of solvent is high leading to hydroxide precipitation, these intermediate reaction products subsequently are likely to undergo into agglomeration. Therefore, longer reaction intervals are required to transform the hydroxides into BF via a dehydration process. When a small volume of an organic compound is mixed with water, i.e. acetone, the surface tension of the solution must decrease, leading to a rapid hydroxide precipitation achieved by the surrounded solvent. The precipitates had a homogeneous dispersion and were easily dehydrated to form BF powders. Another statement proposed to explain the BF synthesis indicates that, (4) the dielectric constant "ε" of the solvent could be lowered due to the organic solvent addition in water to form the hydrothermal media.

Thereby, the reduction on "ε" causes the decrease of the BF solubility, promoting a high-saturation Bi and Fe in the solvent media. When this state is reached, it provides the optimum conditions for a rapid nucleation of the BF phase. The presence of acetone accelerates the dissolution of the reactants, enabling the crystallization to occur in the supersaturated fluid, finally leading to the achievement of pure-phase BF powders at a very low temperature [81].

The synthesis of a perovskite composite with a composition of 0.5BF-0.5BT was satisfactorily obtained at 200°C employing a KOH concentration of 4 M for 6 h [82]. The crystallization of perovskite 0.5BF-0.5BT is achieved by a "dissolution and crystallization process"; this theory has been widely applied to explain the hydrothermal reaction pathways. The relevant chemical reactions that occur during the synthesis that are correlated with the crystallization of BF-BT are as follows (reactions 20 to 24):

$$2Bi\left(NO_3\right)_{3(aq)} + 6KOH_{(aq)} = Bi_2O_{3(s)} + 6K^+ + 6NO^-_3 + 3H_2O_{(l)} \tag{20}$$

$$Fe\left(NO_3\right)_3 + 3KOH = Fe\left(OH\right)_3 + 3K^+ + 3NO^-_3 \tag{21}$$

$$TiCl_4 + H_2O = TiOCl_2 + 2HCl \tag{22}$$

$$TiOCl_2 + 2HCl + 4KOH = Ti\left(OH\right)_{4(aq)} + 4K^+ + 4Cl^- + H_2O_{(l)} \tag{23}$$

$$Ti\left(OH\right)_4 + 2OH^- = \left[Ti\left(OH\right)_6\right]^{2-} \tag{24}$$

When the hydrothermal synthesis takes place, several reactions occurred between the reactants of bismuth oxide, iron III hydroxide, barium hydroxide and titanium hydroxide. To produce the desired crystalline phases following the chemical reactions 21 and 22 must occur in the hydrothermal system at particular conditions of temperature, time and pressure. Additionally, the results demonstrated that the reaction temperature and the KOH concentration must be optimized to limit the presence of reaction by-products. Under these conditions, the formation of perovskite 0.5BF-0.5BT is likely to be conducted following a single-step reaction.

$$Bi_2O_3 + 2Fe\left(OH\right)_3 = 2BiFeO_3 + 3H_2O \tag{25}$$

$$Ba^{2+} + \left[Ti\left(OH\right)_6\right]^{2-} = BaTiO_3 + 3H_2O \tag{26}$$

A reduced number of researchers have carried out the synthesis of RF (R= rare earth elements group or Y) under hydrothermal conditions using rare-earth nitrates as precursor feedstock. The hydrothermal synthesis of RF particles was conducted in highly concentrated alkaline solvents between 2 and 44 M for 48 h, whereas the reaction temperature required to promote the crystallization of the single ABX_3 phase varied between the range of 230 up to 240°C [83–85]. The formation mechanism of the RF crystals was associated with a precipitation process promoted by the precursor dissolution and nucleation processes. The role of alkalinity was a crucial parameter to control the dissolution, nucleation and growth processes because the chemical reagent precursors are chemically stable even in strong alkaline solvents such as NaOH or KOH.

Similar to the BT powders, the microwave hydrothermal synthesis has been used to synthesize perovskite BF to achieve fast crystallization kinetics and control microstructural aspects of the particles. The hydrothermal crystallization process assisted by microwave radiation accomplishes the formation of BF single phase with rhombohedral structure, for very short reaction intervals of 1 h. The precursor selected to perform this reaction were the reagent-grade salts of bismuth ($Bi(NO_3)_3$) and iron ($FeCl_3$, $Fe(NO_3)_3$). The synthesis process can occur at temperatures in the range of 150–200°C in KOH solutions with concentrations varying between 0.05 and 4 M [86, 87]. Experimental results indicate that crystallization of pure-phase BF powders is possible to take place at 200°C for 1 h with a KOH concentration of 0.05 M, due to the fast heating supplied by the microwave radiation with the aid of the precursor feedstock selected [86]. One point that deserves emphasize is related to the abnormal growth of the BF particles; this causes marked agglomeration of the BT particle. This fact is proposed to be the primary reason for the occurrence of a change in the crystal structure because a high OH^- concentration favours the nucleation process. Moreover, Wang et al. [88] studied the synthesis of BF by using a polyanion, poly(methyl vinil ether-alt-maleic acid) (PMVEMA), KOH and Na_2CO_3 solutions with concentrations of 1 and 8 M, respectively. The BF pure phase was synthesized under the same conditions determined elsewhere [87], and a simple ultrasonic purification method was developed to obtain the pure phase. In particular, PMVEMA was selected because it contains a large number of –COOH groups that operate as anions, which avoid bringing other metal ions. Additionally, the presence of these ions also can control the concentration change of metal ions such as Na^+ and K^+ in the hydrothermal reaction system [88].

2.4.3. Synthesis of zirconite-type perovskite ($AZrO_3$)

Another oxide compound of interest due to its potential applications is the perovskite BZ. Continuous synthesis technologies that allowed the researchers to produce tens of grams of powder products per production hour are devoted to developing at an industrial level. The perovskite BZ phase has been obtained by a continuous hydrothermal synthesis proceeding in a single step under supercritical conditions [89, 90]. The cubic BZ pure-phase particles have been produced at higher temperatures around 450–500°C using NaOH as the solvent media. A high temperature is necessary to complete the release of OH– ions coupled with the neutralization of protons, because at severe supercritical conditions the reaction vessel can suffer a rapid acidic corrosion. The reactions that occur in the autoclave vessel are described by chemical reactions 27 to 29. Before the occurrence of these reactions, the dissolution of the precursors must take place, due to rapid ionic mobility enhanced by the dielectric constant of the solvent "ε":

$$Ba(NO_3)_2 \rightarrow Ba^{2+} + 2NO_3^- \tag{27}$$

$$NaOH \rightarrow Na^+ + OH^- \tag{28}$$

$$ZrO(NO_3)_2 \rightarrow ZrO^{2+} + 2NO_3 \tag{29}$$

The crystallization goes further under supercritical when the precursors are mixed because of the high ionic mobility in the solvent phase due to the low dielectric constant. The reaction that occurs is described as follows:

$$ZrO^{2+} + Ba^{2+} + 2OH^- \rightarrow BaZrO_{3(s)} + 2H^+ \tag{30}$$

In contrast, the possibility of preparing BZ powders with cubic structure was investigated under hydrothermal conditions at 130°C for 24 h. In these experiments, the adjustment of hydrothermal media feedstock to a value of 13 occurs by adding a KOH solution with a concentration of 0.5 M [91]. Furthermore, the formation of RBZ perovskite solid solutions with cubic structure required a minimum reaction temperature of 200°C for 24 h. The reaction was conducted using a highly concentrated KOH of 16 M as a solvent; the strong alkaline solutions rapidly dissolved the reactant precursor [92]. The morphology of the RBZ particles obtained under the severe alkaline hydrothermal conditions resembled hollow nanospheres. However, the particle sizes were simply controlled varying the Rb content in the precursor and keeping the KOH concentration constant. It was confirmed that adjusting the KOH concentration would change the inter-particle force, varying the size of the aggregates.

2.4.4. Synthesis of lanthanum chromites and manganites-type perovskites and their solid solutions ($LaCrO_3$, $LaMnO_3$)

In recent years, much work has been conducted for synthesizing perovskite-type $LaCrO_3$ and $LaMnO_3$ (LC and LM) under mild hydrothermal conditions. There are several reports on the preparation of nanoparticles under hydrothermal conditions. Zheng et al. [93] studied the effect of solvent alkalinity by employing KOH with concentrations between 2 and 16 M. Another parameter evaluated in this work was the molar ratio of KOH/Cr = 10–80, the hydrothermal treatments were carried out at a 260°C for a fixed time of 7 days. These authors found that a highly concentrated alkaline solvent is required when the selected B-cation has an amphoteric behaviour, as it is the particular case of Cr^{3+}. Therefore, the formation of pure LC required ≥8 M of KOH in the reaction medium.

A series of solid solutions based on the perovskite-type structure has been hydrothermally prepared. The synthesis of $La_{1-x}Sr_xCrO_3$ (LSC) with orthorhombic structure has been carried out using different reaction media to prepare in a preliminary stage the co-precipitated gel precursor. Perovskite LSC powders were produced using Triton non-ionic surfactant and NH_4OH as co-precipitation medium; the pH of the starting aqueous media was 8. The reaction rate associated with this process was affected by the common ion effect caused by the metal precursor as well as by the steric barrier due to the presence of the surfactant micelles [94]. Also, the synthesis of LSC fine powders in a media with a NaOH concentration of 0.5 M was achieved by the dissolution–crystallization mechanism. The solvent even at the concentration selected enhances a homogeneous nucleation leading to the formation of dispersed particles. The hydrothermal solvent chemically reacts with the solid species incorporated into the vessel, producing the dissolution of all solids and the ionic saturation of the aqueous solvent media. In consequence, a spontaneous precipitation of LC and LSC powders occurs [95]. The synthesis of these powders also proceeded in a reaction medium of KOH solution. In this procedure, the KOH solution was mixed in two steps. Firstly, KOH was mixed with $CrCl_3$ and to produce $Cr(OH)_3$, while in the second step KOH was used to adjust the alkalinity of the reaction medium after the sources of Sr and La were added. Moreover, a higher alkalinity is necessary for the preparation of LSC because Cr^{3+} behaves as an amphoteric element. In fact, the optimum alkalinity range that promotes the crystallization of LSC solid solutions is from 5 up to 8 M [96]. A similar reaction pathway was determined for the synthesis of rare-earth and yttrium orthochromite perovskites. Because when a rare earth element partially substitutes one of the major constituents in the perovskite structure, it provokes that a concentrated alkaline solution must be used as a precursor solvent (i.e. 10–12 M KOH) [97].

Several types of lanthanum manganite solid solutions have been synthesized under hydrothermal conditions. The synthesis of LBM nanowires ($La_{0.5}Ba_{0.5}MnO_3$) with cubic perovskite structure was produced at 270°C for 25 h with a net [OH–] = 10 M; using $La(NO_3)_3$, $Ba(OH)_2$, $KMnO_4$ and $MnCl_4$ as precursors [98]. Likewise, at the similar conditions of synthesis of LCM ($La_{0.5}Ca_{0.5}MnO_3$) solid solution with orthorhombic structure was produced. The reaction associated with the hydrothermal processing investigated is given as follows [99].

$$\begin{aligned} &3KMnO_{4(aq)} + 7MnCl_{2(aq)} + 5La\left(NO_3\right)_{3(aq)} + 5Ca\left(NO_3\right)_{2(aq)} + 36KOH_{(aq)} \rightarrow \\ &10La_{0.5}Ca_{0.5}MnO_{3(s)} + 18\,H_2O_{(aq)} + 25KNO_{3(aq)} + 14KCl_{(aq)} \end{aligned} \tag{31}$$

Recently, perovskite LSM ($La_{1-x}Sr_xMnO_{3+\delta}$) were obtained at 150°C for 20 h by adjusting the suspension (hydrothermal media) pH to a value of 9 with an ammonia solution. Based on the results of preliminary runs, a systematic study was conducted aiming to obtain the desired chemical compositions of solid solutions of manganite [100]. A mechanism of dendrite nucleation leads to the synthesis of this particular perovskite compound. This fact was established based on the crystalline phase evolution analyses of the reaction product produced after each treatment, which was conducted in a hydrothermal media of 4 M NaOH. It was found that the SLM solid solution with hexagonal perovskite structure and with particle morphology of hexagonal platelets was formed at temperatures below 220°C. In contrast, the mechanism of embryo nucleation differs when the reaction temperature was 240°C. At this temperature, the LSM particles were formed via the epitaxial nucleation; this mechanism acts preferentially at the edges of the hexagonal platelet. In consequence, the growth of "tree-like" LCM dendrites is promoted under the hydrothermal conditions investigated [101]. In recent years, Spooren et al. [102] studied the synthesis of LCM, LSM and LBM solid solutions using a KOH reaction media; the use of other solvents was also investigated. However, in all the new solvents tested, the formation of reaction by-products was not avoided. The hydrothermal synthesis of LBM, LSM and LCM were achieved in the KOH solvent solution for different molar compositional ratios of:

Mn^{2+}:MnO_4^{2-}:Ba^{2+}:La^{3+}:KOH:H_2O equivalent to 7:3:5:5:1250:3256 at 240°C for 24 h.

Mn^{2+}:MnO_4^{2-}:Sr^{2+} :La^{3+}:KOH:H_2O equivalent to 7:3:5:5:1250:3256 at 240°C for 24 h.

Mn^{2+}:MnO_4^{2-}:Ca^{2+}:La^{3+}:KOH:H_2O equivalent to 7:3:5:5:1250:3256 at 270°C for 24 h.

At present, there are several authors interested in the study of perovskite solid solutions. [103–104]. The investigations were directed towards to develop a simple route of synthesis for LCSM ($La_{1-x}Ca_xSr_xMnO_3$) solid solutions with orthorhombic structure and cuboidal morphology. The synthesis was conducted at 240°C for 72 h. The preparation of perovskite LAM ($La_{1-x}Ag_xMnO_3$) occurred employing the conventional hydrothermal and microwave-assisted hydrothermal method with a reaction media saturated with NH_4OH solution to adjust the solution pH at 8.5 [105]. Furthermore, orthorhombic perovskite manganates YCM ($YMnO_3$, and $Y_{1-x}Ca_xMnO_3$, x = 0.07, 0.55, 0.65) were prepared in one step by a mild hydrothermal synthesis. The optimum concentration of the alkaline solvent media employed to conduct the hydrothermal crystallization 23 M to stabilize the single phase of YCM [106]. The pH of the solvent solution is a critical parameter to achieve the crystallization of PSM and NSM manganites. The adequate KOH molar concentration that favours reaching the equilibrium between the dissolution crystallization events are between 5.0 and 10 M. Therefore, the synthesis of solid solutions PSM ($Pr_{0.5}Sr_{0.5}MnO_3$), NSM ($Nd_{0.5}Sr_{0.5}MnO_3$) and alkali-earth manganese oxides such as 4H-$SrMnO_3$ and 2H-$BaMnO_3$ were conducted using KOH with concentrations within the abovementioned range [107]. As for the case of perovskite $RMnO_3$ (R = Dy, Ho, La, Pr, Nd, Tb and Bi), a common feature of the mild hydrothermal synthesis is the high alkalinity of the solutions. Such high alkalinity provides a critical condition, which considerably influences the crystallization and composition of the $RMnO_3$ product in the hydrothermal synthesis [108–109]. The high alkalinity of the reaction system also constitutes an optimal condition that considerably influences the crystallization of perovskite manganites. To maintain the suitable alkalinity required for the hydrothermal reaction, it is necessary to use a strong solvent such as KOH [110].

2.4.5. Synthesis of miscellaneous type perovskites and their solid solutions ($ANbO_3$, $ATaO_3$)

The solvent concentration also has an effect on the nucleation and crystal growth of other perovskite compounds, namely NN and KN ($NaNbO_3$, $KNbO_3$). The formation of this particular group of perovskites has been tailored in a media composed of a mixture of KOH and NaOH solutions. The solvent has a composition assigned by the K^+:Na^+ molar ratio of 1:1, while the alkaline solvent solutions concentrations (NaOH or KOH) were 0.6, 0.8, 1.0 and 1.6 M. In the hydrothermal environment, OH^- ions exhibit a catalytic behaviour assisting the mass transport of the ionic solute species during the crystallization process. Also, these ions contribute to the particle nucleation and acceleration of the reaction [111]. However, when only NaOH was used as mineralizer, it was found that the reaction took place by an in situ transformation mechanism. The reaction speed was increased at low concentration of [OH^-] with increasing temperature. A high [OH^-] is not favourable to prepare perovskite NN, and there is an optimum [OH^-]. The concentration of [OH^-] and the reaction speed were correlated, which indicated that a high alkalinity is not useful for the synthesis of niobate [112]. In contrast, when a particular polymorph of Nb_2O_5 was selected as a precursor, the stability of the hexaniobate ion in solution was greater for lower KOH concentration and low temperature. In other words, the induction period related to the conversion of the hexaniobate ion to perovskite powder varied changing the concentration of KOH for a given temperature. Hence, it is possible to produce orthorhombic NN powders at 200°C using solutions of 8.4 M NaOH, 0.25 M Nb_2O_5 and KN powders; the reaction between solutions of different Nb_2O_5 concentrations (0.0015–0.38 M) and KOH (6.7–15 M) was only investigated at 200°C [113]. The reaction mechanism associated with the synthesis of AN is markedly different from that of NN reported in the literature. The heterogeneous chemical reaction between NH_4HF_2, Ag_2O, and Nb_2O_5 gives perovskite AN at pH = 3, adjusting the reaction media adding either an acidic solution (HF) or mild basic solution (NH_4OH). In the hydrothermal processing, water is not only used as the solvent for the crystallization of inorganic substances from solution, but other chemical processes also participate in the energy transfer, catalysis, dissolution and hydrolysis. The experimental results indicate that water solvent, as a solvent, provides a more powerful environment for the hydrothermal reactions for the crystallization of AN perovskites [114].

Cubic perovskite NT ($NaTaO_3$) powders were hydrothermally produced at 200°C for 24 h, employing a reaction medium with a concentration of 7 M and using KOH as mineralizer [115]. Due to the higher OH^- concentration contained in the hydrothermal system, a reaction by-product (pyrochlore) was obtained after the treatment. The pyrochlore crystalline phase subsequently reacted with the solvent media (KOH) to form a perovskite non-stoichiometric phase. In contrast, it was found that the reaction between Ta_2O_5 and NaOH for the synthesis of perovskite NT ($NaTaO_3$) powders follows the dissolution–recrystallization mechanism under microwave-assisted hydrothermal conditions. This reaction occurred without the addition of organic templates compounds and catalysts to the reaction medium. The thermodynamic energy barrier associated with the phase conversions, in this case, was estimated to be in the following order: $Ta_2O_5/Na_2Ta_2O_6 < NaTaO_3/Na_2Ta_2O_6 < Ta_2O_5/NaTaO_3$. A pre-treatment of the Ta_2O_5 raw material by ball milling is a crucial step to obtain pure perovskite NT phase [116].

2.4.6. Synthesis of B site perovskite ($MeBO_3$)

At present, there are a few studies focused on the synthesis of ordered perovskites partially substituted at the B site. The synthesis of a few perovskite $MeBO_3$ group solid solutions have been carried out under mild hydrothermal conditions, where the elements chosen were Me = Pb or Ba; B = Ti, Zr, Zn, or Ta, among others (hereafter these solid solutions are denominated as PZT, BTZ, BZT, respectively). These studies have been oriented to analyze the consolidation and functional properties of the synthesized materials, but any details were considered to clarify the chemical reaction pathways associated synthesis reaction that promotes the crystallization of these perovskite compounds. The present section of the review describes the conditions at which the synthesis of the perovskite solid solutions is possible to progress in a hydrothermal environment.

The hydrothermal processing procedure established for the synthesis of PZT perovskites involves the preparation of a hydrated mixed gel containing the main elements that constitute this compound. In this study, the factor exhaustively investigated was the molar Zr/Ti ratio. The effect of this parameter was considered to determine the optimum molar Zr/Ti ratio that chemically stabilizes the perovskite crystalline structure in different hydrothermal environments. As a consequence, the PZT single phase must belong to one of the three different crystalline structures, namely tetragonal, rhombohedral or pseudo-tetragonal [117]. The synthesis of PZT might occur in the sequence depicted by the chemical reactions that occur during the hydrothermal crystallization of PZT, represented by reactions 32 to 34:

$$PbO + 0.1\ TiO_2 \rightarrow \left(PbO : 0.1\ TiO_2\right)_{(s)} \tag{32}$$

$$2\left(PbO : 0.1\ TiO_2\right)_{ss} + 0.8\ TiO_{2(gel)} + ZrO_{2(gel)} \rightarrow 2Pb\left(Zr_{0.5}Ti_{0.5}\right)O_{3(amorphous)} \tag{33}$$

$$2Pb\left(Zr_{0.5}Ti_{0.5}\right)O_{3(amorphous)} \rightarrow 2Pb\left(Zr_{0.5}Ti_{0.5}\right)O_{3(s)} \tag{34}$$

The presence of H_2O under pressure (40–50 MPa) enhanced the solubility of TiO_2 in PbO, due to rapid transport of the ionic species in the supercritical fluid. In the KOH mild alkaline fluid between 160 and 350°C, PbO partially substituted with Ti rapidly reacts with the homogeneously dispersed precursor gel. The reaction accompanied by a progressive splitting off even large PbO particles (6 μm) occurred nearly the supercritical state of the hydrothermal media at 350°C by adjusting the solvent pH between 12 and 14. At this condition, the crystallization of the PZT tetragonal perovskite preferentially proceeded according to the chemical equilibrium that is given by Eq. (33). Therefore, the successful reactant mixing ratio Pb:Zr:Ti was 1:0.5:0.5, at which thermodynamically the hydrothermal media reach the equilibrium to achieve the crystallization of the tetragonal unit cell. The increase of the reaction time during the hydrothermal treatment provokes an increase in the crystallinity of the PZT product. It is well known that under hydrothermal treatment in an aqueous medium with basic pH (13.5), the hydroxide cation complexes of Pb, Zr and Ti undergoes a transition into PZT [118]. The crystallization of a nanocrystalline perovskite precursor phase was promoted using colloidal wet chemical processing and low-temperature crystallization stages. The common chemistry preparative procedure involves the co-precipitation of the precursor perovskite constituents from a saturated solution. The co-precipitated colloid gel is obtained using an alkaline KOH solution with a concentration of 4 M producing a colloid suspension. The suspension containing the co-precipitated colloid (Pb-Zr-Ti-hydroxide) hydrothermally treated in an autoclave at 250°C for 5 h leads to prepare the tetragonal structured PZT particles exhibiting a cubic morphology and particle size of 600 nm [119]. On the other hand, the production of single crystal tetragonal perovskite PZT ($PbZr_{0.52}Ti_{0.48}O_3$) has been conducted adding a polymer compound. The process denominated as polymer-assisted hydrothermal synthesis was carried out at 200°C for 12 h. Among the polymer additives selected to control the micelles creation were PVA and PAA. The polymer plays an important role in the growth of PZT single crystals with tetragonal structure. The micelles tend to cover the faceted surfaces of the PZT particles, and this event produces a reduction in the superficial energy of the growing particles. Consequently, the PZT particles exhibit a preferential growth along the c axis of the tetragonal structure [120]. Besides the employment of water-soluble organic compounds aiming to control the particle crystal growth. A variety of organic bases provided another advantage for synthesizing PZT perovskites. The possibility to apply the organic bases as potential solid solvents in hydrothermal media was recently investigated in the preparation of PZT perovskite powders. The experiments were carried out in an alkali-free reaction media at 160°C for 72 h. The organic alkaline media selected to conduct the hydrothermal treatments was the tetramethylammonium hydroxide (TMAOH), this also operated as a pH-adjusting agent [121].

Other solid solutions of interest are the BTZ ($BaTi_{0.8}Zr_{0.2}O_3$) and the BZT ($BaZr_{1-x}Ti_xO_3$). The synthesis of these compounds at mild conditions has not been studied in detail yet. In one study conducted to prepare BTZ perovskite, a stoichiometric powdered mixture of Ti and Zr hydroxides, which was added to a 6 M $Ba(OH)_2$ solution. The barium saturated alkaline $Ba(OH)_2$ solution was selected as a solvent for the hydrothermal treatments to adjust the slurry before the hydrothermal synthesis. In this reaction media it was possible to carry out the crystallization of BZT particles with cubic structure, spherical morphology and molecular level homogeneous composition [122]. Likewise, BZT hollow nanospheres with cubic structure were synthesized at 200°C for 24 h using a KOH solution with a concentration of 16 M. The formation of the particles occurred due to Ostwald ripening mechanism, and the excess of Ti in the fluid may change the reaction mechanism that favours the hollowing process [123]. Finally, the synthesis reaction for the formation of BZnT ($BaZn_{1/3}Ta_{1/3}O_3$) was conducted in a reaction medium with a strong organic base (TMAOH), and the raw materials used were barium and zinc acetate, and tantalum oxalate solution. The reactions involved in the hydrothermal synthesis of the BZnT perovskite particles are reactions 35 to 37. Initially, in the first stage of the reaction occurred the decomposition of the organic alkaline produces a reduction in the total concentration of OH^- species at temperatures up to 150°C, simultaneously the nucleation of BZnT embryos occurred in the solution according to the chemical reaction Eq. (35). The decomposition of the TMAH to trimethylamine and methanol (Eq. 36) takes place at temperatures above 220°C in the hydrothermal vessel. The product trimethylamine is a weak alkali, and it has the same chemical reactivity such as that of the ammonia in an aqueous solution (Eq. 37), [124]. Hence, the low chemical stability of the organic alkaline might hinder the crystallization of the desired BZnT perovskite phase because the supersaturation state is difficult to reach in the reaction system.

$$3Ba\left(MeCO_2\right)_{2(aq)} + Zn\left(MeCO_2\right)_{2(aq)} + Ta_2O_5.xH_2O + 8NMe_4^{+}OH^{-}_{(aq)} \rightarrow$$
$$3BaZn_{1/3}Ta_{2/3}O_{3(s)} + 8\left(MeCO_2\right)^{-}NMe_4^{+}{}_{(aq)} + \left(4+x\right)H_2O_{(l)} \quad (35)$$

$$NMe_4^{+}OH^{-}_{(aq)} \rightarrow NMe_{3(aq)} + MeOH_{(aq)} \quad (36)$$

$$NMe_{3(aq)} + H_2O_{(l)} \Leftrightarrow NHMe_3^{+}{}_{(aq)} + OH^{-}_{(aq)} \quad (37)$$

2.4.7. Hydrothermal synthesis of AB site perovskites

The study of the synthesis of solid solutions with simultaneous substitutions in both locations (A and B sites) has been conducted under combined hydrothermal conditions. In a preliminary study, the synthesis of complex perovskite compounds (PBSZ and PBZT) was investigated using a KOH solution with concentrations varying in the range of 1.0–3.0 M. The elements that were partially incorporated into the ABX_3 structure were Pb, Ba and Sr added as A ions and Zr as B ion [125]. In the reaction media of KOH with concentrations of 2.0 and 3.0 M, most of the mixtures containing different contents of Ba and Sr were successfully crystallized and correspond to the tetragonal perovskite pure phase. However, the single perovskite phase was difficult to conduct in the alkaline solution with a concentration below 1 M. In conclusion, a hydrothermal media saturated with OH^- ions is usually required to prepare perovskite ternary compounds under hydrothermal conditions. On the other hand, the synthesis of modified lead titanate powder such as $(Pb_{0.88}Sm0._{08})(Ti_{0.99}Mn0_{.01})O_3$ has been conducted at pH between 8 and 12 using NaOH as mineralizer and metal nitrates precursor solutions. The conditions that favour the crystallization of the perovskite structure are 290°C for 10 h at a pH = 10 [126]. In general, the vast number of studies discussed in the preliminary sections, regarding the synthesis of several groups of perovskite compounds, indicate that the processing parameters (precursor metal salts reagents, solvent media, pH of the precursor solution, reaction time, and temperature) must be carefully selected to carry on the processing perovskite compounds either by hydrothermal microwave-assisted synthesis, hydrothermal surfactant-assisted synthesis or supercritical water conditions.

2.5. Summary

This review describes in detail the state-of-the-art regarding the theoretical and practical aspects associated with the synthesis of a wide number of perovskite-structured compounds. The use of the thermodynamic and kinetic modelling has proved to be a powerful tool to optimize the conditions for perovskite compounds under hydrothermal conditions. These tools allow in a rational way to engineer the processing of smart perovskite compounds, in the most cost-effective approach and an environmental friendly way as well. One of the particular advantages of hydrothermal processing is associated with the preparation of monodispersed nanometer-sized particles exhibiting a control over their shape and size in addition to their chemical homogeneity. However, the models discussed in the present review are only applicable to a very narrow boundary of experimental conditions in which the heterogeneous hydrothermal reaction equilibriums are achieved, namely in the pH range from 0 to 14. The vast practical expertise gained from the past three decades in the synthesis of a wide number of perovskite-structured compounds using the hydrothermal technique indicates that some caution must be executed, when applying the proposed models to diverse perovskite systems. The results derived from the kinetics and thermodynamical models might vary depending on the reaction temperature, concentration of precursors and the alkalinity of the hydrothermal media. Therefore, more conscious analyses based on inorganic chemistry fundamentals coupled with chemical engineering and kinetic modelling might help to derive the adequate conditions to achieve controlled crystallization of particles with specific particle size and morphology. Although the present information regarding the hydrothermal synthesis of perovskite compounds, in terms of the theoretical and practical approaches, has been evaluated with a high statistical reproducibility in small-scale laboratory reactors, there is a potential field to explore when scaling up the process to larger industrial reactors, where mixing problems might frequently raise due to slow and less-efficient agitation systems. Mixing-limited precipitation rate is one of the problems commonly encountered in the scale-up of precipitation processes.

Author details

Juan Carlos Rendón-Angeles[1*], Zully Matamoros-Veloza[2], Karla Lorena Montoya-Cisneros[1], Jorge López Cuevas[1] and Kazumichi Yanagisawa[3]

*Address all correspondence to: jcarlos.rendon@cinvestav.edu.mx

1 Research Institute for Advanced Studies of the NPI, Ceramic Department, Ramos Arizpe, Coahuila, México

2 Technological Institute of Saltillo, Metal-Mechanics Department, Saltillo, Coahuila, México

3 Research Laboratory of Hydrothermal Chemistry, Kochi University, Kochi, Japan

References

[1] Mohammad HH. Characterisation of Mixed-Metal Oxides Prepared by Hydrothermal Synthesis. [thesis]. Coventry CV4 7AL, United Kingdom: The University of Warwick; 2013.

[2] Mats J, Peter L. Crystallography and Chemistry of Perovskites. Handbook of Magnetism and Advanced Magnetic Materials. p. 2007. DOI: 10.1002/9780470022184.hmm411

[3] Kay HF, Bailey PC. Structure and properties of $CaTiO_3$. Acta Crystallogr 1957;10:219–26. DOI:10.1107/S0365110X57000675

[4] Sasaki S, Prewitt CT, Bass JD, Schulze WA. Orthorhombic perovskite $CaTiO_3$ and $CdTiO_3$: structure and space group. Acta Crystallogr 1987;C43:1668–74. DOI:10.1107/S0108270187090620

[5] Verma AS, Jindal VK. ABX_3 type oxides and halides: their structure and physical properties. In: Perovskites: Structure, Properties and Uses. Editors: Maxim Borowski, Editorial: 2010, Chapter: 11. pp. 463–479. DOI: 10.1002/chin.201121201

[6] John BG, Zhou JS. Orbital ordering in orthorhombic perovskites. J Mater Chem 2007;17:2394–405. DOI: 10.1039/b701805c

[7] Nicole AB, Craig JF. Why are there so few perovskite ferroelectrics? J Phys Chem 2013;117:13339–49. DOI.org/10.1021/jp402046t

[8] Ivan AS, Cengiz S, Elbio D. Ferroelectricity in the magnetic *E*-phase of orthorhombic perovskites. Phys Rev Lett 2006;97:227204-1–227204-4. DOI:10.1103/PhysRevLett.97.227204

[9] Goodenough JB, Longo JM. In: Landolt-Cornstein. Numerical data and functional relationships in science and technology, new series, group III: crystal and solid state physics. Magnetic and other properties of oxides and related compounds, part A, vol. 4. Berlin: Springer; 1970. p. 126.

[10] Warren EP, David JS. Electronic structure and half-metallic transport in the $La_{1-x}Ca_xMnO_3$ system. PRB 1996;53,3:1146–60. DOI: 10.1103/PhysRevB.53.1146

[11] Goodenough JB, Longo JM. Crystallographic and magnetic properties of perovskite and perovskite – related compounds. In: LANDOLT-BÖRNSTEIN Zahlenwerte und Funktionen aus Naturwissenschaften und Technik. Neue Serie. Gesamtherausgabe: K.-H. Hellwege. Band 4 Magnetische und andere Eigennschaften von Oxiden und werwandten Verbindungen. Spring-Verla; 1970. pp. 126–129.

[12] Wiebe CR, Greedan JE, Kyriakou PP, Luke GM, Gardner JS, Fukaya A, Gat-Malureanu IM, Russo PL, Savici AT, Uemura YJ. Frustration-driven spin freezing in the S= ½ fcc perovskite Sr_2MgReO_6. PRB 2003;68:134410-1–134410-10. DOI: 10.1103/PhysRevB.68.134410

[13] Cava RJ, Batlogg B, Krajewski JJ, Farrow R, Rupp Jr LW, White AE, Short K, Peck WF, Kometani T. Superconductivity near 30 K without copper: the $Ba_{0.6}K_{0.4}BiO_3$ perovskite. 1988;332:814:814–816. DOI:10.1038/332814a0

[14] Sami V, Maarit K. $A_2B'B''O_6$ perovskites: a review Prog Solid State Chem 2015;43:1–36. http://dx.doi.org/10.1016/j.progsolidstchem.2014.08.001

[15] Eduardo CCdeS, Reginaldo M. Properties and applications of perovskite proton conductors. Mater Res 2010;13(3):385–94. http://dx.doi.org/10.1590/S1516-14392010000300018

[16] Peña MA, Fierro JLG. Chemical structures and performance of perovskite oxides. Chem Rev 2001;101:1981–2017. DOI: 10.1021/cr980129f

[17] Bonanos N, Knight KS, Ellis B. Perovskite solid electrolytes: structure, transport properties and fuel cell applications. Solid State Ionics 1995;79:161–70. DOI 10.1016/0167-2738(95)00056-C

[18] Stephen JS. Recent advances in perovskite-type materials for SOFC cathodes. Fuel Cells Bull 2001;4(33):6–12. DOI: 10.1016/S1464-2859(01)80254-6

[19] Stephen JS. Recent advances in Perovskite-type materials for solid oxide fuel cell cathodes. Int J Inorg Mater 2001;3:113–21. DOI:10.1016/S1466-6049(01)00004-6

[20] Harlan UA. Review of p-type doped perovskite materials for SOFC and other applications. Solid State Ionics 1992;52(1–3):33–41. DOI: 10.1016/0167-2738(92)90089-8

[21] Bangerta U, Falke U, Weidenkaff A. Ultra-high resolution EEL studies of domains in Perovskite. J Phys Conf Ser 2006;26:17–20. DOI:10.1088/1742-6596/26/1/004.

[22] Tapas KM, Gopalakrishnan J. From rocksalt to perovskite: a metathesis route for the synthesis of perovskite oxides of current interest. J Mater Chem 2004;14:1273–80. DOI: 10.1039/b315263d

[23] Hans-Conrad zL, Qingbiao Z, Daniel EB, Michael WC. 2H-perovskite related oxides: Synthesis, structures, and predictions. Cryst Eng Comm 2012;14:23–39. DOI: 10.1039/c1ce05788j

[24] Jesús P-G, Rainer S, Emilio M. Microwave-assisted synthesis and characterization of perovskite oxides. In: Jinghua Z, Huan L, eds. Perovskite: Crystallography, Chemistry and Catalytic Performance. Novascience Publishers, Hauppauge (USA) 2013. pp. 117–140. DOI: 10.1002/chin.201424225

[25] Parkin IP, Solid state metathesis reaction for metal borides, silicides, pnictides and chalcogenides: ionic or elemental pathways. Chem Soc Rev 1996;25:199–207, DOI: 10.1039/CS9962500199

[26] Stanley MW. Hydrothermal synthesis of transition metal oxides under mild conditions. Solid State Mater Sci 1996;1[2]:227–32. DOI: 10.1016/S1359-0286(96)80089-1

[27] Thierry D, Ali A-Z, Berger M-H, Xavier B, Hochepied J-E. Hydrothermal synthesis of ferroelectric mixed potassium niobate–lead titanate nanoparticles. J Am Ceram Soc 2014;97[5]:1456–64. DOI: 10.1111/jace.12778

[28] Stefan GE, Abicht H-P, Dronskowski R, Müller T, Reller A, Weidenkaff A. Perovskite-related oxynitrides – recent developments in synthesis, characterisation and in-

vestigations of physical properties. Prog Solid State Chem 2009;37:173–205. DOI: 10.1016/j.progsolidstchem.2009.11.003

[29] David S. Related chemical themes in the synthesis of advanced ceramic materials. Mater Sci Eng 1989;A109:261–4. ISBN: 9780521424189

[30] Muthurajan H, Kumar HH, Samuel V, Gupta UN, Ravi V. Novel hydroxide precursors to prepare $NaNbO_3$ and $KNbO_3$. Ceram Int 2008;34:671–3. DOI:10.1016/j.ceramint.2006.12.014

[31] Irena P, Darko M, Miha D. Chemical synthesis of $KNbO_3$ and $KNbO_3$–$BaTiO_3$ ceramics. J Eur Ceram Soc 2005;25:2713–7. DOI:10.1016/j.jeurceramsoc.2005.03.128

[32] Bhattacharyya K, Tyagi AK. A novel soft-chemical method for the synthesis of nanocrystalline $MNbO_3$ (M=Na, K). J Alloys Compd 2009;470:580–3. DOI:10.1016/j.jallcom.2008.03.024

[33] Meinan L, Dongfeng X, Shouchen Z, Haiyang Z, Jiyang W, Kenji K. Chemical synthesis of stoichiometric lithium niobate powders. Mater Lett 2005;59:1095–7.

[34] Lu C-H, Lo S-Y, Wang Y-L. Glycothermal preparation of potassium niobate ceramic particles under supercritical conditions. Mater Lett 2002;55:121–5. DOI:10.1016/S0167-577X(01)00633-4

[35] Byrappa K. Hydrothermal Processing. Kirk-Othmer Encyclopedia of Chemical Technology. Wiley; 2005. pp. 1–43. DOI: 10.1002/0471238961.0825041804012319.a01.pub2. ISBN 9780471238966

[36] Byrappa K, Masahiro Y. Hydrothermal technology—principles and applications. In: Gary EM, Stephen MR, Roitan FB, eds. Handbook of Hydrothermal Technology. Technology for Crystal Growth and Materials Processing. William Andrew Publishing/ Noyes Publications. 2001. 11 p. ISBN 0-8155-1445-X

[37] Suchanek WL, Lencka MM, Riman RE. Hydrothermal Synthesis of ceramic materials. In: Donald AP, Roberto F-P, Allan HH, eds. Aqueous System at Elevated Temperature and Pressures: Physical Chemistry in Water, Steam and Hydrothermal Solutions. Elsevier; 2004. pp. 717–744. ISBN: 978-0-12-544461-3

[38] Malgorzata ML, Richard ER. Intelligent synthesis of smarteramic materials. In: Mel S, ed. Encyclopedia of Smart Materials. Volume 1. John Wiley. 2002. pp. 568–580. ISBN 0471177806, 9780471177807

[39] Byrappa K, Adschiri T. Hydrothermal technology for nanotechnology. Prog Crystal Growth Character Mater 2007;53:117–66. DOI:10.1016/j.pcrysgrow.2007.04.001

[40] Malgorzata ML, Richard ER. Thermodynamic modeling of hydrothermal synthesis of ceramic powders. Chem Mater 1993;5(1):61–70. DOI: 10.1021/cm00025a014

[41] Malgorzata ML, Richard ER. Thermodynamics of the hydrothermal synthesis of calcium titanate with reference to other alkaline-earth titanates. Chem Mater 1995;7[1]: 18–25. DOI: 10.1021/cm00049a006

[42] Malgorzata ML, Richard ER. Hydrothermal synthesis of Perovskite materials: Thermodynamic modeling and experimental verification. Ferroelectrics 1994;151:159–64. DOI: 10.1080/00150199408244737

[43] Malgorzata ML, Erik N, Andrzej A, Richard ER. Hydrothermal synthesis of carbonate-free strontium zirconate: thermodynamic modeling and experimental verification. Chem Mater 1997;9[5]:1116–25 DOI: 10.1021/cm960444n

[44] Malgorzata ML, Richard ER. Synthesis of lead titanate: thermodynamic modeling and experimental verification. J Am Ceram Soc 1993;76[10]:2649–59. DOI: 10.1111/j.1151-2916.1993.tb03994.x

[45] Cho S-B, Noh J-S, Malgorzata ML, Richard ER. Low temperature hydrothermal synthesis and formation mechanisms of lead titanate ($PbTiO_3$) particles using tetramethylammonium hydroxide: thermodynamic modelling and experimental verification. J Eur Ceram Soc 2003;23:2323–35. DOI:10.1016/S0955-2219(03)00085-2

[46] Gelabert MC, Laudise RA, Riman RE. Phase stability, solubility and hydrothermal crystal growth of $PbTiO_3$. J Cryst Growth 1999;197[1–2]:195–203. DOI: 10.1016/S0022-0248(98)00914-2

[47] James OE, Jr, Catherine CH-H, Bonnie LG, Malgorzata ML, Richard R. Kinectics and mechanism of hydrothermal synthesis of barium titanate. J Am Ceram Soc 1996;79[11]:2929–39. DOI: 10.1111/j.1151-2916.1996.tb08728.x

[48] Bonnie LG, Malgorzata ML, Richard ER. Low-temperature hydrothermal synthesis of phase-pure (Ba,Sr)TiO_3 perovskite using EDTA. J Am Ceram Soc 2004;87[11]:2025–32. DOI: 10.1111/j.1151-2916.2004.tb06355.x

[49] Malgorzata ML, Andrzej A, Richard ER. Hydrothermal precipitation of lead zirconate titanate solid solutions: thermodynamic modeling and experimental synthesis. J Am Ceram Soc 1995;78[10]:2609–18. DOI: 10.1111/j.1151-2916.1995.tb08030.x

[50] Andrea T, Vincenzo B, Maria TB, Massimo V, Paolo N. Kinetic modeling of aqueous and hydrothermal synthesis of barium titanate ($BaTiO_3$). Chem Mater 2005;17[21]: 5346–56. DOI: 10.1021/cm051119f

[51] Andrea T, Maria TB, Vincenzo B, Massimo V, Carlo B, Paolo N. Kinetics and mechanism of aqueous chemical synthesis of $BaTiO_3$ particles. Chem Mater 2004;16[8]:1536–43. DOI: 10.1021/cm031130k

[52] Daniele LM. On the use of bi-variate population balance equations for modelling barium titanate nanoparticle precipitation. Chem Eng Sci 2009;64:697–708. DOI:10.1016/j.ces.2008.04.052

[53] Ouchiyama N, Rough SL, Bridgwater J. Apopulation balance approach to describing bulk attrition. Chem Eng Sci 2005;60:1429–40. DOI:10.1016/j.ces.2004.08.037

[54] Doraiswami R, Alan WM. Population balance modeling. Promise for the future. Chem Eng Sci 2002;57[4]:595–606. DOI: 10.1016/S0009-2509(01)00386-4.

[55] Lee W, Kim HK, Cho SB. Hydrothermal preparation of $BaTiO_3$ powders from modified hydroxide precursors. Ferroelectrics 2006;333:233–41. DOI: 10.1080/00150190600701319

[56] Changlong C, Yuling W, Xiuling J, Dairong C. Hydrothermal synthesis of $BaTiO_3$: Crystal phase and the Ba^{2+} ions leaching behavior in aqueous médium. Mater Chem Phys 2008;110:186–91. DOI:10.1016/j.matchemphys.2008.01.031

[57] Durrani SK, Naz S, Hayat K. Thermal analysis and phase evolution of nanocrystalline perovskite oxide materials synthesized via hydrothermal and self-combustion methods. J Therm Anal Calorim 2014;115:1371–80. DOI 10.1007/s10973-013-3452-1

[58] Wen C, Tingke R, Aiwu W, Jie H, Junqing W, Jiasong Z, Weidong X. A simple and controllable hydrothermal route for the synthesis of monodispersed cube-like barium titanate nanocrystals. Ceram Int 2015;41:4514–22. http://dx.doi.org/10.1016/j.ceramint.2014.11.146

[59] Jooho M, Jeffrey AK, Henrik K, James HA. Hydrothermal synthesis of ferroelectric perovskites from chemically modified titanium isopropoxide and acetate salts. J Mater Res 1999;14[2]:425–35. DOI: 10.1557/JMR.1999.0061

[60] Lee S-K, Choi G-J, Hwang U-Y, Koo K-K, Park T-J. Effect of molar ratio of KOH to Ti-isopropoxide on the formation of $BaTiO_3$ powders by hydrothermal method. Mater Lett 2003;57:2201–7. DOI:10.1016/S0167-577X(02)01174-6

[61] Xiao W, Gang X, Zhaohui R, Yonggang W, Ge S, Gaorong H. Size-controlled synthesis of $BaTiO_3$ nanocrystals via a hydrothermal route. Mater Lett 2008;62:3666–9. DOI: 10.1016/j.matlet.2008.04.022

[62] Yonggang W, Gang X, Linlin Y, Zhaohui R, Xiao W, Wenjian W, Piyi D, Ge S, Gaorong H. Hydrothermal synthesis of single-crystal $BaTiO_3$ dendrites. Mater Lett 2009;63:239–41. DOI:10.1016/j.matlet.2008.09.050

[63] Xinhua Z, Junyi W, Zhenghai Z, Jianmin Z, Shunhua Z, Zhiguo L, Naiben M. Perovskite nanoparticles and nanowires: microwave–hydrothermal synthesis and structural characterization by high-resolution transmission electron microscopy. J Am Ceram Soc 2008;91:2683–9. DOI: 10.1111/j.1551-2916.2008.02494.x

[64] Lan X, Kongjun Z, Jing W, Qilin G, Yang C, Hongjuan Z, Jinsong L, Jinhao Q. Microwave-assisted sol–hydrothermal synthesis of tetragonal barium titanate nanoparticles with hollow morphologies. J Mater Sci Mater Electron 2015;26:1597–601. DOI: 10.1007/s10854-014-2581-z

[65] Ohara Y, Koumoto K, Shimizu T, Yanagida H. Hydrotherrnal synthesis of fibrous lead titanate Powders. J Mater Sci 1995;30:263–6. DOI 10.1007/BF00352160

[66] Gang X, Xiaoqiang H, Vladimir K, Shuquan C, Xin Y, Chunying C, Ge S, Gaorong H. Hydrothermal synthesis of single-crystalline tetragonal perovskite $PbTiO_3$ nanosheets with dominant (001) or (111) facets. Cryst Eng Comm 2014;16:4373–6. DOI: 10.1039/c4ce00234b

[67] Mario LM, Elaine CP, Gabriela SdN, Valeria ML, Julio RS, Valmor RM, Maria IBB, Juan A, José AV, Elson L. Structural and optical properties of $CaTiO_3$ perovskite-based materials obtained by microwave-assisted hydrothermal synthesis: an experimental and theoretical insight. Acta Mater 2009;57:5174–85. DOI:10.1016/j.actamat.2009.07.019

[68] Tatiana MM, Mario LM, Ivo MP, Francini CP, Edson RL, Ieda LVR, José AV, Leinig AP, Elson L. $CaTiO_3$:Eu^{3+} obtained by microwave assisted hydrothermal method: A photoluminescent approach. Opt Mater 2010;32:990–7. DOI:10.1016/j.optmat.2010.01.039

[69] Yonggang W, Gang X, Linlin Y, Zhaohui R, Xiao W, Wenjian W, Piyi D, Ge S, Gaorong H. Formation of single-crystal $SrTiO_3$ dendritic nanostructures via a simple hydrothermal method. J Cryst Growth 2009;311:2519–23. DOI:10.1016/j.jcrysgro.2009.01.103

[70] Rangel-Hernandez YM, Rendón-Angeles JC, Matamoros-Veloza Z, Pech-Canul MI, Diaz-de la ST, Yanagisawa K. One-step synthesis of fine $SrTiO_3$ particles using $SrSO_4$ ore under alkaline hydrothermal conditions. Chem Eng 2009;155:483–92. DOI: 10.1016/j.cej.2009.07.024

[71] Xiao W, Gang X, Zhaohui R, Ge S, Gaorong H. Effect of KOH concentration on the phase and morphology of hydrothermally synthesized $Pb_{0.70}La_{0.30}TiO_3$ fine powders. Mater Lett 2008;62:3719–21. DOI:10.1016/j.matlet.2008.04.039

[72] Jean-François T, Sophie d'A, Christian C, Philippe C, Mohamed R, Anne L. Influence of hydrothermal synthesis conditions on BNT-based piezoceramics. J Eur Ceram Soc 2011;31:1997–2004. DOI:10.1016/j.jeurceramsoc.2011.04.025

[73] Jianhong P, Mirabbos H, Baowei C, Juan W, Yunhua X. Surfactant-free hydrothermal synthesis of submicron $BiFeO_3$ powders. Appl Phys A 2011;103:511–6. DOI 10.1007/s00339-010-6024-2

[74] Xin Y, Gang X, Zhaohui R, Xiao W, Chunying C, Siyu G, Ge S, Gaorong H. The hydrothermal synthesis and formation mechanism of single-crystalline perovskite $BiFeO_3$ microplates with dominant (012) facets. Cryst Eng Comm 2014;16:4176–82. DOI: 10.1039/c3ce42488j

[75] Xian-Zhi C, Zhong-Cheng Q, Jian-Ping Z, Gangqiang Z, Xiao-Bing B, Peng L. Large-scale growth and shape evolution of bismuth ferrite particles with a hydrothermal method. Mater Chem Phys 2011;126:560–7. DOI:10.1016/j.matchemphys.2011.01.027

[76] Chao C, Jinrong C, Shengwen Y, Lingjuan C, Zhongyan M. Hydrothermal synthesis of perovskite bismuth ferrite crystallites. J Cryst Growth 2006;291:135–9. DOI: 10.1016/j.jcrysgro.2006.02.048

[77] Hong J, Yasutaka M, Nobuhiro K, Yoshinnori Y, Takahiro T, Nobukazu K. Hydrothermal synthesis of perovskite-type $BiFeO_3$. J Ceram Soc Jpn 2008;116[7]:837–9.

[78] Yonggang W, Gang X, Zhaohui R, Xiao W, Wenjian W, Piyi D, Ge S, Gaorong H. Low temperature polymer assisted hydrothermal synthesis of bismuth ferrite nanoparticles. Ceram Int 2008;34:1569–71. DOI:10.1016/j.ceramint.2007.04.013

[79] Chun YS, Jinbao X, Aerpati Y, Lei W. Hydrothermal synthesis of perovskite bismuth ferrite micro/nano powders. Integr Ferroelect 2014;153:33–41. DOI: 10.1080/10584587.2014.902668

[80] Zhiwu C, Yongpeng W, Jianqiang H. Ethanol-assisted hydrothermal synthesis and characterization of $BiFeO_3$ nanopowders. J Am Ceram Soc 2013;96[5]:1345–8. DOI: 10.1111/jace.12296

[81] Zhiwu C, Wuliang J. Low-temperature acetone-assisted hydrothermal synthesis and characterization of $BiFeO_3$ powders. J Mater Sci: Mater Electron 2014;25:4039–45. DOI 10.1007/s10854-014-2126-5

[82] Hongbin Q, Hailong Z, Bo-Ping Z, Lihua X. Hydrothermal synthesis of perovskite $BiFeO_3$–$BaTiO_3$ crystallites. J Am Ceram Soc 2011;94[11]:3671–4. DOI: 10.1111/j.1551-2916.2011.04839.x

[83] Zhiqiang Z, Li G, Haixia Y, Qiang L, Feng Y. Hydrothermal synthesis and magnetic properties of multiferroic rare-earth orthoferrites. J Alloys Compd 2014;583:21–3. http://dx.doi.org/10.1016/j.jallcom.2013.08.129

[84] Youjin Z, Ao Z, Xiaozhi Y, Hongmei H, Yun F, Chengpeng Y. Cubic $GdFeO_3$ particle by a simple hydrothermal synthesis route and its photoluminescence and magnetic properties. Cryst Eng Comm 2012;14:8432–9. DOI: 10.1039/c2ce26233a

[85] Mingyu S, Chenyang Z, Tingsong Z, Lin Y, Lei G, Hongming Y, Shouhua F. The multiferroic perovskite $YFeO_3$. Appl Phys Lett 2013;102:062903-1–062903-4. http://dx.doi.org/10.1063/1.4791697

[86] Guo-Qiang T, Yu-Qin Z, Hong-Yan M, Ao X, Hui-Jun R. Controllable microwave hydrothermal synthesis of bismuth ferrites and photocatalytic characterization. J Am Ceram Soc 2012;95[1]:280–9. DOI: 10.1111/j.1551-2916.2011.04775.x

[87] Biasotto GB, Simoes AZ, Foschini CR, Zaghete MA, Varela JA, Longo E. Microwave-hydrothermal synthesis of perovskite bismuth ferrite nanoparticles. Mater Res Bull 2011;46:2543–7. DOI:10.1016/j.materresbull.2011.08.010

[88] Zhi W, Wenfei X, Hui P(a), Xiaodong T(b). Polyanion modulated evolution of perovskite $BiFeO_3$ microspheres to microcubes by a microwave assisted hydrothermal method. J Mater Res 2013;28[11]:1498–504. DOI: 10.1557/jmr.2013.130

[89] Millot N, Xin B, Pighini C, Aymes D. Hydrothermal synthesis of nanostructured inorganic powders by a continuous process under supercritical conditions. J Eur Ceram Soc 2005;25:2013–6. DOI: 10.1016/j.jeurceramsoc.2005.03.202

[90] Aimable A, Xin B, Millot N, Aymes D. Continuous hydrothermal synthesis of nanometric $BaZrO_3$ in supercritical water. J Solid State Chem 2008;181:183–9. DOI: 10.1016/j.jssc.2007.11.015

[91] Zhouguang L, Yougen T, Limiao C, Yadong L. Shape-controlled synthesis and characterization of $BaZrO_3$ microcrystals. J Cryst Growth 2004;266:539–44. DOI: 10.1016/j.jcrysgro.2004.02.107

[92] Jia G, Guo Z, Jianguo Y, Yongcun Z, Guodong L, Guangshan Z, Yongnan Z. Fabrication and luminescence properties of $Rb_xBa_{1-x}ZrO_3$ hollow nanospheres. Mater Lett 2014;116:164–6. http://dx.doi.org/10.1016/j.matlet.2013.11.002

[93] Wenjun Z, Wenqin P, Guangyao M, Dingkun P. Hydrothermal synthesis and characterization of $LaCrO_3$. J Mater Chem 1999;9:2833–6. DOI: 10.1039/A904399C

[94] Bo HC, Shin-Ae P, Bong KP, Ho HC, Yong-Tae K. Controlled synthesis of $La_{1-x}Sr_xCrO_3$ nanoparticles by hydrothermal method with nonionic surfactant and their ORR activity in alkaline médium. Mater Res Bull 2013;48:3651–6. http://dx.doi.org/10.1016/j.materresbull.2013.04.084

[95] Rendón-Angelesa JC, Yanagisawa K, Matamoros-Veloza Z, Pech-Canula MI, Mendez-Nonell J, Diaz-de la ST. Hydrothermal synthesis of perovskite strontium doped lanthanum chromite fine powders and its sintering. J Alloys Compd 2010;504:251–6. DOI: 10.1016/j.jallcom.2010.05.103

[96] Shan W, Keke H, Beining Z, Jiaqi Z, Shouhua F. Mild hydrothermal synthesis and physical property of perovskite Sr doped $LaCrO_3$. Mater Lett 2013;101:86–9. http://dx.doi.org/10.1016/j.matlet.2013.03.083

[97] Kripasindhu S, Martin RL, Reza JK, Jeremy S, Richard IW. Direct hydrothermal synthesis and physical properties of rare-earth and yttrium orthochromite perovskites. Chem Mater 2011;23:48–56. DOI: 10.1021/cm102925z

[98] Deliang Z, Hong Z, Yuheng Z. Hydrothermal synthesis of $La_{0.5}Ba_{0.5}MnO_3$ nanowires. Appl Phys Lett 2002;80[9]:1634–6. DOI: 10.1063/1.1455690#

[99] Zhang T, Jin CG, Qian T, Lu XL, Bai JM, Li XG. Hydrothermal synthesis of single-crystalline $La_{0.5}Ca_{0.5}MnO_3$ nanowires at low temperature. J Mater Chem 2004;14:2787–9. DOI: 10.1039/b405288a

[100] Bernard C, Dauzet G, Mathieu F, Durand B, Puech-Costes E. Optimization of the hydrothermal synthesis of $La_{0.7}Sr_{0.3}MnO_{3+\delta}$ using an original optimal experimental design. Mater Lett 2005;59:2615–20. DOI: 10.1016/j.matlet.2005.02.063

[101] Darko M, Tanja G, Klementina Z, Darja L. Hydrothermal synthesis of $La_{1-X}Sr_XMnO_3$ dendrites. J Cryst Growth 2013;375:78–83. http://dx.doi.org/10.1016/j.jcrysgro.2013.04.019

[102] Jeroen S, Richard IW, Franck M. A study of the manganites $La_{0.5}M_{0.5}MnO_3$ (M=Ca, Sr, Ba) prepared by hydrothermal synthesis. J Mater Chem 2005;15:1542–51. DOI: 10.1039/b417003b

[103] Sun WA, Li JQ, Ao WQ, Tang JN, Gong XZ. Hydrothermal synthesis and magnetocaloric effect of $La_{0.7}Ca_{0.2}Sr_{0.1}MnO_3$. Powder Technol 2006;166(2):77–80. DOI: 10.1016/j.powtec.2006.05.015

[104] Li JQ, Sun WA, Ao WQ, Tang JN. Hydrothermal synthesis and Magnetocaloric effect of $La_{0.5}Ca_{0.3}Sr_{0.2}MnO_3$. J Magnet Magnet Mater 2006;302:463–6. http://dx.doi.org/10.1016/j.jmmm.2005.10.007

[105] Simon I, Akim K, Patrick G, Didier L. Conventional hydrothermal process versus microwave-assisted hydrothermal synthesis of $La_{1-x}Ag_xMnO_{3+\delta}$ (x=0, 0.2) perovskites used in methane combustion C R Chim 10 2007;10:1216–26. http://dx.doi.org/10.1016/j.crci.2007.08.002

[106] Zhiqiang Z, Li G, Feng Y, Hydrothermal synthesis, magnetism and resistivity of orthorhombic perovskite manganates $Y_{1-x}Ca_xMnO_3$ (x=0, 0.007, 0.55, 0.65). J Alloys Compd 2013;571:123–31.http://dx.doi.org/10.1016/j.jallcom.2013.03.220

[107] Jeroen S, Richard IW. Hydrothermal synthesis of the perovskite manganites $Pr_{0.5}Sr_{0.5}MnO_3$ and $Nd_{0.5}Sr_{0.5}MnO_3$ and alkali-earth manganese oxides $CaMn_2O_4$, 4H-$SrMnO_3$, and 2H-$BaMnO_3$. J Solid State Chem 2005;178:1683–91. http://dx.doi.org/10.1016/j.jssc.2005.03.006

[108] Yongwei W, Xiaoying L, Yan C, Fangli C, Shouhua F, Xiaoyang L, Hydrothermal synthesis of two perovskite rare-earth manganites, $HoMnO_3$ and $DyMnO_3$. J Solid State Chem 2005;178:1317–20. http://dx.doi.org/10.1016/j.jssc.2004.12.039

[109] Yan C, Hongming Y, Ge T, Ganghua Z, Shouhua F. Hydrothermal synthesis and magnetic properties of RMn_2O_5 (R=La, Pr, Nd, Tb, Bi) and $LaMn_2O_{5+\delta}$. J Solid State Chem 2007;180[4]:1340–6. http://dx.doi.org/10.1016/j.jssc.2007.02.005

[110] Yan C, Hongming Y, Guanghua L, Ge T, Shouhua F. Crystal growth and magnetic property of orthorhombic $RMnO_3$ (R=Sm–Ho) perovskites by mild hydrothermal

synthesis. J Cryst Growth 2007;305[1]:242–8. http://dx.doi.org/10.1016/j.jcrysgro.2007.03.052

[111] Kongjun Z, Yang C, Xiaohui W, Lin B, Jinhao Q, Hongli J. Hydrothermal synthesis of sodium niobate with controllable shape and structure. Cryst Eng Comm 2012;14:411–6. DOI: 10.1039/C1CE06100C

[112] Wu SY, Zhang W, Chen XM. Formation mechanism of $NaNbO_3$ powders during hydrothermal synthesis. J Mater Sci Mater Electronics 2010;21[5]:450–5. DOI: 10.1007/s10854-009-9937-9

[113] Gregory KLG, Fred FL, Sossina MH, Carlos GL, Gregory KLG, Fred FL, Sossina MH, Carlos GL. Hydrothermal synthesis of $KNbO_3$ and $NaNbO_3$ powders. J Mater Res 2003;18[2]:338–45. DOI: 10.1557/JMR.2003.0044

[114] Haibo C, Mingyu S, Chenyang Z, Hongming Y, Shouhua F. Hydrothermal syntheses and structural phase transitions of $AgNbO_{3.}$ J Am Ceram Soc 2012;95[11]:3673–77. DOI: 10.1111/j.1551-2916.2012.05392.x

[115] Gregory KLG, Sossina MH, Carlos GL, Fred FL. Hydrothermal synthesis of perovskite and pyrochlore powders of potassium tantalate. J Mater Res 2002;17:3168–76. DOI: 10.1557/JMR.2002.0458

[116] Jingying S, Guiji L, Nan W, Can L. Microwave-assisted hydrothermal synthesis of perovskite $NaTaO_3$ nanocrystals and their photocatalytic properties. J Mater Chem 2012;22:18808–13. DOI: 10.1039/c2jm33470d

[117] Kutty TRN, Balachandran R. Direct precipitation of lead zirconate titanate by the hydrothermal method. Mat Res Bull 1984;19:1479–88.

[118] Anjali AA, Malini SB. Synthesis of nanocrystalline PZT by hydrothermal method. Defence Sci J 2007;57[1]:35–9.

[119] Ralph N, Norbert BH. Hydrothermal synthesis of nanocrystalline perovskite powder systems. Mol Cyst Liq Cfysr 2000;353:329–40.

[120] Gang X, Zhaohui R, Piyi D, Wenjian WE, Ge S, Gaoirong H. Polymer-assisted hydrothermal synthesis of single-crystalline tetragonal perovskite $PbZr_{0.52}Ti_{0.48}O_3$ nanowires. Adv Mater 2005;17[7]:907–10. DOI: 10.1002/adma.200400998

[121] Ivan P, Malgorzata ML, Andrzej A, Richard ER. Hydrothermal synthesis of lead zirconate titanate ($PbZr_{0.52}Ti_{0.48}O_3$) using organic mineralizers. In: Applications of Ferroelectrics, 1996. ISAF '96, Proceedings of the Tenth IEEE International Symposium on (Volume:2); 18–21 Aug 1996; East Brunswick, NJ: ieee; DOI: 10.1109/ISAF.1996.598129

[122] Fukai K, Hidaka K, Aoki M, Abe K. Preparation and properties of uniform fine perovskite powders by hydrothermal synthesis. Ceram Int 1990;16:285–90. DOI: 10.1016/0272-8842(90)90041-D

[123] Xiaojuan W, Juan C, Zhiqiang S, Hua H, Lang L, Jianguo Y, Yongcun Z, Guodong L, Yongnan Z. Hydrothermal synthesis and photoluminescence properties of $BaZr_{1-x}Ti_xO3$ hollow nanospheres. Mater Lett 2012;86:21–4. http://dx.doi.org/10.1016/j.matlet.2012.07.035

[124] Ian ML, Shehan W, Clive BP. Hydrothermal synthesis and characterisation of perovskite $BaZn_{1/3}Ta_{2/3}O_3$. J Mater Chem 1999;9:2663–70.

[125] Wendelbo R, Akporiaye DE, Karlsson A, Plassen M, Olafsen A. Combinatorial hydrothermal synthesis and characterisation of perovskites. J Eur Ceram Soc 2006;26:849–59. DOI: 10.1016/j.jeurceramsoc.2004.12.031

[126] Millar CE, Wolny WW, Pardo L. Hydrothermal synthesis of modified lead titanate powders. In: Applications of Ferroelectrics, 1992. ISAF '92, Proceedings of the Eighth IEEE International Symposium on. 30 Aug–2 Sep 1992. Greenville, SC. DOI: 10.1109/ISAF.1992.300622

Fabrication of Yttrium-Doped Barium Zirconate for High Performance Protonic Ceramic Membranes

W. Grover Coors, Anthony Manerbino, David Martinefski and Sandrine Ricote

Additional information is available at the end of the chapter

http://dx.doi.org/10.5772/61660

Abstract

Barium zirconate has emerged as the leading candidate material for fabricating dense ceramic membranes for hydrogen separation. B-sites in the ABO_3 perovskite are acceptor-doped with a +3 cation – most commonly yttrium – charge-compensated by the formation of oxygen ion vacancies in the lattice. A minor fraction of B-sites can be filled with cerium to give $BaZr_{0.9-x}Ce_xY_{0.1}O_{3-d}$, $x \leq 0.2$. Upon hydration at elevated temperatures, weakly-bound protons are formed in the lattice. This produces a cubic perovskite ceramic proton conductor useful in diverse applications, such as protonic ceramic fuel cells, electrolysers, and catalytic membrane reactors operating at temperatures between 600 and 800 °C. A necessary requirement for fabricating thin ceramic membranes for proton diffusion is to maximize grain size while eliminating percolating porosity. However, high-density, large-grained barium zirconate is a very difficult material to prepare by traditional powder sintering methods. This chapter describes a new methodology for making protonic ceramic membranes with large grains and virtually no residual porosity. This discovery has the potential to have a profound impact on energy conversion efficiency of the various membrane devices envisioned for the coming hydrogen energy economy.

Keywords: Protonic ceramics, Yttrium-doped barium zirconate, Hydrogen separation membranes, Solid-state reactive sintering

1. Introduction

Almost thirty-five years ago, Hiroyasu Iwahara reported high temperature proton conduction in strontium cerate perovskite [1]. In that same year, 1981, there was sufficient interest in the potential for proton conductors to warrant the first international symposium on solid-state proton conductors in Paris (SSPC-I). Interestingly, the possibility of proton transport in perovskite ceramics was not widely recognized at this time. Several more years would pass before Iwahara's ground-breaking discoveries would be appreciated. Even at the fourth international symposium in 1989 (SSPC-IV) there was still general scepticism about proton conduction in perovskites [2]. It was not really until the early 1990's that the commercial importance of ceramic proton conductors sparked widespread interest among researchers. Acceptor-doped barium cerate soon became the model ceramic proton conductor. It exhibited high proton conductivity and was easy to fabricate into testable specimens, but before too long it was realized that barium cerate lacked the necessary chemical stability in the hot, moist, carbon dioxide-containing atmospheres of the typical use environment.

The first significant advancement into the modern era of ceramic proton conductors came from Klaus-Dieter Kreuer at the Max Planck Institute in 1999 [3] with the identification of the ceramic perovskite, barium zirconate, as a better candidate material due to its much greater chemical stability. However, unlike barium cerate, barium zirconate proved to be difficult to sinter by traditional ceramic methods. Some successes were achieved with solid solutions of barium cerate and zirconate [4], but still, it remained a challenge to make dense ceramic with more than 50% zirconium on B-sites, and any less than about 60% zirconium was still chemically unstable.

The second major breakthrough for barium zirconate ceramic processing came from Babilo and Haile at Caltech in 2005 [5] with the discovery of solid-state reactive sintering. Rather than using pre-reacted calcine powder, fully dense, large-grained barium zirconate ceramic membranes were fabricated directly from precursor powders with a small addition of NiO as a sintering additive. In this process, unreacted Ba from $BaSO_4$ and NiO form a binary eutectic glass that simultaneously promotes solid-state reaction and sintering. As the solid-state reaction proceeds, barium is extracted from the glass phase to take up A-sites in the more stable ABO_3 perovskite. Ultimately, the BaO-NiO glass becomes sufficiently barium deficient that the glass freezes out, leaving behind a thin film of NiO, which coats the grain boundaries [6].

In bulk ceramic material prepared this way, a small amount of NiO as a second phase has practically no impact, but when exposed to reducing atmosphere at high temperatures, the NiO nucleates to form Ni nanoparticles that decorate the dihedrals and grain facets. Under extreme operating conditions, depending on size and density, these nanoparticles can result in mechanical strain and fractures along grain boundaries. The formation of Ni nanoparticles hinders the use of membranes made by solid state reactive sintering in reducing atmosphere. Other sintering additives, which do not form nanoparticles, have been tried without much success. The eutectic temperature of BaO-NiO turns out to be almost perfect for the simultaneous solid-state reaction and liquid phase sintering process.

It became necessary to discover a way to remove the residual NiO from ceramic specimens after sintering. Such a process was discovered only last year, and will be described in what follows for barium zirconate-based materials with the nominal formula, $BaZr_{0.9-x}Ce_xY_{0.1}O_{3-d}$, where $x \leq 0.2$. The term BZY will be used to refer to the ceramic formulation with more than 70 mol % Zr.

2. Background on barium zirconate

Over the past ten years, the perovskite ceramic, yttrium-doped barium zirconate has emerged as the most promising ceramic proton conductor [7-14]. In the ABO_3 structure, barium occupies A-sites and zirconium occupies B-sites. Yttrium (Y^{+3}) substitutes for Zr^{+4} on B-sites as an acceptor dopant that is compensated by the formation of oxygen ion vacancies. These vacancies on the oxygen ion sublattice have the remarkable property that water from the gas phase enters the oxygen vacancies at elevated temperatures, donating two quasi-free protons to the lattice (Eq 1) [15-18]. This behaviour requires just the right oxygen ion separation, and is unique to certain oxide ceramics, such as perovskites, and does not occur in more densely packed structures like yttrium-doped zirconia. Hydrogen transport through the ceramic is by means of protons rather than gaseous diffusion. Protons are introduced into the lattice when water vapour reacts with an oxygen ion vacancy at the surface by the hydration reaction (In Kröger-Vink notation),

$$H_2O(g) + V_O^{\bullet\bullet} + O_O^{\times} \rightarrow 2OH_O^{\bullet} \tag{1}$$

One of the necessary requirements for ceramic proton conductors is gas-tightness against hydrogen. This places a new set of constraints on polycrystalline ceramic fabrication that goes well beyond the simple idea of "density". Ceramics processed by conventional means are often called dense, but fail the requirement for gas-tightness, particularly when thin membranes are used at elevated temperatures. This is because traditional sintering involves removal of the pores that form between adjacent ceramic grains during sintering. During intermediate-stage sintering, the dihedral pores actually form a series of interconnected channels that become easy paths for diffusion of molecular hydrogen and other gases. It is only during the final stage of sintering that the continuous pore channels are pinched off leaving a dihedral pore at each vertex of adjoining grains. These ideas are well-treated in many textbooks on ceramic sintering such as Barsoum [19]. Depending on the refractoriness of the ceramic material, these dihedral pores are nearly impossible to eliminate entirely, so the matter of ceramic density is more a qualitative determination of when the ceramic becomes "dense enough". It is the scattering of light from these pores that is the primary reason that oxide ceramics are typically opaque. Ceramic materials, such as magnesium aluminate spinel, can be made transparent with the use of sintering aids and high sintering pressures and temperatures, which effectively squeeze out the dihedral pores [20]. Transparency is a very good indication of high density. Even though "dense enough" in the case of ceramic membranes typically means that there are no pore channels that percolate through the matrix, the best evidence for density is transparency.

Barium zirconate is a very difficult-to-sinter material at reasonable temperatures. The refractoriness of barium zirconate is the main reason why adoption of the material as a ceramic proton conductor has been so slow. Techniques, such as hot isostatic pressing [21] and spark plasma sintering [21-23] have been explored with limited success. The most common approach to processing such materials is to begin with very high purity nanopowders prepared by sol-gel or spray pyrolysis methods [24-25]. This approach has worked well for fabricating membranes dense enough for hydrogen separation applications, but fine-grained ceramic has more grain boundaries perpendicular to the direction of proton transport than coarse-grained ceramic. The core region between ceramic oxide grains in polycrystalline ceramics tend to become positively charged. This is not a problem for anion conductors, such as ZrO_2 and CeO_2 [26], but for proton conductors, the positive grain boundary charge serves to repel the positively charged protons, resulting in significant intrinsic grain boundary impedance [27-30]. Large grain size, on the other hand, extends the low temperature range where suitable operation is possible by decreasing the total grain boundary resistance.

For optimal proton conductivity it is desirable to obtain a ceramic microstructure with the largest possible grain size to reduce the effect of grain boundary impedance. Grain growth in barium zirconate during final-stage sintering is very slow, so simply sintering nanosized material for a longer time is not a viable option. Using micron-size calcined powders is also not an option. Traditionally, ceramic powder with the desired phase is prepared beforehand by solid-state reaction of the precursor oxides. This is how the ceramic powders were prepared for the sintered specimen shown in Fig. 1. Stoichiometric quantities of ZrO_2, Y_2O_3, and $BaSO_4$ were blended and reacted together in a covered crucible to make phase-pure $BaZr_{0.9}Y_{0.1}O_{3-d}$ (BZY10) powder, which was subsequently milled to about one micron size following the traditional method for preparing ceramic materials. Organic binder was added to the powder, pressed into a pellet under high pressure to ensure high green bulk density, and then sintered in air at 1650 °C for up to 24 hours. It is obvious that ceramic prepared this way is not suitable for application as a hydrogen separation membrane.

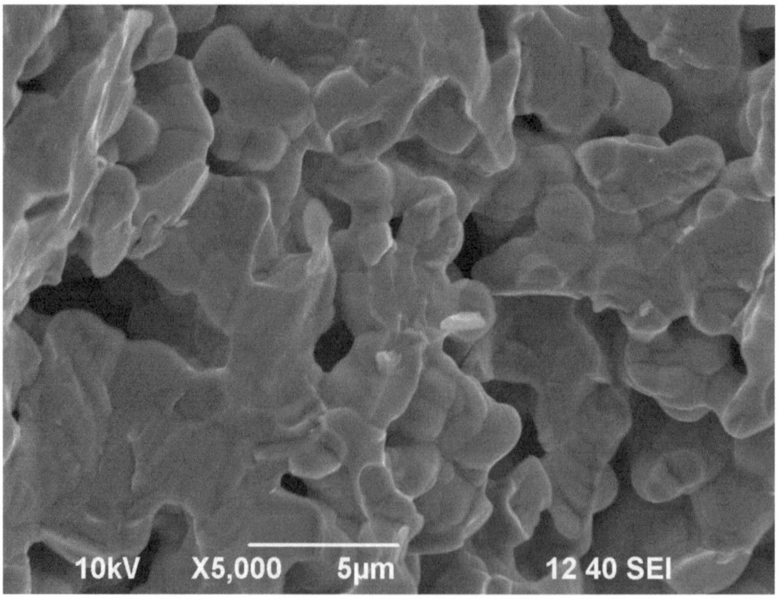

Figure 1. Poor sinterability of micron-size pre-calcined BZY10 powder at 1650 °C for 24 hours in air (Courtesy of Wade Rosensteel, Colorado School of Mines).

http://dx.doi.org/10.5772/61660

An entirely new approach to fabrication of membrane quality barium zirconate ceramic, called solid-state reactive sintering (SSRS), has been developed [5-6, 31-33]. In this method the precursor powders are reacted and sintered in the same operation so that the perovskite phase is produced *in situ* and grain boundary movement never occurs. This results in large grain size and the absence of dihedral pores as shown in Fig. 2. This process will be explained in detail in Section 3.

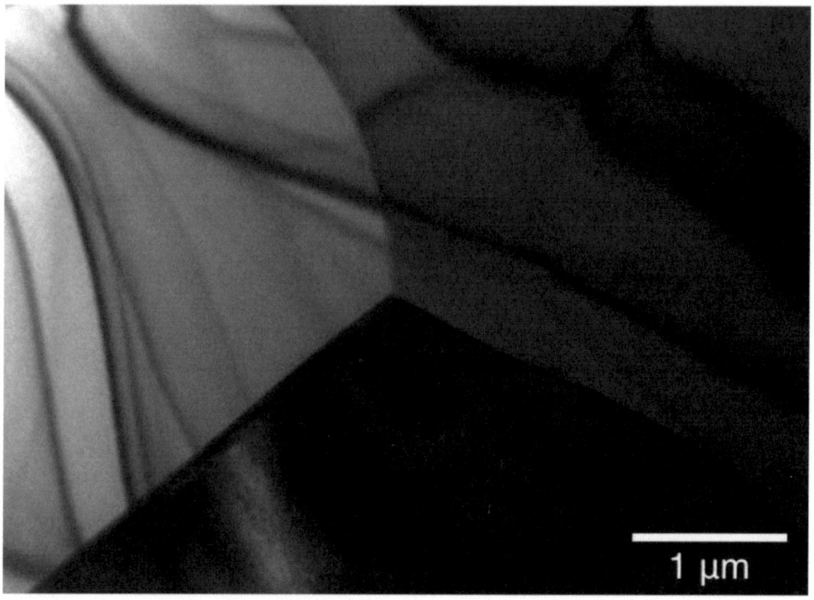

Figure 2. Low magnification bright-field TEM of dihedral in as-fired $BaZr_{0.7}Ce_{0.2}Y_{0.1}O_{3-d}$ (Courtesy of Daniel Clark, Colorado School of Mines).

3. Solid state reactive sintering

3.1. Mechanism of the solid-state reactive sintering

The idea behind reactive sintering is quite simple. However, the ability to execute SSRS depends on identifying a glassy phase that consists of one of the components of the final desired ceramic phase. In the case of BZY, the A-site constituent, BaO, forms a binary eutectic glass with nickel oxide with a low melting point of 1125 °C. The BaO-NiO binary phase diagram is shown in Fig. 3. In solid-state reactive sintering, stoichiometric quantities of $BaSO_4$, ZrO_2, CeO_2 and Y_2O_3 are mixed together with a small amount of NiO – typically 0.1 wt% with organic binders to form a green body. During the initial heating cycle, NiO reacts with $BaSO_4$, releasing SO_2, and forming a liquid at the eutectic composition. As the temperature increases, ZrO_2 and Y_2O_3 begin to react together with Ba from the glass to form the more stable BZY perovskite phase. As more and more of the BZY phase forms, the BaO-NiO glass becomes increasingly

NiO rich, following the composition of the liquidus upwards to the right. Once the final sintering temperature is reached at about 1600 °C, the glass freezes at a composition of about 80-90% NiO and liquid phase sintering ceases. The last remaining Ba is extracted more slowly from the glass by solid-state diffusion and only NiO is left behind as a second phase [6].

As discussed in more details in Section 4.1., $BaSO_4$ is used instead of $BaCO_3$ because it is insoluble in water. Solid state reactive sintering can also be performed with $BaCO_3$.

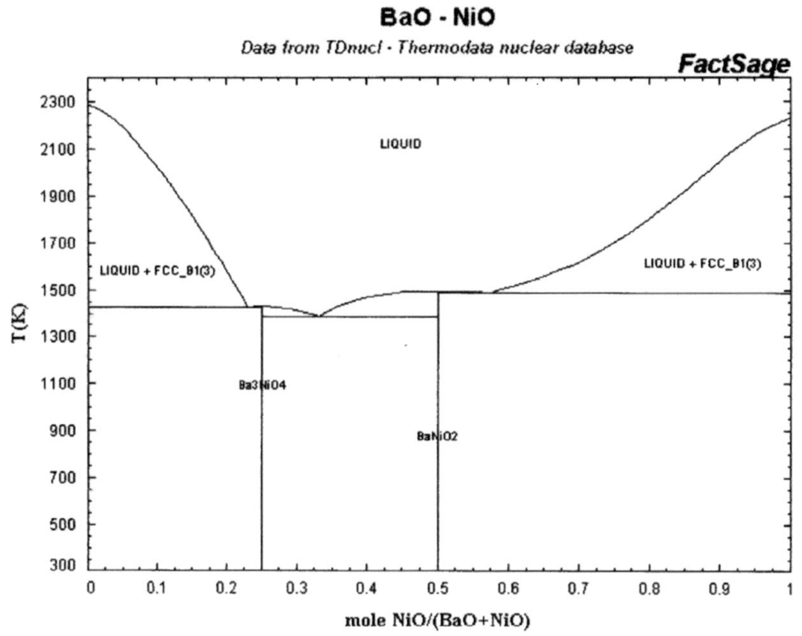

Figure 3. Binary phase diagram for BaO-NiO. (TDnucl – Thermodata nuclear database, FactSage).

The resulting BZY is phase pure, but the amount of residual NiO second phase, depends on the starting amount. Too much NiO leads to inclusions as shown in Fig. 4, but too little leads to incomplete BZY sintering. Ideally, there should be just enough NiO to coat the grain boundaries of the BZY phase with no more than about one monolayer of Ni atoms. Face centered cubic (FCC) NiO has a lattice parameter of 4.218 Å, which is a good match for BZY, 4.240 Å. It is straightforward to estimate the optimal amount of NiO if the BZY grains are treated as volume-filling truncated octahedra. The volume (V) of a 5 μm truncated octahedron is $8\sqrt{2}\,s^3$, where s is the side length ($3s$ is the side length of the regular octahedron). The distance between opposing square faces is $2\sqrt{2}s \approx 5$ μm. Therefore, V = 62.5 μm^3. The surface area of the same truncated octahedron is $A = (6+12\sqrt{2})\,s^2 = 83.7$ μm^2. The thickness of one monolayer of NiO (body diagonal of the fcc lattice) is 0.6 nm, giving a volume of NiO of 0.05 μm^3 per 5 μm BZY grain. The volume ratio is, thus, 0.08%. The density ratio of NiO to BZY is 6.67/6.16 = 1.08, so the amount of NiO required to coat 5 μm BZY gains with a monolayer of NiO is slightly less than 0.1 wt%. Any more than this leads to inclusions, as shown in Fig. 4, and any less is not sufficient to promote complete sintering. This calculation assumes that no NiO dissolves in the BZY crystallites. There is some evidence suggesting that NiO may enter the lattice as interstitial ions or even substitute on B-sites, but the amounts are negligible. Most of the NiO remains at grain boundaries. When just the right amount of NiO is used, the grain boundaries appear very clean, as shown in Fig. 5.

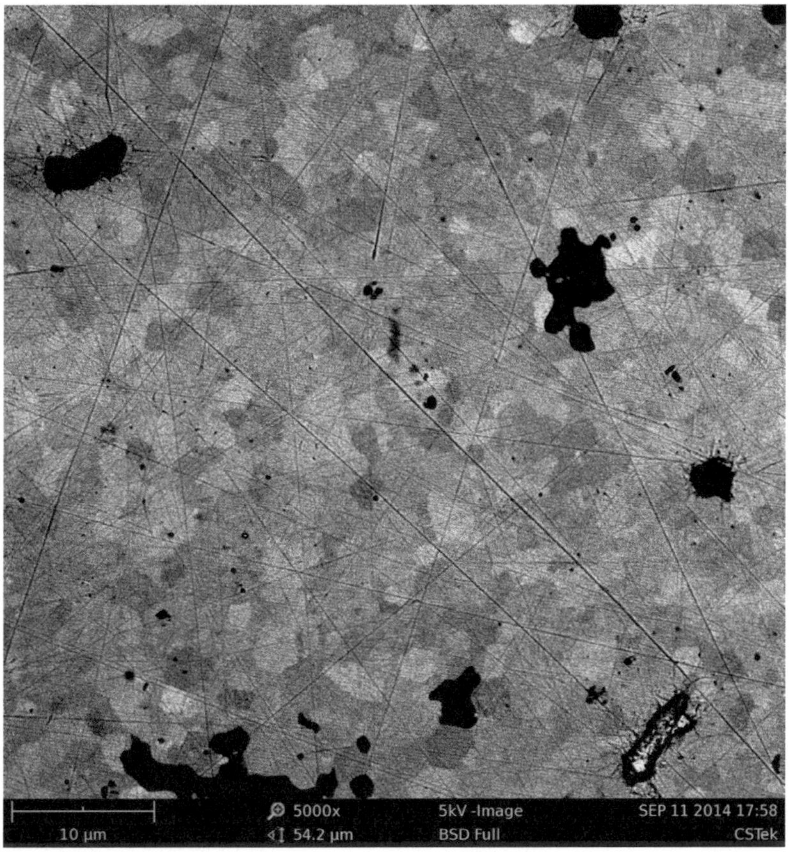

Figure 4. Back-scattered electron micrograph of polished BZY10 prepared by SSRS with 1 wt% NiO after specimen reduction. Dark regions are metallic nickel inclusions.

3.2. Ni nanoparticles

The small amount of NiO left behind at grain boundaries from SSRS has no practical effect as long as the ceramic is only exposed to oxidizing environments. But this defeats the purpose of a ceramic membrane intended for use in reducing atmosphere. When exposed to hydrogen at elevated temperatures, the NiO gets reduced. Upon reduction, metallic nickel nucleates along the grain boundaries and coarsens into nanoparticles that range in size from a few nanometers to 100 nm, depending on time and temperature. Fig. 6 shows the Ni-nanoparticles that form when a specimen of BZY prepared by SSRS with only 0.1 wt% NiO is annealed in hydrogen

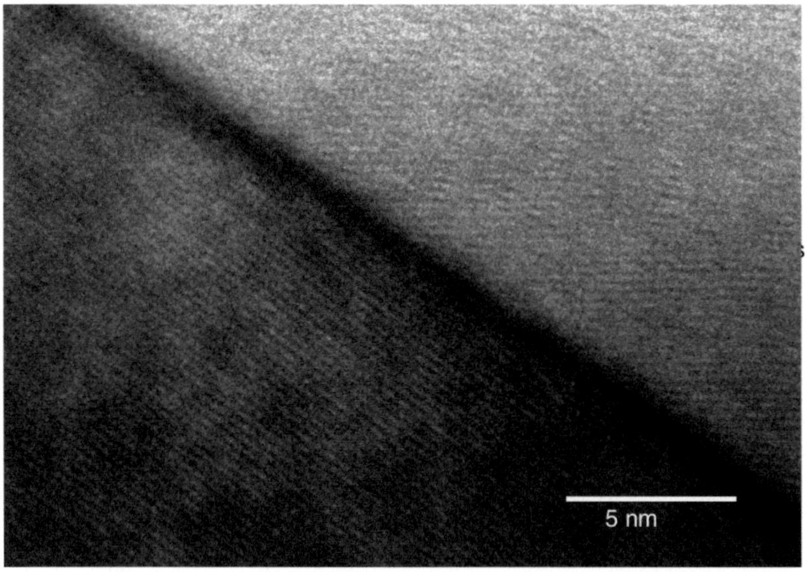

Figure 5. High resolution bright field TEM of as-fired $BaZr_{0.7}Ce_{0.2}Y_{0.1}O_{3-d}$ prepared by SSRS with 0.1 wt% NiO. Grain boundary region ≈ 5.8 Å (Courtesy of Daniel Clark, Colorado School of Mines).

at 1000 °C for 24 hours. It is interesting to note that the Ni-nanoparticles that grow on the crystal facets are all about the same size and uniformly distributed. This is reminiscent of star formation in a galaxy and provides clear evidence that the Ni-nanoparticles nucleated from a uniform coating of NiO between BZY grains. Using the calculation above for the volume of NiO per square micron of grain facet - 0.6×10^6 nm^3 – the volume of equivalent Ni metal is about 75% or about 0.45×10^6 $nm^3/\mu m^2$. A spherical 50 nm Ni particle has a volume of about 65,000 nm^3, so about 7 Ni nanoparticles per square micron of grain facet would be expected, which is about what is observed in Fig. 6.

One of the complications that arises in these protonic ceramic membranes is that the formation of Ni nanoparticles is irreversible. When the ceramic is exposed again to oxidizing conditions, rather than recoating the grain boundaries with a monolayer of NiO, the nanoparticles try to reoxidize in place, with a corresponding volume increase. This introduces local strain, which can lead to catastrophic failure. Fig. 7 shows the fracture that is characteristic of this condition. It is necessary to extract the NiO from the ceramic prior to reduction to levels below which fracture does not occur.

4. Nickel extraction

4.1. Specimen preparation

Barium zirconate and barium cerate can be reacted to make a complete solid solution over the entire composition range [4,6,9,14]. The most common nomenclature in use in the literature is

Figure 6. FESEM of Ni nanoparticles decorating grain boundary after reduction of $BaZr_{0.7}Ce_{0.2}Y_{0.1}O_{3-d}$ without NiO extraction heat-treatment.

BZCY*mn*, where *m* and *n* refer to the mole fractions of Zr and Ce on B-sites (times 10), respectively. For example, the formula for BZCY72 is $BaZr_{0.7}Ce_{0.2}Y_{0.1}O_{3-d}$. When there is no ceria, the common designation is BZY10, signifying 10 mol% yttrium. A small amount of ceria - up to 20 mol% - is advantageous in some cases. The material becomes chemically unstable in harsh environments for Ce > 20 mol% [34-35]. Furthermore, the electronic conductivity in dry hydrogen can be significant for Ce > 10 mol% [36]. In this monograph, the term, BZY, will be used for any ceramic formulation with more than 70 mol% Zr since structurally, there is practically no difference.

Solid state reactive sintering works with almost any ceramic green forming technique, so it is suitable for making everything from extruded tubes to cast tape. The method described here is for making 20 mm diameter by 1 mm thick discs by the time-honored method of slip casting. Slip casting is very easy and does not require expensive tooling. The discs are cast by simply filling PVC rings, arranged on a plaster-of-paris plate, with slip. The slip is prepared by mixing stoichiometric amounts of $BaSO_4$, ZrO_2, CeO_2, and Y_2O_3 with additional 0.1 wt% NiO, DI water and conventional water-soluble, slip-casting binders. The reason for using $BaSO_4$ rather than $BaCO_3$ is that barium sulfate is insoluble in water. It is important in slip casting that no barium (or any cation) diffuse from the slip into the plaster-of-paris mold or base, thus altering the stoichiometry. Also, for health and safety reasons, it is inadvisable to use a water-soluble form of barium due to its toxicity. The slip is mixed in a Nalgene bottle on a jar roller for 2 hours with 10 mm zirconia balls. In reactive sintering, there is no need to adjust the particle size of

Figure 7. FESEM of intragranular fracture due to reoxidation of Ni nanoparticles in BZY10 (Courtesy of Thomas McGilvray, UCSD).

the precursor powders. It is sufficient just to obtain a homogeneous slip. The slip is filtered through a 325 mesh screen to remove the zirconia balls and any large agglomerates or undissolved binder, and continuously stirred prior to use to prevent settling.

Discs are cast by injecting a pre-calibrated amount of slip from a syringe into rings placed on a plaster-of-paris plate, and allowed to dry for several hours. Once dry and hard, the top surfaces of the cast discs are sanded flat to a thickness of about 3.5 mm. The discs are then stacked on top of BZY setters inside of a refractory enclosure to prevent excessive NiO loss during sintering. The initial sintering is carried out in air at 1600 °C for 16 hours. After sintering, the surfaces of the discs are ground with a diamond wheel on both faces to an overall thickness of 1.2 mm.

At this point in the process, the ceramic discs are fully dense and black. A second high temperature heat treatment is required to extract the NiO to a tolerable level. This is done by packing the ceramic in a bed of BZY powder inside of a ceramic retaining ring as shown in Fig. 8. A second firing cycle is carried out, also at 1600 °C. NiO extraction relies on relatively fast diffusion of NiO along grain boundaries until it reaches the surface of the specimen where it sublimes and is captured by the surrounding sacrificial BZY powder. The length of time required depends on the degree of extraction desired and the dimensions of the specimen. In the case of 1.2 mm thick discs, 16 hours is sufficient for achieving NiO level below 100 ppm. Fig. 9 shows a comparison of grain size distribution for the as-fired ceramic and after NiO extraction at 1600 °C for 16 hours. The grain-size measurements were carried out on powder obtained by crushing the specimens, assuming that fracture occurred predominantly at grain boundaries. This is supported by the observation that the typical fracture surfaces of this material are predominantly inter-granular. The as-fired specimen experienced 16 hours at 1600 °C, whereas the NiO-extracted specimen effectively experienced 32 hours at this temperature. It may be observed that the average grain size increased only slightly, from 5 μm to 6.5 μm, demonstrating that grain growth during sintering in BZY is rather slow.

Figure 8. Set up for NiO extraction process. Center photo shows the black sintered disc placed on top of the BZY powder bed. Each part is covered with powder.

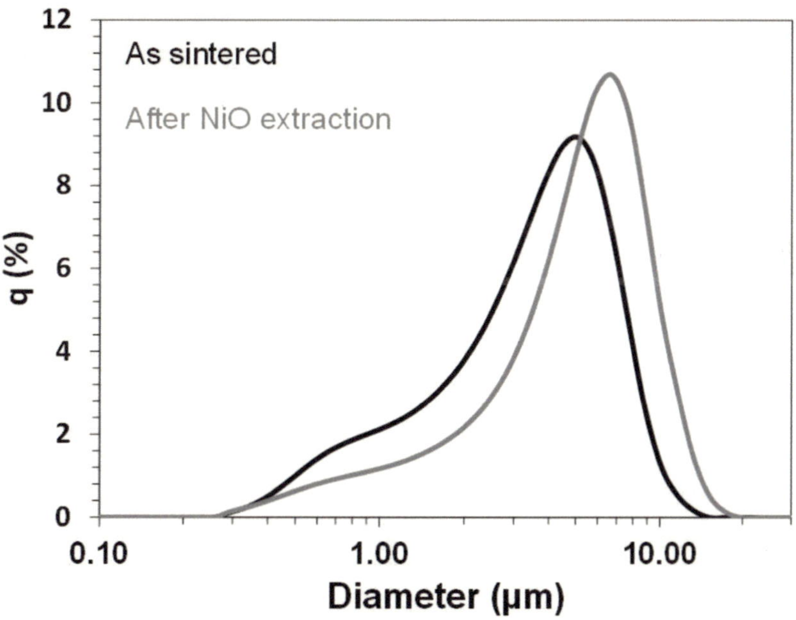

Figure 9. Grain size distributions of BZCY72 for as-fired (black) and after NiO extraction at 1600 °C for an additional 16 hours followed by an annealing treatment (grey) (Obtained with a Horiba Particle Size analyzer).

4.2. Characterization of the specimens after Ni extraction

4.2.1. Structural characterization

After the NiO extraction step, the discs are ground once more on both surfaces to the final 1.0 mm thickness, and ground on the outside diameter if necessary. A final annealing step in 4% H_2-bal Ar at 800 °C for 24 hours is carried out to condition the ceramic discs for the testing environment. Fig. 10 shows the appearance of two specimens, BZCY81 and BZCY72, after NiO extraction at 1600 °C, followed by annealing in H_2. The two specimens are transparent amber in color, and virtually Ni-free. Specimens with nickel extraction at only 1425 °C are dark and almost metallic-looking, demonstrating incomplete NiO extraction.

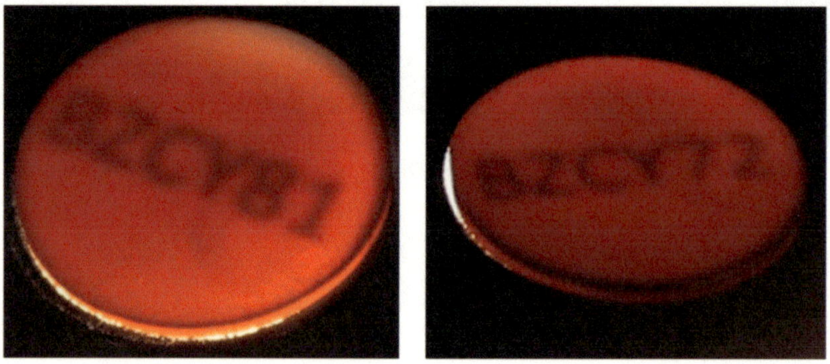

Figure 10. BZCY81 and BZCY72 specimens after NiO extraction at 1600 °C, followed by annealing in H_2. At 800 °C for 24 hrs.

XRD was used to ensure that the process of NiO extraction and subsequent annealing did not modify the BZY phase in any way. Five batches of specimens were prepared for this purpose: 1) As-fired, without NiO extraction or H_2 anneal, 2) NiO extraction only, 1425 °C for 16 hours, 3) NiO extraction, 1425 °C for 16 hours, plus H_2 anneal, 4) NiO extraction only, 1600 °C for 16 hours, and 5) NiO extraction, 1600 °C for 16 hours, plus H_2 anneal. The XRD patterns for 2θ = 20 to 120 degrees are shown in Fig. 11a. It may be observed that all cubic BZY peaks are present. The extra peak at 2θ =26.8, present on all diffractograms, comes from small barium peroxide (ICDD file 00-007-0233). More significant is the finding in Fig. 11b where the (321) peaks are magnified. It may be observed that there is a very slight increase in the peak position with increasing NiO extraction. The cubic lattice parameter for as-fired material is 4.248 Å and that for the material after NiO extraction for 16 hours at 1600 °C is 4.245 Å, or 0.07% contraction. This could suggest that at least some of the NiO may have originated from the perovskite lattice, but it should be kept in mind that this change is much less than the change typically observed due to hydration and dehydration, so this conclusion is tenuous.

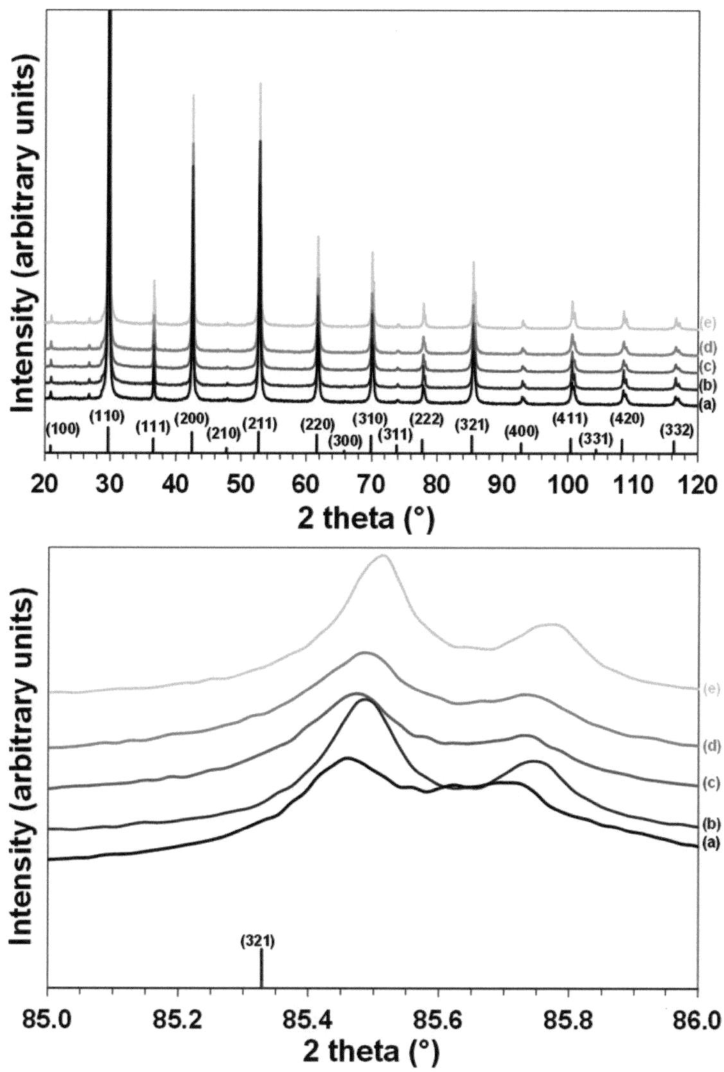

Figure 11. XRD of BZCY72 specimens prepared with different NiO extraction temperature. Full pattern (upper) and magnified (321) peak (lower). (a) as fired, (b) Ni extraction 16h 1475 °C, (c) Ni extraction 16h 1475 °C and annealed, (d) Ni extraction 16h 1600 °C, (e) Ni extraction 16h 1600 °C and annealed. The black bars correspond to the 04-011-7317 ICDD file.

After reduction in H_2 nearly all of the nickel may be assumed to be metallic, and ferromagnetic. This provides a convenient method for quantitative analysis of residual nickel content. It is extremely difficult to measure low levels of NiO reliably in sintered ceramics when the NiO is not uniformly distributed. By reducing the specimens, only metallic nickel remains, which can be quantitatively determined by magnetometry down to the parts per million level. A Quatum Design SQUID-based VSM with EverCool™ at 1.8 K to 400 K, oven temp of 1000 K, uniform field of 0.05 Oe to 70,000 Oe and magnetic moment, 10-8 emu to 10 emu, AC frequency: 0.1 Hz to 1000 H was used. Fig. 12 shows the magnetization plots after reduction in H_2 as a function of NiO extraction temperature (each for 16 hours); 1) 1425 °C, 2) 1550 °C and 3) 1600 °C. The estimate of the amount of Ni in the as-fired specimens was about 0.1 wt%. The NIST standard of 54.888 emu/g-Ni was used to determine the amount of Ni in the specimens at 5000 Oe. For example, the specimen with NiO extraction at 1475 °C had a magnetic moment of 0.022 emu/g-sample. Dividing this by 54.888 emu/g-Ni gives 0.039 wt % (times 58.69 g/mol-Ni/248.06 g/mol BZCY27 = 0.0092 mol%, or 92 ppm). Even at 92 ppm, the specimen was black and opaque. It was necessary to reduce the nickel to less than 0.002%, or about 4 ppm, in order to achieve transparency. It is seen that the extent of nickel extraction is highly dependent on NiO extraction temperature. Longer extraction times may be necessary for thicker specimens, but 1 mm is a good thickness for test specimens hydrogen flux measurements and electrode development.

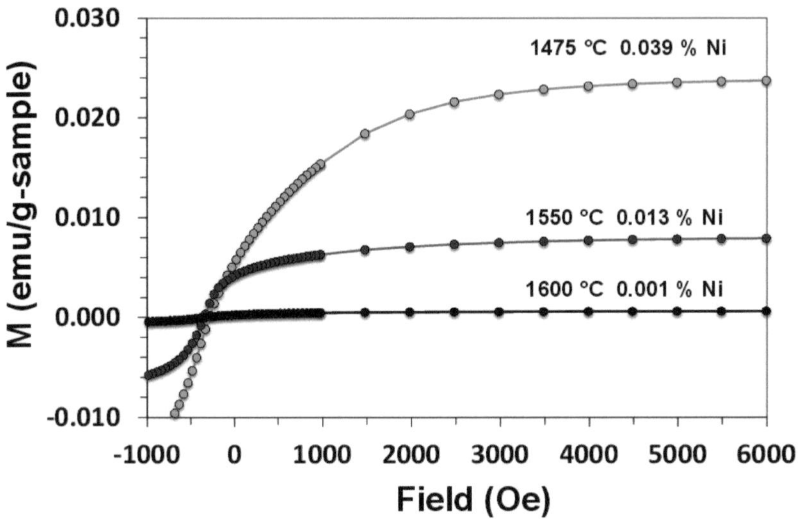

Figure 12. Magnetization curves for different NiO extraction temperatures. Ni concentrations calculated at 5000 Oe (0.5 Tesla) using NIST standard of 54.888 emu/g-Ni (courtesy of Jim O'Brien, UCSD).

4.2.2. Microstructural characterization

Back-scattered electron images on a polished BZCY72 specimen after Ni extraction and annealing in reducing atmosphere reveal the optimal microstructure: large grains (from 2 to 10 microns) with no dihedral pores and no Ni accumulation. An example is shown in Fig. 13. The grain boundaries of the specimen after nickel extraction and annealing in reducing atmosphere are free from nickel nano-particles (Fig. 14).

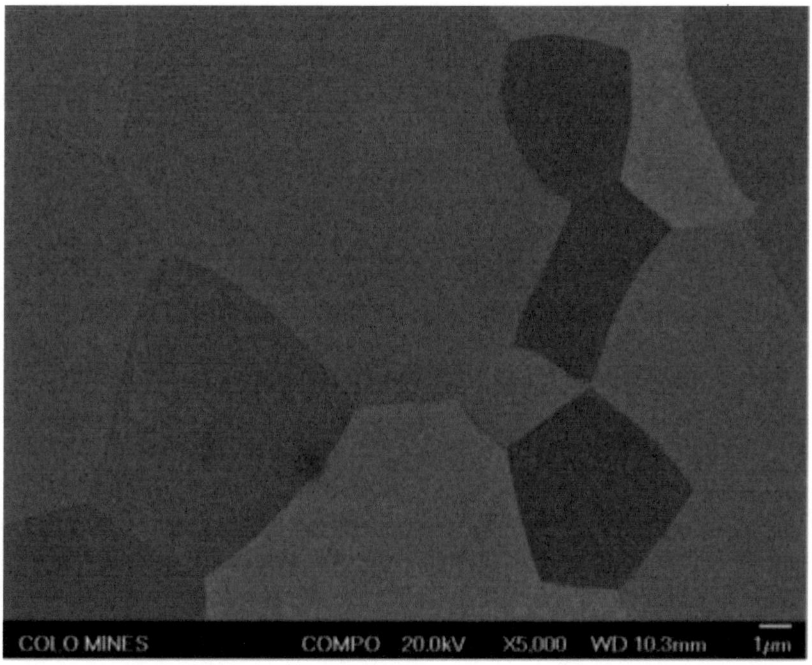

Figure 13. Back-scattered electron micrograph of a polished cross section of a BZCY72 ($BaZr_{0.7}Ce_{0.2}Y_{0.1}O_{3-d}$) after Ni extraction (16h 1600 °C) and annealing in reducing atmosphere.

Figure 14. Secondary electron micrographs of a fractured cross-section of a BZCY72 ($BaZr_{0.7}Ce_{0.2}Y_{0.1}O_{3-d}$) after Ni extraction (16h 1600 °C) and annealing in reducing atmosphere.

4.2.3. Electronic characterization

In moist atmosphere, BZY materials incorporate protonic defects ($OH_O^{\bullet}$) in their lattice according to Eq. 1. For such a reaction to happen, oxygen vacancies are necessary. Some ceramic proton-conductors, such as pyrochlores [37-38], have intrinsic oxygen vacancies. For barium zirconate based materials, the oxygen vacancies are extrinsically created by substituting the B-site of the perovskite (4+) by a trivalent cation (most commonly yttrium). Once in the lattice, the protonic defects diffuse according to the Grotthuss mechanism [39]. The protons are the only mobile species while the oxygen is localized in the vicinity of its crystallographic position [3].

In oxidizing atmosphere, oxygen dissociates in the oxygen vacancies. The corresponding reaction is most of the time reported as equation (2) [40-43], which can be misleading because the electron holes are not valence holes. Indeed, in models, these electron holes are treated as localized, since the Nernst Einstein equation is used [44]. However, no localization is defined. To avoid confusion, equation 2 can be rewritten as equation 3, assuming the electron holes are localized on the O-site [45-46].

$$\tfrac{1}{2}O_2(g) + V_O^{\bullet\bullet} \rightarrow O_O^{\times} + 2h^{\bullet} \qquad (2)$$

$$\tfrac{1}{2}O_2(g) + V_O^{\bullet\bullet} + O_O^{\times} \rightarrow 2O_O^{\bullet} \qquad (3)$$

The conductivity was measured on 1 mm-thick pellets prepared as described in section 3.1 but with a smaller diameter (13 mm). Pt electrodes were painted on both sides of the pellet and fired in air for 30 min at 1000 °C. Impedance spectra were recorded from 20 Hz to 1 MHz using Hewlett Packard 4284A Precision LCR Meter interfaced with Labview and were fitted using the Zsimpwin software. The oxygen partial pressure dependences of the conductivity, with an example for BZCY72 in Fig. 15, exhibit the typical behavior for a BZY material in moist atmosphere:

- A plateau at oxygen partial pressure below 10^{-5} atm, which corresponds to the ionic conductivity. It is agreed that for BZY material, the ionic conductivity is mainly protonic at temperature below 700 °C [47-48],
- An increase of the conductivity with the oxygen partial pressure in oxidizing atmosphere, from the incorporation of the oxygen (Eq. 3).

According to Fig. 15, the protonic conductivity of the BZCY72 specimens after Ni extraction is 2.5 and 1.3 $mS.cm^{-1}$ at 700 and 600 °C respectively. These values are consistent with those reported in the literature for BZCY72 [9,32,34-35,49], indicating that the extraction process did not deteriorate the specimens.

Some electronic conductivity in reducing atmosphere was measured in barium cerate based materials [50-51], generating from the reduction of the cerium cations and the formation of small polarons. In the present work, the conditions were not reducing enough to generate cerium small polarons. However, this phenomenon has been observed on thin BZCY72 membranes [36]. Depending on the desired operation (gas composition, humidification rate and temperature), the percentage of cerium needs to be adjusted.

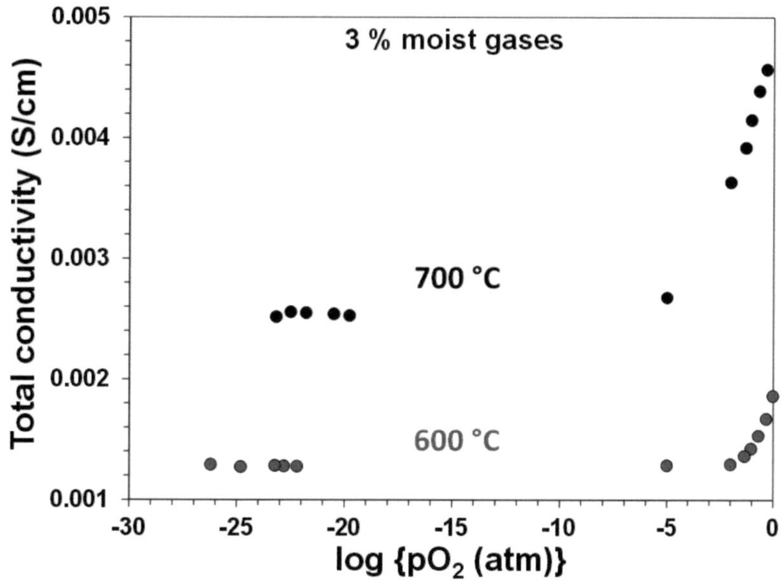

Figure 15. pO_2 dependence of the conductivity measured on a BZCY27 pellet after Ni extraction.

As mentioned in the introduction, the high grain-boundary resistivity in the BZY materials has been explained by the presence of a space-charge layer [27-30]. The width and the magnitude of the space-charge layer depends on the synthesis/sintering processes [52]. The grain-boundary contribution to the conductivity can be decreased by having a large grain microstructure. Indeed, the larger the grains, the smaller the grain boundary density. But other factors, such as the dopant accumulation, the barium evaporation or the segregation of impurities at the grain-boundary, play an important role in the space-charge layer.

The conductivity was recorded as a function of temperature in 3% moist 5% H_2, balance Ar. At high temperature, the spectra were fitted with 2 (RQ) elements in series, with R, a resistance and Q, a constant phase element. The different processes could be analyzed using the pseudo-capacitance. The first (RQ) was assigned to the bulk (~10^{-11} F) and the second one to the Pt electrodes (~10^{-6} F). With decreasing temperature, the grain-boundary contribution becomes visible. Because of the frequency range used for the measurement, the electrode contribution does not appear anymore at low temperature (lower frequencies would be needed). Therefore, at low temperature, the spectra were also fitted with two (RQ) in series, the first one for the bulk and the second one for the grain boundaries (~10^{-9} F). Example of the fits of the impedance spectra at 600 and 400 °C are displayed in Fig. 16. The Arrhenius plot of the conductivity in

3% moist 5% H_2, balance Ar, cf. Fig. 17, gives an activation energy for proton diffusion of 0.47 eV for the bulk and 0.78 eV for the grain-boundaries. Similar values were reported in the literature [13,28,30,52].

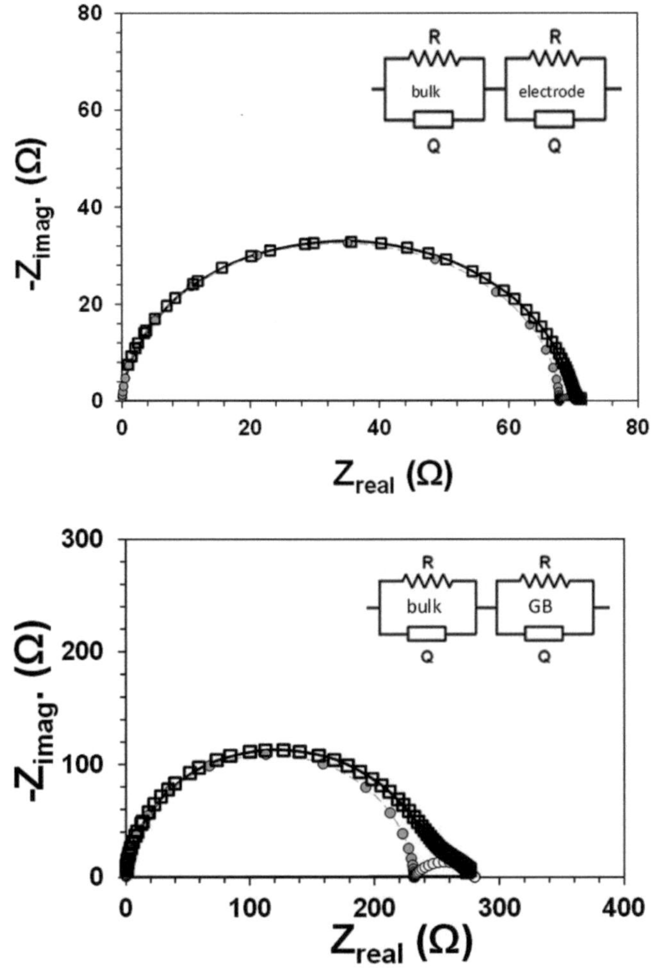

Figure 16. Impedance spectra of BZCY72 after Ni extraction in 3% moist 5% H_2, balance Ar. (upper) 600 °C and (lower) 400 °C. GB stands for grain boundary

5. Summary

A process has been described for making dense, large-grained BZY suitable for use in hydrogen diffusion membranes. It is nearly impossible to make meaningful measurements on membranes films, themselves, on the order of tens of microns thick. The thick specimens shown in Fig. 10 are intended for testing purposes, where it is desirable to have

http://dx.doi.org/10.5772/61660

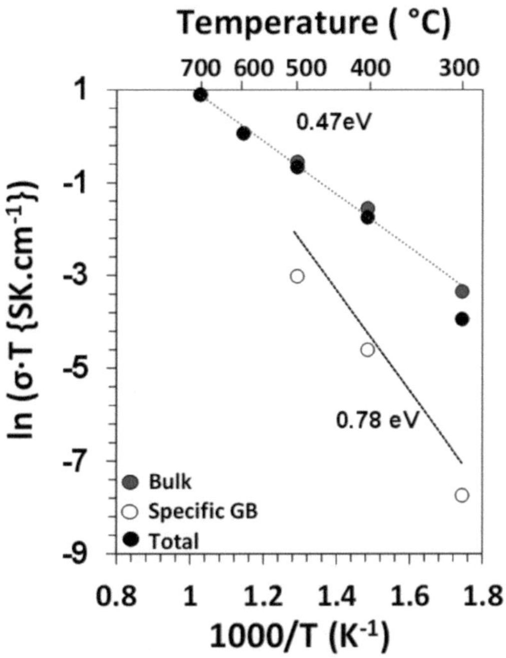

Figure 17. Arrhenius plot for BZCY72 after Ni extraction in 3% moist 5% H_2, balance Ar. The specific grain-boundary (GB) conductivity is obtained by multiplying the grain-boundary conductivity by the ratio of the pseudo-capacitance of the bulk over the pseudo-capacitance of the grain boundaries [53].

ceramic material that is representative of thin membranes for such things as electrode development, measurement of bulk mechanical and electrical properties, etc. The fact that the specimens are transparent is incidental, except that this unique property makes it possible to measure bulk optical properties, such as band gap, by transmission spectra (see Chapter 16 of Fundamentals of Ceramics [19]).

Practical hydrogen diffusion membranes must be as thin as possible in order to maximize the hydrogen flux, but still remain impervious to all other gas species. Fig. 18 shows the cross-section of a typical BZCY72 membrane coating on a porous support tube for a protonic ceramic fuel cell or membrane reactor from a 10 mm diameter production tube from CoorsTek Membranes Sciences. The 20 μm-thick membrane is shown in the upper part of the image coating a 0.8 mm thick Ni/BZCY72 anode-support. The image was obtained on a polished specimen that highlights the microstructure. The fully dense and large-grained microstructure of the membrane is clearly seen. Such a membrane is capable of passing a hydrogen flux of about 5 μmol/cm^2.s (7.3 nml/min/cm^2) from a H_2/He permeant stream at 700 °C while passing negligible detectable helium, as determined by analysing the permeate by mass spectrometry.

Solid-state reactive sintering is a very robust and cost effective method for making such membranes on the scale of hundreds of millions of square meters that will be required for commercialization of this technology.

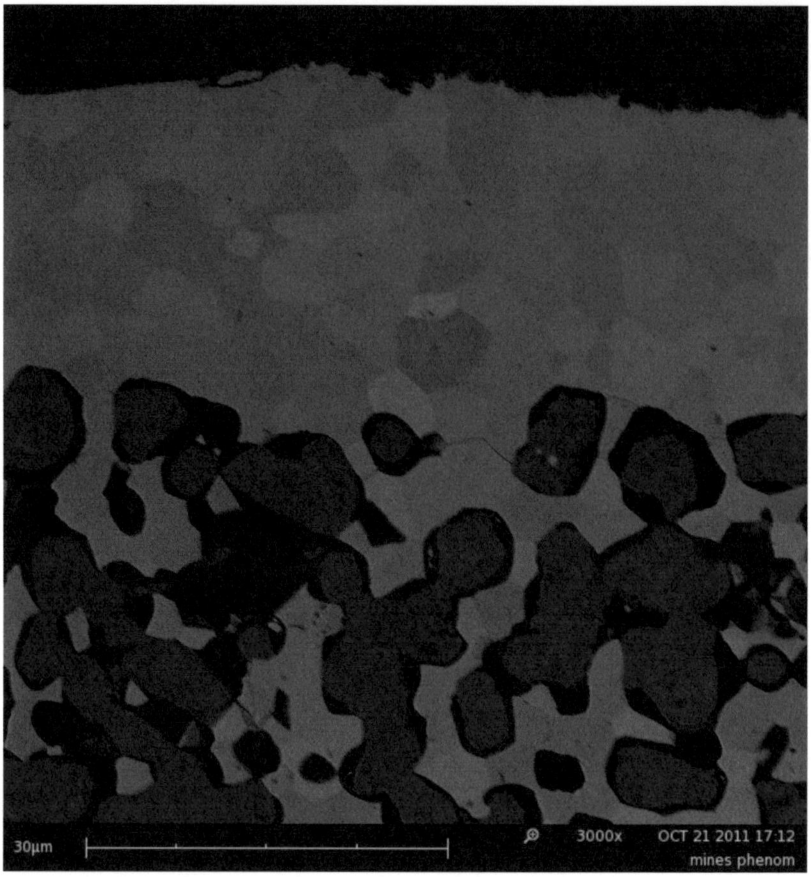

Figure 18. Polished cross-section of BZCY72 membrane on porous Ni/BZCY72 cermet support.

Acknowledgements

The authors acknowledge Daniel Clark from Colorado School of Mines for the transmission electron micrograph, Dr. Jim O' Brien for providing the Squid Data, Mc Gilvray at UCSD for FESEM and Dr. Grant Hudish at CoorsTek for collecting the XRD data.

Author details

W. Grover Coors[1], Anthony Manerbino[1], David Martinefski[1] and Sandrine Ricote[2]

1 CoorsTek, Inc. Golden, Colorado, USA

2 Colorado School of Mines, Golden, Colorado, USA

References

[1] Iwahara H, Eska T, Uchida H, Maeda N (1981) Proton Conduction in Sintered Oxides and its Application to Steam Electrolysis for Hydrogen Production. Solid State Ionics 3/4:359-363.

[2] Slade RT (1989) Preface to the Special Proceedings of the 4th International Conference on Solid State Proton Conductors. Solid State Ionics 35.

[3] Kreuer KD (1999) Aspects of the formation and mobility of protonic charge carriers and the stability of perovskite-type oxides. Solid State Ionics 125:2085-302.

[4] Ryu KH, Haile SM (1999) Chemical stability and proton conductivity of doped $BaCeO_3$–$BaZrO_3$ solid solutions. Solid State Ionics 125:355-367.

[5] Babilo P, Haile SM (2005) Enhanced sintering of yttrium-doped barium zirconate by addition of ZnO. J. Am. Ceram. Soc. 88:2362-2368.

[6] Coors WG (2011) Co-ionic Conduction in Protonic Ceramics of the Solid Solution, $BaCe_xZr_{(y-x)}Y_{(1-y)}O_{3-d}$; Part I: Fabrication and Microstructure. In: Sikalidis C, editor. Ceramic Materials Book 3. Intech. pp.479-500.

[7] Kreuer KD (2003) Proton-conducting oxides. Annu. Rev. Mater. Res. 22:333-359.

[8] Babilo P, Uda T, Haile SM (2007) Processing of yttrium-doped barium zirconate for high proton conductivity. J. Mater. Res. 22:1322-1330.

[9] Katahira K, Kohchi Y, Shimura T, Iwahara H (2000) Protonic conduction in Zr-substituted $BaCeO_3$. Solid State Ionics. 138:91-98.

[10] Schober T, Bohn HG (2000) Water vapor solubility and electrochemical characterization of the high temperature proton conductor $BaZr_{0.9}Y_{0.1}O_{2.95}$. Solid State Ionics 127:351-360.

[11] Tao S, Irvine JTS (2007) Conductivity studies of dense yttrium-doped $BaZrO_3$ sintered at 1325°C. J. Solid State Chem. 180:3493-3503.

[12] Pergolesi D, Fabbri E, Traversa E (2012) Chemically stable anode-supported solid oxide fuel cells based on Y-doped barium zirconate thin films having improved performance. Electrochemistry Communications 12:977-980.

[13] Duval SBC, Holtappels P, Vogt UF, Stimming U, Graule T (2010) Characterisation of $BaZr_{0.9}Y_{0.1}O_{3-d}$ Prepared by Three Different Synthesis Methods: Study of the Sinterability and the Conductivity. Fuel Cells 5:312-321.

[14] Ricote S, Bonanos N, Marco De Luca MC, Caboche G (2009) Structural and conductivity study of the proton conductor $BaCe_{0.9-x}Zr_xY_{0.1}O_{3-d}$. J. Power Sources 193:189-193.

[15] Stotz VS, Wagner C (1960) Die Lőslichkeit von Wasserdampf und Waeserstoff in festen Oxiden. Ber Bunsenges Phys Chem. 70:781–788.

[16] Bonanos N (1992) Transport properties and conduction mechanism in high-temperature protonic conductors. Solid State Ionics. 53-56:967-974.

[17] Kosacki I, Tuller HL (1995) Mixed conductivity in $SrCe_{0.95}Yb_{0.05}O_3$, protonic conductors. Solid State Ionics. 80:223-229.

[18] Krug F, Schober T, Springer T (1995) In situ measurements of the water uptake in Yb doped $SrCeO_3$. Solid State Ionics 81:111-118.

[19] Barsoum MW (2003) Sintering and grain growth. In: Taylor & Francis Group. Fundamentals of Ceramics. LLC. pp. 302-354.

[20] Meir S, Kalabukhov S, Froumin N, Dariel MP, Frage N (2009) Synthesis and Densification of Transparent Magnesium Aluminate Spinel by SPS Processing. J. Am. Ceram. Soc. 92:358-364.

[21] Dahl PI, Lein HL, Yu Y, Tolchard J, Grande T, Einarsrud M-A, Kjølseth C, Norby T, Haugsrud R (2011) Microstructural characterization and electrical properties of spray pyrolyzed conventionally sintered or hot-pressed $BaZrO_3$ and $BaZr_{0.9}Y_{0.1}O_{3-\delta}$. Solid State Ionics 182:32-40.

[22] Anselmi-Tamburini U, Buscaglia MT, Viviani M, Bassoli M, Bottino C, Buscaglia V, Nanni P, Munir ZA (2006) Solid-state synthesis and spark plasma sintering of submicron $BaY_xZr_{1-x}O_{3-x/2}$ (x = 0, 0.08 and 0.16) ceramics. J. European Ceram. Soc. 26:2313-2318.

[23] Ricote S, Bonanos N, Wang HJ, Boukamp BA (2012) Conductivity study of dense $BaZr_{0.9}Y_{0.1}O_{3-\delta}$ obtained by spark plasma sintering. Solid State Ionics 213:36-41.

[24] Barison S, Battagliarin M, Cavallin T, Doubova L, Doubova M, Mortalo C, Boldrini S, Malavasi L, Gerbasi R (2008) High conductivity and chemical stability of $BaCe_{1-x-y}Zr_xY_yO_{3-d}$ proton conductors prepared by a sol–gel method. J. Mater. Chem. 18:5120-5128.

[25] Stuart PA, Unno T, Ayres-Rocha R, Djurado E, Skinner SJ (2009) Synthesis and sintering behaviour of $BaZrYO_3$ powders prepared by spray pyrolysis. J. European Ceram. Soc. 29:697-702.

[26] Kim S, Fleig J, Maier J: Space charge conduction (2003) Simple analytical solutions for ionic and mixed conductors and application to nanocrystalline ceria. Phys. Chem. Chem. Phys. 5:2268-2273.

[27] Iguchi F, Sata N, Yugami H (2010) Proton transport properties at the grain boundary of barium zirconate based proton conductors for intermediate temperature operating SOFC. J. Mater. Chem. 20:6265-6270.

[28] Kjølseth C, Fjeld H, Prytz Ø, Dahl PI, Estournes C, Haugsrud R, Norby T (2010) Space–charge theory applied to the grain boundary impedance of proton conducting $BaZr_{0.9}Y_{0.1}O_{3-\delta}$. Solid State Ionics 181:268-275.

[29] Chen CT, Danel CE, Kim S (2011) On the origin of the blocking effect of grain-boundaries on proton transport in yttrium-doped barium zirconates. J. Mater. Chem. 21:5435-5442.

[30] Shirpour M, Merkle R, Maier J (2012) Evidence for space charge effects in Y-doped $BaZrO_3$ from reduction experiment. Solid State Ionics. 216:1-5.

[31] Tong J, Clark D, Hoban M, O'Hayre R (2010) Cost-effective solid-state reactive sintering method for high conductivity proton conducting yttrium-doped barium zirconium ceramics. Solid State Ionics 181:496-503.

[32] Ricote S, Bonanos N, Manerbino A, Coors WG (2012) Conductivity study of dense $BaCe_xZr_{0.9-x}Y_{0.1}O_{3-d}$ prepared by solid state reactive sintering at 1500°C. International J. Hydrogen Energy 37:7954-7961.

[33] Fang S, Wang S, Brinkman KS, Chen F (2014) A sinteractive Ni–$BaZr_{0.8}Y_{0.2}O_{3-d}$ composite membrane for hydrogen separation. J. Mater. Chem. 2:5825-5833.

[34] Zong Z (2007) Stability and conductivity study of the $BaCe_{0.9-x}Zr_xY_{0.1}O_{2.95}$ systems. Solid State Ionics 178:213-220.

[35] Ricote S, Bonanos N, Caboche G (2009) Water vapour solubility and conductivity study of the proton conductor $BaCe_{(0.9-x)}Zr_xY_{0.1}O_{(3-\delta)}$. Solid State Ionics 180:990-997.

[36] Babiniec SM, Ricote S, Sullivan NP (2015) Characterization of ionic transport through $BaCe_{0.2}Zr_{0.7}Y_{0.1}O_{3-d}$ membranes in galvanic and electrolytic operation. International J. Hydrogen Energy 40: 9278–9286.

[37] Subramanian MA, Aravamudan G, Rao GVS (1983) Oxide pyrochlores –a review. Progess in Solid State Chemistry 15:55-143.

[38] Besikiotis V, Ricote S, Hjorth Jensen M, Norby T, Haugsrud R (2012) Conductivity and hydration trends in disordered fluorite and pyrochlore oxides: A study on lanthanum cerate–zirconate based compounds. Solid State Ionics 279:26-32.

[39] Agmon N (1995) The Grotthuss mechanism. Chem. Phys. Letters 244:456-462.

[40] Norby T, Larring Y (2007) Concentration and transport of protons in oxides. Current Opinion in Solid State & Materials Science 2:593-599.

[41] Poulsen FW (1999) Method for calculating ionic and electronic defect concentrations in proton containing perovskites. J. Solid State Chem. 143:115-121.

[42] Schober T, Schilling W, Wenzl H (1996) Defect model of proton insertion into oxides. Solid State Ionics 86-88:653-658.

[43] Uchida H, Maeda N, Iwahara H (1983) Relation between proton and hole conduction in $SrCeO_3$-based solid electrolytes under water containing atmospheres at high temperatures. Solid State Ionics 11:117-124.

[44] Oishi M, Akoshima S, Yashiro K, Sato K, Mizusaki J, Kawada T (2008) Defect structure analysis of B-site doped perovskite-type proton conducting oxide $BaCeO_3$ Part 2:

The electrical conductivity and diffusion coefficient of $BaCe_{0.9}Y_{0.1}O_{3-\delta}$. Solid State Ionics 179:2240-2247.

[45] Stokes SJ, Islam MS (2010) Defect chemistry and proton-dopant association in $BaZrO_3$ and $BaPrO_3$. J. Mater. Chem. 20:6258-6264.

[46] Hermet J, Bottin F, Dezanneau G, Geneste G (2012) Thermodynamics of hydration and oxidation in the proton conductor Gd-doped barium cerate from density functional theory calculations. Physical review B 85:205137 1-14

[47] Iwahara H (1995) Technological challenges in the application of proton conducting ceramics. Solid State Ionics 77:289-298.

[48] Ricote S, Bonanos N, Wang HJ, Haugsrud R (2011) Conductivity, transport number measurements and hydration thermodynamics of $BaCe_{0.2}Zr_{0.7}Y_{(0.1-\xi)}Ni_{\xi}O_{(3-\delta)}$. Solid State Ionics 185:11-17.

[49] Fish JS, Ricote S, O'Hayre R, Bonanos N (2014) Electrical properties and flux performance of composite ceramic hydrogen separation membranes. J. Mater. Chem. A. 3:5392-5401.

[50] He T, Kreuer KD, Baikov Yu.M, Maier J (1997) Impedance spectroscopic study of thermodynamics and kinetics of a Gd-doped $BaCeO_3$ single crystal. Solid State Ionics 95:301-308.

[51] Bonanos N (1993) Transport study of the solid electrolyte $BaCe_{0.9}Gd_{0.1}O_{2.95}$ at high temperatures. J. Phys. Chem. Solids 54:867-870.

[52] Ricote S, Bonanos N, Manerbino A, Sullivan NP, Coors WG (2014) Effects of the fabrication process on the grain-boundary resistance in $BaZr_{0.9}Y_{0.1}O_{3-d}$. J. Mater. Chem. A. 2:16107-16115.

[53] Haile SM, Staneff G, Ryu KH (2001) Non-stoichiometry, grain-boundary transport and chemical stability of proton conducting perovskites. J. Materials Science 36:1149-1160.

Chapter 4

Perovskite Nanomaterials – Synthesis, Characterization, and Applications

Nada F. Atta, Ahmed Galal and Ekram H. El-Ads

Additional information is available at the end of the chapter

http://dx.doi.org/10.5772/61280

Abstract

Inorganic perovskite-type oxides are fascinating nanomaterials for wide applications in catalysis, fuel cells, and electrochemical sensing. Perovskites prepared in the nanoscale have recently received extensive attention due to their catalytic nature when used as electrode modifiers. The catalytic activity of these oxides is higher than that of many transition metals compounds and even some precious metal oxides. They exhibit attractive physical and chemical characteristics such as electronic conductivity, electrically active structure, the oxide ions mobility through the crystal lattice, variations on the content of the oxygen, thermal and chemical stability, and supermagnetic, photocatalytic, thermoelectric, and dielectric properties.

Nanoperovskites have been utilized as catalysts in oxygen reduction and hydrogen evolution reactions exhibiting high electrocatalytic activity, lower activation energy, and high electron transfer kinetics. In addition, some perovskites are promising candidates for the development of effective anodic catalysts for direct fuel cells showing high catalytic performance. Moreover, they are recently utilized in electrochemical sensing of alcohols, gases, glucose, H_2O_2, and neurotransmitters. They can enhance the catalytic performance in terms of selectivity, sensitivity, unique long-term stability, excellent reproducibility, and anti-interference ability. In addition, organometallic halide perovskites exhibited efficient intrinsic properties for photovoltaic solar cells exhibiting good stability and high efficiency.

This chapter introduces a comprehensive coverage of the progress in perovskites research and their applications. Emphasis is given toward several intrinsic properties of perovskites, namely, electronic conductivity, electrically active structure, and electrochemical performance in terms of synthesis routes and stability. The different synthesis methods of the perovskites (coprecipitation, sol-gel, microwave, citrate/nitrate, etc.) will be summarized in this chapter. The synthesis method affected structural, surface, and catalytic properties of the prepared perovskites to a great extent. Also, this chapter will update the reader with the various applications of nanoperovskites particularly in fuel cells, catalysis, electrochemical sensing, and solar cells.

Keywords: Nanomaterials, Perovskites, Sensors, Catalysis, Fuel cells

1. Introduction

1.1. General introduction to perovskites

The mineral $CaTiO_3$ was discovered by Geologist Gustav Rose in the Ural Mountains in 1839, and it was named perovskite in recognition beholden to Count Lev Alexevich von Perovski, an eminent Russian mineralogist [1–5]. The name perovskite represented any compound that has ABC_3 formula where an octahedron of C ions surrounded the B ion. The Earth's crust contains various types of perovskites and the most abundant ones are $MgSiO_3$ and $FeSiO_3$. Perovskite family includes several types of oxides like transition metal oxides with the formula ABO_3. Some examples of ABO_3 perovskites and their corresponding properties are summarized in Table 1 [1, 2, 5].

Perovskite oxides exhibit an array of electrical properties and a variety of solid-state phenomena from insulating, semiconducting, metallic, and superconducting characters; therefore, they are very fascinating to be studied and applied in a large scale. Many of ABO_3 perovskites are cubic or nearly cubic in structure in their ideal form; however, one or more phase transitions may be achieved particularly at low temperature. In addition, many of them showed magnetic ordering and as a result, large variety of magnetic structures can be found. Some perovskites contained localized electrons, some contained delocalized energy-band states, and the behavior of other perovskites was a transition between these two types. The perovskite structures can incorporate ions of various size and charge showing great flexibility of composition. Moreover, substitutions of ions into the A- and/or B-sites or deviation from ideal stoichiometry resulted in altering the electronic properties of the perovskites. Perovskites exhibit atomic arrangement in the form of 3-dimensional array of corner sharing octahedra. On the other hand, layered perovskites included 2-dimensional layers of corner sharing octahedral separated by cations layers. As a result, the electronic energy bands of perovskites and layered perovskites are very unusual and their structure is unique in properties [1, 2]. Perovskites displayed diversity of electric, optical, and magnetic properties because of the fact that 90% of the elements in the periodic table can be stable in the perovskite structure and the feasibility of partial substitutions of cations in A- and B-sites forming $A_{1-x}A'_xB_{1-y}B'_yO_3$ [6]. Perovskites showed great interest in several applications due to their wide various and useful properties in photochromic, electrochromic, image storage, switching, filtering, and surface acoustic wave signal processing devices. They were utilized as catalytically active catalyst for several reactions like carbon monoxide and hydrocarbons oxidation, hydrogen evolution reaction and nitrogen oxides, and oxygen reduction reactions. They also have a good impact in many electrochemical applications like sensing, biosensing, photoelectrolysis of water-producing hydrogen, and fuel cells [1, 2].

Insulating	Metallic	Magnetic	Superconducting
			$SrTiO_3$ (n-type)
WO_3	ReO_3	$PbCrO_3$	Na_xWO_3 (t)
$NaTaO_3$	$NaWO_3$	$LaCrO_3$	K_xWO_3 (t)
$SrTiO_3$	$KMoO_3$	$CaMnO_3$	K_xWO_3 (h)
$BaTiO_3$	$SrNbO_3$	$LaMnO_3$	Rb_xWO_3 (h)
$KTaO_3$	$LaTiO_3$	$LaCoO_3$	Cs_xWO_3 (h)
$LiNbO_3$	$LaWO_3$	$LaFeO_3$	Li_xWO_3 (h)
t; tetragonal, h; hexagonal			

Table 1. Some perovskites and corresponding properties [1].

1.2. Crystallography of the perovskite structure

In the ABO_3 form, B is a transition metal ion with small radius, larger A ion is an alkali earth metals or lanthanides with larger radius, and O is the oxygen ion with the ratio of 1:1:3. In the cubic unit cell of ABO_3 perovskite, atom A is located at the body center, atom B is located at the cube corner position, and oxygen atoms are located at face-centered positions (Figure 1). The 6-fold coordination of B cation (octahedron) and the 12-fold coordination of the A cation resulted in the stabilization of the perovskite structure. The perfect perovskite structure was described by Hines et al. as corner linked BO_6 octahedra with interstitial A cations [1–10]. Some distortions may exist in the ideal cubic form of perovskite resulted in orthorhombic, rhombohedral, hexagonal, and tetragonal forms (Figure 1) [3–7]. Figure 2 represented the distortion from cubic perovskite to orthorhombic one. In general, all perovskite distortions maintaining the A- and the B-site oxygen coordination was achieved by the tilting of the BO_6 octahedra and an associated displacement of the A cation [4].

V.M. Goldschmidt presented much of the early work on the synthetic perovskites and developed the principle of the tolerance factor *t*, which is applicable to the empirical ionic radii at room temperature [2–9]:

$$t = \left(r_A + r_O\right) \ / \ \left[2^{1/2}\left(r_B + r_O\right)\right],$$

where r_A is the radius of the A-site cation, r_B is the radius of the B-site cation, and r_O is the radius of oxygen ion O^{2-}. The tolerance factor can be used to estimate the suitability of the combination of cations for the perovskite structure [2]. It is a real measure of the degree of the distortion of perovskite from the ideal cubic structure so the value of t tends to unity as the structure approaches the cubic form [4]. From the equation, the tolerance factor will decrease when r_A decreases and/or r_B increases. Based on the analysis of tolerance factor value, Hines et al. solely suggested that the perovskite structure can be estimated. For $1.00 < t < 1.13$, $0.9 < t < 1.0$, and $0.75 < t < 0.9$, the perovskite structure is hexagonal, cubic, and orthorhombic, respectively. For $t < 0.75$, the structure was adopted to hexagonal ilmenite structure ($FeTiO_3$) [4].

Generally, two requirements should be fulfilled for perovskite formation:

1. Electroneutrality; the perovskite formula must have neutral balanced charge therefore the product of the addition of the charges of A and B ions should be equivalent to the whole charge of the oxygen ions. An appropriate charge distribution should be attained in the forms of $A^{1+}B^{5+}O_3$, $A^{4+}B^{2+}O_3$ or $A^{3+}B^{3+}O_3$.

2. Ionic radii requirements; $r_A > 0.090$ nm and $r_B > 0.051$ nm, and the tolerance factor must have values within the range $0.8 < t < 1.0$ [2–8].

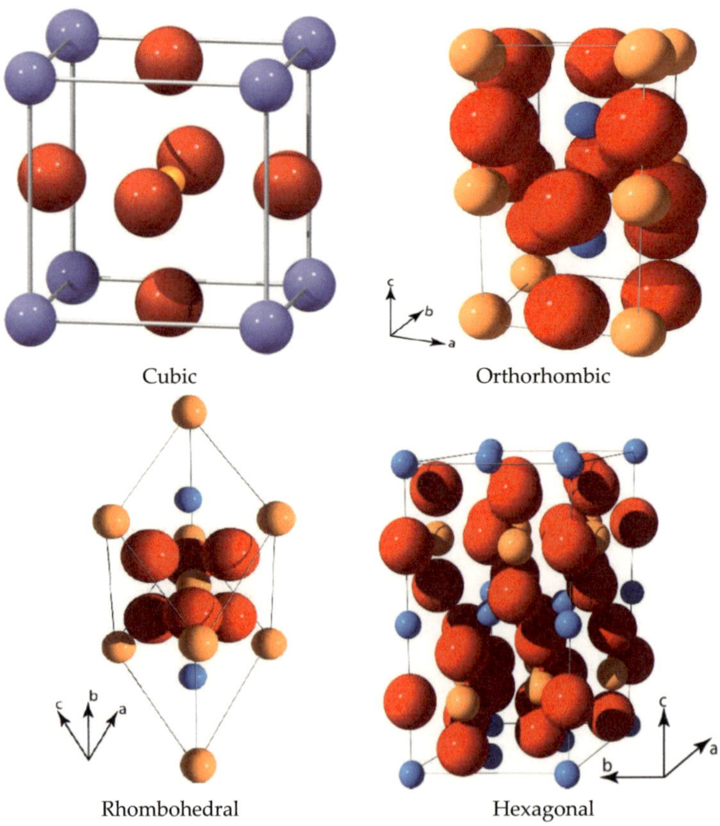

Figure 1. Different perovskite unit cells. Blue spheres represent the A cations, yellow spheres represent the B cations and red spheres represent oxygen anions forming an octahedra [4].

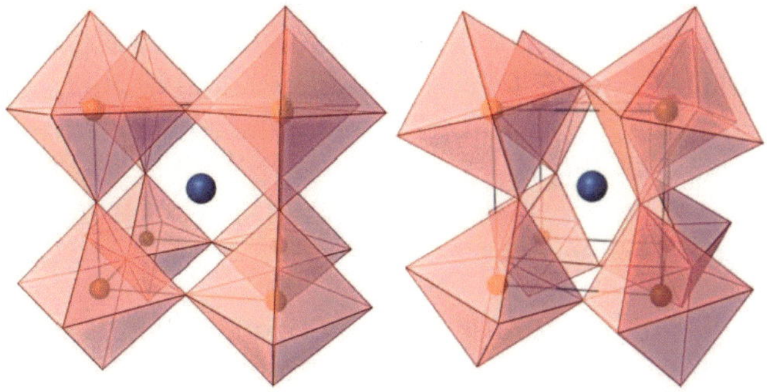

Figure 2. Perovskite distortion from (a) cubic to (b) orthorhombic [4].

1.3. Typical properties of perovskites

Perovskite exhibited a variety of fascinating properties like ferroelectricity as in case of $BaTiO_3$ and superconductivity as in case of $Ba_2YCu_3O_7$. They exhibited good electrical conductivity close to metals, ionic conductivity and mixed ionic and electronic conductivity. In addition, several perovskites exhibited high catalytic activity toward various reactions. Table 2 contains a summary of typical properties of perovskites. Several typical properties will be discussed in this section like ferroelectricity, magnetism, superconductivity, and catalytic activity [9].

Typical property	Typical compound
Ferromagnetic property	$BaTiO_3$, $PdTiO_3$
Piezoelectricity	$Pb(Zr, Ti)O_3$, $(Bi, Na)TiO_3$
Electrical conductivity	ReO_3, $SrFeO_3$, $LaCoO_3$, $LaNiO_3$, $LaCrO_3$
Superconductivity	$La_{0.9}Sr_{0.1}CuO_3$, $YBa_2Cu_3O_7$, $HgBa_2Ca_2Cu_2O_8$
Ion conductivity	$La(Ca)AlO_3$, $CaTiO_3$, $La(Sr)Ga(Mg)O_3$, $BaZrO_3$, $SrZrO_3$, $BaCeO_3$
Magnetic property	$LaMnO_3$, $LaFeO_3$, La_2NiMnO_6
Catalytic property	$LaCoO_3$, $LaMnO_3$, $BaCuO_3$
Electrode	$La_{0.6}Sr_{0.4}CoO_3$, $La_{0.8}Ca_{0.2}MnO_3$

Table 2. Typical properties of perovskite oxides [9].

1.3.1. Dielectric properties

There are some properties inherent to dielectric materials like ferroelectricity, piezoelectricity, electrostriction, and pyroelectricity. One of the important characteristic of perovskites is ferroelectric behavior, which is obvious in $BaTiO_3$, $PdZrO_3$, and their doped compounds. The ferroelectric behavior of $BaTiO_3$ was strongly related to its crystal structure. $BaTiO_3$ was subjected to three phase transitions; as the temperature increases, it was converted from monoclinic to tetragonal then to cubic. At temperature higher than 303 K, $BaTiO_3$ does not show any ferroelectric behavior as it crystallizes into cubic structure. $BaTiO_3$ showed high dielectric constant based on the anisotropy of its crystal structure resulting in large dipole moment generation in $BaTiO_3$ [9].

1.3.2. Electrical conductivity and superconductivity

One of the obvious properties of perovskites is superconductivity. Cu-based perovskites act as high-temperature superconductors, and La-Ba-Cu-O perovskite was first reported. The presence of Cu in B-site is essential for the superconductivity and various superconducting oxides can be manufactured with different A-site ions. Some examples of high temperature superconductors are $YBa_2Cu_3O_7$, $Bi_2Sr_2Ca_2Cu_3O_{10}$, and $HgBa_2Ca_2Cu_3O_{8+\delta}$ with critical temperature of superconducting transition (Tc) of 130–155 K. The superconductivity is associated with the layers of Cu-O in Cu-based perovskites, and the Tc value is associated with the Cu-O layers number in the crystal lattice. The synthesis of 5 or more Cu-O-layered perovskites did not achieve successfully due to the low chemical stability. One of the highly significant superconductors with great Tc value is $YBa_2Cu_3O_7$. In addition, the oxygen nonstiochiometry is one of the most significant reasons for the high value of Tc. In $YBa_2Cu_3O_{7-\delta}$, when the value of $\delta < 0.5$, it crystallizes into orthorhombic structure, which is superconductive. For $\delta > 0.5$, it showed a tetragonal structure that does not show any superconductivity. The crystal structure affected greatly the superconductivity in high Tc oxides, and as a result, high Tc values can be achieved by improving the chemical stability of the perovskite crystal structure. Furthermore, some perovskites exhibited great electronic conductivity similar to that of metals like Cu. $LaCoO_3$ and $LaMnO_3$ are examples of perovskites exhibiting high electronic conductivity, and therefore they are utilized as cathodes in solid oxide fuel cells displaying superior hole conductivity of 100 S/cm. The electronic conductivity of the perovskites can be enhanced by doping the A-site with another cation, which resulted in increasing the quantity of the mobile charge carriers created by the reparations of charge [9].

1.3.3. Catalytic activity

Perovskites showed excellent catalytic activity and high chemical stability; therefore, they were studied in a wide range in the catalysis of different reactions. Perovskites can be described as a model of active sites and as an oxidation or oxygen-activated catalyst. The stability of the perovskite structure allowed the compounds preparation from elements with unusual valence states or a high extent of oxygen deficiency. Perovskites exhibited high catalytic activity, which is partially associated with the high surface activity to oxygen reduction ratio or oxygen activation that resulted from the large number of oxygen vacancies. Perovskites can act as automobile exhaust gas catalyst, intelligent automobile catalyst and cleaning catalyst, etc., for various catalytic environmental reactions. It was reported in the literature that perovskites containing Cu, Co, Mn, or Fe showed excellent catalytic activity toward the direct decomposition of NO at high temperature, which is considered one of the difficult reactions in the catalysis ($2NO \rightarrow N_2+O_2$). Perovskites showed superior activity for this reaction at high

http://dx.doi.org/10.5772/61280

temperatures because of the presence of oxygen deficiency and the simple elimination of the surface oxygen in the form of a reaction product. NO decomposition activity was enhanced upon doping. Also, under an atmosphere that is rich with oxygen up to 5%, $Ba(La)Mn(Mg)O_3$ perovskite exhibited superior activity toward the decomposition of NO [9].

Perovskite showed a great impact as an automobile catalyst; intelligent catalyst. Pd-Rh-Pt catalysts was utilized as an effective catalyst for the removal of NO, CO and uncombusted hydrocarbons. There is another catalyst that consists of fine particles, with high surface-to-volume ratio, and can be utilized to reduce the amount of precious metals used. However, these fine particles exhibited very bad stability under the operation conditions leading to catalyst deactivation. Therefore, perovskite oxides can be used showing redox properties to preserve a great dispersion state. Upon oxidation, Pd is oxidized in the form of $LaFe_{0.57}Co_{0.38}Pd_{0.05}O_3$ and upon reduction; fine metallic particles of Pd were produced with radius of 1–3 nm. This cycle resulted in partial replacement of Pd into and sedimentation from the framework of the perovskite under oxidizing and reducing conditions, respectively, displaying a great dispersion state of Pd. Also, this cycle improved the excellent long-term stability of Pd during the pollutants removal from the exhaust gas. Exposing the catalyst to oxidizing and reducing atmosphere resulted in the recovery of the high dispersion state of Pd. This catalyst is known as intelligent catalyst because of the great dispersion state of Pd and the excellent stability of the perovskite structure [9].

2. Methods of perovskite synthesis

2.1. Solid-state reactions

In solid-state reactions, the raw materials and the final products are in the solid-state therefore nitrates, carbonates, oxides, and others can be mixed with the stoichiometric ratios. Perovskites can be synthesized via solid-state reactions by mixing carbonates or oxides of the A- and B-site metal ions corresponding to the perovskite formula ABO_3 in the required proportion to obtain the final product with the desired composition. They are ball milling effectively in an appropriate milling media of acetone or isopropanol [11, 12]. Then the obtained product is dried at 100 °C and calcined in air at 600 °C for 4–8 h under heating/cooling rates of 2 °C/min. After that, the calcined samples are ground well and sieved. Then it was calcined again at 1300–1600 °C for 5–15 h under the heating/cooling rate of 2 °C/min to confirm the formation of single phase of perovskite. Again grinding and sieving was carried out for the calcined samples [11, 13, 14]. The synthesis of $BaCeO_3$-based proton conductor perovskites [13] and $BaCe_{0.95}Yb_{0.05}O_{3-\delta}$ [11] was achieved through the previous methodology using $BaCO_3$, CeO_2, and Yb_2O_3 as the starting materials and isopropanol as the milling media [10].

2.2. Gas phase preparations

Gas phase reaction or transport can be used for the deposition of perovskite films with a specific thickness and composition. Laser ablation [15], molecular beam epitaxy [16], dc sputtering [17], magnetron sputtering [18], electron beam evaporation [19], and thermal evaporation [20] techniques were developed for gas phase deposition. Gas phase deposition can be categorized into three types: (i) deposition at a low substrate temperature then postannealing at high temperature, (ii) deposition at an intermediate temperature of 873 to 1,073 K then postannealing treatment, and (iii) deposition at the crystallization temperature under suitable atmosphere. $YBa_2Cu_3O_7$ films can be synthesized by the coevaporation of Y, Cu, and BaF_2 then annealing at high temperatures in O_2 atmosphere wet with water vapor to reduce the annealing time and substrate interaction [20].

2.3. Wet chemical methods (solution preparation)

These methods included the sol-gel preparation, coprecipitation of metal ions using precipitating agents like cyanide, oxalate, carbonate, citrate, hydroxide ions, etc., and thermal treatment [21], which resulted in a single-phase material with large surface area and high homogeneity. These methods presented good advantages such as lower temperature compared to the solid-state reactions, better homogeneity, greater flexibility in forming thin films, improved reactivity and new compositions and better control of stoichiometry, particle size, and purity. Therefore, they opened new directions for molecular architecture in the synthesis of perovskites. Solution methods were classified based on the means used for solvent removal. Two classes were identified: (i) precipitation followed by filtration, centrifugation, etc., for the separation of the solid and liquid phases and (ii) thermal treatment such as evaporation, sublimation, combustion, etc., for solvent removal. There are several factors must be taken in consideration in solution methods like solubility, solvent compatibility, cost, purity, toxicity, and choice of presumably inert anions [10].

2.3.1. Precipitation

2.3.1.1. Oxalate-based preparation

This method is built on the assimilation of oxalic acid with carbonates, hydroxides, or oxides producing metal oxalates, water, and carbon dioxide as products [22]. The solubility problem is minimized as the pH of the resulting solution is close to 7. An oxidizing atmosphere like oxygen was used during calcination to avoid the formation of carbide and carbon residues [23]. Clabaugh et al. utilized an aqueous chloride solution with oxalic acid to obtain unique and novel complex compound of $BaTiO(C_2O_4)_2\ 4H_2O$ as a precursor for the preparation of finely divided and stoichiometric $BaTiO_3$ [24].

2.3.1.2. Hydroxide-based preparation

This method is often used due to its low solubility and the possible variety of precipitation schemes. The sol-gel process can be used to produce a wide range of new materials and improve their properties. It presented some advantages over the other traditional methods like chemical homogeneity, low calcination temperature, room temperature deposition, and controlled hydrolysis for thin film formation. $BaZrO_3$ powders in its pure crystalline form can be prepared by the precipitation in aqueous solution of high basicity [25]. $LaCoO_3$ was prepared by the simultaneous oxidation and coprecipitation of a mixture containing equimolar amounts of La(III) and Co(II) nitrates producing a gel containing hydroxide then calcination at 600 °C [26].

2.3.1.3. Acetate-based preparation

Different perovskites were prepared by mixing acetate ions alone or together with nitrate ions with the metal ions salts. $La_{1-x}Sr_xCoO_3$ with $x = 0, 0.2, 0.4, 0.6$ [27] was prepared using acetate precursors then calcination at 1,123 K in air for 5 h. $La_{1-x}Sr_xCo_{1-y}Fe_yO_3$ [28] was prepared using iron nitrate and strontium, cobalt, and lanthanum acetates then calcination at 1,123 K in air between 5 and 10 h.

2.3.1.4. Citrate-based preparation

Citrate precursors can be used and undergo several decomposition steps in the synthesis of perovskite [29]. These steps included the decomposition of citrate complexes and removal of CO_3^{2-} and NO_3^- ions. $LaCo_{0.4}Fe_{0.6}O_3$ can be prepared by this method, and the mechanism was investigated by thermogravimetry, XRD, and IR spectroscopy.

2.3.1.5. Cyanide-based preparation

Rare earth orthoferrites ($REFeO_3$) and cobalt compounds ($RECoO_3$) were prepared using cyanides complexes via thermal decomposition of the rare earth ferricyanide and cobalticyanide compounds [30]. $LaFe(CN)_6$ $6H_2O$, $LaCo(CN)_6$ $5H_2O$, and even ferrocyanides such as $NH_4LaFe(CN)_6$ $5H_2O$ are precipitated from the aqueous solution. This method presented some advantages like control of stoichiometry and low calcination temperature. The same method was used for the preparation of europium and other rare earth hexacyanoferrate compounds [31].

2.3.2. Thermal treatment

2.3.2.1. Freeze-drying

The freeze-drying method can be achieved through the following steps: (i) dissolution of the starting salts in the suitable solvent, water in most cases; (ii) freezing the solution very fast to keep its chemical homogeneity; (iii) freeze-drying the frozen solution to get the dehydrated salts without passing through the liquid phase; and (iv) decomposition of the dehydrated salts to give the desired perovskite powder. The rate of heat loss from the solution is the most important characteristic for the freezing step. This rate should be as high as possible to decrease the segregation of ice-salt. Also, in case of multicomponent solutions, the heat loss rate should be high to prevent the large-scale segregation of the cation components [10, 12, 21].

2.3.2.2. Plasma spray-drying

This method was applicable to various precursors, including gaseous, liquid, and solid materials. It was applied for the preparation of various ceramic, electronic, and catalytic materials. It presented many advantages in terms of economy, purity, particle size distribution, and reactivity. This method was achieved through two steps: (i) injection of the reactants and (ii) generation and interaction of the molten droplets (with substrate or with the previously generated droplets). The thick film of $YBa_2Cu_3O_x$ covering large areas was prepared via this approach, and the optimum superconducting oxide phase was obtained by varying the preparation conditions like plasma parameters, substrate temperatures, and film postdeposition treatment [32].

2.3.2.3. Combustion

A redox reaction, which is thermally induced, occurs between the oxidant and fuel. A homogenous, highly reactive, and nanosized powder was prepared by this method. When compared with the other traditional methods, a single-phase perovskite powder can be obtained at lower calcination temperatures or shorter reaction times. One of the most popular solution combustion methods is citrate/nitrate combustion, where citric acid is the fuel and metal nitrates are used as the source of metal and oxidant. It is similar to the Pechini process "sol-gel combustion method" to a large extent, but in citrate/nitrate combustion, ethylene glycol or other polyhydroxy alcohols are not used. In addition, in citrate/nitrate combustion, the nitrates are not eliminated in the form of NO_x, but they remain in the mixture with the metal-citrate complex facilitating the auto-combustion. Iron, cobalt, and cerium-perovskite can be prepared via citrate/nitrate combustion synthesis [12, 33]. In addition, uniform nanopowder of $La_{0.6}Sr_{0.4}CoO_{3-\delta}$ was prepared by the combined citrate–EDTA method, where the precursor solution was made of metal nitrates, citric acid, and EDTA under controlled pH with ammonia [34]. $La_{0.8}Sr_{0.2}Co_{0.2}Fe_{0.8}O_{3-\delta}$ [35] and Sr- or Ce-doped $La_{1-x}M_xCrO_3$ catalysts [36] were prepared by citrate/nitrate combustion method. Furthermore, the Pechini "citrate gel" process includes two stages: (i) a complex was formed between the metal ions and citric acid, then (ii) the produced complex was polyesterified with ethylene glycol to maintain the metal salt solution in a gel in a homogenous state. This approach presented some advantages like high purity, minimized segregation, and good monitoring of the resulting perovskite composition. $LaMnO_3$ [37, 38], $LaCoO_3$ [39–41], and $LaNiO_3$ [42] were prepared by citric acid gel process producing nanophasic thin films [10].

2.3.2.4. Microwave synthesis

The microwave irradiation process (MIP), evolving from microwave sintering, was applied widely in food drying, inorganic/organic synthesis, plasma chemistry, and microwave-induced catalysis. MIP showed fascinating advantages: (i) fast reaction rate, (ii) regular heating, and (iii) efficient and clean energy. The microwave preparations were achieved in domestic microwave oven at frequency of 2.45 GHz with 1 kW as the maximum output power. Dielectric materials absorbed microwave energy converted directly into heat energy through the polarization and dielectric loss in the interior of materials [43]. The energy efficiency reached 80–90% which is much higher than the conventional routes. MIP was recently utilized to prepare perovskites nanomaterials reducing both the high temperature of calcination (higher than 700 °C) and long time (greater than 3 h) required for pretreatment or sintering [10]. $GaAlO_3$ and $LaCrO_3$ perovskites with ferroelectric, superconductive, high-temperature ionic conductive and magnetic ordering properties, faster lattice diffusion, and grain size with smaller size were prepared in MIP [44–47]. $CaTiO_3$ powders prepared in MIP presented a fast structural ordering than powders dealt in ordinary furnace [48]. Hydrothermal conventional and dielectric heating were utilized to prepare La–Ce–Mn–O catalysts. Hydrothermal MIP leads to formation of $La_{1-x}Ce_xMnO_{3+\varepsilon}CeO_2$ ($x + \varepsilon = 0.2$) with enhanced catalytic activity [49] while using the conventional heating methods lead to formation of $LaMnO_3 + CeO_2$. Moreover, nanosized single-phase perovskite-type $LaFeO_3$ [50], $SmFeO_3$, $NdFeO_3$, $GdFeO_3$, barium iron niobate powders [51], $KNbO_3$ [52], $PbWO_4$ [53], $CaMoO_4$ [54] and MWO_4 (M: Ca, Ni) [55], strontium hexaferrite [56], and $SrRuO_3$ [57] were prepared in MIP showing finer particles, higher specific surface areas and shorter time for synthesis of single crystalline powders.

3. Doping of perovskites

The different properties of perovskites and their catalytic activity are highly affected by the method of synthesis, conditions of calcination (time, atmosphere, fuel, temperature, etc.), and A- and/or B-site substitutions. The catalytic activity of the perovskite is highly affected by partial or total substitutions on A- and/or B-site cations because of the oxidation state modification, the variation of the chemical state of the elements at A- and/or B-site, the generation of oxygen vacancies, the mobility of oxygen lattice, and the formation of structural defects [58–60]. The powerful bond between the B-site metal ions and the oxygen ions can be used to determine the basic characters of perovskites, and as a result, the B-site cation is responsible for the perovskite catalytic activity [61, 62]. Therefore, partial substitution of B-site cation with other metals M in $AB_{1-y}M_yO_3$ will display the properties of both metals: the main metal B and the dopant one M [62]. On the other hand, the cation A can stabilize the unusual oxidation states of B-site cations by the controlled formation of crystal lattice vacancies, which lead to different catalytic performances [61].

Upon doping A- and/or B-sites in ABO_3 perovskite oxides, the catalytic activity, ionic and electronic conductivity, and flexible physical and chemical properties can be altered for utilization in various applications [63–66]. Different cations with different sizes and charges can be hosted in the A- and B-sites of these perovskites; thus, many studies can be performed to utilize doped perovskites in various applications. Multiple cationic substitutions can be accepted in the stable perovskite lattice provided that Goldschmidt tolerance factor ranged between 0.75 and 1 and electroneutrality are preserved [59, 67, 68]. Therefore, variable amounts of different structural and electronic lattice defects can be accommodated in the perovskite structure as a result of their nonstoichiometry. This will further affect the activity of the perovskite and stabilize the unusual valence states of different metal ions [61, 67]. Some physical characteristics of perovskite-type oxides seriously associated with structural characters were affected greatly by the structural deformations from the ideal cubic structure of the perovskite [69].

The type of the metal ion at the B-sites and their partial substitutions can be used to determine the catalytic activity of perovskites. The substitution of B-site metal ion with various metal ions

M in the doped perovskite $AB_{1-y}M_yO_3$ showed a vast spectrum for the alteration of the catalytic and physical properties of the prepared perovskite [64, 67]. There is an effective synergism between the crystal lattice of the perovskite and the metal ions dissolved in the lattice upon doping. This synergism resulted in enhanced redox reaction and better catalytic activity of the prepared perovskite [70]. As well, a dramatic change in the transport and magnetic properties of the ABO_3 perovskite can be achieved upon doping the B-site due to an ionic valence effect and/or an ionic size effect [64]. Furthermore, upon doping the B-site in ABO_3 perovskites with transition metals especially noble metals, the stability of the perovskite was improved and the catalytic activity was enhanced greatly [71]. In addition, the incorporation of two different B ions with appropriate various charge and size may be altered the simple perovskite structure. If the two ions in B-site were used with equimolar amounts, the resulted perovskite can be represented as $AB_{0.5}B'_{0.5}O_3$ with unit cell appearing as doubled along the three axes. In addition, if B and B′ have different charges, there is a slight shift of the oxygen toward the highly charged ion in the ordered structure although the octahedral symmetry of B and B′ cations is maintained [5]. Different B-sites doped perovskites were mentioned in the literature showing enhanced catalytic properties like $LaNi_xCo_{1-x}O_3$ [72], $LaB_{0.9}Pd_{0.1}O_3$ [73], $LaMn_{1-x}PdxO_3$ [74], $BaFe_{1-x}Y_xO_{3-\delta}$ [75], $BaFe_{0.85}Cu_{0.15}O_{3-\delta}$ [76], $LaNi_{1-x}Fe_xO_3$ [77], $LaFe_{0.95-x}Co_xPd_{0.05}O_3$ [78] and $LaCo_{0.95}Pd_{0.05}O_3$ [79]. On the other hand, $(La_{1-x}Sr_x)_yMnO_{3\pm\delta}$ [80], $La_{1-x}Ce_xGaO_3$, $La_{1-x}Pr_xGaO_3$ and $La_{1-x}Nd_xGaO_3$ [81], $La_{1-x}Ca_xMnO_3$ [82], $La_{1-x}Na_xMnO_{3+\delta}$, $La1_{-x}Ca_xMnO_{3+\delta}$ [83], $(Ba_{0.93}Fe_{0.07})TiO_3$ [84], $La_{1-x}Sr_xNiO_3$, and $La_{1-x}Sr_xMnO_3$ [85] are examples of A-site doped perovskites.

4. Characterization of perovskites

X-ray powder diffraction (XRD) can be used to differentiate the different phases of the prepared perovskites. Single-crystal XRD is another analysis used to characterize the structure of the perovskite. Thermal stability of the prepared perovskites can be tested using thermal analysis techniques like TGA, DTA, and DSC. On the other hand, scanning (SEM) and transmission (TEM) electron microscopies can be utilized to identify the different morphological and surface characteristics of the prepared perovskites. Also, BET can be utilized for surface area measurement. In addition, Fourier transform infrared spectroscopy (FTIR) and X-ray photoelectron spectroscopy (XPS) can be used to completely identify the formed phases [10, 86–88].

4.1. XRD

XRD can be used for the phase identification and the relative percents of different phases of the prepared materials. Also, some structural parameters like particle size, lattice parameters (a, b, and c), lattice volume, and theoretical density can be calculated from the XRD data. Also, XRD can be used to optimize the preparation conditions of the different perovskites [3, 87–89]. Galal et al. prepared $SrPdO_3$ by citrate/nitrate combustion method at different pH values; 2, 7, and 10 at calcination temperature 750 °C for 3 h and the XRD patterns of these samples were shown in Figure 3A. The XRD data were compared with the ICCD card of $SrPdO_3$ (card number 00-025-0908). For pH 2, the experimental data and the theoretical one are well matched and supported the formation of the primary orthorhombic perovskite phase of $SrPdO_3$ (the major diffraction peak 110) and the appearance of secondary phase $SrPd_3O_4$ (210). Only $SrPd_3O_4$ phase appeared in case of samples prepared at pH 7 and 10 as (110) peak disappeared. Therefore, pH 2 was the optimal pH for $SrPdO_3$ preparation. Also, the type of fuel (citric acid, urea, and glycine) used in the preparation of $SrPdO_3$ can be optimized using XRD (Figure 3B). $SrPdO_3$ was the primary phase in all cases but with different percents of $SrPdO_3$ (110) with respect to $SrPd_3O_4$ (210). The high percent was in case of urea and the small one in case of citric acid. Some structural parameters were calculated and summarized in Table 3 with good matching with theoretical data.

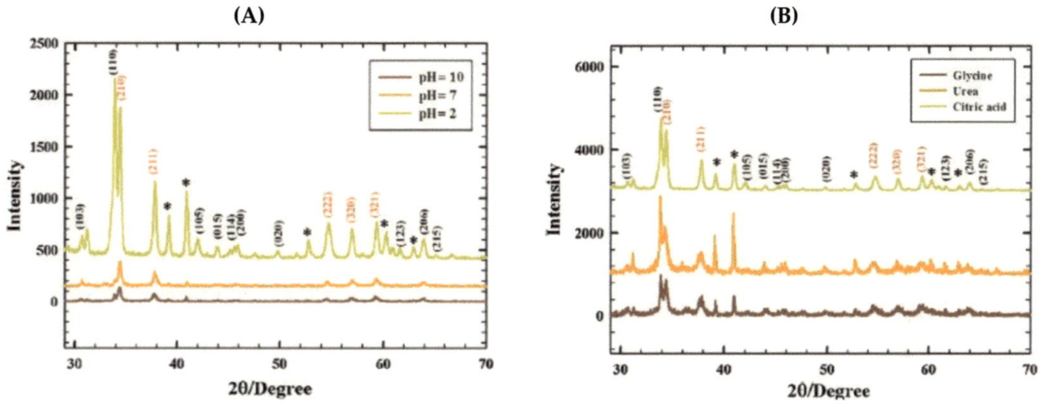

Figure 3. XRD patterns of $SrPdO_3$ prepared by combustion method at different pH values (A) and different fuels (B). Miller indices (h, l,k) are written in black line for $SrPdO_3$, red line for $SrPd_3O_4$ and the symbol (*) for $SrCl_2$. $6H_2O$ [89].

	Lattice Structure	Particle Size (nm)	Lattice Parameters (A°)	Lattice Volume (°A³)	Theoretical density (g/cm³)
Standard $SrPdO_3$ (ICCD card; 00-025-0908)	Orthorhombic		a = 3.977 b = 3.530 c = 12.82	179.98	4.47
citrate-nitrate method	Orthorhombic	34.0	a = 3.983 b = 3.541 c = 12.80	180.59	4.45
urea-nitrate method	Orthorhombic	45.4	a = 3.954 b = 3.563 c = 12.82	180.63	4.45
glycine-nitrate method	Orthorhombic	25.7	a = 3.972 b = 3.527 c = 12.83	179.69	4.47

Table 3. Structural parameters calculated from XRD data [89].

4.2. SEM and TEM

SEM and TEM can be used to study the morphology and surface characteristics of the perovskite nanomaterials. The preparation conditions, synthesis method, type of A- and B-site metal ions, and doping A- and/or B-sites affected greatly the SEM of the prepared perovskites [88–94]. Galal et al. prepared $LaNiO_3$, $LaCoO_3$, $LaFeO_3$, and $LaMnO_3$ by the microwave-assisted citrate method at 720 W as operating power for 30 min under microwave irradiation. The SEM images for the different perovskites were shown in Figure 4 presenting different morphologies depending on the kind of metal ion at B-site, respectively. $LaNiO_3$ showed compact surface with high degree of ordering while $LaCoO_3$ and $LaMnO_3$ showed spherical grains agglomerations with smaller grain size in case of $LaMnO_3$. $LaFeO_3$ showed dissimilar morphology with a porous surface containing particles with bonelike shape. In addition, $LaFeO_3$ presented greater electrocatalytic activity toward hydrogen evolution reaction compared to other types of perovskites [90].

Furthermore, the high-resolution TEM (HRTEM) can be used to show the different morphologies and particle characteristics of the different perovskites [86, 88, 95]. HRTEM images for $LaNiO_3$, $LaCoO_3$, $LaFeO_3$, and $LaMnO_3$ by the microwave-assisted citrate method were shown in Figure 5, respectively. The HRTEM images clearly showed orthorhombic phase with high crystallinity in case of $LaFeO_3$, while HRTEM images of $LaNiO_3$, $LaCoO_3$, and $LaMnO_3$ showed hexagonal distorted rhombohedral phases. The diffraction patterns obtained via HRTEM for the different perovskites were comparable with the XRD data [86, 95].

Figure 4. SEM micrographs of (A) $LaNiO_3$, (B) $LaCoO_3$, (C) $LaFeO_3$ and (D) $LaMnO_3$ prepared by the microwave-assisted citrate method at 720 W for 30 min, with a magnification of 35,000 times [90].

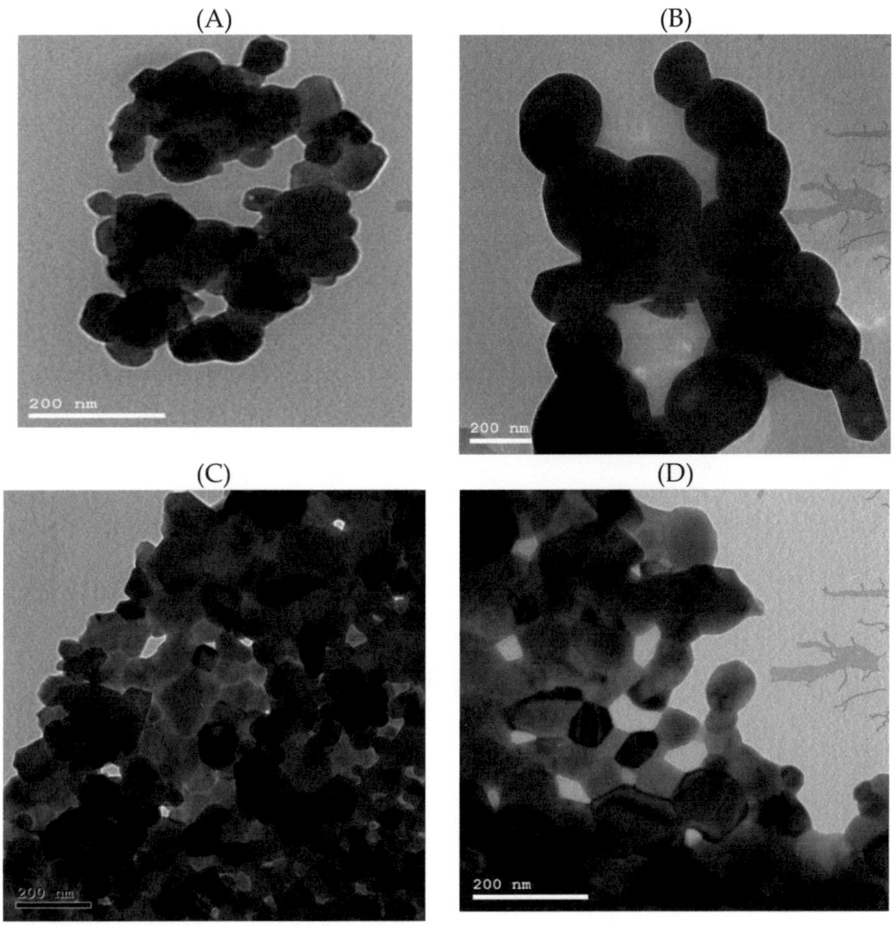

Figure 5. HRTEM micrographs of (A) $LaNiO_3$, (B) $LaCoO_3$, (C) $LaFeO_3$ and (D) $LaMnO_3$ prepared by the microwave-assisted citrate method at 720Wfor 30 min [86].

4.3. BET

The electrochemical performance and electrocatalytic activity of the perovskites are greatly associated with the specific surface area of the materials; therefore, it is necessary to measure the specific surface area of the prepared materials. The surface area values of different perovskites can be measured by Brunauer–Emmett–Teller (BET) nitrogen adsorption. The preparation conditions, synthesis method, type of A- and B-site metals, and presence of different dopants can greatly affect the surface area of the prepared perovskites [86, 87, 95]. Biniwale et al. prepared $LaFeO_3$ via different methods, namely, sol-gel, combustion, and coprecipitation and measured the corresponding surface area and the average pore diameter of the prepared perovskites. The order of decreasing the surface area of the prepared $LaFeO_3$ was sol-gel (16.5 $m^2\ g^{-1}$) > combustion (9.3 $m^2\ g^{-1}$) > coprecipitation method (5.4 $m^2\ g^{-1}$). The order of increasing the average pore diameter of the prepared $LaFeO_3$ was sol-gel (119 °A) <

coprecipitation method (140 °A) < combustion (205 °A). Sol-gel and combustion methods resulted in porous surface with internal pores contributing to higher surface area, while coprecipitation method resulted in less internal pores and lower surface area due to longer calcination time [87].

4.4. Thermal analysis

Thermal analysis can be utilized to identify the thermal stability and the decomposition temperature of the prepared perovskites. The optimum calcination temperature of any perovskite can be identified using thermal analysis [88, 96, 97]. Galal et al. prepared $SrPdO_3$ for the first time, and its optimum calcination temperature was investigated using TGA and DTG of the citrate complex of Sr and Pd (Figure 6A). The breakdown of the citrate complex occurs at ~330 °C through a smooth weight loss step. $SrPdO_3$ formation was achieved at ~750 °C through a sharp weight loss step [96].

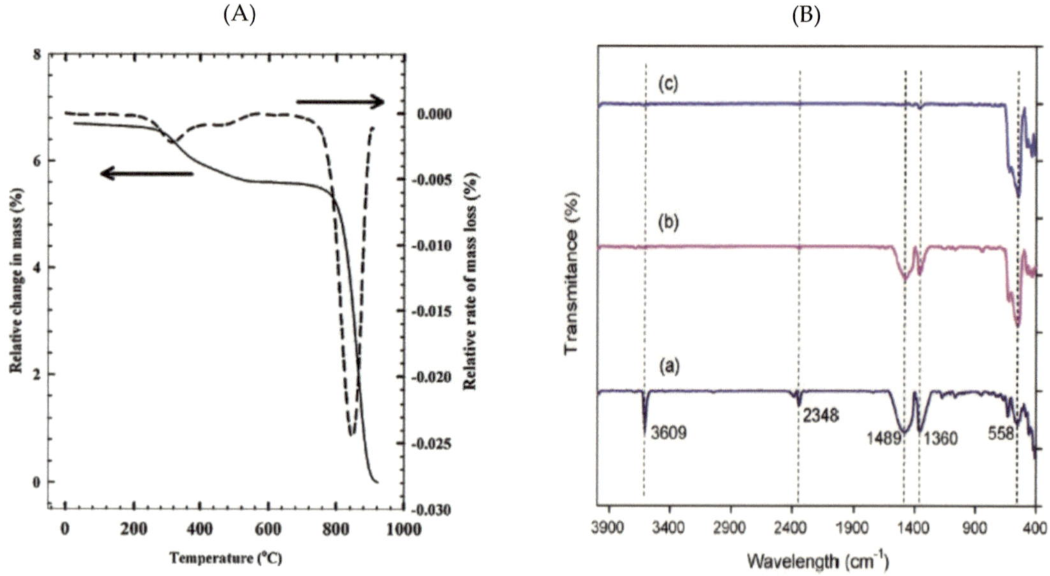

Figure 6. (A) TG spectrum of Sr and Pd mixed citrate complex, heating rate was 10 °C min^{-1} [27] and (B) FTIR spectra for $LaFeO_3$ synthesized using (a) co-precipitation method, (b) combustion method and (c) sol–gel method [87].

4.5. FTIR

The chemical bonding and chemical structure of the prepared perovskites can be investigated via FTIR. The FTIR can give structural confirmation similar to that obtained via XRD [87, 97–100]. Biniwale et al. prepared $LaFeO_3$ via different methods: sol-gel, combustion, and coprecipitation, and the FTIR for them was shown in Figure 6B [87]. The FTIR of $LaFeO_3$ showed an absorption band at 558 cm^{-1} related to the stretching vibration mode of Fe-O. Another band appeared at 430 cm^{-1} was related to the deformation vibration mode of O-Fe-O. $LaFeO_3$

http://dx.doi.org/10.5772/61280

prepared via coprecipitation method showed a sharp band at 3609 cm^{-1}, which is related to La-O in lanthanum oxide. In case of the other two methods, the band at 3600 cm^{-1} disappeared indicating the formation of relatively pure perovskite phase. Other bands appeared at 1360 and 1480 cm^{-1}, indicating other phases in case of coprecipitation method. As a result and as mentioned in literature, the absorption peak around 558 cm^{-1} was related to the stretching modes of metallic oxygen bond [87, 97–100].

4.6. XPS

The surface compositions of the various components of the prepared perovskites can be identified via XPS [101–106]. Lee et al. prepared $La_{0.9}FeO_3$ and $LaFeO_3$ samples and identify their structural composition via XPS analysis [102]. Figure 7 showed the XPS spectra of La (3d), Fe (2p), and O (1s) in $La_{0.9}FeO_3$ and $LaFeO_3$ samples. The binding energy of La ($3d_{5/2}$) was 833.5 eV and 833.8 eV in case of $LaFeO_3$ and $La_{0.9}FeO_3$, respectively, corresponding to the La^{3+} ions in the form of oxide. By contrast, the binding energy of Fe ($2p_{3/2}$) was 710.2 eV for both samples corresponding to Fe^{3+} ions in the form of oxide. The Fe (2p) XPS signal cannot distinguish between Fe^{3+} and Fe^{4+}. The XPS signal of O (1s) was divided into two peaks in case of $La_{0.9}FeO_3$ appearing at 529.9 and 532.1 eV. While for $LaFeO_3$, O (1s) XPS signal was divided into three peaks appearing at 529.4, 531.9, and 534.4 eV. The O (1s) binding energy values at 529.9 and 529.4 eV in both samples are attributed to lattice oxygen species. The peaks at 532.1 and 531.9 eV are ascribed to the chemisorbed oxygen species as OH^- or O^-. The chemisorbed oxygen species appeared at binding energy higher than that of lattice oxygen species by 2.1–2.5 eV. The peak appeared at 534.4 eV in case of $LaFeO_3$ was ascribed to the adsorbed water species associated with the surface lanthanum oxide which is highly hygroscopic [102].

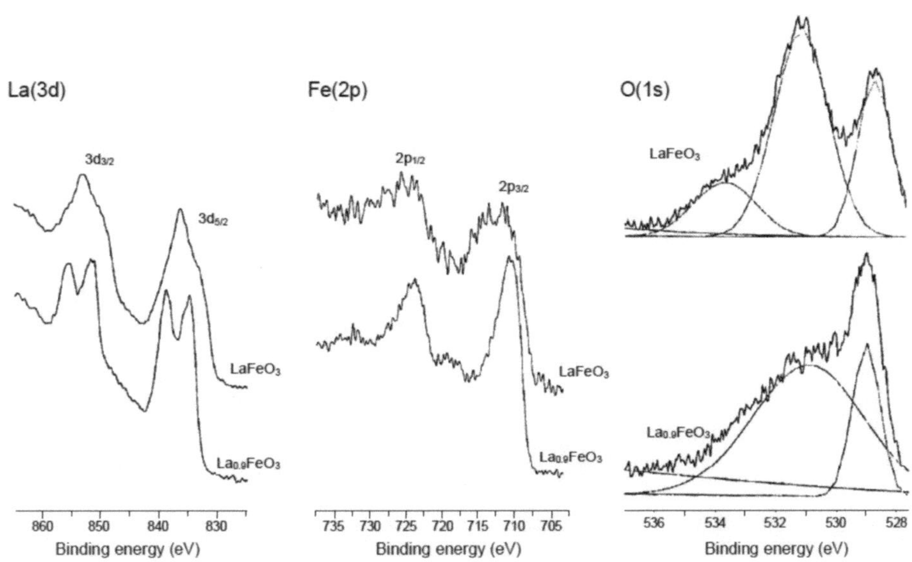

Figure 7. XPS spectra of La(3d), Fe(2p) and O(1s) in $La_{0.9}FeO_3$ and $LaFeO_3$ samples [102].

5. Applications of perovskites

Inorganic perovskite-type oxides exhibited attractive physical and chemical characteristics such as electronic conductivity, electrically active structure, the oxide ions mobility through the crystal lattice, variations on the content of the oxygen, thermal and chemical stability, and supermagnetic, photocatalytic, thermoelectric, and dielectric properties. They are fascinating nanomaterials for wide applications in catalysis, fuel cells, and electrochemical sensing. The catalytic activity of these oxides is higher than that of many transition metals compounds and even some precious metal oxides.

Nanoperovskites are recently utilized in electrochemical sensing of alcohols, gases, amino acids, acetone, glucose, H_2O_2, and neurotransmitters exhibiting good selectivity, sensitivity, unique long-term stability, excellent reproducibility, and anti-interference ability. Moreover, they have been utilized as catalysts in oxygen reduction and hydrogen evolution reactions exhibiting high electrocatalytic activity, lower activation energy, and high electron transfer kinetics. In addition, some perovskites are promising candidates for the development of effective anodic catalysts for direct fuel cells showing high catalytic performance.

5.1. Sensors and biosensors

5.1.1. Gas sensors

There are a number of requirements that the materials utilized as gas sensors must satisfy, namely, good resemblance with the target gases, manufacturability, hydrothermal stability, convenient electronic structure, resistance to poisoning, and adaptation with existing technologies. There is a wide variety of materials that can be used as gas sensors like semiconductors, namely, SnO_2, In_2O_3, and WO_3, and perovskite oxides, namely, $LaFeO_3$ and $SrTiO_3$. Perovskite oxides are fascinating materials as gas sensors due to their perfect thermal stability, ideal band gap "3–4 eV," and difference in size between the cations of A- and B-sites, allowing different dopants addition for controlling semiconducting properties and their catalytic properties. Perovskites including titanates, ferrites, and cobaltates were utilized as gas sensors for detecting CO, NO_2, methanol, ethanol, and hydrocarbons [107–110]. $LaCoO_3$ prepared via high-energy ball milling exhibited the highest amount of grain boundaries, the best CO gas sensing properties, and the smallest crystallite size of 11 nm compared to that prepared via solid-state and sol-gel reactions. The maximum response ratio increased to 26% in case of milling method with maximum response ratio temperature of 125 °C compared to 7% and 17% in case of solid-state reaction and sol-gel method. In addition, the specific surface area increased greatly from 4 $m^2\,g^{-1}$ to 66 $m^2\,g^{-1}$ by extra milling step, and the mobility of the oxygen was enhanced by growing the extra milling step and surface area [107]. A summary of various perovskite oxides for different gas sensing was given in Table 4.

5.1.2. Glucose sensor

It is very important to analytically determine H_2O_2 and glucose in many fields like food, clinic, and pharmaceutical analyses. H_2O_2 is considered one of the most important oxidizing agents

Perovskite	Sensing for	Response ratio %	Oxide type	Reference
$LaCoO_3$	CO	75% for 100 ppm CO in dry air at 125 ∘C	p-type	[107]
$La_{0.9}Ce_{0.1}CoO_3$	CO	240 % with respect to 100ppm CO in air	p-type	[108]
$NdFe_{1-x}Co_xO_3$	CO	1215% at 170 ºC for 0.03% CO gas	When x<0.3; p-type conduction behavior, when x > 0.3; n-type.	[109]
$La_{0.8}Pb_{0.2}Fe_{0.8}Cu_{0.2}O_3$	CO	Not reported	p-type	[111]
$La_{0.8}Pb_{0.2}FeO_3$	methanol	146.6 at 230 ∘C for 400 ppm methanol	p-type	[112]
$LaFeO_3$	ethanol	Not reported	p-type	[113]
$LaMg_{0.1}Fe_{0.9}O_3$	ethanol	Not reported	p-type	[114]
$SrFeO_3$	ethanol	Not reported	p-type	[115]
$SmFe_{0.9}Mg_{0.1}O_3$	ethanol	Not reported	p-type	[116]
$LaMnO_3$	ethanol	Not reported	Not reported	[117]
$La_{0.875}Ba_{0.125}FeO_3$	ethanol	sensitivity to 500 ppm ethanol is 58 at 170 °C	p-type	[118]
$Ca_xLa_{1-x}FeO_3$	ethanol	Not reported	Not reported	[119]
$LaCo_{0.1}Fe_{0.9}O_3$	ethanol	Not reported	p-type	[120]
$LaFeO_3$	ethanol	Not reported	p-type	[121]
$La_{1-x}Pb_xFeO_3$	ethanol	80% for 100 ppm in a wide temperature range of 140–240 ∘C	x = 0.30; p-type, x = 0.32–0.50; n-type	[122]
$LaFeO_3$ and $SmFeO_3$	NO_2, CO	Not reported	p-type	[123]
$LaFeO_3$	NO_2	Not reported	Not reported	[124]
$LaFeO_3$	NO_x	Not reported	Not reported	[125]
$SrTi_{1-x}Fe_xO_{3-\delta}$	hydrocarbons	Not reported	p-type	[126]
$LnFeO_3$ (Ln = La, Nd and Sm)	hydrocarbons	Not reported	p-type	[127]

Table 4. A summary of different perovskites for gas sensing.

in chemical and food industries. Glucose is a fundamental metabolite for most of the living organisms and for the clinical examination of diabetes mellitus, a worldwide health problem.

As a result, it is very important to construct biosensors for the sensitive determination of H_2O_2 and glucose [128–137]. Different types of enzymatic glucose sensors were constructed and used in the literature exhibiting the advantages of simplicity and sensitivity. However, enzymatic glucose sensors suffered from the lack of stability and the difficult procedures required for the effective immobilization of enzyme on the electrode surface. The lack of enzyme stability was attributed to its intrinsic nature because the enzyme activity was highly affected by poisonous chemicals, pH, temperature, humidity, etc. As a result, most attention was given for sensitive, simple, stable, and selective nonenzymatic glucose sensor. Different novel materials were proposed for the electrocatalytic oxidation of glucose like noble nanometals, nanoalloys, metal oxides, and inorganic perovskite oxides. Inorganic perovskite oxides as nanomaterials exhibited fascinating properties for glucose sensing like ferroelectricity, superconductivity, charge ordering, high thermopower, good biocompatibility, catalytic activity, and the ability of the perovskite structure to accommodate different metallic ions [128–137]. Zhen Zhang et al. utilized carbon paste electrode modified with $Co_{0.4}Fe_{0.6}LaO_3$ as a promising nonenzymatic H_2O_2 and glucose sensor. This sensor displayed perfect electrocatalytic activity toward H_2O_2 and glucose oxidation in alkaline medium due to the presence of large amount of active sites in the modifier. The linear dynamic range for H_2O_2 at this surface was 0.01 to 800 μM with low detection limit of 2.0 nM. For glucose, two ranges were obtained from 0.05 to 5 μM and from 5 to 500 μM with detection limit of 10 nM. The proposed sensor exhibited rapid response, excellent long-term stability, and anti-interference ability toward ascorbic acid, uric acid, and dopamine [132]. Furthermore, Atta et al. have recently modified $SrPdO_3$ perovskite with a film of gold nanoparticles to be utilized as a nonenzymatic voltammetric glucose sensor. The studied sensor exhibited high electrocatalytic activity toward glucose oxidation exploring the effective synergism between $SrPdO_3$ and gold nanoparticles. $SrPdO_3$ perovskite facilitated the charge transfer process and acted as an effective supporting substrate for gold nanoparticles. The catalytic activity of $SrPdO_3$ was attributed to the deficiency of the surface for oxygen which resulted in enhanced intrinsic reactivity toward glucose oxidation. Another reason for the catalytic activity of $SrPdO_3$ was the matrix effect induced by the stable crystal structure of the perovskite where there is a homogenous distribution of Pd^{4+} cations in the inert matrix of the perovskite during the reaction. This nanocomposite showed good performance toward glucose sensing in terms of highly reproducible response, high sensitivity, wide linearity, low detection limit, good selectivity, long-term stability, and applicability in real urine samples and blood serum [137]. A summary of different types of perovskites used for enzymatic and nonenzymatic H_2O_2 and glucose sensing was given in Table 5, exhibiting high sensitivity, wide linear range, low detection limit, anti-interference ability, applicability in real samples, and long-term stability.

5.1.3. Neurotransmitters sensor

Dopamine (DA) is an essential catecholamine neurotransmitter that exists in the mammalian central nervous system. The depletion of DA can lead to Parkinson's disease; therefore, its determination is very crucial. The interference of ascorbic acid (AA) and uric acid (UA) with DA is the major problem in DA detection [89, 140–143]. Therefore, it is very important to present a modified surface which can be sensitively and selectively detect DA even in presence

Perovskite	Sensing for	Sensor type	Detection limit	Sensitivity	Reference
graphite/$SrPdO_3$/Au_{nano}	glucose	Non-enzymatic	10.1 μmol L^{-1}	422.30 μA/ mmolL^{-1}	[137]
$LaTiO_3$-Ag0.1	glucose	Non-enzymatic	2.50×10^{-9} M	7.80×10^2 μA $mM^{-1}cm^{-2}$	[128]
$LaNi_{0.5}Ti_{0.5}O_3/CoFe_2O_4$	H_2O_2	Non-enzymatic	23 nM	3.21 μA μM^{-1} cm^{-2}	[129]
$LaNi_{0.6}Co_{0.4}O_3$	H_2O_2 and glucose	Non-enzymatic	1 nM H_2O_2 8 nM glucose	1812.84 μA mM^{-1} cm^{-2} for H_2O_2 and 643.0 μA mM^{-1} cm^{-2} for glucose	[130]
CPE–$La_{0.66}Sr_{0.33}MnO_3$–GO_x (glucose oxidase)	glucose	Enzymatic	Not reported	158.1 μAmol^{-1} L	[131]
$Co_{0.4}Fe_{0.6}LaO_3$	H_2O_2 and glucose	Non-enzymatic	2 nM H_2O_2 and 10 nM glucose	Not reported	[132]
$LaTiO_3$–Ag0.2	glucose	Non-enzymatic	2.1×10^{-7} M	784.14 μA mM^{-1} cm^{-2}	[133]
$LaNi_{0.5}Ti_{0.5}O_3$–$NiFe_2O_4$	glucose	Enzymatic	0.04 mM	Not reported	[134]
$La_{0.6}Ca_{0.4}Ni_{0.7}Fe_{0.3}O_3$	H_2O_2	Non-enzymatic	Not reported	Not reported	[135]
$La_{0.5}Sr_{0.5}CoO_{3-\delta}$	H_2O_2	Non-enzymatic	Not reported	Not reported	[136]
$La_{0.7}Sr_{0.3}NiO_3$ /chitosan/GCE	H_2O_2	Enzymatic	9.0×10^{-8} mol/L	Not reported	[138]
$LaNiO_3$	H_2O_2 and glucose	Non-enzymatic	33.9 nM H_2O_2	Not reported	[139]

Table 5. A summary of different perovskites for H_2O_2 and glucose sensing.

of high concentration of AA and UA. Atta et al. presented carbon paste electrode modified with $SrPdO_3$ (CpE/$SrPdO_3$) as a promising electrochemical DA sensor in biological fluids with unique long-term stability and low detection limit of 9.3 nmol L^{-1}. CpE/$SrPdO_3$ can simultaneously detect DA in the presence of high concentrations of AA and UA and can successfully detect DA in human urine samples with excellent recovery results in terms of selectivity, accuracy, precision, and detection limit. The proposed sensor showed high sensitivity, good selectivity, and anti-interference ability [89]. Moreover, higher response toward DA oxidation was achieved at CpE/$SrPdO_3$ compared to electrodeposited palladium nanoparticles modified CpE (CpE/Pd) with equivalent loading of Pd^{4+} salt. The higher catalytic activity at CpE/$SrPdO_3$ was explained in terms of the oxygen–surface interaction between the oxygen atoms of the hydroxyl groups and the transition element in the perovskite. One characteristic of perovskite is the deficiency of its surface for oxygen. As a result, the "dihydroxy"-oxygen adsorb onto perovskite surface with the formation of a "moderate" bond between oxygen atoms and the transition element in the oxide [89]. Furthermore, the descriptor that controls the catalytic process in perovskites is the type of transition metal in the perovskite, which is related to the number of occupied d orbital states of a specific symmetry, for example, of the active metal. This is associated with the surface ability to bond oxygen on the basis of the calculations of the density functional theory. Therefore, the oxygen adsorption energy represented a perfect descriptor for the oxidation of DA at CpE/$SrPdO_3$. A summary of different perovskites used for neurotransmitters sensing was given in Table 6.

Perovskite	Sensing for	Linear dynamic range	Detection limit (nM)	Reference
$LaFeO_3$ microspheres made up of nanospheres	DA	$2 \times 10^{-8} - 1.6 \times 10^{-6}$ M	59	[140]
CpE/$SrPdO_3$	DA	7 – 70 μmol L^{-1}	9.3	[89]
Graphite/$SrPdO_3$	L-dopa EP, NE, DA, DOPAC and ST	0.6 – 9 μmol L^{-1}	1.63	[141]
$LaFeO_3$ nanostructure dendrites	DA	$8.2 \times 10^{-8} - 1.6 \times 10^{-7}$ M	62	[142]
$LaFeO_3$ nanoparticles	DA	$1.5\times10^{-7} - 8.0\times10^{-4}$ M	30	[143]

Table 6. A summary of different perovskites for neurotransmitters sensing.

5.2. Solid oxide fuel cells

Fuel cells have come into view as efficient alternatives to combustion engines due to their potential to minimize the environmental influence of the use of fossil fuels. A fuel cell uses certain type of chemical fuel as its energy source, and there is a direct transformation of chemical energy into electrical energy like a battery. Fuel cells are attractive because of their great efficiency, modular and distributed nature, low emissions, zero noise pollution, and role in any future hydrogen fuel economy. There are several types of fuel cells depending on operating temperature, fuel type, electrolyte type, and mobile ions. Polymer electrolyte membrane fuel cells, molten carbonate, phosphoric acid or alkali fuel cells, and solid oxide fuel cell are the most common examples of fuel cells [144]. Table 7 contained some fuel cells types and some selected features [144]. Here we will concern on solid oxide fuel cells. Solid oxide fuel cells (SOFCs), based on conducting electrolyte in the form of an oxide-ion, can generate electricity and heat and they are considered as energy-saving, environment-friendly, and effective energy conversion devices. SOFCs exhibited several features compared to the other types of fuel cells like high-energy conversion efficiency, cheap materials, low sensitivity to the fuel impurities, low pollution emissions, environmental compatibility, and excellent fuel flexibility [145–160]. Figure 8 showed the working principle of a solid oxide fuel cell [157]. The high temperature of SOFC operation resulted in the difficult choice of the proper materials and the decreased cell durability. Thus, providing materials for SOFCs with good performance at intermediate temperatures (500–800 °C) is very essential so that the cell cost and the startup and shut down time can be reduced and its long-term stability can be improved [148, 155]. Perovskite oxides exhibited fascinating properties like good electrical conductivity similar to that of metals, high ionic conductivity, and perfect mixed ionic and electronic conductivity. Depending on the differences in the electrical conductive characteristics of perovskites, they are chosen as an effective component in SOFC [9]. In addition, mixed-conduction perovskite oxides possess beneficial electrochemical reaction; structural, thermal, and chemical stabilities; high electrical conductivity; high catalytic activity toward the oxygen reduction; and ideal mixed electronic and ionic conductivities to be used as effective component for intermediate temperatures SOFC (IT-SOFC) [147–160]. Shao and Haile utilized $Ba_{0.5}Sr_{0.5}Co_{0.8}Fe_{0.2}O_{3-\delta}$ as an effective cathode for intermediate SOFC with the fuel of humidified H_2 and the cathode gas of air. This cathode exhibited 1010 and 402 mW cm^{-2} as the maximum power density at 600 and 500 °C, respectively [145]. On the other hand, Goodenough reported the use of double perovskite $Sr_2MgMnMoO_{6-\delta}$ as an anode material for SOFC with dry methane as the fuel and 438 mW cm^{-2} as the maximum power density at 800 °C. This anode material showed long-term stability, stability in reducing atmosphere, tolerance to sulfur, and characteristic of oxygen deficiency [161]. Table 8 contained a summary of different perovskites used as anode and cathode for SOFC illustrating the fuel type, the operating temperature, and the maximum power density.

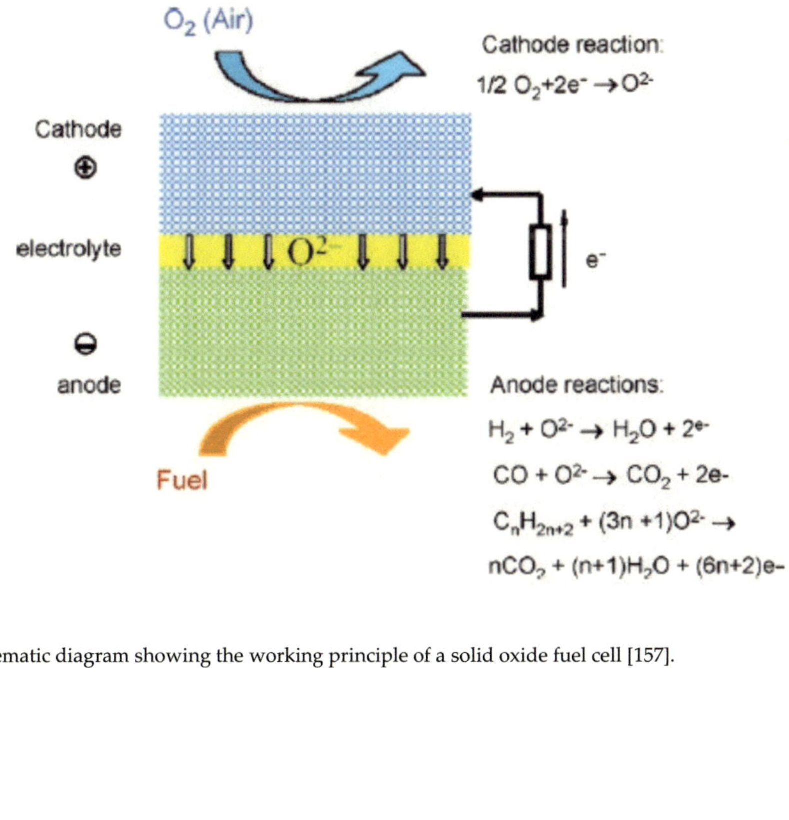

Figure 8. Schematic diagram showing the working principle of a solid oxide fuel cell [157].

Type	Temperature °C	Fuel	Electrolyte	Mobile ion
PEM: polymer electrolyte membrane or proton exchange membrane	70–110	H_2, methane	Sulphonated polymer (Nafion)	$(H_2O)_nH^+$
AFC: Alkali fuel cell	100–250	H_2	Aqueous KOH	OH^-
PAFC: Phosphoric acid fuel cell	150–250	H_2	H_3PO_4	H^+
MCFC: Molten carbonate fuel cell	500–700	Hydrocarbons, CO	$(Na, K)_2CO_3$	$CO_3^{\ 2-}$
SOFC: Solid oxide fuel cell	700–1000	Hydrocarbons, CO	$(Zr, Y)O_{2-\delta}$	O^{2-}

Table 7. Fuel cell types and selected features [144].

Perovskite	Anode/cathode in the cell	Fuel used	Operating temperature °C	Maximum power density mW cm^{-2}	Reference
$Ba_{0.5}Sr_{0.5}Co_{0.8}Fe_{0.2}O_{3-\delta}$	Cathode	Humidified H_2 (~3% H_2O)	600 500	1010 402	[145]
$Ba_{0.5}Sr_{0.5}Co_{0.2}Fe_{0.8}O_{3-\delta}$	Cathode	Humidified H_2	800	266	[147]
$La_{0.7}Sr_{0.3}Co_{0.5}Fe_{0.5}O_3$	Cathode	Not reported	Not reported	Not reported	[148]
$La_{0.6}Sr_{0.4}Fe_{0.8}Co_{0.2}O_3$	Cathode	glycerol	800	Not reported	[149]
$LaBaCuFeO_{5+x}$	Cathode	Humidified H_2	700 550	327 105	[150]
$LaBaCuCoO_{5+x}$	Cathode	Humidified H_2	700 550	432 171	[150]
$Y_{0.8}Ca_{0.2}BaCoFeO_{5+\delta}$	Cathode	Humidified H_2	650	426	[151]
$NdFeO_3$	anode	sulfur vapor or SO_2	620 650	0.154 0.265	[153]
$Sm_{0.5}Sr_{0.5}CoO_{3-\delta}$	cathode	Not reported	700	936	[158]
$Sr_2Fe_{1.4}Mo_{0.6}O_{6-\delta}$	cathode	H_2 or methanol	Not reported	Not reported	[159]
La0.75Sr0.25Cr0.5Mn0.5O3	anode	methane	Not reported	Not reported	[162]
$Sr_2MgMnMoO_{6-\delta}$	anode	dry methane	800	438	[161]
$La_{0.8}Sr_{0.2}Cr_{0.97}V_{0.03}O_3$	anode	dry methane	800	Not reported	[163]

Table 8. A summary of different perovskites for SOFC.

5.3. Catalyst

Perovskite oxides can be widely used as catalyst in modern chemical industry, exhibiting appropriate solid-state, surface, and morphological properties [6]. Several perovskites exhibited enhanced catalytic activity toward different reactions like hydrogen evolution and oxygen evolution and reduction reactions [9].

5.3.1. Hydrogen evolution reaction

Because of the advantages of high heat of combustion, abundant sources, and no pollution, hydrogen is considered as an ideal fuel. Hydrogen evolution reaction (HER) is a fascinating reaction in the renewable energy field. This reaction is very important in (i) metal electrodeposition and corrosion in acids, (ii) storage of energy through production of hydrogen, and (iii) as the hydrogen oxidation reaction reverse in low-temperature fuel cells. One of the most studied reactions in electrochemistry is the electrocatalysis in HER. The material used for HER should have (i) intrinsic electrocatalytic activity, (ii) considerable active surface area per unit volume, and (iii) good stability. To reduce the cost of electrolytic HER, the overpotential required for the operation of the electrolyzer at considerable current densities should be reduced. The overpotential reduction can be achieved through the electrode active surface area enhancement or by the selection of electrode materials of high catalytic activity. The steps of the reaction in acidic solutions are as follows:

$$M + H_3O^+ + e^- \leftrightarrow MH_{ads} + H_2O \quad (1)$$

$$MH_{ads} + H_3O^+ + e^- \leftrightarrow H_2 + M + H_2O \quad (2)$$

$$MH_{ads} + MH_{ads} \leftrightarrow H_2 + 2M \quad (3)$$

The first step in HER is the proton discharge (volume reaction, Eq. (1)), which is followed by electrodesorption step (Heyrovsky reaction, Eq. (2)) or proton recombination step, physical desorption, (Tafel reaction, Eq. (3)) [93]. Galal et al. confirmed the high catalytic activity of different perovskite oxides toward hydrogen evolution reaction [90–94, 96]. $LnFeO_3$ perovskites (Ln= Gd, La, Sm, and Nd) were prepared by the microwave assistant-citrate method, and single-phase perovskites were formed with uniform distribution of small average particle size. Tafel and electrochemical impedance measurements were used to study the catalytic activity of $LnFeO_3$ toward HER showing the effect of the type of the lanthanide ion on HER and the partial substitution effect at the La-site in $La_{1-y}Sm_yFeO_3$. The order of decreasing the catalytic activity toward HER was $NdFeO_3 > LaFeO_3 > SmFeO_3 > GdFeO_3$ based on activation energies calculations and the strength of Fe-O bond, which is related to A-type metal ion. Furthermore, the order of decreasing the catalytic activity in case of doped samples was $La_{0.75}Sm_{0.25}FeO_3 > La_{0.5}Sm_{0.5}FeO_3 > La_{0.25}Sm_{0.75}FeO_3 > LaFeO_3 > SmFeO_3$ displaying greater catalytic activity of ternary perovskites compared to that of binary ones [94]. On the other hand,

Galal et al. prepared $SrPdO_3$ by the citrate method for the first time showing enhanced catalytic activity toward HER up to 100 times with respect to the unmodified surface with 27.9 kJ mol $^{-1}$ as the calculated activation energy. The rate-determining step was the hydrogen adsorption on the catalyst and the order of the reaction at the catalyst surface was 0.86 [96]. Table 9 contained a summary of different perovskites used as catalysts for HER with the values of exchange current density at constant overpotential, activation energy, reaction order, and the rate-determining step.

Perovskite	Exchange current density j/ μA.cm^{-2}	Activation energy kJ mol^{-1}	Reaction order	Rate determining step	Reference
$NdFeO_3$	-107.82	24.68	0.72	adsorption of hydrogen on the catalyst	[94]
$LaFeO_3$	-367.37	41.15	0.37	adsorption of hydrogen on the catalyst	[94]
$SmFeO_3$	-147.23	57.89	0.41	adsorption of hydrogen on the catalyst	[94]
$GdFeO_3$	-451.86	81.37	0.72	adsorption of hydrogen on the catalyst	[94]
$La_{0.75}Sm_{0.25}FeO_3$	Not reported	Not reported	Not reported	adsorption of hydrogen on the catalyst	[94]
$SrPdO_3$	−946.3	27.9	0.86	adsorption of hydrogen on the catalyst	[96]
$LaFeO_3$	−105.6	51.61	0.37	adsorption of hydrogen	[90]
$LaCoO_3$	−348.9	45.37	0.74	adsorption of hydrogen	[90]
$LaNiO_3$	−158.5	41.15	0.94	adsorption of hydrogen	[90]
$LaMnO_3$	−41.9	55.05	0.87	adsorption of hydrogen	[90]
$SrRuO_3$ (microwave)	−5326.0	6.67	1.14	adsorption of hydrogen	[92]
$SrRuO_3$ (sol-combustion)	−425.5	49.45	0.98	adsorption of hydrogen	[92]
$SrRuO_3$ (coprecipitation)	−298.0	86.32	0.88	adsorption of hydrogen	[92]
$CaRuO_3$ (coprecipitation)	-1064.9	42.60	0.94	adsorption of hydrogen	[93]
$BaRuO_3$ (coprecipitation)	-3687.2	17.68	1.03	adsorption of hydrogen	[93]

Table 9. A summary of different perovskites for HER catalysis.

5.3.2. Oxygen reduction and oxygen evolution reactions

Oxygen reduction reaction (ORR) and oxygen evolution reaction (OER) are considered one of the most important electrode reactions in many industrial processes like fuel cells, metal electrowinning, water electrolysis, electro-organic synthesis, cathodic protection, and rechargeable metal air batteries [95, 164–166]. Platinum-based catalysts and precious metal oxides are the most common catalysts for ORR or OER, but they are expensive and scarce. Therefore, it is very important to develop other catalysts for ORR or OER. Mixed metal perovskite oxides of transition and rare earth metals are promising low-cost alternatives to precious metal catalysts for both ORR and OER [165]. Perovskite oxide exhibited unique electronic and magnetic properties, defective structure, and good cation ordering resulting in disorder-free channels of oxygen vacancies and enhanced mobility of oxygen ions [164]. Some perovskite oxides were reported as electrocatalysts for ORR and OER and summarized in Table 10 [95, 164–166]. Ruizhi Yang et al. prepared $Ba_{0.5}Sr_{0.5}Co_{0.8}Fe_{0.2}O_3$ by sol-gel method and utilized it as ORR or OER catalyst in basic medium of KOH. The proposed catalyst exhibited higher catalytic activity toward OER than the unmodified electrode [164]. Furthermore, Galal et al. prepared $LaFeO_3$ by microwave-assisted citrate method and studied its catalytic activity toward OER in acid medium of $HClO_4$. $LaFeO_3$ exhibited greater electrocatalytic activity toward OER by about 100-folds compared to the unmodified electrode. The current density at 1.5 V increased from 3.6×10^{-5} in case of unmodified electrode to 1.2×10^{-3} A/cm^2 in case of modified one. The calculated activation energy was 20 kJ/mol, which is much lower than that reported for other iron compounds and even some precious metal oxides like RuO_2. This was attributed to the matrix effect induced by the stable crystal structure of the perovskite [95]. In addition, $La_{0.6}Ca_{0.4}CoO_3$, prepared by the sol-gel method, showed high catalytic activity and relative stability toward oxygen electrochemistry in basic medium of KOH. $La_{0.6}Ca_{0.4}CoO_3$ exhibited single-phase perovskite structure, high conductivity, and large surface area [165, 166].

Perovskite	**Medium**	**Exchange current density mA.cm^{-2}**	**Activation energy kJ mol^{-1}**	**Reaction order**	**Rate determining step**	**Reference**
$LaFeO_3$	$HClO_4$	1.24×10^{-5}	20.13	0.82	Difficult to be determined	[95]
$Ba_{0.5}Sr_{0.5}Co_{0.8}Fe_{0.2}O_3$	KOH	Not reported	Not reported	Not reported	Not reported	[164]
$La_{0.6}Ca_{0.4}CoO_3$	KOH	Not reported	Not reported	Not reported	Not reported	[165]
$La_{1-x}Ca_xCoO_3$ series $(0 < x < 0.6)$	KOH	Not reported	Not reported	Not reported	Not reported	[166]

Table 10. A summary of different perovskites for ORR or OER catalysis.

5.4. Solar cells

Solar energy is a green source of energy that can be used instead of energy sources based on fossil fuels. Solar radiation can be directly converted into electrical energy in a suitable way creating various applications for solar energy. Solar energy can be efficiently converted into electricity using photovoltaic solar cells based on silicon. The obvious disadvantage of silicon-based solar cell is the high price of electricity generated from it so that there is a potential need to develop solar cell with low cost. Recently, attention was paid to solar cells based on organic/inorganic solid-state methylammonium lead halide (CH3NH3PbX3, X=Br, I) hybrid perovskite. This type of solar cells presented effective points such as a conversion efficiency of about 20%; its cost is lower than that of conventional silicon solar cells and the availability of the raw materials. These 3D organometal halide perovskite exhibited excellent intrinsic properties for photovoltaic applications like excellent stability, appropriate band gap (~1.55 eV), high absorption coefficient (1.5×10^4 cm^{-1} at 550 nm), long hole-electron diffusion length (~100 nm for $CH_3NH_3PbI_3$ and ~1 μm for $CH_3NH_3PbI_{3-x}Cl_x$), high carrier mobility and transport, charge carriers with small effective mass, low temperature of processing, and easy processing steps [167–179]. Figure 9A showed the structure of ABX_3 perovskite (X = oxygen, carbon, nitrogen or halogen), where A and B cations are placed in a cubo-octahedral and an octahedral site, respectively. A is usually divalent and B is tetravalent when O^{2-} anion is used. On the other hand, Figure 9B showed the structure of $CH_3NH_3PbI_3$ halide perovskite where the A-site is occupied by $CH_3NH_3^+$ (organic component) and B-site cation is occupied by Pb^{2+}. As indicated by the example, halide perovskites contain monovalent and divalent cations in A- and B-sites, respectively, to maintain electrical neutrality [169, 171]. Like oxide perovskite, the tolerance factor of halide perovskite should be as close to one to maintain a stable and symmetrical crystal structure [169].

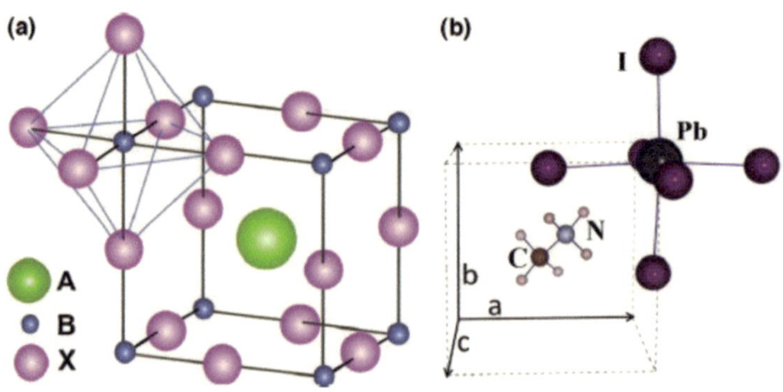

Figure 9. (a) ABX_3 perovskite structure showing BX_6 octahedral and larger A cation occupied in cubo-octahedral site. (b) Unit cell of cubic $CH_3NH_3PbI_3$ perovskite [171].

The quality of the perovskite film is very crucial for solar cells. Several methods have been used to form perovskite films with high quality such as single step solution method, vapor assistant solution process, sequential deposition of inorganic and organic precursor, and coevaporation of the precursors [167]. $CH_3NH_3PbI_3$ perovskite film was prepared with high quality by adding small amounts of N-methyl-2-pyrrolidone and a mixture of g-butyrolactone and dimethylsulfoxide via a solution method. A power conversion efficiency of 11.77% with fill factor of 80.52% was obtained based on the structure of ITO/PEDOT:PSS (poly(3,4-

ethylenedioxythiophene):polystyrenesulfonate)/perovskite/ PCBM (fullerene-derivative phenyl-C61-butyric acid methyl ester)/Ca/Al under one sun illumination (100 mW cm^{-2}) [167]. Table 11 contained a summary of different models based on perovskites used for solar cells applications with the values of power conversion efficiency, fill factor, method of perovskite formation, solar cell composition, cost, and stability.

Solar cell composition	Perovskite formation	Power conversion efficiency	Fill factor	Cost	Stability	Reference
ITO/PEDOT:PSS/ $CH_3NH_3PbI_3$ perovskite/ PCBM/Ca/Al	Solution method	11.77 %	80.52 %	low	highly stable	[167]
conducting polymer (poly-aniline) $CH_3NH_3PbI_3$ perovskite	Solution method	21 %	Not reported	low	Not reported	[168]
Mesoporous TiO_2/ $CH_3NH_3PbI_3$/ graphite paste	Solution method	1.11 %	44.5 %	low	slightly stable	[174]
Ti foil/TiO_2 nanotubes (TNTs) with organic– inorganic halide perovskite absorber "$CH_3NH_3PbI_3$" and transparent carbon nanotube electrode	Solution method	8.31 %	64 %	low	highly stable	[175]
ITO/conductive-polymer/ $CH_3NH_3PbI_3$/C_{60}/BCP/Ag	Solution method	16.5 %	75 %	low	highly stable	[177]
glass/ITO/RGO/ $CH_3NH_3PbI_3$/$PC_{61}BM$/ bathocuproine (BCP)/Ag	Solution method	10.8 %	71.6 %	low	highly stable	[179]

Table 11. A summary of different models based perovskites for Solar cells applications

6. Conclusions

Inorganic perovskite-type oxides are excellent nanomaterials for wide applications in catalysis, fuel cells, and electrochemical sensing, exhibiting attractive physical and chemical characteristics. They showed electronic conductivity, electrically active structure, the oxide ions mobility through the crystal lattice, variations on the content of the oxygen, thermal and chemical stability and supermagnetic, photocatalytic, thermoelectric, and dielectric properties. Nanoperovskites have been utilized as catalysts in oxygen reduction and hydrogen evolution reactions exhibiting high electrocatalytic activity, lower activation energy and high electron transfer kinetics. In addition, some perovskites are promising candidates for the development of effective anodic catalysts for direct fuel cells showing high catalytic performance. Moreover, they are recently utilized in electrochemical sensing of alcohols, gases, glucose, H_2O_2, and neurotransmitters. They can enhance the catalytic performance in terms of unique long-term stability, sensitivity, excellent reproducibility, selectivity, and anti-interference ability. In addition, organometallic halide perovskites exhibited efficient intrinsic properties to be utilized as a photovoltaic solar cell with good stability and high efficiency.

Acknowledgements

The authors would like to acknowledge the financial support from Cairo University through the Vice President Office for Research Funds.

Author details

Nada F. Atta*, Ahmed Galal and Ekram H. El-Ads

*Address all correspondence to: Nada_fah1@yahoo.com

Department of Chemistry, Faculty of Science, Cairo University, Giza, Egypt

References

[1] Wolfram T, Ellialtioğlu S. Electronic and Optical Properties of d-band Perovskites. 1st ed. Cambridge University Press: New York; 2006. 1 p.

[2] Galasso FS. Structure, Properties and Preparation of Perovskite-Type Compounds. In: Smoluchowski R, Kurti N, editors. 1st ed. Pergamon Press: New York; 1969. p. 3–49. Chapter 2.

[3] Ali SMM. Synthesis of Nano-particles Using Microwave Technique, the Study of their Physical Properties and Some Applications [PhD thesis]. Faculty of Science Cairo University; 2009.

[4] Perovskite Perfect Lattice. Chapter 3. p. 79–114.

[5] Johnsson M, Lemmens P. Crystallography and Chemistry of Perovskites. 2005. p. 1–11.

[6] Peňa MA, Fierro JLG. Chemical structures and performance of perovskite oxides. Chemical Reviews. 2001;101:1981–2017. DOI: 10.1016/j.apsusc.2014.10.168.

[7] Fu1 D, Itoh M. Ferroelectrics—Material Aspects. In: Lallart M, editor. InTech: Shizuoka University, Tokyo Institute of Technology, Japan; 2011. p. 413–442. Chapter 20.

[8] Porta P, De Rossi S, Faticanti M, Minelli G, Pettiti I, Lisi L, Turco M. Perovskite-type oxides: I. Structural, magnetic, and morphological properties of $LaMn_{1-x}Cu_xO_3$ and $LaCo_{1-x}Cu_xO_3$ solid solutions with large surface area. Journal of Solid State Chemistry. 1999;146:291–304. DOI: 10.1006/jssc.1999.8326.

[9] Ishihara T. Perovskite Oxide for Solid Oxide Fuel Cells, Fuel Cells and Hydrogen Energy. In: Ishihara T, editor. Springer Science Business Media, LLC; 2009. p. 1–16. Chapter 1.

[10] Lozano-Gorrín AD. Polycrystalline Materials—Theoretical and Practical Aspects. In: Zakhariev Z, editor. Universidad de La Laguna. Spain: InTech; 2012. p 107–124. Chapter 5.

[11] Nieto S, Polanco R, Roque-Malherbe R. Absorption kinetics of hydrogen in nanocrystals of $BaCe_{2.95}Yb_{0.05}O_3$-proton conducting perovskite. Journal of Physical Chemistry C. 2007;111:2809–2818. DOI: 10.1021/jp067389i.

[12] Ecija A, Vidal K, Larrañaga A, Ortega-San-Martín L, Arriortua MI. Advances in Crystallization Processes. In: Mastai Y, editor. InTech; 2012. p. 486–506. Chapter 19.

[13] Wu J. [PhD thesis]. California Institute of Technology, Pasadena, CA; 2005.

[14] Martinez E, Viches ES, Beltran-Porter A, Beltran-Porter D. Low temperature synthesis of Ba-Fe mixed oxides having perovskite structures. Materials Research Bulletin. 1985;21:511–514. DOI: 10.1016/0025-5408(86)90103-0

[15] Dijkkamp D, Venkatesan T, Wu XD, Shaheen SA, Jisrawi N, Min-Lee YH, McLean WL, Croft M. Preparation of Y-Ba-Cu-O oxide superconductor thin films using pulsed laser evaporation from high Tc bulk material. Applied Physics Letters. 1987;51(8):619–621. DOI: 10.1063/1.98366.

[16] Webb C, Weng SL, Eckstein JN, Missert N, Char K, Schlom DG, Hellman ES, Beasley MR, Kapitulink A, Harris JS. Applied Physics Letters 1988;51:12.

[17] Croteau A, Matsubara S, Miyasaka Y, Shohata N, Ferroelectric lead zirconate titanate ($Pb(Zr, Ti)O_3$) thin films prepared by metal target sputtering. Japanese Journal of Applied Physics, Part 1. 1990;26:18. DOI: 10.7567/JJAPS.26S2.18.

[18] Han Z, Bourget L, Li H, Ulla M, Millman WS, Baun HP, Xu MF, Sarma BK, Levy M, Tonner BP. Single target deposition, post proceeding and electron spectroscopy of perovskite superconductor thin films. AIP Conference Proceedings. 1988;165:66. DOI: 10.1063/1.37082.

[19] Naito M, Hammond RH, Oh B, Hahn MR, Hsu JWP, Rosenthal P, Marshall AF, Beasley MR, Geballe TH, Kapitulnik A. Thin film synthesis of the high Tc oxide supercon-

ductor yttrium barium copper oxide ($YBa_2Cu_3O_7$) by electron-beam codeposition. Journal of Materials Research. 1987;2:713–725. DOI: 10.1557/JMR.1987.0713.

[20] Mankiewich PM, Scofield JH, Skocpol WJ, Howard RE, Dayem AH. Good E., Reproducible technique for fabrication of thin films of high transition temperature superconductors. Applied Physics Letters. 1987;51(21):1753–1755.

[21] Tejuca LG, Fiero JLG. Properties and Applications of Perovskite Type-Oxides. Marcel Dekker Inc. Ed. Chemical Industries. 1993;50.

[22] Gallagher PK, Fleming DA. Influence of oxygen partial pressure on the synthesis of barium yttrium copper oxide ($Ba_2YCu_3O_7$) from a novel oxalate precursor. Chemistry of Materials. 1989;1:659–664. DOI: 10.1021/cm00006a019.

[23] Gallagher PK, Schrey F. The thermal decomposition of freeze-dried tantalum and mixed lithium–niobium oxalate. Thermochimica Acta. 1970;1:465–476. DOI: 10.1016/0040-6031(70)85017-1.

[24] Clabaugh WS, Swiggard EM, Gilchrist RJ. Preparation of barium titanyl oxalate tetrahydrate for conversion to barium titanate of highly purity. Journal of Research of the National Bureau of Standards. 1956;56:289–291.

[25] Boschini F, Rulmont A, Cloots R, Vertruyen B. Rapid synthesis of submicron crystalline barium zirconate $BaZrO_3$ by precipitation in aqueous basic solution below 100 °C. Journal of the European Ceramic Society 2009;29(8):1457–1462. DOI: 10.1016/j.jeurceramsoc.2008.09.001.

[26] Barnard KR, Foger K, Turney TW, Williams RD. Lanthanum cobalt oxide oxidation catalysts derived from mixed hydroxide precursors. Journal of Catalysis. 1990;125(2):265–275. DOI: 10.1016/0021-9517(90)90302-Z.

[27] Nakamura T, Misono M, Yoneda Y. Reduction and oxidation and catalytic properties of perovskite type mixed oxide catalysts. Chemistry Letters. 1981;1589.

[28] Zhang HM, Shimizu Y, Teraoka Y, Miura N, Yamazoe N. Oxygen sorption and catalytic properties of $La_{1-x}Sr_xCo_{1-y}Fe_yO_3$ Perovskite-type oxides. Journal of Catalysis. 1990;121(2):432–440. DOI: 10.1016/0021-9517(90)90251-E.

[29] Zhang HM, Teraoka Y, Yamazoe N. Preparation of perovskite type oxides with large surface area by citrate process. Chemistry Letters. 1987:665–668.

[30] Gallagher PK. A simple method for the preparation of (RE)FeO_3 and (RE)CoO_3. Materials Research Bulletin. 1967; 3,225–232. DOI: 10.1016/0025-5408(68)90123-2.

[31] Gallagher PK, Prescott B. Further studies of the thermal decomposition of europium hexacyanoferrate(III) and ammonium europium hexacyanoferrate(II). Inorganic Chemistry. 1970;9:2510–2512. DOI: 10.1021/ic50093a026.

[32] Cuomo JJ, Guarnieri C, Richard M, Shivashankar SA, Roy RA, Yee DS, Rosenberg R. Large area plasma spray deposited superconducting $YBa_2Cu_3O_7$ thick films. Advanced Ceramic Materials. 1989;2(3B):422–429.

[33] Deganello F, Marcì G, Deganello G. Citrate–nitrate auto-combustion synthesis of perovskite-type nanopowders: a systematic approach. Journal of the European Ceramic Society. 2009;29(3):439–450. DOI: 10.1016/j.jeurceramsoc.2008.06.012.

[34] Tao Y, Shao J, Wang J, Wang WG. Synthesis and properties of La0.6Sr0.4CoO3–δ nanopowder. Journal of Power Sources. 2008;185(2):609–614. DOI: 10.1016/j.jpowsour.2008.09.021.

[35] Park JH, Kim JP, Kwon HT, Kim J. Oxygen permeability, electrical property and stability of $La_{0.8}Sr_{0.2}Co_{0.2}Fe_{0.8}O_{3-\delta}$ membrane. Desalination. 2008;233(1–3):73–81. DOI: 10.1016/j.desal.2007.09.045.

[36] Rida K, Benabbas A, Bouremmad F, Peña MA, Sastre E, Martínez-Arias A. Effect of strontium and cerium doping on the structural characteristics and catalytic activity for C_3H_6 combustion of perovskite $LaCrO_3$ prepared by sol-gel. Applied Catalysis B: Environmental. 2008;84(3–4):457–467.

[37] Hammami R, Aïssa SB, Batis H. Effects of thermal treatment on physico-chemical and catalytic properties of lanthanum manganite $LaMnO_{3+y}$. Applied Catalysis A: General. 2009;353(2):145–153. DOI: 10.1016/j.apcata.2008.10.048.

[38] Li Y, Xue L, Fan L, Yan Y. Journal of Alloys and Compounds. (2008).

[39] Popa M, Calderón-Moreno JM. Lanthanum cobaltite thin films on stainless steel. Thin Solid Films. 2009;517(5):1530–1533. DOI: 10.1016/j.tsf.2008.08.187.

[40] Ivanova S, Senyshyn A, Zhecheva E, Tenchev K, Nickolov V, Stoyanova R, Fuess H. Mechanochemical synthesis and characterization of nanodimensional iron–cobalt spinel oxides. Journal of Alloys and Compounds. 2009;485(1):356–361. DOI: 10.1016/j.jallcom.2009.05.107.

[41] Wang J, Manivannan A, Wu N. Sol-gel derived $La_{0.6}Sr_{0.4}CoO_3$ nanoparticles, nanotubes, nanowires and thin films. Thin Solid Films 2008;517(2):582–587. DOI: 10.1016/j.tsf.2008.06.095.

[42] Rivas ME, Fierro JLG, Guil-López R, Peña MA, La Parola V, Goldwasser MR. Preparation and characterization of nickel-based mixed-oxides and their performance for catalytic methane decomposition. Catalysis Today. 2008;133–135:367–373. DOI: 10.1016/j.cattod.2007.12.045.

[43] Jin Q. Microwave Chemistry. China Science Press: Beijing; 1999.

[44] Selvam MP, Rao K. microwave synthesis and consolidation of gadolinium aluminum perovskite: a ceramic extraordinaire. Advanced Materials. 2000;12:1621–1624. DOI: 10.1002/1521-4095(200011)12:21<1621::AID-ADMA1621>3.0.CO;2-X.

[45] Gibbons KE, Blundell SJ, Mihut AI, Gameson I, Edwards PP, Miyazaki Y, Hyatt NC, Jones MO, Porch A. Rapid synthesis of colossal magnetoresistance manganites by microwave dielectric heating. Chemical Communications. 2000;1:159–160. DOI: 10.1039/A907677H.

[46] Selvam MP, Rao KJ. Microwave preparation and sintering of industrially important perovskite oxides: $LaMO_3$ (M = Cr, Co, Ni). Journal of Materials and Chemistry. 2003;13:596–601. DOI: 10.1039/B211602B.

[47] Yan H, Huang X, Lu Z, Hu H, Xue R, Chen L. Microwave synthesis of $LiCoO_2$ cathode materials. Journal of Power Sources. 1997;68:530–532. DOI: 10.1016/S0378-7753(96)02565-7.

[48] Cavalcante LS, Marques VS, Sczancoski JC, Escote MT, Joya MR, Varela JA, Santos MRMC, Pizani PS, Longo E. Synthesis, structural refinement and optical behavior of $CaTiO_3$ powders: a comparative study of processing in different furnaces. Chemical Engineering Journal. 2008;143(1–3):299–301.

[49] Kaddouri A, Gelin P, Dupont N. Methane catalytic combustion over La–Ce–Mn–O–perovskite prepared using dielectric heating. Catalysis Communications. 2009;10(7):1085–1089. DOI: 10.1016/j.catcom.2008.12.063.

[50] Farhadi S, Momeni Z, Taherimehr M. Rapid synthesis of perovskite-type $LaFeO_3$ nanoparticles by microwave-assisted decomposition of bimetallic $La[Fe(CN)_6]$ $5H_2O$ compound. Journal of Alloys and Compounds. 2009;471(1–2):L5–L8. DOI: 10.1016/j.jallcom.2008.03.113.

[51] Charoenthai N, Traiphol R, Rujijanagul G. Microwave synthesis of barium iron niobate and dielectric properties. Materials Letters. 2008;62(29):4446–4448. DOI: 10.1016/j.matlet.2008.07.047.

[52] Paula AJ, Parra R, Zaghete MA, Varela JA. Synthesis of $KNbO_3$ nanostructures by a microwave assisted hydrothermal method. Materials Letters. 2008;62(17–18):2581–2584. DOI: 10.1016/j.matlet.2007.12.059.

[53] Ryu JH, Koo S, Chang DS, Yoon J, Lim CS, Shim KB. Microwave-assisted synthesis of $PbWO_4$ nano-powders via a citrate complex precursor and its photoluminescence. Ceramics International. 2006;32(6):647–652. DOI: 10.1016/j.ceramint.2005.04.025.

[54] Ryu JH, Koo S, Chang DS, Yoon J, Lim CS, Shim KB. Microwave-assisted synthesis of $CaMoO_4$ nano-powders by a citrate complex method and its photoluminescence property Journal of Alloys and Compounds. 2005;390(1–2);245–249. DOI: 10.1016/j.jallcom.2004.07.064.

[55] Ryu JH, Yoon J, Lim CS, Oh W, Shim KB. Microwave-assisted synthesis of nanocrystalline MWO_4 (M: Ca, Ni) via water-based citrate complex precursor. Ceramics International. 2005;31(6):883–888. DOI: 10.1016/j.ceramint.2004.09.015.

[56] Fu Y, Lin C. Fe/Sr ratio effect on magnetic properties of strontium ferrite powders synthesized by microwave-induced combustion process. Journal of Alloys and Compounds 2005;386(1–2):222–227. DOI: 10.1016/j.jallcom.2004.04.148.

[57] Galal A, Darwish SA, Atta NF, Ali SM, Abd El Fatah AA. Synthesis, structure and catalytic activity of nano-structured Sr–Ru–O type perovskite for hydrogen production. Applied Catalysis A: General. 2010;378:151–159. DOI: 10.1016/j.jallcom.2004.04.148.

[58] Li Z, Meng M, Li Q, Xie Y, Hu T, Zhang J. Fe-substituted nanometric $La_{0.9}K_{0.1}Co_{1-x}Fe_xO_{3-\delta}$ perovskite catalysts used for soot combustion, NO_x storage and simultaneous catalytic removal of soot and NOx. Chemical Engineering Journal. 2010;164:98–105. DOI: 10.1016/j.cej.2010.08.036.

[59] Silva GRO, Santos JC, Martinelli DMH, Pedrosa AMG, Souza MJBd, Melo DMA. Synthesis and characterization of $LaNi_xCo_{1-x}O_3$ perovskites via complex precursor methods. Materials Sciences and Applications. 2010;1:39–45. DOI: 10.4236/msa.2010.12008.

[60] Jahangiri A, Aghabozorg H, Pahlavanzadeh H. Effects of Fe substitutions by Ni in La–Ni–O perovskite-type oxides in reforming of methane with CO_2 and O_2. International Journal of Hydrogen Energy. 2013;38:10407–10416. DOI: 10.1016/j.ijhydene.2013.05.080.

[61] Russo N, Palmisano P, Fino D. Pd substitution effects on perovskite catalyst activity for methane emission control. Chemical Engineering Journal. 2009;154:137–145. DOI: 10.1016/j.cej.2009.05.015.

[62] Oishi M, Yashiro K, Sato K, Mizusaki J, Kawad T. Oxygen nonstoichiometry and defect structure analysis of B-site mixed perovskite-type oxide (La, Sr)(Cr, M)$O_{3-\delta}$ (M=Ti, Mn and Fe). Journal of Solid State Chemistry.2008;181:3177–3184. DOI: 10.1016/j.jssc.2008.08.015.

[63] Liu X, Zhao H, Yang J, Li Y, Chen T, Lu X, Ding W, Li F. Lattice characteristics, structure stability and oxygen permeability of $BaFe_{1-x}Y_xO_{3-\delta}$ ceramic membranes. Journal of Membrane Science. 2011;383:235–240. DOI: 10.1016/j.memsci.2011.08.059.

[64] Dho J, Hur NH. Magnetic and transport properties of lanthanum perovskites with B-site half doping. Solid State Communications. 2006;138:152–156. DOI: 10.1016/j.ssc.2006.02.008.

[65] Suntivich J, May KJ, Gasteiger HA, Goodenough JB, Shao-Horn Y. A Perovskite Oxide Optimized for Oxygen Evolution Catalysis from Molecular Orbital Principles. Science. 2011;334:1383–1385.

[66] Vojvodic A, Nørskov JK. Optimizing Perovskites for the Water-Splitting Reaction. Science. 2011;334:1355–1356. DOI: 10.1126/science.1215081.

[67] Alifanti M, Auer R, Kirchnerova J, Thyrion F, Grange P, Delmon B. Activity in methane combustion and sensitivity to sulfur poisoning of $La_{1-x}Ce_xMn_{1-y}Co_yO_3$ perovskite oxides. Applied Catalysis, B. 2003;41:71–81. DOI: 10.1016/S0926-3373(02)00194-7.

[68] Kida T, Yamasaki A, Watanabe K, Yamazoe N, Shimanoe K. Oxygen-permeable membranes based on partially B-site substituted $BaFe_{1-y}M_yO_{3-\delta}$ (M=Cu or Ni). Journal of Solid State Chemistry. 2010;183:2426–2431. DOI: 10.1016/j.jssc.2010.08.002.

[69] Sora IN, Caronna T, Fontana F, Fernàndez CdJ, Caneschi A, Green M. Crystal structures and magnetic properties of strontium and copper doped lanthanum ferrites. Journal of Solid State Chemistry. 2012;191:33–39. DOI: 10.1016/j.jssc.2012.02.020.

[70] Rodríguez GCM, Saruhan B. Effect of Fe/Co-ratio on the phase composition of Pd-integrated perovskites and its H_2-SCR of NO_x performance. Applied Catalysis, B. 2010;93:304–313. DOI: 10.1016/j.apcatb.2009.10.004.

[71] Ke-bin Z, Hong-de C, Qun T, Bao-wei Z, Di-xin S, Xiao-bai X. Synergistic effect of palladium and oxygen vacancies in the Pd/perovskite catalysts synthesized by the SPC method. Journal of Environmental Sciences. 2005;17(1):19–24. PMID: 15900751.

[72] Silva GRO, Santos JC, Martinelli DMH, Pedrosa AMG, de Souza MJB, Melo DMA. Synthesis and characterization of $LaNi_xCo_{1-x}O_3$ perovskites via complex precursor methods. Materials Sciences and Applications, 2010;1:39–45. DOI: 10.4236/msa.2010.12008.

[73] Russo N, Palmisano P, Fino D. Pd substitution effects on perovskite catalyst activity for methane emission control. Chemical Engineering Journal. 2009;154:137–141. DOI: 10.1016/j.cej.2009.05.015.

[74] Li C, Jiang B, Fanchiang W, Lin Y. The effect of Pd content in $LaMnO_3$ for methanol partial oxidation. Catalysis Communications. 2011;16:165–169. DOI: 10.1016/j.catcom.2011.09.028.

[75] Liu X, Zhao H, Yang J, Li Y, Chen T, Lu X, Ding W, Li F. Lattice characteristics, structure stability and oxygen permeability of $BaFe_{1-x}Y_xO_{3-\delta}$ ceramic membranes. Journal of Membrane Science. 2011;383:235– 240. DOI: 10.1016/j.memsci.2011.08.059.

[76] Kida T, Yamasaki A, Watanabe K, Yamazoe N, Shimanoe K. Oxygen-permeable membranes based on partially B-site substituted $BaFe_{1-y}M_yO_{3-\delta}$ (M= Cu or Ni). Journal of Solid State Chemistry. 2010;183:2426–2431. DOI: 10.1016/j.jssc.2010.08.002.

[77] Jahangiri A, Aghabozorg H, Pahlavanzadeh H. Effects of Fe substitutions by Ni in LaNiO perovskite-type oxides in reforming of methane with CO_2 and O_2. International of Hydrogen Energy. 2013;38:10407–10416. DOI: 10.1016/j.ijhydene.2013.05.080.

[78] Rodrìguez GCM, Saruhan B. Effect of Fe/Co-ratio on the phase composition of Pd-integrated perovskites and its H_2-SCR of NO_x performance. Applied Catalysis B: Environmental. 2010;93:304–313. DOI: 10.1016/j.apcatb.2009.10.004.

[79] Ke-bin Z, Hong-de C, Qun T, Bao-wei Z, Di-xin S, Xiao-bai X. Synergistic effect of palladium and oxygen vacancies in the Pd/perovskites catalysts synthesized by SPC method. Journal of environmental sciences. 2005;17(1):19–24. PMID: 15900751.

[80] Poulsen FW. Defect chemistry modelling of oxygen-stoichiometry, vacancy concentrations, and conductivity of $(La_{1-x}Sr_x)MnO_{3\pm\delta}$. Solid State Ionics. 2000;129:145–162. DOI: 10.1016/S0167-2738(99)00322-7.

[81] Parveen A, Gaur NK. Effect of A-site doping on thermal properties of $LaGaO_3$. Solid State Sciences. 2012;14:814–819. DOI: 10.1016/j.solidstatesciences.2012.03.032.

[82] Srivastava A, Gaur NK, Kaur N, Singh RK. Effect of cation doping on low-temperature specific heat of $LaMnO_3$ manganite. Journal of Magnetism and Magnetic Materials. 2008;320:2596–2601. DOI: 10.1016/j.jmmm.2008.05.042.

[83] Malavasi L, Ritter C, Mozzati MC, Tealdi C, Islam MS, Azzoni CB, Flor G. Effects of cation vacancy distribution in doped $LaMnO_{3+\delta}$ perovskites. Journal of Solid State Chemistry. 2005;178:2042–2049. DOI: 10.1016/j.jssc.2005.04.019.

[84] Lin F, Shi W. Effects of doping site and pre-sintering time on microstructure and magnetic properties of Fe-doped $BaTiO_3$ ceramics. Physica B. 2012;407:451–456. DOI: 10.1016/j.physb.2011.11.013.

[85] Wei HJ, Cao Y, Ji WJ, Au CT. Lattice oxygen of $La_{1-x}Sr_xMO_3$ (M = Mn, Ni) and $LaMnO_{3-\alpha}F_\beta$ perovskite oxides for the partial oxidation of methane to synthesis gas. Catalysis Communications. 2008;9:2509–2514. DOI: 10.1016/j.catcom.2008.06.019.

[86] Abd Al-Rahman YM. Characterization and some applications of nano-inorganic oxides synthesized by microwave technique [MSc thesis]. Faculty of Science Cairo University; 2013.

[87] Gosavi PV, Biniwale RB. Pure phase $LaFeO_3$ perovskite with improved surface area synthesized using different routes and its characterization. Materials Chemistry and Physics. 2010;119:324–329. DOI: 10.1016/j.matchemphys.2009.09.005.

[88] Vijayakumar C, Kumar HP, Solomon S, Thomas JK, Warriar PRS, Koshy J. Synthesis, characterization, sintering and dielectric properties of nanostructured perovskite-type oxide, Ba_2GdSbO_6. Bulletin Materials Science. 2008;31(5):719–722. DOI: 10.1007/s12034-008-0113-2.

[89] Atta NF, Ali SM, El-Ads EH, Galal A. Nano-perovskite carbon paste composite electrode for the simultaneous determination of dopamine, ascorbic acid and uric acid. Electrochimica Acta. 2014;128:16–24. DOI: 10.1016/j.electacta.2013.09.101.

[90] Galal A, Atta NF, Ali SM. Investigation of the catalytic activity of $LaBO_3$ (B = Ni, Co, Fe or Mn) prepared by the microwave-assisted method for hydrogen evolution in acidic medium. Electrochimica Acta. 2011;56:5722–5730. DOI: 10.1016/j.electacta.2011.04.045.

[91] Galal A, Atta NF, Ali SM. Optimization of the synthesis conditions for $LaNiO_3$ catalyst by microwave assisted citrate method for hydrogen production. Applied Catalysis A: General. 2011;409–410:202–208. DOI: 10.1016/j.apcata.2011.10.005.

[92] Galal A, Darwish SA, Atta NF, Ali SM, Abd El Fatah AA. Synthesis, structure and catalytic activity of nano-structured Sr–Ru–O type perovskite for hydrogen production. Applied Catalysis A: General. 2010;378:151–159. DOI: 10.1016/j.apcata.2010.02.015.

[93] Atta NF, Galal A, Ali SM. The catalytic activity of ruthenates $ARuO_3$ (A= Ca, Sr or Ba) for the hydrogen evolution reaction in acidic medium. International Journal of Electrochemical Science. 2012;7:725–746. DOI: 71772857.

[94] Atta NF, Galal A, Ali SM. The effect of the lanthanide ion-type in $LnFeO_3$ on the catalytic activity for the hydrogen evolution in acidic medium. International Journal of Electrochemical Science. 2014;9:2132–2148. DOI: 94967196.

[95] Ali SM, Abd Al-Rahman YM, Galal A. Catalytic activity toward oxygen evolution of $LaFeO_3$ prepared by the microwave assisted citrate method. Journal of Electrochemical Society. 2012;159(9):F600–F605. DOI: 10.1149/2.063209jes.

[96] Galal A, Atta NF, Darwish SA, Abd El Fatah AA, Ali SM. Electrocatalytic evolution of hydrogen on a novel $SrPdO_3$ perovskite electrode. Journal of Power Sources. 2010;195:3806–3809. DOI: 10.1016/j.jpowsour.2009.12.091.

[97] Ghosh S, Dasgupta S. Synthesis, characterization and properties of nanocrystalline perovskite cathode materials. Materials Science—Poland. 2010;28(2):427–438.

[98] Viruthagiri G, Praveen P, Mugundan S, Gopinathan E. Synthesis and characterization of pure and nickel doped $SrTiO_3$ nanoparticles via solid state reaction route. Indian Journal of Advances in Chemical Science. 2013;1(3):132–138.

[99] Li S. Preparation and characterization of perovskite structure lanthanum gallate and lanthanum aluminate based oxides [PhD thesis]. Royal Institute of Technology Stockholm, Sweden; 2009.

[100] Pecchi G, Campos CM, Jiliberto MG, Delgado EJ, Fierro JLG. Effect of additive Ag on the physicochemical and catalytic properties of $LaMn_{0.9}Co_{0.1}O_{3.5}$ perovskite. Applied Catalysis A: General. 2009;371:78–84. DOI: 10.1016/j.apcata.2009.09.031.

[101] Lakshminarayanan N, Choi H, Kuhn1 JN, Ozkan US. Effect of additional B-site transition metal doping on oxygen transport and activation characteristics in $La_{0.6}Sr_{0.4}(Co_{0.18}Fe_{0.72}X_{0.1})O_{3-\delta}$ (where X= Zn, Ni or Cu) perovskite oxides. Applied Catalysis B: Environmental. 2011;103:318–325. DOI: 10.1016/j.apcatb.2011.01.038.

[102] Cho Y, Choi K, Kim Y, Jung J, Lee S. Characterization and catalytic properties of surface La-rich $LaFeO_3$ perovskite. Bulletin of the Korean Chemical Society. 2009;30(6):1368–1372.

[103] Pawlak DA, Ito M, Oku M, Shimamura K, Fukuda T. Interpretation of XPS O (1s) in mixed oxides proved on mixed perovskite crystals. Journal of Physical Chemistry B. 2002;106:504–507. DOI: 10.1021/jp012040a.

[104] Ito A, Masumoto H, Goto T. Microstructure and electrical conductivity of $SrRuO_3$ thin films prepared by laser ablation. Materials Transactions. 2006;47(11):2808–2814.

[105] Pradier CM, Hinnen C, Jansson K, Dahl L, Nygren M, Flodstrom A. Structural and surface characterization of perovskite-type oxides; influence of A and B substitutions upon oxygen binding energy. Journal of Materials Science. 1998;33:3187–3191. DOI: 10.1023/A:1004312326617.

[106] Wu Q, Liu M, Jaegermann W. X-ray photoelectron spectroscopy of $La_{0.5}Sr_{0.5}MnO_3$. Materials Letters. 2005;59:1980–1983. DOI: 10.1016/j.matlet.2005.01.038.

[107] Ghasdi M, Alamdari H. CO sensitive nanocrystalline $LaCoO_3$ perovskite sensor prepared by high energy ball milling. Sensors and Actuators B. 2010;148:478–485. DOI: 10.1016/j.snb.2010.05.056.

[108] Ghasdi M, Alamdari H, Royer S, Adnot A. Electrical and CO gas sensing properties of nanostructured $La_{1-x}Ce_xCoO_3$ perovskite prepared by activated reactive synthesis. Sensors and Actuators B. 2011;156:147–155. DOI: 10.1016/j.snb.2011.04.003.

[109] Ru Z, Jifan H, Zhouxiang H, Ma Z, Zhanlei W, Yongjia Z, Hongwei Q. Electrical and CO-sensing properties of $NdFe_{1-x}Co_xO_3$ perovskite system. Journal of Rare Earths. 2010;28(4):591–595. DOI: 10.1016/S1002-0721(09)60160-5

[110] Fergus JW. Perovskite oxides for semiconductor-based gas sensors. Sensors and Actuators B. 2007;123:1169–1179. DOI: 10.1016/j.snb.2006.10.051.

[111] Song P, Wang Q, Yang Z. The effects of annealing temperature on the CO-sensing property of perovskite $La_{0.8}Pb_{0.2}Fe_{0.8}Cu_{0.2}O_3$ nanoparticles. Sensors and Actuators B. 2009;141:109–115. DOI: 10.1016/j.snb.2009.05.040.

[112] Doroftei C, Pop PD, Iacomi F. Synthesis of nanocrystalline La–Pb–Fe–O perovskite and methanol-sensing characteristics. Sensors and Actuators B. 2011;161(1):977–981. DOI: 10.1016/j.snb.2011.11.078

[113] Benali A, Azizi S, Bejar M, Dhahri E, Graça MFP. Structural, electrical and ethanol sensing properties of double-doping $LaFeO_3$ perovskite oxides. Ceramics International. 2014;40(9):14367–14373. DOI: 10.1016/j.ceramint.2014.06.029.

[114] Liu X, Cheng B, Hu J, Qin H, Jiang M. Semiconducting gas sensor for ethanol based on $LaMg_xFe_{1-x}O_3$ nanocrystals. Sensors and Actuators B. 2008;129:53–58. DOI: 10.1016/j.snb.2007.07.102.

[115] Wang Y, Chen J, Wu X. Preparation and gas-sensing properties of perovskite-type $SrFeO_3$ oxide. Materials Letters. 2001;49:361–364. DOI: 10.1016/S0167-577X(00)00400-6.

[116] Liu X, Hu J, Cheng B, Qin H, Jiang M. Preparation and gas sensing characteristics of p-type semiconducting $LnFe_{0.9}Mg_{0.1}O_3$ (Ln = Nd, Sm, Gd and Dy) materials. Current Applied Physics. 2009;9:613–617. DOI: 10.1016/j.cap.2008.05.014.

[117] Najjar H, Batis H. La–Mn perovskite-type oxide prepared by combustion method: Catalytic activity in ethanol oxidation. Applied Catalysis A: General. 2010;383:192–201. DOI: 10.1016/j.apcata.2010.05.048.

[118] Sun L, Qin HW, Cao E, Zhao M, Gao F, Hu J. Gas-sensing properties of perovskite $La_{0.}875Ba_{0.125}FeO_3$ nanocrystallinepowders. Journal of Physics and Chemistry of Solids. 2011;72:29–33. DOI: 10.1016/j.jpcs.2010.10.074.

[119] Kong L, Shen Y. Gas-sensing property and mechanism of $Ca_xLa_{1-x}FeO_3$ ceramics. Sensors and Actuators B. 1996;30:217–221. DOI: 10.1016/0925-4005(96)80052-9.

[120] Feng C, Ruan S, Li J, Zou B, Luo J, Chen W, Dong W, Wu F. Ethanol sensing properties of $LaCo_xFe_{1-x}O_3$ nanoparticles: effects of calcination temperature, Co-doping, and carbon nanotube-treatment. Sensors and Actuators B. 2011;155:232–238. DOI: 10.1016/j.snb.2010.11.053.

[121] Khetre SM. Ethanol gas sensing properties of nano-porous $LaFeO_3$ thick film. Sensors and Transducers. 2013;149(2):13–19.

[122] Zhang L, Hu J, Song P, Qin H, Jiang M. Electrical properties and ethanol-sensing characteristics of perovskite $La_{1-x}Pb_xFeO_3$. Sensors and Actuators B. 2006;114:836–840. DOI: 10.1016/j.snb.2005.08.002.

[123] Martinelli G, Carotta MC, Ferroni M, Sadaoka Y, Traversa E. Screen-printed perovskite-type thick films as gas sensors for environmental monitoring. Sensors and Actuators B. 1999;55:99–110. DOI: 10.1016/S0925-4005(99)00054-4.

[124] Carotta MC, Butturi MA, Martinelli G, Sadaoka Y, Nunziante P, Traversa E. Microstructural evolution of nanosized $LaFeO_3$ powders from the thermal decomposition of a cyano-complex for thick film gas sensors. Sensors and Actuators B. 1997;44:590–594. DOI: 10.1016/S0925-4005(97)00177-9.

[125] Di Bartolomeo E, Grilli ML, Yoon JW, Traversa E. NO_x sensors based on interfacing nano-sized $LaFeO_3$ perovskite-type oxide and ionic conductors. Project of Special Materials for Advanced Technologies—MSTA II. Via della Ricerca Scientifica, 00133 Rome, Italy.

[126] Sahner K, Moos R, Matam M, Tunney JJ, Post M. Hydrocarbon sensing with thick and thin film p-type conducting perovskite materials. Sensors and Actuators B. 2005;108:102–112. DOI: 10.1016/j.snb.2004.12.104.

[127] Giang HT, Duy HT, Ngan PQ, Thai GH, Thu DTA, Thu DT, Toan NN. Hydrocarbon gas sensing of nano-crystalline perovskite oxides $LnFeO_3$ (Ln = La, Nd and Sm). Sensors and Actuators B. 2011;158:246– 251. DOI: 10.1016/j.snb.2011.06.013.

[128] Jia F, Zhong H, Zhang W, Li X, Wanga G, Songa J, Cheng Z, Yin J, Guo L. A novel nonenzymatic ECL glucose sensor based on perovskite $LaTiO_3$-Ag0.1 nanomaterials. Sensors and Actuators B. 2015;212:174–182. DOI: 10.1016/j.snb.2015.02.011.

[129] Ye D, Xu Y, Luo L, Ding Y, Wang Y, Liu X, Xing L, Peng J. A novel nonenzymatic hydrogen peroxide sensor based on $LaNi_{0.5}Ti_{0.5}O_3/CoFe_2O_4$ modified electrode. Colloids and Surfaces B: Biointerfaces. 2012;89:10–14. DOI: 10.1016/j.colsurfb.2011.08.014.

[130] Zhang Z, Gu S, Ding Y, Jin J. A novel nonenzymatic sensor based on $LaNi_{0.6}Co_{0.4}O_3$ modified electrode for hydrogen peroxide and glucose. Analytica Chimica Acta. 2012;745:112–117. DOI: 10.1016/j.aca.2012.07.039.

[131] Luque GL, Ferreyra NF, Leyva AG, Rivas GA. Characterization of carbon paste electrodes modified with manganese based perovskites-type oxides from the amperometric determination of hydrogen peroxide. Sensors and Actuators B. 2009;142:331–336. DOI: 10.1016/j.snb.2009.07.038.

[132] Zhang Z, Gu S, Ding Y, Zhang F, Jin J. Determination of hydrogen peroxide and glucose using a novel sensor platform based on $Co_{0.4}Fe_{0.6}LaO_3$ nanoparticles. Microchim Acta. 2013;180:1043–1049. DOI: 10.1007/s00604-013-1012-9.

[133] Wang Y, Zhong H, Li X, Jia F, Shi Y, Zhang W, Cheng Z, Zhang L, Wang J. Perovskite $LaTiO_3$–Ag0.2 nanomaterials for nonenzymatic glucose sensor with high performance. Biosensors and Bioelectronics. 2013;48:56–60. DOI: 10.1016/j.bios.2013.03.081.

[134] Wang Y, Xu Y, Luo L, Ding Y, Liu X. Preparation of perovskite-type composite oxide $LaNi_{0.5}Ti_{0.5}O_3$–$NiFe_2O_4$ and its application in glucose biosensor. Journal of Electroanalytical Chemistry. 2010;642:35–40. DOI: 10.1016/j.jelechem.2010.02.001.

[135] Shimizu Y, Komatsu H, Michishita S, Miura N, Yamazo N. Sensing characteristics of hydrogen peroxide sensor using carbon based-electrode loaded with perovskite-type oxide. Sensors and Actuators B. 1996;34:493–498. DOI: 10.1016/S0925-4005(97)80021-4.

[136] Anh DTV, Olthuis W, Bergveld P. Sensing properties of perovskite oxide $La_{1-x}Sr_xCoO_{3-\delta}$ obtained by using pulsed laser deposition. MESA+ Research Institute, University of Twente: Enschede, The Netherlands. p. 618–620.

[137] El-Ads EH, Galal A, Atta NF. Electrochemistry of glucose at gold nanoparticles modified graphite/$SrPdO_3$ electrode—towards a novel non-enzymatic glucose sensor. Journal of Electroanalytical Chemistry. 2015;749:42–52. DOI: 10.1016/j.jelechem.2015.04.033.

[138] Wen CY. Study and Application of Composite Materials Based on Perovskite Nanoparticles [MSc thesis]. Jiangsu University of Science and Technology; 2011.

[139] Wang B, Gu S, Ding Y, Chu Y, Zhang Z, Ba X, Zhang Q, Li X. A novel route to prepare $LaNiO_3$ perovskite-type oxide nanofibers by electrospinning for glucose and hydrogen peroxide sensing. Analyst. 2013;138(1):362–367. DOI: 10.1039/c2an35989h.

[140] Thirumalairajan S, Girija K, Mastelaro VR, Ganesh V, Ponpandian N. Detection of the neurotransmitter dopamine by a glassy carbon electrode modified with self assembled perovskite $LaFeO_3$ microspheres made up of nanospheres. RSC Advances. 2014;4:25957–25962. DOI: 10.1039/C4RA03467H.

[141] Atta NF, Ali SM, El-Ads EH, Galal A. The electrochemistry and determination of some neurotransmitters at $SrPdO_3$ modified graphite electrode. Journal of the Electrochemical Society. 2013;160(7):G3144-G3151. DOI: 10.1149/2.022307jes.

[142] Thirumalairajan S, Girija K, Ganesh V, Mangalaraj D, Viswanathan C, Ponpandian N. Novel synthesis of $LaFeO_3$ nanostructure dendrites: a systematic investigation of growth mechanism, properties, and biosensing for highly selective determination of neurotransmitter compounds. Crystal Growth and Design. 2013;13:291–302. DOI: 10.1021/cg3014305

[143] Wang G, Sun J, Zhang W, Jiao S, Fang B. Simultaneous determination of dopamine, uric acid and ascorbic acid with $LaFeO_3$ nanoparticles modified electrode. Microchim Acta. 2009;164:357–362. DOI: 10.1007/s00604-008-0066-6.

[144] Haile SM. Fuel cell materials and components. Acta Materialia. 2003;51:5981–6000. DOI: 10.1016/j.actamat.2003.08.004.

[145] Shao Z, Haile SM. A high-performance cathode for the next generation of solid-oxide fuel cells. Nature. 2004;431:170–173. DOI: 10.1038/nature02863.

[146] Ding X, Kong X, Wang X, Jiang J, Cui C. Characterization and optimization of $Ln_{1.7}Sr_{0.3}CuO_4$ (Ln = La, Nd)-based cathodes for intermediate temperature solid oxide fuel cells. Journal of Alloys and Compounds. 2010;502:472–476. DOI: 10.1016/j.jallcom.2010.04.199.

[147] Song K, Lee K. Characterization of $Ba_{0.5}Sr_{0.5}M_{1-x}Fe_xO_{3-\delta}$ (M = Co and Cu) perovskite oxide cathode materials for intermediate temperature solid oxide fuel cells. Ceramics International. 2012;38:5123–5131. DOI: 10.1016/j.ceramint.2012.03.015.

[148] Conceição LD, Silva AM, Ribeiro NFP, Souza MMVM. Combustion synthesis of $La_{0.7}Sr_{0.3}Co_{0.5}Fe_{0.5}O_3$ (LSCF) porous materials for application as cathode in IT-SOFC. Materials Research Bulletin. 2011;46:308–314. DOI: 10.1016/j.materresbull.2010.10.009.

[149] Faro ML, Minutoli M, Monforte G, Antonucci V, Aricó AS. Glycerol oxidation in solid oxide fuel cells based on a Ni-perovskite electrocatalyst. Biomass and Bioenergy. 2011;35:1075–1084. DOI: 10.1016/j.biombioe.2010.11.018.

[150] Ling Y, Lin B, Zhao L, Zhang X, Yu J, Peng R, Meng G, Liu X. Layered perovskite $LaBaCuMO_{5+x}$ (M= Fe, Co) cathodes for intermediate-temperature protonic ceramic membrane fuel cells. Journal of Alloys and Compounds. 2010;493:252–255. DOI: 10.1016/j.jallcom.2009.12.072.

[151] Lianghao Y, Yonghong C, Qingwen G, Dong T, Xiaoyong L, Guangyao M, Bin L. Layered perovskite oxide $Y_{0.8}Ca_{0.}2BaCoFeO_{5+\delta}$ as a novel cathode material for inter-

mediate-temperature solid oxide fuel cells. Journal of Rare Earths. 2015;33(5):519–523. DOI: 10.1016/S1002-0721(14)60450-6.

[152] Chen Y, Wei Y, Zhong H, Gao J, Liu X, Meng G. Synthesis and electrical properties of $Ln_{0.6}Ca_{0.4}FeO_{3-\delta}$ (Ln Pr, Nd, Sm) as cathode materials for IT-SOFC. Ceramics International. 2007;33:1237–1241. DOI: 10.1016/j.ceramint.2006.03.035.

[153] Tongyun C, Liming S, Feng L, Weichang Z, Qianfeng Z, Xiangfeng C. $NdFeO_3$ as anode material for S/O_2 solid oxide fuel cells. 2012;30(11):1138–1141. DOI: 10.1016/S1002-0721(12)60194-X.

[154] Takahashl T, Iwahara H. Ionic conduction in perovskite-type oxide solid solution and its applications to the solid electrolyte fuel cell. Energy Conversion. 1971;11:105–111. DOI: 10.1016/0013-7480(71)90121-5.

[155] Bonanos N, Knight KS, Ellis B. Perovskite solid electrolytes: Structure, transport properties and fuel cell applications. Solid State Ionics. 1995;79:161–170. DOI: 10.1016/0167-2738(95)00056-C.

[156] Skinner SJ. Recent advances in perovskite-type materials for solid oxide fuel cell cathodes. International Journal of Inorganic Materials. 2001;3:113–121. DOI: 10.1016/S1466-6049(01)00004-6.

[157] Sun C, Stimming U. Recent anode advances in solid oxide fuel cells. Journal of Power Sources. 2007;171:247–260. DOI: 10.1016/j.jpowsour.2007.06.086.

[158] Wang F, Chen D, Shao Z. $Sm_{0.5}Sr_{0.5}CoO_{3-\delta}$-infiltrated cathodes for solid oxide fuel cells with improved oxygen reduction activity and stability. Journal of Power Sources. 2012;216:208–215. DOI: 10.1016/j.jpowsour.2012.05.068.

[159] Li H, Zhao Y, Wang Y, Li Y. $Sr_2Fe_{2-x}Mo_xO_{6-\delta}$ perovskite as an anode in a solid oxide fuel cell:Effect of the substitution ratio. Catalysis Today. DOI: 10.1016/j.cattod.2015.04.025.

[160] Raza MA, Rahman IZ, Beloshapkin S. Synthesis of nanoparticles of $La_{0.75}Sr_{0.25}Cr_{0.5}Mn_{0.5}O_{3-\delta}$ (LSCM) perovskite by solution combustion method for solid oxide fuel cell application. Journal of Alloys and Compounds. 2009;485:593–597. DOI: 10.1016/j.jallcom.2009.06.059.

[161] Huang YH, Dass RI, Xing ZL, Goodenough JB. Double perovskites as anode materials for solid-oxide fuel cells. Science. 2006;312:254–257. DOI: 10.1126/science.1125877

[162] Tao SW, Irvine JTS. A redox-stable efficient anode for solid-oxide fuel cells. Nature Materials. 2003;2:320–323. DOI: 10.1038/nmat871.

[163] Vernoux P, Guillodo M, Fouletier J, Hammou A. Alternative anode material for gradual methane reforming in solid oxide fuel cells. Solid State Ionics. 2000;135:425–431. DOI: 10.1016/S0167-2738(00)00390-8.

[164] Jin C, Cao X, Lu F, Yang Z, Yang R. Electrochemical study of $Ba_{0.5}Sr_{0}.5Co_{0.8}Fe_{0.2}O_{3}$ perovskite as bifunctional catalyst in alkaline media. International Journal of Hydrogen Energy. 2013;38:10389–10393. DOI: 10.1016/j.ijhydene.2013.06.047.

[165] Malkhandi S, Manohar AK, Yang B, Prakash GKS, Narayanan SR. Properties of calcium-doped lanthanum cobalt oxide perovskite electrocatalysts for oxygen evolution in alkaline medium. The Electrochemical Society. Abstract 292, 220th ECS Meeting. 2011.

[166] Kahoul A, Hammouche A, Nâamoune F, Chartier P, Poillerat G, Koenig JF. Solvent effect on synthesis of perovskite-type $La_{1-x}Ca_{x}CoO_{3}$ and their electrochemical properties for oxygen reactions. Materials Research Bulletin. 2000;35:1955–1966. DOI: 10.1016/S0025-5408(00)00395-0.

[167] Bao X, Wang Y, Zhu Q, Wang N, Zhu D, Wang J, Yang A, Yang R. Efficient planar perovskite solar cells with large fill factor and excellent stability. Journal of Power Sources. 2015;297:53–58. DOI: 10.1016/j.jpowsour.2015.07.081.

[168] Bhatt P, Pandey K, Yadav P, Tripathi B, KanthP C, Pandey MK, Kumar M. Investigating the charge carrier transport within the hole-transport material free perovskite solar cell processed in ambient air. Solar Energy Materials and Solar Cells. 2015;140:320–327. DOI: 10.1016/j.solmat.2015.04.028.

[169] Chen Q, Marco ND, Yang YM, Song T, Chen C, Zhao H, Hong Z, Zhou H, Yang Y. Under the spotlight: the organic-inorganic hybrid halide perovskite for optoelectronic applications. Nano Today. DOI: 10.1016/j.nantod.2015.04.009.

[170] Green MA, Ho-Baillie A, Snaith HJ. The emergence of perovskite solar cells. Nature Photonics. 2014;8:506–514. DOI: 10.1038/NPHOTON.2014.134.

[171] Park N. Perovskite solar cells: an emerging photovoltaic technology. Materials Today. DOI: 10.1016/j.mattod.2014.07.007.

[172] Schoonman J. Organic-inorganic lead halide perovskite solar cell materials: a possible stability problem. Chemical Physics Letters. DOI: 10.1016/j.cplett.2014.11.063.

[173] Sun H, Ruan P, Bao Z, Chen L, Zhou X. Shell-in-Shell TiO_2 hollow microspheres and optimized application in light-trapping perovskite solar cells. Solid State Sciences. 2015;40:60–66. DOI: 10.1016/j.solidstatesciences.2014.12.015.

[174] Tripathi B, Bhatt P, KanthP C, Yadav P, Desai B, Pandey MK, Kumar M. Temperature induced structural, electrical and optical changes in solution processed perovskite material: Application in photovoltaics. Solar Energy Materials and Solar Cells. 2015;132:615–622. DOI: 10.1016/j.solmat.2014.10.017.

[175] Wang X, Li Z, Xu W, Kulkarni SA, Batabyal SK, Zhang S, Cao A, Wong LH. TiO_2 nanotube arrays based flexible perovskite solar cells with transparent carbon nanotube electrode. Nano Energy. DOI: 10.1016/j.nanoen.2014.11.042.

[176] Werner J, Dubuis G, Walter A, Löper P, Moon S, Nicolay S, Morales-Masis M, Wolf SD, Niesen B, Ballif C. Solar Energy Materials and Solar Cells. 2015;141:407–413. DOI: 10.1016/j.solmat.2015.06.024.

[177] Yan W, Li Y, Li Y, Ye S, Liu Z, Wang S, Bian Z, Huang C. High-performance hybrid perovskite solar cells with open circuit voltage dependence on hole transporting materials. Nano Energy. DOI: 10.1016/j.nanoen.2015.07.024.

[178] Yang Z, Cai B, Zhou B, Yao T, Yu W, Liu FS, Zhang W, Li C. An up-scalable approach to $CH_3NH_3PbI_3$ compact films for high-performance perovskite solar cells. Nano Energy. DOI: 10.1016/j.nanoen.2015.05.027.

[179] Yeo J, Kang R, Lee S, Jeon Y, Myoung N, Lee C, Kim D, Yun J, Seo Y, Kim S, Na S. Highly efficient and stable planar perovskite solar cells with reduced graphene oxide nanosheets as electrode interlayer. Nano Energy. DOI: 10.1016/j.nanoen.2014.12.022.

Chapter 5

Perovskite Oxide Nanocrystals — Synthesis, Characterization, Functionalization, and Novel Applications

Heng Wu and Xinhua Zhu

Additional information is available at the end of the chapter

http://dx.doi.org/10.5772/61640

Abstract

Perovskite oxide nanocrystals exhibit a wide spectrum of attractive properties such as ferroelectricity, piezoelectricity, dielectricity, ferromagnetism, magnetoresistance, and multiferroics. These properties are indispensable for applications in ferroelectric random access memories, multilayer ceramic capacitors, transducers, sensors and actuators, magnetic random access memories, and the potential new types of multiple-state memories and spintronic devices controlled by electric and magnetic fields. In the past two decades, much effort has been made to synthesize and characterize the perovskite oxide nanocrystals. Various physical and chemical deposition techniques and growth mechanisms are explored and developed to control the morphology, identical shape, uniform size, perfect crystalline structure, defects, and homogenous stoichiometry of the perovskite oxide nanocrystals. This chapter provides a comprehensive review of the state-of-the-art research activities that focus on the rational synthesis, structural characterization, functionalization, and unique applications of perovskite oxide nanocrystals in nanoelectronics. It begins with the rational synthesis of perovskite oxide nanocrystals, and then summarizes their structural characterizations. Fundamental physical properties of perovskite oxide nanocrystals are also highlighted, and a range of novel applications in nanoelectronics, information storages, and spintronics are discussed. Finally, we conclude this review with some perspectives/outlook and future researches in these fields.

Keywords: Perovskite oxide nanocrystals, synthesis, structural characterization, functionalization, applications

1. Introduction

Perovskite oxide nanocrystals have important properties in ferroelectricity, piezoelectricity, dielectricity, ferromagnetism, magnetoresistance, and multiferroics. The perovskite structure is named for the prototype $CaTiO_3$ mineral called perovskite, which is generally a metal oxide with the formula ABO_3, where B is a small transition metal cation and A is a larger s-, d-, or f-block cation. In a cubic perovskite, the larger cation A resides on the corners of the unit cell, the smaller cation B is in the center of the unit cell, and the oxygen ions (O_2^-) are on the centers of the faces [1]. The perovskite structure can also be built from three-dimensional corner-sharing BO_6 octahedra that are connected through B–O–B linkages. The A-site cation fits in the large cavity at the center of eight corner-sharing BO_6 octahedra, and the B-site cation resides in the interstitial site of an octahedron of oxygen anions [1]. Interestingly, and of technological importance, a variety of crystallize in the perovskite structure. Typical perovskite materials of technological importance are ferroelectric $BaTiO_3$, $PbTiO_3$, dielectric $(Ba,Sr)TiO_3$, piezoelectric $Pb(Zr,Ti)O_3$, electrostrictive $Pb(Mg,Nb)O_3$, magnetoresistant $(La,Ca)MnO_3$, and multiferroic $BiFeO_3$. They have attracted interest for several decades, with tremendous applications including ferroelectric random access memories, multilayer ceramic capacitors, transducers, sensors and actuators, magnetic random access memories, and the potential new types of multiple-state memories and spintronic devices controlled by electric and magnetic fields [1–8]. The major challenge in manufacturing these materials is the processing of the materials with reliable and reproducible properties [9, 10]. Following a similar trend to the miniaturization as the conventional CMOS (complementary metal oxide semiconductor) devices, the down-sized electronic devices based on perovskite electronic ceramic materials have also been developed. Advances toward nanoscale electronics have additionally increased interest in this field of perovskite oxide nanocrystals [11–13]. For example, to develop high volume efficient multilayered ceramic capacitors (MLCCs), the sizes of $BaTiO_3$ nanoparticles with high purity and uniform shape used for fabricating the next generation of MLCCs will be lowered down to tens of nanometers. Therefore, synthesis of high-purity, ultra-fine, and agglomerate-free perovskite nanoparticles with controlled particle size, morphology, and stoichiometry is the critical step in processing of perovskite ceramics with desirable properties. Perovskite oxide nanocrystals with the general formula of ABO_3 play a very important role in today's technological advances, which can be found in the application fields of passive dielectric materials, piezoelectric actuators and transducers, sensors, and micro/nano electromechanical systems and many more.

The evolution of a method to produce perovskite oxide nanocrystals with precise stoichiometry and desired properties is much complex. Conventionally, perovskite nanoparicles are prepared by solid-state reactions between the corresponding oxides or oxides and carbonates at temperatures above 1000°C [14, 15]. However, the resulting microstructures of perovskite nanoparticles obtained from this method are not suitable for the miniaturization of electronic devices, due to their significant particle agglomeration, poor chemical homogeneity, and coarse large particle sizes. To resolve the problems arising from the conventional ceramic techniques and to produce homogeneous and stoichiometric perovskite nanocrystals, in recent years, wet chemical routes have been developed [12, 16–18]. They can be better controlled from the molecular precursor to the final material to give highly pure and homogeneous products, allowing for the low reaction temperatures used. The size and morphology of the nanocrystals can be controlled, and metastable phases could be prepared [18].

The objective of this chapter is to provide an overview of the state-of-the-art in perovskite nanocrystals, which covers their synthesis, characterization, functionalization, and novel applications. First, we review the synthetic methods for perovskite nanocrystals, which include the syntheses using solid and liquid precursors. The second section deals with the electron microstructural characterization of perovskite nanocrystals. In the context of functionalization, we discuss the unique properties of perovskite oxide nanocrystals (e.g., ferroelectric and dielectric, electrical, magnetic, and multiferroic properties), and the size effects for these unique properties are also discussed. And then a broad range of novel applications of perovskite oxide nanocrystals is addressed. Finally, we provide a perspective on the future outlook of perovskite nanocrystals.

2. Synthesis of perovskite oxide nanocrystals

Due to the particle size, dimensionality, and composition governing the resultant properties of the nanostructured perovskite materials that are assembled from nanocrystals as building blocks to achieve certain desired properties, the synthesis of high-purity, ultra-fine, and agglomerate-free perovskite oxide nanocrystals with controlled particle size, morphology, and stoichiometry is the first and perhaps the most crucial step in processing of perovskite ceramics with desirable properties. The major issues for the synthesis of perovskite oxide nanocrystals include: (a) the control of particle size and composition and (b) the control of the interfaces and distributions of the nanobuilding blocks within the fully formed nanostructured perovskite compounds. Over the past several decades, various methods have been developed to prepared perovskite oxide nanocrystals and the related nanostructured perovskite compounds. In the subsequent sections, some important methods for the preparation of perovskite oxide nanocrystals are described.

2.1. Solid-state reaction method

The solid-state reaction method is the most traditional one for preparing perovskite oxide nanocrystals (e.g., $BaTiO_3$, $PbTiO_3$, $Pb(Zr,Ti)O_3$, etc.) [14, 15]. This process includes weighting starting materials (the corresponding oxides or oxides and carbonates), mixing, milling, and calcining them at elevated temperatures to form the perovskite phase. For example, in synthesis of $BaTiO_3$ nanocrystals by solid-state reaction method, the reaction process in air has been proposed to take place in at least three stages and relies on the diffusion of Ba^{2+} ions into TiO_2 [19]. Firstly, $BaCO_3$ reacts with the outer surface region of TiO_2 to form a surface layer of $BaTiO_3$ on individual TiO_2 grains. Further diffusion of Ba^{2+} ions into TiO_2 necessitates the formation of Ba_2TiO_4 between the unreacted $BaCO_3$ and the previously formed $BaTiO_3$. After prolonged sintering periods, the intermediate Ba-rich phase Ba_2TiO_4 reacts with the remaining TiO_2 in the core-regions of the TiO_2 grains to form $BaTiO_3$. The high-temperature calcination produces an agglomerated powder with a coarse particle size which requires additional milling process. However, contamination and other undesirable features during the milling process can create defects in the manufactured products. Furthermore, the more components in the ceramic powders, the more difficult it may be to achieve the desired homogeneity, stoichiometry, and phases. By using nanocrystalline $BaCO_3$ and TiO_2 as starting materials, Buscaglia et al. [20] have recently synthesized the perovskite $BaTiO_3$ nanocrystals with size of ~100 nm and narrow particle size distribution, *via* a solid-state reaction at calcination temperatures as low as 800°C. The average particle size of powders obtained *via* this method is essentially determined by the particle size of the used TiO_2 because the reaction rate is controlled by the diffusion rate of barium ions into the TiO_2 lattice [21]. Similar reaction mechanism was also found in the synthesized process of $BaZrO_3$ powders [22]. The morphology of $BaZrO_3$ nanoparticles was dependent upon the initial size and shape of the used starting ZrO_2 particles. Therefore, fine $BaZrO_3$ powders with particle size of 70–100 nm composing of crystallites of ~ 20–30 nm can be synthesized by using very fine (70–90 nm) starting ZrO_2 particles and coarse (~ 1 μm) $BaCO_3$ particles commercially available and calcination at ~1000°C. Higher calcination temperatures accelerate the initial stage of reaction but often lead to coarser and more-agglomerated powders.

2.2. Molten-Salt Method (MSS)

Molten-salt synthesis (MSS) is one of the methods for preparing perovskite oxide nanopowders, which involves the use of a molten salt as the medium for preparing perovskite oxides from their constituent materials (oxides and carbonates) [23]. This method allows melt-solid reacting much faster due to the small diffusion distances and higher mobility of oxides in the melt. Many molten salt solvents such as alkali chlorides, sulphates, carbonates, and hydroxides are used in the past acting as a medium of reaction for the constituent oxides. MSS attracted much attention due to its advantages such as one-step, rapid, and environmentally friendly for synthesizing perovskite oxide functional materials. The melt increases the reaction rate due to the small diffusion distances and high mobility of oxides in it. In addition, as a medium of the reaction, the choice of the molten salt system is variety, e.g., alkali chlorides, sulphates, carbonates, and hydroxides. The features of this synthetic method are the simplest, versatile, and cost-effective approaches available for obtaining a pure perovskite phase at a relatively low temperature for a shorter soaking time. Generally, the procedure of synthesize perovskite oxide powders by MSS is shown below. First, heating the raw materials consisting of reactants and salt system above the melting temperature of the salt, so that the produced particles are synthesized at the melted salt solvent. The characteristics of the powders can be controlled by tuning the heating temperature and duration time. After cooling down to room temperature naturally, the product is washed by appropriate solvent (typically, deionized water) and dried subsequently, and the complex perovskite oxide powder is obtained. In the past decade, MSS has been widely used to synthesize a range of perovskite oxide nanocrystals. For example, perovskite $SrFeO_3$ nanocrystals were successfully synthesized in molten $NaNO_3$–KNO_3 eutectic with Na_2O_2 from a mixture of strontium nitrate and ferric nitrate [24]. The effects of metal precursors, salt medium, annealing temperature, and oxidizing properties of the melt on the phase compositions, crystallite sizes, and morphology of the resulting metal oxides were systematically investigated. It was found that the formation of the $SrFeO_3$ phase was mainly dependent upon the nature of the metal precursor and salt medium. Metal nitrates were the suitable precursors and $NaNO_3$–KNO_3 eutectic with Na_2O_2 was the suitable salt medium, which resulted in the formation of pure $SrFeO_3$ nanocrystals at a much lower temperature of 400°C. Pure cubic phase $SrTiO_3$ nanocrystals were synthesized in the eutectic NaCl–KCl molten salts at 700°C by heating TiO_2 and $Sr(NO_3)_2$ powders for 6 h [25]. The sizes of the as-synthesized $SrTiO_3$ nanocrystals were dependent upon the kinds of TiO_2 precursors, indicating that the formation process of $SrTiO_3$ in the molten salts was mainly controlled by the template formation mechanism. Single-crystalline perovskite $BaZrO_3$ nanocrystals were also obtained using BaC_2O_4 and ZrO_2 as precursors and NaOH/KOH as the molten salts at 700°C [26]. Rare earth orthoferrites with the general formula of $LnFeO_3$ (Ln = La, Pr, Nd) were synthesized in molten NaOH flux at 400°C [27], and $LaMO_3$ (M = Mn, Fe, Co, Ni) were obtained in molten nitrates or nitrites [28]. Recently, perovskite Pb-based relaxors [29], Ba-based dielectric oxides [30], and perovskite multiferroic bismuth ferrites [31–35] were also synthesized by MSS process.

2.3. Wet chemical methods

Since perovskite oxide nanocrystals synthesized *via* conventional solid-state reactions have uncontrolled and irregular morphologies, the electrical properties of the sintered ceramics are very poor. To solve this problem, various wet chemical methods have been developed to synthesize perovskite nanocrystals. The popular wet chemical methods for the preparation of perovskite nanocrystals, include sol–gel method [36–38], alkoxide–hydroxide sol-precipitation method [39–41], hydrothermal method [42–44], microwave-hydrothermal [45–47], and solvothermal syntheses [48–50]. The most important advantages of the wet chemical methods include easy controlling of the chemical stoichiometry, producing nanocrystals with narrow size distribution, and low crystallization temperature due to the constituents mixed at the quasi-atomic level in a solution system. Due to the wet chemical solution process, a dopant such as paramagnetic ions or rare-earth ions could be readily introduced during the preparation of the precursor solution. In the following subsequent sections, various wet chemical methods used for preparation of perovskite oxide nanocrystals are introduced.

2.3.1. Sol–gel (colloidal) processing

Sol–gel process is a popular processing route for the synthesis of perovskite oxide nanocrystals (e.g., $BaTiO_3$, $PbTiO_3$, $BiFeO_3$) [36–38]. This process involves the formation of a sol by dissolving the metal aloxide, metal-organic, or metal-inorganic salt precursors in a suitable solvent, subsequent drying of the gel followed by calcination and sintering at high temperature to form perovskite nanocrystals. Due to the reacting species homogenized at the atomic level in a sol–gel process, the diffusion distances are considerably reduced compared to a conventional solid-state reaction; therefore, the product can be formed at much lower temperatures. In this process, the selection of starting materials, concentration, pH value, and heat treatment schedule play an important role in affecting the properties of perovskite nanocrystals. This has been demonstrated in the case of $BaTiO_3$

perovskite oxide nanocrystals [38, 51, 52]. Barium acetate and titanium isopropoxide are often used as starting materials to synthesize $BaTiO_3$ nanocrystals. However, the different rates in the hydrolysis and condensation of Ba and Ti precursors often give rise to the chemical component segregation in the obtained gels. To solve this problem, acetic acid or acetylaceton was often used to control the hydrolysis rate of the Ti precursor, since these complexing agents acts as chelating agents to coordinate with Ti species [53,54].

For the obtained gels, a heat treatment at high temperature over 600°C is required to remove the unreacted organics and to crystallize the powders. Several steps involve in the transformation from the precursor to the crystalline $BaTiO_3$ nanocrystals, including the transformation from the precursor to the amorphous $BaTiO_3$, and then to the three-dimensional nucleation of the crystalline $BaTiO_3$ in the amorphous matrix, and finally to the nanocrystal growth of $BaTiO_3$ *via* a solid-state reaction [55]. To better control the grain size and its distribution, the heat treatment process parameters of the gels (e.g., post-annealing temperature, time and atmosphere, heating rate) must be optimized [56]. Normally, higher annealing temperature or longer annealing time can lead to larger grain size of the powders, while slow heating rate and inert annealing atmosphere can inhabit the aggregated behavior of nanocrystals in comparison to air or oxygen atmosphere. That was demonstrated in the synthesis of $Pb(Zr,Ti)O_3$ nanocrystals [56]. By using these techniques, monodispersed perovskite oxide nanocrystals and related nanostructured materials have been successfully fabricated. The particle size can be adjusted from a few nanometers to micrometers *via* controlling the solid-state polymerization and the heat treatment process [36, 55].

2.3.2. Hydrothermal process

Hydrothermal synthesis involves heating an aqueous suspension of insoluble salts in an autoclave at a moderate temperature and pressure where the crystallization of a desired phase is taking place. As a powerful method for synthesis of very fine and homogeneous perovskite powders with a controllable size distribution and morphology, its application to the growth of $BaTiO_3$ powders with the desired size and particle morphology has been widely investigated [57–60]. Based on the high-resolution transmission electron microscopy (HRTEM) observations on the incompletely and fully reacted powders, Pinceloup et al. [57] proposed a dissolution–precipitation model for hydrothermal synthesis of $BaTiO_3$ nanocrystals using $Ba(OH)_2$ and TiO_2 as precursors. In this model, TiO_2 particles are first dissolved to form hydroxyl titanium complexes $[Ti(OH)^{n-}]$ and then react with barium ions in the solution to precipitate $BaTiO_3$. On the other hand, Hertl [58] and Hu et al. [61] proposed another in situ heterogeneous transformation model, in which TiO_2 particles react initially with the dissolved barium to produce a continuous layer of $BaTiO_3$, and the additional barium must diffuse through this layer and reacts with TiO_2 until the supply of TiO_2 is exhausted. This model was supported experimentally by the hydrothermal conversion from TiO_2 microspheres to nanocrystalline $BaTiO_3$ [59]. Eckert et al. [60] also reported on a mechanism evolution from a dissolution–precipitation process at the early stage of the reaction to an in situ mechanism for the longer reaction times. Recently, Walton et al. [62] investigated the hydrothermal crystallization of $BaTiO_3$ by time-resolved powder neutron diffraction methods in situ, using the newly developed Oxford/ISIS hydrothermal cell. They directly observed that the rapid dissolution of the barium source was followed by dissolution of the titanium source before the onset of crystallization of $BaTiO_3$. These qualitative observations strongly suggest that a homogeneous dissolution–precipitation mechanism dominates in the hydrothermal crystallization of $BaTiO_3$ rather than other possible mechanisms proposed in the literatures [81–83]. These contradictive experimental observations reported previously probably result from the different hydrothermal conditions.

2.3.3. Microwave-hydrothermal process

The microwave-hydrothermal process is a rapid process, which has the potential to enhance the crystallization kinetics of hydrothermal process. The term microwave-hydrothermal process was named by Komarneni et al. [63] in 1992, and this process has been used for the rapid synthesis of numerous ceramic oxides, hydroxylated phases, porous materials, and hematite powders [64–67]. It offers many distinct advantages over the conventional hydrothermal synthesis, such as cost savings, rapid internal heating, and synthesis of new materials.

Up to date, numerous reports have been published on the synthesis of $BaTiO_3$ nanocrystals by microwave-hydrothermal process below 200°C, and these processes were found to be very rapid but they all yielded cubic phase [45–47, 68]. For example, Khollam et al. [47] obtained submicron-sized $BaTiO_3$ powders (0.1–0.2 μm) at holding time of 30 min. One of the first approaches on the synthesis of the nanosized $BaTiO_3$ powders (about 30 nm) at 30 min was reported by Jhung et al. [69]. Recently, tetragonal $BaTiO_3$ powders are synthesized by microwave-hydrothermal method at a typical temperature of 240°C from hydrous titanium oxide and barium hydroxide, in the absence of chloride ions and alkali metal ions to avoid contaminations. The effects of synthesis conditions, including reaction temperature and time, and reactant composition, on the formation of tetragonal structure and particle size of $BaTiO_3$ powders, have been systematically investigated [70]. The results have shown that the amount of the tetragonal phase and the particle size increased quickly with reaction time, whereas the content of lattice hydroxyl groups decreased. Tetragonal $BaTiO_3$ powder with nearly full tetragonality (c/a ratio = 1.010) was obtained *via* the microwave-hydrothermal process performed at 240°C for 20 h [70]. As the reaction temperature was lowered down to 220°C, the formation of tetragonal structure and the growth of particles slowed down substantially, showing a critical effect of the reaction temperature on the microwave-hydrothermal processing of tetragonal $BaTiO_3$. Higher $Ba(OH)_2$/Ti mole ratio enhanced the formation of tetragonal $BaTiO_3$ and so did higher initial concentration of Ti with fixed $Ba(OH)_2$/Ti ratio. Besides the $BaTiO_3$ nanocrystals, Ba_1–xSr$x$$TiO_3$ (x = 0.1–0.4) nanocrystals with the average size about 20 nm were also prepared at a relatively short period of time (10 min) *via* microwave-hydrothermal synthesis [71]. The structure and the average size of BST were determined to be in the range of 20–50 nm depending on the synthesis time (10–90 min).

Perovskite $BiFeO_3$ nanocrystals exhibit multiferroic properties at room temperature (ferroelectric and magnetic order coexisting in the same phase) and is, therefore, a good candidate for potential multiferroic application in information technology. Recently, perovskite $BiFeO_3$ nanocrystals with diameters of 10–50 nm are synthesized by microwave-hydrothermal synthesis [72,73]. The starting reactants are used as $Fe(NO_3)_3{\bullet}9H_2O$ and $Bi(NO_3)_3{\bullet}5H_2O$ together with KOH as the mineralizer. Figure 2 shows the XRD pattern of the perovskite $BiFeO_3$ nanocrystals synthesized by microwave-hydrothermal process. As shown in Figure 2, all the reflection peaks can be readily indexed as a rhombohedrally distorted perovskite BFO (JCPDS card No. 86-1518) with space group R3c and lattice parameters of a = 5.582 Å and c = 13.876 Å. No peaks from other phase were detected. In addition, the sharp diffraction peaks indicate that highly crystallized and phase-pure BFO nanocrystals can be obtained under the present synthesis conditions. The particle size and morphology of the as-obtained perovskite $BiFeO_3$ nanocrystals were revealed by TEM images. Figure 3(a) represents a typical low-magnitude TEM image, in which the BFO nanocrystals exhibit a spherical morphology with particle sizes of 15–55 nm. No agglomerated particles were observed, and nearly monodispersive behavior was observed in these BFO nanocrystals. Their statistic particle size distribution is shown in Figure 3(b), which clearly demonstrates that the average particle size of the BFO nanoparticle is ~35 nm, this value is much smaller than the previously reported ones [61, 74–77]. A high-magnification TEM image is shown in Figure 3(c), which clearly demonstrates the well-dispersed BFO nanocrystals with a spherical morphology. The selected area electron diffraction (SAED) pattern of the as-obtained BFO nanocrystals is shown in Figure 3(d), which exhibits polycrystalline diffraction rings consisting of discrete diffraction spots. Based on the analysis of the diffraction rings, the first six diffraction rings can be indexed as (012), (104)/(110), (006)/(202), (024), (116)/(122), and (214), which is in well agreement with the XRD results. The high crystallinity nature of the BFO nanocrystals is also proven by the lattice fringes observed in the HRTEM image. Figure 3(e) shows an HRTEM image taken from a single BFO nanocrystal with a particle size of ~12 nm, and the lattice fringes of (202) and (113) crystal planes are clearly resolved. The Fourier filtered HRTEM image of the single BFO nanocrystal is shown in Figure 3(f), and the inset is a fast Fourier transform (FFT) pattern of the nanocrystal obtained by Gatan Digital Micrography software. The angle between the (202) and (113) crystal planes measured from Figure 3(f) is 58°, which is very close to the theoretical value of 57.99° (the angle $\angle(202):(113)$ = 57.99° for BFO). Therefore, the single-crystalline nature of BFO nanocrystals is confirmed by the SAED pattern and HRTEM image.

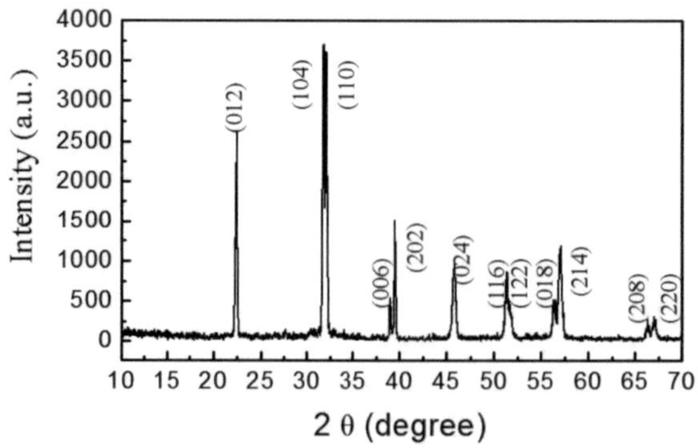

Figure 1. XRD pattern of the as-synthesized perovskite $BiFeO_3$ nanocrystals by microwave-hydrothermal reaction. Reproduced with permission from [73].

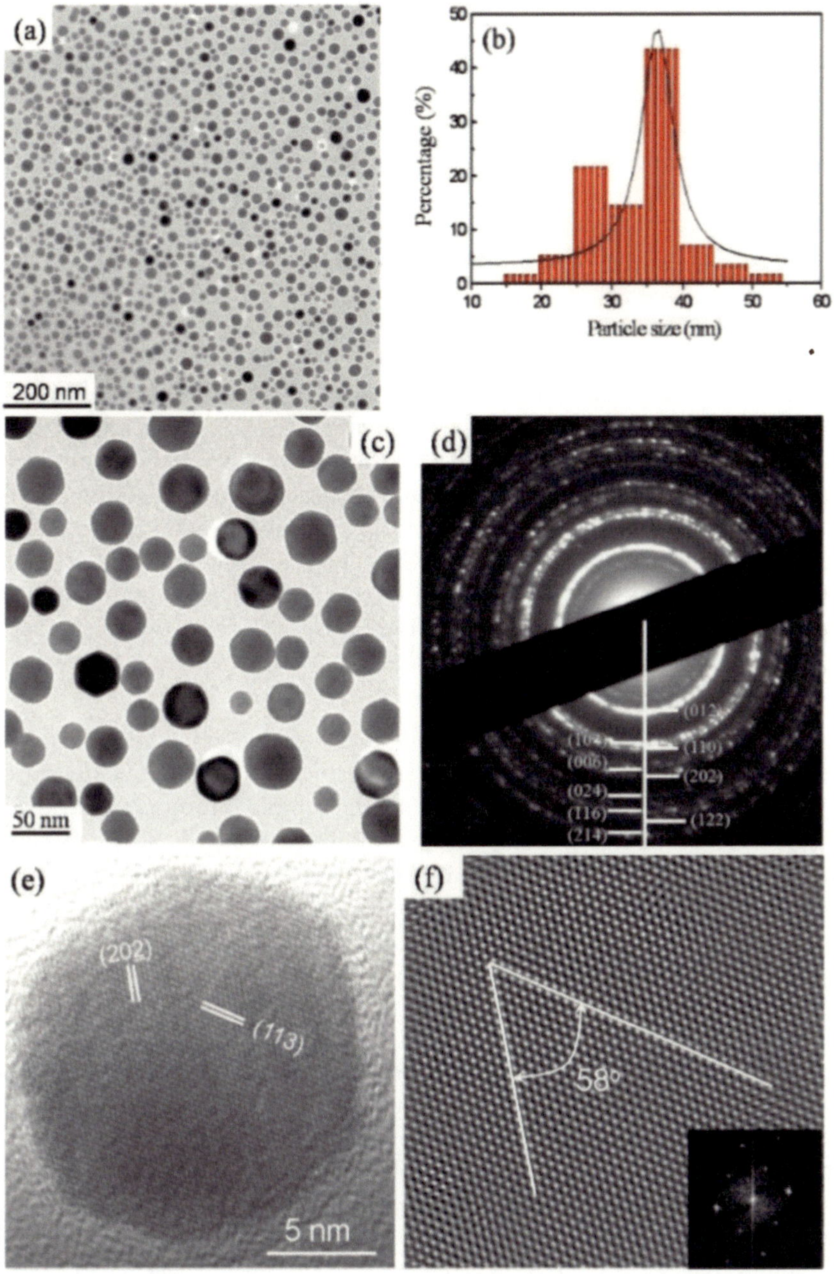

Figure 2. (a) Typical low-magnitude TEM image of the perovskite $BiFeO_3$ nanocrystals, (b) statistic particle size distribution obtained from low-magnitude TEM image, (c) high-magnitude TEM image, (d) selected area electron diffraction pattern, in which the first six diffraction rings are indexed as (012), (104)/(110), (006)/(202), (024), (116)/(122), and (214). (d) HRTEM image of a single perovskite BFO nanocrystal with a diameter of ~12 nm, and (e) Fourier filtered HRTEM image. Inset is a fast Fourier transform (FFT) pattern of the nanocrystal. Reproduced with permission from [73].

3. Structural characterization of perovskite oxide nanocrystals

Up to date, numerous TEM investigations have been carried out to investigate the microstructural features of perovskite $BaTiO_3$ nanocrystals [79–83]. For example, the internal pore structures in the hydrothermal $BaTiO_3$ nannopowders were examined by TEM [79]. Figure 3 shows TEM images of (a) as-received $BaTiO_3$ powder (particle size ~60 nm) and (b) $BaTiO_3$ powder annealed at 673 K. It was observed that some particles had internal pores of various sizes in as-received $BaTiO_3$ powders, as indicated by the arrows in Figure 3(a). A large pore with a cubic-shaped was observed in the particle annealed at 673 K in Figure 3(b). Lattice images were observed around the particle, and these indicated that the pore existed inside the particle. A three-dimensional structure of the internal pore was successfully observed and constructed by TEM tomography system. The results showed that no inclusion was found in the pores, and such internal pores were not lattice defects at the atomic level because their sizes were a few tens of nanometers. Large pores were involved in as-received $BaTiO_3$ powders, and their numbers decreased at >1,128 K. Some of the internal pores were released from the particle's surface and/or during the grain growth. The presence of the pores affected the density of the $BaTiO_3$ particle. The behavior of the internal pore was observed in situ with increasing temperature on the thermal stage of a TEM device. The results showed that at >1,128 K, some pores moved out from the particle's surface during TEM observation. This temperature roughly agreed with the temperature at which the density of $BaTiO_3$ powder sharply increases. During observation with increasing temperature, a thin layer appeared on the particle's surface at temperature over 573 K and then disappeared at 1,193 K.

The hydrothermal $BaTiO_3$ powder with a small particle size is stabilized in a cubic phase at room temperature [80–83], which implies that the distortion of the $[TiO_6]$ structure resulting in a cubic-to-tetragonal phase transition as cooling the sample through the Curie temperature is not taking place. A plausible reason is that the small size of the $BaTiO_3$ nanocrystals, which are so small that the structural defects in the particles prevent the completion of the structural transition, leading to high strains within the crystals. The high strains inside the nanoparticles introduced by structural defects (e.g., lattice defects) would make the unit cell distortion (c/a ratio) much smaller than that in the standard $BaTiO_3$. To reveal the high strains in the hydrothermal $BaTiO_3$ nanoparticles by TEM images, Zhu et al. [81] recorded both bright- and dark-field TEM images from the hydrothermal $BaTiO_3$ nanoparticles. Figure 4(a) is a bright-field TEM image recorded by using a small objective aperture that selects only the (000) central transmitted beam, which shows narrow-distribution spherical nanoparticles. The dark-field image shown in Figure 4(b) was recorded by using a smaller objective aperture that selects the part of the {100} and {110} reflections, as indicated by a circle in Figure 4(c). The dark-field image displayed in Figure 4(b) clearly shows high strains in some $BaTiO_3$ nanoparticles. By using the bright- and dark-field TEM images, Lu et al. [82] also reported several types of TEM contrast variations in an individual $BaTiO_3$ nanocrystal synthesized *via* hydrothermal method at a temperature of 230°C. It is believed that the different types of variations of TEM contrast indicate the existence of different strains in $BaTiO_3$ nanograins. Therefore, in a TEM image, large strain is indicated by a contrast variation across a particle. If a particle is single crystalline and has no strain, it should be uniform in contrast. However, for a single crystalline particle, if the TEM image shows dark-bright variation in contrast, it is likely to have a high strain within the grain. Strain affects the diffraction behavior of the electrons, resulting in dramatic contrast change. The hydrothermal $BaTiO_3$ nanoparticles exhibit a cubic structure (a high-temperature phase) at room temperature; such an abnormal crystallographic phenomenon is closely related to the existence of high strains in these $BaTiO_3$ nanoparticles. The strains introduced by a high concentration of lattice defects such as OH^- ions and barium vacancies can make the unit cell distortion (*c*/*a* ratio) much smaller compared with that of the standard $BaTiO_3$. As a result, no peak splitting was detected in the XRD patterns of the hydrothermal $BaTiO_3$ powders even though they belong to the tetragonal phase.

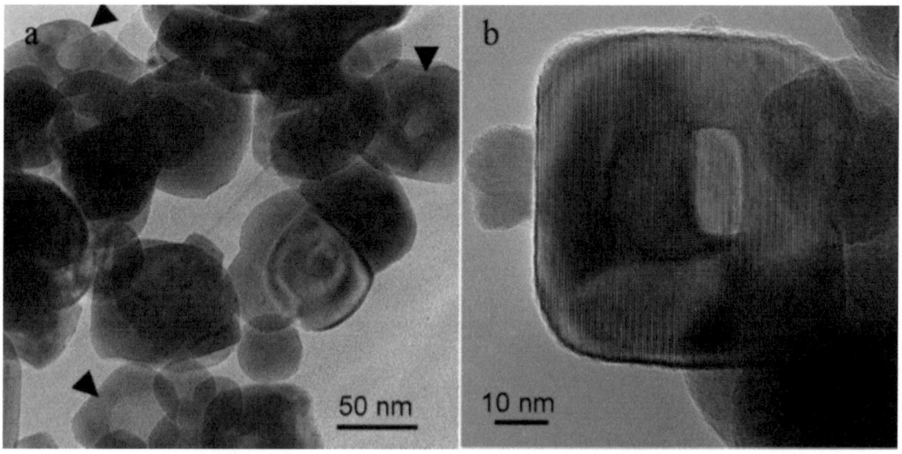

Figure 3. Transmission electron microscopy images of (a) as-received $BaTiO_3$ powder (particle size ~60 nm) and (b) $BaTiO_3$ powder annealed at 673 K. Figures reproduced with permission from [79].

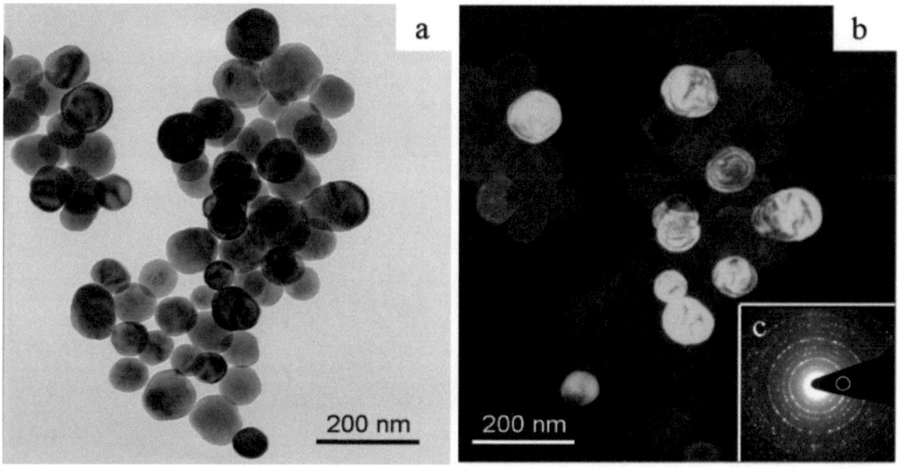

Figure 4. (a) Bright-field and (b) dark-field TEM images recorded from the hydrothermal $BaTiO_3$ nanoparticles. (c) A selected area electron diffraction pattern from the $BaTiO_3$ particles showing a perovskite structure. The circle indicates the size and position of the objective aperture used to record the dark-field image displayed in (b). Figures reproduced with permission from [80].

Due to the ability of revealing the local atomic structures, HRTEM image is the most useful and appropriate technique for identifying the local structures at the edges of perovskite nanocrystals. For example, a terrace-ledge-kink (TLK) surface structure is frequently observed at the edges of the hydrothermal perovskite $BaTiO_3$ [84,85] and $ZnZrO_3$ [86] nanocrystals with rough surface morphology, and in most cases the terraces and ledges lie on the {100} planes, as shown in Figures 5 and 6, respectively. Such a TLK surface structure can be explained by the periodic bond chain theory, which was originally developed by Hartmann and Perdok [87]. The rarely seen {110} surface in the perovskite $BaTiO_3$ and $ZnZrO_3$ nanoparticles were found to be reconstructed so that the surface was composed of corners bound by {100} mini-faces like the triangular small islands.

Internal defect textures, such as nanoscale multiple (111) twining and complicated (111) intergrowth defects, were also observed in the $BaTiO_3$ nanocrystals synthesized by sol–gel and stearic acid-gel (SAG) methods. They were identified as hexagonal-type $BaTiO_3$ structure [88, 89]. Complex arrangements of defects lying on the (111) plane were observed in the SAG-derived $BaTiO_3$ nanocrystal with particle size of 10 nm. The density of the small defects was estimated to be on the order of $10^{27}/m^3$ in the SAG-derived $BaTiO_3$ nanocrystals. These high density of defects could result in the cubic phase structure of SAG-derived $BaTiO_3$ powders even with grain size large up to 3.50 μm [88].

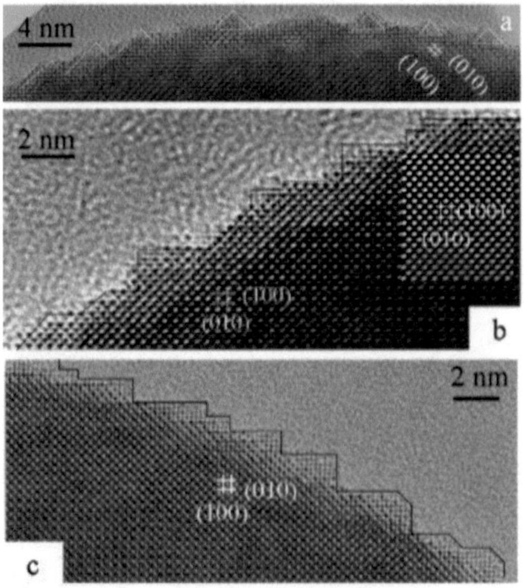

Figure 5. HRTEM images of the surface structures at the edges of $BaTiO_3$ nanoparticles viewed from the [001] direction. (a) Both a terrace-ledge-kink (TLK) surface structure and small nucleated and triangular islands with two to three atomic layer thickness are observed. (b) and (c) TLK surface structure with both terraces and ledges lying on the {100} planes; only a small amount of ledges lie on the (110) plane. The inset in (b) is a Fourier-filtered image of the corresponding position, which clearly demonstrates two perpendicular sets of (100) and (010) planes. Reproduced with permission from [84].

http://dx.doi.org/10.5772/61640

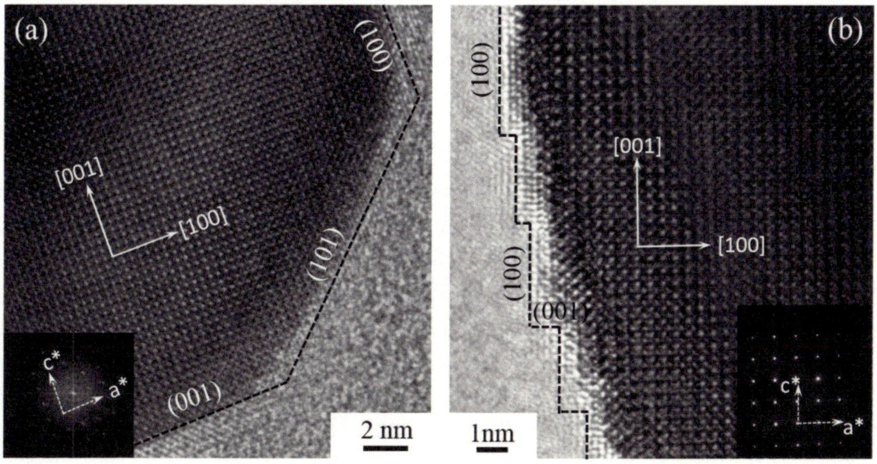

Figure 6. HRTEM images of the surface structures at the edges of perovskite $ZnZrO_3$ nanocrystals synthesized at different Zn/Zr molar ratios in the precursors. (a) Zn/Zr = 3.0, and (b) Zn/Zr = 4.0. Inset in Figure a is the FFT patterns of the corresponding HRTEM image, and inset in Figure d is the SAED pattern taken from the [010]-zone axis. The {100} and (101) facets are indicated in Figure a, and surface steps lying on the {100} planes are indicated in Figure b. Reproduced with permission from [86].

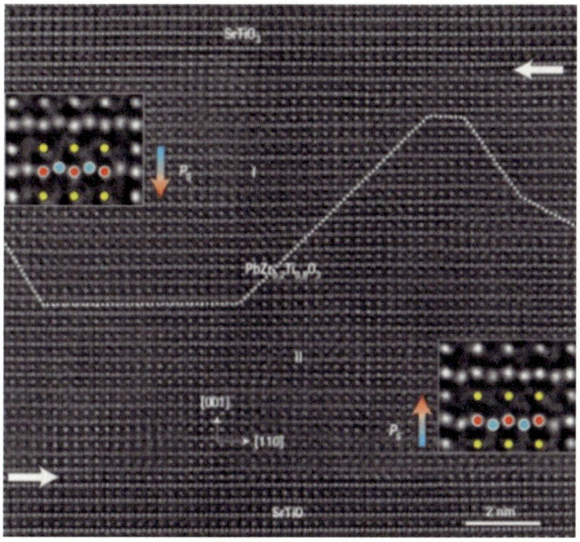

Figure 7. Atomic-scale imaging of the electric dipoles formed by the relative displacements of the Zr/Ti cation columns and the O anion columns in the approximately 10-nm-thick $PbZr_{0.2}Ti_{0.8}O_3$layer sandwiched between two $SrTiO_3$ layers. The image was viewed along the [] direction and recorded under negative spherical-aberration imaging conditions. The atom columns appear bright on a dark background. The horizontal arrows denote the horizontal interfaces between the $PbZr_{0.2}Ti_{0.8}O_3$ film and the top and the bottom $SrTiO_3$ film layers. The dotted line traces the 180° domain wall between the domain I and domain II. The arrows denoted by 'P_S' show the directions of the polarization in the 180° domains. Two insets show higher magnifications of the dipoles formed by the displacements of ions in the unit cells. Yellow symbols denote PbO atom columns seen end-on, red symbols for Zr/Ti columns, and blue symbols for oxygen. Reproduced with permission from [92].

Traditionally, spherical aberration (Cs) of magnetic lenses limits the resolutions of HRTEM and STEM images. In recent years, spherical aberration correctors (e.g., hexapole type Cs-correctors proposed by Rose [90]) have been developed to substantially reduce the effective value of Cs of the objective lens. The Cs-corrected HRTEM mode offers a tunable spherical aberration coefficient from negative to positive values. Properly combining a negative Cs with a positive defocus, at no cost to point resolution, an HRTEM image with bright-contrast of atoms on dark background can be obtained, which can be directly interpreted without image simulation, and light elements such as oxygen atoms and even their vacancies can also be imaged [91,92]. For example, by using the Cs-corrected imaging technique, Jia et al. [92] first performed the atomic-scale investigations of the electric dipoles near (charged and uncharged) 180° domain walls in thin epitaxial $PbZr_{0.2}Ti_{0.8}O_3$ film sandwiched between two $SrTiO_3$ layers. Figure 7 is an atomic-scale image of the electric dipoles formed by the relative displacements of the Zr/Ti cation columns and the O anion columns in $PbZr_{0.2}Ti_{0.8}O_3$ film, viewed from the direction and recorded under negative spherical-aberration imaging conditions. The local tetragonality c/a and spontaneous polarization inside the domains and across the domain wall were calculated. For the first time, a large difference in atomic details between charged and uncharged domain walls was reported. Such breakthrough would improve our ability to see and thoroughly explore the properties of perovskite nanocrystals. We can foresee that the new Cs-corrected HRTEM and STEM will benefit perovskite nanopowder materials research in the new era.

4. Properties of perovskite nanocrystals

4.1. Ferroelectric and dielectric properties

Perovskite $BaTiO_3$ nanocrystals can be used as initial building blocks to fabricate thin films, which exhibit highly uniform nanostructured texture and grain sizes. Recently, well-isolated $BaTiO_3$ nanocrystals smaller than 10 nm with control over aggregation and crystal densities have been synthesized and used to construct films with a uniform nanocrystalline grain texture [93]. The ferroelectric behavior was found in these $BaTiO_3$ nanocrystallined films with grain sizes in the range of 10–30 nm. Their relative dielectric constants were in the range of 85–90 over the 1–100 KHz with low dielectric loss of 0.03–0.04, representing a promising application in thin-film capacitance [93]. The nanometer-scale ferroelectric property of tetragonal $BaTiO_3$ particles with sizes of 6–12 nm is also reported by Nuraje et al. [94]. The ferroelectric polarization of these nanoparticles can be manipulated by electrostatic force microscopy (EFM), as demonstrated in Figure 8. First, the electric polarization of the $BaTiO_3$ nanoparticles was manipulated by applying a voltage, V_{write}, to the conductive atomic force microscopy(AFM) tip that gently contacts the nanoparticles (Figure 8a). After the local electric polarization is written onto the nanoparticles, the resulting polarization is probed using EFM with a lower voltage, V_{probe}, by measuring the shift in the resonance frequency of the AFM tip. As shown in Figure 8(a), during the probing process the AFM tip is raised at a constant height above the nanoparticles in order to avoid interference between the manipulated polarization and V_{probe}. The raised distance of 50 nm enables one to image only the contribution from the surface charges associated with the local electric polarization of the $BaTiO_3$ nanoparticles. After a V_{write} of +12 V was applied to $BaTiO_3$ nanoparticles with an average diameter of 12 nm (Figure 8b),

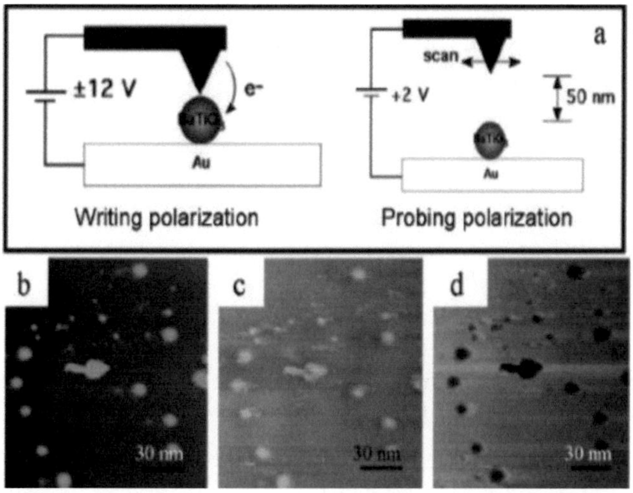

Figure 8. (a) Schematic diagram of manipulating and probing the electric polarization of $BaTiO_3$ nanoparticles with EFM. (b) Topological AFM image of $BaTiO_3$ nanoparticles. (c) and (d) EFM images of $BaTiO_3$ nanoparticles with V_{probe} = +2 V after V_{write} = ± 12 V was applied on the nanoparticles across a conductive AFM tip and a gold substrate. Reproduced with permission from [95].

the EFM image of those nanoparticles to which a V_{probe} of +2 V was applied, shown in Figure 8(c), appears in a brighter contrast compared with the background due to the repulsive electrostatic interaction between the tip and the nanoparticles. After a V_{write} of –12 V applied to the same $BaTiO_3$ nanoparticles, the EFM image of those nanoparticles to which a V_{probe} of –2 V was applied, shown in Figure 8(d), appears in a darker contrast compared with the background due to the attractive electrostatic interaction. It should be noted that the experiments that involved scanning the manipulated nanoparticles with V_{probe} = –2 V resulted in reverse EFM images, which confirms that the probe voltage did not interfere significantly with the written polarization. These EFM images indicate that the $BaTiO_3$ nanoparticles synthesized in the peptide nanorings at room temperature possess a ferroelectric property with spontaneous electric polarization, which can be reoriented by an external electric field.

4.2. Magnetic properties

The magnetic properties of perovskite oxide nanocrystals are dependent upon the particle size or large surface-area-to-volume ratio. For most biomedical and magnetofluidic applications, magnetic nanoparticles with fairly uniform size and having a Curie temperature above room temperature are highly required. In this application, a well-controlled, reproducible, and narrow distribution of ferromagnetic nanoparticle sizes is important in addition to having a large magnetic moment per particle. Lipham et al. [99] have synthesized polycrystalline, nanometer-sized powders of $La_{1-x}SrxMnO_3$ by a citrate gel technique. The particle sizes of which were in the range of 11–17 nm, their saturated magnetizations at 300 K in the range of 7–26 emu/g, and magnetic transition temperature from 275 to 350K. The saturated magnetization and magnetic transition temperature increased linearly with increasing the average particle size.

The magnetic properties of single-crystalline multiferroic $BiFeO_3$ nanoparticles were also reported by Park et al. [100] and Selbach et al. [101]. Their results demonstrate that the $BiFeO_3$ nanoparticles exhibit strong size-dependent magnetic properties that correlate with: (a) increased suppression of the known spiral spin structure (period length of ~62 nm) with decreasing nanoparticle size and (b) uncompensated spins and strain anisotropies at the surface. Both zero-field-cooled and field-cooled magnetization curves exhibit spin-glass freezing behavior due to a complex interplay between finite size effects, inter-particle interactions, and a random distribution of anisotropy axes in the nanoparticle assemblies.

4.3. Multiferroic properties

Multiferroic materials are characterized by the coexistence of spin, orbital, and/or electric-dipole orders and cross-term effects between the corresponding different degrees of freedom [102,103]. Magnetoelectric (ME) multiferroics are an example of materials that combine simultaneously ferromagnetic and ferroelectric in the same phase, with coupling between the two orders. That implies they possess spontaneous magnetization which can be reoriented by an applied magnetic field, and spontaneous polarization which can be reoriented by an applied electric field. Up to date, large ME effects have been observed in the form of ferroelectric phase transitions induced by magnetic fields in perovskite manganites [104] and switching local ferromagnetism by an electric field through the coupling between multiferroic $BiFeO_3$ and a ferromagnet [105,106]. Magnetoelectric memory effects and magnetic switching of ferroelectric domains (and the converse process) have been demonstrated in many multiferroic materials. By using an optical second harmonic generation technique ferrotoroidic domains are spatially revealed in $LiCoPO_4$ multiferroics, where independent antiferromagnetic domains exist; with this method, the coupling between magnetic and electric domains is also confirmed [107]. Due to the weak coupling behaviors between the ferroelectric and the magnetic order parameters in the single-phase multiferroics and small values of the electric and magnetic polarizations at room temperature, the applications of single-phase multiferroics are not very attractive in the near future. However, for the single-phase multiferroics, there still exist some basic questions to be answered such as the origin of the ferroelectricity in some unusual multiferroics [104]. Up to date, several coupling mechanisms between the magnetic and ferroelectric ordering in the single-phase multiferroics have been proposed and extensively investigated; it is still unclear how to obtain high-temperature single-phased ferromagnetic–ferroelectric multiferroicity. From a viewpoint of the practical applications, it is very important and is still a major challenge to develop robust room-temperature ferromagnetic ferroelectrics that are sufficiently insulating to sustain a large macroscopic polarization [108].

5. Applications of perovskite oxide nanocrystals

Due to their high dielectric, ferroelectric, piezoelectric, pyroelectric, and electro-optic properties, perovskite oxide nanocrystals have wide ranges of applications, such as multilayered ceramic capacitors, ferroelectric memories, voltage tunable capacitors, surface acoustic wave devices, microactuators, and IR detectors. An emerging application seeks to exploit the multiferroic properties of perovskite oxide nanocrystals to develop novel multifunctional devices controlled by magnetic and electric fields. This magnetoelectric coupling enables the manipulation of the ferroelectric polarization by a magnetic field [104] or the control of the antiferromagnetic vector orientation by an electric field [105]. This latter opportunity is very appealing for spintronics as it may allow to control the magnetization of a ferromagnet, exchange coupled to a ferroelectric antiferromagnet, through an electric field. In other words, it may enable electrical writing of magnetic information with a low power consumption, like an magnetic random access memory(MRAM) cell [109]. In the subsequent sections, typical applications of perovskite oxide nanocrystals are introduced.

5.1. Nanoelectronics

An important application of perovskite oxide nanocrystals in electronics is the multilayered ceramic capacitors(MLCCs). The MLCCs based on $BaTiO_3$ powders have used in many electronic devices such as video camera, cell phone, laptop computers, and automobiles. Two key features of the MLCCs have attracted high degree of interest in them. One is the discovery of the ferroelectric properties of perovskite $BaTiO_3$ and its high relative dielectric constant. The other is the technical breakthrough for making multilayer ceramic units in a small volume, thus satisfying the need for an economical manufacture of systems, and the requirements for printed circuit boards and for hybrid circuits on ceramic substrates. With increasing the volumetric efficiency of electronic devices, MLCCs are needed to use much thin dielectric layers. Recently, it is expected that the dielectric layers as thin as 1 μm or less will be available in MLCCs for the next generation of electronic components. Besides the thin dielectric layers, other key factors for the development of future highly volume-efficient and high-capacitance MLCCs should be considered, such as the use of high dielectric films, large numbers of active dielectric layers, improvement in the overlap area, and stacking precision of the electrodes. For ideal $BaTiO_3$ nanopowders used for the next generation MLCCs, they should have high purity, homogeneous compositions and cation distributions, uniform sizes and shapes, and weak agglomeration. Generally, solution-based techniques are used to prepare such very fine powders, since these techniques could synthesize homogeneous, phase-pure, and stoichiometric $BaTiO_3$ nanocrystals with finer particle sizes due to the finer scale of mixing and subsequently lower processing temperature.

5.2. Information storage devices

Ferroelectric materials possess an electric dipole moment even in the absence of an external electric field. Such a spontaneous polarization is caused by the positional bistability of constituent ions in the crystal and its direction can be adjusted by an external field. Since the response time of the ion displacement is the order of ns or less, so non-volatile random access memory called FeRAM (ferroelectric random access memory) can be realized using ferroelectric capacitors, in which two states of "0" and "1" in the binary logic are represented by the direction of the spontaneous polarization. Two different types of FeRAM cells can be used to achieve data storage and read operations, which are named as 1T-1C and 2T-2C configurations [110]. Because the 2T-2C cell occupies a large area, it is only used for FeRAM memories with densities of 256 kbit or less. A significantly smaller cell size can be achieved by eliminating the complementary reference capacitor in each cell and accessing a reference signal from a single reference capacitor placed outside the cell array. The cell size in 1T-1C configuration is reduced because only one transistor and one capacitor are required to form the popular cell. In the FeRAM cell, the polarization direction is set by a positive or negative voltage pulse to the ferroelectric capacitor defining logical "0" or "1". For readout, another voltage pulse is applied and the stored bit configuration determines whether or not the polarization switched direction. The ferroelectric capacitor suffers from the fact that data are destroyed during readout and bit reprogramming is required after each read cycle, i.e., the data read process is destructive. As a result, switching the polarization during both write and read operations would cause reduced endurance, resulting in short lifetime of FeRAM cells because of the fatigue problem of the ferroelectric material. In addition, during the read-processing data storage is volatile and data could be erased if powder supply is lost during the reading process. To overcome the destructive readout scheme of FeRAM, recently considerable efforts have focused on the development of a so-called ferroelectric gate field-effect transistor (FeFET). The principle of a ferroelectric-gate field-effect transistor is based on a conventional Si MOSFET (metal oxide semiconductor ferroelectric-gate field effect transistor) whose gate dielectrics is a ferroelectric materials. This FET-type FeRAMs have such unique features as non-volatile data storage, non-destructive data readout, and the single-transistor-type cell structure. Further advantages are the reduced power consumption and better scaling properties as in the case of usual FeRAM cells. Although the FET-type FeRAMs have been studied since the 1950s [111], however, up to date the commercially available devices have not been fabricated, mainly due to the interfacial structures between the perovskite ferroelectric and a semiconductor. Any imperfections at the interface, such as the formation of undesirable phases or electronic trapping states, will seriously degrade the performance of the device. Essential challenges for the FeFET are the improvement of the retention time and the suppression of serious parasitic effects such as the charge traps at the Si–ferroelectric interface. One possible solution in the case of a FeFET is the incorporation of one insulating buffer layer between the Si and the ferroelectric, which is composed of either a dielectric material (MFIS structure) or a stacked structure of conductive and dielectric materials (MFMIS structure). Several buffer layers such as SiO_2, CeOx, and Si_3N_4 have been investigated. The most recent results by using alternative gate oxides such as HfO_2 or HfAlO show encouraging results, and the retention time for the gate layer sequence of Pt/SBT/HfO_2/Si is up to 30 days [112]. However, insertion of the buffer layer causes new problems such as current injection that lead to short data retention time. Another considered factor is the leakage current of both the ferroelectric film and the buffer layer. If the charge neutrality at a node between the two capacitors is destroyed by the leakage current, electric charges on the electrodes of the buffer layer capacitor disappear, which means that carriers on the semiconductor surface disappear and the stored data cannot be readout by drain current of the FET, even if the polarization of the ferroelectric film is retained. FeFET based on epitaxial perovskite heterostructures was also demonstrated, in which doped rare-earth manganates such as $La_{0.7}Ca_{0.3}MnO_3$ were used as the semiconductor channel material, and $Pb(Zr_{0.2}Ti_{0.8})O_3$ as the ferroelectric gate in the prototypical epitaxial field effect device [112]. The carrier concentration of the semiconductor channel can be tuned by varying the manganate stoichiometry. The enhanced interface characteristics associated with the ferroelectric-manganate interface allows for the fabrication of field effect devices with channel resistance modulation of at least a factor of 3 and retention on the order of hours.

Perovskite oxide nanocrystals also exhibit good dielectric properties, which can be used for dynamic random access memory (DRAM). The DRAM is the primary working medium for information storage in the microelectronic devices that comprise the entire litany of electronic systems. DRAM works by using a submicron-sized capacitor, representing one bit of memory, to store a given amount of electrical charge: if the charge is present, it represents a digital "1", if not then the bit is a "0". Each bit is addressed using a complementary metal oxide semiconductor field effect transistor, which acts as a valve for adding to or removing charge from the capacitor upon application of a voltage. Conventional DRAMs employ either SiO_2 capacitors, or a combination of SiO_2 and Si_3N_4 nitride, which is termed ONO (oxy-nitride). The next generation of DRAMs will utilize Ta_2O_5 replacing ONO, but the tantalum oxide relative dielectric constant is only about 25, and it appears that the DRAM evolution will skip this intermediate stage and pass directly to very high dielectric materials ($\varepsilon_r = 500 - 1500$) that are ferroelectric or nearly ferroelectric. Many of these materials are perovskite oxides with ABO_3 perovskite, or closely related variations of perovskites. Several good reviews of this technology are given by Gnade et al. [113], Tasch and Parker [114], and Scott [115]. In going from 64 Mbit to 4 Gbit DRAM, the most promising dielectric material is perovskite barium strontium titanate (BST), which can be processed by physical deposition, especially sputtering with good results [116]. A dielectric material should have a low leakage current and high dielectric constant, and it is desirable to use the paraelectric BST for DRAM. Among the $Pb(Zr,Ti)O_3$ (PZT), $SrTiO_3$, and BST ferroelectric materials, $(Ba,Sr)TiO_3$ is the most promising material for DRAM capacitors because it is a better insulator with a higher dielectric constant than PZT and $SrTiO_3$, and it can be controlled to be the paraelectric phase with an appropriate ratio of Ba/Sr composition [117].

5.3. Spintronics

Applications of the magnetoelectric coupling and multiferroics in the fields of spintronics are increasing rapidly, and numerous possible device architectures have been proposed and fabricated [118]. For example, in the ferroelectric antiferromagnets, such as the multiferroic perovskites, the magnetic structure could be modulated or controlled by the application of an electric field [119,120]. For example, non-volatile control of the orientation of the antiferromagnetic axis can be achieved by using the coupling between ferroelectricity and antiferromagnetism that can provide permanent [121]. Similarly, the coupling between the antiferromagnetic component and an adjacent ferromagnet also allows to switch the ferromagnetism by the application of an electric field. Therefore, potential new types of multiple-state memories and spintronic devices controlled by electric and magnetic fields can be developed. One specific demonstration is a multiferroic tunnel junction used as a spin filter device with the potential to control both electrically and magnetically [122,123]. In such a spin filter device, the multiferroic tunnel barrier should be as thin as 1–2 nm thickness. Experiments and theoretical calculations show that a critical thickness of a few unit cells is closely related to the screening properties of the electrodes [124,125]. Another potential application of multiferroics in spintronics is the spin wave devices controlled by electric field. Smolenskii and Chupis theoretically analyzed the excitation of spin waves by an alternating electric field, and of ferroelectric oscillations by a magnetic field [126], and concluded that such excitations should be strongest in ferroelectric ferrimagnets, which could be used to produce magneto-electric generators or spin wave amplifiers driven by electric field or current. Up to now, few works on the influence of electric fields on spin waves are reported. Recently, Pimenov et al. first demonstrated such hybrid excitations in multiferroic manganites 125]. It is expected that this field should be a hot topic in near future.

6. Future outlook of perovskite nanocrystals

Perovskite nanopowdes (e.g., $BaTiO_3$) are widely used in MLCCs in the electronic industry due to their excellent dielectric properties. In recent years, miniaturization of ferroelectric $BaTiO_3$ powders to the nanometer scale is very desirable for their applications in the next generation of electronics; however, the main challenge lies in the synthesis of barium titanate nanocrystals at room temperature with a tetragonal crystalline structure, which induces the ferroelectric property. Much effort has been concentrated on synthesizing the high-purity, homogeneous, weakly agglomerated nanocrystals with tetragonal structure. However, it is known that the ferroelectricity becomes weaker with a decrease in particle size and disappears below a certain critical size, known as the size effect. Lots of fundamental and experimental studies are needed to understand the size effect of ferroelectricty for nanosized perovskite powders, to develop future high volume-efficient and high-capacitance MLCCs.

Perovskite oxide nanocrystals also exhibit multiferroic behavior, which is the fundamental of giant ME effects and ME phase control. However, these phenomena are now primarily investigated in the viewpoint of basic research rather than the practical applications. The main reason is that the number of multiferroic compounds with perovskite structure is very small, in many cases use of bulk single crystals is necessary and only very few multiferroics exhibit ME behavior at room temperature. Several microscopic physical mechanisms of ME behaviours in multiferroics have been revealed and the precise criteria searching for new multiferroics have been given out. More and more perovskite multiferroics will be developed and the improvement on tuning the ME performance will drive multiferroics much closer to practical applications in the near future. Now, the major challenge is developing the room-temperature perovskite multiferroics insulators and expanding their applications in the fields of micro-electronics.

7. Conclusions

In this chapter, we have reviewed various processing routes, characterization, functionalization, and novel application areas of perovskite oxide nanocrystals. Significant progress has been achieved in the development of processing routes for perovskite oxide nanocrystals, which ranges from the solid-state reaction to chemical solution depositions. By introducing

http://dx.doi.org/10.5772/61640

the $BaTiO_3$ nanoparticles into the multilayer dielectrics, much advancement has been achieved in reducing dielectric thickness and increasing the volumetric efficiency of $BaTiO_3$-based MLCCs, to meet the miniaturization demand for electronic devices. At the nanoscale, highly accurate microstructural characterizations are usually required for investigating the microstructures of perovskite nanocrystals. The unique properties of perovskite oxide nanocrystals (e.g., ferroelectric and dielectric, electrical, magnetic, and multiferroic properties) are addressed based on the selected recent literature. An important conclusion from the extensive review is that there exists size effects for these unique properties. Better understanding of the size effects in perovskite oxide nanocrystals will be helpful in selecting the critical size and dimension for the purpose of implementing perovskite oxide nanocrystals in various devices. Applications of perovskite oxide nanocrystals have been identified with a major focus in areas such as nanoelectronics, information storage devices, and their potential applications in spintronics are also briefly introduced. As research into perovskite oxide nanocrystals spreads its wings, becoming more extensive, a complete review on this subject has become an arduous task. However, in this chapter a modest attempt is made to analyze recent significant developments in researches of perovskite oxide nanocrystals and their possible applications in various industries.

Acknowledgements

This work is partially supported by National Natural Science Foundation of China (grant nos. 11174122 and 11134004), National Basic Research Program of China (grant no. 2015CB654900), the open project from National Laboratory of Solid State Microstructures, Nanjing University (grant no. M26012), and six big talent peak project from Jiangsu province (grant no. XCL-004).

Author details

Heng Wu and Xinhua Zhu*

*Address all correspondence to: xhzhu@nju.edu.cn

National Laboratory of Solid State of Microstructures, School of Physics, Nanjing University, Nanjing, China

References

[1] Schaak RE, Mallouk TE. Perovskites by design: a toolbox of solid-state reactions. Chem Mater 2002;14(4):1455–71. DOI: 10.1021/cm010689m

[2] Bhalla A, Guo R, Roy R. The perovskite structure–a review of its role in ceramic science and technology. Mater Res Innovat 2000;4(1):3–26. DOI: 10.1007/s100190000062

[3] Dawber M, Rabe K, Scott J. Physics of thin-film ferroelectric oxides. Rev Mod Phys 2005;77(4):1083–130. DOI: 10.1103/RevModPhys.77.1083

[4] Gruverman A, Kholkin A. Nanoscale ferroelectrics: processing, characterization and future trends. Rep Prog Phys 2006;69(8):2443–74. DOI: 10.1088/0034-4885/69/8/R04

[5] Hill NA. Why are there so few magnetic ferroelectrics. J Phys Chem B 2000;104(29): 6694–709. DOI: 10.1021/jp000114x

[6] Fiebig M, Lottermoser T, Frohlich D, Goltsev AV, Pisarev RV. Observation of coupled magnetic and electric domains. Nature 2002;419(6909):818–20. DOI: 10.1038/nature01077

[7] Wang J, Neaton JB, Zheng H, Nagarajan V, Ogale SB, Liu B, et al. Epitaxial $BiFeO_3$ multiferroic thin film heterostructures. Science 2003;299(5613):1719–22. DOI: 10.1126/science.1080615

[8] Hur N, Park S, Sharma PA, Ahn JS, Guha S, Cheong SW. Electric polarization reversal and memory in a multiferroic material induced by magnetic fields. Nature 2004;429(2990):392–5. DOI: 10.1038/nature02572

[9] Haertling GH. Ferroelectric ceramics: history and technology. J Am Ceramic Soc 1999;82(4):797–818. DOI: 10.1111/j.1151-2916.1999.tb01840.x

[10] Moulson AJ, Herbert JM. Electroceramics: Materials, Properties, Applications. 1st ed. New Jersey, USA: John Wiley & Sons; 2003. 557 p. DOI: 10.1002/0470867965

[11] Yoon DH, Lee BI. $BaTiO_3$ properties and powder characteristics for ceramic capacitors. J Ceramic Process Res 2002;3(2):41–7.

[12] Yoon D. Tetragonality of barium titanate powder for a ceramic capacitor application. J Ceramic Process Res 2006;7(4):343.

[13] Pithan C, Hennings D, Waser R. Progress in the synthesis of nanocrystalline $BaTiO_3$ powders for MLCC. Int J Appl Ceramic Technol 2005;2(1):1–14. DOI: 10.1111/j.1744-7402.2005.02008.x

[14] Chu MS, Rae AW. Manufacturing dielectric powders. Am Ceramic Soc Bull 1995;74(6):69–72.

[15] Hu YH, Harmer MP, Smyth DM. Solubility of BaO in $BaTiO_3$. J Am Ceramic Soc 1985;68(7):372–6. DOI: 10.1111/j.1151-2916.1985.tb10145.x

[16] Phule P, Risbud S. Low-temperature synthesis and processing of electronic materials in the BaO-TiO_2 system. J Mater Sci 1990;25(2):1169–83. DOI: 10.1007/BF00585422

[17] Joshi UA, Lee JS. Template-free hydrothermal synthesis of single-crystalline barium titanate and strontium titanate nanowires. Small 2005;1(12):1172–6. DOI: 10.1002/smll.200500055

[18] Niederberger M, Pinna N, Polleux J, Antonietti M. A general soft-chemistry route to perovskites and related materials: synthesis of $BaTiO_3$, $BaZrO_3$, and $LiNbO_3$ nanoparticles. Angewandte Chemie International Edition 2004;43(17):2270–3. DOI: 10.1002/anie.200353300

[19] Templeton LK, Pask JA. Formation of $BaTiO_3$ from $BaCO_3$ and TiO_2 in air and in CO_2. J Am Ceramic Soc 1959;42(5):212–6. DOI: 10.1111/j.1151-2916.1959.tb15455.x

[20] Buscaglia MT, Bassoli M, Buscaglia V, Alessio R. Solid-state synthesis of ultrafine $BaTiO_3$ powders from nanocrystalline $BaCO_3$ and TiO_2. J Am Ceramic Soc 2005;88(9):2374–9. DOI: 10.1111/j.1551-2916.2005.00451.x

[21] Beauger A, Mutin JC, Niepce JC. Synthesis reaction of metatitanate $BaTiO_3$. J Mater Sci 1983;18(10):3041–6. DOI: 10.1007/BF00700786

[22] Ubaldini A, Buscaglia V, Uliana C, Costa G, Ferretti M. Kinetics and mechanism of formation of barium zirconate from barium carbonate and zirconia powders. J Am Ceramic Soc 2003;86(1):19–25. DOI: 10.1111/j.1151-2916.2003.tb03271.x

[23] Kimura T. Molten salt synthesis of ceramic powders. In: Sikalidis C. (Ed.) Advances in Ceramics-Synthesis and Characterization, Processing and Specific Applications. 1st ed. Rijeka, Croatia: INTECH Open Access Publisher; 2011. pp. 75–100. DOI: 10.5772/20472

[24] Yang J, Li RS, Li XC, Long YL, Zhou JY, Zhang YM. Molten salt synthesis of $SrFeO_3$ nanocrystals. J Ceramic Soc Jpn 2011;119(1394):736–9. DOI: 10.2109/jcersj2.119.736

[25] Li H-L, Du Z-N, Wang G-L, Zhang Y-C. Low temperature molten salt synthesis of $SrTiO_3$ submicron crystallites and nanocrystals in the eutectic NaCl–KCl. Mater Lett 2010;64(3):431–4. DOI: 10.1016/j.matlet.2009.11.040

[26] Zhou H, Mao Y, Wong SS. Probing structure–parameter correlations in the molten salt synthesis of $BaZrO_3$ perovskite submicrometer-sized particles. Chem Mater 2007;19(22):5238–49. DOI: 10.1021/cm071456j

[27] Shivakumara C. Low temperature synthesis and characterization of rare earth orthoferrites $LnFeO_3$ (Ln=La, Pr and Nd) from molten NaOH flux. Solid State Commun 2006;139(4):165–9. DOI: 10.1016/j.ssc.2006.05.030

[28] Fossdal A, Einarsrud M-A, Grande T. Phase equilibria in the pseudo-binary system SrO–Fe_2O_3. J Solid State Chem 2004;177(8):2933–42. DOI: 10.1016/j.jssc.2004.05.007

[29] Yoon K, Cho Y, Kang D. Molten salt synthesis of lead-based relaxors. J Mater Sci 1998;33(12):2977–84. DOI: 10.1023/A:1004310931643

[30] Thirumal M, Jain P, Ganguli AK. Molten salt synthesis of complex perovskite-related dielectric oxides. Mater Chem Phys 2001;70(1):7–11. DOI: 10.1016/S0254-0584(00)00384-9

[31] Chen J, Xing X, Watson A, Wang W, Yu R, Deng J, et al. Rapid synthesis of multiferroic $BiFeO_3$ single-crystalline nanostructures. Chem Mater 2007;19(15):3598–600. DOI: 10.1021/cm070790c

[32] Chen J, Yu R, Li L, Sun C, Zhang T, Chen H, et al. Structure and shape evolution of $Bi_{1-x}La_xFeO_3$ perovskite microcrystals by molten salt synthesis. Eur J Inorg Chem 2008;2008(23):3655–60. DOI: 10.1002/ejic.200800263

[33] He X, Gao L. Synthesis of pure phase $BiFeO_3$ powders in molten alkali metal nitrates. Ceramics Int 2009;35(3):975–8. DOI: 10.1016/j.ceramint.2008.04.013

[34] Zheng XH, Chen PJ, Ma N, Ma ZH, Tang DP. Synthesis and dielectric properties of $BiFeO_3$ derived from molten salt method. J Mater Sci: Mater Electronics 2012;23(5): 990–4. DOI: 10.1007/s10854-011-0533-4

[35] Zhu X, Zhou J, Jiang M, Xie J, Liang S, Li S, et al. Molten salt synthesis of bismuth ferrite nano- and microcrystals and their structural characterization. J Am Ceramic Soc 2014;97(7):2223–32. DOI: 10.1111/jace.12897

[36] Frey MH, Payne DA. Synthesis and processing of barium titanate ceramics from alkoxide solutions and monolithic gels. Chem Mater 1995;7(1):123–9. DOI: 10.1021/cm00049a019

[37] Hwang U-Y, Park H-S, Koo K-K. Low-temperature synthesis of fully crystallized spherical $BaTiO_3$ particles by the gel–sol method. J Am Ceramic Soc 2004;87(12):2168–74. DOI: 10.1111/j.1151-2916.2004.tb07486.x

[38] Brutchey RL, Morse DE. Template-free, low-temperature synthesis of crystalline. Angewandte Chemie 2006;118(39):6714–6. DOI: 10.1002/ange.200602571

[39] Kiss K, Magder J, Vukasovich MS, Lockhart RJ. Ferroelectrics of ultrafine particle size: I, synthesis of titanate powders of ultrafine particle size. J Am Ceramic Soc 1966;49(6):291–5. DOI: 10.1111/j.1151-2916.1966.tb13265.x

[40] Yoon S, Baik S, Kim MG, Shin N. Formation mechanisms of tetragonal barium titanate nanoparticles in alkoxide–hydroxide sol-precipitation synthesis. J Am Ceramic Soc 2006;89(6):1816–21. DOI: 10.1111/j.1551-2916.2006.01056.x

[41] Yoon S, Baik S, Kim MG, Shin N, Kim I. Synthesis of tetragonal barium titanate nanoparticles via alkoxide–hydroxide sol-precipitation: effect of water addition. J Am Ceramic Soc 2007;90(1):311–4. DOI: 10.1111/j.1551-2916.2006.01361.x

[42] Dutta PK, Gregg JR. Hydrothermal synthesis of tetragonal barium titanate ($BaTiO_3$). Chem Mater 1992;4(4):843–6. DOI: 10.1021/cm00022a019

[43] Clark I, Takeuchi T, Ohtori N,C, Sinclair D. Hydrothermal synthesis and characterisation of $BaTiO_3$ fine powders: precursors, polymorphism and properties. J Mater Chem 1999;9(1):83–91. DOI: 10.1039/A805756G

[44] Lee HS, Lee JJ, Chang TS, Kim JW, Koo SM. Hydrothermal synthesis for large barium titanate powders at a low temperature: effect of titania aging in an alkaline solution. J Am Ceramic Soc 2007;90(9):2995–7. DOI: 10.1111/j.1551-2916.2007.01828.x

[45] Choi GJ, Kim HS, Cho YS. $BaTiO_3$ particles prepared by microwave-assisted hydrothermal reaction using titanium acylate precursors. Mater Lett 1999;41(3):122–7. DOI: 10.1016/S0167-577X(99)00117-2

[46] Newalkar BL, Komarneni S, Katsuki H. Microwave-hydrothermal synthesis and characterization of barium titanate powders. Mater Res Bull 2001;36(13–14):2347–55. DOI: 10.1016/S0025-5408(01)00729-2

[47] Khollam YB, Deshpande AS, Patil AJ, Potdar HS, Deshpande SB, Date SK. Microwave-hydrothermal synthesis of equi-axed and submicron-sized $BaTiO_3$ powders. Mater Chem Phys 2001;71(3):304–8. DOI: 10.1016/S0254-0584(01)00286-3

[48] Chen D, Jiao X. Solvothermal synthesis and characterization of barium titanate powders. J Am Ceramic Soc 2000;83(10):2637–9. DOI: 10.1111/j.1151-2916.2000.tb01606.x

[49] Kwon S-G, Choi K, Kim B-I. Solvothermal synthesis of nano-sized tetragonal barium titanate powders. Mater Lett 2006;60(7):979–82. DOI: 10.1016/j.matlet.2005.10.089

[50] Kolen'ko YV, Kovnir KA, Neira IS, Taniguchi T, Ishigaki T, Watanabe T, et al. A novel, controlled, and high-yield solvothermal drying route to nanosized barium titanate powders. J Phys Chem C 2007;111(20):7306–18. DOI: 10.1021/jp0678103

[51] Viviani M, Lemaitre J, Buscaglia MT, Nanni P. Low-temperature aqueous synthesis (LTAS) of $BaTiO_3$: a statistical design of experiment approach. J Eur Ceramic Soc 2000;20(3):315–20. DOI: 10.1016/S0955-2219(99)00150-8

[52] Ciftci E, Rahaman MN, Shumsky M. Hydrothermal precipitation and characterization of nanocrystalline $BaTiO_3$ particles. J Mater Sci 2001;36(20):4875–82. DOI: 10.1023/A:1011828018247

[53] Thomas R, Varadan VK, Komarneni S, Dube DC. Diffuse phase transitions, electrical conduction, and low temperature dielectric properties of sol–gel derived ferroelectric barium titanate thin films. J Appl Phys 2001;90(3):1480–8. DOI: 10.1063/1.1367318

[54] Zhang HX, Kam CH, Zhou Y, Han XQ, Buddhudu S, Xiang Q, et al. Green upconversion luminescence in Er^{3+}:$BaTiO_3$ films. Appl Phys Lett 2000;77(5):609–11. DOI: 10.1063/1.127060

[55] Cho W-S. Structural evolution and characterization of BaTiO3 nanoparticles synthesized from polymeric precursor. J Phys Chem Solids 1988;59(5):659–66. DOI: 10.1016/S0022-3697(97)00227-8

[56] Liu C, Zou B, Rondinone AJ, Zhang ZJ. Sol–gel synthesis of free-standing ferroelectric lead zirconate titanate nanoparticles. J Am Chem Soc 2001;123(18):4344–5. DOI: 10.1021/ja001893y

[57] Pinceloup P, Courtois C, Vicens J, Leriche A, Thierry B. Evidence of a dissolution–precipitation mechanism in hydrothermal synthesis of barium titanate powders. J Eur Ceramic Soc 1999;19(6–7):973–7. DOI: 10.1016/S0955-2219(98)00356-2

[58] Hertl W. Kinetics of barium titanate synthesis. J Am Ceramic Soc 1988;71(10):879–83. DOI: 10.1111/j.1151-2916.1988.tb07540.x

[59] Hu MZC, Kurian V, Payzant EA, Rawn CJ, Hunt RD. Wet-chemical synthesis of monodispersed barium titanate particles -hydrothermal conversion of TiO_2 microspheres to nanocrystalline $BaTiO_3$. Powder Technol 2000;110(1–2):2–14. DOI: 10.1016/S0032-5910(99)00262-4

[60] Eckert JO, Hung-Houston CC, Gersten BL, Lencka MM, Riman RE. Kinetics and mechanisms of hydrothermal synthesis of barium titanate. J Am Ceramic Soc 1996;79(11):2929–39. DOI: 10.1111/j.1151-2916.1996.tb08728.x

[61] Gao F, Chen XY, Yin KB, Dong S, Ren ZF, Yuan F, et al. Visible-light photocatalytic properties of weak magnetic $BiFeO_3$ nanoparticles. Adv Mater 2007;19(19):2889–92. DOI: 10.1002/adma.200602377

[62] Walton RI, Millange F, Smith RI, Hansen TC, O'Hare D. Real time observation of the hydrothermal crystallization of barium titanate using in situ neutron powder diffraction. J Am Chem Soc 2001;123(50):12547–55. DOI: 10.1021/ja011805p

[63] Komarneni S, Roy R, Li QH. Microwave-hydrothermal synthesis of ceramic powders. Mater Res Bull 1992;27(12):1393–405. DOI: 10.1016/0025-5408(92)90004-J

[64] A. Katsuki H, Komarneni S. Microwave-hydrothermal synthesis of monodispersed nanophase α-Fe_2O_3. J Am Ceramic Soc 2001;84(10):2313–7. DOI: 10.1111/j.1151-2916.2001.tb01007.x

[65] Komarneni S, Li Q, Stefansson KM, Roy R. Microwave-hydrothermal processing for synthesis of electroceramic powders. J Mater Res. 1993;8(12):3176–83. DOI: 10.1557/JMR.1993.3176

[66] Komarneni S, Li QH, Roy R. Microwave-hydrothermal processing for synthesis of layered and network phosphates. J Mater Chem 1994;4(12):1903–6. DOI: 10.1039/JM9940401903

[67] Komarneni S, Li QH, Roy R. Microwave-hydrothermal processing of layered anion exchangers. J Mater Res 1996;11(8):1866–9. DOI: 10.1557/JMR.1996.0236

[68] Liu S-F, Abothu IR, Komarneni S. Barium titanate ceramics prepared from conventional and microwave hydrothermal powders. Mater Lett 1999;38(5):344–350. DOI: 10.1016/S0167-577X(98)00187-6

[69] Jhung SH, Lee J-H, Yoon JW, Hwang YK, Hwang J-S, Park S-E, et al. Effects of reaction conditions in microwave synthesis of nanocrystalline barium titanate. Mater Lett 2004;58(25):3161–5. DOI: 10.1016/j.matlet.2004.06.006

[70] Sun W, Li C, Li J, Liu W. Microwave-hydrothermal synthesis of tetragonal $BaTiO_3$ under various conditions. Mater Chem Phys 2006;97(2–3):481–7. DOI: 10.1016/j.matchemphys.2005.08.051

[71] Pązik R, Hreniak D, Stręk W. Microwave driven hydrothermal synthesis of $Ba_{1-x}Sr_xTiO_3$ nanoparticles. Mater Res Bull 2007;42(7):1188–94. DOI: 10.1016/j.materresbull.2006.10.021

[72] Zhu X, Hang Q, Xing Z, Yang Y, Zhu J, Liu Z, et al. Microwave hydrothermal synthesis, structural characterization, and visible-light photocatalytic activities of single-crystalline bismuth ferric nanocrystals. J Am Ceramic Soc 2011;94(8):2688–93. DOI: 10.1111/j.1551-2916.2011.04430.x

[73] Zhu X, Yang Y, He K, Zhu J, Ye S, Zhou S, et al. Microwave-hydrothermal synthesis and structural characterization of multiferroic bismuth ferrite nanostructures. Ferroelectrics 2010;409(1):204–10. DOI: 10.1080/00150193.2010.486248

[74] Chen C, Cheng J, Yu S, Che L, Meng Z. Hydrothermal synthesis of perovskite bismuth ferrite crystallites. J Crystal Growth 2006;291(1):135–9. DOI: 10.1016/j.jcrysgro.2006.02.048

[75] Han JT, Huang YH, Wu XJ, Wu CL, Wei W, Peng B, et al. Tunable synthesis of bismuth ferrites with various morphologies. Adv Mater 2006;18(16):2145–8. DOI: 10.1002/adma.200600072

[76] Joshi UA, Jang JS, Borse PH, Lee JS. Microwave synthesis of single-crystalline perovskite $BiFeO_3$ nanocubes for photoelectrode and photocatalytic applications. Appl Phys Lett 2008;92(24):242106. DOI: 10.1063/1.2946486

[77] Li S, Lin Y-H, Zhang B-P, Wang Y, Nan C-W. Controlled fabrication of $BiFeO_3$ uniform microcrystals and their magnetic and photocatalytic behaviors. J Phys Chem C 2010;114(7):2903–8. DOI: 10.1021/jp910401u

[78] Hennings D, Schreinemacher S. Characterization of hydrothermal barium titanate. J Eur Ceramic Soc 1992;9(1):41–6. DOI: 10.1016/0955-2219(92)90075-O

[79] Nakano H, Urabe K, Oikawa T, Ikawa H. Characterization of internal pores in hydrothermally synthesized $BaTiO_3$ particle by transmission electron microscopy. J Am Ceramic Soc 2004;87(8):1594–7. DOI: 10.1111/j.1551-2916.2004.01594.x

[80] Sakabe Y. Grain size effect on dielectric properties and crystal structure of fine-grained $BaTiO_3$ ceramics. J Korean Phys Soc 1998;32:S260–4.

[81] Zhu X, Zhu J, Zhou S, Liu Z, Ming N. Hydrothermal synthesis of nanocrystalline $BaTiO_3$ particles and structural characterization by high-resolution transmission elec-

tron microscopy. J Crystal Growth 2008;310(2):434–41. DOI: 10.1016/j.jcrysgro.2007.10.076

[82] Wang ZL, Cowley JM. Simulating high-angle annular dark-field stem images including inelastic thermal diffuse scattering. Ultramicroscopy 1989;31(4):437–53. DOI: 10.1016/0304-3991(89)90340-9

[83] Pennycook SJ, Jesson DE. High-resolution incoherent imaging of crystals. Phys Rev Lett 1990;64(8):938–41. DOI: 10.1103/PhysRevLett.64.938

[84] Zhu X, Wang J, Zhang Z, Zhu J, Zhou S, Liu Z, et al. Atomic-scale characterization of barium titanate powders formed by the hydrothermal process. J Am Ceramic Soc 2008;91(3):1002–8. DOI: 10.1111/j.1551-2916.2007.02227.x

[85] Zhu X, Zhu J, Zhou S, Liu Z, Ming N, Hesse D. $BaTiO_3$ nanocrystals: hydrothermal synthesis and structural characterization. J Crystal Growth 2005;283(3–4):553–62. DOI: 10.1016/j.jcrysgro.2005.05.080

[86] Zhu X, Zhou J, Zhu J, Liu Z, Li Y, Al-Kassab T. Structural characterization and optical properties of perovskite $ZnZrO_3$ nanoparticles. J Am Ceramic Soc 2014;97(6): 1987–92. DOI: 10.1111/jace.12883

[87] Hartman P, Perdok WG. On the relations between structure and morphology of crystals. I. Acta Crystallograph 1955;8(1):49–52. DOI: 10.1107/S0365110X55000121

[88] Jiang B, Peng JL, Bursill LA, Ren TL, Zhang PL, Zhong WL. Defect structure and physical properties of barium titanate ultra-fine particles. Phys B: Condens Matt 2000;291(1–2):203–12. DOI: 10.1016/S0921-4526(99)01874-8

[89] Jiang B, Peng JL, Bursill LA. Intergrowth defects and dielectric response of $BaTiO_3$ nano-particles. Ferroelectrics 1998;207(1):587–610. DOI: 10.1080/00150199808217271

[90] Rose H. Outline of a spherically corrected semiaplanatic medium-voltage transmission electron-microscope. Optik 1990;85(1):19–24.

[91] Urban KW. Studying atomic structures by aberration-corrected transmission electron microscopy. Science 2008;321(5888):506–10. DOI: 10.1126/science.1152800

[92] Jia C-L, Mi S-B, Urban K, Vrejoiu I, Alexe M, Hesse D. Atomic-scale study of electric dipoles near charged and uncharged domain walls in ferroelectric films. Nature Mater 2008;7(1):57–61. DOI: 10.1038/nmat2080

[93] Huang L, Chen Z, Wilson JD, Banerjee S, Robinson RD, Herman IP, et al. Barium titanate nanocrystals and nanocrystal thin films: synthesis, ferroelectricity, and dielectric properties. J Appl Phys 2006;100(3):034316. DOI: 10.1063/1.2218765

[94] Nuraje N, Su K, Haboosheh A, Samson J, Manning EP, Yang Nl, et al. Room temperature synthesis of ferroelectric barium titanate nanoparticles using peptide nanorings as templates. Adv Mater 2006;18(6):807–11. DOI: 10.1002/adma.200501340

[95] Markovic D, Kusigerski V, Tadic M, Blanusa J, Antisari MV, Spasojevic V. Magnetic properties of nanoparticle $La_{0.7}Ca_{0.3}MnO_3$ prepared by glycine–nitrate method without additional heat treatment. Scripta Materialia 2008;59(1):35–8. DOI: 10.1016/j.scriptamat.2008.02.020

[96] Moreira ML, Soares JM, de Azevedo WM, Rodrigues AR, Machado FLA, de Araújo JH. Structural and magnetic properties of nanoparticles of $La_{2/3}Sr_{1/3}MnO_3$. Phys B: Condens Matt 2006;384(1–2):51–3. DOI: 10.1016/j.physb.2006.05.044

[97] Uskoković V, Košak A, Drofenik M. Silica-coated lanthanum–strontium manganites for hyperthermia treatments. Mater Lett 2006;60(21–22):2620–2. DOI: 10.1016/j.matlet.2006.01.047

[98] Lipham ND, Tsoi GM, Wenger LE. Synthesis and characterization of Sr-doped lanthanum manganite nanoparticles. Magnetics, IEEE Transact 2007;43(6):3088–90. DOI: 10.1109/TMAG.2007.893850

[99] Park T-J, Papaefthymiou GC, Viescas AJ, Moodenbaugh AR, Wong SS. Size-dependent magnetic properties of single-crystalline multiferroic $BiFeO_3$ nanoparticles. Nano Lett 2007;7(3):766–72. DOI: 10.1021/nl063039w

[100] Selbach SM, Tybell T, Einarsrud M-A, Grande T. Size-dependent properties of multiferroic $BiFeO_3$ nanoparticles. Chem Mater 2007;19(26):6478–84. DOI: 10.1021/cm071827w

[101] Joshi PC, Desu SB. Structural, electrical, and optical studies on rapid thermally processed ferroelectric $BaTiO_3$ thin films prepared by metallo-organic solution deposition technique. Thin Solid Films 1997;300(1–2):289–94. DOI: 10.1016/S0040-6090(96)09468-0

[102] Cheong S-W, Mostovoy M. Multiferroics: a magnetic twist for ferroelectricity. Nat Mater 2007;6(1):13–20. DOI: 10.1038/nmat1804

[103] Rao CNR, Serrao CR. New routes to multiferroics. J Mater Chem 2007;17(47):4931–8. DOI: 10.1039/B709126E

[104] Zhao T, Scholl A, Zavaliche F, Lee K, Barry M, Doran A, et al. Electrical control of antiferromagnetic domains in multiferroic $BiFeO_3$ films at room temperature. Nat Mater 2006;5(10):823–9. DOI: 10.1038/nmat1731

[105] Chu Y-H, Martin LW, Holcomb MB, Gajek M, Han S-J, He Q, et al. Electric-field control of local ferromagnetism using a magnetoelectric multiferroic. Nat Mater 2008;7(6):478–82. DOI: 10.1038/nmat2184

[106] Fiebig M. Revival of the magnetoelectric effect. J Phys D: Appl Phys 2005;38(8):R123–52. DOI: 10.1088/0022-3727/38/8/R01

[107] Ederer C, Spaldin NA. Magnetoelectrics: a new route to magnetic ferroelectrics. Nat Mater 2004;3(12):849–51. DOI: 10.1038/nmat1265

[108] Yu J, Chu J. Progress and prospect for high temperature single-phased magnetic ferroelectrics. Chin Sci Bull 2008;53(14):2097–112. DOI: 10.1007/s11434-008-0308-3

[109] Setter N, Damjanovic D, Eng L, Fox G, Gevorgian S, Hong S, et al. Ferroelectric thin films: review of materials, properties, and applications. J Appl Phys 2006;100(5): 051606. DOI: 10.1063/1.2336999

[110] Kohlstedt H, Ishiwara H. Nanoelectronics and Information Technology - Advanced Electronic Materials and Novel Devices. 1st ed. Wienheim, Germany: WILEY-VCH Verlag; 2003. 1002 p.

[111] Martinelli G, Carotta MC, Traversa E, Ghiotti G. Thick-film gas sensors based on nano-sized semiconducting oxide powders. MRS Bull 1999;24(06):30–6. DOI: 10.1557/ S0883769400052489

[112] Mathews S, Ramesh R, Venkatesan T, Benedetto J. Ferroelectric field effect transistor based on epitaxial perovskite heterostructures. Science 1997;276(5310):238–40. DOI: 10.1126/science.276.5310.238

[113] Tasch AF, Jr, Parker LH. Memory cell and technology issues for 64- and 256-Mbit one-transistor cell MOSD DRAMs. Proc IEEE 1989;77(3):374–88. DOI: 10.1109/5.24125

[114] Scott J. High-dielectric constant thin films for dynamic random access memories (DRAM). Annu Rev Mater Sci 1998;28(1):79–100. DOI: 10.1146/annurev.matsci.28.1.79

[115] Hwang CS, Park SO, Cho HJ, Kang CS, Kang HK, Lee SI, et al. Deposition of extremely thin (Ba, Sr) TiO_3 thin films for ultra-large-scale integrated dynamic random access memory application. Appl Phys Lett 1995;67(19):2819–21. DOI: 10.1063/1.114795

[116] Horikawa T, Mikami N, Makita T, Tanimura J, Kataoka M, Sato K et al. Dielectric properties of (Ba, Sr)TiO_3 thin films deposited by RF sputtering. Jpn J Appl Phys 1993;32(9S):4126–30. DOI: 10.1143/JJAP.32.4126

[117] Yoshino K, Shimoda Y, Kawagishi Y, Nakayama K, Ozaki M. Temperature tuning of the stop band in transmission spectra of liquid-crystal infiltrated synthetic opal as tunable photonic crystal. Appl Phys Lett 1999;75(7):932–4. DOI: 10.1063/1.124558

[118] Martí X, Sánchez F, Hrabovsky D, Fàbrega L, Ruyter A, Fontcuberta J, et al. Exchange biasing and electric polarization with $YMnO_3$. Appl Phys Lett 2006;89(3):032510. DOI: 10.1063/1.2234285

[119] Béa H, Bibes M, Cherifi S, Nolting F, Warot-Fonrose B, Fusil S, et al. Tunnel magnetoresistance and robust room temperature exchange bias with multiferroic $BiFeO_3$ epitaxial thin films. Appl Phys Lett 2006;89(24):242114. DOI: 10.1063/1.2402204

[120] Žutić I, Fabian J, Das Sarma S. Spintronics: fundamentals and applications. Rev Mod Phys 2004;76(2):323–410. DOI: 10.1103/RevModPhys.76.323

[121] Kimura T, Kawamoto S, Yamada I, Azuma M, Takano M, Tokura Y. Magnetocapacitance effect in multiferroic $BiMnO_3$. Phys Rev B 2003;67(18):180401. DOI: 10.1103/PhysRevB.67.180401

[122] Shaw TM, Trolier-McKinstry S, McIntyre PC. The properties of ferroelectric films at small dimensions. Ann Rev Mater Sci 2000;30(1):263–98. DOI: 10.1146/annurev.matsci.30.1.263

[123] Tsymbal EY, Kohlstedt H. Tunneling across a ferroelectric. Science 2006;313(5784):181–3. DOI: 10.1126/science.1126230

[124] Junquera J, Ghosez P. Critical thickness for ferroelectricity in perovskite ultrathin films. Nature 2003;422(6931):506–9. DOI: 10.1038/nature01501

[125] Pimenov A, Mukhin AA, Ivanov VY, Travkin VD, Balbashov AM, Loidl A. Possible evidence for electromagnons in multiferroic manganites. Nat Phys 2006;2(2):97–100. DOI: 10.1038/nphys212

[126] Eerenstein W, Mathur ND, Scott JF. Multiferroic and magnetoelectric materials. Nature 2006;442(7104):759–65. DOI: 10.1038/nature05023

Chapter 6

Synthesis, Crystal Structure, and Physical Properties of the Perovskite Iridates

Yunqi Cai, Yan Li and Jinguang Cheng

Additional information is available at the end of the chapter

http://dx.doi.org/10.5772/61281

Abstract

Perovskite iridates have emerged as a new paradigm for studying the strongly correlated electron physics with strong spin-orbit coupling. The "113" alkaline-earth iridates $AIrO_3$ (A = Ca, Sr, Ba) display a rich variety of crystallographic and electronic states and are now attracting growing research interest. This chapter aims to provide an overview for these "113" iridates, including the materials' synthesis, crystal structure, major physical properties, and other interesting results such as the effects of pressure and chemical substitutions, as well as theoretical perspectives.

Keywords: Perovskite iridates, Spin-orbit coupling, Post-perovskite, Polytype, Semimetal

1. Introduction

The discoveries of high-transition-temperature superconductivity in cuprates and the colossal magnetoresistance in manganites made the first-row (3d) transition-metal oxides (TMOs) with perovskite-related structures the central topics of condensed matter physics over the past four decades. The strong electron–electron correlations intrinsic for these narrow-band 3d-electron systems are believed to be at the heart of rich physics. Following the general wisdom based on the 3d TMOs, the third-row (5d) counterparts having a spatially much extended 5d orbitals were expected to have much reduced electron–electron correlations, U, and broaden bandwidth, W, i.e. U << W, leading to a Pauli paramagnetic metallic ground state, Figure 1(a). Such an expectation, however, was recently found to be violated in many 5d-electron iridium oxides (iridates), such as Sr_2IrO_4 [1], in which an antiferromagnetic insulating ground state was instead observed. Recent studies have revealed that such discrepancy originates from the inherently strong spin-orbit coupling (SOC) for these heavy 5d elements, which have a typical value of SOC, $\zeta_{SO} \approx 0.3$–0.5 eV, comparable with the magnitude of U and W, and thus cannot be treated as a negligible perturbation as in the 3d TMOs.

Since an unrealistically large U is required to open a Mott gap in Sr_2IrO_4, Figure 1(b), Kim et al. [2] proposed that the strong SOC splits the otherwise broad t_{2g} band of the octahedral-site, low-spin Ir^{4+}($5d^5$) array into a filled, low-energy $J_{eff} = 3/2$ quartet band and a half-filled, high-energy $J_{eff} = 1/2$ doublet band, Figure 1(c, e). A moderate Hubbard U can then open a Mott gap, leading to the SOC-driven $J_{eff} = 1/2$ Mott insulating state, Figure 1(d). Subsequent experimental [3] and theoretical [4] investigations have confirmed such a novel $J_{eff} = 1/2$ state in the strong SOC limit. Since then, the 5d TMOs have emerged as a new paradigm for studying the strongly correlated electron physics with strong SOC. In particular, the iridates have attracted special attention in that the combination of relativistic SOC and electron–electron correlations has been proposed to generate more exotic, unprecedented quantum states of matters, such as the strong topological insulators, Weyl semimetal, quantum spin liquids, and even unconventional superconductors [5].

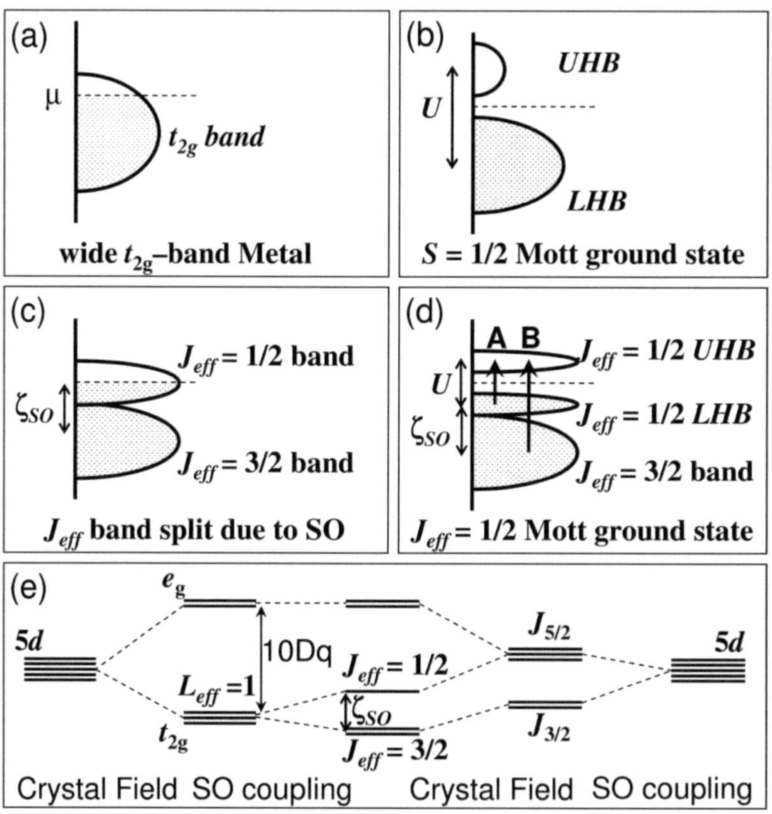

Figure 1. Schematic energy diagrams for the $5d^5$ (t_{2g}^5) configuration: (a) without SOC and U, (b) with an unrealistically large U but no SOC, (c) with SOC but no U, and (d) with SOC and U, (e) 5d level splitting by the crystal field and SOC. Adapted from Reference [2].

Since the importance of SOC was first recognized in Sr_2IrO_4, which is the $n = 1$ member of the Ruddlesden–Popper series $Sr_{n+1}Ir_nO_{3n+1}$ ($n = 1, 2, \infty$), these perovskite (Pv) iridates are currently archetypal systems for studying the interplay of SOC and electron–electron correlations. During the past few years, numerous studies have been performed on the single-layer Sr_2IrO_4 ($n = 1$) and bilayer $Sr_3Ir_2O_7$ ($n = 2$); for a review, see Reference [6]. With increasing the number of Pv layers n, the bandwidth of the $J_{eff} = 1/2$ band is expected to increase, and a dimensionality controlled insulator-to-metal transition is eventually realized in the orthorhombic Pv $SrIrO_3$ ($n = \infty$) [7]. Recent advances in this emergent field have turned much attention to the "113" alkaline-earth iridates $AIrO_3$ (A = Ca, Sr, Ba) with the Pv-related structures. However, in-depth investigations on these compounds are hindered to a great extent by the harsh synthesis conditions as well as their complex structural variations. For example, the above-mentioned orthorhombic Pv $SrIrO_3$ can only be stabilized in the bulk form under high-pressure and high-temperature (HPHT) conditions [8], or in the form of thin films by applying epitaxial strain [9]. Recent theoretical investigations proposed an intriguing topological semimetal state for the orthorhombic Pv phase [10]. When synthesized at ambient pressure, on the other hand, $SrIrO_3$ adopts a so-called six-layer (6H) polytype [8], which has been characterized as a non-Fermi-liquid metal approaching a ferromagnetic quantum critical point [11]. As the sister compounds of $SrIrO_3$, both $CaIrO_3$ and $BaIrO_3$ also display multiple structural polymorphs with interesting structural–property relationships. Depending on the synthesis conditions, $CaIrO_3$ can be stabilized in either the Pv or the post-perovskite (pPv) structure, which displays, respectively, a paramagnetic metal and an antiferromagnetic insulator ground states [12, 13]. Although the pPv $CaIrO_3$ was regarded as the $J_{eff} = 1/2$ Mott insulator [14], recent studies revealed a clear deviation from the ideal $J_{eff} = 1/2$ state due to the pronounced structural distortions [15]. In addition, in the field of geosciences, $CaIrO_3$ has been studied extensively as an analogy of $MgSiO_3$ to elucidate the mechanism of Pv to pPv transition at the boundary of Earth's lowermost mantle, or the D'' layer [16]. In the case of $BaIrO_3$, it also exhibits multiple structural polytypes with interesting structural–property relationship. At ambient pressure, $BaIrO_3$ adopts a nine-layer structure (9R). It is the first known ferromagnetic insulator among the 5d TMOs and exhibits intriguing charge-density-wave formation accompanying the ferromagnetic order [17, 18]. When treating the 9R phase under different pressures, three more polytypes, i.e. 5H, 6H, and 3C, have been identified [19]. Following the sequence of 9R → 5H → 6H → 3C, their ground states change progressively from a weak ferromagnetic insulator with $T_c = 180$ K for 9R, through a ferromagnetic metal with $T_c = 50$ K for 5H, and an exchange enhanced non-Fermi-liquid metal for 6H approaching a ferromagnetic quantum critical point, finally to a Pauli paramagnetic Fermi-liquid metal for 3C [20, 21]. These results demonstrate an intimate structure–property relationship that has been well documented in the 3d TMOs. A brief summary of the structural types and interesting physical properties for these "113" alkaline-earth iridates $AIrO_3$ (A = Ca, Sr, Ba) is given in Table 1.

Although there are many publications dealing with an individual compound, a monograph that provides a comprehensive overview for these "113" alkaline-earth iridates is still lacking to our knowledge. Taking into account the growing research interests on these iridates, it is imminent to summarize the currently available knowledge in a single chapter. Thus, this chapter aims to bring together the available information in literature for these "113" iridates.

In the following, we will give a comprehensive literature survey for each $AIrO_3$, covering the materials' synthesis, crystal structure, and major physical properties, as well as other interesting results such as the effects of chemical substitutions and theoretical investigations. Finally, we will give a brief concluding remark on the current research status and provide an outlook on the future research directions on these iridates.

Compound	Structure type	Interesting physical properties
$CaIrO_3$	pPv	AF insulator with $T_N = 110$ K, stripe-type AF order with spin canting;
	Pv	PM semimetal with possible Dirac node protected by reflection symmetry
$SrIrO_3$	6H	Exchange enhanced PM metal with nFL behaviors due to proximity to a FM QCP
	Pv	PM semimetal with possible Dirac node protected by reflection symmetry
$BaIrO_3$	9R	Weak FM insulator with a simultaneous CDW formation below $T_c \approx 180$ K
	5H	Weak FM metal with $T_c \approx 50$ K
	6H	Exchange enhanced PM metal with nFL behaviors due to proximity to a FM QCP
	3C	FL PM metal

AF: Antiferromagnetic; PM: Paramagnetic; FM: Ferromagnetic; FL: Fermi liquid;

nFL: non Fermi liquid; CDW: Charge density wave; QCP: Quantum critical point

Table 1. A summary of the "113" Alkaline-earth iridates $AIrO_3$ (A=Ca, Sr, Ba)

2. $CaIrO_3$

$CaIrO_3$ has two different orthorhombic polymorphs, i.e. the layered pPv structure with space group *Cmcm* and the $GdFeO_3$-type Pv structure with space group *Pbnm*. These two compounds have been known since 1960s [22, 23] and received significant attention from geologists since 2004 as an analogy of $MgSiO_3$, the main constituent mineral of the Earth's lower mantle [16, 24]. More recently, they have emerged as important correlated 5d-electron systems with strong SOC [14, 25]; the strong local distortion in pPv $CaIrO_3$ makes it a model system to investigate the interplay of non-cubic crystal field splitting and SOC [15], while the orthorhombic Pv $CaIrO_3$ might be considered as an intriguing semimetal with symmetry-protected Dirac points [26].

2.1. Synthesis

There are some discrepancies in literature regarding the synthesis of pPv $CaIrO_3$ at ambient pressure. In the earlier studies [12, 22], it was reported that single-phase pPv phase cannot be obtained at ambient pressure through a solid-state reaction from $CaCO_3$ and IrO_2 in air. Recently, Harai et al. [27] reported that pure pPv $CaIrO_3$ can be prepared by heating the stoichiometric mixture of CaO and IrO_2 powders sealed in an evacuated silica tube at 1000°C over 20 h. On the other hand, since the pPv structure is a high-pressure phase, pPv $CaIrO_3$ can be readily obtained by utilizing HPHT synthesis. For example, Ohgushi et al. [25] reported the synthesis of single-phase pPv $CaIrO_3$ at 4 GPa and 1150°C.

Needle-shaped pPv $CaIrO_3$ single crystals have been reported to grow out of the $CaCl_2$ flux. By adopting a tenfold flux and a relatively low soaking temperature of 836 and 950°C, respectively, Sugahara et al. [28] and Hirai et al. [29] obtained tiny single crystals for the purpose of crystal-structure refinements. On the other hand, Ohgushi et al. [14] seems to grow sizable pPv $CaIrO_3$ single crystals for anisotropic magnetic property measurements by employing a higher flux molar ratio (16:1) and a higher soaking temperature of 1200°C. However, our attempts by using the latter approach ended up with Ca_2IrO_4 rather than the pPv $CaIrO_3$.

Because Pv $CaIrO_3$ is a metastable phase, it cannot be synthesized via a solid-state reaction route at ambient pressure. Alternatively, Sarkozy et al. [12] reported the preparation of pure Pv phase by thermal decomposition at 650–700°C in air of the hydroxide intermediate $CaIr(OH)_6$, which can be obtained according to the following wet-chemical reaction scheme:

$$K_2IrCl_6 \xrightarrow{H_2O} IrCl_6^{\ 2-} \xrightarrow{KOH,pH10-12} Ir(OH)_6^{\ 2-} \xrightarrow{Ca^{2+}} CaIr(OH)_6 \downarrow .$$

By following this approach, we obtained nearly single-phase Pv $CaIrO_3$ with a trace amount of IrO_2 (0.2 wt.%) and Ca_2IrO_4 (1.3 wt.%) [30]. Recently, Kojitani et al. [31] determined a large positive Clapeyron slope for the pPv/Pv transition of $CaIrO_3$, i.e. Pv structure is the high-temperature phase of pPv. Thus, Pv $CaIrO_3$ can be obtained by transforming pPv phase at higher temperature under given pressures. For example, Ohgushi et al. [13] have reported the synthesis of single-phase Pv $CaIrO_3$ at 1 GPa and 1450°C. In addition, thin films of Pv $CaIrO_3$ have recently been epitaxially stabilized on various substrates [26, 32].

2.2. Crystal structure

pPv $CaIrO_3$ As shown in Fig. 2(a), the crystal structure of pPv $CaIrO_3$ can be described as a pseudo-2D layered structure having IrO_6 octahedral sheets stacked alternatively with the Ca^{2+} ions along the crystallographic *b* axis. Within the octahedral sheets, IrO_6 octahedra share edges along the *a* axis to form rutile-type chains; these chains are then interconnected with each other via apical oxygen atoms along the *c* axis. Because of the significant interest in geosciences, the crystal structure of pPv $CaIrO_3$ have been thoroughly studied by several groups [28, 29]. To illustrate the peculiar features of pPv $CaIrO_3$, here we adopted the results given by Sugahara *et al.* [28] who refined the crystal structure based on the single-crystal X-ray diffraction (XRD) technique. The crystal structure was refined in an orthorhombic space group *Cmcm* (No. 63) with Ca at *4c* (0, y, 1/4), Ir at *4a* (0, 0, 0), O1 at *4c* (0, y, 1/4), and O2 at *8f* (0, y, z) sites, respectively. The obtained unit cell parameters are a = 3.147 Å, b = 9.866 Å, c = 7.302 Å, and V = 226.7 Å^3 at room temperature. The refined positional parameters and selected bond lengths and bond angles after Reference [28] are listed in Table 2. Within the buckled octahedral layer, the IrO_6 octahedral chains display an alternative rotation about the *a* axis, resulting in Ir-O1-Ir bond angles of 134.3°. As a result, for a given octahedron the local *z* axis that is along the Ir-O1 bond deviates from the crystallographic *c* axis by about 23°. In addition, IrO_6 octahedra show a significant tetragonal compression, with two short Ir-O1 (1.978 Å) and four long Ir-O2 (2.066 Å) bonds. Octahedral-site distortions can be generally described by the orthorhombic vibrational modes $Q_2 = l_x - l_y$ and $Q_3 = (2l_z - l_x - l_y)/\sqrt{3}$, where l_x, l_y, l_z are the bond lengths for bonding along local *x, y, z* directions. In pPv $CaIrO_3$, the octahedral-site distortion corresponds to a negative mode of $Q_3 = -0.102$. For comparison, the PtO_6 octahedra in the pPv $CaPtO_3$ with a filled t_{2g}^6 manifold show a negligible $Q_3 = -0.001$. Such a comparison highlights a strong orbital-lattice coupling in pPv $CaIrO_3$ with low-spin t_{2g}^5 configuration for Ir^{4+} ions, for which the single hole would be expected to have a dominant *yz* ± *izx* orbital character [30], which has been confirmed recently by the resonant inelastic X-ray spectroscopy [15].

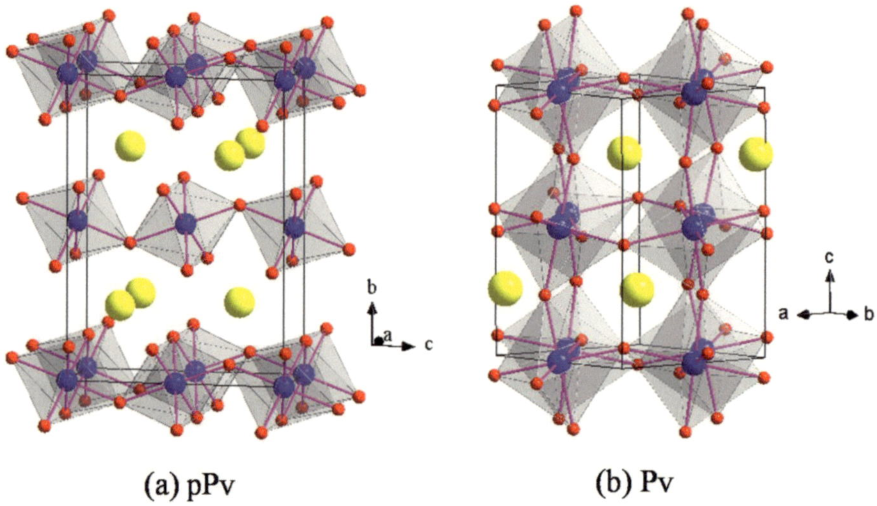

Figure 2. Crystal structure of $CaIrO_3$ polymorphs: (a) pPv and (b) Pv.

Pv $CaIrO_3$ As shown in Fig. 2(b), the crystal structure of Pv $CaIrO_3$ is built up from corner-shared IrO_6 octahedra in three dimensions with Ca cations in the interstitial positions. It has been known over 40 years that Pv $CaIrO_3$ adopts the $GdFeO_3$-type structure; however, structural refinements have not been performed until recently. We present here our Rietveld refinement results [30] based on the high-resolution synchrotron XRD on polycrystalline Pv $CaIrO_3$ prepared with the wet-chemical method mentioned above. The crystal structure was refined in space group *Pbnm* (No. 62) with Ca at *4c* (*x, y*, 1/4), Ir at *4b* (0.5, 0, 0), O1 at *4c* (*x, y*, 1/4) and O2 at *8d* (*x, y, z*) sites, respectively. The lattice parameters at room temperature are determined as a = 5.35046 Å, b = 5.59291 Å, c = 7.67694 Å, and V = 229.73 Å^3. The obtained positional parameters and selected bond lengths and bond angles after Reference [30] are listed in Table 3. In comparison with the pPv phase, the IrO_6 octahedra are less distorted with three sets of Ir-O distances of 2.006 Å, 2.020 Å, and 2.038 Å; the average Ir-O distance of 2.021 Å is consistent with the ionic radii sum for Ir^{4+} (0.625 Å) and O^{2-} (1.40 Å). The averaged Ir-O-Ir bond angle is about 145.5° in the Pv phase.

Atom	Site	*x*	*y*	*z*	B_{iso} (Å^2)
Ca	4c	0	0.7492	1/4	0.40
Ir	4a	0	0	0	0.32
O1	4c	0	0.0779	1/4	0.70
O2	8f	0	0.3658	0.4452	0.79
Ir-O1 (×2)	1.978		Ir-O1-Ir	134.3	
Ir-O2 (×4)	2.066		O1-Ir-O2	86.3	
<Ir-O>	2.037		O1-Ir-O2	93.7	
Ir-Ir (×2)	3.1472				
Ir-Ir (×2)	3.651				

Table 2. Refined positional parameters and selected bond lengths (Å) and bond angles (°) for pPv $CaIrO_3$ from single-crystal XRD [28]: space group *Cmcm* (No. 63), *a* = 3.1472 Å, *b* = 9.8655 Å, *c* = 7.3018 Å, *V* = 226.71 Å^3, Z = 4.

Atom	Site	*x*	*y*	*z*	B_{iso} (Å^2)
Ca	4c	-0.01403	0.05962	1/4	0.71
Ir	4b	0.5	0	0	0.27
O1	4c	0.10487	0.47110	1/4	0.92
O2	8d	0.69257	0.30488	0.05602	1.07
Ir-O1 (×2)	2.006		Ir-O1-Ir	146.15	
Ir-O2 (×2)	2.020		Ir-O2-Ir	144.95	
Ir-O2 (×2)	2.038				
<Ir-O>	2.021				

Table 3. Refined positional parameters and selected bond lengths (Å) and bond angles (°) for Pv $CaIrO_3$ from powder XRD[30]: space group *Pbnm* (No. 62), *a* = 5.35046 Å, *b* = 5.59291 Å, *c* = 7.67694 Å, *V* = 229.73 Å^3, Z = 4.

2.3. Physical properties

pPv $CaIrO_3$ It is an antiferromagnetic insulator with $T_N \approx 110$ K. Due to the difficulty in obtaining single-phase samples, the physical properties of pPv $CaIrO_3$ were not characterized until 2006 by Ohgushi et al. [25], who first reported its electrical transport and magnetic properties on polycrystalline samples synthesized under HPHT conditions. As shown in Figure 3, its resistivity $\varrho(T)$ increases quickly upon cooling, following the Arrhenius-type behaviour, i.e. $\varrho(T) = \varrho_0 \exp(\Delta/T)$, with the activation energy Δ = 0.17 eV; magnetic susceptibility χ = M/H exhibits a sharp transition at T_N = 115 K, below which a weak ferromagnetic moment of ~ 0.04 μ_B/Ir was observed from the isothermal magnetization curve. In addition, a huge coercive field

$H_c \approx 4$ T was evidenced at low temperatures. Density functional calculations by Subedi [33] demonstrated that the inclusion of SOC can split the t_{2g} bands into fully filled $J_{eff} = 3/2$ bands and half-filled $J_{eff} = 1/2$ bands, as shown schematic in Figure 1(c), and that both SOC and moderate U are required to reconcile the experimentally observed Mott insulating behaviour. By performing the resonant X-ray diffraction at the L absorption edges of pPv $CaIrO_3$ single crystals, Ohgushi et al. [14] determined its magnetic structure as a stripe-type antiferromagnetic order, i.e. the Ir moments are aligned parallel along the *a* axis and antiparallel along the *c* axis with a canted ferromagnetic component along the *b* axis. Bogdanov et al. [34] carried out *ab initio* quantum chemical calculations and reproduced such a striped antiferromagnetic structure. Moreover, their calculations predicted a strong antiferromagnetic exchange interaction of $J_c = 121$ meV through the corner-shared path along the *c* axis, and a weak nearest-neighbour ferromagnetic coupling of $J_a \approx -7.3$ meV within the edge-shared chains along the *a* axis. In this regard, pPv $CaIrO_3$ can be regarded as a $J_{eff} = 1/2$ quasi-1D antiferromagnet. Although the above results suggested that a $J_{eff} = 1/2$ ground state is realized in pPv $CaIrO_3$, first-principles calculations [33, 34] evidenced significant deviations from the ideal $J_{eff} = 1/2$ state with highly uneven admixture of the t_{2g} components due to the pronounced tetragonal distortion. In agreement with these calculations, a very recent resonant inelastic X-ray scattering (RIXS) study by Sala et al. [15] confirmed the departure from the $J_{eff} = 1/2$ state. By analyzing the RIXS spectrum, they estimated the effective tetragonal crystal field splitting $\Delta = -0.71$ eV and the SOC $\zeta_{SO} = 0.52$ eV, from which a ground state wave function $|\mathbf{0}, \pm\rangle = \mp \mathbf{0.32}|xy, \mp\rangle + \mathbf{0.67}(|yz, \pm\rangle \mp i|zx, \pm\rangle)$ with a dominant yz±izx orbital character was derived.

The Mott insulating nature of quasi-2D pPv $CaIrO_3$ have motivated Ohgushi et al. [25] to metallize it via the carrier doping. They successfully prepared a series of hole-doped $Ca_{1-x}Na_xIrO_3$ ($0 \leq x \leq 0.37$) with pPv structure under HPHT conditions and realized a filling-control antiferromagnetic insulator to paramagnetic metal transition around $x = 0.3$, near which, however, no superconductivity was observed. Nevertheless, anomalous properties such as the non-Fermi-liquid behaviour and positive magnetoresistance violating Kohler's rule were observed for the metallic samples. In the same report, attempt of electron doping via Y^{3+} substitutions for Ca^{2+} was unsuccessful due to the formation of pyrochlore phase. In a recent work by Gunasekera et al. [35], however, Y-substituted $Ca_{1-x}Y_xIrO_3$ ($0 \leq x \leq 0.5$) samples with pPv structure were successfully fabricated via a conventional solid-state reaction route at ambient pressure. Similar with the Na^+-doped case, about 30%-Y^{3+} doping can also drive an insulator-to-metal transition; superconductivity was not observed either in this case. In striking contrast with the gradual suppression of T_N by Na doping, surprisingly, T_N remains unchanged upon Y substitutions up to 50%, except that the magnitude of ac susceptibility peak decreases about one order. Further experiments are needed to exclude the possibility that the magnetic order arises from the minor $CaIrO_3$ phase. Anyhow, metallization of the quasi-2D pPv $CaIrO_3$ represents an interesting direction to pursue exotic electronic state in the vicinity of metal–insulator transition.

Pv $CaIrO_3$ In sharp contrast with the antiferromagnetic insulating ground state of pPv $CaIrO_3$, the Pv phase has been reported as a Pauli paramagnetic metal by Sarkozy et al. [12] in 1974.

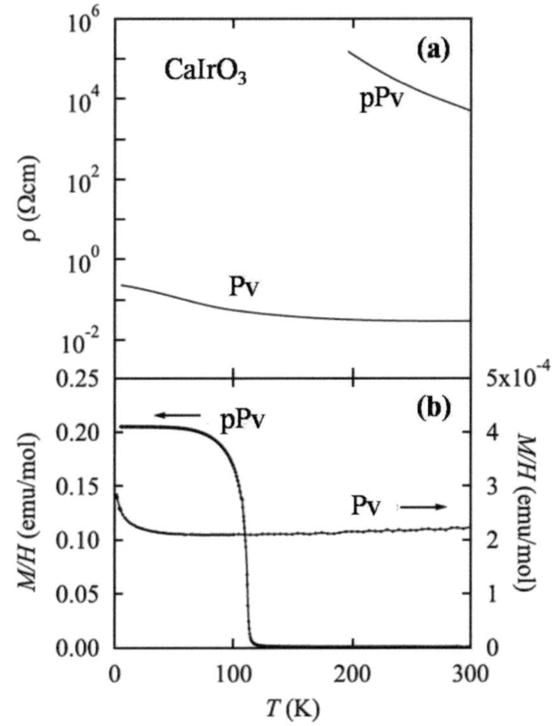

Figure 3. Temperature dependence of (a) resistivity ϱ(T) and (b) magnetic susceptibility M/H for the two polymorphs of $CaIrO_3$, pPv for post-perovskite and Pv for perovskite. Adapted from Reference [13].

Recent characterizations by Ohgushi et al. [13] on the dense pellets obtained via HPHT synthesis confirmed the paramagnetic nature, but the temperature dependence of resistivity shows bad metal behaviour with a non-diverge upturn at low temperatures, Figure 3. Such a bad metallic behaviour was reproduced on the epitaxially stabilized thin films grown on various substrates [26, 32], and has been ascribed to a semimetallic ground state with the conduction and valence band touching at Fermi level. The observations of a sign change and a nonlinear magnetic-field dependence of the Hall resistance are consistent with the coexistence of electron and hole charge carriers [26]. As discussed below for Pv $SrIrO_3$, the semimetallic state might originate from the symmetry-protected Dirac nodes around the Fermi level due to a combined effect of SOC and reflection symmetry of the *Pbnm* orthorhombic lattice. In light of the recent theoretical proposals for the orthorhombic Pv iridates discussed below, further experimental studies on the semimetallic Pv $CaIrO_3$ are highly desirable.

pPv versus Pv $CaMO_3$ (M = Ir, Rh, Ru) The distinct ground states of pPv and Pv $CaIrO_3$ reflect the intimate structural–property relationships. In addition to $CaIrO_3$, both $CaRuO_3$ [36] and $CaRhO_3$ [37] have also been reported to possess quenchable Pv and pPv polymorphs. Besides the importance in geosciences as analogy materials of $MgSiO_3$, these compounds with partially filled d-electron shells are important correlated electron systems with intriguing physical properties [37, 38]. In a similar manner as $CaIrO_3$, their ground states differ sharply as the

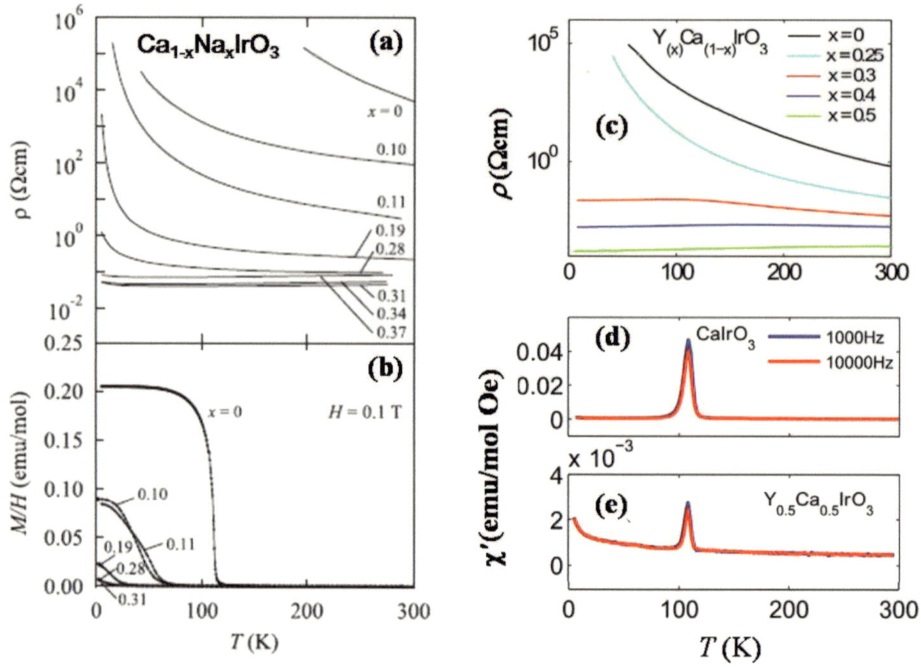

Figure 4. Insulator–metal transition in pPv $CaIrO_3$ induced by hole (Na^+) and electron (Y^{3+}) doping. (a, b) shows the temperature dependence of resistivity and magnetic susceptibility of $Ca_{1-x}Na_xIrO_3$. Adapted from Reference [25]; (c, d, e) show the temperature dependence of resistivity and ac magnetic susceptibility of $Ca_{1-x}Y_xIrO_3$. Adapted from Reference [35].

structure changes: Pv $CaRuO_3$ is a well-known exchanged-enhanced paramagnetic metal on the verge of a ferromagnetic instability, whereas the pPv phase is an antiferromagnetic insulator with $T_N = 270$ K [38]; Pv $CaRhO_3$ is a Pauli paramagnetic metal while the pPv phase is insulating and undergoes a canted antiferromagnetic transition below $T_N = 90$ K [37]. Current first-principles calculations for the pPv compounds failed to capture the correct ground state; it seems that in addition to electron–electron correlations, SOC also plays an essential role in producing the insulating ground state for these 4d and 5d-electron systems [39].

In addition to the interest in fundamental physics, the $CaIrO_3$ ceramics have also been investigated by Keawprak et al. [40] for the potential thermoelectric applications. They prepared both phases of $CaIrO_3$ with spark plasma sintering technique and evaluated their thermoelectric properties from room temperature up to 1023 K. The highest dimensionless figure of merit (ZT) reaches 0.02 and 0.003 for Pv and pPv phase, respectively.

3. $SrIrO_3$

Depending on the synthesis conditions, $SrIrO_3$ can form in two different structures, i.e. the monoclinically distorted 6H polytype and the orthorhombic $GdFeO_3$-type Pv structure [8]. The former is a rare stoichiometric oxide exhibiting non-Fermi-liquid behaviours near a ferromagnetic quantum critical point [11]. The latter was recently found to be an exotic narrow-band semimetal that may harvest many topological and magnetic insulating phases [10, 41, 42].

3.1. Synthesis

The 6H phase can be readily prepared in the polycrystalline form at ambient pressure by sintering the stoichiometric mixture of $SrCO_3$ and IrO_2 (or Ir) at 900–1,100°C in air [8]. Single crystals of 6H phase with dimensions ~0.4 × 0.4 × 0.6 mm^3 have been grown in Pt crucibles with the $SrCl_2$ self-flux techniques [11]. The Pv phase is a HP form of $SrIrO_3$. Longo et al. [8] performed the first HPHT syntheses and established the temperature–pressure phase diagram for the 6H-Pv transformation of $SrIrO_3$. It was found that the 6H phase transforms to the Pv structure above 1,650°C at 2 GPa and above 700°C at 5 GPa. Recent HPHT syntheses of Pv $SrIrO_3$ were usually performed at 1,000–1,100°C and 5–6 GPa [43, 44]. For these samples, Rietveld refinements on the powder XRD patterns evidenced the presence of ~3–4 wt.% IrO_2 impurity. Since the Pv phase is metastable, it remains a challenge to obtain sizable bulk single crystals under HP conditions. However, Pv $SrIrO_3$ films and superlattices have been stabilized at ambient pressure via applying the epitaxial strain with various techniques, including the metalorganic chemical vapour deposition [9], pulsed laser deposition [45], and reactive oxide molecular-beam expitaxy [42]. As discussed below, given the tolerance factor $t < 1$, it is unusual for $SrIrO_3$ to adopt the 6H structure at ambient pressure. It was recently reported [46, 47] that the 6H structure of $SrIrO_3$ can be destabilized by partial substitution of M = Li, Fe, Co, Ni, Zn for Ir in $SrIr_{1-x}M_xO_3$ and converted to the Pv structure within a narrow composition range around $x = 0.2$. In these cases, it was suggested that the presence of e_g type orbitals on the M ions contributes to the breakdown of face-sharing octahedral dimmers in the 6H structure.

3.2. Crystal structure

6H $SrIrO_3$ The crystal structure of 6H $SrIrO_3$ was first determined by Longo et al. [8] as a monoclinic distortion of the hexagonal $BaTiO_3$ structure. The 6H hexagonal structure can be described as close-packed SrO_3 layers stacked perpendicular to the c axis in the sequence hcchcc, where h and c refer to hexagonal (ABAB...) and cubic (ABCABC...) close packing, respectively. The Ir atoms occupy the oxygen octahedra formed by the SrO_3 layers, and the IrO_6 octahedra share common faces across an h layer and common corners across a c layer. As a result, the above hcchcc stacking sequence results in two independent positions for the Ir atoms. As shown in Figure 5(a), two $Ir2O_6$ octahedra form pairs of face-shared octahedra that are joined by common corners to a plane of corner-sharing $Ir1O_6$ octahedra. Therefore, the 6H structure can be alternatively depicted as a stacking of layers of corner- (C) and face (F)-sharing IrO_6 octahedra in the sequence FCCFCC along the c axis.

In the original work by Longo et al., the oxygen positional parameters were not refined due to the low scattering of oxygen relative to Ir and Sr. Based on the neutron diffraction data, Qasim et al. [46] recently provided a full refinement on the crystal structure of 6H $SrIrO_3$ with $a = 5.6040$ Å, $b = 9.6256$ Å, $c = 14.1834$ Å, and $\beta = 93.202°$ in space group *C2/c* (No. 15). The refined positional parameters and selected bond lengths and bond angles after Reference [46] are listed in Table 4. In this structure, the Ir1O_6 octahedron has an average Ir-O distance of 2.006 Å typical of Ir^{4+} with the individual distances in a narrow range 1.987–2.038 Å. In contrast, the Ir2O_6 octahedra in the Ir_2O_9 dimers are not regular with a longer average Ir-O distance of 2.030 Å. The Ir2–Ir2 distance, 2.770 Å, is relatively short and close to the separation of 2.72 Å found in Ir metal, suggesting a strong Ir–Ir bonding across the common faces. Variable temperature XRD measurements on the 6H phase confirmed that the monoclinic structure is stable without any structural transition up to 1,000°C in air.

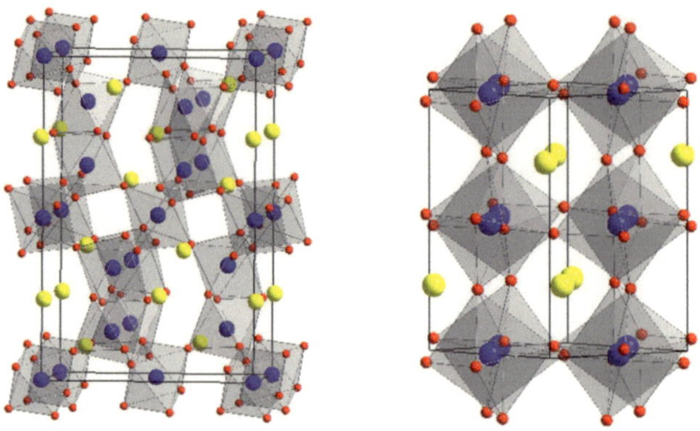

Figure 5. Crystal structure of $SrIrO_3$ polytypes: (a) 6H and (b) Pv.

Pv SrIrO₃ Similar as Pv $CaIrO_3$, it consists of a 3D network of corner-sharing IrO_6 octahedra that are cooperatively rotated and tilted about the pseudocubic [110] and [001] axes, Figure 5(b). Blanchard et al. [44] recently performed a thorough study on the crystal structure of Pv $SrIrO_3$ based on the synchrotron and neutron diffraction data. The crystal structure was refined in space group *Pbnm* (No. 62) with Sr at *4c* (*x*, *y*, 1/4), Ir at *4a* (0, 0, 0), O1 at *4c* (*x*, *y*, 1/4), and O2 at *8d* (*x*, *y*, *z*) sites, respectively. The lattice parameters at room temperature are determined as a = 5.60075 Å, b = 5.57115 Å, and c = 7.89601 Å. The obtained positional parameters and selected bond lengths and bond angles after Reference [44] are listed in Table 5. As can be seen, the individual IrO_6 octahedron is relatively rigid with three Ir–O distances being approximately equal. The average Ir–O distance at 300 K of 2.016 Å is consistent with the ionic radii sum for Ir^{4+} and O^{2-}. From the refined atomic coordinates, Blanchard et al. also estimated the two independent octahedral tilt angles, i.e. $\psi = 11.5°$ for out-of-phase tilt about the pseudocubic [110] axis, and $\varphi = 8.7°$ for in-phase tilt about the pseudocubic [001] axis, respectively. These tilting angles were found to be nearly temperature-independent below room temperature. The orthorhombic *Pbnm* structure was shown to persist over the temperature range 3–1,070 K.

6H-Pv transformation As pointed out by Longo et al. [8], $SrIrO_3$ and $SrMnO_3$ are the only $SrBO_3$ (B = Ti, Zr, Hf, Cr, Mo, Tc, Fe, Ru, Sn, Pb, Ce, Th) compounds that do not adopt the Pv structure at ambient pressure. Given the tolerance factor $t \leq 1$, the Pv structure would be stabilized for these compounds. However, $SrMnO_3$ has the 4H polytype structure with a stacking sequence of hchc along the *c* axis, while $SrIrO_3$ crystallizes in the 6H polytype as mentioned above. It would appear that the hexagonal polytypes with their face-shared octahedra and trigonal crystal fields are stabilized by the outer electron configurations that allow for metal–metal bonding along the *c* axis. For example, in the case of low-spin Ir^{4+}, the trigonal crystal field of the hexagonal polytype should split the $t_{2g}{}^5$ orbitals to $e_{g\sigma}{}^0 e_{g\pi}{}^4 a_{1g}{}^1$, which allow for metal–metal bonding along the *c* axis via the half-filled a_{1g} orbitals. Thus, the 6H $SrIrO_3$ and 4H $SrMnO_3$ can be regarded as a compromise between the continuous face-shared chains of the 2H polytype and the geometrically favoured Pv structure. Since high pressure prefers the cubic close packing with a higher density than the hexagonal close packing, 6H $SrIrO_3$ transforms to the Pv structure under high-pressure conditions with a ~3% volume reduction.

Atom	Site	*x*	*y*	*z*	B_{iso} (Å^2)
Sr1	4e	0	0.0092	1/4	0.0285
Sr2	8f	0.0122	0.6667	0.0957	0.0482
Ir1	4a	0	0	0	0.0478
Ir2	8f	0.9820	0.6660	0.84698	0.0459
O1	4e	0	0.4981	1/4	0.0584
O2	8f	0.2411	0.2649	0.2603	0.0287
O3	8f	0.8112	0.4077	0.0474	0.0572
O4	8f	0.9407	0.1544	0.4087	0.0535
O5	8f	0.3238	0.4204	0.1058	0.0586

Ir1-O3 (×2)	2.038	Ir1-O3-Ir2	149.6
Ir1-O4 (×2)	1.987	Ir1-O4-Ir2	158.8
Ir1-O5 (×2)	1.994	Ir1-O5-Ir2	149.3
<Ir1-O>	2.006		
Ir2-O1	2.100	Ir2-O1-Ir2	82.5
Ir2-O2	2.055	Ir2-O1-Ir2	85.1
Ir2-O2	2.040		
Ir2-O3	1.974		
Ir2-O4	1.957		
Ir2-O5	2.051		
<Ir2-O>	2.030		
Ir2-Ir2	2.770		

Table 4. Refined positional parameters and selected bond lengths (Å) and bond angles (°) for 6H $SrIrO_3$ from neutron diffraction [46]: space group *C2/c* (No. 15), *a* = 5.60401 Å, *b* = 9.6256 Å, *c* = 14.1834 Å, β = 93.202°, *V* = 763.89 Å^3, Z = 12.

Atom	Site	x	y	z	B_{iso} (Å^2)
Sr	4c	-0.0068	0.4687	1/4	0.019
Ir	4a	0	0	0	0.017
O1	4c	0.0718	0.0049	1/4	0.019
O2	8d	0.2126	0.2877	-0.0369	0.022
Ir-O1 (×2)	2.015		Ir-O1-Ir	156.92	
Ir-O2 (×2)	2.018		Ir-O2-Ir	156.22	
Ir-O2 (×2)	2.018				
<Ir-O>	2.017				

Table 5. Refined positional parameters and selected bond lengths (Å) and bond angles (°) for Pv $SrIrO_3$ from neutron diffraction [44]: space group *Pbnm* (No. 62), a = 5.60075 Å, b = 5.57115 Å, c = 7.89601 Å, V = 246.376 Å^3, Z = 4.

3.3. Physical properties

6H $SrIrO_3$ Although the 6H $SrIrO_3$ has been synthesized more than 50 years ago, its physical properties were not characterized in detail until 2007 by Cao et al. [11], who reported the magnetic, electrical transport, and calorimetric properties of 6H $SrIrO_3$ single crystals grown out of the $SrCl_2$ flux. The primary results are summarized in Figure 6. Magnetic susceptibility $\chi(T)$ measurements evidenced no long-range magnetic order down to 1.7 K, but exhibited at low temperatures strong enhancements that diverge as $\chi \propto T^{\gamma}$ with $1/2 < \gamma < 1$, suggesting the proximity to a ferromagnetic instability. The isothermal magnetization $M(H)$ at 1.7 K indeed displays a saturation behaviour at $H \sim 3$ T, yet the saturation moment is very small, being less than 0.03 μ_B/Ir. The low-temperature specific heat $C(T)$ exhibits a pronounced $-T\log T$ dependence, which is characteristic of non-Fermi-liquid systems. Such a $C/T \sim -\log T$ behaviour can be readily enhanced in low applied fields up to 1.1 T, vanishes for $H > 2$ T, and eventually changes to a $T^{3/2}$ power law expected for a ferromagnetically ordered state at H = 8 T. In accordance with the $C(T)$ results, both the c-axis resistivity, ϱ_c, and the ab-plane resistivity, ϱ_{ab}(T), follow a non-Fermi-liquid $T^{3/2}$ dependence over a wide temperature range up to 120 K under zero field, while a Fermi-liquid T^2 behaviour is restored upon applying an external field $H \geq 5$ T. Taking into account all these observations, 6H $SrIrO_3$ can be regarded a rare example of stoichiometric oxide that exhibits non-Fermi-liquid behaviours near a ferromagnetic quantum critical point. As will be shown explicitly in $BaIrO_3$, such a quantum critical point can be realized via subtle structural variations.

Pv $SrIrO_3$ In the original work by Longo et al. [8], Pv $SrIrO_3$ has been described as a Pauli paramagnetic metal. In 2008, Zhao et al. [43] reinvestigated the physical properties of Pv $SrIrO_3$ bulk sample prepared under 5 GPa and 1,000°C. They observed two characteristic temperatures $T^* \approx 170$ K and $T_{MI} \approx 44$ K: at T^*, the paramagnetic susceptibility $\chi(T)$ starts to increase with temperature, and the resistivity $\varrho(T)$ exhibits a slope change, followed below T^* by the presence of unusual linear field dependence positive magnetoresistance (MR) that reaches about 12% at 5 K and 7 T; a broad metal–insulator transition was observed at T_{MI}. However, these observations are largely disapproved by the very recent work of Blanchard et

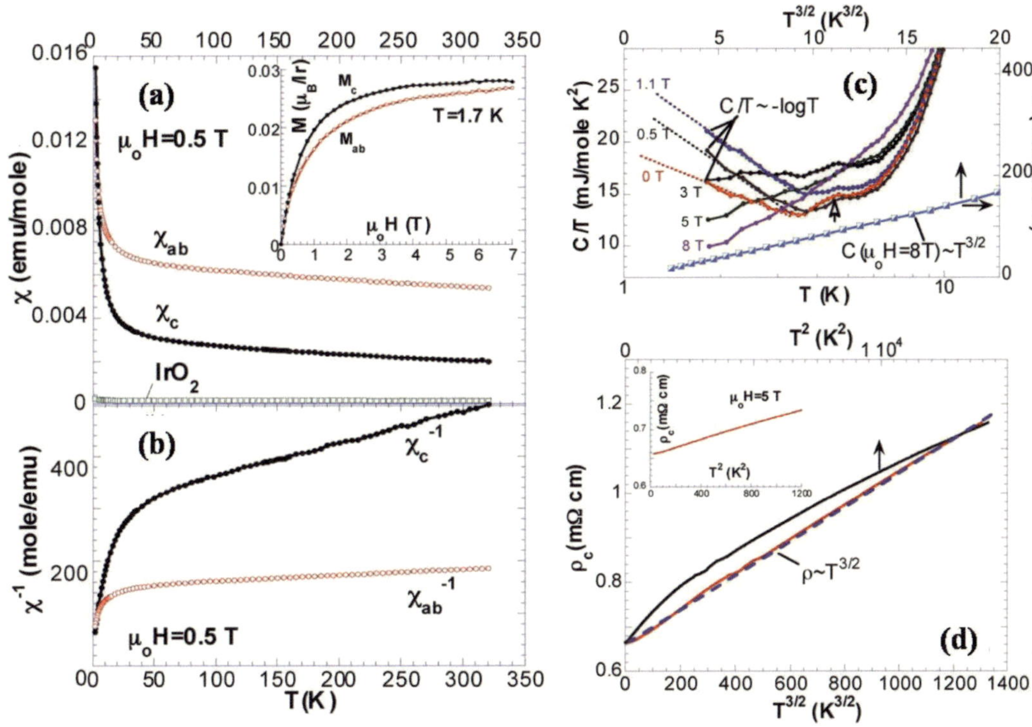

Figure 6. Physical properties of 6H $SrIrO_3$ single crystal. Adapted from Reference [11]. (a, b) show the temperature dependence of magnetic susceptibility measured along the *c* axis and *ab* plane, inset of (a) shows the magnetization curve at 1.7 K. (c) shows the specific heat C(T) data illustrating the $C/T \sim -\log T$ behaviour. (d) shows the resistivity following the $T^{3/2}$ behaviour.

al. [44], who studied the bulk samples prepared under 6 GPa and 1,100°C. As shown in Figure 7, it was found in the latter work that $\varrho(T)$ exhibits metallic conductivity down to 2 K, following Fermi-liquid T^2 dependence between 2 and 30 K, without showing an upturn at low temperature. In addition, they observed a smaller positive MR up to 2% at 7 *T* and 2–50 K. These discrepancies might originate from the polycrystalline nature of the studied samples, in which the extrinsic effects such as the grain boundary and impurities can largely influence the transport properties. As mentioned above, Pv $SrIrO_3$ films of single-crystal quality can be stabilized via applying epitaxial strain. The resistivity upturn at low temperatures have been frequently observed in these thin films [9, 48], but the upturn temperature displays a broad distribution, in support of an extrinsic property due to weak Anderson localization. Nevertheless, such sensitivity to defects reflects the bad metal character of semimetallic Pv $SrIrO_3$ approaching the boundary of metal–insulator transition.

As the end member of the Ruddlesden–Popper series $Sr_{n+1}Ir_nO_{3n+1}$ (n = 1, 2, ∞), Pv $SrIrO_3$ has recently attracted much attention due to the presence of nontrivial features within the J_{eff} = 1/2 bands. Density-functional theory first-principles calculations by Carter et al. [10] found that in the strong SOC limit the bands near the Fermi energy are mostly composed of J_{eff} = 1/2 states. Interestingly, they found a node near the U point, Figure 8(a), thus revealing the semimetallic

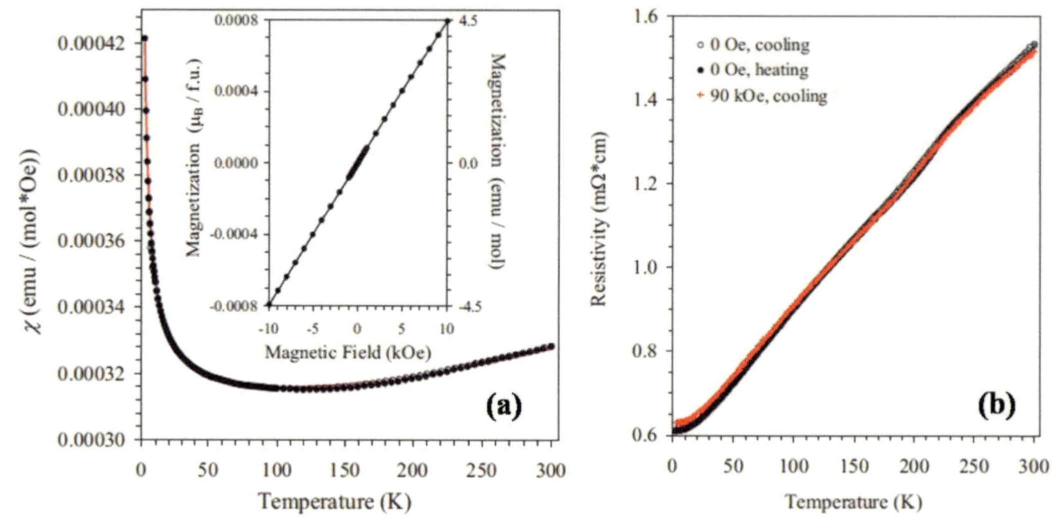

Figure 7. Temperature dependence of (a) magnetic susceptibility and (b) resistivity of Pv $SrIrO_3$. Adapted from Reference [44].

nature of Pv $SrIrO_3$. By constructing a tight-binding model, they confirmed the presence of a line node near the U point in the Brilloiun zone, and further shown that the line node originates from the reflection symmetry of the crystal structure at the $z = 1/4$ and 3/4 planes presented in the orthorhombic *Pbnm* space group. Since the line node is protected by the underlying lattice symmetry, it has been further proposed that perturbations breaking the sublayer reflection symmetry can lift the line node and convert the system into an insulating phase; the system may become a strong topological insulator at a certain point. In addition, as shown in Figure 8(b), magnetically ordered metallic and insulating phases have also been proposed to arise in the U versus SOC phase diagram of Pv $SrIrO_3$ [41]. Moreover, Chen et al. [49] further proposed that the presence of reflection symmetry in orthorhombic Pv iridates may realize a novel class of topological crystalline metals with zero-energy surface states at certain planes.

Recent angle-resolved photoemission spectroscopy on Pv $SrIrO_3$ films by Nie et al. [42] has uncovered such an exotic semimetallic state with very narrow bands near the Fermi surface consisting of heavy hole-like pockets around ($\pm\pi$, 0) and (0, 0) and light electron-like pockets at ($\pm\pi/2$, $\pm\pi/2$). Surprisingly, the bandwidth of Pv $SrIrO_3$ is found to be narrower than that of Sr_2IrO_4, in contrary to the general expectations of broaden bandwidth with increasing dimensionality [7]. Since the semimetallic ground state has been confirmed experimentally, it is of particular interest to achieve the proposed topological and/or magnetic states via tuning the SOC, U, and/or lattice symmetry. In this regard, Matsuno et al. [45] have made an important step towards these exotic phases; they tailored a spin-orbit magnetic insulator out of the semimetallic state via controlling the dimensionality of [$(SrIrO_3)_m$, $SrTiO_3$] superlattices. By utilizing HPHT synthesis, we prepared a series of Sn-doped $SrIr_{1-x}Sn_xO_3$ orthorhombic perovskites. We found that substitutions of isovalent, nonmagnetic Sn^{4+} for Ir^{4+} ions lead to a breakdown of the semimetallic state, and convert the paramagnetic, semimetallic ground state

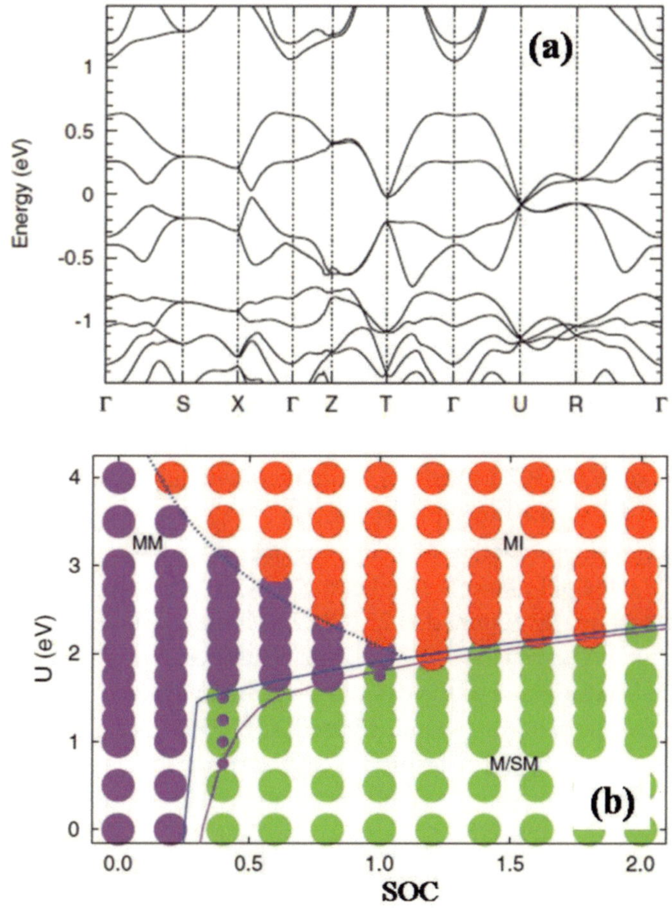

Figure 8. (a) LDA band structure of Pv $SrIrO_3$ with Hubbard U = 2 eV and SOC, demonstrating the presence of the node near the U point of J_{eff} = 1/2 band near the Fermi level; (b) the phase diagram of Pv $SrIrO_3$ in the U-SOC plane containing three phases: magnetic metal (MM), nonmangetic metal or semimetal (M/SM), and magnetic insulator (MI). Adapted from Reference [41].

of Pv $SrIrO_3$ to an antiferromagnetic insulator with a concomitant metal–insulator transition at T_N [50]. These recent experimental efforts demonstrated that semimetallic Pv $SrIrO_3$ is a promising candidate for realizing distinct topological and magnetic insulating states that deserve further investigations in the near future. On the other hand, the Pv $SrIrO_3$ film has also been regarded as potential electrode material for microelectronic devices [9].

4. $BaIrO_3$

At ambient pressure, $BaIrO_3$ crystallizes in the nine-layer (9R) polytype. It is the first known ferromagnetic insulator with $T_c \approx 180$ K among the 5d TMOs [17]. Detailed studies on single crystals revealed a charge-density-wave (CDW) formation below the ferromagnetic order [18, 51]. Recent experimental [52] and theoretical [53] investigations further revealed it as an exotic spin-orbit Mott insulator that is of great current research interest. Following the general trend of perovskite hexagonal polytypes, we have explored the high-pressure sequences of $BaIrO_3$ and found three more polytypes, i.e. 5H, 6H, and 3C [19, 20, 21]. Their ground states exhibit an interesting evolution from a ferromagnetic insulator to a Pauli paramagnetic metal passing through a ferromagnetic quantum critical point tuned by the gradual structural changes as detailed below.

4.1. Synthesis

The ambient-pressure 9R phase can be readily obtained by sintering the stoichiometric mixtures of $BaCO_3$ and Ir at 1,000°C in air. The sample should be cooled down slowly for the last sintering in order to ensure an oxygen stoichiometry [54]. Single crystals have been reported to grow out of the $BaCl_2$ flux at a relatively low temperature of 1,000 K [18]. HPHT synthesis is needed for all the other polytypes [19, 21, 55, 56]. For the HP syntheses around 1,000°C, the 9R polytype is stable up to 3 GPa, the 5H phase exists only in a narrow pressure range around 4 GPa, the 6H phase is stabilized in a wide pressure range from 5 to ~20 GPa, and the 3C phase was finally obtained at 25 GPa. We have employed the two-stage (Walker- or Kawai-type) multianvil systems for the HPHT syntheses. During the HPHT experiments, the sample was first compressed to the desired pressure by eight truncated tungsten carbide anvils, and then the temperature was increased to ~1,000°C and kept for 30 min before quenching to room temperature. The resultant samples were recovered after releasing pressure and then subjected to various characterizations at ambient pressure.

4.2. Crystal structure

9R $BaIrO_3$ As shown in Figure 9(a), the crystal structure of the 9R phase consists of Ir_3O_{12} trimers of face-sharing octahedra that are linked by their vertices to form columns parallel to the *c*-axis, with a stacking of layers of corner-sharing (C) and face-sharing (F) IrO_6 octahedra in the order FFCFFCFFC along the *c* axis. Except for the monoclinic distortion, it is isostructural with the 9R $BaRuO_3$. The monoclinic distortion generates twisting and buckling of the Ir_3O_{12} trimers that are tilted ~12° relative to each other. Here, we adopted the crystal structure of 9R $BaIrO_{2.94}$ obtained by Powell et al. [54] from the NPD data, which were refined in a structural model defined in the *C2/m* space group, with three kinds of Ba atoms at 4*i* (*x*, 0, *z*) positions, four types of unequivalent Ir atoms at 4*i*, 2*a* (0, 0, 0) and 2*d* (0.5, 0, 0.5) sites, and six types of oxygen atoms at 4*i* and 8*j* (*x*, *y*, *z*) positions. The obtained unit-cell parameters are $a = 9.9992$ Å, $b = 5.7490$ Å, $c = 15.1707$ Å, and $\beta = 103.27°$. The final positional parameters and the selected bond lengths and bong angles after Reference [54] are listed in Table 6. Ir1–Ir2 and Ir3–Ir4 distances, of 2.618 Å and 2.627 Å respectively, are even smaller than the separation of 2.72 Å found in Ir metal, which indicates significant interactions between iridium cations at the centre of face-shared pairs of octahedra. It is important to note that, although this polytype has been compared to the ambient 9R $BaRuO_3$, with rhombohedral (*R-3m*) symmetry and a stacking sequence $(FFC)_3$, the monoclinic distortion described for ambient 9R $BaIrO_3$ actually involves a shorter periodicity, with a stacking sequence $(FFC)_2$ along the *c* axis, as shown in Figure 9(a).

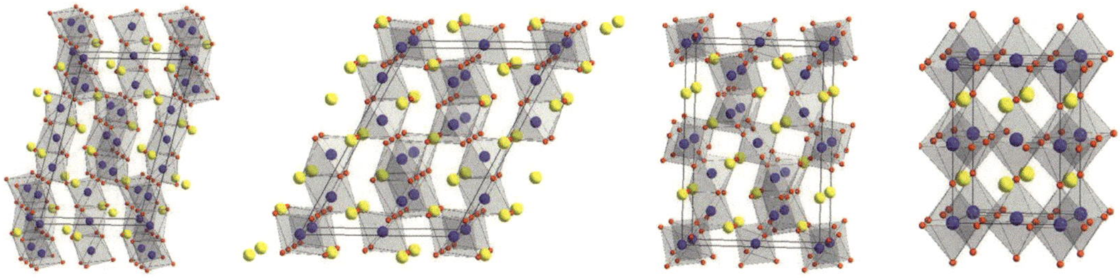

Figure 9. Crystal structure of the $BaIrO_3$ polytypes: (a) 9R, (b) 5H, (c) 6H, and (d) 3C.

Atom	Site	*x*	*y*	*z*	B_{iso} (Å^2)
Ba1	4i	0.7779	0	0.2501	0.42
Ba2	4i	0.3686	0	0.0720	0.42
Ba3	4i	0.1515	0	0.4224	0.42
Ir1	4i	0.0845	0	0.1766	0.16
Ir2	2a	0	0	0	0.16
Ir3	4i	0.4657	0	0.3230	0.16
Ir4	2d	0.5	0	0.5	0.33
O1	4i	0.2926	0	0.2287	0.33
O2	8j	0.0507	0.2421	0.2617	0.33
O3	4i	0.8931	0	0.0994	0.33
O4	8j	0.1164	0.2362	0.0839	0.33
O5	8j	0.4036	0.2291	0.4040	0.33
O6	4i	0.6427	0	0.4240	0.33

Ir1-O1	2.049	Ir1-Ir2	2.618
Ir1-O2 (×2)	1.979	Ir3-Ir4	2.627
Ir1-O3	2.001		
Ir1-O4 (×2)	2.032	Ir1-O1-Ir3	157.3
<Ir1-O>	2.01	Ir1-O2-Ir3	164.0
Ir2-O3 (×2)	2.038	Ir1-O3-Ir2	80.8

Atom	Site	x	y	z	B_{iso} (Å^2)
Ir2-O4 (×4)		2.034	Ir1-O4-Ir2	80.1	
<Ir2-O>		2.04	Ir3-O5-Ir4	81.4	
Ir3-O1		1.978	Ir3-O6-Ir4	80.0	
Ir3-O2 (×2)		2.037			
Ir3-O5 (×2)		1.955			
Ir3-O6		2.057			
<Ir3-O>		2.02			
Ir4-O6 (×2)		2.030			
Ir4-O5 (×4)		2.035			
<Ir4-O>		2.03			

Table 6. Refined positional parameters and selected bond lengths (Å) and bond angles (°) for 9R $BaIrO_3$ from neutron diffraction [54]: space group *C2/m* (No. 12), $a = 9.9992$ Å, $b = 5.7490$ Å, $c = 15.1707$ Å, $\beta = 103.27°$, $V = 848.81$ Å^3, Z = 12.

5H $BaIrO_3$ The 5H phase was discovered as a new perovskite polytype [19]. As shown in Figure 9(b), its crystal structure contains chains of double dimer units that are corner-connected via oxygen atoms. These clusters of four octahedra are interleaved with single layers of vertex-sharing IrO_6 octahedra, forming infinite chains along the *c* axis. Adjacent chains are interlinked along the *a* and *b* directions via Ir–O–Ir vertex-sharing bridges. Alternatively, the structure can be described as stacking of layers of corner-sharing (C) and face-sharing (F) IrO_6 octahedra along the sequence FCFCC. The crystal structure was refined in the monoclinic *C2/m* (No. 12) space group, with three kinds of Ba atoms at 2*c* (0.5, 0.5, 0.5) and 4*i* (x, 0, z), three types of Ir atoms at 2*a* (0, 0, 0) and 4*i* sites, and six unequivalent oxygen atoms at 8*j* (x, y, z), 4*i*, 2*d* (0.5, 0, 0.5), and 4*f* (0.75, 0.75, 0) positions. The lattice parameters at room temperature are determined as $a = 9.9554$ Å, $b = 5.7434$ Å, $c = 13.8049$ Å, and $\beta = 119.23°$. The final positional parameters and the selected bond lengths and bond angles after Reference [19] are listed in Table 7. As can be seen, Ir–O distances vary in the range 1.90 Å for Ir1–O4 to 2.23 Å for Ir2–O3. The average value, of 2.03 Å, is consistent with the ionic radii sum for Ir^{4+} and O^{2-}. It is noteworthy that the structure contains three kinds of octahedra with rather distinct average sizes: <Ir–O> are 1.985 Å, 2.072 Å, and 2.017 Å for Ir1, Ir2, and Ir3 octahedra. The two largest octahedra, Ir2 and Ir3, are those forming dimers, where the Ir–O bonds are weakened by the Ir–Ir bonds. According to these bond distances, the bond valences for the three types of octahedra are 4.26(8)+, 3.35(8)+, and 4.06(9)+, indicating that Ir1 and Ir2 are under certain compressive and tensile stresses, respectively.

6H $BaIrO_3$ Same as the 6H $SrIrO_3$, the crystal structure of 6H $BaIrO_3$ consists of dimers of face-sharing octahedra separated by single corner-sharing octahedron, showing the sequence FCCFCC along the *c* axis. Based on the XRD data, we have refined its crystal structure in the monoclinic *C2/c* space group with two kinds of Ba atoms at 4*e* (0, y, ¼) and 8*f* (x, y, z) positions, Ir1 at 4*a* (0, 0, 0) and Ir2 at 8*f* sites, and four independent oxygen atoms, O1 at 4*e*, O2, O3, and O4 at 8*f* positions. The obtained unit-cell parameters are $a = 5.7483$ Å, $b = 9.9390$ Å, $c = 14.3582$

Å, and $\beta = 91.319°$. The final positional parameters and selected bond lengths and bond angles after Reference [19] are listed in Table 8. As can be seen, the $Ir2O_6$ octahedra within the face-sharing dimmers are considerably more expanded than the $Ir1O_6$ octahedra, with average Ir–O distances of 2.16 and 1.99 Å, respectively. This is probably a consequence of the metal–metal bond linking the couples of Ir2 atoms in the dimmers, with Ir2–Ir2 distances of 2.710 Å.

Atom	Site	*x*	*y*	*z*	B_{iso} (Å^2)
Ba1	2c	0.5	0.5	0.5	1.2
Ba2	4i	-0.191	0	0.7139	0.1
Ba3	4i	-0.247	0.5	0.8886	0.9
Ir1	2a	0	0	0	3.2
Ir2	4i	-0.4529	0	0.8215	1.0
Ir3	4i	-0.6046	0	0.5920	2.4
O1	8j	0.053	0.72	0.7002	2.3
O2	8j	-0.0093	-0.770	0.1085	1.2
O3	4i	0.703	0	0.303	4.0
O4	4i	0.217	0	0.0607	0.6
O5	2d	0.5	0	0.5	1.5
O6	4f	0.75	0.75	0.5	3.9
Ir1-O2 (×4)		2.029	Ir2-Ir3	2.735	
Ir1-O4 (×2)		1.898			
<Ir1-O>		1.985	Ir2-O1-Ir3	84.0	
Ir2-O1 (×2)		2.12	Ir1-O2-Ir2	165.1	
Ir2-O2 (×2)		2.23	Ir2-O3-Ir3	79.1	
Ir2-O3		2.038	Ir1-O4-Ir2	160.3	
Ir2-O4		2.105	Ir3-O5-Ir3	180	
<Ir2-O>		2.072	Ir3-O6-Ir3	180	
Ir3-O1 (×2)		2.004			
Ir3-O3		2.11			
Ir3-O5		1.994			
<Ir3-O>		2.017			

Table 7. Refined positional parameters and selected bond lengths (Å) and bond angles (°) for 5H $BaIrO_3$ from neutron diffraction [19]: space group *C2/m* (No. 12), $a = 9.9554$ Å, $b = 5.7434$ Å, $c = 13.8049$ Å, $\beta = 119.231°$, $V = 688.8$ Å^3, Z = 10.

3C BaIrO$_3$ A single-phase Pv $BaIrO_3$ was finally obtained at 25 GPa [21]. Instead of the simple cubic phase with space group *Pm-3m*, it was found that the XRD pattern of the Pv phase can be refined excellently in the tetragonal *I4/mcm* (No. 140) space group with the Ba atom at *4b* (0, ½, ¼), the Ir atom at *4c* (0, 0, 0), and two kinds of O atoms at *4a* (0, 0, ¼) and *8h* (*x*, *x*+½, 0) sites. The obtained unit-cell parameters are $a = b = 5.7044$ Å and $c = 8.0926$ Å. The final positional parameters and the main bond distances and bond angles after Reference [21] are listed in Table 9. It should be noted that we denoted this phase as "3C" in order to follow the conventional notations.

Atom	Site	*x*	*y*	*z*	B_{iso} (Å^2)
Ba1	4e	0	-0.0052	1/4	0.3
Ba2	8f	0.0078	0.3349	0.0912	0.25
Ir1	4a	0	0	0	0.4
Ir2	8f	0.9936	0.3323	0.8442	0.27
O1	4e	0	0.499	1/4	-0.2
O2	8f	0.2180	0.2390	0.2427	-0.2
O3	8f	0.036	0.846	0.0852	-0.2
O4	8f	0.286	0.087	0.049	-0.2
O5	8f	0.809	0.090	0.103	-0.2
Ir1-O3 (×2)		1.93	Ir1-O3-Ir2	164.4	
Ir1-O4 (×2)		2.02	Ir1-O4-Ir2	151.4	
Ir1-O5 (×2)		2.01	Ir1-O5-Ir2	153.6	
<Ir1-O>		1.99			
Ir2-O1		2.19	Ir2-O1-Ir2	76.4	
Ir2-O2		2.22	Ir2-O1-Ir2	75.1	
Ir2-O2		2.23			
Ir2-O3		2.10			
Ir2-O4		2.09			
Ir2-O5		2.11			
<Ir2-O>		2.16			
Ir2-Ir2		2.710			

Table 8. Refined positional parameters and selected bond lengths (Å) and bond angles (°) for 6H $BaIrO_3$ from powder XRD [19]: space group *C2/c* (No. 15), $a = 5.7483$ Å, $b = 9.9390$ Å, $c = 14.3582$ Å, $\beta = 91.319°$, $V = 820.12$ Å^3, Z = 12.

The small tetragonal distortion of the 3C $BaIrO_3$ phase is unexpected; we should have a cubic phase as found for $BaRuO_3$ formed under high pressure. Such a distortion to tetragonal symmetry by cooperative rotations of the $IrO_{6/2}$ octahedra about the *c* axis is typical of an $A^{2+}B^{4+}O_3$ perovskite with a tolerance factor $t \equiv (A-O)/[\sqrt{2}\,(B-O)]$ a little smaller than unity. However, stabilization of $BaIrO_3$ in the 9R polytype at ambient pressure is consistent with $t >$ 1 obtained from tabulated equilibrium ionic radii. The larger compressibility of the Ba–O bond makes it possible to stabilize the 3C phase of $BaIrO_3$ under 25 GPa pressure, but compression of the Ba–O bond by cubic symmetry should not reduce the tolerance factor below $t = 1$. Retention of the cubic symmetry of the IrO_6 octahedra shows that the threefold degeneracy of the 5d π^* bands is not a factor. In fact, the tolerance factor calculated from the measured <Ba –O> and <Ir–O> bond lengths in Table 9 gives a $t = 0.998$ consistent with the tetragonal structure observed. Therefore, we conclude that at 25 GPa there has been a first-order transition of the Ba–O equilibrium bond length to give a $t < 1$, which indicates that the 5d π^* bands of the IrO_6 array may also have transitioned for $J_{eff} = 1/2$ and 3/2 bands as a result of a reduction of the orbital angular momentum where the bandwidth is broadened. The high-pressure equilibrium (Ba–O) bond length is retained as a metastable bond length on removal of the pressure, and the Ir–O bonds are not under a tensile stress.

Atom	Site	x	y	z	B_{iso} (Å^2)
Ba	4b	0	0.5	1/4	0.72
Ir	4c	0	0	0	0.49
O1	4a	0	0	1/4	0.8
O2	8h	0.2313	0.7313	0	0.8
Ir-O1 (×2)	2.023		Ir-O1-Ir	180	
Ir-O2 (×2)	2.023		Ir-O2-Ir	171.1	
<Ir-O>	2.023				

Table 9. Refined positional parameters and selected bond lengths (Å) and bond angles (°) for 3C $BaIrO_3$ from powder XRD [21]: space group *I4/mcm* (No. 140), $a = b = 5.7044$ Å, $c = 8.0916$ Å, $V = 263.30$ Å^3, Z = 4.

Polytype structures The polytype structures of the ABO_3 oxides and the phase transformation under high pressure were established during the 1960–1970s, see the Review [57]. As a general trend, the number of the hexagonal close packing along the *c* axis in a unit cell is reduced as the synthesis pressure increases, which led to a decrease (increase) of face(corner)-sharing octahedra. This is consistent with that fact that pressure stabilizes preferentially the denser phase. The observed crystallographic densities of the 9R, 5H, 6H and 3C phases of $BaIrO_3$ are 8.84, 9.08, 9.17, and 9.36 g/cm^3, respectively. They progressively increase as expected, since these phases have been stabilized at increasing pressures. This sequence corresponds, therefore, to more dense packing of the BaO_3 layers along the *c* axis, showing an evolution to structures with more corner (C) sharing and fewer face (F) sharing octahedra, i.e. C:F ratios increase in the order 1:2 (9R), 3:2 (5H), 2:1 (6H), and ∞ (3C). As detailed below, the physical properties exhibit interesting evolution in response to these systematic structural variations.

4.3. Physical properties

9R $BaIrO_3$ As mentioned above, 9R $BaIrO_3$ is the first known ferromagnet among the 5d TMOs [17]. Cao et al. [18] performed the first detailed experimental study on the single-crystal samples and uncovered a CDW formation accompanying the ferromagnetic order at $T_c \approx 180$

K. The experimental evidences in support of the CDW formation included [18]: (1) a sudden increase of resistivity at T_c, (2) the presence of non-linear conductivity with negative differential resistivity below T_c, (3) an optical gap formation at ~ 1,200 $cm^{-1} \approx 9\kappa_B T_c$ in the electron excitation spectrum and a splitting of a phonon mode at 350 cm^{-1} for $T < T_c$, and (4) the emergent X-ray satellite structure below T_c. Besides the transitions at $T_c \approx 180$ K, two additional anomalies have also been observed [18] upon cooling on the c-axis resistivity ϱ_c, which first changes to a metallic behaviour below $T_1 = 80$ K and then suddenly enters a Mott-like insulating state below $T_2 = 26$ K, Figure 10. These two additional transitions are absent for the resistivity within the *ab* plane. The simultaneous occurrence of ferromagnetic order and CDW formation is quite unusual, and 9R $BaIrO_3$ has thus been the subject of extensive investigations since then. Later on, Nakano and Terasaki [51] carried out similar current-voltage (I-V) measurements on their single crystals by using a pulsed current in order to exclude the self-heating effects. Their observations of the giant nonlinear conduction only appearing below 30 K, well below $T_c \approx$ 180K, questioned the above scenario of a simultaneous onset of a CDW and a ferromagnetic transition. Instead of the sliding motion of CDW, they proposed an interplay between two different bands is likely the origin of the nonlinear conduction observed in $BaIrO_3$ [51]. Such discrepancy might arise from the different sample quality. Nevertheless, a clear gap opening is unambiguously evidenced at T_c by other experimental probes, such as the Seebeck coefficient [51, 58] and the high-resolution photoemission spectroscopy [59]. Currently, it remains elusive whether the gap opening is driven by the magnetic order or the Fermi surface nesting.

The observation of weak ferromagnetism and insulating ground state in the 9R $BaIrO_3$ has attracted renewed interest in recent years in light of the SOC-driven Mott insulating state for iridates. As for the nature of the weak ferromagnetism, there also exist long-standing discrepancies. Experimentally, a tiny Ir moment of ~0.03 μ_B/Ir was observed below T_c. In addition, a modified Curie-Weiss fitting to the inverse susceptibility also evidenced a small effective moment of ~ 0.13 μ_B [18]. Originally, a spin canting from a localized full-moment antiferromagnetic configuration had been invoked to explain the tiny ordered moment [17]. In contrast, Cao et al. [18] proposed a model of band magnetism with intrinsically small Ir moment due to d–p hybridization and small exchange splitting. Indeed, the muon-spin relaxation measurements by Brooks et al. [60] provided direct experimental evidences in support of a small Ir moment, i.e. they observed clear oscillations below T_c and found an extremely small internal field at the muon site. Such an itinerant picture of band magnetism, however, is incompatible with the observation of high coercive force and anisotropy in magnetization measurements. By employing the X-ray absorption spectroscopy (XAS) and X-ray magnetic circular dichroism (XMCD) techniques, Laguna-Marco et al. [52] recently elucidated an atomic-like nature of the Ir moment with the orbital moment being ~1.5 times larger than the spin moment, thus highlighting the importance of SOC in addressing the magnetic order of 9R $BaIrO_3$. After taking into account both SOC and moderate on-site coulomb interactions, first-principles calculations by Ju et al. [53] identified 9R $BaIrO_3$ as an exotic spin-orbit Mott insulator with multiple J_{eff} = 1/2 states associated with the unique face-sharing Ir_3O_{12} octahedral units within the structure.

Although the atomic-like nature of Ir local moment in 9R $BaIrO_3$ was found to be extremely stable against temperature, pressure, and chemical substitutions [52, 61], these external stimuli

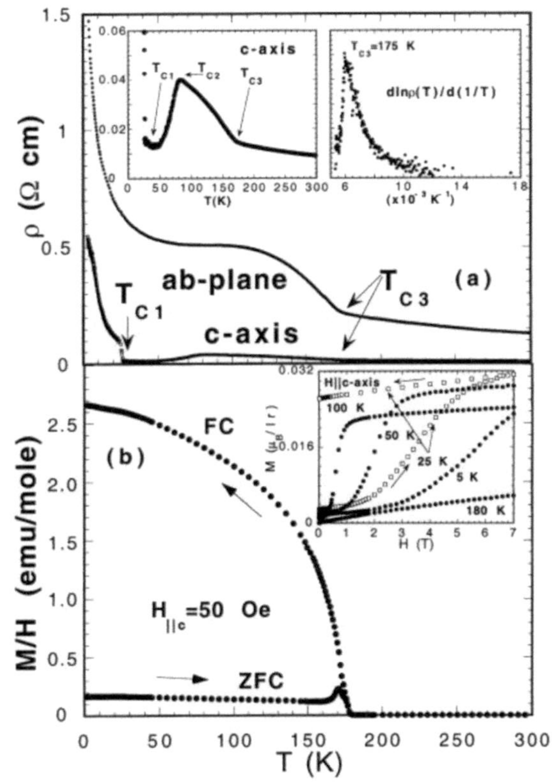

Figure 10. (a) Temperature dependence of resistivity for 9R $BaIrO_3$ for two major crystallographic directions. The first inset shows details of c-axis conductivity and the second the sharp peak in dlnρ/d(1/*T*) denotes the onset of ferromagnetism. (b) Field-cooled (FC) ad zero field-cooled (ZFC) magnetization showing the ferromagnetic transition at T_c. The inset shows isothermal magnetization at several temperatures, illustrating a large hysteresis. Adapted from Reference [18].

can easily lead to a breakdown of the weak ferromagnetism and nonmetallic ground state. For example, Cao et al. [62] grown a series of Sr-doped $Ba_{1-x}Sr_xIrO_3$ single crystals and found that the chemical pressure applied via Sr doping drastically suppresses T_c and immediately leads to a non-metal to metal transition at high temperatures. On the other hand, although the application of external pressure of ~4.5 GPa can also quench the weak ferromagnetism as Sr doping, $BaIrO_3$ becomes more insulating under pressure [61]. Such a disparate response of transport and magnetic properties to the chemical and physical pressure has been ascribed to the different compression rates of the lattice parameters *a* and *c* upon Sr doping and external pressure. Interestingly, Korneta et al. [63] found that a dilute rare-earth R^{3+} doping (~ 4%) of $BaIrO_3$ can also suppress the weak ferromagnetism and lead to a metallic state, whereas the application of modest external pressure readily restores the insulating state. Further studies are needed to clarify whether the weak ferromagnetism is also recovered in the pressure-induced insulating state. All these above results demonstrate a delicate interplay between structural and electronic degrees of freedom in 9R $BaIrO_3$.

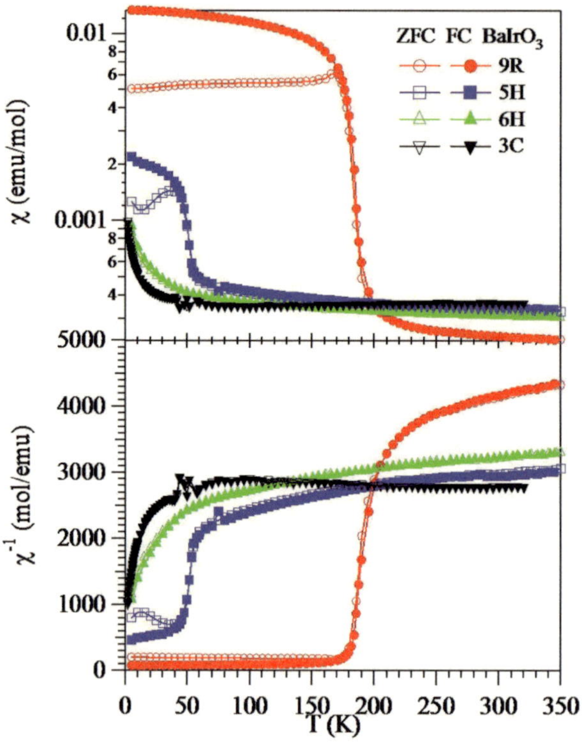

Figure 11. Temperature dependence of magnetic susceptibility $\chi(T)$ and its inverse $\chi^{-1}(T)$ for the $BaIrO_3$ polytypes, illustrating the evolution of the magnetic ground state. Adapted from Reference [21].

5H $BaIrO_3$ The 5H phase is a weak ferromagnetic metal with $T_c \approx 50$ K, Figure 11. Cheng et al. [20] and Zhao et al. [56] have characterized this compound through measurements of magnetic susceptibility $\chi(T)$, resistivity $\varrho(T)$, thermoelectric power $S(T)$, and specific heat $C(T)$. A weak ferromagnetic transition at $T_c \approx 50$ K was clearly observed in $\chi(T)$, and well reflected as a kink in the plots of ρ vs T, S/T vs ln$T$, and C/T vs T. *In situ* high-pressure resistivity measurements show that T_c decreases gradually with pressure, and reaches about 40 K under 1.5 GPa.

6H $BaIrO_3$ The 6H phase has been independently identified and characterized by Zhao et al. [55] and Cheng et al. [20]. Similar with the 6H $SrIrO_3$, it is an exchange-enhanced paramagnetic metal with non-Fermi-liquid behaviours. Zhao et al. [55] reported that its resistivity $\rho(T)$ follows a linear T dependence below 20 K, whereas a $T^{5/3}$ dependence was observed for $T < 60$ K by Cheng et al. [20]. Such a discrepancy should arise from the polycrystalline nature of the studied samples. As mentioned above, a non-Fermi-liquid $\rho \sim T^{3/2}$ behaviour has also been found in the 6H $SrIrO_3$ single crystals due to the proximity to a ferromagnetic quantum critical point. In order to verify similar situation taking place in 6H $BaIrO_3$, we measured the thermopower $S(T)$ that is insensitive to grain boundaries. We indeed found a linear relationship in the plot of S/T versus –ln$T$ over a wide temperature range, in strong support of the realization of ferromagnetic quantum critical point [20]. Based on the low-temperature specific heat and

http://dx.doi.org/10.5772/61281

magnetic susceptibility, the obtained Sommerfeld–Wilson ratio $R_W = \frac{\pi^2}{3}\left(\frac{k_B}{\mu_B}\right)^2 \frac{\chi_0}{\gamma} = 2.14(3)$ provides further evidence for strong electron-electron correlations.

3C $BaIrO_3$ 3C $BaIrO_3$ is characterized as a Pauli paramagnetic metal with a Fermi-liquid behaviour [21]. Its resistivity $\rho(T)$ displays a metallic behaviour down to at least 1.8 K and follows the Fermi-liquid behaviour, i.e. $\rho(T) = \rho_0 + AT^2$ below 6 K with $\rho_0 = 0.0584(1)$ Ω cm and $A = 8.1(1)$ μΩ cm K^{-2}, respectively. The magnetic susceptibility $\chi(T)$ exhibits a nearly temperature-independent Pauli paramagnetism with a shallow minimum around 85 K as observed around $T^* \approx 170$ K in Pv $SrIrO_3$. Such an upturn with temperature has been ascribed to the higher-order temperature-dependent term in the Pauli paramagnetism. Low-temperature specific heat C(T) analysis yields an electronic specific-heat coefficient $\gamma = 6.84(6)$ mJ/mol K^2 and a Debye temperature $\Theta_D = 335$ K.

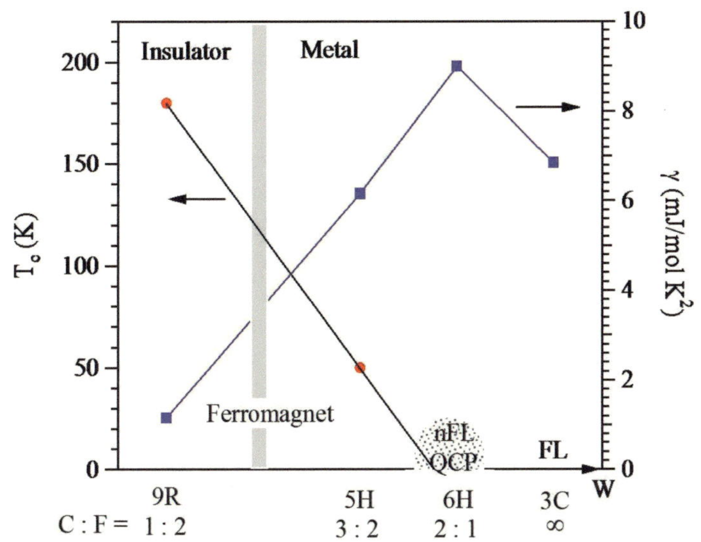

Figure 12. A schematic phase diagram of the $BaIrO_3$ polytypes showing the evolution of magnetic transition temperautre T_c (left) and the electronic specific-heat coeffcieint γ (right) as a function of the corner-to-face sharing octahedral C:F ratio. Adapted from Reference [21].

Structure–property evolutions in the $BaIrO_3$ polytypes As shown in Figure 9, the major change of the crystal structures for these polytypes can be described by the stacking sequence of IrO_6 octahedra that evolves from 9R(CFFCFFCFF) → 5H(CFCCF) → 6H(CCFCCF) → 3C(CCC), where C and F stands for corner- and face-sharing, respectively. Figure 11 displays the temperature dependence of magnetic susceptibility $\chi(T)$ and its inverse $\chi^{-1}(T)$ for the $BaIrO_3$ polytypes, illustrating the evolution of the magnetic ground state. Figure 12 shows a schematic phase diagram of the $BaIrO_3$ polytypes. With increasing C:F ratio in the sequence 9R(1:2) → 5H(3:2) → 6H(2:1) → 3C(∞), the ground states of $BaIrO_3$ evolve from a ferromagnetic insulator with $T_c = 180$ K in the 9R phase to a ferromagnetic metal with $T_c = 50$ K in the 5H phase, then to an exchange-enhanced paramagnetic metal with non-Fermi-liquid behaviour near a ferromagnetic quantum critical point in the 6H phase, and finally to a Fermi-liquid metal in the 3C phase. Such a structure–physical property evolution has been ascribed to a progressive bandwidth broadening in the sense that the corner-shared arrangement of IrO_6 octahedron can facilitate the Ir overlap integral mediated via O-2p orbitals. Since the electronic specific-heat coefficient γ is proportional to the density of states at the Fermi energy, the gradual enhancement of γ from 9R to 6H phase shown in Figure 12 is in agreement with the bandwidth argument. Both the 6H and 3C phases are metallic. Since the 6H phase is close to a ferromagnetic quantum critical point, the γ is much enhanced due to critical fluctuations relative to that in the 3C phase with an even broader bandwidth. From this systematic study on $BaIrO_3$ polytypes, we can understand why the 6H $SrIrO_3$ is a non-Fermi-liquid metal near a ferromagnetic quantum critical point [11]. Moreover, the study on $BaIrO_3$ demonstrated that HPHT synthesis of structurally closely related perovskite polytypes represents an effective approach to fine tune the physical properties of interest via modifying the octahedral arrangement.

5. Conclusions

We have summarized in this chapter the current knowledge on the materials' synthesis, crystal structure, and physical properties of the "113" alkaline-earth iridates $AIrO_3$ (A = Ca, Sr, Ba), which display a rich variety of crystallographic and electronic states that are of great current research interest. For $CaIrO_3$, it can form in either the layered pPv or the orthorhombic Pv structure, and thus serves as an important analogue of $MgSiO_3$ to investigate the Pv/pPv transformation in the Earth's lowermost mantle in geosciences. Corresponding to different crystal structures, their electronic ground states differ sharply: the pPv phase is an antiferromagnetic Mott insulator with $T_N = 110$ K while the Pv phase is a paramagnetic semimetal with possible Dirac nodes protected by the lattice symmetry. The presence of strong structural distortion in pPv $CaIrO_3$ makes it a model system to investigate the interplay of non-cubic crystal field and SOC in iridates. On the other hand, metallization of the pPv phase via electron or hole doping represents an important approach to realize the exotic electronic states on the verge of insulator–metal transition. For $SrIrO_3$, it crystallizes in the 6H polytype at ambient pressure and transforms to the orthorhombic Pv structure under high-pressure conditions. The 6H phase is an exchange enhanced paramagnetic metal with non-Fermi-liquid behaviour due to the proximity of ferromagnetic quantum critical point, while the Pv phase is revealed as an exotic narrow-band semimetal with symmetry-protected Dirac nodes within the $J_{eff} = 1/2$ band near the Fermi level. The presence of nontrivial features in the low energy electronic states makes these "113" orthorhombic Pv iridates $AIrO_3$ (A = Ca, Sr) promising candidates for realizing various topological and magnetic insulating phases via tuning the SOC, Hubbard interactions, and/or lattice symmetry. In this regard, epitaxial growth of superlattices and high-pressure synthesis of bulk materials with proper chemical design are currently important approaches to tailor the proposed quantum phases out of the semimetallic state. For $BaIrO_3$, it adopts a nine-layer 9R polytype at ambient pressure, and can be transformed to 5H, 6H, and 3C phases under different high pressure conditions. The 9R phase is a weak ferromagnetic insulator with $T_c = 180$ K, and can be regarded as an exotic SOC Mott insulator with multiple $J_{eff} = 1/2$ states associated with the unique Ir_3O_{12} structural units. An atomic-like nature of the

Ir moment driven by strong SOC is rather stable against external perturbations, but the weak ferromagnetism can be easily suppressed by applying chemical and physical pressures or dilute rare-earth substitutions for Ba^{2+}. In contrast, the nonmetallic ground state displays distinct response to the chemical and physical pressure, highlighting the delicate interplay of crystal structure and electronic degrees of freedom for this quasi-1D compound. With increasing the ratio of corner-to-face sharing octahedra in the sequence 9R(1:2) → 5H(3:2) → 6H(2:1) → 3C(∞), the ground states of $BaIrO_3$ evolve from a ferromagnetic insulator with T_c = 180 K in the 9R phase to a ferromagnetic metal with T_c = 50 K in the 5H phase, then to an exchange-enhanced paramagnetic metal with non-Fermi-liquid behaviour near a ferromagnetic quantum critical point in the 6H phase, and finally to a Fermi-liquid metal in the 3C phase. Such a structure–physical property evolution demonstrated that HPHT synthesis of structurally closely related perovskite polytypes represents an effective approach to fine tune the physical properties of interest via modifying the octahedral arrangement.

Acknowledgements

We are grateful to J.-Z. Zhou, J. B. Goodenough, José Alonso, Y. Uwatoko, and M. Akaogi for collaborations on work related to this review. This work is supported by the National Basic Research Program of China (Grant No. 2014CB921500), the National Science Foundation of China (Grant Nos. 11304371, 51402019), and the strategic Priority Research Program of the Chinese Academy of Sciences (Grant No. XDB07020100).

Author details

Yunqi Cai[1], Yan Li[2] and Jinguang Cheng[1*]

*Address all correspondence to: jgcheng@iphy.ac.cn

1 Beijing National Laboratory for Condensed Matter Physics and Institute of Physics, Chinese Academy of Sciences, Beijing, China

2 College of Materials Science and Engineering, Beijing Institute of Petrochemical Technology, Beijing, China

References

[1] Cao G, Bolivar J, McCall S, Crow JE, Guertin RP. Weak ferromagnetism, metal-to-nonmetal transition, and negative differential resistivity in single-crystal Sr_2IrO_4. Phys Rev B 1998;57:R11039.

[2] Kim BJ, Jin H, Moo SJ, Kim JY, Park BG, Leem CS, Yu J, Noh TW, Kim C, Oh SJ, Park JH, Durairaj V, Cao G, Rotenberg E. Novel J_{eff}=1/2 Mott state induced by relativistic spin-orbit coupling in Sr_2IrO_4. Phys Rev Lett 2008;101:076402.

[3] Kim BJ, Ohsumi H, Komesu T, Sakai F, Morita T, Takagi H, Arima T. Phase sensitive observation of a spin-orbital Mott state in Sr_2IrO_4. Science 2009;323:1329.

[4] Jackeli G, Khaliullin G. Mott insulators in the strong spin-orbit coupling limit: from Heisenberg to a quantum compass and Kitaev models. Phys Rev Lett 2009;102:017205.

[5] Witczak-Krempa W, Chen G, Kim YB, Balents L. Correlated quantum phenomena in the strong spin-orbit regime. Annu Rev Condens Matter Phys 2014;5:57–82.

[6] Cao G, Delong LE. Frontiers of 4d- and 5d-Transition Metal Oxides. World Scientific Publishing Co. Pte. Ltd.; 2013.

[7] Moon SJ, Jin H, Kim KW, Choi WS, Lee YS, Yu J, Cao G, Sumi A, Funakubo H, Bernhard C, Noh TW. Dimensionality-controlled insulator-metal transition and correlated metallic state in 5d transition metal oxides $Sr_{n+1}Ir_nO_{3n+1}$ (n = 1, 2, and infinite). Phys Rev Lett 2008;101:226402.

[8] Longo JM, Kafalas JA, Arnott RJ. Structure and properties of the high and low pressure forms of $SrIrO_3$. J Solid State Chem 1971;3:174–9.

[9] Kim YK, Sumi A, Takahashi K, Yokoyama S, Ito S, Watanabe T, Akiyama K, Kaneko S, Saito K, Funakubo H. Metalorganic chemical vapour deposition of epitaxial perovskite $SrIrO_3$ films on (100) $SrTiO_3$ substrates. J Appl Phys 2006;45:L36–8.

[10] Carter JM, Vijay Shankar V, Ahsan Zeb M, Kee HY. Semimetal and topological insulator in perovskite iridates. Phys Rev B 2012;85:115105.

[11] Cao G, Durairaj V, Chikara S, DeLong LE, Parkin S, Schlottmann P. Non-Fermi-liquid behaviour in nearly ferromagnetic $SrIrO_3$ single crystal. Phys Rev B 2007;76:100402(R).

[12] Sarkozy RF, Moeller CW, Chamberland BL. The characterization of calcium iridium oxides. J Solid State Chem 1974;9:242–6.

[13] Ohgushi K, Yagi T, Gotou H, Kiuchi Y, Ueda Y. Metallization of quasi-two-dimensional Mott insulator $CaIrO_3$ with S = 1/2 spins. Physica B 2009;404:3261–3.

[14] Ohgushi K, Yamaura J, Ohsumi H, Sugimoto K, Takeshita S, Tokuda A, Takagi H, Takata M, Arima T. Resonant X-ray diffraction study of the strongly spin-orbit-coupled Mott insulator $CaIrO_3$. Phys Rev Lett 2013;110:217212.

[15] Moretti Sala M, Ohgushi K, Al-Zein A, Hirata Y, Monaco G, Krisch M. $CaIrO_3$: a spin-orbit Mott insulator beyond the j_{eff}=1/2 ground state. Phys Rev Lett 2014;112:176402.

[16] Shim S-H. The postperovskite transition. Annu Rev Earth Planet Sci 2008;36:569.

[17] Lindsay R, Strange W, Chamberland BL, Moyer JR. Weak ferromagnetism in $BaIrO_3$. Solid State Comm 1993;86:759–63.

[18] Cao G, Crow JE, Guertin RP, Henning PF, Homes CC, Strongin M, Basov DN, Lochner E. Charge density wave formation accompanying ferromagnetic ordering in quasi-one-dimensional $BaIrO_3$. Solid State Comm 2000;113:657–62.

[19] Cheng JG, Alonso JA, Suard E, Zhou JS, Goodenough JB. A new perovskite polytype in the high-pressure sequence of $BaIrO_3$. J Am Chem Soc 2009;131:7461–9.

[20] Cheng JG, Zhou JS, Alonso JA, Goodenough JB, Sui Y, Matsubayashi K, Uwatoko Y. Transition from a weak ferromagnetic insulator to an exchange-enhanced paramagnetic metal in the $BaIrO_3$ polytypes. Phys Rev B 2009;80:104430.

[21] Cheng JG, Ishii T, Kojitani H, Matsubayashi K, Matsuo A, Li X, Shirako Y, Zhou JS, Goodenough JB, Jin CQ, Akaogi M, Uwatoko K. High-pressure synthesis of the $BaIrO_3$ perovskite: a Pauli paramagnetic metal with a Fermi liquid ground state. Phys Rev B 2013;88:205114.

[22] McDaniel CL, Schneider SJ. Phase relations in the CaO-IrO_2-Ir system in air. J Solid State Chem 1972;4:275–80.

[23] Rodi VF, Babel D. Erdalkaliiridium (IV) oxide: Kristallstruktru von $CaIrO_3$. Z. Anorg Allg Chem 1965;336:17–23.

[24] Oganov AR, Ono S. Theoretical and experimental evidence for a post-perovskite phase of $MgSiO_3$ in Earth's D'' layer. Nature 2004;430:445–8.

[25] Ohgushi K, Goto H, Yagi T, Kiuchi Y, Sakai F, Ueda Y. Metal-insulator transition in $Ca_{1-x}Na_xIrO_3$ with post-perovskite structure. Phys Rev B 2006;74:241104(R).

[26] Hirai D, Matsuno J, Takagi H. Semimetallic transport properties of epitaxially stabilized perovskite $CaIrO_3$ films. ArXiv:1501.01433. 2015.

[27] Hirai S, Sanehira T, Nishiyama N, Irifune T, Klemme S, Bromiley G, Attfield JP. Tuning of structure, morphology, and magnetism in postperovskite oxide solid solutions. Chem Mater 2011;23:114–21.

[28] Sugahara M, Yoshiasa A, Yoneda A, Hashimoto T, Sakai S, Okube M, Nakatsuka A, Ohtaka O. Single-crystal X-ray diffraction study of $CaIrO_3$. Am Miner 2008;93:1148–52.

[29] Hirai S, Welch MD, Aguado F, Redfern SAT. The crystal structure of $CaIrO_3$ post-perovskite revisited. Z Kristallogr 2009;224:345–50.

[30] Cheng JG, Zhou JS, Goodenough JB, Sui Y, Ren Y, Suchomel MR. High-pressure synthesis and physical properties of perovskite and post-perovskite $Ca_{1-x}Sr_xIrO_3$. Phys Rev B 2011;83:064401.

[31] Kojitani H, Furukawa A, Akaogi M. Thermochemistry and high-pressure equilibria of the post-perovskite phase transition in $CaIrO_3$. Am Miner 2007;92:229–32.

[32] Jang SY, Kim H, Moo SJ, Choi WS, Jeon BC, Yu J, Noh TW. The electronic structure of epitaxially stabilized 5d perovskite $Ca_{1-x}Sr_xIrO_3$ (x = 0, 0.5, and 1) thin films: the role of strong spin-orbit coupling. J Phys: Condens Matter 2010;22:485602.

[33] Subedi A. First-principles study of the electronic structure and magnetism of $CaIrO_3$. Phys Rev B 2012;85:020408(R).

[34] Bagdanov NA, Katukuri VM, Stoll H, Brink J, Hozoi L. Post-perovskite $CaIrO_3$: A j=1/2 quasi-one-dimensional antiferromagnet. Phys Rev B 2012;85:235147.

[35] Gunasekera J, Chen Y, Kremenak JW, Miceli PF, Singh DK. Mott insulator-to-metal transition in yttrium doped $CaIrO_3$. J Phys: Condens Matter 2015;27:052201.

[36] Kojitani H, Shirako Y, Akaogi M. Post-perovskite phase transition in $CaRuO_3$. Phys Earth Planet Inter 2007;165:127–34.

[37] Yamaura K, Shirako Y, Kojitani H, Arai M, Young DP, Akaogi M, Nakashima M, Katsumata T, Inaguma Y, Takayama-Muromachi E. Synthesis and magnetic and charge-transport properties of the correlated 4d post-perovskite $CaRhO_3$. J Am Chem Soc 2009;131:2722–6.

[38] Shirako Y, Satsukawa H, Wang XX, Li JJ, Guo YF, Arai M, Yamaura K, Yoshida M, Kojitani H, Katsumata T, Inaguma Y, Hiraki K, Takahashi T, Akaogi M. Integer spin-chain antiferromagnetism of the 4d oxide $CaRuO_3$ with post-perovskite structure. Phys Rev B 2011;83:174411.

[39] Zhong GH, Li YL, Liu Z, Zeng Z. Ground state properties of perovskite and post-perovskite $CaRuO_3$: ferromagnetism reduction. Solid State Sci 2010;12:2003–9.

[40] Keawpark N, Tu R, Goto T. Thermoelectricitiy of $CaIrO_3$ ceramics prepared by spark plasma sintering. J Cera Soc Jpn 2009;117(4):466–9.

[41] Ahsan Zeb M, Kee HY. Interplay between spin-orbit coupling and Hubbard interactions in $SrIrO_3$ and related Pbnn perovskite oxides. Phys Rev B 2012;86:085149.

[42] Nie YF, King PDC, Kim CH, Uchida M, Wei HI, Faeth BD, Ruf JP, Ruff JPC, Xie L, Pan X, Fennie CJ, Schlom DG, Shen KM. Interplay of spin-orbit interactions, dimensionality, and octahedral rotations in semimetallic $SrIrO_3$. Phys Rev Lett 2015;114:016401.

[43] Zhao JG, Yang LX, Yu Y, Li FY, Ru RC, Fang Z, Chen LC, Jin CQ. High-pressure synthesis of orthorhombic $SrIrO_3$ perovskite and its positive magnetoresistance. J Appl Phys 2008;103:103706.

[44] Blanchard PER, Reynolds E, Kennedy BJ, Kimpton JA, Avdeev M, Belik AA. Anomalous thermal expansion in orthorhombic perovskite $SrIrO_3$: interplay between spin-orbit coupling and the crystal lattice. Phys Rev B 2014;89:214106.

[45] Matsuno J, Ihara K, Yamamura S, Wadati H, Ishii K, Vijay Shankar V, Kee HY, Takagi H. Engineering spin-orbital magnetic insulator by tailoring superlattice. ArXiv: 1401.1066. 2014.

[46] Qasim I, Kennedy BJ, Avdeev M. Synthesis, structures and properties of transition metal doped $SrIrO_3$. J Mater Chem A 2013;1:3127–32.

[47] Bremholm M, Yim CK, Hirai D, Climent-Pascual E, Xu Q, Zandberen HW, Ali MN, Cava RJ. Destabilization of the 6H-$SrIrO_3$ polymorph through partial substitution of zinc and lithium. J Mater Chem 2012;22:16431–6.

[48] Wu FX, Zhou J, Zhang LY, Chen YB, Zhang ST, Gu ZB, Yao SH, Chen YF. Metal-insulator transition in $SrIrO_3$ with strong spin-orbit interaction. J Phys: Condens Matter 2013;25:125604.

[49] Chen Y, Lu YM, Kee HY. Topological crystalline metal in orthorhombic perovskite iridates. Nat Commun 2015;6:6593.

[50] Cheng JG, Zhou JS, Goodenough JB, Matsubayashi K, Uwatoko Y. Weak ferromagnetism in the orthorhombic perovskite $SrIr_{0.8}Sn_{0.2}O_3$. JPS Conf Proc 2014;3:013014.

[51] Nakano T, Terasaki I. Giant nonlinear conduction and Thyristor-like negative differential resistance in $BaIrO_3$ single crystal. Phys Rev B 2006;73:195106.

[52] Laguna-Marco MA, Haskel D, Souza-Neto N, Lang JC, Krishnamurthy VV, Chikara S, Cao G, Van Veenendaal M. Orbital magnetism and spin-orbit effects in the electronic structure of $BaIrO_3$. Phys Rev Lett 2010;105:216407.

[53] Ju WW, Liu GQ, Yang ZQ. Exotic spin-orbital Mott insulating states in $BaIrO_3$. Phys Rev B 2013;87:075112.

[54] Powell AV, Battle PD. A time-of-flight powder neutron diffraction study of non-stoichiometry in barium iridates $BaIrO_{3-\delta}$. J Alloy Comp 1996;232:147.

[55] Zhao JG, Yang LX, Yu Y, Li FY, Yu RC, Jin CQ. Structural and physical properties of the 6M $BaIrO_3$: a new metallic iridate synthesized under high pressure. Inorg Chem 2009;48:4290–4.

[56] Zhao JG, Yang LX, Yu Y, Li FY, Yu RC, Jin CQ. Physical properties of the 5M $BaIrO_3$: a new weak ferromagnetic iridate synthesized under high pressure. Solid State Commun 2010;150:36–9.

[57] Goodenough JB, Kafalas JA, Longo JM. High-Pressure Synthesis: Preparative Methods in Solid State Chemistry. New York and London: Academic Press, Inc.; 1972.

[58] Kini NS, Bentien A, Ramakrishnan S, Geibel C. Specific heat and transport study of the co-existence of chage-density-wave and weak ferromagnetism in $BaIrO_3$. Physica B 2005;359–361:1264.

[59] Maiti K, Singh RS, Medicherla VRR, Rayaprol S, Sampathkumaran EV. Origin of charge density wave formation in insulators from high resolution photoemission study of $BaIrO_3$. Phys Rev Lett 2005;95:016404.

[60] Brooks ML, Blundell SJ, Lancaster T, Hayes W, Pratt FL, Frampton PPC, Battle PD. Unconventional magnetic properties of the weakly ferromagnetic metal $BaIrO_3$. Phys Rev B 2005;71:220411(R).

[61] Laguna-Marco MA, Fabbris G, Souza-Neto N, Chikara S, Schilling JS, Cao G, Haskel D. Different response of transport and magnetic properties of $BaIrO_3$ to chemical and physical pressure. Phys Rev B 2014;90:014419.

[62] Cao G, Lin XN, Chikara S, Durairaj V, Elhami E. High-temperature weak ferromagnetism on the verge of a metallic state: impact of dilute Sr doping on $BaIrO_3$. Phys Rev B 2004;69:174418.

[63] Korneta OB, Chikara S, Parkin S, DeLong LE, Schlottmann P, Cao G. Pressure-induced insulating state in $Ba_{1-x}R_xIrO_3$ (R=Gd, Eu) single crystals. Phys Rev B 2010;81:045101.

Section 2

Perovskite Materials: Properties

Chapter 7

Metal–Insulator Transitions and Non-Fermi Liquid Behaviors in 5d Perovskite Iridates

Abhijit Biswas, Ki-Seok Kim and Yoon Hee Jeong

Additional information is available at the end of the chapter

http://dx.doi.org/10.5772/61285

Abstract

Transition metal oxides, in particular, 3d or 4d perovskites, have provided diverse emergent physics that originates from the coupling of various degrees of freedom such as spin, lattice, charge, orbital, and also disorder. 5d perovskites form a distinct class because they have strong spin-orbit coupling that introduces to the system an additional energy scale that is comparable to bandwidth and Coulomb correlation. Consequent new physics includes novel J_{eff} = 1/2 Mott insulators, metal–insulator transitions, spin liquids, and topological insulators. After highlighting some of the phenomena appearing in the Ruddlesden–Popper iridate series $Sr_{n+1}Ir_nO_{3n+1}$ (n = 1, 2, and ∞), we focus on the transport properties of perovskite $SrIrO_3$. Using epitaxial thin films on various substrates, we demonstrate that metal–insulator transitions can be induced in perovskite $SrIrO_3$ by reducing its thickness or by imposing compressive strain. The metal–insulator transition driven by thickness reduction is due to disorder, but the metal–insulator transition driven by compressive strain is accompanied by peculiar non-Fermi liquid behaviors, possibly due to the delicate interplay between correlation, disorder, and spin-orbit coupling. We examine various theoretical frameworks to understand the non-Fermi liquid physics and metal–insulator transition that occurs in $SrIrO_3$ and offer the Mott–Anderson–Griffiths scenario as a possible solution.

Keywords: Transition metal oxides, iridates, metal–insulator transitions, non-Fermi liquid behaviors, localization

1. Introduction

Transition metal oxides (TMOs), ranging from simple binary oxides to more complex ternary or quaternary compounds, have been a subject of intense activities in condensed matter physics and materials science [1–5]. Of various transition metal oxides, perovskites and their variations are a particularly important class because they display a rich spectrum of various competing phases and physical properties. Structurally, in perovskites with the chemical formula ABO_3, large rare-earth or alkali-metal cations occupy the A-sites (tetrahedral oxygen interstices) and transition metal cations occupy the B-sites (octahedral oxygen interstices). Each transition metal ion is thus surrounded by six oxygen anions to form a BO_6 octahedron, and the octahedra are connected to each other three-dimensionally. This metal–oxygen network transition mostly determines the electronic properties, while the close-packed A cations and oxygen anions provide structural stability. One characteristic feature of the perovskite family is that numerous transition metal ions of different size and valence are allowed in the same structure. Along with this elemental diversity, mutually interacting quantum degrees of freedom such as lattice, spin, charge, and orbital lead to numerous emergent physical properties in perovskites. For example, the conducting properties of perovskites range from insulating to semiconducting to metallic and finally to superconducting. The electrical properties of perovskites are equally diverse; they also display various magnetic and multifunctional properties [6–14]. These unparalleled varieties in the electronic properties of perovskites hold tremendous application potential and thus provide a continuous impetus for research in the field.

The network of BO_6 octahedra in transition metal perovskites strongly influences their electronic properties because *d* orbitals form bands with characteristic effective masses and different strengths of on-site Coulomb interactions. Indeed, diverse phenomena have been treated in terms of various relevant energy scales such as bandwidth (W) and Coulomb repulsion (U), and the relevant efforts can be termed *d* orbital physics [15,16]. Disorder (D), which is unavoidable in real solids also, has an important influence on the physics of transition metal oxides.

In materials science, numerous studies of 3*d*- and 4*d*-based transition metal perovskites have been conducted to find ways to exploit their functionalities that depend sensitively on structural distortion and crystal chemistry. However, perovskite studies have been limited mainly to those that contain 3*d* or 4*d* transition metal elements. In this regard, 5*d* perovskites form a special class because the strong spin-orbit coupling (SOC, Λ) introduces to the system an additional energy scale that is comparable to bandwidth and Coulomb correlation [17]. For example, metal oxide compounds containing 5*d* iridium (e.g., Sr_2IrO_4, $Sr_3Ir_2O_7$, $SrIrO_3$, $BaIrO_3$, Na_2IrO_3, $Eu_2Ir_2O_7$, $Na_4Ir_3O_8$, and Sr_2GdIrO_6) display a variety of emerging properties (Table 1). Of the various exotic properties of 5*d* perovskites, the transport properties and, in particular, the metal–insulator transitions (MIT) in iridate perovskites are the subject of the present chapter.

Simple extrapolation of the characteristics of 3*d* or 4*d* systems fails to predict the characteristics of 5*d* systems. Moving down the periodic table from 3*d* → 4*d* → 5*d*, the orbitals in the solids that contain the corresponding *d* orbitals become increasingly extended and so does the bandwidth ($W_{3d} < W_{4d} < W_{5d}$). Along with an increase in the bandwidth, the associated on-site Coulomb repulsion would decrease correspondingly ($U_{3d} > U_{4d} > U_{5d}$). From these considerations, one may expect higher metallicity and less magnetic instability for the materials with more extended 5*d* orbitals, compared with the systems containing 3*d* or 4*d* elements, because the Stoner criterion in these compounds favors paramagnetic metallic states. Surprisingly, however, the properties of the 5*d*-based transition metal oxides exhibit extremely rich behaviors (Table 1). The Ruddlesden–Popper (RP) series of iridium-based $Sr_{n+1}Ir_nO_{3n+1}$ (n = 1, 2, and ∞) as well as other structural families such as pyrochlores, kagome-type lattices, honeycomb-type structures, and double perovskites are mostly insulating (few are metallic) but may have exotic physical properties, including novel Mott insulator, lattice-driven magnetoresistance, giant magnetocaloric effect, quantum criticality, charge density wave, geometrically frustrated magnetism, possible topological insulator, and Weyl semimetal [17–19]. These diverse behaviors in iridates require that a new area of physics must be considered; this is SOC ($\Lambda_{3d} < \Lambda_{4d} < \Lambda_{5d}$). In iridates, the energy scales of W, U, and Λ are all comparable and thus compete with each other. In addition, D is also important, and the emergent properties are the results of strong interplay among Coulomb interaction, SOC, and disorder.

Structure	Compound	Properties
Layered	Sr_2IrO_4 $Sr_3Ir_2O_7$ Ba_2IrO_4	J_{eff} = 1/2 Mott insulator, complex magnetism, lattice magnetoresistance
Hexagonal	$BaIrO_3$ $Ca_5Ir_3O_{12}$ Sr_3NiIrO_6	J_{eff} = 1/2 Mott insulator, non-Fermi liquid, magnetism, geometric frustration
Pyrochlore	$Y_2Ir_2O_7$ $Bi_2Ir_2O_7$ $Eu_2Ir_2O_7$	Topological insulator, strong magnetic instability, metal-insulator transition
Honeycomb	Na_2IrO_3 Li_2IrO_3	Topological insulator, zig-zag magnetic order
Orthorhombic	$SrIrO_3$ $CaIrO_3$	Metal-insulator transition, non-Fermi liquid, semi-metal
Double perovskite	Sr_2YIrO_6 Sr_2GdIrO_6 La_2MgIrO_6	Correlated insulator, exotic magnetism, spin wave excitations
Kagome	$Na_4Ir_3O_8$	Spin liquid

Table 1. 5d transition metal iridium (Ir)-based oxides, which are mostly insulators with exotic magnetic states.

Electrical conductivity is one of the properties that used to characterize or to classify solids, so a transition from a metal to insulator or vice versa as a function of a control parameter such as temperature, pressure, or magnetic field has been a topic of interest to the condensed matter physics community for several decades [20–24]. However, despite enormous efforts, understanding of the MIT at the microscopic level is still under debate. In transition metal oxides and strongly correlated systems, the MIT is often accompanied by a change in structural or magnetic symmetry. In contrast, in the presence of a sufficient amount of disorder, MIT is often not associated with any uniform ordering or change in symmetry. Perhaps the most important mechanisms that underlay MITs in correlated systems are interaction-driven or correlation-driven Mott localization, magnetic-order-driven Slater insulator, and disorder-driven Anderson localization. In perovskites that involve heavy 5*d* transition metal elements (e.g., Re, Os, Ir), SOC comes as another parameter and becomes comparable in strength to other relevant energy scales. All of these different parameters in the system compete with each other so their interplay stabilizes new exotic ground states. In this chapter, our foremost aim is to provide a brief description of current research on MITs that occur in Ir-based perovskites. Our emphasis will be on a model system ($SrIrO_3$) to provide experimental evidence that the interplay among correlation, SOC, and disorder is important in achieving different MITs. We also wish to present our understanding about the observed non-Fermi liquid physics and MITs to provide insight and motivation for further activities in this rapidly developing, yet poorly understood, field of strong spin-orbit-coupled 5*d*-based oxide physics.

2. Metal–insulator transitions and representative types of insulators

2.1. Metal–insulator transitions

Electrical resistance R (or resistivity ρ if intrinsic quantity is used) is a key property that characterizes materials, which are typically classified as metals or insulators. A very good metal (e.g., Cu) can have an electrical resistivity as small as $\rho \sim 10^{-10}$ Ω cm, whereas a good electrical insulator (e.g., quartz) has a resistivity as high as $\rho \sim 10^{20}$ Ω cm. The causes of this huge difference in resistivity are now well understood in modern solid-state physics. The difference between the resistivities of metals and those of insulators implies that MITs may be accompanied by a large change in ρ (up to several orders of magnitude). However, many MITs of current interest are often accompanied not by a large change in magnitude but by a qualitative change in conducting behaviors.

Metals can be distinguished from insulators by their different responses of R to temperature T. Metals are defined as materials in which R decreases as T decreases ($dR/dT > 0$), whereas insulators are materials in which R increases as T decreases ($dR/dT < 0$) (Figure 1). Strictly, the ultimate distinction between the metal and the insulator can be made at absolute zero; a metal would continue to be conductive, whereas an insulator would lose its conductivity. MITs of fundamental interest change a system from a phase with $dR/dT < 0$ to another with $dR/dT > 0$, or vice versa.

The conventional theory of electrical transport was first formulated by Drude [25], immediately after the discovery of the electron. The semiclassical Drude theory of electronic conductivity was built on the idea of the kinetic theory of gases, considering a metal as a gas of electrons. The key concepts in that description are the mean free path ℓ (i.e., the average length that an electron travels between successive collisions) and relaxation time τ (i.e., the average time between successive collisions). According to the Drude theory, the electronic conductivity σ ($= 1/\rho$) is directly proportional to τ as $\sigma = \frac{ne^2\tau}{m}$, where n is the density of electrons, e is the electron charge, and m is electron mass. For good metals, $\ell \sim 100$ nm and $\tau \sim 10^{-14}$ s. This semiclassical Drude model is still used today as a quick way to estimate a material's property.

http://dx.doi.org/10.5772/61285

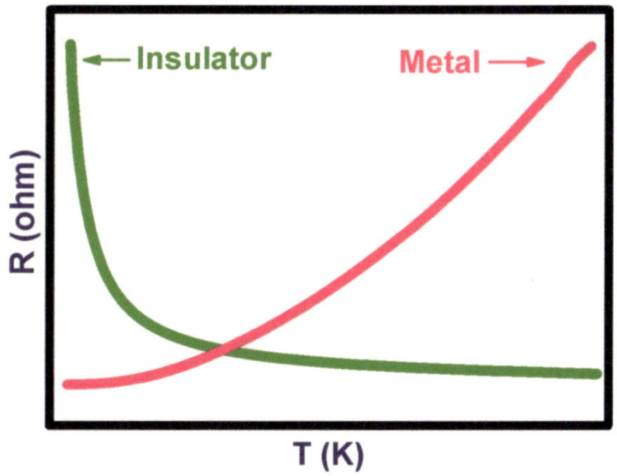

Figure 1. Metals are broadly defined as materials with d*R*/d*T* > 0, whereas insulators are as ones with d*R*/d*T* < 0. *R*: resistance; and *T*: temperature. At absolute zero, the resistance of an insulator would reach infinity with zero conductivity while a metal would still possess finite conductivity.

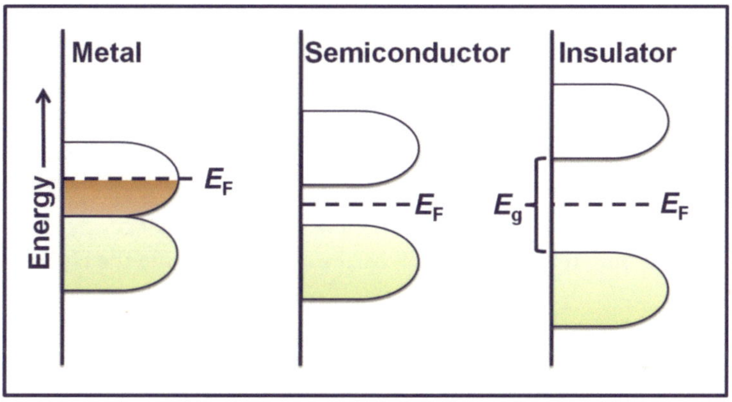

Figure 2. Schematic band diagram of metal, semiconductor, and insulator. E_F, and E_g are the Fermi energy and band gap. A semiconductor is an insulator with a small energy gap. Upper one: conduction band (CB); lower one: valence band (VB).

Quantum mechanics has been used to clarify the transport properties of solids, and now metals, semiconductors, and insulators are classified according to the band theory of solids (Figure 2). In metals, Fermi energy E_F is within the conduction band, whereas in insulators, E_F separates the electronic band into an empty upper and filled lower band. For example, quartz is an insulator with a band gap E_g ~8.9 eV, whereas Cu is a metal. Si is also an insulator but has a small E_g ~1.1 eV and thus is called a semiconductor. Although band theory successfully describes the conducting nature of many materials, it fails to account for the behaviors of many transition metal oxides that have relatively narrow separation between conduction and valence band. Many materials with partially filled *d* orbitals and an odd number of electrons per lattice site, which should be metals according to band theory, are actually insulators (e.g., NiO,

V_2O_3, and Fe_3O_4). The band theory considers single electron and thus may not adequately consider many-body effects such as electron–electron interaction that occur in these complex materials. In addition, *D*, which is unavoidable in real solids, must also be considered. Many theoretical explanations have been proposed to account for the metallic and insulating properties of so-called correlated materials; we briefly summarize these ideas in this section.

2.2. Mott insulator

In 1937, de Boer and Verwey found that many metal–oxides (e.g., NiO, MnO, and FeO) show insulating features despite the fact that these oxides have partially filled *d*-bands [26]. Soon after their discovery, Peierls suggested that strong Coulomb repulsion between electrons could be the origin of the insulating behaviors [27]. He remarked that "it is quite possible that the electrostatic interaction between the electrons prevents them from moving at all. At low temperatures the majority of the electrons are in their proper places in the ions. The minority which have happened to cross the potential barrier find therefore all the other atoms occupied, and in order to get through the lattice have to spend a long time in ions already occupied by other electrons. This needs a considerable addition of energy and so is extremely improbable at low temperatures" [20]. Peierl's speculation aroused interest in so-called strongly correlated systems. In 1949, Mott offered a theoretical explanation of how electron–electron correlation could yield an insulating state, now called a Mott Insulator [21].

Consider a lattice model with a single electron orbital at each site. If the electron–electron interactions are not considered, a single band would be formed from the overlap of the atomic orbitals: when two electrons, one with spin up (↑) and the other with spin down (↓), occupy each site, the band becomes full. However, when two electrons occupy the same site, they would feel a large Coulomb repulsion, which can be explained using the Hubbard Hamiltonian (Eq. (1)):

$$H = -t\sum_{ij\sigma}\left(c_{i\sigma}^{\dagger}c_{j\sigma} + h.c.\right) + U\sum_{i}n_{i\uparrow}n_{i\downarrow} - \mu\sum_{i,\sigma}n_{i\sigma} \tag{1}$$

where $c_{i,\sigma}^{\dagger}$ and $c_{j,\sigma}$ ($\sigma = \uparrow or \downarrow$) are the creation and the annihilation operators, respectively, for electrons on the site *i* and *j* with spin σ. Electron hopping between nearest neighbor sites occurs with hopping constant $-t$. *U* is the amount of energy for each pair of electrons that occupy the same lattice site and represents on-site Coulomb correlation. $n_{i\sigma}$ is the number operator and μ is the chemical potential.

The ratio between *t* and *U* determines whether the system is a metal or an insulator. When electron–electron correlation is negligible compared to hopping, $t/U >> 1$, the electrons tunnel between the sites without hindrance and the system is metallic. When $t/U << 1$, electron–electron correlation is strong, the electrons are localized due to Coulomb interaction, and the system becomes an insulator. In the Hubbard model, at half filling $\langle n_{i\sigma} \rangle = \frac{1}{2}$, the Mott–Hubbard insulating phase with one electron per site would appear.

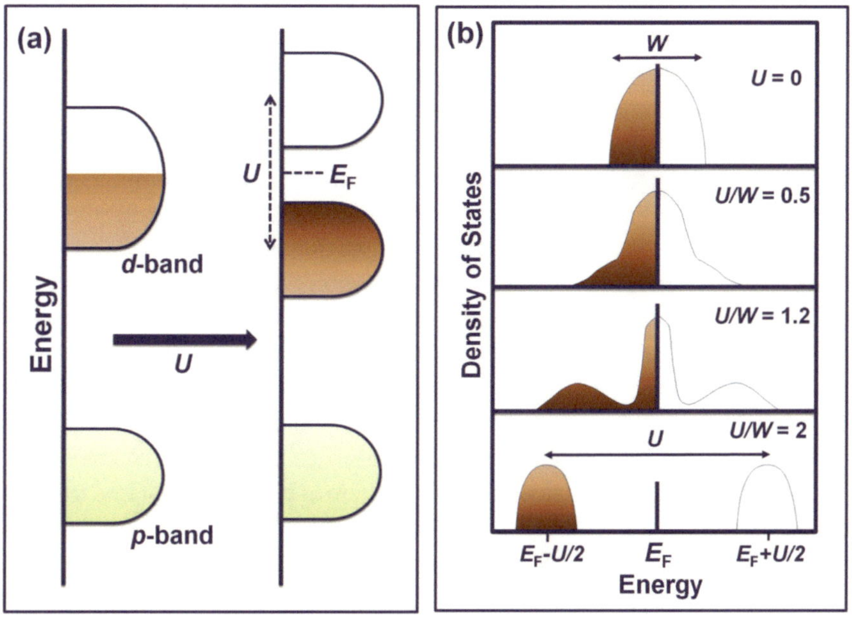

Figure 3. (a) Schematic representation of the energy levels for a Mott–Hubbard insulator where on-site Coulomb interaction *U* splits the *d*-band into lower Hubbard band and upper Hubbard band. (b) Evolution of the density of states (DOS) of electrons as a function of *U/W* (*W* = bandwidth) as the system evolves from a metal to an insulator. Adapted with permission from Kotliar and Vollhardt [28]. © 2009, AIP Publishing LLC

In transition metal oxides, often oxygen *p*-bands remain unchanged when interaction (*U*) occurs (Figure 3(a)). In contrast, the *d*-band splits into two subbands (upper Hubbard band and lower Hubbard band), and the half-filled state becomes an insulator with the opening of a charge gap (Figure 3(b)). The Hubbard model allows two types of MIT: the filling-control metal–insulator transition (FC-MIT), which originates from variation of electron concentration or chemical potential (μ/U); and a bandwidth-control metal–insulator transition (BC-MIT), which originates from variation of hopping energy or bandwidth (t/U).

2.3. Slater insulator

Mott insulators in the previous section are electron systems with an odd number of electrons per unit cell; despite the fact that although band theory suggests that these systems would be metallic, they are insulators. Mott insulators belong to the $U >> t$ regime in the Hubbard model. However, another aspect must be considered to explain the insulating property even in the Hubbard model. In fact, many insulating systems, especially transition metal oxides (e.g., Na_2IrO_3 [19]), often have antiferromagnetic ground states. Slater focused on this point. He proposed that formation of spin density waves (long-range magnetic order) due to electron–electron interaction may be an origin of the insulating phase itself [29]. The difference between a Mott insulator and a Slater insulator is that for a Mott insulator, the system remains insulating even above Néel temperature T_N, whereas a Slater insulator the system should be metallic above T_N.

The principle behind a magnetic-order-driven Slater insulator may be easily understood from the band picture. Slater insulators belong to the $t >> U$ regime, which is opposite to the case for Mott insulators. Suppose for simplicity that the lattice is half filled and thus on average each site holds one electron. A low-energy state for a half-filled system ensues if a periodicity doubling of the lattice occurs (i.e., the Brillouin zone is halved). At the new Brillouin zone boundary, an energy gap for charge excitations occurs and the system becomes insulating. According to Slater, such an insulating behavior is closely connected with the appearance of magnetic order at T_N, for example, electrons in a bipartite lattice [30], i.e., one that consists of two interpenetrate sublattices A and B in such a way that the nearest neighbors of any site are members of the opposite sublattice, for example, a rock salt arrangement in a simple cubic lattice. Thus, the nearest neighbors of an electron from A are those from B, and vice versa. Because ↑ and ↓ electrons mutually repel due to Coulomb interaction, they become preferentially arranged on alternating A and B sublattice sites. Hence, spins form a spin-density wave (SDW) whose wave vector is commensurate with the lattice. Because the electrons tend to avoid each other, the potential energy increases, and this gain is balanced by a loss in kinetic energy because of localization of electrons. The spin-density wave provides a necessary periodicity-doubling potential. With an increase in temperature, thermal fluctuations affect the ordering and narrow the energy gap. Eventually, at T_N, which is typically ~10^2 K, the ordering is destroyed and the insulating property disappears.

2.4. Anderson insulator

Defects and impurities are unavoidable in real materials, and disorder can never be neglected in reality. Thus, disorder-induced MIT is a subject of continuing interest in condensed matter physics. In 1958, Anderson initiated the field of so-called localization by arguing that sufficiently strong randomness will localize all the electronic states within a given band and that diffusion may be completely suppressed, thereby leading to an MIT at $T = 0$ K [31]. In the absence of disorder, even a small amount of hopping can delocalize the electrons. However, in the presence of sufficient disorder, the hopping process only spreads an initially localized state over a finite distance; defined as the localization length ξ_0. As a result, the states at the band tail become localized and sometimes even a whole band becomes localized. Mott argued that there must exist a critical energy E_C called mobility edge, which separates the localized states from the extended ones (Figure 4(a)) [21]. If $E_F < E_C$, the system is an insulator; otherwise, it is a metal.

The basic idea behind the localization phenomenon is rather straightforward. Suppose an electron propagates in a disordered medium from point A to point B. To obtain the probability that an electron reaches point B, the probabilities of all possible paths from A to B must be summed. In a disordered medium, the phases of the interference terms may be so random that on average they mutually cancel; the resultant state can be explained by the diffusion model of conductivity, but this simple argument may not be always applicable. Imagine a wave that travels from point A along a random path to point B and then goes back to A. Two possible paths are illustrated (Figure 4(b)): a randomly chosen path and the same path traversed in time reversed sense. The two paths interfere constructively and should be treated coherently as long as time reversal symmetry is preserved. Then the probability of the electron's return to A is twice as large as it would have been if probabilities were added (first squared and then added). The enhanced back-scattering, known as weak localization, reduces the conductance between A and B by increasing the electron's likelihood of returning to its starting point, and eventually leads to localization [32]. At high temperatures, coherence is lost due to thermal vibrations and back-scattering effects diminish. However, at low temperatures, thermal vibrations and inelastic scattering cease and other channels through which electrons can exchange energy become interrupted, so quantum interference immobilizes electrons to induce localization transitions. In 1979, Abrahams et al. developed the phenomenological scaling theory of localization [33, 34]. In metallic systems in two dimensions, resistivity often increases at low temperatures when T is decreased. This σ_{2D} ln $\propto T$ variation is due to weak localization and was first experimentally demonstrated by Dolan and Osheroff [35].

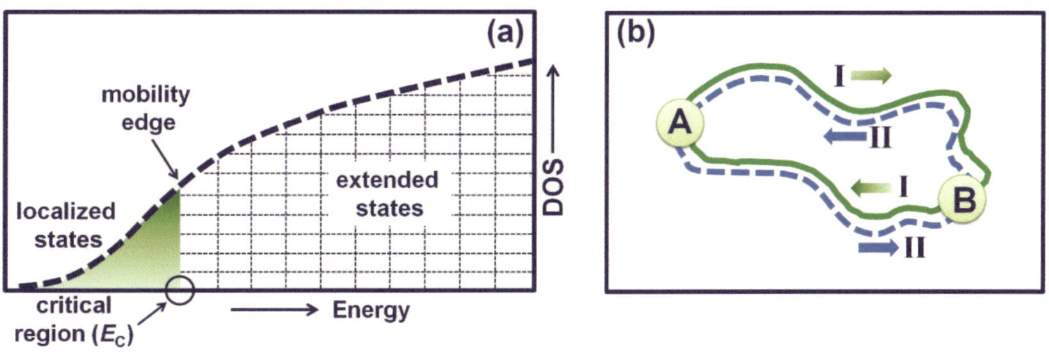

Figure 4. (a) Concept of the mobility edge E_C; electronic states > E_C are extended and those states < E_C are localized. If the Fermi energy is in the localized state, then the system is insulating; otherwise, it is metallic. DOS: density of states. (b) Propagation of electronic waves in a disordered medium. If two waves follow the same path from A to B and then to A, one in a clockwise direction and the other in a time-reversed direction, then they interfere constructively on returning to A.

In 1960, Ioffe and Regel realized that the electron mean free path ℓ can never be shorter than the lattice spacing a, because at $\ell \leq a$, coherent quasi-particle motion would vanish and the system would be an insulator [36]. In 1972, Mott proposed a similar argument and formed a criterion $\ell_{min} \approx a$ for minimum metallic conductivity [37]. This criterion was named the Mott–Ioffe–Regel (MIR) limit for resistivity saturation [38, 39]. Unfortunately, the MIR limit is not universally observed and disagrees with the now-widely accepted scaling theory of localization. Nevertheless, the MIR limit is still quoted frequently and is now generalized to more complex media; the criteria ranging from $k_F\ell_{min} \approx 1$ through $\ell_{min} \approx a$ to $k_F\ell_{min} \approx 2\pi$, where the wave vector $k_F = 2\pi/\lambda$. In some materials MIT occurs at the MIR limit. For two-dimensional materials, $k_F\ell$ is related to the ratio of quantum resistance h/e^2 to sheet resistance R_{sheet} as $k_F l = \frac{h\,\ell e^2}{R_{sheet}} \approx \frac{26\ \mathrm{k\Omega}\ell\square}{R_{sheet}}$ [40].

3. Recent results on 5*d* perovskite iridates

5*d* perovskites, in particular iridates, show many new interesting phenomena; examples include novel insulating states, exotic magnetism, spin-liquid behaviors, Weyl semimetals, and topological insulators. In this regard, RP series $Sr_{n+1}Ir_nO_{3n+1}$ have been widely investigated and are probably the most widely studied iridate system. In this section, we review some of the recent results on 5*d* perovskite iridates.

3.1. Perovskites

The term *perovskite* was named for the discovery of $CaTiO_3$ in honor of Count Lev Alexander Von Perovski, a Russian mineralogist. An ideal perovskite structure has an ABX_3 stoichiometry. Although most of the perovskite are oxides (X = Oxygen), other forms like fluorides, halides, and sulfides also exist in literature. Perovskite ABO_3 and, more generally, perovskite families including its variations accept a large number of transition metal elements of various size and valence into the B-sites (e.g., $A^{1+}B^{5+}O_3^{2-}$, $A^{2+}B^{4+}O_3^{2-}$, and $A^{3+}B^{3+}O_3^{2-}$), so the variety of transition metal oxides is unlimited: examples include cuprates, manganites, ruthenates, nickelates, titanates, and recently iridates [41, 42]. Along with this elemental diversity, mutually interacting quantum degrees of freedom such as lattice, spin, charge, and orbital leads to numerous emergent physical properties in perovskite families [1–14]. The perovskite families of current interest, iridates, can be classified as follows:

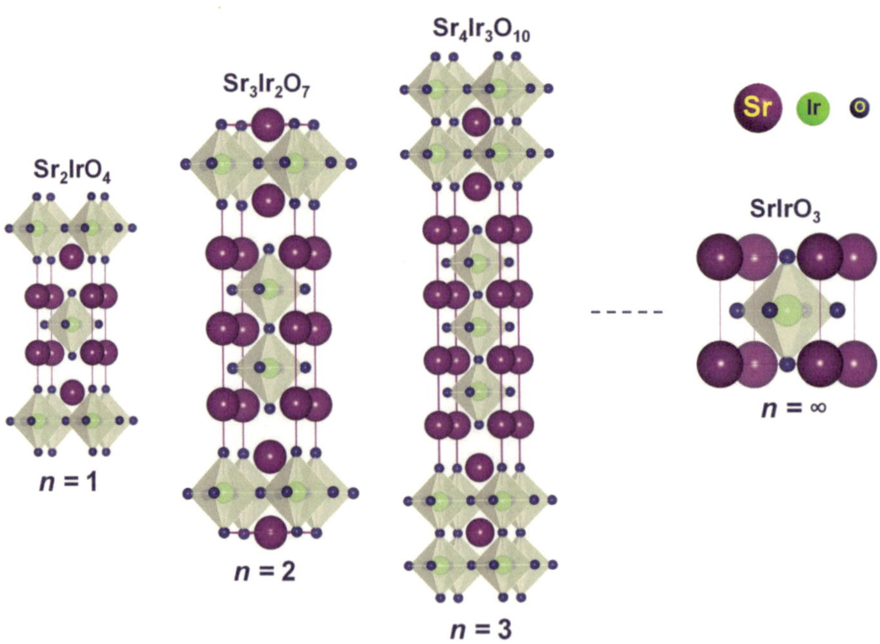

Figure 5. The $n = 1$ (Sr_2IrO_4), $n = 2$ ($Sr_3Ir_2O_7$), $n = 3$ ($Sr_4Ir_3O_{10}$), and $n = \infty$ ($SrIrO_3$) members of the homologous Ruddlesden–Popper series $Sr_{n+1}Ir_nO_{3n+1}$. Structures were drawn using VESTA software.

http://dx.doi.org/10.5772/61285

a. Perovskites

General formula of ABO_3 where A or B are metal cations and O as anions. B cations are surrounded by six oxygen anions forming BO_6 octahedra (e.g., $SrIrO_3$) (Figure 5).

b. Double perovskites

General formula of $A_2BB'O_6$ (ABO_3+$AB'O_3$), where the BO_6 and $B'O_6$ octahedra are alternatively arranged in two sublattices. For double perovskites, alkaline or rare earth ions occupy the A-sites, while the B-sites represent transition metal ions (e.g., Sr_2RuIrO_6 constitutes alternative unit cells of $SrRuO_3$ and $SrIrO_3$).

c. Layered perovskites

Layered perovskites consist of stacked two-dimensional slabs of the ABO_3 layer. Three subcategories have been recognized.

i. Ruddlesden–Popper series

The general formula of Ruddlesden–Popper (RP) series is $A_{n+1}B_nO_{3n+1}$, where n represents the number of octahedral layers in the repeating unit and can be visualized as repeated stacking of $AO(ABO_3)_n$ [43, 44]. n = 1 corresponds to one BO_6 octahedron (Sr_2IrO_4), and n = 2 to two BO_6 octahedra ($Sr_3Ir_2O_7$) (Figure 5). The differentiating characteristics for the layered perovskites are the motif (Sr^{2+}) that separates the layers, and the offsetting of the layers from each other. As n increases from $n = 1$ to $n = \infty$, the dimensionality of the compounds changes from two to three.

ii. Aurivillius phase

The general formula of the Aurivillius phase is $(Bi_2O_2)(A_{n+1}B_nO_{3n+1})$ [45]. This structure consists of layers of Bi_2O_2 separated by n layers of perovskites.

iii. Dion–Jacobson phases

The general formula of the Dion–Jacobson phase is $M(A_{n+1}B_nO_{3n+1})$, where M is a cation with valence +1, usually an alkali metal [46, 47].

Ideal perovskite ABO_3 has symmetric cubic structure with a lattice constant of ~4 Å. However, most perovskites, deviate from the cubic structure if the Goldschmidt tolerance factor t_f, given by $t_f = \frac{r_A + r_O}{\sqrt{2}(r_B + r_O)}$ (r_A, r_B, and r_O represents the ionic radii of ions A, B, and O, respectively), deviates from 1 [48]. When $t_f < 1$, (i.e., radius of cation A is small), the O anions move toward the A cation, so BO_6 octahedra tilt to shrink the available volume for A cations that is empty. Because B-O-B bonds are highly flexible and BO_6 octahedra are flexible in shape and size depending upon the cationic size, valence, and position, the overall deformation reduces the cubic symmetry, and result in structural transitions to orthorhombic, tetragonal, or hexagonal states that have lower symmetry than the cubic state.

3.2. *d* orbitals

The critical factor in the orbital physics involving *d* orbitals is their anisotropic charge distributions that arise from the wave functions, which take different shapes depending on energy when electrons are bound to atomic nuclei by Coulomb force. Electronic properties of perovskites would be severely affected by the chemistry of the transition metals at the center of BO_6 octahedra that have corner-sharing oxygen anions [15, 16]. Consider a transition metal element which is surrounded by six O^{2-} in the BO_6 octahedron. This configuration gives rise to a crystal field which hinders the free motion of *d*-electrons; consequently, orbital angular momentum is usually quenched, and the *d* orbitals due to the crystal field energy Δ_{oct} split in energy into (1) $d_{x2\text{-}y2}$ and d_{z2} states, which form twofold degenerate e_g orbitals with higher energy, and (2) threefold degenerate t_{2g} orbitals d_{xy}, d_{yz}, and d_{zx} states at lower energy (Figure 6). As the degeneracy of spherical symmetry of an isolated atom is removed, the *d* orbitals begin to fill, starting from a low-energy state and continuing to higher energy states. The actual filling arrangement depends on the competition between the crystal field and on-site exchange interaction described by Hund's rule [49]. For example, for 5*d* perovskite $SrIrO_3$, Ir^{4+} has five electrons in the *d* orbitals ($5d^5$); the electrons distributions from basic viewpoint are illustrated (Figure 7). For transport properties, the band structure would depend sensitively on an overlap between the *d* orbital of the B-site transition metal element and the *p* orbitals of the surrounding oxygen's.

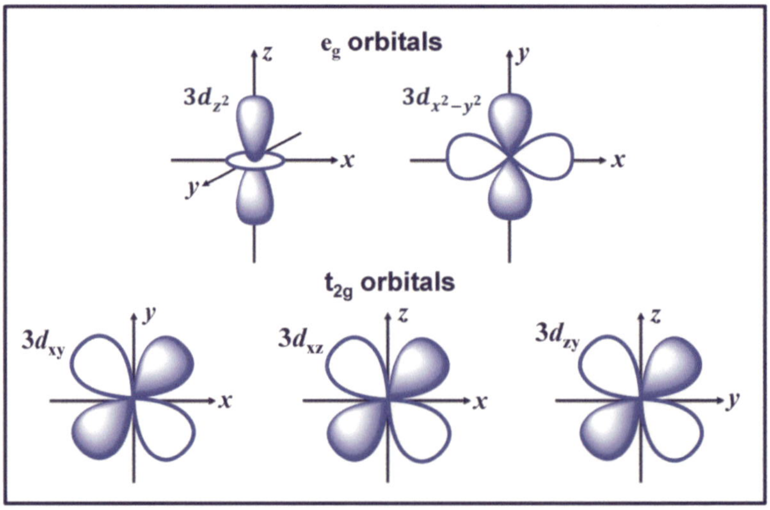

Figure 6. Five *d* orbitals in a cubic crystal field which split into two e_g orbitals and three t_{2g} orbitals.

Generally, in transition metal oxides, the electronic properties are further complicated by the interaction of various degrees of freedom surrounding the *d*-electrons, i.e., charge, orbital, spin, and lattice. These degrees of freedom would produce relevant energy scales of similar magnitude such as bandwidth, Coulomb repulsion, and SOC. Because of the large number of transition metals, the number of possible metal-based perovskite transition is enormous. These

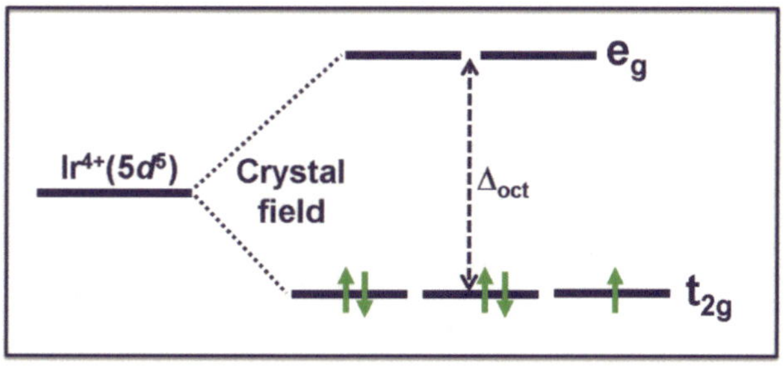

Figure 7. Simple view of distribution of five *d*-electrons of Ir^{4+} (d^5) in e_g *and* t_{2g} orbitals split by octahedral crystal field energy Δ_{oct}.

transition metals include those of which the order orbital is 3*d* (e.g., Fe, Co, Ni), 4*d* (e.g., Mo, Ru, Rh), or 5*d* (e.g., W, Re, Ir). 3*d* orbitals are well localized and thus form a narrow band (*W*) with a large on-site Coulomb interaction (*U*). 4*d* orbitals are spatially more extended than their 3*d* counterparts. As 5*d* orbitals are more spatially extended than 3*d* or 4*d* ones, as a result, nearest-neighbor orbitals overlap significantly, and therefore *W* is wider in 5*d* orbitals than in 3*d* or 4*d* cases, i.e., $W_{3d} < W_{4d} < W_{5d}$.

Because 5*d* orbitals are extended, on-site Coulomb repulsion or correlation *U* is weaker for 5*d* orbitals than form 3*d* or 4*d* ones, i.e., $U_{5d} < U_{4d} < U_{3d}$. Thus, naively one would expect 5*d* systems to be more metallic and less magnetic than those based on 3*d* and 4*d* transition metal oxides. Indeed, 5*d* perovskite $SrIrO_3$ is a correlated paramagnetic metal according to the expectation, but this is an exception. Surprisingly, many other 5*d* perovskites such as Sr_2IrO_4, and $Sr_3Ir_2O_7$ are insulators. This unexpected fact can be explained by the high SOC in 5*d* systems. In general, 5*d* systems have larger SOC than do 3*d* or 4*d* systems, i.e., $\Lambda_{3d} < \Lambda_{4d} < \Lambda_{5d}$ because SOC is proportional to the fourth power of the atomic number Z (i.e., $\Lambda_{soc} \propto Z^4$) [49]. In iridate compounds, $Z = 77$ for Ir, so the SOC strength is very high, even comparable to on-site Coulomb repulsion (~0.5 eV). This high SOC strength leads to modification of the electronic structure and gives rise to novel emergent phenomena.

3.3. Ruddlesden–Popper series $Sr_{n+1}Ir_nO_{3n+1}$

3.3.1. *Spin-orbit coupling and band structure evolution*

In this section, we present some unusual phenomena that are observed in RP series $Sr_{n+1}Ir_nO_{3n+1}$ ($n = 1, 2,$ and ∞). Because the SOC strength is so much larger in Ir than in a typical 3*d* system, SOC contributes to lifting the fivefold degeneracy of the atomic *d*-levels. In crystals, the crystal field and SOC act together to split the t_{2g} levels into $J_{eff} = 3/2$ and $J_{eff} = 1/2$ levels (Figure 8). Two equivalent views of the splitting of 5*d* orbitals due to Δ_{oct} and Λ_{soc} are illustrated [17].

Now, considering electron hopping in solids, the J_{eff} levels would become bands. Ir^{4+} has five 5*d* electrons; four of them fill the lower $J_{eff} = 3/2$ band and one partially fills the $J_{eff} = 1/2$ band with the Fermi level in the $J_{eff} = 1/2$ band. The band structure evolves across the RP series $Sr_{n+1}Ir_nO_{3n+1}$ (Figure 9). In the $n = 1$ case, Sr_2IrO_4, SOC splits the t_{2g} band into two J_{eff} bands, so the

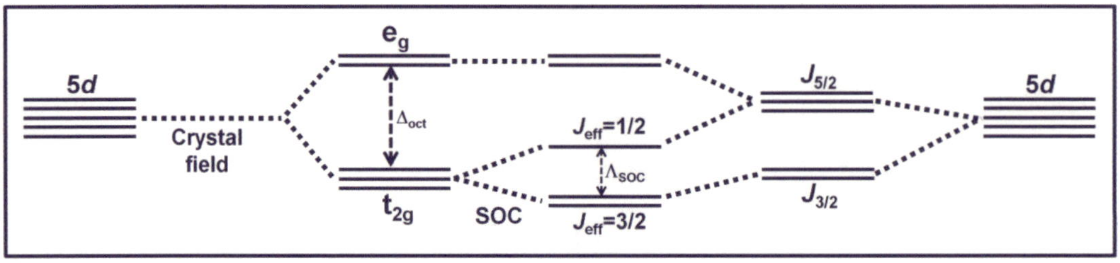

Figure 8. Two equivalent views of the splitting of 5*d* orbitals due to crystal field Δ_{oct} and strong SOC Λ_{soc}. Horizontal lines in the center column represent the energy levels of the final configuration. The final configuration can be reached from the left side that depicts the crystalline view: crystal field splitting occurs first and then SOC generates J_{eff} = 3/2 and 1/2 states. This result can also be viewed from the atomic viewpoint: the atomic fivefold degeneracy is lifted due to SOC, and then the crystal field turns quantum number *J* into $J_{eff.}$

bandwidth of the conduction band (J_{eff} = 1/2) is effectively reduced and correlation *U* can split the conduction band again into upper Hubbard bands (UHB) and lower Hubbard bands (LHB). Thus, Sr_2IrO_4 becomes a Mott insulator (Figure 9(a)). As *n* increases in the RP series, the bandwidth of the J_{eff} = 1/2 band also increases, possibly due to increase in the coordination number. $Sr_3Ir_2O_7$, a barely insulator, is the intermediate case (Figure 9(b)). In $SrIrO_3$ with $n = \infty$, *U* cannot split the conduction band with relatively wide bandwidth, so this compound is a correlated metal (Figure 9(c)).

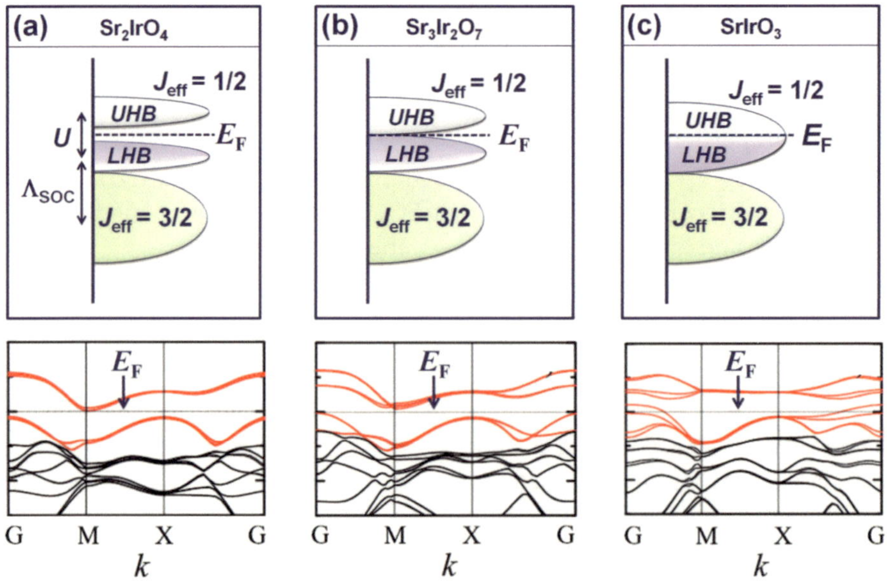

Figure 9. Evolution of the band structure across Ruddlesden–Popper series $Sr_{n+1}Ir_nO_{3n+1}$. Schematic diagrams (upper panel) and calculated results (lower panel, LDA+U+SOC). Sr_2IrO_4 (n = 1) is a Mott insulator; $SrIrO_3$ ($n = \infty$) is a correlated metal. Barely insulating $Sr_3Ir_2O_7$ (n = 2) is in intermediate. UHB: upper Hubbard; LHB: lower Hubbard band, respectively. In band structure calculation, the red and dark lines represent the J_{eff} =1/2 and J_{eff} =3/2 bands. The lower panels are reprinted with permission from Moon et al. [50]. © American Physical Society

3.3.2. Sr_2IrO_4

The first material in the RP series of $Sr_{n+1}Ir_nO_{3n+1}$ is a layered compound Sr_2IrO_4 (n = 1). The crystal structure of Sr_2IrO_4 consists of alternating layered stacking of SrO-IrO_2-SrO perovskite units with ideal K_2NiF_4-type tetragonal cell (Figure 5). The lattice constants are ~5.494 Å in the *ab*-plane and ~25.796 Å along the *c*-axis with space group $I4_1/acd$ [51]. Most importantly, it has a highly insulating nature (Figure 10(a)) [52] in contrast to a naive expectation that the extended nature of 5*d* orbitals would lead to a significant overlap of the nearest-neighbor orbitals and thus a broad electronic bandwidth. This insulating state arises from strong SOC and subsequent Coulomb repulsion (Figure 9(a)) and is termed a "Mott J_{eff} = 1/2 insulating state" of the 5*d* electron system [53, 54]. Magnetically, Sr_2IrO_4 exhibits weak ferromagnetism (canted antiferromagnetism) with T_N ~ 240 K (Figure 10(b)) with very small ferromagnetic moment (0.023 μ_B/Ir) [52]. The Mott insulating state due to strong SOC was first confirmed by Kim et al. by x-ray absorption (Figure 10(c)), based on the selection rules associated with the 2*p* to 5*d* transitions. Kim et al. also used theoretical calculations to identify the unusual nature of the Ir^{4+} state in accord with [53]. Angle Resolved Photoemission Spectroscopy (ARPES) (Figure 10(d)) also confirmed the Mott J_{eff} = 1/2 state [55].

Considering the origin of the Mott J_{eff} = 1/2 insulating state, it should be very sensitive to external perturbations; if so, the relevant energy parameters can be tuned to modify the electronic states. In Sr_2IrO_4 thin films under tensile (compressive) strain, the correlation energy is affected by in-plane lattice strain with increase (decrease) in bandwidth [56]. Sr_2IrO_4 remains an insulator even under pressure up to 55 GPa [57]; this stability illustrates the robustness of this insulating state. Also, a pressure-induced, fully reversible, giant piezoresistance was detected at room temperature [58]. The electronic band gap could be tuned electrically, and this characteristic demonstrates potential application in next-generation electronic devices [59]. Alkali-metal doping induces a close resemblance of the electronic state to that of high-temperature cuprates and therefore represents a step toward high-temperature superconductivity [60]. Many other exciting properties have been observed in this compound, including magnetic structural change, spin-orbit tuned MITs, lattice-driven magnetoresistivity, electron-doped tuned electronic structure, anisotropic magnetoresistance, and excitonic quasi-particle [61–66]. The interplay of crystal field splitting, SOC, and correlation effects in layered Sr_2IrO_4 determines the 5*d* electronic structure and leads to realization of a completely new class of materials with a novel quantum state.

3.3.3. $Sr_3Ir_2O_7$

The crystal structure of $Sr_3Ir_2O_7$ (n = 2) in this RP series of $Sr_{n+1}Ir_nO_{3n+1}$, is of tetragonal cell with a = 3.896 Å and c = 20.879 Å and space group *I4/mmm* [67]. It consists of strongly coupled bilayers of Ir-O octahedra which are separated by Sr-O interlayer (Figure 5). With the increase of number of octahedral layers (n), the electronic bands progressively broaden, and in particular, the bandwidth of the J_{eff} = 1/2 band increases from 0.48 eV for n = 1 to 0.56 eV for n = 2 [68]. Still the transport measurement shows a well-defined, barely insulating J_{eff} = 1/2 states (Figure 11(a)) [69]. Theoretical calculations based on LDA + U + SOC also provide evidence for the existence of the barely insulating band structure in which Fermi energy is between the J_{eff} = 1/2

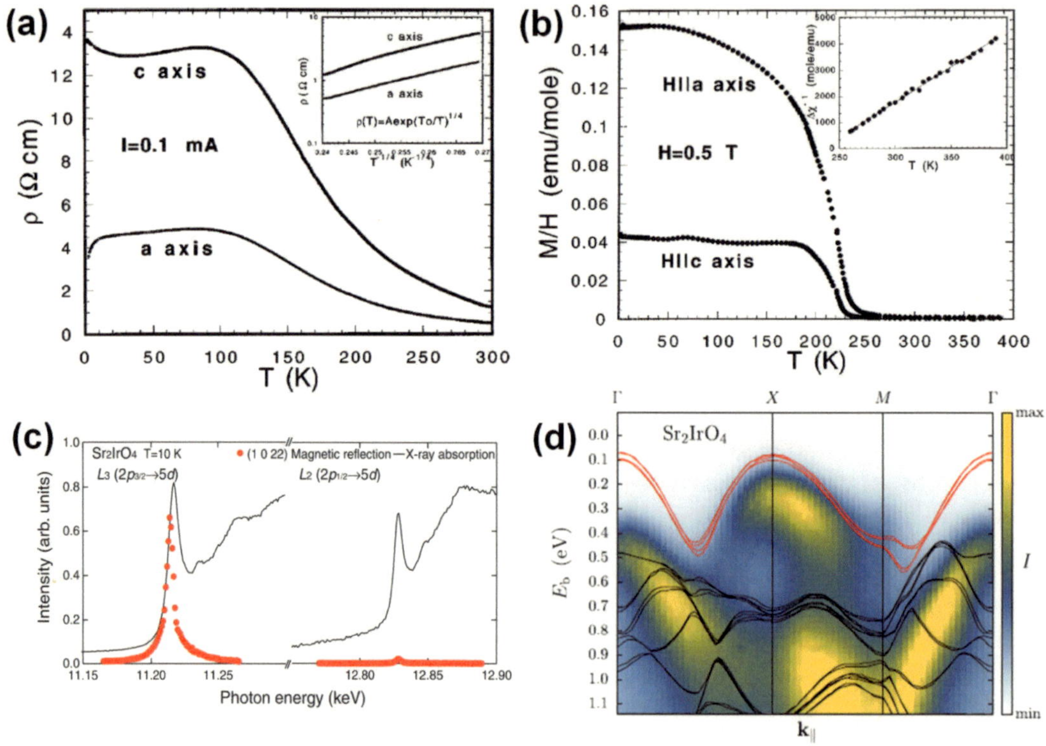

Figure 10. Characteristic properties of Sr_2IrO_4. (a) Resistivity along *ab*-plane and *c*-axis showing insulating nature with activation energy of ≈70 meV. (b) Magnetic measurements showing weak ferromagnetism with T_N ~ 240 K. (c) Black lines: X-ray absorption spectra indicating the presence of Ir L_3 ($2p_{3/2}$) and L_2 ($2p_{1/2}$) edges around 11.22 and 12.83 keV. Red dots: intensity of the magnetic (1 0 22) peak. (d) Experimental ARPES spectra for Sr_2IrO_4 (hν = 85 *eV*; *T* = 100 K). J_{eff} = 1/2 band (red line) and J_{eff}=3/2 band (black line) are shown. Reprinted with permission from Cao et al. [52], Kim et al. [54], and Wojek et al. [55]. © American Physical Society, © AAAS and © IOP Publishing.

bands (Figure 9(b)) [50]. The onset of weak ferromagnetism occurs at T_C ~ 285 K and is closely associated with the rotation of IrO_6 octahedra about the *c*-axis. Indeed, the temperature dependence M(*T*) of magnetization closely tracks the rotation of the octahedra, as characterized by Ir-O-Ir bond angle. $Sr_3Ir_2O_7$ also exhibits an intriguing *M* reversal for in-plane magnetization below 20 K with the onset of a rapid reduction at T_D ~ 50 K (Figure 11(b)). The barely insulating nature related to splitting of the J_{eff} = 1/2 band was again observed with x-ray scattering and absorption (Figure 11(c)) [70] and supported by ARPES measurements (Figure 11(d)) [55].

The ground state of bilayered $Sr_3Ir_2O_7$ is highly sensitive to small external perturbations such as chemical doping, high pressures, and magnetic field. By replacing Sr^{2+} with La^{3+}, electrons can be doped into bulk samples and can lead to an insulator-to-metal transition [71]. Also, the application of a high hydrostatic pressure leads to a drastic reduction in the electrical resistivity; this observation suggests that the system is near an MIT [57]. Although this system has not been fully explored yet, some studies such as resonant inelastic x-ray scattering, scanning tunneling spectroscopy, and optical conductivity have been performed [72–74]. These obser-

http://dx.doi.org/10.5772/61285

vations indicated that $Sr_3Ir_2O_7$ is a good model system to explore the mechanism for novel Mott states near an MIT boundary at which competitive interplay between SOC and Coulomb interactions persists.

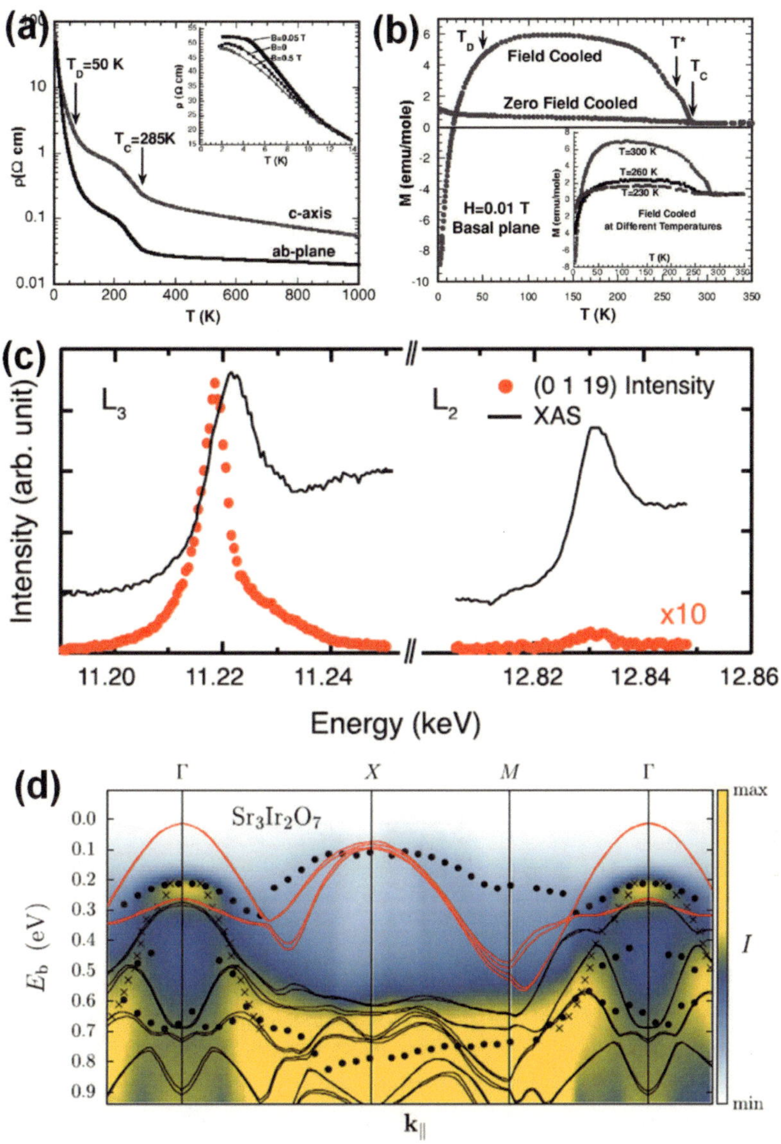

Figure 11. Characteristic properties of $Sr_3Ir_2O_7$. (a) Resistivity as a function of *T* for the basal plane and along the *c*-axis. (b) In-plane magnetization *M* vs. *T* at magnetic field of 100 Oe. (c) Energy scan of (0 1 19) reflection scanned around Ir L_3 and L_2 resonances. Red dots: scattering intensity; black lines: x-ray absorption spectra. (d) Experimental ARPES spectra (hν = 10.5 *eV*; *T* = 9 K). Calculated band structure: $J_{eff} = 1/2$ (red line) and $J_{eff} = 3/2$ bands (black line). Reprinted with permission from Wokec et al. [55], Cao et al. [69], and Kim et al. [70]. © American Physical Society, and © IOP Publishing.

3.3.4. $SrIrO_3$

The last compound in the RP series ($n = \infty$) is $SrIrO_3$; it is a correlated metal. $SrIrO_3$ is a three-dimensional system and has the largest coordination number in the RP series. An increase in the coordination number would lead to the increase of the bandwidth, and thus correlation-driven band splitting would not occur (Figure 9(c)). In fact, in $SrIrO_3$, the bandwidth of the $J_{eff} = 1/2$ band reaches 1.01 eV, so no gap appears within the $J_{eff} = 1/2$ band or between the $J_{eff} = 1/2$ band and $J_{eff} = 3/2$ band [50].

In 1971, Longo et al. first synthesized polycrystalline $SrIrO_3$ [75]. The stable ambient structure of $SrIrO_3$ is a monoclinic distortion of hexagonal $BaTiO_3$ structure [75, 76]. However, the structure transforms to an orthorhombic perovskite (space group *Pnma*) at 40 kbar and T = 1000 °C (Table 2). If a perovskite sample obtained at high temperature and high pressure is quenched, it remains a perovskite at room temperature. Because this book is about perovskites, our main focus remains on the properties of perovskite $SrIrO_3$. Zhao et al. [77] and Blanchard et al. [78] again synthesized orthorhombic perovskite samples under high pressure and performed electric and magnetic measurements which showed that perovskite $SrIrO_3$ is truly a paramagnetic metal (Figure 12).

Synthesis conditions	Structure	Lattice parameters
Atmospheric pressure, T = 900 °C	Monoclinic distortion of hexagonal $BaTiO_3$ structure	a = 5.604 Å b = 9.618 Å c = 14.17 Å β = 93.26°
P_{O2} = 40 kbar, T = 1000 °C	Orthorhombic	a = 5.60 Å b = 5.58 Å c = 7.89 Å

Table 2. Bulk $SrIrO_3$ can assume two forms depending on synthesis conditions. Corresponding lattice constants are summarized [75].

In situ ARPES showed that perovskite $SrIrO_3$ is an exotic narrow-band semimetal [79]. The bandwidth was surprisingly narrower than other two-dimensional RP phases, and the semimetallic nature is caused by the unusual coexistence of heavy hole-like and light electron-like bands contrary to the coordination number argument given previously. The observed unusual property may originate from the interplay of strong SOC, dimensionality, and both in- and out-of-plane IrO_6 octahedral rotations. Recent theoretical calculations also suggest an extremely interesting possibility that the interplay of the lattice structure and large SOC produces Dirac nodes in the $J_{eff} = 1/2$ band, and engineering topological phases at interfaces and in superlattices would alter the system to be close to a topological crystalline metal [80–82]. It is obvious that $SrIrO_3$ is an intriguing system in its own right and further studies are warranted.

http://dx.doi.org/10.5772/61285

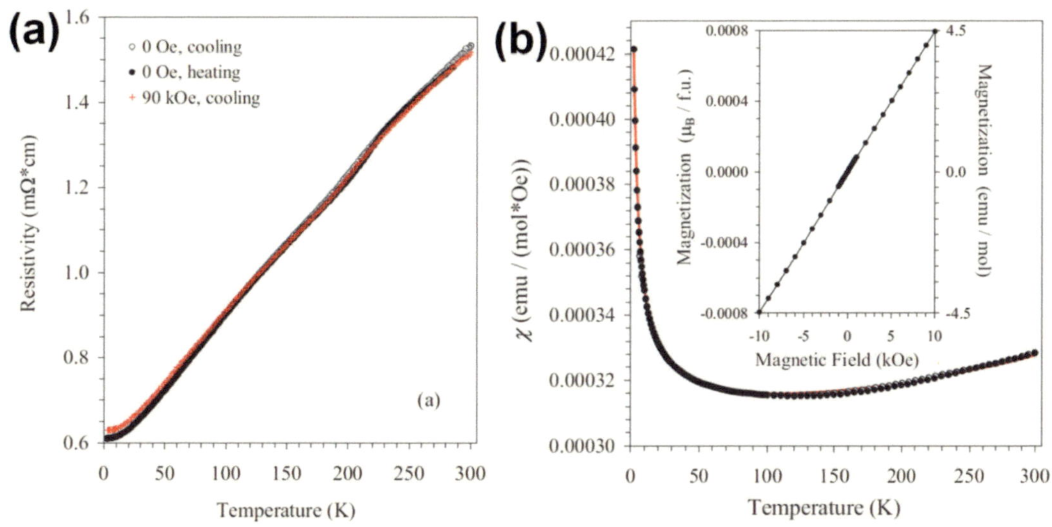

Figure 12. Characteristic properties of perovskite $SrIrO_3$. (a) Temperature dependence of electrical resistivity measured under different conditions. It remains metallic down to low temperatures. (b) It remains paramagnetic down to low temperatures. Magnetic susceptibility gives no sign of long range ordering. Inset confirms paramagnetism at $T = 5$ K. Reprinted with permission from Blanchard et al. [78]. © American Physical Society.

4. Non-Fermi liquid physics and metal–insulator transitions in $SrIrO_3$ films

Among RP series compounds $Sr_{n+1}Ir_nO_{3n+1}$, the end member $SrIrO_3$ is of particular interest because it is a correlated metal that exhibits unusual electronic transport properties that deviate from the normal Fermi liquid behaviors. The MITs and associated non-Fermi liquid physics that occur in $SrIrO_3$ with strong SOC would provide an opportunity to extend the limit of our current knowledge of the physics of MITs as presented in the previous section. Thus, the present section constitutes the core of this chapter, and we first summarize salient features of the transport properties of $SrIrO_3$, particularly in thin film form. Then we proceed to present the relevant theoretical frameworks to understand the non-Fermi liquid physics that underlay the MITs in the system.

4.1. Salient features of the metal–insulator transitions in $SrIrO_3$

$SrIrO_3$ is believed to be close to an MIT as evidenced by the evolution of the RP series $Sr_{n+1}Ir_nO_{3n+1}$ from being an insulator to a correlated metal with increasing $n = 1 \rightarrow 2 \rightarrow \infty$. The transport properties of correlated $SrIrO_3$ can be anticipated to be susceptible to external perturbations and MITs, and that the associated unusual properties could be induced in $SrIrO_3$ if, for example, an external stress is applied to the system. As introduced in the earlier section, the two most important mechanisms for MITs in correlated transition metal oxides are correlation-driven Mott localization and disorder-driven Anderson localization. When the system is under variable external stress, one may be able to tune W, U, and D to some extent and expose the interesting physics that is controlled by the parameters known as effective correlation (U/W)

and effective disorder (*D/W*). For these reasons, we attempted to synthesize perovskite $SrIrO_3$. Perovskite $SrIrO_3$, however, is metastable at room temperature and is obtainable only by applying an elevated pressure (~40 kbar) at high temperature (~1,000 °C) and subsequent quenching. While it is not easy to obtain single crystals of perovskite $SrIrO_3$ due to technical difficulties dealing with high pressures, the crystals can be stabilized by using thin film synthesis to produce them. In this case, the underlying substrates provide compressive strain, which replaces pressure, and epitaxial perovskite $SrIrO_3$ thin films are easily obtained. More importantly, compressive strains can be imposed on $SrIrO_3$ films by choosing substrates with appropriate lattice parameters. By depositing $SrIrO_3$ films on $GdScO_3$, $DyScO_3$, $SrTiO_3$, and $NdGaO_3$, one can impose progressively larger compressive strain in the films (Figure 13). Biswas et al. [83] provides further details on thin-film synthesis.

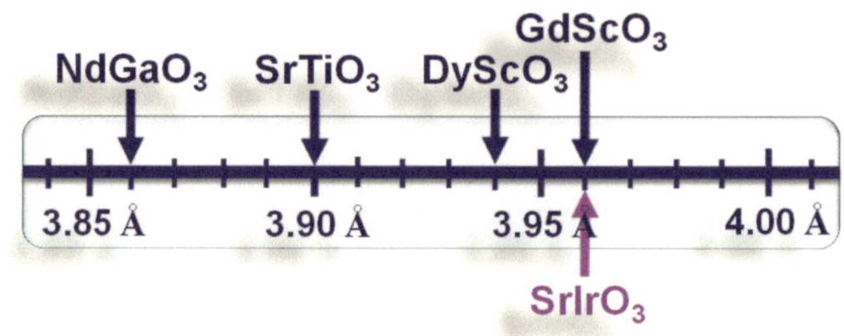

Figure 13. Pseudocubic lattice parameters of $SrIrO_3$ and various substrates available. $SrIrO_3$ and the substrates $GdScO_3$, $DyScO_3$, and $NdGaO_3$ are orthorhombic; $SrTiO_3$ is cubic.

The various scenarios on MITs suggest that an MIT may be obtained by varying the thickness of $SrIrO_3$ films. Indeed in films deposited on $GdScO_3$, which has a lattice constant well matched with that of $SrIrO_3$ (Figure 13), an MIT occurs as the thickness is reduced from 4 nm to 3 nm [83, 84]; a 4-nm film is metallic (with a resistivity upturn at low temperatures), but a 3-nm film is insulating. The resistivity upturn at low temperatures in the 4-nm film is well described by the weak localization in two dimensions, and the insulating behavior of the 3-nm film can be explained by the variable-range hopping. Thus, the thickness-driven MIT for $SrIrO_3$ (Figure 14(a)) is due to disorder and falls in the class of Anderson localization [83]. Reducing the thickness of the film seems to increase the effectiveness of given disorder and of the grain size effect to scatter the charge carriers. The transition from being metallic with a low temperature upturn in resistivity for small disorder to fully insulating over the whole temperature range with the increase of disorder is a realization of the disorder-driven MIT. However, the temperature variation of the resistance of all the metallic films (thickness 4 nm, 10 nm, and 35 nm) follows $\rho \propto T^{4/5}$. This nontrivial exponent indicates that although the MIT itself is driven by disorder, the underlying transport mechanism not so simple. Also, in some materials (e.g., $SrVO_3$), thickness-dependent MIT is caused by a decrease of the coordination number or by an increase in the correlation effect [85].

MIT also occurs in perovskite $SrIrO_3$ thin films when compressive strain is imposed on the films when the thickness is kept constant (Figure 14(b)) [83]. The imposed strain changes the

http://dx.doi.org/10.5772/61285

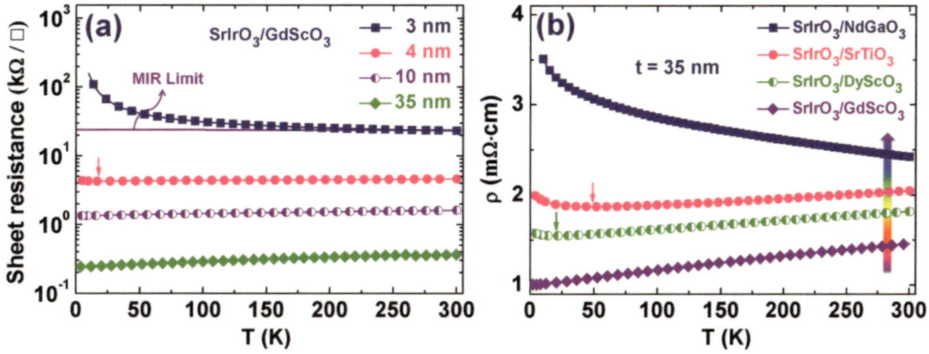

Figure 14. Metal–insulator transitions in perovskite $SrIrO_3$. (a) Sheet resistance for films of varying thickness and the thickness-driven metal–insulator transition. Mott–Ioffe–Regel (MIR) limit corresponds to the two-dimensional quantum sheet resistance (h/e^2) ~26 $k\Omega/\square$. (b) Resistivity for films of constant thickness 35 nm on various substrates and compressive-strain-driven metal–insulator transitions. The vertical arrow indicates an increase of compressive strain. Small arrows in both figures indicate the temperature below which resistivity upturns show up. Reprinted with permission from Biswas et al. [83]. © 2014, AIP Publishing LLC.

Ir-O bond length d and the Ir-O-Ir bond angle θ; these changes affect the bandwidth as $W \propto \frac{\cos\psi}{d^{3.5}}$, where $\psi = (\pi - \theta)/2$ is the buckling deviation of the *Ir-O-Ir* bond angle θ from π [83]. d is relatively difficult to change, but θ can be readily affected by lattice strain. When θ is decreased by compressive strain, the electronic hopping integral between Ir 5*d* orbitals is reduced, so the buckling drives the system toward the insulating state. In addition, compression would also slightly increase correlation (*U*). Thus, the overall change in effective correlation (*U/W*) would seem to induce the MIT; however, as compressive strain increases, the effective disorder (*D/W*) in the system also increase. Disorder was already shown to play a decisive role in the thickness-driven MIT and has a strong influence in the strain-driven MIT, as exposed by optical conductivity measurements [86]. Optical absorption spectroscopy provides insight into electronic band structure and free-carrier dynamics; optical absorption spectroscopy measurements of compressively strained $SrIrO_3$ films showed Drude-like, metallic responses without an optical gap opening. This result indicates that localization has a measurable effect on strain-induced MIT in perovskite $SrIrO_3$ thin films. In fact, extensive transport measurements in the compressively strained $SrIrO_3$ revealed is that the MIT is not simply due to either disorder or correlation [83, 87]. Thus, $SrIrO_3$ seems to provide a rare example in which the interplay of correlation and disorder in the presence of SOC causes the MIT.

The most remarkable feature of the electrical transport of perovskite $SrIrO_3$ under compression is that the temperature variation of resistivity deviates from the Fermi liquid behavior in a peculiar way. For $SrIrO_3$ films of 35-nm thickness on $GdScO_3$, $DyScO_3$, $SrTiO_3$, and $NdGaO_3$ substrates, the electrical resistivity not only shows non-Fermi liquid behaviors ($\rho \propto T^{\varepsilon}$ with $\varepsilon \neq 2$), but ε evolves from 4/5 to 1 to 3/2 as the compressive strain is increased, specifically, $\rho \propto T^{4/5}$ for films on $GdScO_3$, $\rho \propto T$ for films on $DyScO_3$, and $\rho \propto T^{3/2}$ for films on $SrTiO_3$ (Figure 15) [83]. Films on $NdGaO_3$ are subjected to the largest strain and become insulating (Figure 14(b)). The present strain-driven MIT is clearly contrasted to the thickness-driven MIT, during which ε remains constant at 4/5 as the thickness is reduced and the system approaches the MIT [83].

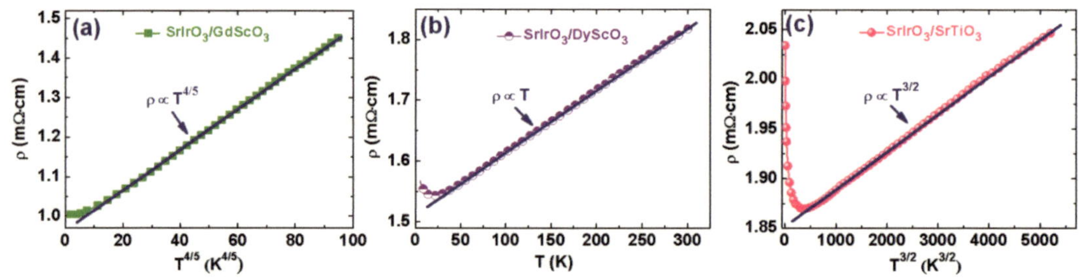

Figure 15. Temperature-dependent resistivity of $SrIrO_3$ thin films of thickness 35 nm on various substrates. The films show distinctly non-Fermi liquid behaviors as $\rho \propto T^{\varepsilon}$ with (a) $\varepsilon = 4/5$ for films on $GdScO_3$, (b) $\varepsilon = 1$ for films on $DyScO_3$, and (c) $\varepsilon = 3/2$ for films on $SrTiO_3$. The films on $DyScO_3$ and $SrTiO_3$ show resistivity upturns at low temperatures. Adapted with permission from Biswas et al. [83]. © 2014, AIP Publishing LLC.

The cause of these peculiar non-Fermi liquid behaviors and consequent MIT in the strain-driven case is not clear. Arguably, because $SrIrO_3$ is paramagnetic without long-range magnetic ordering, localized states might induce the formation of local magnetic moments without collective magnetic fluctuations because they occur near the MIT. Such localized moments or small magnetic clusters can influence the electronic transport significantly in the presence of disorder. Indeed, the presence of disorder in the strained films is indicated by the increases in resistivity at low temperatures (Figure 14). Also, disorder and the evolution of non-Fermi liquid physics are possibly inter-connected. In a correlated metal, when disorder is sufficiently high, the system can enter a so-called Griffiths phase, which consists of a mixture of islands of Fermi liquid and Mott insulating regions, and that has non-Fermi liquid behaviors. In the following sections, we will try to propose a model that incorporates all these factors and that may explore a new paradigm for non-Fermi liquid physics.

4.2. UV physics: Emergence of localized magnetic moments and dynamical mean-field theory

Essential experimental features for $SrIrO_3$ films of thickness 35-nm grown on $GdScO_3$, $DyScO_3$, $SrTiO_3$, and $NdGaO_3$ are non-Fermi liquid transport phenomena near MIT, where electrical resistivity $\rho \propto T^{\varepsilon}$ shows anomalous temperature dependences with $\varepsilon = 4/5$, 1, 3/2, and −1/4 respectively. (The resistivity for the insulating phase can be described with a negative exponent.) The continuous change of the temperature exponent ε implies that a particular type of interplay between correlations of electrons and disorders is expected to have an important influence on physics near the MIT and raises fundamental questions about the nature of the non-Fermi liquid. The first question is whether this non-Fermi liquid physics is involved with either UV (ultraviolet) or IR (infrared) physics. Here, UV physics means that localized magnetic moments appear near an MIT and are regarded to be the source of strong inelastic scattering

http://dx.doi.org/10.5772/61285

events due to their extensive entropy and to be responsible for non-Fermi liquid transport phenomena. The theoretical framework of dynamical mean-field theory was designed to simulate this local-moment physics quite well [88].

Dynamical mean-field theory describes Mott quantum criticality successfully and suggests that it appears at high temperatures, where not only MITs but also the so-called bad metal physics have been revealed [89, 90]. A noticeable point is that this local-moment UV physics seems to be universal, regardless of IR physics in which such local moments are expected to disappear by forming either singlets or magnetic orders and thereby reducing the entropy dramatically at low temperatures. Because incompletely screened local moments have important effects in non-Fermi liquid physics, the appearance of negative magnetoresistance (MR) (MR $=\frac{\varrho(B)-\varrho(0)}{\varrho(0)}$) can be expected. However, the experiment on 35-nm $SrIrO_3$ films on various substrates confirms positive MR, which is less than 1% up to magnetic field strength of 9 Tesla (Figure 16) [83]. Although this result does not necessarily mean that the local-moment physics may not be important in the MIT of $SrIrO_3$ thin films, the positive MR leads us to focus on IR physics.

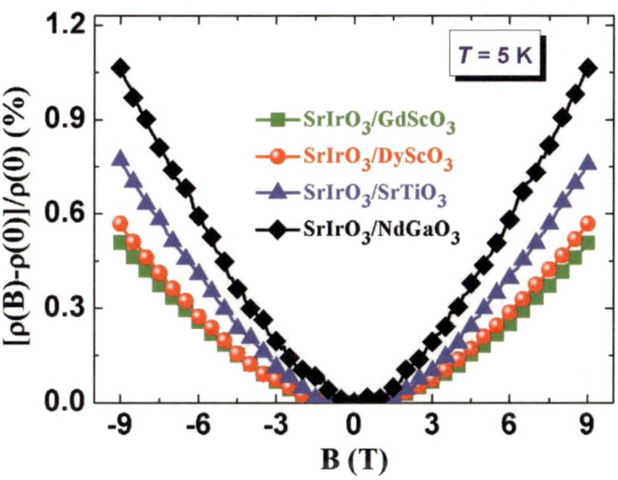

Figure 16. Magnetoresistance (MR) at $T = 5$ K of $SrIrO_3$ thin films of thickness 35 nm grown on various substrates. MR was positive in all films regardless of compressive strain which determines the magnitude of MR

Another important phenomenon is that spectral weight transfer in optical conductivity from the Drude part to a mid-infrared region as temperature increases [38]. This change is a characteristic feature of bad metals and occurs at finite temperatures in the metallic side near an MIT. Dynamical mean-field theory reveals that local-moment UV physics describes this bad-metal Mott physics nicely. If local-moment physics are not considered, the spectral weight transfer in the optical conductivity cannot be easily reproduced without symmetry breaking. Unfortunately, no experiment has been performed yet in determining the importance of UV local-moment physics on non-Fermi liquid states of $SrIrO_3$ thin films. With the contradictory picture of UV physics and observed positive MR, in the next section, we focus on IR physics, which assumed to be responsible for the observed non-Fermi liquid physics near the MIT of $SrIrO_3$ thin films.

4.3. IR physics: Slater quantum criticality vs. Mott–Anderson–Griffiths scenario

Here, IR physics means that long-wave length and low-energy fluctuations determine the non-Fermi liquid physics near the MIT of $SrIrO_3$ thin films. Then, quantum criticality would be the first choice, where quantum critical fluctuations involved with symmetry breaking, which cause to strong inelastic scattering between low-energy electrons. This scattering is responsible for non-Fermi liquid transport phenomena [91]. Unfortunately, we failed to figure out quantum criticality involved with any kinds of orders, in particular, those associated with magnetism, although we do not exclude more-delicate symmetry breaking near the MIT of $SrIrO_3$ thin films. Furthermore, quantum criticality itself cannot readily explain the continuous change of ε in the relationship $\rho \propto T^{\varepsilon}$ of electrical resistivity to temperature on the metallic side of the MIT. One way to understand this continuous change is to consider the effect of disorder on quantum criticality. The Harris criterion is on the stability of quantum criticality against weak randomness, in the sense of average, involved with a space dimension and a critical exponent of correlation length [92]. When the Harris criterion is violated, the clean quantum critical point becomes destabilized. As a result, a novel disordered quantum critical point can emerge, respecting the Harris criterion. However, the continuous evolution of the non-Fermi liquid physics is difficult to understand even in this situation. However, the strength of randomness can grow indefinitely, resulting in the so-called infinite randomness fixed point; when this happens, samples become extremely inhomogeneous due to effectively enhanced disorders, with ordered regions coexisting with disordered islands. The statistical distribution of the ordered islands shows a power-law tail, which implies that rare events that correspond to the power-law tail of the distribution function have important influences in non-Fermi liquid physics, particularly thermodynamics [93]. For example, an antiferromagnetic Heisenberg model with random exchange coupling lies at an infinite randomness fixed point, at which the antiferromagnetic quantum critical point disappears and is replaced by a random singlet state [94]. An attractive feature of the infinite randomness fixed point is that non-Fermi liquid physics still survive near it, i.e., away from the quantum critical point, where dynamics of rare regions associated with the long tail part of the distribution function governs singular behaviors of the extremely inhomogeneous state. These phenomena are referred to as quantum Griffith effects [95]. The continuous change of non-Fermi liquid transport exponents reminds us of the quantum Griffith phase.

We suspect that the MIT in 35-nm $SrIrO_3$ films on $GdScO_3$, $DyScO_3$, $SrTiO_3$, and $NdGaO_3$ can be achieved by electron correlations, in which lattice mismatches between substrates and $SrIrO_3$ thin films are expected to control the ratio between interactions and hopping integrals. An essential question is on the role of disorder in this Mott transition, combined with electron correlations. Considering that the films show positive MR at magnetic field strength up to 9 Tesla, we suggest that magnetic correlations may not have important influence in this MIT. In fact, the positive MR suggests that the resulting insulating phase may be paramagnetic.

http://dx.doi.org/10.5772/61285

Furthermore, the residual resistivity is close to the Mott–Ioffe–Regel (MIR) limit; this similarity implies that the concentration of disorders is not low. Therefore, we conjecture that the interplay between electron correlations and disorders can give rise to a Griffith-type phase between Landau's Fermi liquid state and the Mott–Anderson insulating phase; the Griffith-type phase allows the continuous change of transport exponents. We call this physics the Mott–Anderson–Griffith scenario. Until now, the Griffith scenario has been realized near the infinite randomness fixed point [93], at which extreme inhomogeneity and associated rare events are responsible for non-Fermi liquid physics that have varying critical exponents. Although the mechanism by which such an infinite randomness fixed point can appear in the Mott–Anderson transition has not been identified, fluctuations between metallic and insulating islands as rare events are expected to allow development of the Mott–Anderson–Griffith phase.

4.4. Model Hamiltonian and tentative Global phase diagram

Recalling the interplay between the spin-orbit coupling and the Hubbard interaction, we start from an Anderson–Hubbard model with one band (Eq. (2)):

$$H = -t\sum_{ij\sigma} c_{i\sigma}^{\dagger} c_{j\sigma} + U\sum_{i}\left(\sum_{\sigma} c_{i\sigma}^{\dagger} c_{i\sigma}\right)^{2} - \sum_{i} v_{i}\left(\sum_{\sigma} c_{i\sigma}^{\dagger} c_{i\sigma}\right) \tag{2}$$

where $c_{i,\sigma}$ represents an electron field at site i, and σ expresses a Kramers doublet state given by total angular momentum, t is a hopping integral, U is the strength of on-site Coulomb interaction, and v_i is a random potential introduced by disorder. When electron correlation becomes negligible, the Hamiltonian reduces to the Anderson model, showing a continuous phase transition from a diffusive Fermi liquid state to an Anderson insulating phase in three dimensions. Local density of states (LDOS) occurs (Figure 17), in which localized eigenstates are given by discrete energy levels. This condition occurs below the mobility edge in the diffusive Fermi liquid state, and over the whole range of energy in the insulating phase. When contributions of random impurities can be neglected, the Hamiltonian becomes the Hubbard model, which shows a Mott transition from Landau's Fermi liquid state to a paramagnetic Mott insulating phase. Regarding this Mott transition based on the UV local-moment physics yields first-order MIT [88]. In contrast, using IR physics to explain the Mott transition yields a continuous MIT, which will be discussed below. In the Landau's Fermi liquid state, a coherent peak occurs at the Fermi energy in the electron spectral function besides incoherent humps, and it disappears to transfer into an incoherent background, resulting in upper and lower Hubbard bands in the paramagnetic Mott insulating phase. A further question is whether the Griffith-type phase appears near the Mott–Anderson transition in the middle region of the phase diagram (Figure 17).

As discussed, we try to describe IR physics to explain the Mott–Anderson MIT in $SrIrO_3$ thin films. Incompletely screened local moments can be expected to be screened completely at low temperatures. One mechanism for this screening is the Kondo effect, in which localized magnetic moments form singlets with itinerant electrons, well described by dynamical mean-

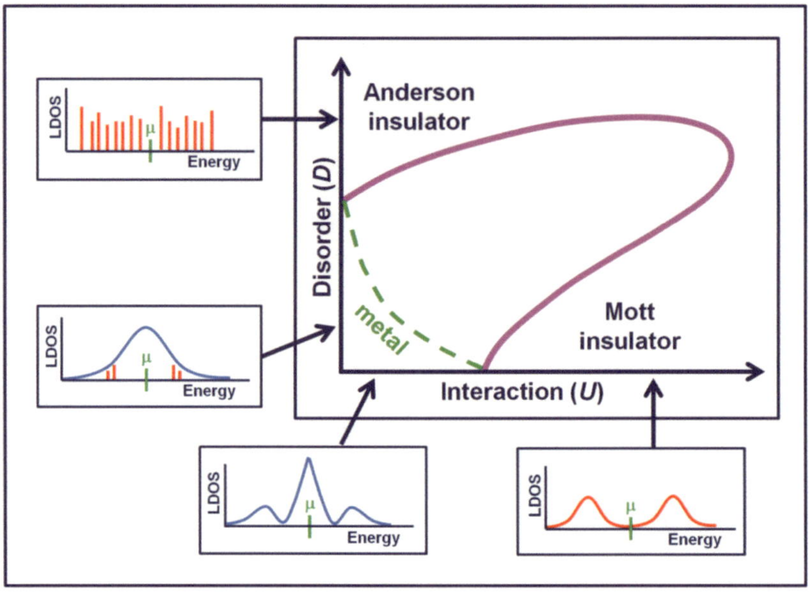

Figure 17. A schematic phase diagram of Anderson–Hubbard model with one band and local density of states (LDOS) of each phase. Adapted with permission from Byczuk [96]. © 2005 American Physical Society.

field theory. The other mechanism entails localized magnetic moments form singlets with themselves as an RKKY-type (Ruderman, Kittel, Kasuya, Yosida) interactions [97–99]. The resulting paramagnetic Mott insulating phase is called a spin-liquid state, in which charge-neutral spinons may emerge as low-energy elementary excitations and may interact among themselves through abundant singlet fluctuations, referred to as gauge fields [100]. Recalling the positive MR, we discuss the Mott–Anderson transition based on the spin-liquid physics.

4.5. Generalization of Finkelstein's nonlinear sigma model approach near the Mott–Anderson transition

To describe the Mott transition involved with the spin-liquid physics, we take the U(1) slave-rotor representation [101] as $C_{i\sigma} = e^{-i\theta_i} f_{i\sigma}$, in which an electron field is expressed as a composite operator of charge and spin degrees of freedom. θ_i accounts for the dynamics of collective charge fluctuations (sound modes), and $f_{i\sigma}$ expresses a charge-neutral spinon field for collective dynamics of spin degrees of freedom. Then, an effective mean-field theory for such variables of θ_i and $f_{i\sigma}$ can be easily formulated from the Hubbard model without randomness; in this model, the Hubbard Hamiltonian is decomposed into two sectors describing the dynamics of spinons and zero-sound modes, respectively, given by (Eq. (3) and Eq. (4)):

$$S_F = \int_0^\beta d\tau \left[\sum_{i,\sigma} f_{i\sigma}^\dagger (\partial_\tau - \mu) f_{i\sigma} - t\chi_f \sum_{ij,\sigma} \left(f_{i\sigma}^\dagger f_{j\sigma} + h.c. \right) \right] \tag{3}$$

http://dx.doi.org/10.5772/61285

$$S_B = \int_0^{\beta} d\tau \left[\frac{1}{2U} \sum_i \left(\partial_\tau b_i^\dagger\right)\left(\partial_\tau b_i\right) - t\chi_\theta \sum_{ij} \left(b_{i\sigma}^{+} b_{j\sigma} + H.c.\right) + \lambda \sum_i \left(\left|b_i\right|^2 - 1\right) + 2L^2 zt\chi_f \chi_\theta \right] \tag{4}$$

Here, the conventional saddle-point approximation has been performed for a spin-liquid-type Mott insulating phase. χ_f describes band renormalization for electrons, and χ_θ approximately expresses the width of incoherent bands. λ is a Lagrange multiplier field to control the spin-liquid to Fermi liquid phase transition and is regarded to be the chemical potential of bosons. This results from a nonlinear $\sigma-$ model description, in which the rotor variable $e^{-i\theta_i}$ is replaced with b_i and the unimodular constraint $|b_i|^2=1$ is considered. z is the nearest coordinate number of our lattice, and L^2 is the size of the system.

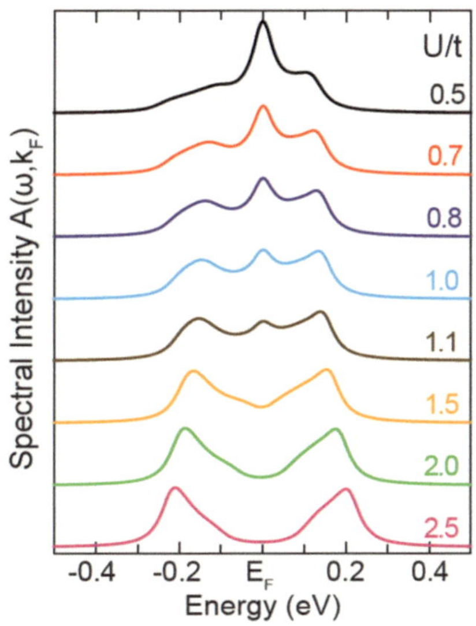

Figure 18. Evolution of the electron spectral function from Landau's Fermi liquid state to a spin-liquid Mott insulating phase [102].

The Mott transition in this spin-liquid approach is realized by the condensation transition of b_i, by which zero-sound modes are gapless in the Fermi liquid state, i.e., $b_i \neq 0$ but become gapped in the spin-liquid phase, i.e., $b_i=0$. As a result, the height of the coherent peak in the electron spectral function, proportional to the condensation amplitude $|b_i|^2$, decreases gradually to disappear toward the spin-liquid Mott insulating phase; during this process, the spectral weight is transferred to the incoherent background of upper and lower Hubbard bands (Figure 18). Our problem is to introduce a random potential into this spin-liquid Mott transition. As long as the strength of the random potential remains smaller than the spinon bandwidth, we are allowed to deal with the role of the random potential perturbatively. This renormalization group analysis has been performed to reveal that the clean spin-liquid Mott critical point becomes unstable as soon as the random potential is turned on. As a result, a disorder critical point appears to be identified with a transition from diffusive spin-liquid glass insulator to diffusive Fermi liquid metal, in which the diffusive spin-liquid glass state consists of a diffusive spin-liquid phase of spinons and a charge glass phase of sound modes (Figure 19) [103].

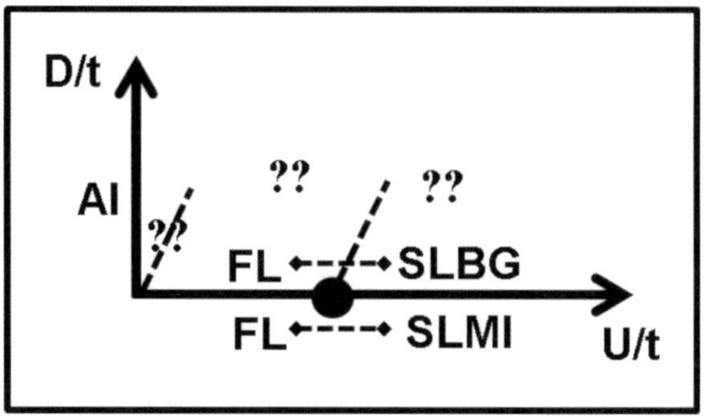

Figure 19. A schematic phase diagram for the spin-liquid Mott–Anderson transition based on weak-coupling renormalization group analysis. FL: Fermi liquid metal, SLMI: spin liquid Mott insulator, SLBG: spin liquid Bose glass, AI: Anderson insulator. Bose glass means that charge dynamics described by sound modes displays glassy behaviors. *U*: strength of local interactions, *D*: disorder, and *t*: hopping integral of electrons. Adapted with permission from Kim [103]. © 2006 American Physical Society.

In this renormalization group analysis, the spinon conductivity has been used as an input parameter, rather than being self-consistently determined, and the diffusive dynamics of spinons has been assumed. As a result, the renormalization group study reached the conclusion that the weak-disorder quantum critical point depends on the residual spinon conductivity, which means that the universality may not appear around this disorder quantum critical point. Suppose that $SrIrO_3$ thin films on the substrate of $GdScO_3$ are near this disordered MIT. Since $SrIrO_3$ thin films on $DyScO_3$, $SrTiO_3$, and $NdGaO_3$ substrates are also near this quantum critical point and their spinon conductivities differ from each other, we may observe continuous change of the temperature exponent ε of the electrical resistivity. Unfortunately, however, this previous study does not evaluate the spinon conductivity self-consistently, so the discussion cannot be beyond our speculation. Therefore, a theoretical framework must be developed to determine both transport coefficients of spinons and sound modes self-consistently.

One problem is that the previous renormalization group analysis does not consider effective interactions between diffusions and Cooperons, i.e., weak-localization corrections [34]. The replica nonlinear $\sigma-$ model approach serves a natural theoretical framework to introduce such quantum corrections [104, 105]. By introducing the replica trick into the effective action for spinons and sound modes, and taking the diffusive spin-liquid fixed point as a saddle point for the spinon dynamics, one can derive an effective field theory as follows (Eq. (5)) [106]:

$$S_{eff} = \int_0^\beta d\tau \int d^d r \; \{ \frac{\pi N_F^f}{4} \sqrt{-\partial_r^2} tr(KQ) + \frac{\pi N_F^f}{4} D_c tr\left(\nabla Q - i a^c \left[Q, \tau^3\right]\right)^2 +$$
$$+ \frac{1}{U}\left\{(\partial_\tau + i\varphi) b^\dagger\right\} \left(1 - U \frac{\sqrt{-\partial_\tau^2}}{\sqrt{-\partial_\tau^2 + D_c(-\nabla^2)}} \right) \left\{(\partial_\tau - i\varphi) b\right\} + t\eta_\theta \left|\left(\nabla - i a^c\right) b\right|^2 + \lambda_c b^\dagger b\} \quad (5)$$
$$+ \frac{1}{\beta} \sum_{i\nu} \sum_q a_i^c (q, i\nu) \left(\gamma_a |\nu| + v_a^2 |q|^2\right) P_{ij}^T a_j^c (-q, -i\nu)$$

Dynamics of diffusions and Cooperons of the spinon sector are described by the replica nonlinear $\sigma-$ model of the first line. Q is a $4N \times 4N$ matrix field, located on the Grassmannian manifold of $\frac{Sp(4N)}{Sp(2N) \times Sp(2N)}$, which spans retarded and advanced, Kramers doublet, and N replica spaces. K is a constant diagonal $4N \times 4N$ matrix, given by $I_{2N \times 2N}$ for the retarded part and $-I_{2N \times 2N}$ for the advanced sector. N_F^f is the spinon density of states at the spinon Fermi-energy and D_c is the diffusion coefficient. a^c a gauge field to describe singlet excitations, more precisely, spin-chirality fluctuations that are neglected in the previous mean-field analysis but introduced here beyond the saddle-point approximation [100], in which the dynamics of gauge fluctuations is described by the last term, which is given by the polarization function of the diffusive dynamics of spinons. When the effects of gauge fluctuations in the dynamics of diffusions and Cooperons can be neglected, the replica nonlinear $\sigma-$ model reduces to that for the Anderson localization [104, 105]. The boson sector of sound modes is essentially the same as before, but gauge fluctuations are introduced beyond the mean-field analysis and the time sector is modified by the renormalization of the diffusive dynamics of spinons. Nonlocal derivative terms for both space and time should be defined in the momentum and frequency space, where the above expression is just for our formal writing. As discussed before, the boson sector describes the spin-liquid Mott transition, whereas the nonlinear $\sigma-$ model part describes Anderson localization.

We claim that this replica nonlinear $\sigma-$ model approach [107–110] generalizes the Finkelstein's nonlinear $\sigma-$ model approach into the region of Mott transitions. Finkelstein's approach does not incorporate Mott physics associated with strong correlations of electrons. Instead, this approach considers the influences of effective interactions of both singlet and triplet channels in the dynamics of diffusions and Cooperons, considers instabilities of interacting diffusive Fermi liquids, and reveals the nature of the Anderson MIT. In contrast, the existing nonlinear $\sigma-$ model approach considers instabilities of interacting diffusive spin-liquids and the nature of the Mott–Anderson MIT. In diffusive Fermi liquids, effective interactions of the triplet channel have an important influence on the Anderson MIT of two spatial dimensions [107–110]. In our problem, we do not consider triplet-channel interactions because the positive magnetoelectrical resistivity suggests that the magnetic correlations may not be important. However, we speculate that gauge fluctuations, regarded to be singlet-channel interactions, affect the dynamics of Cooperons seriously, thereby suppressing weak-localization corrections and stabilizing the metallic state of spinons. This would be a novel mechanism to produce metallicity near the Mott–Anderson transition. Renormalization group analysis has not been performed yet for this effective field theory.

5. Summary

Transition metal perovskites have been the platform for numerous emergent physics that originate from the coupling of the fundamental degrees of freedom such as spin, lattice, charge, and orbital as well as of disorder, which is unavoidable in solids. The physics of 5*d* perovskite iridates, in particular, has attracted considerable attention after the discovery of the novel $J_{eff} = 1/2$ Mott insulating state and the evolution of dimensionality controlled MIT in the RP series $Sr_{n+1}Ir_nO_{3n+1}$, originating due to strong SOC of 5*d* element which is comparable to its Coulomb correlation or bandwidth. $SrIrO_3$, the end member of the RP series, shows many unusual phenomena, including different kinds of MITs. We fabricated epitaxial thin films of various thickness on various substrates to induce MIT in perovskite $SrIrO_3$ by thickness reduction or imposed compressed strain. The MIT driven by thickness reduction occurs due to disorder, but the MIT driven by compressive strain in the films on different substrates is accompanied by peculiar non-Fermi liquid behaviors with an evolving temperature exponent in the electrical resistivity relationship. The latter MIT and associated non-Fermi liquid behaviors are probably due to the delicate interplay between correlation, SOC, and disorder, and thus pose a theoretical challenge to our understanding of non-Fermi liquid physics and MIT.

To reveal the nature of this non-Fermi liquid physics near the MIT of $SrIrO_3$ thin films on various substrates, we first discussed the influence of emergent localized magnetic moments, referred to as UV physics. The observed positive MR in the whole temperature range led us to pursue another direction, referred to as IR physics, in which long-wave length and low-energy fluctuations would be important. Because quantum criticality itself cannot explain the continuous evolution of the non-Fermi liquid physics, we speculate that the interplay between strong correlation of electrons and not-so-weak disorder may contribute to the continuously varying non-Fermi liquid physics. One possible scenario within IR physics is to consider quantum Griffiths effects with extreme inhomogeneity, in which local fluctuations between metallic and insulating phases, referred to as rare events, and may dominate the non-Fermi liquid physics. To realize this so-called Mott–Anderson–Griffiths scenario, we try to combine spin-liquid Mott physics with Anderson localization that describes a Mott transition from Landau's Fermi liquid state to a spin-liquid Mott insulating phase. Theoretical understanding of this special class of materials is in its early stages, and many new emergent phenomena in iridates are yet to be explored theoretically or experimentally.

Acknowledgements

YHJ acknowledges the support by NRF (grant no. 2011-0009231). KSK was supported by NRF (grant no. 2012R1A1B3000550) and also by TJ Park Science Fellowship of the POSCO TJ Park

Foundation. YHJ and KSK were jointly supported by CTM at POSTECH (grant nos. 2011-0030785 and 2011-0030786).

Author details

Abhijit Biswas[1], Ki-Seok Kim[1,2] and Yoon Hee Jeong[1*]

*Address all correspondence to: yhj@postech.ac.kr

1 Department of Physics, POSTECH, Pohang, South Korea

2 Institute of Edge of Theoretical Science, POSTECH, Pohang, South Korea

References

[1] Rao C N R, Raveau B. Transition Metal Oxides: Structure, Properties, and Synthesis of Ceramic Oxides. 2nd Edition. Wiley–VCH; 1998. 392 pp.

[2] Maekawa S, Tohyama T, Barnes S E, Ishihara S, Koshibae W, Khaliullin G. Physics of Transition Metal Oxides. Springer Science & Business Media; 2004. 337 pp.

[3] Tokura Y. Correlated-electron physics in transition-metal oxides. Phys. Today. 2003; 56: 50–55. (DOI: http://dx.doi.org/10.1063/1.1603080)

[4] Goodenough J B. Perspective on engineering transition-metal oxides. Chem. Mater. 2014; 26: 820–829. (DOI: 10.1021/cm402063u)

[5] Dagotto E. Complexity in strongly correlated electronic systems. Science. 2005; 309: 257–262. (DOI: 10.1126/science.1107559)

[6] Johnsson M, Lemmens P. Perovskite and thin films- crystallography and chemistry. J. Phys. Condens. Matter. 2007; 20: 264001-6. (DOI: 10.1088/0953-8984/20/26/264001)

[7] Schlom D G, Chen L-Q, Pan X, Schmehl A, Zurbuchen M A. A thin film approach to engineering functionality into oxides. J. Am. Ceram. Soc. 2008; 91: 2429–2454. (DOI: 10.1111/j.1551-2916.2008.02556.x)

[8] Martin L W, Chu Y-H, Ramesh R. Advances in the growth and characterization of magnetic, ferroelectric, and multiferroic oxide thin films. Mater. Sci. Eng. R. 2010; 68: 89–133. (DOI:10.1016/j.mser.2010.03.001)

[9] Zubko P, Gariglio S, Gabay M, Ghosez P, Triscone J M. Interface physics in complex oxide heterostructures. Annu. Rev. Condens. Matter Phys. 2011; 2: 141–165. (DOI: 10.1146/annurev-conmatphys-062910-140445)

[10] Granozio F M, Koster G, Rijnders G. Functional oxide interfaces. MRS Bull. 2013; 38: 1017–1023. (DOI: 10.1557/mrs.2013.282)

[11] Ngai J H, Walker F J, Ahn C H. Correlated oxide physics and electronics. Annu. Rev. Mater. Res. 2014; 44: 1–17. (DOI: 10.1146/annurevmatsci 070813-113248)

[12] Sulpizio J A, Ilani S, Irvin P, Levy J. Nanoscale Phenomena in oxide Heterostructures. Annu. Rev. Mater. Res. 2014; 44: 117–149. (DOI: 10.1146/annurev-matsci-070813-113437)

[13] Chakhalian J, Freeland J W, Millis A J, Panagopoulos, Rondinelli J M. Colloquium: emergent properties in plane view: strong correlations at oxide interfaces. Rev. Mod. Phys. 2014; 86: 1189–1202. (DOI: http://dx.doi.org/10.1103/RevModPhys.86.1189)

[14] Schlom D G, Chen L-Q, Fennie C J, Gopalan V, Muller D A, Pan X, Ramesh R, Uecker R. Elastic strain engineering of ferroic oxides. MRS Bull. 2014; 39: 118–130. (DOI: 10.1557/mrs.2014.1)

[15] Tokura Y, Nagaosa N. Orbital physics in transition-metal oxides. Science. 2000; 288: 462–468. (DOI: 10.1126/science.288.5465.462)

[16] Cheong S W. Transition metal oxides: the exciting world of orbitals. Nat. Mater. 2007; 6: 927–928. (DOI:10.1038/nmat2069)

[17] Cao G, De-Long L. Frontiers of 4*d*- and 5*d*-Transition Metal Oxides. 1st edition. World Scientific; 2013. 328 pp.

[18] Krempa W W, Chen G, Kim Y B, Balents L. Correlated quantum phenomena in the strong spin-orbit regime. Annu. Rev. Condens. Matter Phys. 2014; 5: 57–82. (DOI: 10.1146/annurev-conmatphys-020911-125138)

[19] Rau J G, Lee E K-H, Kee H-Y. Spin-Orbit Physics Giving Rise to Novel Phases in Correlated Systems: Iridates and Related Materials. arXiv:1507.06323

[20] Imada M, Fujimori A, Tokura Y. Metal–insulator transitions. Rev. Mod. Phys. 1998; 70: 1039–1263. (DOI: 10.1103/RevModPhys.70.1039)

[21] Mott N F. Metal–Insulator Transitions. Taylor & Francis, London; 1990.

[22] Dobrosavljević V, Trivedi N, Valles J M Jr. Conductor insulator quantum phase transitions. Oxford; 2012.

[23] Miranda E, Dobrosavljević V. Disorder-driven non-Fermi liquid behavior of correlated electrons. Rep. Prog. Phys. 2005; 68: 2337–2408. (DOI:10.1088/0034-4885/68/10/R02)

[24] Edwards P P, Johnston R L, Rao C N R, Tunstall D P, Hensel. The metal–insulator transition: a perspective. Phil. Trans. R. Soc. Lond. A. 1998; 356: 5–22. (DOI: 10.1098/rsta.1998.0146)

[25] Drude P. Zur Elektronentheorie der metalle. Annalen der Physik. 1900; 306: 566–613. (DOI: 10.1002/andp.19003060312)

[26] Boer J H, Verwey E J W. Semi-conductors with partially and with completely filled 3*d*-lattice bands. Proc. Phys. Soc. 1937; 49: 59–71. (DOI:10.1088/0959-5309/49/4S/307)

[27] Mott N F, Peierls R. Discussion of the paper by de Boer and Verwey. Proc. Phys. Soc. 1937; 49: 72–73. (DOI:10.1088/0959-5309/49/4S/308)

[28] Kotliar G, Vollhardt D. Strongly correlated materials: insights from dynamical mean-field theory. Phys. Today. 2004; 57: 53–59. (DOI: http://dx.doi.org/10.1063/1.1712502)

[29] Slater J C. Magnetic effects and the Hartree–Fock equation. Phys. Rev. 1951; 82: 538–541. (DOI: 10.1103/PhysRev.82.538)

[30] Gebhard F. The Mott Metal–Insulator Transition: Models and Methods. Springer; 1997.

[31] Anderson P W. Absence of diffusion in certain random lattices. Phys. Rev. 1958; 109: 1492–1505. (DOI:10.1103/PhysRev.109.1492)

[32] Lagendijk A, Tiggelen B V, Wiersma D S. Fifty years of Anderson localization. Phys. Today. 2009; 62: 24–29. (DOI: http://dx.doi.org/10.1063/1.3206091)

[33] Abrahams E, Anderson P W, Licciardello D C, Ramakrishnan T V. Scaling theory of localization: absence of quantum diffusion in two dimensions. Phys. Rev. Lett. 1979; 42: 673–676. (DOI:10.1103/PhysRevLett.42.673)

[34] Lee P A, Ramakeishnan T V. Disordered electronic systems. Rev. Mod. Phys. 1985; 57: 287–337. (DOI: 10.1103/RevModPhys.57.287)

[35] Dolan G J, Osheroff D D. Nonmetallic conduction in thin metal films at low temperatures. Phys. Rev. Lett. 1979; 43: 721–724. (DOI:10.1103/PhysRevLett.43.721)

[36] Ioffe A F, Regel A R. Semiclassical concept of transport. Prog. Semicond.1960; 4: 237

[37] Mott N F. Conduction in non-crystalline systems IX. The minimum metallic conductivity. Phil. Mag. 1972; 26: 1015–1026. (DOI:10.1080/14786437208226973)

[38] Hussey N E, Takenaka K, Takagi H. Universality of the Mott–Ioffe–Regel limit in metals. Phil. Mag. 2004; 84: 2847–2864. (DOI:10.1080/14786430410001716944)

[39] Gunnarsson O, Calandra M, Han J E. Colloquium: saturation of electrical resistivity. Rev. Mod. Phys. 2003; 75: 1085–1099 (DOI:10.1103/RevModPhys.75.1085)

[40] Licciardello D C, Thouless D J. Constancy of minimum metallic conductivity in two dimensions. Phys. Rev. Lett. 1975; 35: 1475–1478. (DOI:10.1103/PhysRevLett.35.1475)

[41] Bhalla A S, Guo R, Roy R. The perovskite structure—a review of its role in ceramic science and technology. Mat. Res. Innovations, 2000; 4: 3–26. (DOI: 10.1007/s100190000062)

[42] Li C, Soh K C K, Wu P. Formability of ABO_3 perovskite. J. Alloys Compd. 2004; 372: 40–48. (DOI: 10.1016/j.jallcom.2003.10.017)

[43] Ruddlesden S N, Popper P. New compounds of the K_2NiF_4 type. Acta Crystallogr. 1957; 10: 538–539. (DOI:10.1107/S0365110X57001929)

[44] Ruddlesden S N, Popper P. The compound $Sr_3Ti_2O_7$ and its structure. Acta Crystallogr. 1958; 11: 54–55. (DOI:10.1107/S0365110X58000128)

[45] Aurivillius B. Ark. Kemi 1949; 1: 463–480, 499–512.

[46] Dion M, Ganne M, Tournoux M. Nouvelles familles de phases $M_IM_{II2}Nb_3O_{10}$ a feuillets "perovskites." Mater. Res. Bull. 1981; 16: 1429–1435. (DOI: 10.1016/0025-5408(81)90063-5)

[47] Jacobson A J, Johnson J W, Lewandowski J T. Interlayer chemistry between thick transition-metal oxide layers: synthesis and intercalation reactions of $K[Ca_2Na_{n-3}Nb_nO_{3n+1}]$ ($3 \leq n \leq 7$). Inorg. Chem. 1985; 24: 3727–3729. (DOI: 10.1021/ic00217a006)

[48] Goldschmidt V M. Die gesetze der krystallochemie. Naturwissenschaften, 1926; 14: 477–485. (DOI: 10.1007/BF01507527)

[49] Cullity B D. Introduction to Magnetic Materials. Addison-Wesley; 1972.

[50] Moon S J, Jin H, Kim K W, Choi W S, Lee Y S, Yu J, Cao G, Sumi A, Funajubo H, Bernhard C, Noh T W. Dimensionality-controlled insulator-metal transition and correlated metallic state in 5d transition metal oxides $Sr_{n+1}Ir_nO_{3n+1}$ (n = 1, 2, and ∞). Phys. Rev. Lett. 2008; 101: 226402-4. (DOI:10.1103/PhysRevLett.101.226402)

[51] Crawford M K, Subramanian M A, Harlow R L, Fernandez-Baca, Wang Z R, Johnston D C. Structural and magnetic studies of Sr_2IrO_4. Phys. Rev. B. 1994; 49: 9198–9201. (DOI:10.1103/PhysRevB.49.9198)

[52] Cao G, Bolivar J, McCall S, Crow J E, Guertin R P. Weak ferromagnetism, metal-to-nonmetal transition, and negative differential resistivity in single-crystal Sr_2IrO_4. Phys. Rev. B (R). 1998; 57: 11039–11042. (DOI:http://dx.doi.org/10.1103/PhysRevB.57.R11039)

[53] Kim B J, Jin H, Moon S J, Kim J-Y, Park B-G, Leem C S, Yu J, Noh T W, Kim C, Oh S-J, Park J-H, Durairaj V, Rotenberg E. Novel J_{eff} = 1/2 Mott state induced by relativistic spin-orbit coupling in Sr_2IrO_4. Phys. Rev. Lett. 2008; 101: 076402-4. (DOI:10.1103/PhysRevLett.101.076402)

[54] Kim B J, Ohsumi H, Komesu T, Sakai S, Morita T, Takagi H, Arima T. Phase-sensitive observation of a spin-orbital mott state in Sr_2IrO_4. Science. 2009; 323: 1329–1332. (DOI: 10.1126/science.1167106)

[55] Wojek B M, Bernsten M H, Boseggia S, Boothroyd A T, Prabhakaran D, Mcmorrow D F, Ronnow H M, Chang J, Tjernberg O. The J_{eff} = 1/2 insulator $Sr_3Ir_2O_7$ studied by means of angle-resolved photoemission spectroscopy. J. Phys. Condens. Matter. 2012; 24: 415602-4. (DOI:10.1088/0953-8984/24/41/415602)

[56] Nichols J, Terzic J, Bittle E G, Korneta O B, Long L E D, Brill J W, Cao G, Seo S S A. Tuning electronic structure via epitaxial strain in Sr_2IrO_4 thin films. Appl. Phys. Lett. 2013; 102: 141908-4. (DOI:http://dx.doi.org/10.1063/1.4801877)

[57] Zocco D A, Hamlin J J, White B D, Kim B J, Jeffries J R, Weir S T, Vohra Y K, Allen J W, Maple M B. Persistent non-metallic behavior in Sr_2IrO_4 and $Sr_3Ir_2O_7$ at high pressures. J. Phys. Condens. Matter. 2014; 26: 255603-4. (DOI: 10.1088/0953-8984/26/25/255603)

[58] Domingo N, López-Mir L, Paradinas M, Holy V, Zelezny J, D, Suresha S J, Liu J, Rayan-Serrao C, Ramesh R, Ocal C, Martí X, Catalan G. Giant reversible nanoscale piezoresistance at room temperature in Sr_2IrO_4 thin films. Nanoscale. 2015; 7: 3453–3459. (DOI: 10.1039/C4NR06954D)

[59] Wang C, Seinige H, Cao G, Zhou J-S, Goodenough J B, Tsoi M. Electrically tunable transport in the antiferromagnetic Mott insulator Sr_2IrO_4. Phys. Rev. B. 2015; 92: 115136-6 (DOI:10.1103/PhysRevB.92.115136).

[60] Kim Y K, Krupin O, Denlinger J D, Bostwick A, Rotenberg E, Zhao Q, Mitchell J F, Allen J W, Kim B J. Science. Fermi arcs in a doped pseudospin-1/2 Heisenberg antiferromagnet. 2014; 345: 187–190. (DOI:10.1126/science.1251151)

[61] Lupascu A, Clancy J P, Gretarsson H, Nie Z, Nichols J, Terzic J, Cao G, Seo S S A, Islam Z, Upton M H, Kim J, Casa D, Gog T, Said A H, Katukuri V M, Stoll H, Hozoi L, van der Brink J, Kim Y-J. Tuning magnetic coupling in Sr_2IrO_4 thin films with epitaxial Strain. Phys. Rev. Lett. 2014; 112: 147201-5. (DOI: 10.1103/PhysRevLett. 112.147201)

[62] Ge M, Qi T F, Korneta O B, De Long D E, Schlottmann P, Crummett W P, Cao G. Lattice-driven magnetoresistivity and metal–insulator transition in single-layered iridates. Phys. Rev. B (R). 2011; 84: 100402-5. (DOI:10.1103/PhysRevB.84.100402)

[63] Fujiyama S, Ohsumi H, Ohashi K, Hirai D, Kim B J, Arima T, Takata M, Takagi H. Spin and orbital contributions to magnetically ordered moments in 5d layered perovskite Sr_2IrO_4. Phys. Rev. Lett. 2014; 112: 016405-5. (DOI:10.1103/PhysRevLett. 112.016405)

[64] Qi T F, Korneta O B, Li L, Butrouna K, Cao V S, Wan X, Schlottmann P, Kaul R K, Cao G. Spin-orbit tuned metal–insulator transitions in single-crystal $Sr_2Ir_{1-x}Rh_xO_4$ ($0 \leq x \leq 1$). Phys. Rev. B. 2012; 86: 125105-6. (DOI:10.1103/PhysRevB.86.125105)

[65] Wang C, Seinige H, Cao G, Zhou J-S, Goodenough J-B, Tsoi M. Anisotropic magnetoresistance in antiferromagnetic Sr_2IrO_4. Phys. Rev. X. 2014; 4: 041034-5. (DOI:10.1103/PhysRevX.4.041034)

[66] Kim J, Daghofer M, Said A H, Gog T, van den Brink J, Khaliullin G, Kim B J. Excitonic quasiparticles in a spin-orbit Mott insulator. Nat. Commun. 2014; 5: 4453–4456. (DOI:10.1038/ncomms5453)

[67] Subramanian M A, Crawford M K, Harlow R L. Single crystal structure determination of double layered strontium iridium oxide [$Sr_3Ir_2O_7$]. Mat. Res. Bull. 1994; 29: 645–650. (DOI: 10.1016/0025-5408(94)90120-1)

[68] Li L, Kong P P, Qi T F, Jin C Q, Yuan S J, DeLong L E, Schlottmann P, Cao G. Tuning the J_{eff}=1/2 insulating state via electron doping and pressure in the double-layered iridates $Sr_3Ir_2O_7$. Phys. Rev. B. 2013; 87: 235127-6. (DOI:10.1103/PhysRevB.87.235127)

[69] Cao G, Xin Y, Alexander C S, Crow J E, Schlottmann P, Crawford M K, Hatlow R L, Marshall W. Anomalous magnetic and transport behavior in the magnetic insulator $Sr_3Ir_2O_7$. Phys. Rev. B. 2002; 66: 214412-7. (DOI:10.1103/PhysRevB.66.214412)

[70] Kim J W, Choi Y, Kim J, Mitchell J F, Jackeli G, Daghofer M, van der Brink J, Khaliullin G, Kim B J. Dimensionality driven spin-flop transition in layered iridates. Phys. Rev. Lett. 2012; 109: 037204-5. (DOI:10.1103/PhysRevLett.109.037204)

[71] Li L, Kong P P, Qi T F, Jin C Q, Yuan S J, DeLong L E, Schlottmann P, Cao G. Tuning Jeff=1/2 insulating state via electron doping and pressure in double-layered iridate $Sr_3Ir_2O_7$. Phys. Rev. B. 2013; 87: 235127-6. (DOI: 10.1103/PhysRevB.87.235127)

[72] Kim J, Said A H, Casa D, Upton M H, Gog T, Daghofer M, Jackeli G, van den Brink J, Khaliullin G, Kim B J. Large spin-wave energy gap in the bilayer iridate $Sr_3Ir_2O_7$: evidence for the enhanced dipolar interactions near the Mott metal–insulator transitions. Phys. Rev. Lett. 2012; 109: 157402-5. (DOI:10.1103/PhysRevLett.109.157402)

[73] Okada Y, Walkup D, Lin H, Dhital C, Chang T-R, Khadka S, Zhou W, Jeng H-T, Bansil A, Wang Z, Wilson S, Madhavan V. Imaging the evolution of metallic states in a correlated iridates. Nat. Mater. 2013; 12: 707–713. (DOI:10.1038/nmat3653)

[74] Park H J, Sohn C H, Jeong D W, Cao G, Kim K W, Moon S J, Jin H, Cho D-Y, Noh T W. Phonon-assisted optical excitation in the narrow bandgap Mott insulator $Sr_3Ir_2O_7$. Phys. Rev. B. 2014; 89: 155115-6. (DOI:10.1103/PhysRevB.89.155115)

[75] Longo J M, Kafalas J A, Arnott R J. Structure and properties of the high and low pressure forms of $SrIrO_3$. J. Solid State Chem. 1971; 3: 174–179. (DOI: 10.1016/0022-4596(71)90022-3)

[76] Cao G, Durairaj V, Chikara S, DeLong L E, Parkin S, Schlottmann P. Non-Fermi liquid behavior in nearly ferromagnetic $SrIrO_3$ single crystals. Phys. Rev. B. (R), 2007; 76: 100402-4. (DOI: 10.1103/PhysRevB.76.100402)

[77] Zhao J G,Yang L X, Yu Y, Li F Y, Yu R C, Fang Z, Chen L C, Jin C Q. High-pressure synthesis of orthorhombic $SrIrO_3$ perovskite and its positive magnetoresistance. J. Appl. Phys. 2008; 103: 103706-5. (DOI:10.1063/1.2908879)

[78] Blanchard P E R, Reynolds E, Kennedy B J, Kimpton J A, Avdev M, Belik A A. Anomalous thermal expansion in orthorhombic peovskite $SrIrO_3$: interplay between spin-orbit coupling and the crystal lattice. Phys. Rev. B. 2014; 89: 214106-8. (DOI: 10.1103/PhysRevB.89.214106)

[79] Nie Y F, King P D C, Kim C H, Uchida M, Wei H I, Faeth B D, Ruf J P, Ruff J P C, Xie L, Pan X, Fennie C J, Schlom D G, Shen K M. Interplay of spin-orbit interactions, dimensionality, and octahedral rotations in semimetallic $SrIrO_3$. Phys. Rev. Lett. 2015; 114: 016401-6. (DOI:10.1103/PhysRevLett.114.016401)

[80] Carter J-M, Shankar V V, Zeb M A, Kee H-Y. Semimetal and topological insulator in perovsakite iridates. Phys. Rev. B. 2012; 85: 115105-6. (DOI:10.1103/PhysRevB.85.115105)

[81] Chen Y, Lu Y-M, Kee H-Y. Topological crystalline metal in orthorhombic perovskite iridates. Nat. Commun. 2015; 6: 6593–6597. (DOI:10.1038/ncomms7593)

[82] Zeb M A, Kee H-Y. Interplay between spin-orbit coupling and Hubbard interaction in $SrIrO_3$ and related Pbnm perovskites. Phys. Rev. B. 2012; 86: 085149-7. (DOI: 10.1103/PhysRevB.86.085149)

[83] Biswas A, Kim K-S, Jeong Y H. Metal insulator transitions in perovskite $SrIrO_3$ thin films. J. Appl. Phys. 2014; 116: 213704-10. (DOI: http://dx.doi.org/10.1063/1.4903314)

[84] Wu F X, Zhou J, Zhang L Y, Chen Y B, Zhang S-T, Gu Z-B, Yao S-H, Chen Y-F, Metal–insulator transition in $SrIrO_3$ with strong spin-orbit interaction. J. Phys. Condens. Matter. 2013; 25: 125604-8. (DOI:10.1088/0953-8984/25/12/125604)

[85] Yoshimatsu K, Okabe T, Kumihashira H, Okamoto S, Aizaki S, Fujimori A, Oshima M. Dimensional-crossover-driven metal–insulator transition in $SrVO_3$ ultrathin Films. Phys. Rev. Lett. 2010; 104: 147601-4. (DOI:10.1103/PhysRevLett.104.147601)

[86] Gruenewald J H, Nichols J, Terzic J, Cao G, Brill J W, Seo S S A. Compressive strain-induced metal–insulator transition in orthorhombic $SrIrO_3$ thin films. J. Mater. Res. 2014; 29: 2491–2496. (DOI:10.1557/jmr.2014.288)

[87] Zhang L, Liang Q, Xiong Y, Zhang B, Gao L, Handong Li H, Chen Y B, Zhou J, Zhang S-T, Gu Z-B, Yao S, Wang Z, Lin Y, Chen Y-F. Tunable semimetallic state in compressive-strained $SrIrO_3$ films revealed by transport behaviors. Phys. Rev. B. 2015; 91: 035110-9. (DOI:10.1103/PhysRevB.91.035110)

[88] Georges A, Kotliar G, Krauth W, Rozenberg M J. Dynamical mean-field theory of strongly correlated fermion systems and the limit of infinite dimension. Rev. Mod. Phys. 1996; 68: 13–125. (DOI:http://dx.doi.org/10.1103/RevModPhys.68.13)

[89] Vucicevic J, Terletska H, Tanaskovic D, Dobrosavljevic V. Finite-temperature crossover and the quantum Widom line near the Mott transition. Phys. Rev. B. 2013; 88: 075143-12. (DOI: 10.1103/PhysRevB.88.075143)

[90] Deng X, Mravlje J, Žitko R, Ferrero M, Kotliar G, Georges A. How bad metal turn good: spectroscopic signature of resilient quasiparticles. Phys. Rev. Lett. 2013; 110: 086401-5. (DOI:10.1103/PhysRevLett.110.086401)

[91] Löhneysen H, Rosch A, Vojta M, Wölfle P. Fermi-liquid instabilities at magnetic quantum phase transition. Rev. Mod. Phys. 2007; 79: 1015–1075. (DOI:10.1103/RevModPhys.79.1015)

[92] Harris A B. Nature of the "Griffiths" singularity in dilute magnets. Phys. Rev. B. 1975; 12: 203–207. (DOI: 10.1103/PhysRevB.12.203)

[93] Vojta T. Phase and phase transitions in disordered quantum systems. AIP Conf. Proc. 2013; 1550: 188–247. (DOI: 10.1063/1.4818403)

[94] Bhatt R N, Lee P A. Scaling studies of highly disordered spin-1/2 antiferromagnetic systems. Phys. Rev. Lett. 1982; 48: 344–347. (DOI:10.1103/PhysRevLett.48.344)

[95] Griffiths R B. Nonanalytic behavior above the critical point in a randomising ferromagnet. Phys. Rev. Lett. 1969; 23: 17–19. (DOI:10.1103/PhysRevLett.23.17)

[96] Byczuk K. Metal–insulator transitions in the Falicov–Kimball model with disorder. Phys. Rev. B. 2005; 71: 205105-8. (DOI:10.1103/PhysRevB.71.205105)

[97] Ruderman M A, Kittel C. Indirect exchange coupling of nuclear magnetic moments by conduction electrons. Phys. Rev. 1954; 96: 99–102. (DOI:http://dx.doi.org/10.1103/PhysRev.96.99)

[98] Kasuya T. A theory of metallic ferro-and antiferromagentism on Zener's model. Prog. Theor. Phys. 1956; 16: 45–57. (DOI: 10.1143/PTP.16.45)

[99] Yosida K. Magnetic properties of Cu-Mn alloys. Phys. Rev.1957; 106: 893–898. (DOI: 10.1103/PhysRev.106.893)

[100] Lee P A, Nagaosa N, Wen X-G. Doping a Mott insulator: physics of high-temperature superconductivity. Rev. Mod. Phys. 2006; 7: 17–85. (DOI: http://dx.doi.org/10.1103/RevModPhys.78.17)

[101] Florens S, Georges A. Slave-rotor mean-field theories of strongly correlated systems and the Mott transition in finite dimensions. Phys. Rev. B 2004; 70: 035114-15. (DOI: 10.1103/PhysRevB.70.035114)

[102] Cho D, Cheon S, Kim K-S, Lee S-H, Cho Y-H, Cheong S-W, Yeom H W. Nanoscale manipulation of the Mott insulating state coupled to charge order in 1T-TaS_2. arXiv: 1505.00690

[103] Kim K-S. Role of disorder in the Mott–Hubbard transition. Phys. Rev. B. 2006; 73: 235115-10. (DOI: 10.1103/PhysRevB.73.235115)

[104] Wegner F. The mobility edge problem: continuous symmetry and a conjecture. Z. Phys. B. 1979; 35: 207–210. (DOI: 10.1007/BF01319839)

[105] Schäfer L, Wegner F. Disordered system with n orbitals per site: Lagrange formulation, hyperbolic symmetry, and goldstone modes. Z. Phys. B. 1980; 3: 113–126. (DOI: 10.1007/BF01598751)

[106] Kim K-M, Kim K-S. in preparation

[107] Finkel'stein A M. Influence of coulomb interaction on the properties of disordered metals. Sov. Phys. JETP 1983; 57: 97–108

[108] Finkel'stein. A M. Electron liquid in disordered conductors. Sov. Sci. Rev. London; 1990; 14: 1–101.

[109] Punnoose A, Finkelstein A M. Metal–insulator transition in disordered two-dimensional electron systems. Science 2005; 310: 289–291. (DOI:10.1126/science.1115660)

[110] Anissimova S, Kravchenko S V, Punnoose A, Finkel'stein A M, Klapwijk T M. Flow diagram of the metal–insulator transition in two dimensions. Nat. Phys. 2007; 3: 707–710. (DOI: 10.1038/nphys685)

Chapter 8

Structural, Magnetic and Transport Properties of B-Site Substituted Perovskite $La_{0.7}Sr_{0.3}MnO_3$

J.B. Yang, M.S. Kim, T. F. Creel, H. Zhao, X.G. Chen, W.B. Yelon and W.J. James

Additional information is available at the end of the chapter

http://dx.doi.org/10.5772/61770

Abstract

In this chapter, in order to understand the structural related magnetic and transport properties of B site substituted perovskites $La_{0.7}Sr_{0.3}MnO_3$ (LSMO), we have systematically investigated the effects of replacing some of the Mn with nonmagnetic elements Ti, Zr, Cu, Al, Zn and magnetic elements Co, Ni, Cr, Fe. The structural, magnetic and electrical phase transitions and transport properties of these compounds were investigated by neutron diffraction, magnetization and electric resistivity measurements.

The abnormal behaviors relative to the parent manganite perovskite are explained by the competition between the double exchange and super exchange interactions, Mn-O bond distance, Mn-O-Mn bond angle, local Jahn-Teller distortion, the dilution of magnetization, the frustration of spins, and the change of valence states.

Keywords: Pervoskite, crystal structure, magnetic structure, neutron diffraction, magnetic properties

1. Introduction

Perovskite oxides have been an interesting research area for scientists due to their promising physical properties including colossal magnetoresistance (CMR), superconductivity, multiferroelectricity, metal-insulator transition (MIT), charge/orbital ordering, etc. Among various perovskite oxides, the manganite is a representative one with fascinating physical properties [37]. The manganite materials such as $La_{1-x}Sr_xMnO_3$, $Nd_{1-x}Sr_xMnO_3$, and $Pr_{1-x}Ca_xMnO_3$ exhibit rich phase diagram involving spin-charge-orbital ordering, canted antiferromagnetic, antiferromagnetic/ferromagnetic ordering, and electronic phase separation [14, 17, 27, 39]. In addition to the abundant magnetic behavior, the MIT often occurs coincidently with structural or magnetic transition [17]. Pure $LaMnO_3$ is an A-type antiferromagnetic insulator. La^{3+} can be partially substituted by a divalent cation such as Sr^{2+} or Ca^{2+}, and $La_{1-x}(Sr,Ca)_xMnO_3$ can become a metallic and ferromagnetic material. The Mn ions are in Mn^{3+} and Mn^{4+} states, which both have a local spin (S=3/2) from their t_{2g}^3 orbitals, and Mn^{3+} ion provides an extra electron from the e_g orbital responsible for conduction. The magnetic and electronic properties in these compounds can be explained using Zener's double exchange (DE) interaction [4, 11, 20, 40]. There is ferromagnetic interaction between the spin of the e_g^1 electron in Mn^{3+} and the local spin of t_{2g}^3. By Sr or Ca doping, the holes were introduced into the e_g band near the Fermi energy, which leads to mobile holes and conduction under an electric field. It was revealed that both DE interactions and the strong electron-phonon coupling should be considered to understand the Sr(Ca)-doped systems [10, 19, 30, 31, 35]. Polaron hopping was also proposed as the dominant conduction mechanism below T_C [3, 12, 21, 32, 41].

In the past decades, the A-site doped manganites $A_{1-x}D_xMnO_3$ have been extensively studied with various attractive properties [14, 17, 27, 29, 33, 39]. In contrast, the B-site doped manganites have not been well studied. The substitution for the Mn (B site) has shown dramatic effect on the magnetic and transport properties of the perovskites [1, 2, 5, 6, 36, 38]. Generally, the B-site doping with 3d ions would destroy the ferromagnetic ordering of the Mn network, leading to the changes in the magnetic and electrical properties of manganites. The reentrant spin glass behavior has been observed in the Cr-doped A-type antiferromagnetic $La_{0.46}S_{r0.54}Mn_{1-x}Cr_xO_3$ due to the competing interaction between the FM and the A-type AFM coupling. The charge-orbital ordered $Nd_{0.5}Ca_{0.5}Mn_{1-x}Cr_xO_3$ is a relaxor ferromagnet [26]. The Fe-doped $La_{1-x}Ca_xMnO_3$ has gone through the localization-delocalization transition as the increase of the dopant concentration [1, 34]. Two ferromagnetic phases appeared in the $LaMn_{0.5}Ni_{0.5}O_3$ sample, which is critically related to the preparation process [18].

Therefore, in this chapter, the samples of $La_{0.7}Sr_{0.3}Mn_{1-x}T_xO_3$ (T= Ti, Zr, Cu, Co and Cr) were prepared, and the effects of substitution on Mn were studied using neutron diffraction(ND), X-ray photoelectron spectra (XPS), magnetic and electric resistivity measurements. The relationship between structure and physical properties are explained by the competition between the DE and super exchange interactions, bandwidth W, bond angle, bond length and the frustration of spins and the change of valence states.

2. Experiments

Samples of $La_{0.7}Sr_{0.3}Mn_{1-x}T_xO_3$ (T=Ti, Zr, Cu, Co, Cr) were synthesized using the standard solid-state reaction method, starting with the high purity La_2O_3, MnO_2, TiO_2, CuO, Cr_2O_3, Co_3O_4 and $SrCO_3$ powders. Appropriate amounts of these powders were weighed and mixed according to the desired stoichiometry for each sample, then sintered in air for one day at 800℃, and cooled naturally to room temperature as the raw material. The raw materials were ground and sintered again in air for one day at 1350℃ with a room-air quench. The reacted powders were ground and cold pressed into disks with the thickness of ~2 mm under a pressure of ~10 MPa. These disks were sintered in air for one more day at 1350℃ and cooled naturally to room temperature. X-ray diffraction of the powders was performed at room temperature using with Cu-Kα radiation. Powder neutron diffraction experiments were performed at the University of Missouri-Columbia Research Reactor (MURR, λ = 1.4875Å) and high resolution powder diffractometer at HZB Germany using neutrons of wavelength (λ = 1.79821Å). The patterns were collected at the temperature range from 5K to 300K. Refinement of the XRD and ND data were carried out using the FULLPROF program. Magnetic measurements were conducted with a SQUID magnetometer (MPMS, Quantum design). The zero-field cooling (ZFC) and field cooling (FC) magnetization curves were measured under applied magnetic field of 50Oe. Magnetoresistance data were collected using a physical properties measurement system (PPMS, Quantum design) with a standard four-point probe method.

3. Results and discussion

3.1. Ti-substituted perovskites, $La_{0.7}Sr_{0.3}Mn_{1-x}Ti_xO_3$ [22]

Ti-substituted perovskites $La_{0.7}Sr_{0.3}Mn_{1-x}Ti_xO_3$, with $0 \leq x \leq 0.20$ were studied using XRD, ND, magnetizatic and magnetoresistance (MR) measurements [22]. Typical ND patterns of the $La_{0.7}Sr_{0.3}Mn_{1-x}Ti_xO_3$ (x=0.05, 0.1 and 0.2) were shown in Fig. 1. All samples show a rhombohedral structure (space group $R\bar{3}c$) from RT to 10 K. The crystal structure of the samples can be well fitted with the rhombohedral space group (No. 167), and the atomic positions of La(Sr): 6a (0,0,1/4), Mn(Ti): 6b (0,0,0); O18e (x,0,1/4). The magnetic structure was refined with P1 space group. It was confirmed that the Ti ions replaced the Mn ions at B sites, since the ionic radius of Ti^{4+} (0.605Å) lies between the ionic radius of Mn^{4+} (0.530 Å) and Mn^{3+} (0.645 Å)[38]. Fig. 2 plots the lattice parameter of the samples at different temperatures. It can be seen that the different behaviours are observed at RT and 10K. The lattice parameter *a* and unit cell volume reach a maximum value for x = 0.10, and then decrease for x > 0.10, while lattice parameter *c* increases with increasing x at 10 K. However, at RT, the lattice parameters *a*, *c* and the unit cell volume *V* all show similar increasing trend with increasing Ti content x. This difference should be related to the different magnetic ordering temperatures due to Ti substitution. In this system, as Mn ions are substituted by Ti ions, a lattice distortion may occur. We have calculated the so-called tolerance factor *t* of the $La_{0.7}Sr_{0.3}Mn_{1-x}Ti_xO_3$, which can be used to indicate the geometric measure of size mismatch of perovskites. The calculated values of *t* are about 0.928 and 0.921 for $La_{0.7}Sr_{0.3}MnO_3$ and $La_{0.7}Sr_{0.3}Mn_{0.8}Ti_{0.2}O_3$, respectively. These values are in the stable range of the $0.89 < t < 1.02$ for the rhombohedral structure. The decrease of the *t* will lead to the decrease of Mn-O-Mn bond angle from 166.5 º to 165 º for x=0 and x=0.2 without a structure change.

Fig. 3 plots the average Mn-O bond length and Mn-O-Mn bond angle of $La_{0.7}Sr_{0.3}Mn_{1-x}Ti_xO_3$ obtained from the ND refinement at 10 K and RT. At 10 K, the Mn-O bond length increases up to *x = 0.10* and keeps constant for *x>0.10*, while the Mn-O-Mn bond angle decreases with increasing x. At RT, the bond length of Mn-O increases up to x = 0.15 and keeps constant for *x≥0.15*, while the Mn-O-Mn bond angle decreases and reaches a minimum value for x = 0.15. The changes in the bond length and bond angle of MnO_6 octahedral may help to diminish the internal strain induced by Ti substitution. The Ti substitution will change the oxygen positions

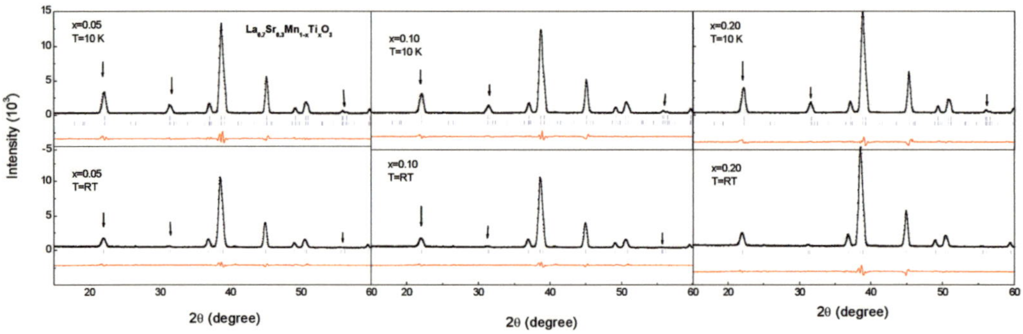

Figure 1. ND patterns of $La_{0.7}Sr_{0.3}Mn_{1-x}Ti_xO_3$ (x = 0.05, 0.10, 0.20) at RT and 10 K. (The bottom curves (red line) are the difference between experimental data and the refinement data. The vertical bars (blue line) represent the magnetic (bottom) and Bragg (top) peak positions). Arrows shows some of the major magnetic diffraction peaks.

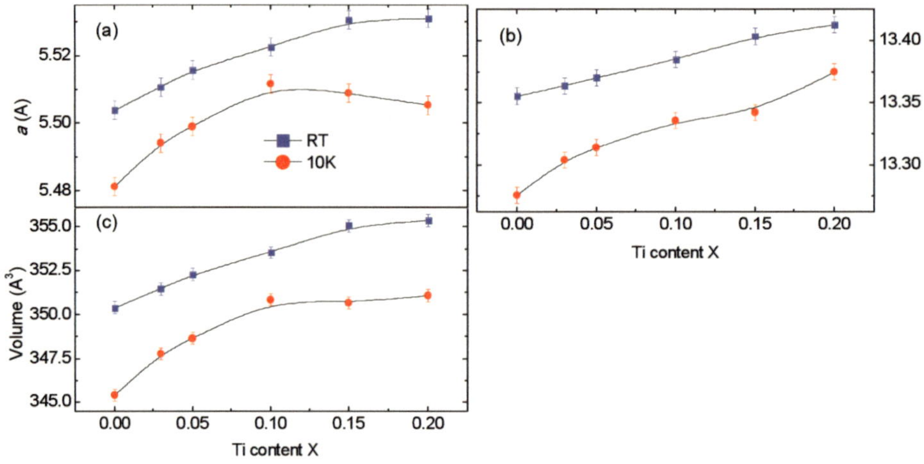

Figure 2. Lattice parameters a, c, and unit cell volumes of $La_{0.7}Sr_{0.3}Mn_{1-x}Ti_xO_3$ with different Ti content x at 10 K and RT.

and affect the Mn-O bond length and the Mn-O–Mn bond angle. This is consistent with the change of the tolerance factor, indicating the increase of Mn-O bond length and the decrease of Mn-O-Mn bond angle are related.

The temperature dependent resistivities for $La_{0.7}Sr_{0.3}Mn_{1-x}Ti_xO_3$ compounds (x = 0.0(a), 0.05(b), 0.10(c), and 0.15(d)) under magnetic fields H = 0, 1, 3, and 5 T were plotted in Fig. 4.

As a comparison, the Curie temperatures *Tc* of the samples are also shown in the figure 4. It can be seen that the resistivity for the x ≤ 0.05 sample shows a metallic-like behavior below the T_C. A MIT is observed for all the x ≥ 0.10 samples at low temperature. A maximum peak in the resistivity is observed below T_C for all samples, and shifts to a lower temperature when x increases. The one-electron bandwidth *W* is one of the fundamental parameters for controlling

http://dx.doi.org/10.5772/61770

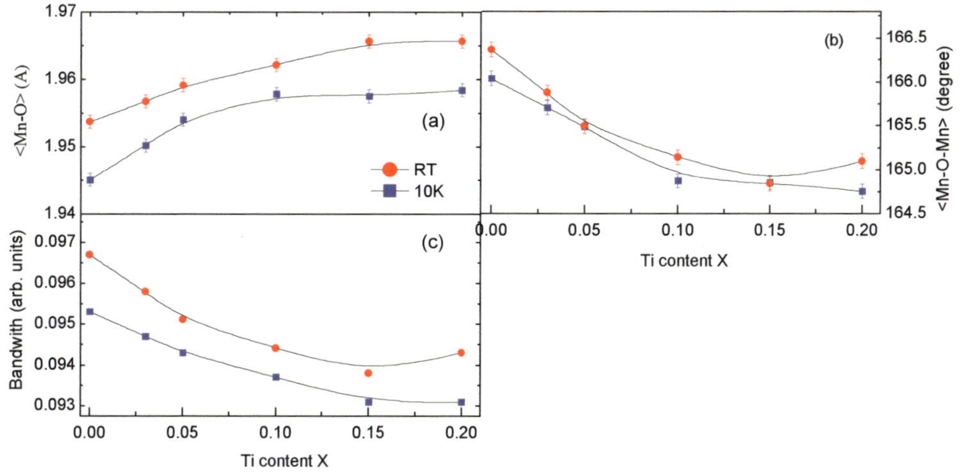

Figure 3. Average Mn-O bond lengths (a), Mn-O-Mn bond angles (b), and electronic bandwidth parameter *W* (c), of $La_{0.7}Sr_{0.3}Mn_{1-x}Ti_xO_3$ at room temperature and at 10K.

the magnetic and electric behavior of correlated electrons system such as perovskite [15, 35]. Using the tight binding approximation, the empirical formula of the *W* for ABO_3-type perovskites is [29].

$$W = \frac{\cos\beta}{(d_{Mn-O})^{3.5}} \tag{1}$$

where $\beta=(\pi-\theta_{<Mn-O-Mn>})/2$, $d_{Mn\text{-}O}$ is the average Mn-O bond length, and $\theta_{<Mn-O-Mn>}$ is the average Mn-O-Mn bond angle. By using the data obtained from ND refinements, the calculated values of the *W* are plot in Fig. 3 (c). It is obvious that *W* decreases with the increasing Ti content x, which will reduce the 2p-3d hybridization between O and Mn ions and increases the electron-phonon coupling. Therefore, it leads to a lower magnetic ordering temperature T_C and higher resistivity with increasing Ti content. Since the double exchange interaction between Mn-Mn ions is strongly dependent on both bond angle and bond distance, the substitution of Mn by Ti will decrease the exchange interaction between Mn-Mn ions.

It is obvious that a field-induced shift of the resistivity maximum occurs for x> 0.05 samples. The MR ratio increases with the Ti content x, and reaches to about 70% for $La_{0.7}Sr_{0.3}Mn_{0.8}Ti_{0.2}O_3$, which is related to the weaker magnetic interaction between Mn-Mn ions. The separation of T_C and the resistivity maximum temperature $T_{\varrho,max}$ becomes wider as Ti content increases due to the weak coupling between the magnetic ordering and the resistivity as compared with $La_{0.7}Sr_{0.3}MnO_3$.

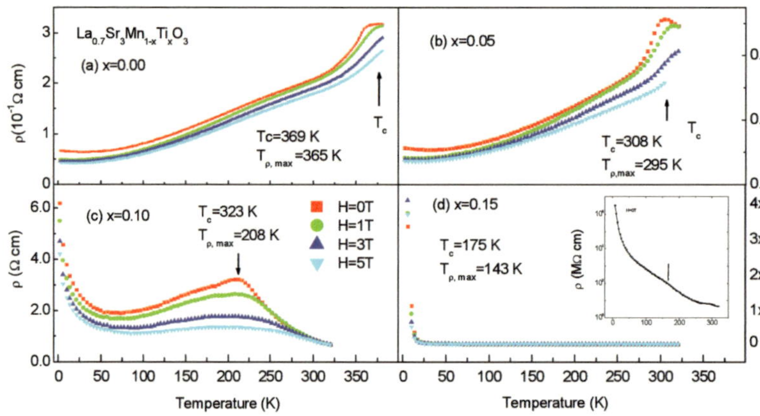

Figure 4. The temperature dependent resistivity for $La_{0.7}Sr_{0.3}Mn_{1-x}TixO_3$ compounds (x = 0.0(a), 0.05(b), 0.10(c), and 0.15(d)) under magnetic field H = 0, 1, 3, and 5T. Arrows indicate the resistivity maximum temperature Tϱ,max. The inset in (d) is the plot of resistivity of x = 0.15 compound (with log scale) in H = 0T.

3.2. Zr-substituted perovskites $La_{0.7}Sr_{0.3}Mn_{1-x}Zr_xO_3$[23]

We have tried to synthesize the $La_{0.7}Sr_{0.3}Mn_{1-x}Zr_xO_3$ compounds with different Zr contents. However, it was found that solubility limit of Zr is about x ~ 0.10, due to the large size (0.72 Å) of Zr^{4+}. Fig. 5 is the ND patterns of $La_{0.7}Sr_{0.3}Mn_{1-x}Zr_xO_3$. It reveals that Zr goes only to the Mn-site. A single phase of $La_{0.7}Sr_{0.3}Mn_{1-x}Zr_xO_3$ was obtained for x≤0.1, which exhibits a rhombohedral structure from 10 K to RT. An impurity $La_2Zr_2O_7$ phase was found for x>0.1 samples. The refined lattice parameters *a, c* and unit cell volume increase with Zr content due to the large ionic radius of the Zr ions as compared to that of the Mn ions.

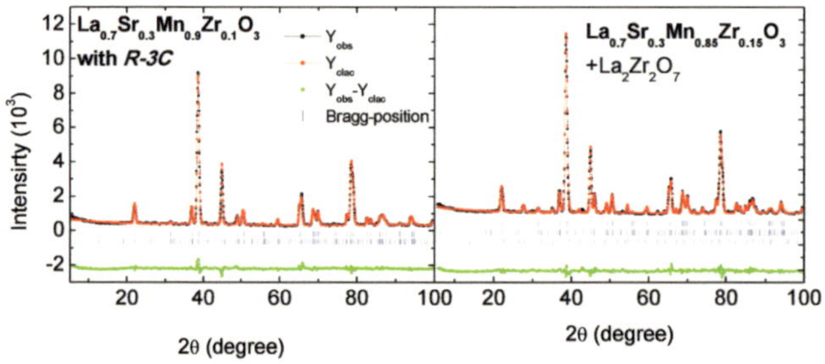

Figure 5. Neutron diffraction patterns of $La_{0.7}Sr_{0.3}Mn_{0.9}Zr_{0.1}O_3$ and $La_{0.7}Sr_{0.3}Mn_{0.85}Zr_{0.15}O_3$ at room temperature.

Fig. 6 plots the average Mn-O bond length,Mn-O-Mn bond angle and bandwidth W obtained from the refined ND patterns at RT and 10K. Similar to those of the Ti substituted samples, the Mn-O bond length increases, while Mn-O-Mn bond angle and band width W decrease with the increase of Zr content. The decrease of the bandwidth W will reduce the overlap between the O-2p and the Mn-3d orbitals, which will reduce the exchange interactions between Mn-

http://dx.doi.org/10.5772/61770

Mn in this system. This is confirmed by the results that the reduction in magnetic moments and the Curie temperature with increased Zr content. A metallic-like behavior was observed for the $La_{0.7}Sr_{0.3}Mn_{1-x}Zr_xO_3$ at low temperature. The contribution from the two-magnon scattering in resistivity becomes larger with increasing Zr content.

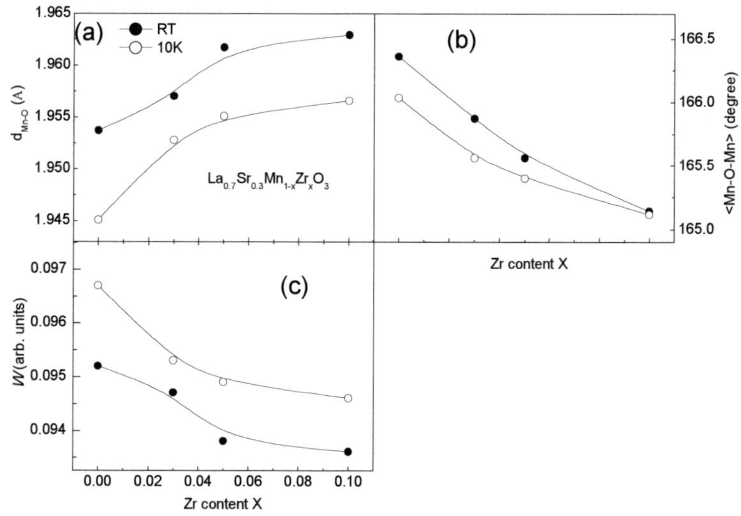

Figure 6. Average Mn-O bond lengths (a), Mn-O-Mn bond angles (b), and electronic bandwidth parameter *W*, of $La_{0.7}Sr_{0.3}Mn_{1-x}Zr_xO_3$ (x = 0.0, 0.03, 0.05, 0.10) at room temperature and at 10K.

3.3. Cu-substituted perovskites $La_{0.7}Sr_{0.3}Mn_{1-x}Cu_xO_3$ [24, 25]

$La_{0.7}Sr_{0.3}Mn_{1-x}Cu_xO_3$ samples with 0 n_p x $z_{i\alpha}$ 0.20 were prepared. XRD and ND patterns indicate that Cu goes into the Mn sites and all samples can be refined with the rhombohedral structure with the R $\bar{3}$ c space-group from 10 K to RT [24, 25]. It was noticed that the lattice parameters a,c and the unit cell volumes decrease with increasing Cu content at RT, while they remain nearly constant with increasing Cu content at 10 K. Since Cu^{2+} has a larger radius (0.73Å) than those of Mn^{3+} and Mn^{4+}, both Cu^{2+} and Cu^{3+} states may appear in these compounds accounting for the abnormal changes of the unit cell volume and the lattice parameters with Cu content. The refined magnetic moments of Mn atoms at 10 K and RT decrease with increasing Cu content. These results agree well with the values obtained from magnetic measurements. There is no sign of antiferromagnetic couplings between Cu and Mn ions.

Fig. 7 plots the Mn-O bond length and Mn-O-Mn bond angle obtained from the ND data. The changes of the Mn-O bond length vs. Cu content show similar trend corresponding to the changes of lattice parameters at 10 K and RT. The average bond angle increases and reaches to a maximum at x = 0.10, then slightly decreases for x > 0.1 at 10K an RT. The bandwidth *W* of the Cu substituted samples shows different effects as compared those of Ti and Zr substituted samples. At 10 K, an increase of the bandwidth *W* with Cu content *x* was observed, which may be related to the Cu valence states(See later discussion). This will affect the exchange mechanisms and magnetic interactions in this system and is observed through the reduction in ferromagnetic moments with increased Cu content. This indicates a likely reduction in the Mn^{3+}-O-Mn^{4+} ferromagnetic double exchange interactions, a direct result of the reduction in Mn^{3+} ions.

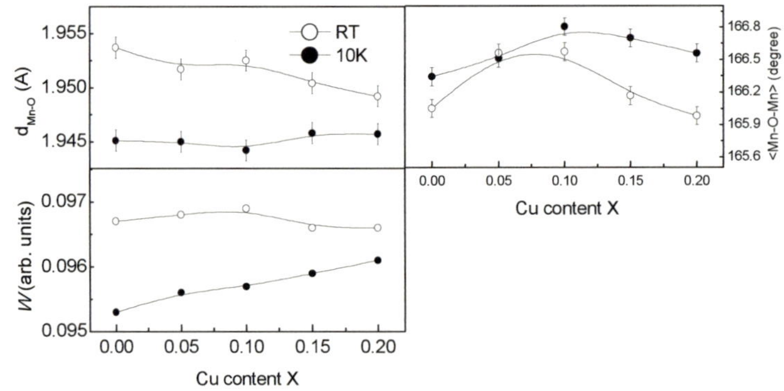

Figure 7. The average Mn-O bond lengths (a), Mn-O-Mn bond angles (b) and band width (c) of $La_{0.7}Sr_{0.3}Mn_{1-x}Cu_xO_3$ at RT and at 10 K.

X-ray photoelectron spectra (XPS) was used to determine the Cu valence states in these compounds. Fig. 8 plots the mole percent of different Cu ions in the Mn sites obtained by fitting the XPS data. The Two kinds of Cu ions (Cu^{3+} and Cu^{2+}) were observed when $x > 0.1$, unlike the other metal substituted systems. The binding energies of both Cu^{2+} and Cu^{3+} states shift to the lower BE energy region, which suggests a strong hybridization between the Cu-2*p* state and the O-*1s* state.

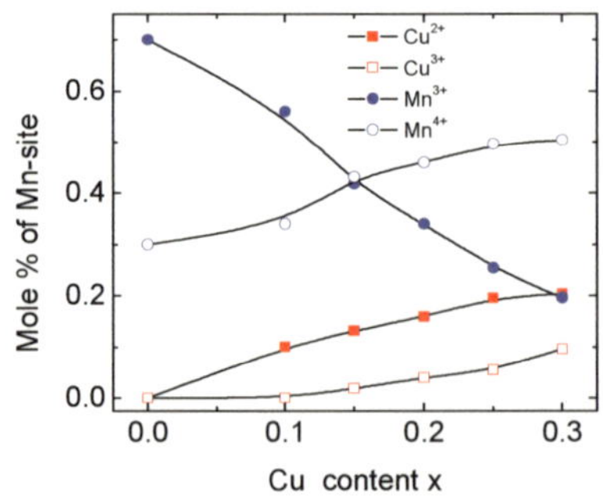

Figure 8. The mole percent of different Cu ions in the Mn sites obtained from XPS

Fig. 9 plots the Curie temperatures for the Cu substituted samples. The T_C decreases with increased Cu content. It was found that the low Cu-doped samples (x= 0.05, Tc=365 K) show almost no decrease in T_C as compared to the parent LSMO. This suggests that the Cu-doping could enhance the exchange coupling of Mn^{3+}-Mn^{4+} due to the Cu^{2+}, and increase the magnetic ordering temperature T_C at a low Cu-doping ratio. At a high doping ratio, the magnetic dilution effect of Cu is predominant, which gives rise to a sharp drop in the Curie temperature (Tc=250 K) for x=0.15. This is also related to the Cu ionic states. While the signature separation between the zero field cooling(ZFC) and field cooling (FC) curves is present, the complex curves observed in the other 3d-substituted systems is only weakly observable in the x=0.15 sample. This indicates there may be antiferromagnetic ordering up to 50 K that is too small to be observable within the resolution of the neutron diffraction analysis.

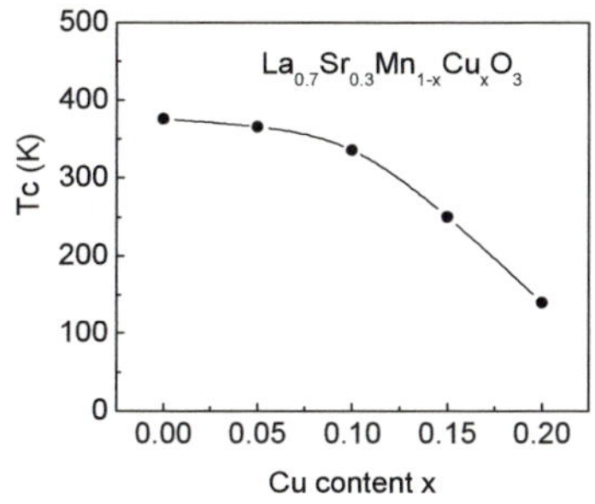

Figure 9. The Curie temperature of $La_{0.7}Sr_{0.3}Mn_{1-x}Cu_xO_3$.

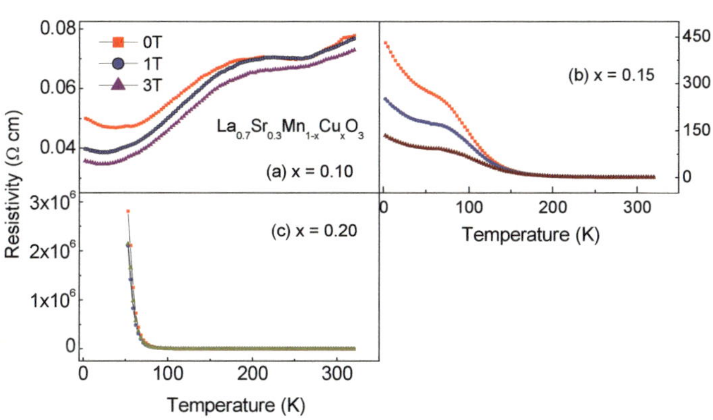

Figure 10. Temperature dependent resistivities of $La_{0.7}Sr_{0.3}Mn_{1-x}Cu_xO_3$ (x = 0.10(a), 0.15(b), 0.20(c)) under applied magnetic fields of H = 0, 1, and 3T.

The temperature dependence of resistivity under various applied fields was measured using PPMS and shown in Fig. 10. With increasing Cu content, the resistivity of the compound increases, while the resistivity decreases with increasing magnetic field. This is ascribed to a reduction of the Mn^{3+}/Mn^{4+} ratio to account for the DE interaction and a reduction in the number of hopping electrons and hopping sites by Cu substitution. The resistivity shows a metal-like behavior with decreasing temperature when x is less than 0.10 samples. A MIT occurs for the $x \geq 0.15$ samples (Fig. 10). A resistivity peak corresponding to the magnetic transition is present.

The suppression of the resistivity by the applied magnetic field occurs over the entire temperature range for all samples. The highest MR ratio of about 80% was obtained for x = 0.15 sample, which might result from the co-existence of Cu^{3+}/ Cu^{2+} and the dilution effect of Cu-doping on the double exchange interaction [25].

3.4. Co-substituted $La_{0.7}Sr_{0.3}Mn_{1-x}Co_xO_3$

Typical ND patterns for the $La_{0.7}Sr_{0.3}Mn_{1-x}Co_xO_3$ samples at RT are shown in Figure 11. It is obvious that the peak intensity of (012) decreases and those of (110) and (104) increase with the increase of the cobalt content x, which is due to the different scattering lengths of Mn and Co ions. Since these peaks are related to magnetic scattering, the changes correspond to the decrease of the magnetic contribution. A Rietveld refinement for all the polycrystalline samples was carried out to understand the detailed crystal and magnetic properties. Figure 12 displays the refined lattice parameters for the $La_{0.7}Sr_{0.3}Mn_{1-x}Co_xO_3$ samples at RT. It shows that the lattice parameters and unit cell volume decrease with increasing the Co content due to the fact that the radius of Co ions (0.55 Å for Co^{3+} and 0.40 Å for Co^{4}+) is smaller than that of Mn ions (0.65 Å for Mn^{3+} and 0.53 Å for Mn^{4+}). Generally speaking, the DE interaction strength among Mn^{3+} and Mn^{4+} can be estimated using the transfer integral, $t \propto \cos(\theta/2)$ and thus strongly depends on the Mn-O-Mn bond angle. The changes in θ value also have strong influence on the effective bandwidth *W*. For a charge-transfer insulator, the band gap energy *Eg* in the insulating phase can be written as $E_g = \Delta - W$, where Δ is the charge-transfer energy and W is the O-2p-like bandwidth (Kim et al., 2005). In fact, Δ changes little in the $La_{1-x}Sr_xMnO_3$ system thus the bandwidth *W* becomes a main factor in turning the band gap energy [16]. The decrease of bandwidth for Co-substituted compounds reduces the overlap between the O-2p and the Mn-3d orbitals. It also increases the band gap energy *Eg*, which in turn decreases the fragile double exchange coupling of Mn^{3}+-O-Mn^{4+}, as well as the magnetization and Curie temperature Tc (375 K for $La_{0.7}Sr_{0.3}MnO_3$, 226 K for $La_{0.7}Sr_{0.3}CoO_3$, and 184, 193, 180 K for x=0.4, 0.5 and 0.6, respectively), with the transition from the metal to insulator.

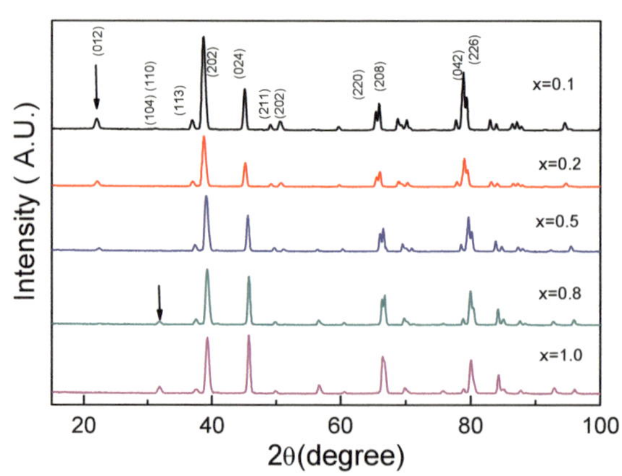

Figure 11. Neutron diffraction patterns of the $La_{0.7}Sr_{0.3}Mn_{1-x}Co_xO_3$ sample at RT.

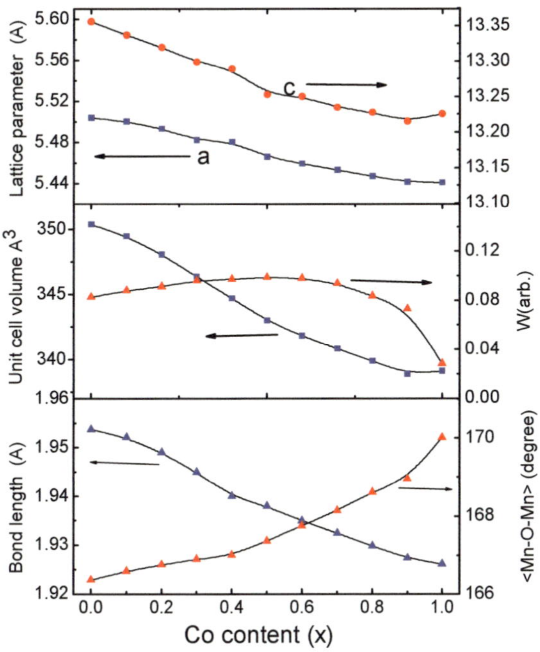

Figure 12. The lattice parameters a, c, unit cell volume, bandwidth *W*, bond length d_{Mn-O}, and bond angle <Mn-O-Mn> for $La_{0.7}Sr_{0.3}Mn_{1-x}Co_xO_3$ at RT.

The temperature dependent ND patterns of $La_{0.7}Sr_{0.3}Mn_{1-x}Co_xO_3$ (x=0.4, 1.0) samples are shown in Figure 13(a). The representative Bragg reflections of neutron diffraction prior to and with the addition of magnetic phase are shown in Figure 13(b). The misfits indicate the magnetic contributions. It is obvious that the (012) reflection has both nuclear and magnetic intensities and the (104), (110) reflections show little magnetic intensity for x=0.4. However, for x=1.0 sample, the (012) peak has magnetic intensity only and (104), (110) has both nuclear and magnetic intensities. There is almost no change of the intensity for x=0.5 and x=0.6 samples whether to add magnetic phase or not. The intensity of the magnetic peak (012) for LSCO decreases with the increase of temperature until vanishes finally, which is similar to LSMO-Co0.4 (i.e. the magnetic peaks (104), (110)).

The Curie temperature (T_C), the coercivity ($_iH_C$), the magnetization and the resistivity of $La_{0.7}Sr_{0.3}Mn_{1-x}Co_xO_3$ are shown in Figure 1. The critical Co-doping contents for the onset/ disappearance of the glassy behavior are x=0.3 and 0.8. The $_iH_C$ and resistivity show a maximum value, while the T_C and magnetization show a minimum value at the critical Co-doping point (x=0.3, 0.8). The ferromagnetic ordered $La_{0.7}Sr_{0.3}MnO_3$ gradually turns into disordered glassy system by the B site Co-doping, which is attributed to the break of the double exchange interaction between Mn-Mn ions and random substitution of the Mn ions. At the intermediate Co-doping region, the ferromagnetic ordered clusters embed in the antiferromagnetic ordering cluster matrix, forming the superparamagnetic-like free spin and reentrant spin glassy states. The resistivity increases due to the break of the double exchange interaction and the phase

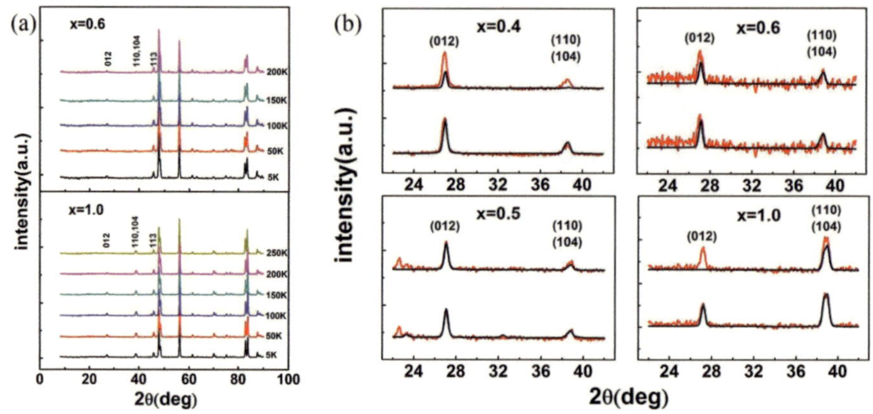

Figure 13. (a) ND patterns for the $La_{0.7}Sr_{0.3}Mn_{1-x}Co_xO_3$ (x=0.4, 1.0) samples at different temperature. (b) Representative Bragg reflections of neutron diffraction data for the $La_{0.7}Sr_{0.3}Mn_{1-x}Co_xO_3$ (x=0.4, 0.5, 0.6, 1.0) samples. Each data graphic contains two groups of plot (red for the experimental data and black for the refined data), the under one with the nuclear and magnetic phase refined and the upper one with only the nuclear phase.

separation. And the destruction of the long range exchange coupling leads to the decrease of the Curie temperature. The sparsely alignment of the cluster makes the decrease of the magnetization

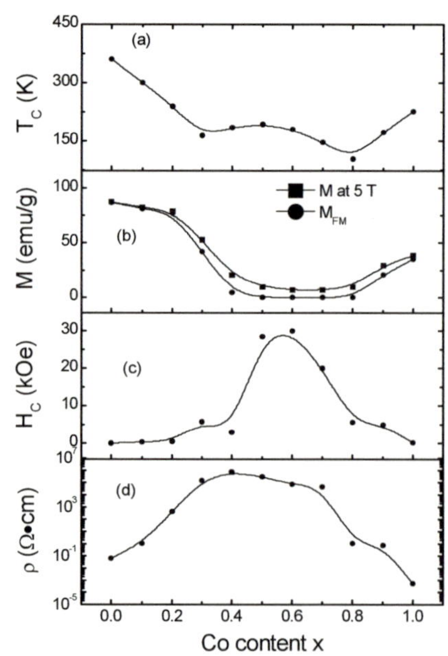

Figure 14. (a) Curie temperature (TC), (b) the magnetization at the field of 5 T and the ferromagnetic component extracted from the M-H curve, (c)the coercivity at 5 K, (d) the resistivity at 70 K of $La_{0.7}Sr_{0.3}Mn_{1-x}Co_xO_3$.

The typical hysteresis loops at various temperatures for $La_{0.7}Sr_{0.3}Mn_{0.5}Co_{0.5}O_3$ were plotted in figure 15. This is a representative one for the intermediate Co-doping samples (0.2<x<0.8), which is consistent with the simple cluster model (where the Co^{3+}-Co^{4+} or Mn^{3+}-Mn^{4+} ferromagnetic clusters exist in the Co^3+-Co^3+(Mn^{3+}) or Co^{4+}-Co^{4+}(Mn^{4+}) antiferromagnetic cluster). It should be noted that the hysteresis loop at 5K shows a jump at the vicinity of 0 T. The jump disappears when the temperature is just above the Curie temperature. This suggests that the observed phenomenon is related to the competition between the ferromagnetic double exchange interaction and the antiferromagnetic superexchange interaction. The $_iH_{C\ i}$s are significantly enhanced due to the freezing and pinning of the domain wall. The hysteresis loops do not show saturation under a magnetic field of 5T, which is consistent with the cluster glass behaviour. It is proposed that under a high magnetic field, the ferromagnetic clusters and partial antiferromagnetic clusters are forced to align along the direction of magnetic field. But the magnetization could not be saturated on account of the existence of antiferromagnetic clusters, so partial soft ferromagnetic clusters with small coercivity are demagnetized easily around H=0. As a result, a sudden drop of the magnetization has occurred, giving rise to a jump in the hysteresis curve.

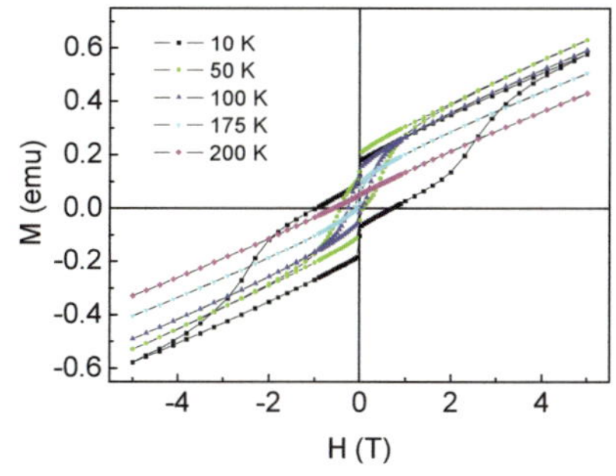

Figure 15. The hysteresis loops of $La_{0.7}Sr_{0.3}Mn_{0.5}Co_{0.5}O_3$ at different temperatures.

3.5. Cr-doped $La_{0.7}Sr_{0.3}Mn_{1-x}Cr_xO_3$ [7, 8, 9]

The $La_{0.7}Sr_{0.3}Mn_{1-x}Cr_xO_3$ (0<x<0.6) have been prepared and influence of the Cr^{3+} substitution for Mn^{3+} was investigated [7, 8, 9]. Figure 14 is the ND patterns of $La_{0.7}Sr_{0.3}Mn_{1-x}Cr_xO_3$ (0<x<0.5) at 10 K. The magnetic contributions to the (012) and (110)+(104) peaks are evident. For x<0.2, the samples are simple ferromagnetic with magnetic moments decreasing with increasing Cr content. For 0.2≤x≤0.4, the (104) + (110) reflections become weaker, but two new peaks ((003) + (011)), inconsistent with a simple ferromagnetic solution emerge. The ((003) + (011)) peak is purely magnetic, while the (113) peak has nuclear and magnetic components. For x=0.5, the (104) + (110) reflections are present but weak, while the (003) + (011) reflections are now dominant. A single, homogeneous, long-range magnetically ordered state with compositionally-dependent charge ordering was proposed to fit the ND patterns [8, 9]. The magnetic structures are related to the competition between Mn-Mn, Mn-Cr and Cr-Cr interactions (double-exchange and superexchange). The metal to semi-metal and semi-metal to insulator transitions can be quantitatively described as due to the localization effect of superexchange. The presence of charge ordered states above the M-I transition concentration (x) arises from the favourable energetics of Mn^{4+}-O-Cr^{3}+ superexchange bonds relative to Mn^{3}+-O-Cr^{3+} bonds.

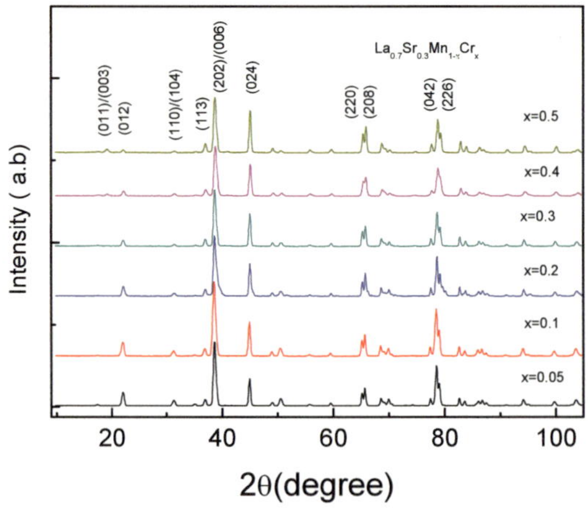

Figure 16. The neutron diffraction patterns of the $La_{0.7}Sr_{0.3}Mn_{1-x}Cr_xO_3$ sample at 10 K.

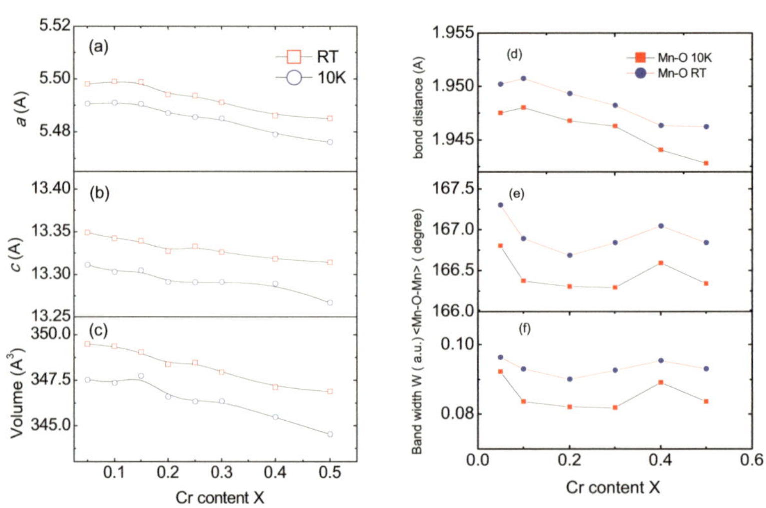

Figure 17. The lattice parameters a, c, unit cell volume, bandwidth W, bond length d_{Mn-O}, and bond angle <Mn-O-Mn> for $La_{0.7}Sr_{0.3}Mn_{1-x}Cr_xO_3$ at RT and 10 K.

At low temperature with small x, the net ferromagnetic behavior of the system is due to the large quantities of Mn^{3+}-O-Mn^{4+} ferromagnetic double exchanges taking place while the system is being driven towards a layered ferromagnetic structure by the antiferromagnetic Cr^{3+}-O-Mn^{4+} superexchanges. In the intermediate region (0.2<x<0.4) charge ordering creates a layered structure and the antiferromagnetic Cr^{3+}-O-Mn^{4+} superexchange continues to drive the system towards an antiferromagnetic state. As x>0.4, the antiferromagnetic Cr^{3+}-O-Cr^{3+} and Cr^{3+}-O-Mn^{4+} superexchange mechanisms become dominate, with charge order persisting, producing a ferrimagnetic structure in lieu of an antiferromagnetic one.

Figure 17 displays the refined lattice parameters, unit cell volume, bond length d_{Mn-O}, bond angle <Mn-O-Mn>, and band width W of $La_{0.7}Sr_{0.3}Mn_{1-x}Cr_xO_3$ samples at RT and 10 K. It can be seen that that the lattice parameters, cell volume and bond distance slightly decrease with increasing the Cr content x, which originates from the fact that the radius of Cr ions (0.615 Å for Cr^{3+}) is little bit smaller than that of Mn ions (0.65 Å for Mn^{3+}). The Mn-O-Mn bond angle decreases to a minimum for x=0.2 and then increases to a maximum for x=0.4. A similar trend can be observed for the band width W. This is consistent with the fact that a charge ordering creates a layered structure and the antiferromagnetic Cr^{3+}-O-Mn^{4+} superexchange continues to drive the system towards an antiferromagnetic state.

4. Conclusion

The B-site substituted LSMOs can be divided into following two groups, (1) those made with the replacement of Mn by other 3d transition metal ions and (2) those made with the replacement of Mn by non-magnetic, closed shell, metal ions such as Ti, Zr. The ionic radii of the substituted elements, the Mn-O-Mn bond angles, the Mn-O bond length, the calculated bandwidths W, and the corresponding T_C's are given in Table I. It should be pointed out that neutron diffraction scattering lengths of the 3d elements are sufficiently different, and uniquely, the scattering length of Mn is negative. This allows relatively small amounts of other elements substituted into the manganites to be accurately located in the unit cell structure by employing neutron diffraction.

Substituted atom	**Electronic configuration**	**Ionic radius (Å)**	Mn-O bond length (Å)	**Mn-O-Mn bond angle(degree)**	T_C (K) (x=0.1)
Ti^{4+}	[Ar]	0.605	1.9621	165.142	231
Zr^{4+}	[Kr]	0.72	1.9566	166.031	318
Zn^{2+}	[Ar]	0.74	1.9538	166.41	340
Cu^{2+}/Cu^{3+}	$[Ar]3d^9/3d^8$	0.73/0.54	1.9507	166.733	300
Fe^{3+}/Fe^{4+}	$[Ar]3d^5/3d^4$	0.645/0.585	1.9568	166.179	310
Cr^{3+}/Cr^{4+}	$[Ar]3d^3/3d^2$	0.615/0.55	1.9507	166.896	327
Co^3+/Co^{4+}	$[Ar]3d^6/3d^5$	0.61/0.53	1.9515	166.591	324
Ni^{2+}/Ni^{3+}	$[Ar]3d^8/3d^7$	0.69/0.6	1.9506	166.986	289
Mn^{3+}/Mn^{4+}	$[Ar]3d^4/3d^3$	0.64/0.53	1.9451	166.03	370

Table 1. The electronic configuration, ionic radii, Mn-O bond length and Mn-O-Mn bond angle at room temperature, and T_C for $La_{0.7}Sr_{0.3}Mn_{1-x}M_xO_3$ (x=0.1, M = Ti, Zr, Zn, Cr, Fe, Co, Ni, and Cu).

4.1. Substitution the Mn-site by 3d-transition metals (TM); Cr, Fe, Co, Ni, Cu

For the TM-substituted LSMOs, they show the same crystal structure with space group $R\bar{3}c$. There are small changes in the lattice parameters, and almost the same valence-band structures near the Fermi level are observed. A decrease in T_C is observed for all elements and a metal to insulator (MIT) transition occurs with increasing substitution content around x=0.2. The decrease in T_C of the parent LSMO with substitution by other 3d transition metals is strongly related to the following two factors; (i) the magnetic moments of the substituting ions and (ii) the ionic radius mismatch between the substituting ion and Mn ions. For the TM-substituted LSMO, the degree of reduction in T_C and saturation magnetization is related to the valence states of the substituting ions and the ionic size mismatch. The highest T_C is obtained from the Cr- and Co-substituted LSMO. This is mainly due to the large magnetic moments of Cr and Co ions and also to the formation of the additional DE interactions between Cr and Mn (or between Co ions). Of the six transition metal-substituted systems, the Cr-based system presented some of the most interesting magnetic properties, followed by the Ni-based system and, finally, the Cu-based system. All systems exhibit layered magnetic behavior, a reduction in ferromagnetism with increasing transition metal content, complex magnetic interactions well below T_C, and a metal to insulator transition around x~0.2 (a value also near the percolation threshold described in many works of around x~0.16). Our data for the Ni [7, 9] and Cr [8] systems strongly suggest the onset of charge ordering occurs coincident with the metal to insulator transition, and not only at specific nodal quantities. We suggest this type of ordering is highly likely in the remaining transition metal-substituted systems, but that additional analysis with neutron data is needed.

4.2. Substitution of the Mn-site by closed shell ions: Ti, Zr or Zn

There are several advantages in using closed shell ions to investigate metal substituted-LSMOs. First, the closed shell ions normally do not affect the magnetic interactions between the Mn ions due to their having no magnetic moment. Second, they have inert gas configurations, and therefore do not contribute to the electron charge density. But there still remains the possibility of secondary effects such as a disturbance of the magnetic ordering and a redistribution of electron charge density by a large ionic size mismatch at the B-site, as with, e.g. Zr^{4+}. In general a decrease in T_C and M_S with increasing substitution is observed for the closed shell ion-substituted LSMOs. This is attributed to the dilution of the magnetic ions and the weakening of the ferromagnetic DE interaction between them. Substitution with Ti is selective because Ti^{4+} substitutes for Mn^{4+}. It would be expected as well that Zr^{4+} or Zn^{2+} would substitute for Mn^{4+}. However the severe ionic size mismatch of $Zr^{4+}$and Zn^{2+} may not allow the substitution of Mn^{4+} by Zr^{4+} or Zn^{2+}. Therefore mixed-valent Mn ions in LSMO are selectively diluted by partial substitution of Mn by these ions. LSMO has the highest T_C when the ratio of Mn^{3+}/Mn^{4+} has an optimal value of 7/3. Upon substitution of the Mn ions with closed shell ions, the ratio of Mn^{3+}/Mn^{4+} deviates from the optimal value of the parent compound. For Ti and Zr substitutions, the Mn^{3+}/Mn^{4+} ratio increases with increasing Ti and Zr content. It should be noted that Zn^{2+} replaces Mn^{3+} which has a larger magnetic moment than Mn^{4+}, and the difference of the magnetic moments in the Mn-sites of the Zn-substituted LSMO and Ti- or Zr-

http://dx.doi.org/10.5772/61770

substituted LSMO is only $0.1\mu_B$ per Mn-site. In this case, the competition between DE and SE interactions is a more important control factor for predicting T_C and M_S. Therefore, substitution onto the Mn-site with Zn^{2+} produces more DE couplings and less SE couplings. In turn one would expect the Zn-substituted LSMO to have a larger M_S and a higher T_C than with Ti or Zr substitution, which produces less DE couplings and more SE couplings.

Acknowledgements

This work was supported by the National Natural Science Foundation of China (Grant Nos 51371009, 50971003 and 51171001), the National Basic Research Program of China (No 2010CB833104, MOST of China), the National High Technology Research and Development Program of China (No 2011AA03A403).

Author details

J.B. Yang[1,2*], M.S. Kim[3], T. F. Creel[3], H. Zhao[1], X.G. Chen[1], W.B. Yelon[3] and W.J. James[3]

*Address all correspondence to: jbyang@pku.edu.cn

1 State Key Laboratory for Mesoscopic Physics, School of Physics, Peking University, Beijing, P. R. China

2 Collaborative Innovation Center of Quantum Matter, Beijing, P. R. China

3 Materials Research Center, Missouri University of Science and Technology, USA

References

[1] Ahn, K. H. ; Wu, X. W. ; Liu, K. ; & Chien, C. L. (1996). Magnetic properties and colossal magnetoresistance of La(Ca)MnO_3 materials doped with Fe, Phys. Rev. B 54, 15299.

[2] Ahn, K. H. ; Wu, X. W. ; Liu, K. ; & Chien, C. L. (1997). Effects of Fe doping in the colossal magnetoresistive $La_{1-x}Ca_xMnO_3$, J. Appl. Phys., 81, 5505.

[3] Alexandrov, A. S. ; & Bratkovsky, A. M. (1999). Carrier Density Collapse and Colossal Magnetoresistance in Doped Manganites, Phys. Rev. Letter., 82, 141.

[4] Anderson, P. W.; & Hasegawa, H. (1955). Considerations on Double Exchange, Phys. Rev., 100, 675.

[5] Blasco, J. ; Garcia, J. ; de Teresa, J. M. ; Ibarra,M. R. ; Perez, J. ; Algarabel, P. A. ; Marquina, C. ; & Ritter, C. (1997). Structural, magnetic, and transport properties of the giant magnetoresistive perovskites $La_{2/3}Ca_{1/3}Mn_{1-x}Al_xO_{3-\delta}$ Phys. Rev. B, 8905.

[6] Cao, D. ; Bridges, F. ; Anderson, M. ; Ramirez, A. P. ; Olapinski, M. ; Subramanian, M. A. ; Booth, C. H. ; and Kwei, G. H. (2001). Local distortions in $La_{0.7}Ca_{0.3}Mn_{1-b}A_bO_3$ (A=Ti and Ga) colossal magnetoresistance samples: Correlations with magnetization and evidence for cluster formation, Phys. Rev. B. , 64, 184409.

[7] Creel, Thomas F.; Yang, Jinbo; Kahveci, Mehmet; Malik, Satish K.; Quezado, S.; Pringle, O.A.; Yelon, William B.; & James, William J. (2013a). Structural and magnetic properties of $La_{0.7}Sr_{0.3}Mn_{1-x}Ni_xO_3$ (x<=0.4), Journal of Applied Physics, 114(1), 013911.

[8] Creel, Thomas F.; Jinbo Yang, Kahveci, Mehmet; Lamsal, Jagat; Malik, S. K.; Quezado, Sylvio; Pringle, Oran A.; Yelon, William B.; & James, William J. (2010). Structural and Magnetic Properties of $La_{0.7}Sr_{0.3}Mn_{1-x}Cr_xO_3$ (x<= 0.5), IEEE Transactions on Magnetics 46(6), 1832-1835.

[9] Creel, Thomas F. (2013b). Structural and magnetic properties of Ni and Cr substituted $La_{0.7}Sr_{0.3}MnO_3$, Doctoral Dissertations. Paper 2031.

[10] Dai, P. ; et al., (1996). Experimental Evidence for the Dynamic Jahn-Teller Effect in $La_{0.65}Ca_{0.35}MnO_3$, Phys. Rev. B 54, R3694.

[11] De, Gennes P. G. (1960). Effects of Double Exchange in Magnetic Crystals, Phys. Rev., 118, 141.

[12] Dessau, D. S. ; Saitoh, T.; Park, C. H.; Shen, Z. X. ; Villella, P. ; Hamad,a N. ; Moritomo, Y. ; and Tokura, Y. (1998). k-dependent electronic structure, a large "ghost" fermi surface, and a pseudogap in a layered mangnetoresistive oxide, Phys. Rev. Lett., 81, 192.

[13] Dho, J. ; Kim, W. S.; & Hur, N. H. (2002). Reentrant Spin Glass Behavior in Cr-Doped Perovskite Manganite, Phys. Rev. Lett. 89, 027202.

[14] Geck, J.; et al., (2001). Evidence for canted antiferromagnetism in lightly doped $La_{1-x}Sr_xMnO_3$, Phys. Rev. B 64, 144430.

[15] Hwang, H. Y.; Cheong, S-W.; Radaelli, P. G.; Marezio, M.; & Batlogg, B. (1995). Lattice Effects on the Magnetoresistance in Doped $LaMnO_3$, Phys. Rev. Lett. 75, 914.

[16] Harrison, W. A. (1980) The electronic structure and properties of solids (Freeman, San Francisco, 1980).

[17] Imada, M. ; Fujimori, A.; & Tokura, Y. (1998). Metal-insulator transitions, Rev. Mod. Phys. 70, 1039.

[18] Joly, V. L. J. ; Joy, P. A. ; Date, S. K. ; & Gopinath, C. S. (2002). Two ferromagnetic phases with different spin states of Mn and Ni in $LaMn_{0.5}Ni_{0.5}O_3$, Phys. Rev. B 65, 184416.

[19] Ju, H. L. ; Sohn, H. C. ; & Krishnan, K. M. (1997). Evidence for O-2p hole-driven conductivity in $La_{1-x}Sr_xMnO3$ (0≤ x≤ 0.7) and $La_{1-x}Sr_xMnOz$ thin films, Phys. Rev. Lett., 74, 3230 (1997).

[20] Kubo, K.; & Ohata, N. (1972). A Quantum Theory of Double Exchange, J. Phys. Soc. Japan ,33, 21-32.

[21] Kim, K. H. ; Jung, J. H. ; & Noh, T. W. (1998). Polaron Absorption in a Perovskite Manganite $La_{0.7}Ca_{0.3}MnO_3$, Phys. Rev. Lett., 81, 1517.

[22] Kim, M. S. ; Yang, J. B. ; Cai, Q.; Zhou, X. D.; James, W. J.; Yelon, W. B.; Parris, P. E.; Buddhikot, D.; & Malik, S. K. (2005a). Structure, magnetic and transport properties of Ti-substituted $La_{0.7}Sr_{0.3}MnO_3$, Phys. Rev. B. 71, 14433.

[23] Kim, M. S.; Yang, J. B.; Cai, Q.; James, W. J.; Yelon, W. B.; Parri,,P. E.;& Malik, S. K. (2007). Structural, magnetic, and transport properties of Zr-substituted $La_{0.7}Sr_{0.3}Mn_{1-x}O_3$, J. Appl. Phys. 102, 013531.

[24] Kim, M. S.; Yang, J. B.; Cai, Q.; Zhou, X.D.; James, W. J.; Yelon, W. B.; Parri, P. E.; & Malik, S. K. (2005b). The effect of Cu-doping on the magnetic and transport properties of $La_{0.7}Sr_{0.3}MnO_3$, J. Appl. Phys. 97, 10H714.

[25] Kim, M.S. ; Yang, J.B. ; Cai, Q. ; Zhou, X.D.; James ,W.J.; Yelon,W.B.; & Parris, P.E., (2008). Electronic structure of $La_{0.7}S_{0.3}Mn_{1-x}Cu_xO_3$, J. Phys. C: Condensed Matter 20, 255228/1.

[26] Kimura, T. ; Tomioka, Y. ; Kumai, R. ; Okimoto, Y. ; & Tokura, Y. (1999). Diffuse Phase Transition and Phase Separation in Cr-Doped $Nd_{1/2}Ca_{1/2}MnO_3$: A Relaxor Ferromagnet, Phys. Rev. Lett. 83, 3940

[27] Klingeler, R. ; Geck J. ; Gross, Pinsard-Gaudart L. ; Revcolevschi, A. ; Uhlenbruck S.; & Büchner, B. (2002). Magnetism and the charge order transition in lightly doped $La_{1-x}Sr_xMnO_3$, Phys. Rev. B 65, 174404.

[28] Liu, X. ; Xu, X. ; & Zhang, Y. (2000). Effect of Ti dopant on the carrier density collapse in colossal magnetoresistance material $La_{0.7}Ca_{0.3}Mn_{1-y}Ti_yO_3$ Phys. Rev. B 62, 15112.

[29] Medarde, M. ; Mesot, J. ; Lacorre, P. ; Rosenkranz, S. ; Fischer, P. ; & Gobrecht, K. (1995). High-pressure neutron-diffraction study of the metallization process in PrNiO3, Phys. Rev. B 52, 9248.

[30] Millis, A. J. ; Littlewood, P. B. ; & Shraiman, B. I. (1995). Double Exchange Alone Does Not Explain the Resistivity of $La_{1-x}Sr_xMnO_3$, Phys. Rev. Lett., 74, 5144.

[31] Millis, A. J. ; Shraiman, B. I.; & Mueller, R. (1996). Dynamic Jahn-Teller Effect and Colossal Magnetoresistance in $La_{1-x}Sr_xMnO_3$, Phys. Rev. Lett., 77,175.

[32] Neumeier, J. J. ; Andres, K. ; and McClellan, K. J. (1999). Thermal expansion of single-crystalline $La_{0.83}Sr_{0.17}MnO_3$: The importance of temperature-induced strain for the electrical resistivity, Phys. Rev. B , 59, 1701.

[33] Okada, K. ; & Yamada, S. (2012). Anisotropic magnetic properties of P $_{1-x}Ca_xMnO_3$, Phys. Rev. B 86, 064430

[34] Ogale, S. B. ; Shreekala, R.; Bathe, R. ; Date, S. K. ; Patil, S. I. ; Hannoyer, B. ; Petit, F. ; & Marest, G. (1998). Transport properties, magnetic ordering, and hyperfine interactions in Fe-doped $La_{0.75}Ca_{0.25}MnO_3$: Localization-delocalization transition, Phys. Rev. B 57, 7841.

[35] Radaelli, P. G. ; Iannone, G., Marezio, M. ; Hwang, H. Y. ; Cheong, S-W.; Jorgensen, J. D. ; & Argyriou, D. N. (1997). Structural effects on the magnetic and transport properties of perovskite $A_{1-x} A'_x MnO_3$ (x=0.25, 0.30), Phys. Rev. B 56, 8265 (1997).

[36] Raveau, B. ; Maignan, A. ; & Martin, C. (1997). Insulator–Metal Transition Induced by Cr and Co Doping in $Pr_{0.5}Ca_{0.5}MnO_3$, Journal of Solid State Chemistry, 130, 162–166

[37] Salamon, Myron B. ; & Jaime, M. (2001). The physics of manganites: Structure and transport, Rev. Mod. Phys. 73, 583.

[38] Shannon, R. D. (1976). Revised effective ionic radii and systematic studies of interatomic distances in halides and chalcogenides, Acta Crystallogr. Sect. A: Cryst. Phys., Diffr., Theor. Gen. Crystallogr., 751.

[39] Tokura, Y.; Kuwahara, H.; Moritomo, Y.; Tomioka, Y.; & Asamitsu, A. (1996). Competing Instabilities and Metastable States in $(Nd,Sm)_{1/2}Sr_{1/2}MnO_3$, Phys. Rev. Lett. 76, 3184.

[40] Zener C. (1951). Interaction between the d -Shells in the Transition Metals. II. Ferromagnetic Compounds of Manganese with Perovskite Structure, Phys. Rev., 82, 403.

[41] Zhao, G. M. ; Smolyaninova, V. ; Prellie,r W. ; & Keller, H. (2000). Electrical transport in the ferromagnetic state of manganites: Small-polaron metallic condition at low temperatures, Phys. Rev. Lett. 84, 6086.

Chapter 9

Microwave Dielectrics with Perovskite-Type Structure

Hitoshi Ohsato

Additional information is available at the end of the chapter

http://dx.doi.org/10.5772/61718

Abstract

Most electroceramics are ferroelectrics, but microwave dielectrics are mostly paraelectrics with a center of symmetry *i*. Microwave dielectrics should possess a perfect crystal structure with neither defects nor internal strain in order to be microwave friendly. They have been used in resonators and filters in mobile telecommunications devices. Perovskite and related compounds are also mostly ferroelectrics, but paraelectrics with a perovskite structure also exist, and are used in microwave dielectrics. Owing to the flexibility of the perovskite structure, many kinds of microwave dielectrics with a perovskite structure have been designed for microwave dielectrics. In this chapter, simple and complex perovskite, and perovskite related materials such as pseudo-tungsten-bronze solid solutions and homologous compounds are introduced for microwave dielectrics. The microwave dielectric properties are revealed through the crystalline structure of the material. Therefore, the relationship between the crystalline structure and properties of the material is presented, and is expected to be of use in the design of novel dielectrics. As many superior materials for microwave dielectrics have been developed and are expected to be used in new applications such as wireless sensors and wireless power transfer by resonant coupling, wave absorption by interference and transparent ceramics with no birefringence, these new applications are also discussed.

Keywords: Microwave dielectrics, Complex Perovskite, Ordering, Tungsten- bronze compounds, Homologous series

1. Introduction

Perovskite and related compounds are the main materials used in microwave dielectrics, as shown in Fig. 1. The data are listed in a database proposed by M. T. Sebastian and published in the book "*Dielectric materials for wireless communication*" [1]. The book cites about 2,300 compounds with about 750 references making it an excellent publication for material scientists and researchers, particularly with respect to microwave technology. The data for dielectric materials is now being updated with about 3,000 compounds and 1,600 references which has now been published [2]. A sizeable amount (about 46%) includes rare-earth (*R*) ions with colors as shown in Fig. 2(a), and these were reviewed in a chapter of the book [3] and various papers [4, 5]. The largest amount of compounds (45%) are of the perovskite-type, known as the 'King' of electroceramics, and the second largest group of compounds, at 21%, are the pseudo-tungsten-bronze solid solutions, also related to perovskite compounds, as shown in Fig. 1.

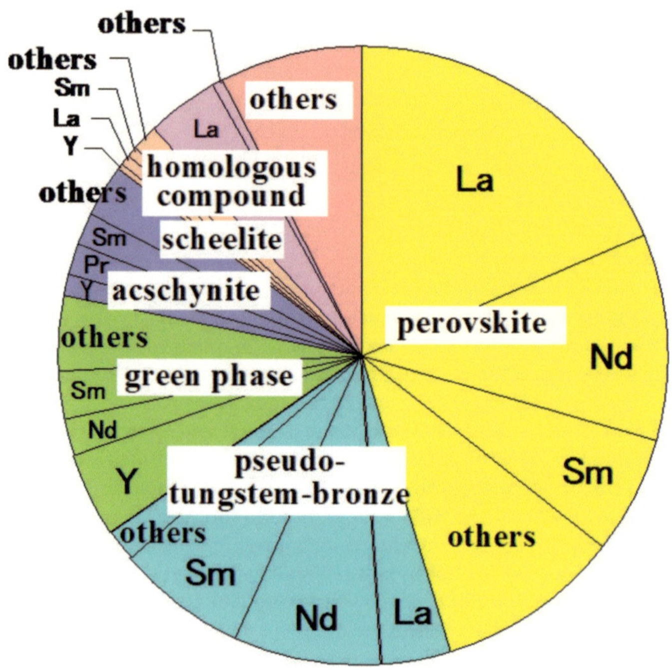

Figure 1. Ratio of microwave dielectric compounds with different crystal structures.

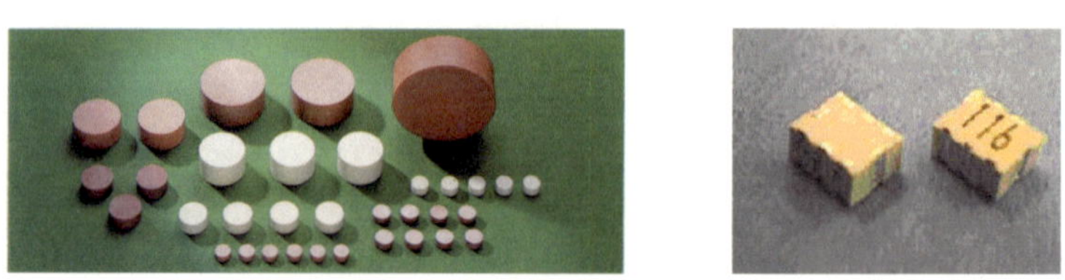

Dielectrics
(a)

LTCC LC filter
(b)

Figure 2. (a) Dielectric resonators. (b) LTCC for LC filter. (NTK/NGK)

Microwave dielectrics have been used as a key constituent of wireless communications [6–9]. Microwave dielectrics are used in resonators, filters and temperature stable capacitors with a near zero temperature coefficient of resonate frequency (*TCf*) / temperature coefficient of

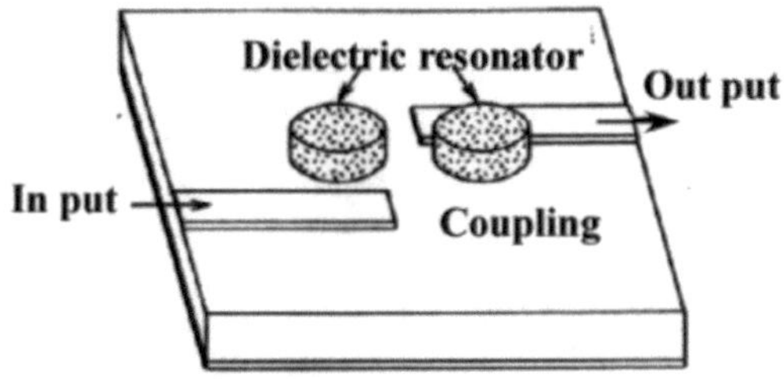

Figure 3. Dielectric resonator with resonate coupling.

dielectric constant ($TC\varepsilon_r$) and so on (Fig. 3). Originally, microwave dielectrics were developed from the temperature stable capacitor which shows a near zero $TC\varepsilon_r$ as explained in section 2. Resonators and filters are used in mobile communication technology. In the early days of information technology (IT), microwave dielectrics replaced cavity resonators and worked extremely well in reducing the size of the equipment used — from the car telephone to the shoulder phone by 1987 for example. Moreover, microwave dielectrics have been developed for a wide range of telecommunication applications, such as mobile phones, wireless LAN, and intelligent transport system (ITS). Fig. 4 shows three directions of development of microwave dielectrics, which presented $Q{\cdot}f$ as a function of the dielectric constant ε_r [10]. The curve in the figure shows the outline of the upper limit of $Q{\cdot}f$ obtained for a given ε_r. The first direction with a high ε_r is mainly used in the miniaturization of mobile phone components. The second one, with a high Q and a high ε_r is in demand for use in mobile phone base stations. The third direction, with a high Q and a low ε_r is for devices working in the millimeter-wave range — the new frontiers of microwave dielectrics because the utilizable frequency region is expanding towards the millimeter-wave due to a shortage of conventional radio frequency (RF) regions. The three important microwave dielectric properties of ε_r, $Q{\cdot}f$ and TCf are explained in section 2.

These compounds are friendly with electromagnetic waves. When irradiated with an electromagnetic wave, the materials should be resonating owing to dielectric polarization changing under an alternating electromagnetic field as shown in Fig. 5(a). The direction of the dielectric polarization should be easily changeable to the opposite direction depending on the electric field. If the material has spontaneous polarization as in ferroelectrics, then inversion losses become large. As a result most microwave dielectrics are paraelectrics with a center of symmetry *i*. The structure of perovskite is flexible, and as a result perovskite shows many kinds of structure, such as cubic, tetragonal, orthorhombic, trigonal, and monoclinic, depending on the particular *A* and *B* cations in ABO_3. The author recommends referring to some reports written by the author himself [11-16].

Microwave dielectrics have been studied for more than a half of a century now. Many materials with suitable properties have been identified and should be used in new applications to develop new technologies to aid the survival of humans on the Earth. The next generation of functional advances in microwave dielectrics are presented in a chapter of the "Handbook of Multifunctional Ceramics" [17].

In this chapter, perovskite and related materials used in microwave dielectrics are presented and the relationships between the crystal structure and the properties of the materials are discussed. Moreover, new applications for microwave dielectrics developed up to date are also are presented.

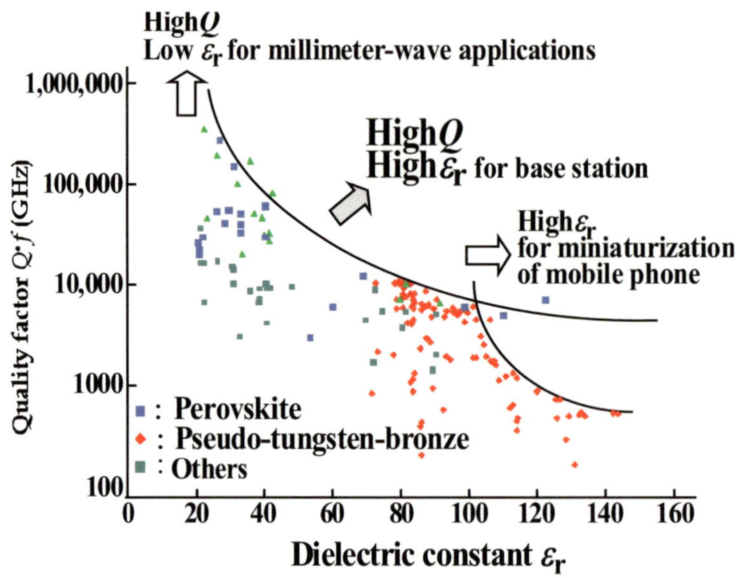

Figure 4. Three directions of R&D for microwave dielectrics. The $Q{\cdot}f$ of microwave dielectrics is shown as a function of ε_r.

2. Three important microwave dielectric properties

There are three important properties of microwave dielectrics: the quality factor Q, the dielectric constants ε_r, and the temperature coefficient of resonant frequency TCf [18].

2.1. Quality factor

Firstly, dielectric materials placed in an electromagnetic field should resonate easily with the electromagnetic waves. In other words, they should have a high quality factor. The quality factor Q is the inverse of the dielectric loss (tanδ) and is presented as follows:

$$Q = 1/\tan\delta$$

However, upon measurement, it is usual to obtain a so-called unloaded quality factor Qu. This is the sum of the reciprocals of the other factors and depends on the dielectric loss of the materials Qd, conduction loss Qc, and radiation loss Qr.

http://dx.doi.org/10.5772/61718

$$1/Qu = 1/Qd + 1/Qc + 1/Qr$$

The losses are generated by dielectric polarization in the presence of an electromagnetic wave. Ferroelectric materials with spontaneous polarization have large dielectric losses because of the large movement of cations. So, paraelectric materials with a center of symmetry *i* are suitable for microwave and millimeter-wave dielectrics. Dielectric loss increases with an increase in frequency as shown in Fig. 5 (b). In the case of ultra-high frequencies, the number of polarity changes increases with frequency. Therefore, dielectric materials with a high *Q* value are desirable.

2.2. Dielectric constant ε_r

The dielectric constant ε_r causes a shortening of wave length λ in dielectrics as shown in Fig. 5(c) according to the following equation:

$$\lambda = \lambda_0 / \sqrt{\varepsilon_r}$$

Here, λ_0 is the wave length in a vacuum. In the microwave region, the ε_r value is expected to be large for the miniaturization of mobile communication equipment. In the millimeter-wave region, the ε_r value is expected to be small. As the wave length is in the millimeter order, miniaturization is not needed.

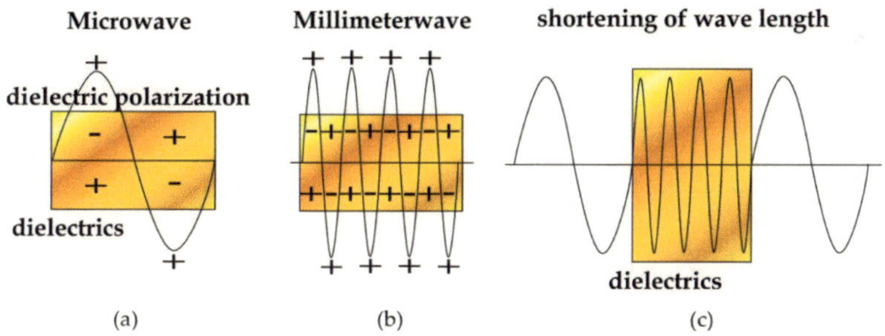

Figure 5. (a) When irradiated by electromagnetic waves, the materials should resonate due to changing dielectric polarization under alternating electromagnetic fields. (b) Dielectric losses increase with an increase in frequency. (c) ε_r causes a shortening of wavelength λ in dielectrics.

There are other more important phenomena such as the time delay T_{PD} according to the following equation:

$$T_{PD} = \sqrt{\varepsilon_r} / c$$

Here, ε_r is the dielectric constant and c is the velocity of light. The time delay is desirable in order to improve the speed of the signal.

The origin of ε_r was considered by difference of crystal structure as shown in Fig. 6 [19]. Silicates with a low ε_r are formed by the tetrahedral framework of SiO_4, with 45% ionic bonds and 55% covalent bonds. Covalent bonds reduce ε_r, because the rattling effect of the cations in a polyhedron should be reduced as a result of the high bond strength. On the other hand, titanates (Fig. 6(c)) with a large ε_r such as $SrTiO_3$, are formed by a TiO_6 octahedral framework, which is almost of ionic bond and has space for ionic displacement. In the case of aluminates, although Al ions also occupy an octahedral framework, the Al ions located in the paired octahedral on the threefold axis repulse each other as shown in Fig. 6(b). The Al ions are immovable in the octahedron and produce a medium ε_r. The order of ε_r is as follows:

$$\varepsilon_{r\,\text{silicate}} < \varepsilon_{r\,\text{aluminate}} < \varepsilon_{r\,\text{titanate}}$$

Silicates with a low ε_r are good candidates for millimeter-wave dielectrics [19].

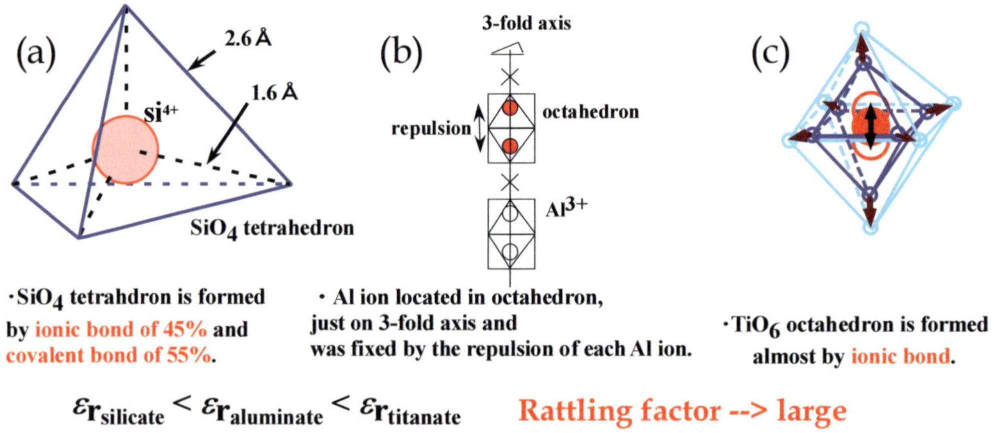

Figure 6. Dielectric constants due to crystal structure: (a) SiO_4 tetrahedron, (b) Al_2O_3 and (c) TiO_6 octahedron.

2.3. Temperature coefficient of resonant frequency *TCf*

The *TCf* is required to be near 0 ppm/°C for global usage in different environmental temperatures. The *TCf* has a relationship with the temperature coefficient of dielectric constant $TC\varepsilon_r$ as follows:

$$TCf = -\left(\alpha + TC\varepsilon_r/2\right).$$

Where α is the thermal expansion coefficient.

Most millimeter-wave compounds with a low ε_r have a large negative *TCf* such as alumina: -65, and forsterite: -70 ppm/°C. The *TCf* of these millimeter-wave dielectrics was improved by two different methods. The first requires the addition of materials with the opposite charge (*i.e.* a positive *TCf*). The addition of rutile with a *TCf* = +450 ppm/°C can adjust the *TCf* of the compound in question [20–23]. The second method is to adjust the *TCf* to near 0 ppm/°C by the formation of the solid solution phases [24]. This is the preferred method because of the small degradation of $Q \cdot f$.

3. Specialized study

3.1. Perovskite-type compounds

3.1.1. Simple perovskite

Although perovskite compounds commonly used in ferroelectrics shouldn't be used for microwave dielectrics as described in the previous section, perovskite compounds can be flexibly applied to microwave dielectrics. This flexibility is due to the depth of the crystal structure. Table 1 shows the polymorphism of $BaTiO_3$ — a representative perovskite-type structure. At room temperature, it is stable due to the tetragonal system, but is ferroelectric without a center of symmetry *i*. It transforms to a cubic structure with paraelectricity above a Curie point of 120 °C. In the near feature, if devices capable of operating under temperatures of more than 120 °C appear, then it may be applied to microwave devices. At low temperatures, all structures of $BaTiO_3$ are ferroelectrics without *i*.

Crystal system	Stable temperature region	Space group	Center of symmetry	Lattice parameters Å
Hexagonal	Above 1460 °C	$P6_3/mmc$ (No.194)	*i*	a = 5.72481 c = 13.9673
Cubic	120~1460 °C	$Pm\bar{3}m$ (No.221)	*i*	a = 4.038
Tetragonal	5~120 °C	$P4mm$ (No.99)	non	a = 3.994 c = 4.038
Orthorhombic	-90~5 °C	$Amm2$ (No.38)	non	a = 3.990 b = 5.669 c = 5.682
Trigonal (Rhombohedral)	Under -90 °C	$R3m$ (No.160)	non	a = 5.6560 c = 6.9509 (a = 4.004, α = 89.870 °)

Table 1. Crystal data for $BaTiO_3$

Currently, microwave dielectrics of perovskite-type and related compounds are detailed in the database created by Sebastian [1, 2]. Some simple perovskite-type compounds are $ATiO_3$,

$AZrO_3$ (A^{2+} = Ba, Sr and Ca) and RBO_3 (R^{3+} = rare earth, B^{3+} = Al, Ga). $MgTiO_3$ and $ZnTiO_3$ with their small ionic cations of Mg and Zn are not perovskite-type structures, but are of the ilmenite-type similar to the structure of Al_2O_3 with oxygen closest packing structure. Table 2(a) shows three microwave properties of simple perovskite-type compounds. These have different crystal structures such as cubic, orthorhombic and hexagonal, but qualify as microwave dielectrics because they have a center of symmetry *i*. $SrTiO_3$ has the crystal structure closest to $BaTiO_3$. It is expected to be a microwave dielectric due to the cubic structure of the paraelectric at room temperature. However, one disadvantageous point is that the temperature coefficient of resonant frequency *TCf* is too large at 1,200 ppm/°C. $CaTiO_3$ with the mineral name "perovskite" is orthorhombic in the space group *Pnma* (No.62) with *i* [25]. The characteristic structure of $CaTiO_3$ is a tilting octahedral. This compound also has a large *TCf* of over 859 ppm/°C, so it could not be used by itself as a microwave dielectrics. Nonetheless, this compound has been used as a stabilizer of *TCf* against microwave compounds with a negative *TCf*, as most useful microwave dielectrics have a positive *TCf*. $MgTiO_3$ with *TCf* = -45 ppm/°C was improved to a near zero *TCf* by adding $CaTiO_3$. This compound with $\varepsilon_r = 21$ and $Q{\cdot}f$ = 8,000 GHz was the first one used in practice in microwave dielectrics. Recently, in an ilmenite system, a Co doped $MgTiO_3$ dielectric with a high $Q{\cdot}f$ (864,000 GHz) was found, and its *TCf* was improved to near zero by the addition of $CaTiO_3$ [28].

(a) Simple perovskite			
Compound	ε_r	$Q{\bullet}f$ (10^3*GHz)	*TCf* (ppm/°C)
$SrTiO_3$	304	3.3	1700
$CaTiO_3$	162	12.96	859
$BaZrO_3$	35	8.8	-
$SrZrO_3$	30	13.6	-60
$CaZrO_3$	30	26.4	-27
$NdGaO_3$	22	85	-
$LaAlO_3$	23.4	68	-44
$SmAlO_3$	20.4	65	-74
$NdAlO_3$	22.3	58	-33
$YAlO_3$	15.7	58	-59
$PrAlO_3$	23.2	51	-25
(b) Modified simple perovskite			
$0.2SrTiO_3{\cdot}0.8LaAlO_3$	26.7	139	-50
$0.64CaTiO_3{\cdot}0.34LaGaO_3$	46.5	48	-2.9
$0.7CaTiO_3{\cdot}0.3NdAlO_3$	43	47	0
$0.35CaTiO_3{\cdot}0.65LaAlO_3$	37	47	-2

$0.7CaTiO_3 \cdot 0.3SmAlO_3$	45	42	1
$0.7CaTiO_3 \cdot 0.3(La_{1/2}Nd_{1/2})(Ga_{1/2}\ Al_{1/2})O_3$	45.2	43	9.3
$NdAlO_3$ +0.25 wt% V_2O_5	21.5	64	-30
$NdAlO_3$+0.25wt%CuO	22.4	63	-35
$LaAlO_3$ +0.25 wt%CuO	20.7	48	-80
$0.95(Mg_{0.95}Co_{0.05})TiO_3 \cdot 0.05CaTiO_3$	20.3	107	-22.8
$0.91(MgZn)_2TiO_4 \cdot 0.09CaTiO_3$	22.5	86	3
$0.94MgTiO_3 \cdot 0.36SrTiO_3$	20.8	71	-1
$0.95MgTiO_3 \cdot 0.05CaTiO_3$	21	56	0
(c) 1:2 type complex perovskite			
$Ba_{0.9925}(Mg_{00.33}Ta_{0.67})O_3$	24.7	152	1.2
$Ba(Mg_{0.3183}Ta_{0.67})O_3$	25.1	120.5	3.3
$Ba(Mg_{1/3}Ta_{2/3})O_3$	24	100.5	8
$Ca(Mg_{1/3}Ta_{2/3})O_3$	21	78	-61
$Sr(Mg_{1/3}Ta_{2/3})O_3$	22	5.6	-50
$Ba(Zn_{1/3}Ta_{2/3})O_3$	30	135	2
$Sr(Zn_{1/3}Ta_{2/3})O_3$	28	55	-62
$Ca(Zn_{1/3}Ta_{2/3})O_3$	25	25	-66
$Ba(Zn_{1/3}Nb_{2/3})O_3$	41.1	86.9	31
$Ba(Mn_{1/3}Ta_{2/3})O_3$	32	58.2	34
$La(Mg_{2/3}Ta_{1/3})O_3$	24.7	65.5	-65
(d) 1:1 type complex perovskite			
$Ba(Mg_{1/2}W_{1/2})O_3$	19	100	-34
$Ca(Ga_{1/2}Ta_{1/2})O_3$	25	100	-81
$Sr(Ga_{1/2}Ta_{1/2})O_3$	27	91	-51
$Sr(Sm_{1/2}Ta_{1/2})O_3$	27.7	59	-63
$La(Mg_{1/2}Ti_{1/2})O_3$	27.6	114.3	-81
$Sm(Co_{1/2}Ti_{1/2})O_3$	25.5	76	-16
$La(Co_{1/2}Ti_{1/2})O_3$	30	67	-64
(e) Solid solutions of 1:2 and 1:1 types of complex perovskite			
$Ba(Mg_{1/2}Ta_{2/3})O_3$:0.5mol% $Ba(Mg_{1/2}W_{1/2})O_3$	24.2	400	-
$0.95Ba(Zn_{1/2}Ta_{2/3})O_3 \cdot 0.05(Sr_{1/4}Ba_{3/4})(Ga_{1/2}Ta_{1/2})O_3$	31	210	-

$0.7Ba(Mg_{1/3}Ta_{2/3})O_3{\cdot}0.3Ba(Co_{1/3}Nb_{2/3})O_3$	27	165	-1.3
$Ba(ZnTa)O_3{\cdot}Ba(ZnNb)O_3$	30	164	0
$Ba(Mg_{1/3}Ta_{2/3})O_3{\cdot}Ba(Zn_{1/3}Ta_{2/3})O_3$	27	150	0
$0.5Ba(MgTa)O_3{\cdot}0.5Ba(ZnTa)O_3$	27	135	1.95
$0.95Ba(Zn_{1/3}Nb_{2/3})O_3{\cdot}0.05Ba(Ga_{1/2}Ta_{1/2})O_3$	38	102.96	19
$Ba(Ni_{1/3}Ta_{2/3})O_3{\cdot}Ba(ZrZnTa)O_3$	30	100	0
(f) Modified complex perovskite			
$Ba(Mg_{1/3}Ta_{2/3})O_3$: $BaSnO_3$, $BaWO_4$,	24	430	5
$Ba[(Zn_{0.6}Co_{0.4})_{1/3}Nb_{2/3}]O_3$	35.6	351.95	-
$Ba[(Mg_{1-x}Zn_x)_{1/3}Ta_{2/3}]O_3$	24-26	200-300	-0.5-1.7
$Ba(Zn_{1/3}Ta_{2/3})O_3$: Ga, Zr	30	165.4	0
$Ba(SnMgTa)O_3$	24.2	120	-
$Ba[(Mg_{0.4}Zn_{0.6})Ta_{2/3}]O_3$	27.7	109.9	6.3
$Ba(Zr_{0.05}Zn_{0.32}Ta_{0.63})O_3$	30.4	105	8
$Ba(M_{0.33}Ta_{0.63}Ti_{0.017}W_{0.017})O_3$	24.5	100.7	12.6
$Ba(Mg_{0.30}Ta_{0.60}Ti_{0.10})O_3$	26.3	100	14.4
$(Ba_{1-z}Sr_z)[Zn_{1/3}(Ta_pNb_{1-p})_{2/3}](Sr_{1-x}Ca_x)\ (Ga_{1/2}Ta_{1/2})O_3$	32-34	180-80	0-10
$Ba(SnMgTa)O_3$	24.2	120	-
(g) Pseudo-tungsten-bronze solid solutions (x: $Ba_{6-3x}\,R_{8+2x}Ti_{18}O_{54}$, R: Rare earth)			
$Ba_4Sm_{9.33}Ti_{18}O_{54}$ (x=2/3)	80	10.7	-15
$Ba_4Nd_{9.33}Ti_{18}O_{54}$ (x=2/3)	82.5	10.1	71.1
$Ba_4Sm_{8.33}EuTi_{18}O_{54}$ (x=2/3)	78.7	9.56	-10.5
$Ba_4Nd_{5.33}Eu_4Ti_{18}O_{54}$ (x=2/3)	78	10.46	10.4
$Ba_4Nd_{8.33}DyTi_{18}O_{54}$ (x=2/3)	78.6	10.04	33.8
$Ba_4Sm_{8.08}Li_{0.25}Ti_{18}O_{54}$ (x=2/3)	82.1	5.62	-2
$Ba_{4.2}(Sm_{0.9}La_{0.1})_{9.2}Ti_{18}O_{54}$ (x=0.6)	84	9.05	1.6
$Ba_{4.5}(Nd_{0.8}Bi_{0.2})_9Ti_{18}O_{54}$ (x=0.5)	106	4.2	8
$Ba_4(La_{1-y-z}Sm_yBi_z)_{9.33}Ti_{18}O_{54}$ (x=2/3, y=0.7, z=0.04)	88.4	6.69	1
$Ba_4(Sm_{1-y}Nd_y)_{9.33}(Ti_{9.95}Sn_{0.05})O_{54}$ (x=2/3, y=0.8, z=0.05)	80	10.6	11
$Ba_4Sr_2Nd_8Ti_{18}O_{54}$ (x= 0)	98.0	6	20
$(Ba_{1-\alpha}Sr_\alpha)_6Sm_8Ti_{18}O_{54}$ (x= 0, α = 0.32)	91.3	8.02	61
$(Ba_{1-\alpha}Sr_\alpha)_{5.7}Sm_{8.2}Ti_{18}O_{54}$($x$=0.1,$\alpha$= 0.298)	85.3	8.71	24
(h) (111) type layered perovskite ($Ba_nLa_4Ti_{3+n}O_{12+3n}$-type homologous series)			

$Ba_xLa_4Ti_{3+x}O_{12+3x}$ (x=0.2)[55]	42	86	-17
$CaLa_4Ti_4O_{15}$ (n=1)	41.1	50.2	-25
$SrLa_4Ti_4O_{15}$ (n=1)	43.8	50.2	-14
$BaLa_4Ti_4O_{15}$ (n=1)	45	47	-11
$Ba(La_{1-y}Al_y)_4Ti_4O_{15}$ (n=1, y=0.11)	44	47	1.3
$Ca(La_{0.875}Nd_{0.125})_4Ti_4O_{15}$ (n=1)	43.4	32.9	-13
$Ba_2La_3Ti_3NbO_{15}$ (n=1)	42.8	21.7	-8
$Ba_3La_2Ti_2Nb_2O_{15}$ (n=1)	49.4	20.2	4
$Ba_3Nd_2Ti_2Nb_2O_{15}$ (n=1)	46.8	19.5	28
$Ba_4NdTiNb_3O_{15}$ (n=1)	38.2	18.7	12
$Ba_2La_4Ti_5O_{18}$ (n=2)	46	31.85	-36
$Ca_2La_4Ti_5O_{18}$ (n=2)	44.7	20.1	6
(i) (100) type layered perovskite ($A_{n+1}B_nO_{2n+1}$ Ruddlesden-Popper phase)			
Sr_2TiO_4 (n=1) No.323[1]	15	1,600	-
Sr_2TiO_4 (n=1) No.1273[1]	37.4	8,160	137
$Sr_3Ti_2O_7$ (n=2) No.785[1]	26	2,400	-
$Sr_3Ti_2O_7$ (n=2) No.1738[1]	57.9	18,850	317
$Sr_4Ti_3O_{10}$ (n=3) No.1422[1]	42	960	-
$Sr_4Ti_3O_{10}$ (n=3) No.1880[1]	76.1	12,700	576
(j) (110) type layered perovskite ($A_nB_nO_{3n+2}$ type homologous series)			
$La_2Ti_2O_7$ (n=4)	47	8,500	-10
$Nd_2Ti_2O_7$ (n=4)	36.5	16,400	-118
$CaLa_4Ti_5O_{17}$ (n=5)	53.7	17,400	-26

Table 2. Microwave dielectric properties of perovskite and perovskite related compounds. No. in (i) list are cited from Sebastian's data base (Book) [1].

$R^{3+}B^{3+}O_3$ compounds containing rare-earth ions (R) in the A-site of the perovskite structure are one of simplest perovskite-type compounds [29]. As the R ion is trivalent, the B ion in the B-site should also be trivalent. Almost all rare earth ions (that is Y to Er) can occupy the A-site. In some compounds, the A-site can be occupied by two or more R ions. Compounds including Sc, Yb and Lu ions have not been reported because of their small radius size. The B-site is occupied by a single ion such as Al, Ga and B, and by a pair of ions that are either divalent or tetravalent such as $Mg^{2+}Ti^{4+}$ [30, 31]. These RBO_3-type compounds as shown in Table 2(a) are preferred for microwave dielectrics because of their small dielectric losses. The crystal structure changes from trigonal to orthorhombic depending on the tolerance factor, as shown

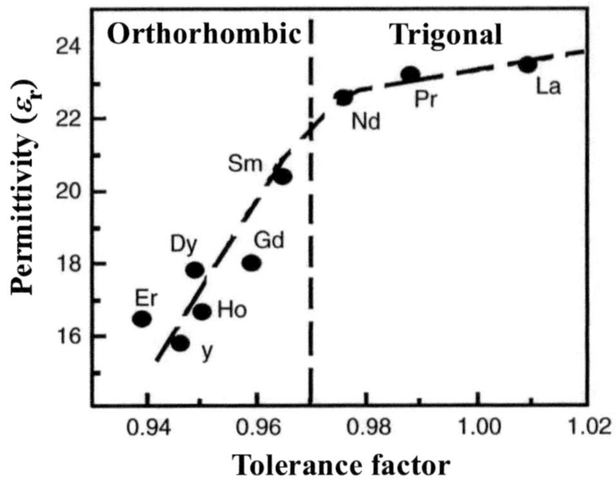

Figure 7. Variation of ε_r as a function of the tolerance factor t in $RAlO_3$.

in Fig. 7 [29]. The compounds with a larger size of ion (La to Nd) are trigonal ($R\bar{3}c$ No.167), and those with a smaller size (Sm to Er) are orthorhombic (*Pnma*, No.62).

The single crystals of $LaAlO_3$ which can be grown easily from melts [32] are used as substrates for superconductor materials such as $YBa_2Cu_3O_{7-x}$, because of their low dielectric losses and their small mismatch for epitaxial growth. It is noticed that strip-line resonators formed by superconductors grown epitaxially on the $LaAlO_3$ single crystal substrate are used in the band-pass filter of base stations in microwave mobile communications. The low dielectric losses come from the low conductivity, based on zero electrical resistivity. Although the *TCf* of $LaAlO_3$ is above -60 ppm/°C [29], this is not an issue whenever it is used as the substrate for a superconductor at a fixed low temperature.

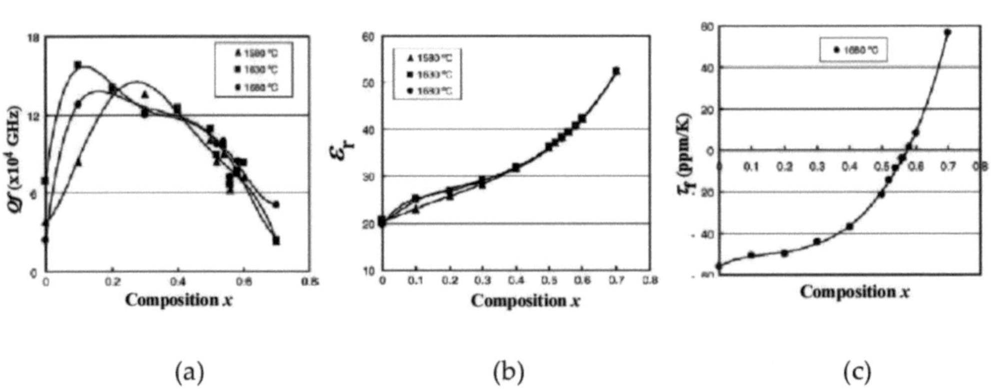

Figure 8. $Q{\cdot}f$ value (a), ε_r (b) and *TCf* (c) of (1-x)$LaAlO_3$-$x$$SrTiO_3$ as a function of composition x.

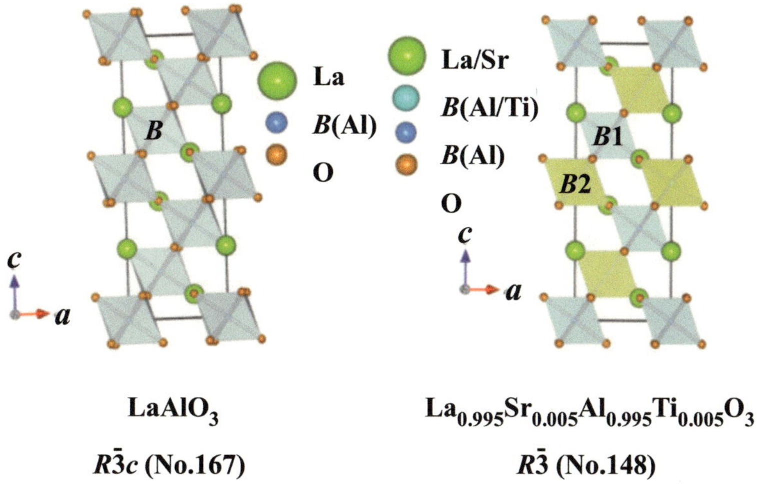

Figure 9. Crystal structure of $LaAlO_3$ (a) with Space Group $R\bar{3}c$ (167), and of $SrTiO_3$ doped $LaAlO_3$ (b) with S.G. $R\bar{3}$ (148).

The *TCf* value is problematic whenever it is used as a resonator or filter at room temperature. In such cases, it is proposed that the *TCf* value is suppressed to near zero ppm/°C by the doping of $SrTiO_3$ or $CaTiO_3$ as shown in Fig. 8(c) [33, 34]. As these solid solutions show a high Q and a high ε_r as shown in Fig. 8(a) and (b), the reason for the improved high Q value is seen through the study and analysis of a single crystal structure. Inagaki *et al.* [35] showed that the crystal system changed from $R\bar{3}c$ (No. 167) to $R\bar{3}$ (No. 148), thereby creating a new position for the Sr ion, as shown in Fig. 9(b) [35–36], and the observed disappearance of the polysynthetic twin. These facts suggest the improvement of $Q{\cdot}f$. Moreover, a $NdTiO_3$-$CaTiO_3$ solid solution system is used for microwave dielectrics with a higher ε_r instead of $LaAlO_3$-$SrTiO_3$ solid solutions. The properties are as follows: $0.2SrTiO_3$-$0.8LaAlO_3$ [34]: $\varepsilon_r = 26.7$, $Q{\cdot}f = 139{,}000$ GHz and $TCf = -50$ ppm/°C ; $0.67CaTiO_3$-$0.33NdAlO_3$ [37]: $\varepsilon_r = 41.98$, $Q{\cdot}f = 42{,}900$ GHz and $TCf = 45$ ppm/°C.

3.1.2. Complex perovskite

3.1.2.1. Types of complex perovskite and their properties

Complex perovskite differs from simple perovskite by a single ion in both the *A*- and *B*-site. It is constituted maintaining the charge valance through the differently charged ions in each site, and is distinguished from the substance containing the substituted different plural ions. Complex perovskite compounds reported so far are as follows [38]:

1:1 type in B-site

$$A^{2+}\left(B^{3+}{}_{1/2}B^{5+}{}_{1/2}\right)O_3, A^{2+}\left(B^{2+}{}_{1/2}B^{6+}{}_{1/2}\right)O_3, A^{2+}\left(B^{1+}{}_{1/2}B^{7+}{}_{1/2}\right)O_3, A^{3+}\left(B^{2+}{}_{1/2}B^{4+}{}_{1/2}\right)O_3,$$

1:2 type in B-site

$$A^{2+}\left(B^{2+}{}_{1/3}B^{5+}{}_{2/3}\right)O_3, A^{2+}\left(B^{3+}{}_{2/3}B^{6+}{}_{1/3}\right)O_3,$$

1:3 type in B-site

$$A^{2+}\left(B^{1+}{}_{1/4}B^{5+}{}_{3/4}\right)O_3,$$

1:1 type in A-site

$$\left(A^{1+}{}_{1/2}A^{3+}{}_{1/2}\right)TiO_3, \left(A^{2+}{}_{2/5}A^{3+}{}_{2/5}\right)TiO_3,$$

In the 1:2 type complex perovskite, many compounds exist with suitable properties. Tables 2(c), (d), (e) and (f) show characteristic microwave dielectrics with good properties selected from Sebastian's database, as referred to above [1, 2]. In the data, the microwave dielectric with the highest $Q{\cdot}f$ value of 430,000 GHz is $Ba(Mg_{1/3}Ta_{2/3})O_3$ (BMT) — the 'king' of microwave dielectrics [39]. The *TCf* also has a desirable value, being near to zero at 3.3 ppm/°C. The $Q{\cdot}f$ value of $Ca(Mg_{1/3}Ta_{2/3})O_3$ when Ca was substituted for Ba decreased to 78,000 GHz [40], and when Sr was substituted for Ba, it decreased even more, to 5,600 GHz [41]. Kageyama [42] showed the $Q{\cdot}f$ values of 1:2 type complex perovskites as a function of the tolerance factor in the Ba and Sr-system as shown in Fig. 10. It brings a high $Q{\cdot}f$ so that the tolerance factors of the Ba-system with large size ions in the *A*-site are large, and the electronic structure of the *B*-site ions is a closed shell. In the case of 1:1 type compounds, $La(Mg_{1/2}Ti_{1/2})O_3$ (LMT) [43] shows the highest $Q{\cdot}f$ of 114,000 GHz. The *A*-site of this compound is occupied by the trivalent rare earth La ion, and the valence of the *B*-site is trivalent and composed of the 1:1 ratio of Mg^{2+} and Ti^{4+}. However, the *TCf* of -81 ppm/°C is not a desirable value. Kageyama [44] systematically studied 1:1-type compounds and clarified that $Ca(Ga_{1/2}Ta_{1/2})O_3$ (CGT) and $Sr(Ga_{1/2}Ta_{1/2})O_3$ (SGT) show high a $Q{\cdot}f$. In this system, though the correlation with the tolerance factor is small, Ga with a closed shell electronic structure contributed to the improvement in $Q{\cdot}f$ values. Wakino *et al.* [45] reported $Ba(Mg_{1/2}W_{1/2})O_3$ (BMW) with a high $Q{\cdot}f$, composed of divalent Mg and six valenced W. These compounds also have the disadvantage of a large *TCf*. One of compounds with a near zero *TCf* is $Ba(Tb_{1/2}Nb_{1/2})O_3$ (BTN) [46] with high $\varepsilon_r = 39$, $Q{\cdot}f = 52{,}400$ GHz and $TCf = -2$ ppm/°C.

3.1.2.2. Is ordering a necessary condition for a high Q value?

The origin of a high *Q* value, especially the relationship between a high *Q* value and ordering based on an order-disorder transition, has been under discussion for a long time [47, 51]. The feature of complex perovskite $A(B_{1/3}B'_{2/3})O_3$ exhibits the phenomenon of the ordering of *B* cations. The ordered phase appeared at low temperature is low symmetry trigonal (rhombo-

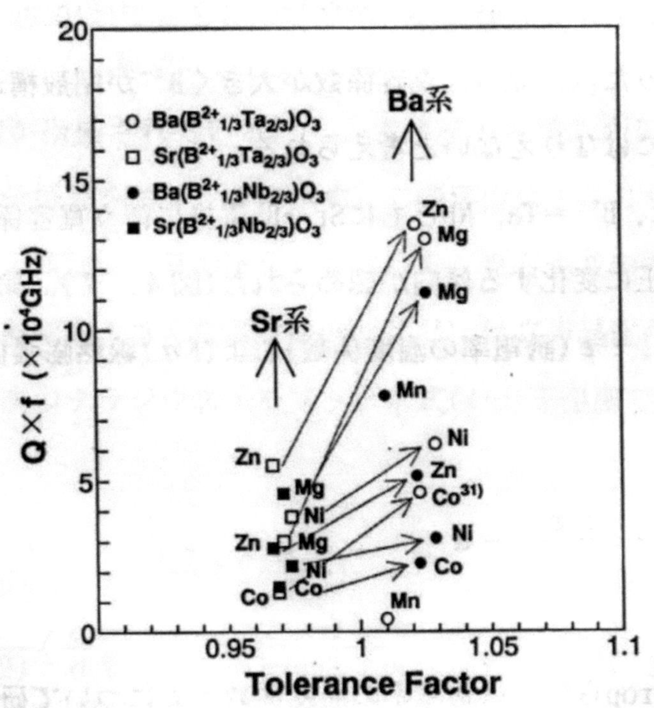

Figure 10. $Q{\cdot}f$ of $A(B^{2+}{}_{1/3}B^{5+}{}_{2/3})O_3$ as a function of the tolerance factor t.

hedral) structure of space group $P\bar{3}m1$ (No. 164) and the disordered phase appeared at high temperature is high symmetry cubic structure of $Pm\bar{3}m$ (No. 221) as shown in Fig. 11. Kawashima *et al.* [52] reported that $Ba(Zn_{1/3}Ta_{2/3})O_3$ (BZT) has a high Q. BZT shows ordering of B cations, as revealed by the splitting and super structure lines on the X-ray powder diffraction (XRPD) patterns for a long sintering time. When the Zn and Ta ions occupy the same position, the structure is a disordered cubic one. On the other hand, if both ions occupy different independent sites, that is ordering, the structure becomes trigonal. This transition is sluggish and the temperature of transition is not clear in some compounds. The relationship between cubic and trigonal crystal structures is shown in Fig. 11. The B cations occupy the octahedra located between the hexagonal closed packing layers composing BaO_3. The ordering is apparent by the periodic arrangement of Zn-Ta-Ta along the c-axis of the trigonal. Though it is believed that ordering brings a high Q, some examples contradicting this have arisen, such as BMT-$Ba(Co_{1/3}Ta_{2/3})O_3$ [53] and $Ba(Mg_{1/3}Ta_{2/3}Sn)O_3$ [54]. Recently, Koga *et al.* [55–59] presented the quantification of the ordering ratio using the Rietveld method and the ordering state in the vicinity of BZT. Kugimiya [60] reported that the composition which deviated from BMT has a high Q because of the high density composition. More recently Surendran *et al.* [61] showed that Ba and Mg deficient BMT compositions have a high Q. In this section, the author presents the primary factors for a high value of Q instead of ordering based on Koga's data [55–59, 61, 62].

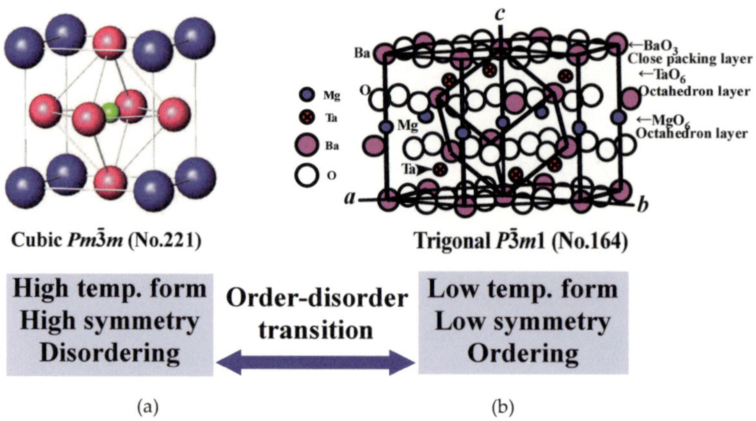

Figure 11. Order-disorder transition of perovskite. (a) High temperature and high symmetry phase with cubic, (b) Low temperature and low symmetry phase with trigonal.

- **Ordering ratio and $Q{\cdot}f$ [55]**

Fig. 12 shows the XRPD patterns (a) and the high angle diffraction peaks (b) of BZT ceramics as a function of sintering time at 1,350 °C. According to sintering time, superlattice lines (asterisked) became clear and the 420 cubic diffraction peak splits gradually into two peaks, namely 226 and 422, in the trigonal system. It is considered that ordered and disordered structures coexist and ordered peaks become intense on sintering of 80 hours or more. These results are consistent with the report by Kawashima *et al* [52].

Fig. 13 shows $Q{\cdot}f$ as functions of ordering ratio (a) obtained by the Rietveld method [63], density (b) and grain size (c). The ordering ratio saturates at about 80 % but the $Q{\cdot}f$ varies from 40,000 to 100,000 GHz. However, the $Q{\cdot}f$ increases with density and grain size. This indicates that the effect of ordering on the Q value is not so important.

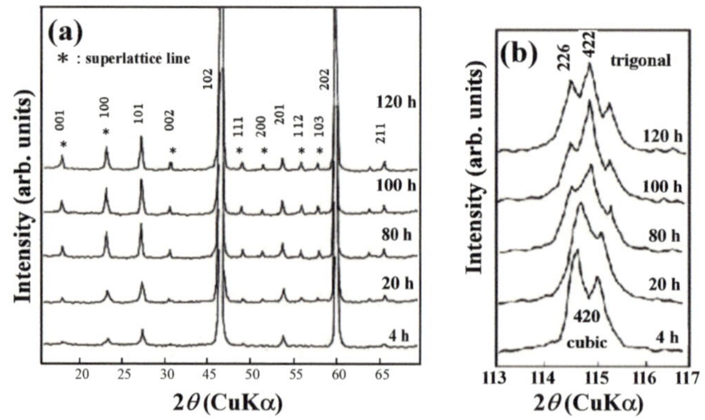

Figure 12. (a) XRPD patterns of BZT ceramics with different sintering time at 1,350 °C. Asterisks are superlattice reflections. (b) Magnified XRPD patterns around $2\theta = 115°$ in which 420 diffraction peak split to 226 and 422.

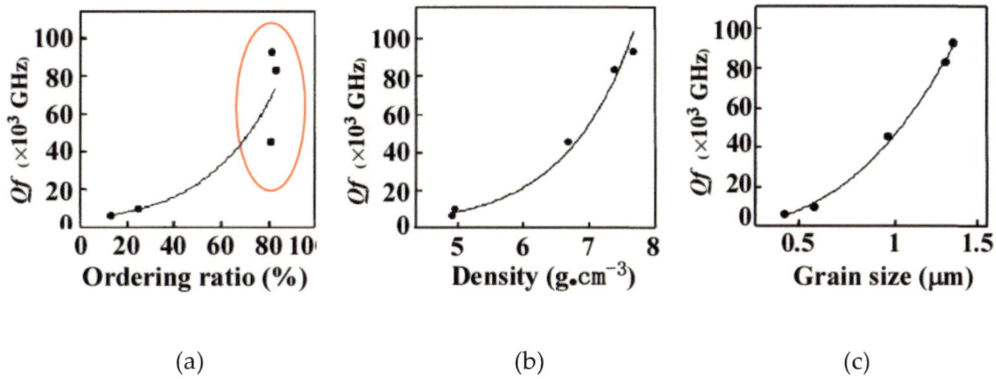

Figure 13. The $Q{\cdot}f$ of BZT ceramics as functions of ordering ratio (a), density (b), and grain size (c).

- **Disordered BZT with a high $Q{\cdot}f$ sintered by SPS** [57]

As both ordered and disordered BZT — with similar microstructures — can be obtained by various heat treatments using a conventional solid state reaction (SSR) as described in the previous section, the effects of the crystal-structural ordering and ceramic microstructure were discussed independently. In the case of BZT, although the high density and high Q ceramics of ordered BZT were synthesized by SSR with a long sintering time of over 80 hours, the dense ceramics of disordered BZT have not been obtained by SSR. Koga *et al.* [57] created high density disordered BZT ceramics with a short sintering time by spark plasma sintering (SPS). In this section, the effects of crystal-structural ordering and ceramic microstructure on the high Q are discussed.

Fig. 14 shows the $Q{\cdot}f$ as a function of the densities in BZT fabricated using SSR and SPS. The samples obtained by SPS were of the disordered cubic type of perovskite as shown in the XRPD pattern (Fig. 15) with a lone 422 reflection compared with the ordered trigonal type with peak separations of 422 and 226 when sintered using SSR (1400 °C 100 h). The SPS samples with high densities were obtained using an extremely short sintering time of 5 mins between 1150 and 1300 °C under 30 Mpa [57]. The short time sintering when using SPS may result in the disordered BZT with a high density of 7.62 g/cm^3, which is approximately 50% higher than that of low density samples of 5.0-6.0 g/cm^3 synthesized by conventional SSR. The full width at half maximum (FWHM) of the 420 peak became narrower with an increase in the temperature from 1,100 to 1,300 °C (Fig. 15). This indicates that the degree of crystallization of the disordered cubic phase is improved without the need for conversion to the ordered trigonal phase. Regardless of the method of synthesis, $Q{\cdot}f$ is strongly dependent on density, and $Q{\cdot}f$ values were improved with density as shown in Fig. 14. The highly crystallized dense disordered BZT ceramics synthesized by SPS showed a significantly high $Q{\cdot}f$ (= 53,400 GHz) similar to that of the ordered BZT sample with the same density (= ca. 7.5 g/cm^3) synthesized by SSR. The crystallization with densification of BZT ceramics should play a more important role in the improvement of the Q factor in the BZT system than structural ordering and grain size. In the high density region (> 7.5 g/cm^3), the variation of the $Q{\cdot}f$ should be clarified.

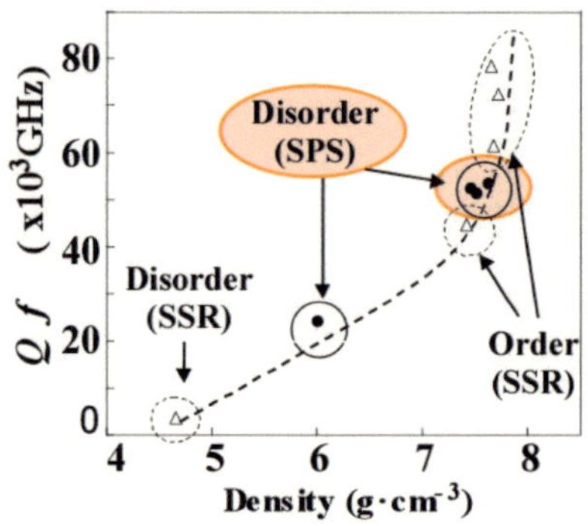

Figure 14. $Q{\cdot}f$ of BZT by solid state reaction (SSR) and spark plasma sintering (SPS) as a function of density. Order: ordered perovskite, Disorder: disordered perovskite.

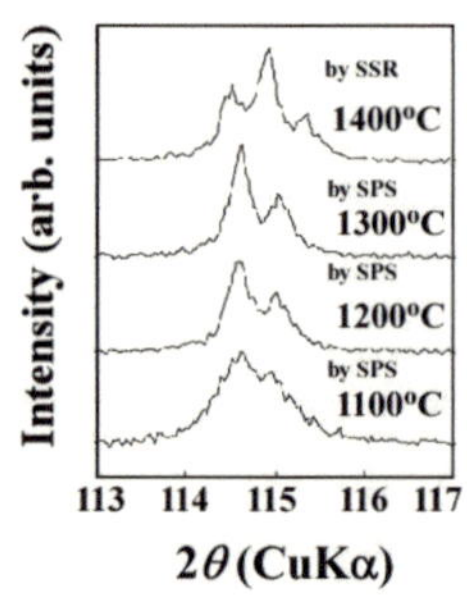

Figure 15. XRPD patterns around 420 diffraction of BZT sintering by SPS for 5 min under 30 MPa with different sintering temperature.

- **$Ba(Zn_{1/3}Nb_{2/3})O_3$ (BZN) with clear order-disorder transition** [58]

Ordering based on the order-disorder transition brings low symmetry, and disordering brings high symmetry as described above. Usually, high symmetry also brings a high Q, similar to ordering. We present an example showing that high symmetry is more influential in bringing about a high Q than ordering is.

BZN clearly shows an order-disorder transition temperature at 1,350 °C as shown in Fig. 16 (a). The transition temperatures of BMT and BZT are un clear because of the high transition temperature. The ordering was confirmed using X-ray diffraction patterns. Fig. 16 shows $Q{\cdot}f$, grain size and density as a function of the sintering temperature of BZN. The disordered sample sintered at 1,400 °C shows a drastic increase of $Q{\cdot}f$, grain size and density when compared with ordered samples sintered at 1,200 and 1,300 °C. As a result of the post-annealing at 1,200 °C over 100h for the disordered sample sintered at 1,400 °C, the structure transformed to order, but the $Q{\cdot}f$ did not improve and instead it decreased slightly in an inverse manner. The grain sizes and densities were not changed by the annealing, as shown in Figs. 16(b) and (c).

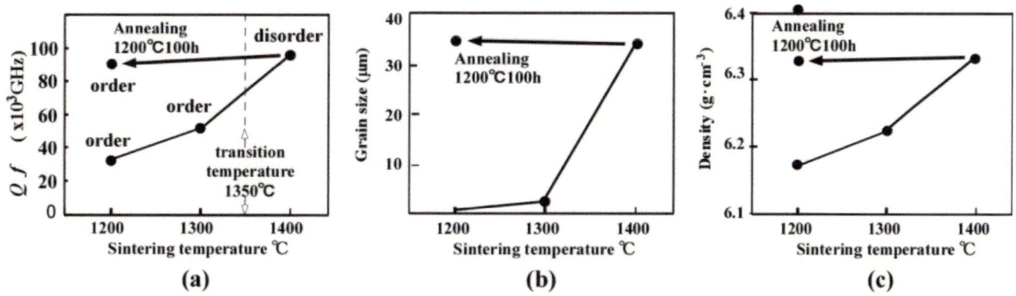

Figure 16. $Q{\cdot}f$ (a), grain size (b) and density (c) of BZN with transition temperature at 1350 °C as a function of sintering temperature. Although the disorder phase with a high Q sintered at 1400 °C annealed at 1200 °C, the $Q{\cdot}f$ did not improve.

Therefore, we can conclude that the crystal-structural ordering in the ceramic BZN system has no significant effect on the improvement of the Q factor. The Q factor strongly depends on the density and grain size, but not on the crystal-structure order. The decrease in $Q{\cdot}f$ as a result of annealing might be dependent on the low symmetry that accompanies ordering.

3.1.2.3. Phase relations and $Q{\cdot}f$ in the vicinity of BZT [56, 59]

- **Koga's research on BZT**

Koga *et al.* [56, 59] studied the phase relation in the vicinity of BZT in the BaO-ZnO-$TaO_{5/2}$ ternary system as shown in Fig. 17. These samples were sintered at 1,400 °C for 100 hours as reported in Koga's paper [56]. These diffraction patterns fit the Rietveld method well [63]. Ordering ratios obtained are shown in Fig. 18(a). Three areas in the vicinity of BZT are presented as shown in Fig. 17.

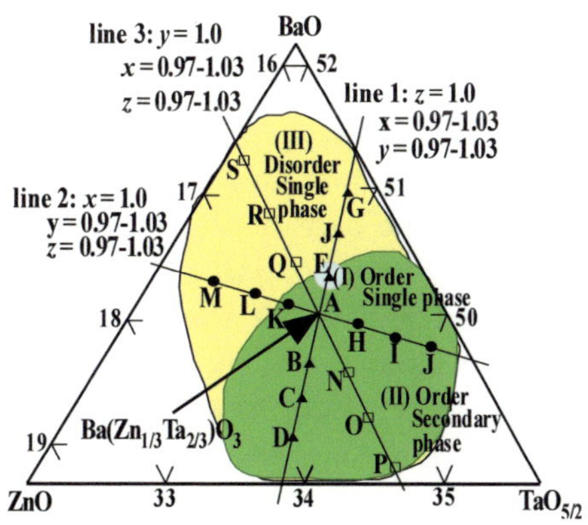

Figure 17. BaO-ZnO-$TaO_{5/2}$ partial ternary system in the vicinity of BZT. Synthesized compositions are shown by the letters A to S. The A point is pure BZT. Three areas are shown and these are (I) for order/single phase, (II) for order/secondary phase, and (III) for disorder/single phase.

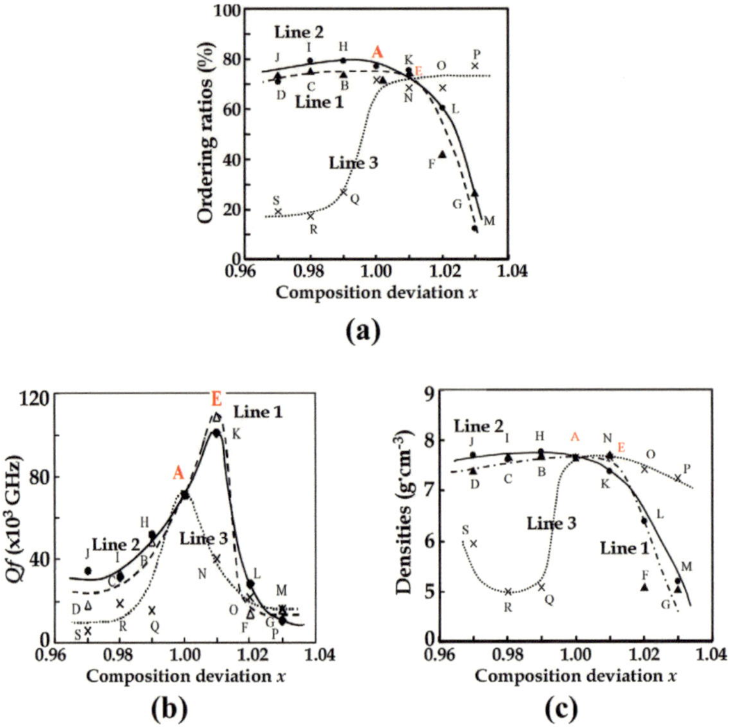

Figure 18. Ordering ratio (a), $Q{\cdot}f$ (b) and density (c) as a function of composition deviation from pure BZT.

(I): Ordering area with BZT single phase

(II): Ordering area with secondary phase

(III): Disordering area with BZT single phase

The first area (I) is composed of a single phase of BZT with an ordered structure and a high $Q{\cdot}f$. The compositions E and K have a $Q{\cdot}f$ about 50 % higher than that of the pure BZT composition A. Composition K is located on the boundary area (I) and has a minor secondary phase as revealed by the SEM figure reported in a previous paper [59]. The ordering ratio in composition E is lower than A, and the density of composition E is same as that of A. The second area (II) is an ordered BZT with a secondary phase $BaTa_2O_6$ with a specific amount of Zn [59] analyzed by X-ray microanalyzer (XMA). The ordering ratio of compounds located in this area is high at about 70 to 8 0 % as shown in Fig. 18(a). Although the structure is ordered, the $Q{\cdot}f$ values decrease according to the deviation from pure BZT as shown in Fig. 18(b). The composition of the ordered BZT compounds is located on Ta_2O_5 rich side, which is precipitated with secondary phase as a eutectic phase diagram system. The third area (III) is precipitated as a single phase BZT solid solution with a disordered structure. The $Q{\cdot}f$ values degrade with a decrease in the ordering ratio and density as shown in Fig. 18(c). The lower density comes from the existence of numerous pores due to hard sintering. The single phase in this area is originated by a solid solution accompanying defects in *B*- and O-sites, which causes degrada-

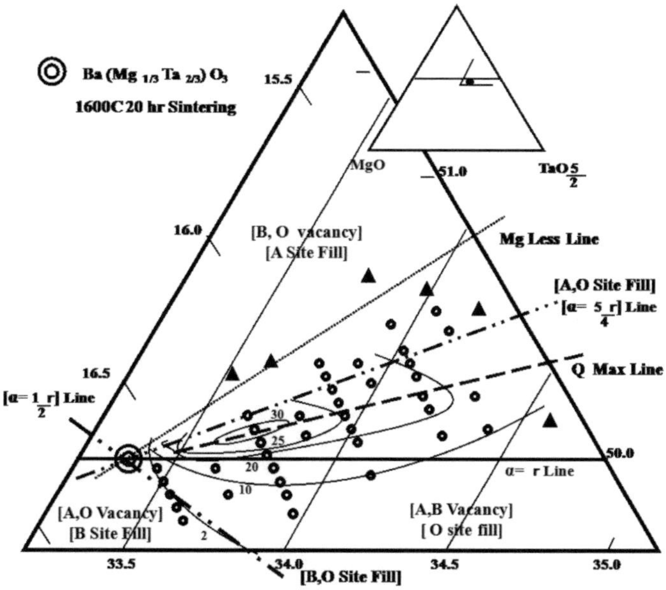

Figure 19. Partial BaO-MgO-$TaO_{5/2}$ ternary system in the vicinity of BMT. On the tie line BMT-$BaTa_{4/5}O_3$, $Ba(Mg_{1/3-\alpha/3}Ta_{2/3+2\alpha/15}V_{\alpha/5})O_3$ solid solutions are formed with high densities and high *Q* values, in which *A*- and O-sites are filled, and the *B*-site has vacancies without charge. Three areas are divided by two lines: $\alpha = 5\gamma/4$ and $\alpha = \gamma/2$. The first one is *B*- and the O-site is vacant although the *A*-site is filled. The second one is *A*- and the *B*-site is vacant although the O-site is filled. The third one is *A*- and the O-site has vacancies, although the *B*-site is filled.

tion of $Q{\cdot}f$. The pores and defects were examined by SEM [59] and Raman scattering spectra [62] respectively.

- **Kugimiya's research on BMT/BMT** [60, 61]

Kugimiya [60] presented the highest $Q{\cdot}f$ composition at the Ta and Ba rich side in a BMT system as shown in Fig. 19. The synthesized samples are precise compositions formed by master batches which mixed samples from the four master batch method. Here, chemical formulae in the vicinity of BMT are reported as follows: Kugimiya presented three areas divided by drawing two lines as shown in Table 3 and Fig. 19.

α	Chemical formula	Vacancy
$\alpha > 5\gamma/4$	$Ba_{1+\alpha}(Mg_{1/3}Ta_{2/3+\gamma}V_{\alpha-\gamma})O_{3+\alpha+5\gamma/2}V_{2\alpha-5\gamma/2}$	A: fill, B, O: vacancy
$\alpha = 5\gamma/4$	$Ba_{1+\alpha}(Mg_{1/3}Ta_{2/3+4\alpha/5}V_{\alpha/5})O_{3+3\alpha}$	A, O: fill, B: vacancy
$5\gamma/4 > \alpha > \gamma/2$	$Ba_{1+\alpha}V_{5\gamma/6-2\alpha/3}(Mg_{1/3}Ta_{2/3+\gamma}V_{\alpha/3-\gamma/6})O_{3+\alpha+5\gamma/2}$	A, B: vacancy, O: fill
$\alpha = \gamma/2$	$Ba_{1+\alpha}V_{\alpha}(Mg_{1/3}Ta_{2/3+\gamma})O_{3+6\alpha}$	A: vacancy, B, O: fill
$\alpha < \gamma/2$	$Ba_{1+\alpha}V_{\gamma-\alpha}(Mg_{1/3}Ta_{2/3+\gamma})O_{3+\alpha+5\gamma/2}V_{\gamma/2-\alpha}$	A, O: vacancy, B: fill

Table 3. Chemical formula for three areas divided by two lines: $\alpha = 5\gamma/4$ and $\alpha = \gamma/2$, here, α and γ are in $Ba_{\alpha}Ta_{\gamma}O_{\alpha+5\gamma/2}$ and vacancies are on the *A*-, *B*- and O-sites.

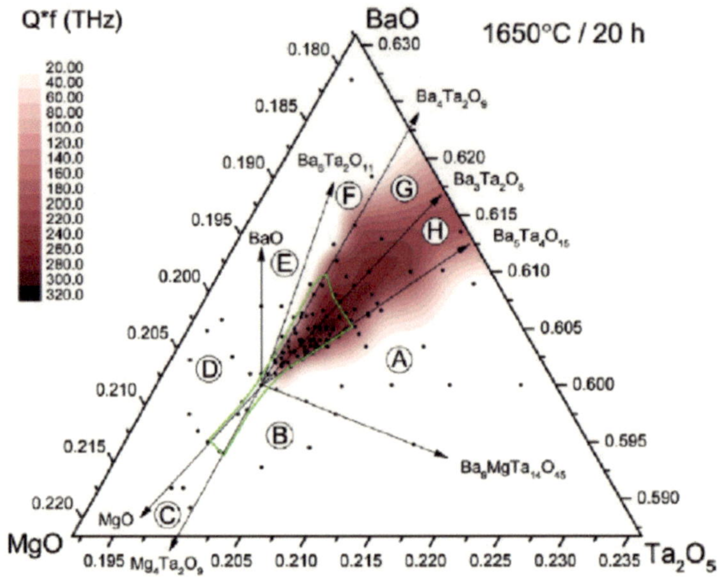

Figure 20. Part of the BaO–MgO–Ta_2O_5 phase diagram in the vicinity of the BMT phase indicating a composition dependence of the $Q{\cdot}f$ for samples sintered at 1650 °C for 20 h. Small black dots indicate the target sample compositions. Green line indicates an approximate boundary of the single-phase BMT.

Here, α and γ are in $Ba_{\alpha}Ta_{\gamma}O_{\alpha+5\gamma/2}$. In the region $\alpha > 5\gamma/4$, the composition denoted by $Ba_{1+\alpha}(Mg_{1/3}Ta_{2/3+\gamma}V_{\alpha-\gamma})O_{3+\alpha+5\gamma/2}V_{2\alpha-5\gamma/2}$ has *B*- and O-site vacancies with holes and electrons. In the $\alpha = 5\gamma/4$ line, the compositions denoted by $Ba_{1+\alpha}(Mg_{1/3}Ta_{2/3+4\alpha/5}V_{\alpha/5})O_{3+3\alpha}$ are the ideal ones without vacancies in the *A*- and O- sites. The *B*-site vacancy is neutralized without charge. The highest $Q{\cdot}f$ composition is located near the line $\alpha = 5\gamma/4$ as shown in Fig. 19. The compositions in the line are ideal for microwave dielectrics because there are no oxygen defects and the density is high due to the substitution of Ta for Mg. In the region $5\gamma/4 > \alpha > \gamma/2$, the composition denoted by $Ba_{1+\alpha}V_{5\gamma/6-2\alpha/3}(Mg_{1/3}Ta_{2/3+\gamma}V_{\alpha/3-\gamma/6})O_{3+\alpha+5\gamma/2}$ has a defect in the *A*- and *B*-sites filled with holes and electrons. In the region at $\alpha = \gamma$, the composition denoted by $Ba_{1+\alpha}V_{\alpha/6}(Mg_{1/3}Ta_{2/3+\alpha}V_{\alpha/6})O_{3+7\alpha/2}$ has the same amount of vacancies in both *A*- and *B*-sites filled with the same holes and electrons. In the region $\alpha = \gamma/2$, the composition denoted by $Ba_{1+\alpha}V_{\alpha}(Mg_{1/3}Ta_{2/3+\gamma})O_{3+6\alpha}$ only has vacancies in the *A*-site with holes and in the *B*-site with excess electrons which introduced instability. In the region $\alpha < \gamma/2$, the composition denoted by $Ba_{1+\alpha}V_{\gamma-\alpha}(Mg_{1/3}Ta_{2/3+\gamma})O_{3+\alpha+5\gamma/2}V_{\gamma/2-\alpha}$ has holes in both the *A*- and O-sites with electrons and excess electrons in the *B*-site, which leads to an unstable crystal structure.

The contour lines in Fig. 19 show *Q* values from 2,000 in the outer area to 30,000 in the center. The highest *Q* value of 50,000 was obtained in the center. The contour is elongated parallel to the *Q* max line as drawn in Fig. 19 and it changes steeply on the perpendicular to the line.

- **Kolodiazhnyi's research on BMT** [64]

The author presented a part of the BaO-MgO-Ta_2O_3 phase diagram in the vicinity of the BMT phase as shown in Fig. 20 [64]. Ceramic samples whose chemical composition falls within the

A, B and C compositional triangles (CTs) in Fig. 20 reach a relative density of 96–98% after sintering at 1,550–1,580 °C for 20 hours. Samples located in the H and G CTs required temperatures of 1,630–1,650 °C to reach a relative density of 96-98 %. The specimens located in the D, E and F CTs retained a density of ≦80% after heat treatment at 1,680 °C. The pure BMT sintered at 1,650 °C for 20 h shows a poor dielectric performance with a $Q{\cdot}f \approx$ 20,000-40,000 GHz. A very large variation in the dielectric properties and density of ceramics was found upon a slight deviation from pure BMT composition. The tendency of the variation was similar to Koga's results as shown in Fig. 17 [56]. Significant improvement in the $Q{\cdot}f$ vales is seen in samples with a slight Mg deficiency, which are located in the G and H CTs. The highest $Q{\cdot}f$ compositions of 330,000–340,000 GHz were found within the H CT close to the BMT-$Ba_3Ta_2O_8$ tie line. Upon approaching the BMT-$Ba_5Ta_4O_{15}$ tie line from the H CT, the $Q{\cdot}f$ starts to decrease and then drops sharply after crossing into the A CT. Mg-rich BMT with a high density and a high degree of 1:2 cation order within B and C CTs showed low $Q{\cdot}f$ values (e.g. $Q{\cdot}f <$ 20,000 GHz). The dominant source of the extrinsic dielectric loss is identified as Mg occupation substituted for Ba in the *A*-site (Mg_{Ba}) which improves 'rattling' inside the dodecahedral position. Ta-poor, non-pure BMT samples located in the D, E and F CTs showed a very low density and high dialectric losses after sintering at 1,650 °C for 20 h.

- **High *Q* by high density composition** [60, 61]

Koga's data [56] and Kolodiazhnyi's [64] data are comparable with Kugimiya's BMT data [60]. The area (I) and the H CT with the highest $Q{\cdot}f$ as shown in Figs 17 and 20 respectively, are located on the opposite side of Kugimiya's data against the BMT-$Ba_5Ta_4O_{15}$ tie line (Fig. 19). These compositions will be comparable with the completed ideal crystal structure $Ba_{1+\alpha}(Mg_{1/3}Ta_{2/3+4\alpha/5}V_{\alpha/5})O_{3+3\alpha}$ reported by Kugimiya [60]. The formula is rewritten as $Ba(Mg_{1/3-\alpha/3}Ta_{2/3+2\alpha/15}V_{\alpha/5})O_3$ solid solutions on the tie-line BMT-$Ba_5Ta_4O_3$. The crystal structure in the composition region is perfect, without defects, and with a high density. The density of BMT increases with the introduction of the $Ba_5Ta_4O_{15}$ phase, because Mg ions are substituted by heavy Ta ions.

Surendran *et al.* [61] also presented compositions with high *Q* values in the two kinds of magnesium and barium deficient nonstoichiometric compositions $Ba(Mg_{1/3-x}Ta_{2/3})O_3$ [x=0.015] and $Ba_{1-x}(Mg_{1/3}Ta_{2/3})O_3$ [x=0.0075] as shown in Fig. 21(a). The microwave dielectric properties of $Ba_{0.9925}(Mg_{0.33}Ta_{0.67})O_3$ [ε_r = 24.7, *TCf* = 1.2 ppm/°C, $Q{\cdot}f$ = 152,580 GHz] and $Ba(Mg_{0.3183}Ta_{0.67})O_3$ [ε_r = 25.1, *TCf* = 3.3 ppm/°C and $Q{\cdot}f$ = 120,500 GHz] were found to be better than stoichiometric BMT [ε_r = 24.2, *TCf* = 8 ppm/°C and $Q{\cdot}f$ = 100,500 GHz]. The important difference from Kugimiya's results [60] is standing on the nonstoichiometry with a barium or magnesium deficiency. We consider that Surendran's data [61] is based on Kugimiya's results [60]. In the case of Mg-deficient BMT, as the composition is located near Kugimiya's area with a high $Q{\cdot}f$, the composition of the main compound must be $Ba(Mg_{1/3-\alpha/3}Ta_{2/3+2\alpha/15}V_{\alpha/5})O_3$ solid solutions on the tie-line BMT-$Ba_5Ta_4O_3$. As shown in Fig. 21(b), in the solid solution area, the Mg deficiencies are filled with Ta and create vacancies in the *B*-site, so that density and the ordering ratio are maintained. On the other hand, the existing area of Ba-deficient BMT is included in Koga's (II) area as shown in Fig. 17, composed of ordered BMT and secondary phase. The ordered BMT will have a similar composition with a high density and a high $Q{\cdot}f$ on the BMT-$Ba_5Ta_4O_3$ tie-line presented by Kugimiya [60]. The compound by Surendran *et al.* [61] may be located in the eutectic phase diagram region accompanying the secondary phase. However, as the amount of secondary phases is small, detection may be difficult. Though the density and ordering ratio are maintained at a high level as shown in Fig. 21(c), $Q{\cdot}f$ values degrade steeply according to the secondary phase. The compound should be stoichiometric and complete, because microwave dielectrics with a high Q are usually free of defects.

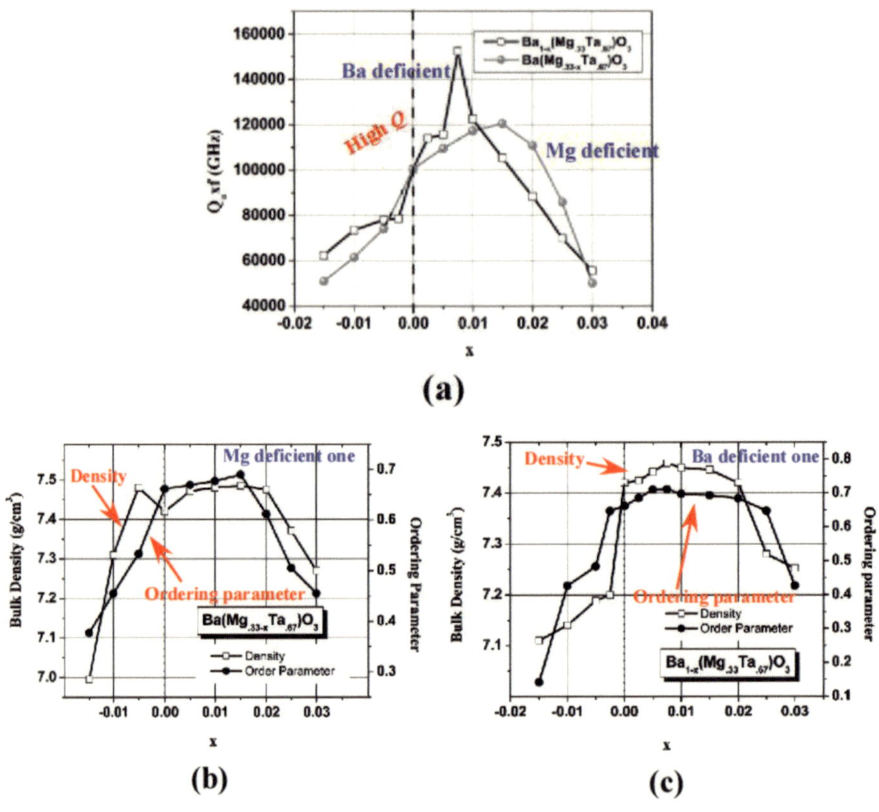

Figure 21. (a) $Q{\cdot}f$ for $Ba(Mg_{1/3-x}Ta_{2/3})O_3$ and $Ba_{1-x}(Mg_{1/3}Ta_{2/3})O_3$ as a function of composition deviation (x), (b) Bulk density and ordering parameter for $Ba(Mg_{1/3-x}Ta_{2/3})O_3$ as a function of x, (c) Bulk density and ordering parameter for $Ba_{1-x}(Mg_{1/3}Ta_{2/3})O_3$ as a function of x.

3.1.2.4. Important points concerning complex perovskite

- A complex perovskite is composed of different ions with different charges such as A^{2+} $(B^{2+}_{1/3}B^{5+}_{2/3})$ O_3, thereby maintaining the charge valance.
- A complex perovskite usually has an order-disorder phase transition. The order phase is a low temperature phase with low crystallographic symmetry, while the disorder phase is a high temperature phase with high symmetry. In the case of a 1:3 type complex perovskite,

the ordered phase is a trigonal (rhombohedral) $R\bar{3}c$ (No. 167), and the disorder phase is a cubic $Pm\bar{3}m$ (No. 221) [55, 65].

- The $Q{\cdot}f$ is dependent on the density instead of on ordering based on SPS sintering [57].
- In the case of a compound with an order-disorder phase transition, the disorder phase with high symmetry might show a high $Q{\cdot}f$ instead of the order phase, based on the results of BZN [58, 65].
- In the vicinity of BZT in the BaO-ZnO-Ta_2O_5 system, there are three areas, namely (I) an ordering area with a single phase; (II) an ordering area with a secondary phase and (III) a disordering area with a single phase [56, 66].
- The composition with the highest $Q{\cdot}f$ should ideally be located on the tie-line BZT/BMT-$BaTa_{4/5}O_3$, on which $Ba(Mg_{1/3-\alpha/3}Ta_{2/3+2\alpha/15}V_{\alpha/5})O_3$ solid solutions are formed. The composition shows a higher density than that of pure BZT/BMT [56,60, 67].

3.2. Perovskite related compounds

3.2.1. Pseudo-tungsten-bronze solid solutions

- **Crystal structure of pseudo-tungsten-bronze solid solutions**

The pseudo-tungsten-bronze $Ba_{6-3x}R_{8+2x}Ti_{18}O_{54}$ (R = rare earth) solid solutions [68, 69] are located on the perovskite-type compound tie-line of $BaTiO_3$ and $R_2Ti_3O_9$ compositions on the BaO-R_2O_3-TiO_2 ternary phase diagram as shown in Fig. 22. The crystal structure contains perovskite blocks of 2 x 2 unit cells (2x2), and pentagonal (A_2) sites as shown in Fig. 23, which are named from similar tetragonal tungsten-bronze structure with 1x1 perovskite blocks [70–72]. These compounds contain two ions with different atomic sizes. The larger Ba ions are found mainly in the pentagonal A_2-site and the smaller rare-earth (R) ions in the rhombic A_1-site. This structure has two more sites, B and C. The B-site is same as perovskite octahedral sites, and the C-site is a triangular site which is usually empty. This structure has a close relationship to the structure of perovskite. If the two ions become the same size, the structure changes to perovskite with only cubic A_1-sites owing to the combination of the A_2 and C-sites as described later at section 4 (Fig. 40). The crystal data are as follows: orthorhombic crystal system of space group *Pbnm* (No.62), point group *mmm*, lattice parameter a = 12.13, b = 22.27, c = 7.64 Å, Z = 2, Dx = 5.91 g/cm^3. This structure has a super lattice along the c-axis of twice [73]. As the space group has a center of symmetry of i as do paraelectrics, it qualifies for microwave dielectrics. The chemical formula of all occupied sites is $Ba_6R_8C_4Ti_{18}O_{54}$ and the structural formula is $[R_8Ba_2]_{A1}[Ba_4]_{A2}[V]_C[Ti_{18}]_BO_{54}$, where V is vacancy. As this chemical formula is x = 0, the chemical formula of the solid solutions is $Ba_{6-3x}R_{8+2x}Ti_{18}O_{54}$, and the structural formula is $[Ba_4]_{A2}[Ba_{2-3x}R_{8+2x}]_{A1}Ti_{18}O_{54}$. Here, the amount of Ba in the A_1-site becomes zero if 2-3x = 0, that is, x = 2/3. This composition is special due to one factor : the structure formula is $[Ba_4]_{A2}[R_{8+4/3}]_{A1}Ti_{18}O_{54}$ and is occupied separately by Ba in A_2 and by R in A_1 as shown in Fig. 24. This special composition is called "compositional ordering".

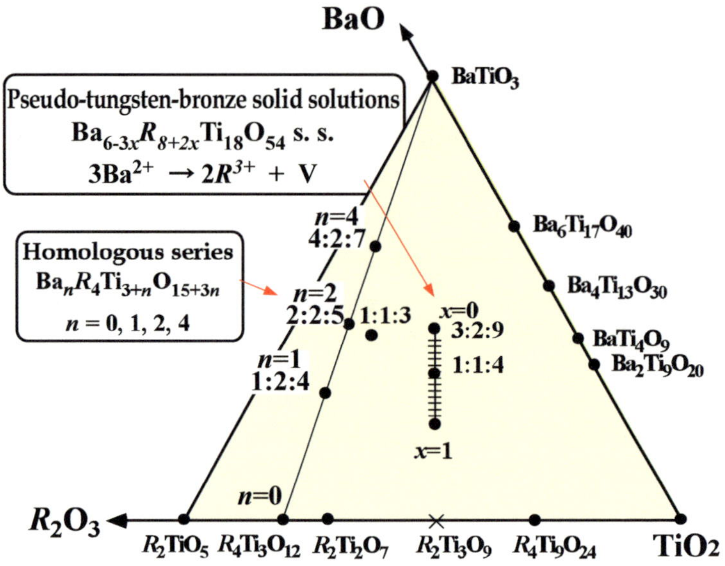

Figure 22. Part of the BaO-R_2O_3-TiO_2 ternary phase diagram with pseudo-tungsten-bronze type solid solutions and homologous compounds.

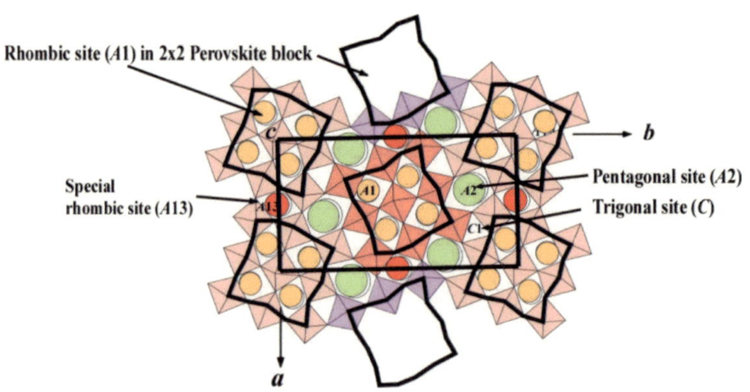

Figure 23. Crystal structure of the pseudo-tungsten-bronze solid solutions viewed in projection along [001]. Pentagonal sites (A_2) are located among 2x2 perovskite blocks with rhombic sites (A_1).

- **Microwave dielectric properties of pseudo-tungsten-bronze solid solutions**

The quality factor $Q{\cdot}f$ of the $x = 2/3$ composition, in which R and Ba ions separately occupy the rhombic site A_1 and the pentagonal site A_2 respectively, show the highest $Q\,f$ values: 10,549 GHz in the Sm system, 10,010 GHz in the Nd system, and 2,024 GHz in the La system[4)] as shown in Fig.25 (a) [74]. The highest quality factor is based on the compositional ordering of R and Ba ions in the A_1 and A_2 sites respectively, as shown in Fig. 24. The ordering distribution of the ions reduces the internal strain and results in the non-linear variation in quality factor.

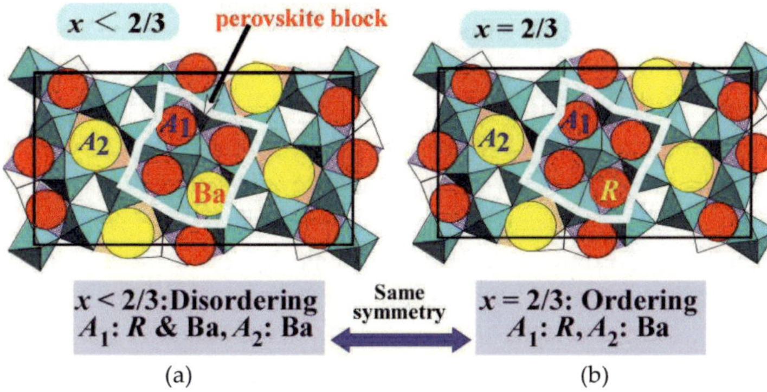

Figure 24. Crystal structure of disordering (a) and compositional ordering (b) on pseudo-tungsten-bronze solid solutions.

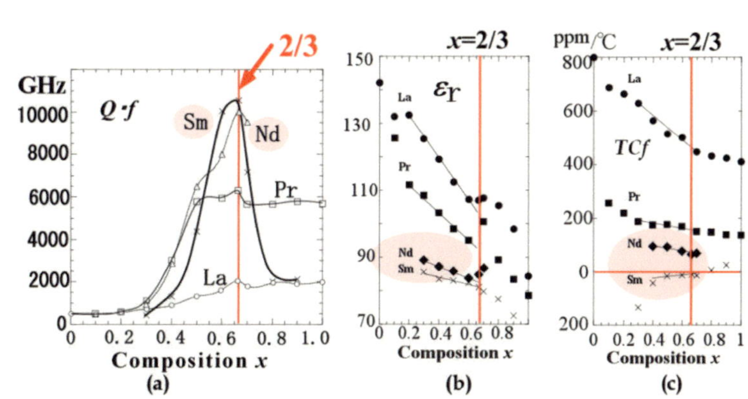

Figure 25. $Q \cdot f$ values (a), ε_r (b) and TCf (c) of pseudo-tungsten-bronze type solid solutions as a function of x in $Ba_{6-3x}R_{8+2x}Ti_{18}O_{54}$ solid solutions.

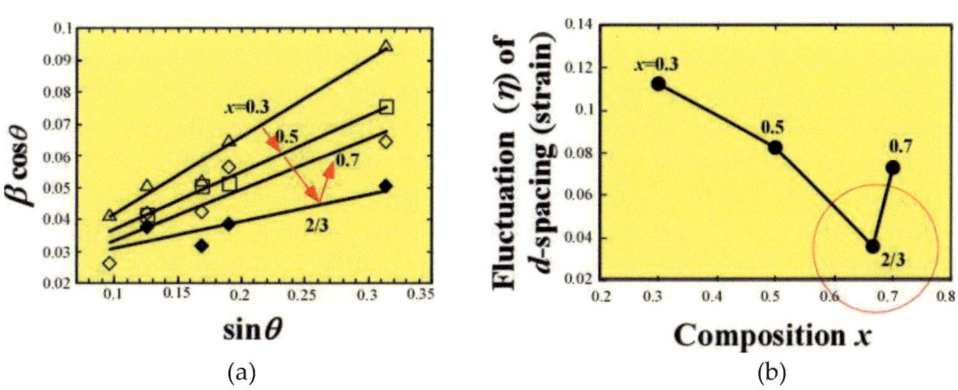

Figure 26. (a) Internal strain η obtained from the slope of equation $\beta \cos\theta = \lambda/t + 2\eta\sin\theta$. (b)

Internal strain η values for $x = 0.3$, 0.5, 2/3 and 0.7 are shown in Fig. 26 [74]. It should be noted that the internal strain for $x = 2/3$ is the lowest in the series of $Ba_{6-3x}Sm_{8+2x}Ti_{18}O_{54}$ solid solutions. This low internal strain comes from the compositional ordering. As the x-values decrease according to the structural formula $[R_{8+2x}Ba_{2-3x}V_x]_{A1}[Ba_4]_{A2}Ti_{18}O_{54}$ in the range of $0 \leq x \leq 2/3$, Ba ions with their larger ionic radii will also occupy a part of the rhombic sites with their smaller size. The location of Ba ions in the A_1- site leads to internal strain around the ions themselves, lowering the $Q\,f$ values. Moreover, the vacancies generated in the A_{1-} sites by the substitution of 3Ba by 2R might be the second reason for the lowering of the internal strain and may lead to the high $Q\,f$ values. On the other hand, as the x-values increase according to the structural formula $[R_{9.33+2(x-2/3)}V_{0.66-(x-2/3)}]_{A1}[Ba_{4-3(x-2/3)}V_{3(x-2/3)}]_{A2}Ti_{18}O_{54}$ in the range of $2/3 \leq x \leq 0.7$, then Ba ions in pentagonal A_2- sites are substituted with vacancies and R ion occupy the vacancies in A_1-site. The decrease in Ba ions produces vacancies in A_2-sites and may lead to unstable crystal structures. Moreover, the decrease in the number of vacancies in the rhombic A_1-sites, accompanied by the decrease of Ba ions in the pentagonal sites might lead to an additional internal strain. These strains are the reason for the lower quality factor at $x = 0.7$. The internal strain around $x = 0.7$ might lead to the limit of solid solutions as shown in Fig. 27 [75]. The solid solution area is different based on the R ions: the region is $0.3 \leq x \leq 0.7$ in the case of Sm and $0.0 \leq x \leq 0.7$ in Pr, Nd and La with inflection points at $x = 0.2$, which may be based on the different substitution sites.

On the other hand, the $Q{\cdot}f$ values of each R analogue with $x = 2/3$ in the $Ba_{6-3x}R_{8+2x}Ti_{18}O_{54}$ solid solutions increase according to a decrease in the rare-earth ion size (lanthanide contraction) as shown in Fig. 28. The Sm analogue has a better $Q{\cdot}f$ than the La analogue, at ca. 10,000 GHz. This crystal structure is maintained by the size difference of the cations between the Ba and R ions. It was revealed that the crystal structure with the largest size difference between Ba and Sm ions shows an excellent quality factor as it has low internal strain. On the other hand, the La analogue shows a low $Q{\cdot}f$ of ca. 2,000 GHz. Though the $Q{\cdot}f$ values of the Pr, Nd and Sm analogues show a linear relationship, that of the La analogue deviates from the linear relationship as shown in Fig. 28 [76]. If the changes in $Q{\cdot}f$ are affected only by ionic radius, then the relationship should be linear. The reason for the deviation might be internal stress depending on the stability of the crystal structure. There are two different cation sites: the A_1-site in the perovskite block and the A_2-site in the differently sized pentagonal columns as described before, which are occupied by differently sized cations. As the difference in ionic radius between Ba and La is not large in comparison with other R ions, the crystal structure is not stable, and shows a tendency of changing toward a perovskite structure which has only a single site for large cations. So, in the case of the La-ion, the internal stress always exists as an intrinsic quality, and the internal stress might cause the deviation of $Q{\cdot}f$ from the expected linear relationship. The ε_r and TCf lines against ionic radius of R increase according to the increasing size of ionic radius. The parameters are not affected by the crystal structure. The reason why the ε_r and TCf lines are proportional has not yet been clarified [76].

The dielectric constant ε_r is affected by the following three factors: (I) volume of TiO_6 octahedra; (II) tilting of octahedra strings; and (III) polarizabilities of R and Ba ions [77]. The dielectric constants ε_r of the solid solutions are proportional to lattice parameters or cell volumes as shown in Fig. 29. As x increased, ε_r decreased linearly (Fig. 25(b)), and lattice parameters or cell volumes also decreased linearly (Fig. 27). Usually, in the perovskite structure, the polarity of the Ti ion in the octahedra is produced as a result of the large octahedral volume. Thus, as the mean value of the volume decreased from 9.946 Å^3 at x=0.5 to 9.925 Å^3 at x=0.7, this volume change is considered to have decreased ε_r. However, the volume change is very small, thus other effects should be examined such as tilting of the TiO_6 octahedra strings as suggested by Valant *et al.* [78]. The tilting angle, which is that between the *c*-axis and the central axis of the octahedra as shown in Fig. 30, is inversely proportional to lattice parameters: the mean tilting angle is 9.99° at x=0.5 and 10.63° at x=0.7, based on the refined crystal structure of the Sm solid solution series [13]. From Fig. 29, it was also deduced that the polarizabilities of *R* ions affect ε_r and *TCf*. In the table of polarizabilities derived by Shannon [79], the La ion, which gives the largest ε_r in the series, also has the largest polarizability among these *R* ions: 6.03 for La, 5.31 for Pr, 5.01 for Nd and 4.74 Å^3 for Sm. The ε_r values decrease with the polarizabilities. On the other hand, the ε_r values also vary linearly as a function of cell volume in each *R*-system as shown in Fig. 29 (a). The variations in ε_r are also affected by the polarizabilities of *R* and Ba ions. The substitution is performed according to the following equation:

$$3\mathrm{Ba} \Leftrightarrow 2R + V$$

The total polarizabilities due to the substitution equation are reduced from 3×6.40 to 2×6.03 Å^3 for the La system. Here, the value of 6.40 Å^3 for the Ba^{+2} ion is larger than that for the La^{+3} ion.

The *TCf* is also plotted as a function of cell volume in Fig. 29 (b). Though a similar tendency to ε_r is observed, the mechanism of *TCf* has not yet been clarified. The *TCf* values of the Sm system are usually negative but close to zero as shown in Fig. 25(c). As *TCf* obeys additional rules, we could easily get a material with *TCf* = 0 ppm/°C. Outstanding materials with *TCf* = 0 ppm/°C have been realized by adding Nd or La to Sm-systems, which are composed of a solid solution with a single phase of x = 2/3 [80]. So, *TCf* is improved to near zero ppm/°C without the degradation of the $Q \cdot f$ value. Usually, as doped materials with different sigh *TCf* located as secondary phase, the $Q \cdot f$ values are degraded.

- Design of outstanding materials based on the crystal structure

In this section, some cases concerning material designs based on the crystal structure are presented. Ohsato *et al.* [72, 81] have researched the crystal structure of microwave materials and clarified the relationship between material properties and crystal structure to aid the design of new outstanding materials.

- **Case 1: Design by the distribution of cations for the improvement of properties when x = 0** [82, 83]

Sr ions are introduced in to this system, in which the ionic size is located between Ba and Sm ions. As mentioned above, $Q \cdot f$ values of $Ba_{6-3x}R_{8+2x}Ti_{18}O_{56}$ solid solutions have the maximum value at x=2/3. In the region x < 2/3, the structural formula of the solid solutions is $[R_{8+2x}Ba_{2-3x}$

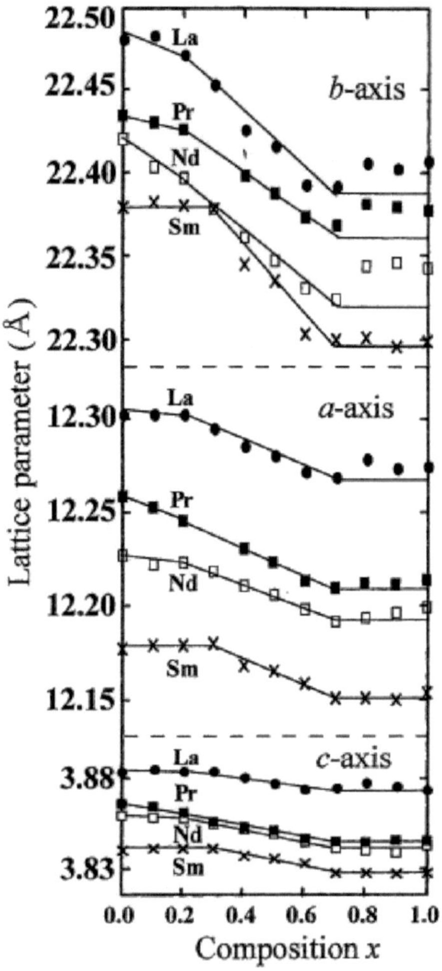

Figure 27. Lattice parameters of $R_{6-3x}R_{8+2x}Ti_{18}O_{54}$ (R = La, Pr, Nd and Sm) solid solutions.

$V_x]_{A1}[Ba_4]_{A2}Ti_{18}O_{54}$. In this region, Ba ions located in A_1-sites result in a deterioration of the quality factor. In the case of x = 0, $Q{\cdot}f$ values are very low as shown in Fig. 31 (a). When Ba ions are substituted by Sr ions such as in $[R_8Sr_2]_{A1}[Ba_4]_{A2}Ti_{18}O_{54}$, $Q{\cdot}f$ values improved markedly from 206 to 5,880 GHz in the case of R = Nd as shown in Fig. 31 (b) [82]. The introduction of Sr ions into A_1-sites may reduce the internal strain / fluctuation of d-spacing, due to the reduction in ionic size in A_1-sites. Mercurio *et al.* [84] reported that the Sr ions occupy A_{13} special sites (Fig. 23), which have a medium size between that of A_1- and A_2-sites. Hence it is expected that R, Sr and Ba ions are ordering in A_1-, A_{13}- and A_2-sites respectively [83].

- **Case 2: Substituting Sr for Ba in A_1-sites when x = 0.6** [85]

The effects of substituting Sr for Ba in the A_1-sites of $Ba_{6-3x}Sm_{8+2x}Ti_{18}O_{54}$ solid solutions were studied in terms of the lattice parameters and microwave dielectric properties as shown in

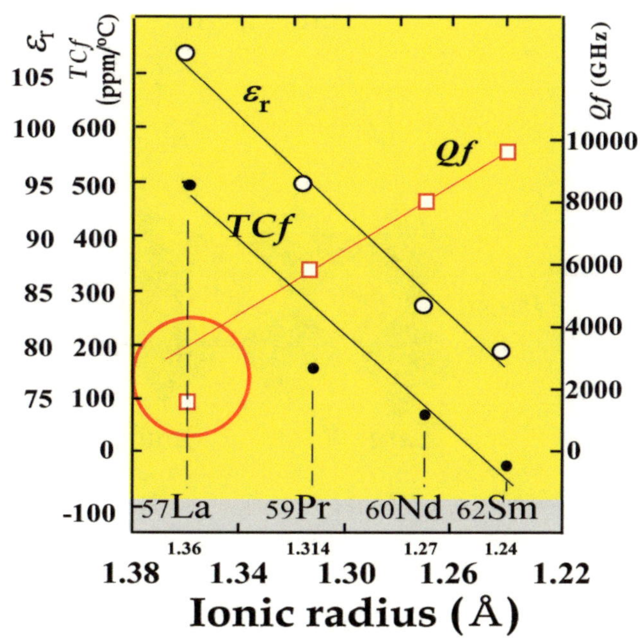

Figure 28. Microwave dielectrics properties as a function of ionic radius of *R* ion.

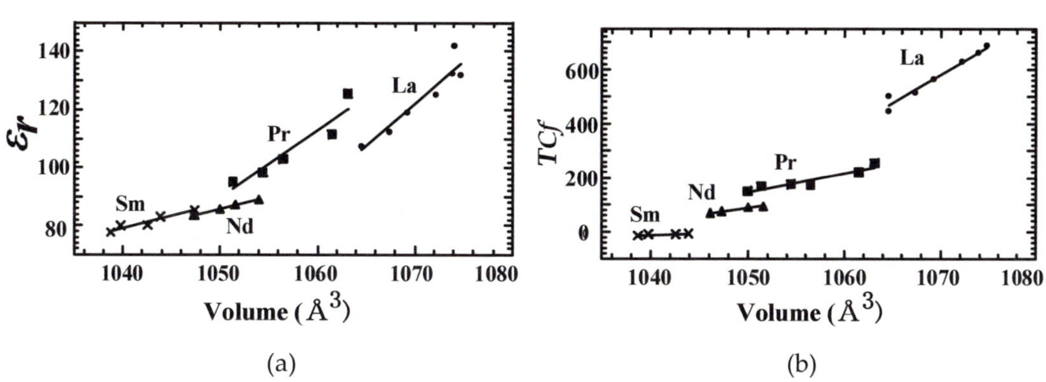

Figure 29. ε_r (a) and *TCf* (b) are shown as a function of unit cell volume.

Figs 32 and 33 respectively [85]. The compositions of the compounds in which Sr is substituted for Ba are as follows: x in $(Ba_{1-\alpha}Sr_{\alpha})_{6-3x}Sm_{8+2x}Ti_{18}O_{54}$ is fixed at 0.6, at which point the $Ba_{6-3x}Sm_{8+2x}Ti_{18}O_{54}$ solid solution has excellent properties, and composition α, in which Sr is substituted for Ba, ranges from 0.0 to 0.2. The properties are concerned with the strain in the crystal structure due to Sr substitution. We derived a structural formula $[Sm_{8+2x}Ba_{2-3x}]_{A1}[Ba_4]_{A2}Ti_{18}O_{54}$, in the range $0 \leq x \leq 2/3$ and another $[Sm_{9.33+2(x-2/3)}]_{A1}[Ba_{4-3(x-2/3)}]_{A2}Ti_{18}O_{54}$ in the range $2/3 \leq x \leq 1$. In the composition $x = 0.6$, the formula is $[Sm_{9.2}Ba_{0.2}]_{A1}[Ba_4]_{A2}Ti_{18}O_{54}$, which includes four Ba ions in A_2-sites on the pentagonal columns,

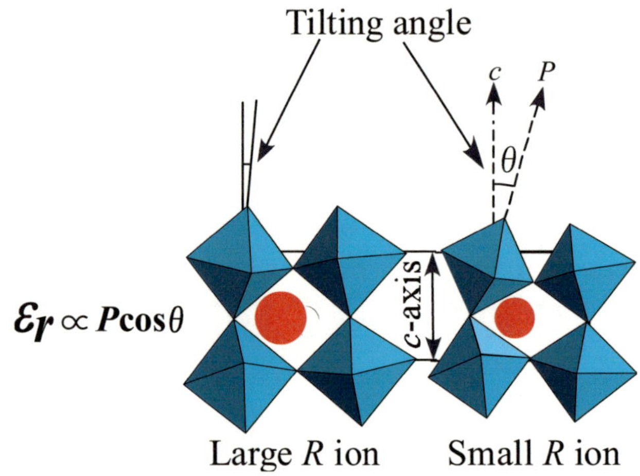

Figure 30. Correlationship between dielectric constant and the tilting angle of the octahedral.

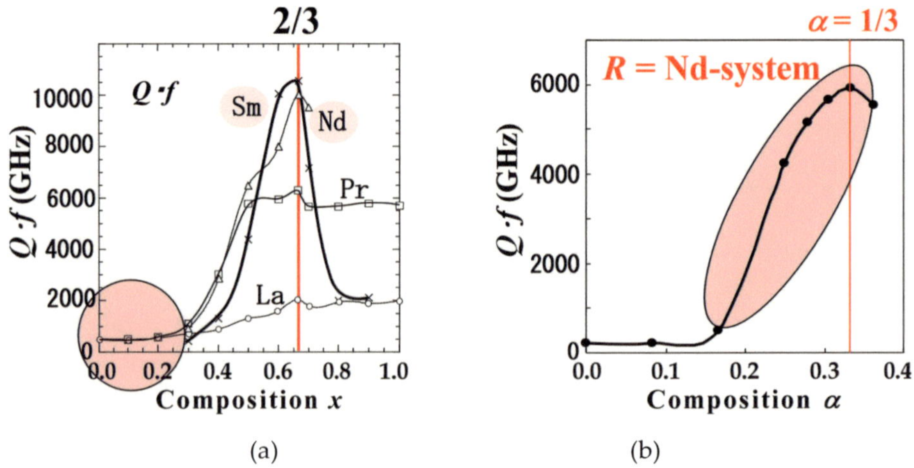

Figure 31. (a) Low $Q{\cdot}f$ around $x = 0$ on the $Q{\cdot}f$ composition figure. (b) $Q{\cdot}f$ values of Nd-system with $x = 0$ improved from 200 to 6000 GHz by substitution Sr for Ba.

and 0.2 Ba ions in A_1-sites on the perovskite blocks with 9.2 Sm. The 0.2 Ba ions in A_1-sites produce the internal strain because the size of the Ba ions is fairly large for the A_1-sites. When 0.2 Ba ions are completely substituted by Sr ions, then the $Q{\cdot}f$ values improve to 10,205 GHz, which shows that the strain in the crystal structure has relaxed somewhat. The composition in which Sr is substituted for 0.2 Ba is $\alpha = 0.048$ in the $(Ba_{1-\alpha}Sr_{\alpha})_{6-3x}Sm_{8+2x}Ti_{18}O_{54}$ substitutional formula. The dielectric properties depend on the lattice parameters, the values of which change at the composition $\alpha = 0.048$ due to the change in the substitution mode of the Sr ions. The temperature coefficient of the resonate frequency *TCf* was changed in the same manner as the dielectric constant [86].

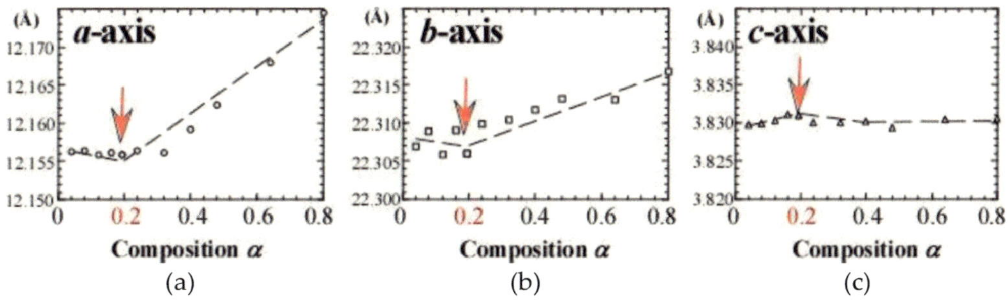

Figure 32. Lattice parameters of $(Ba_{1-\alpha}Sr_{\alpha})_{4.2}Sm_{9.2}Ti_{18}O_{54}$ solid solutions.

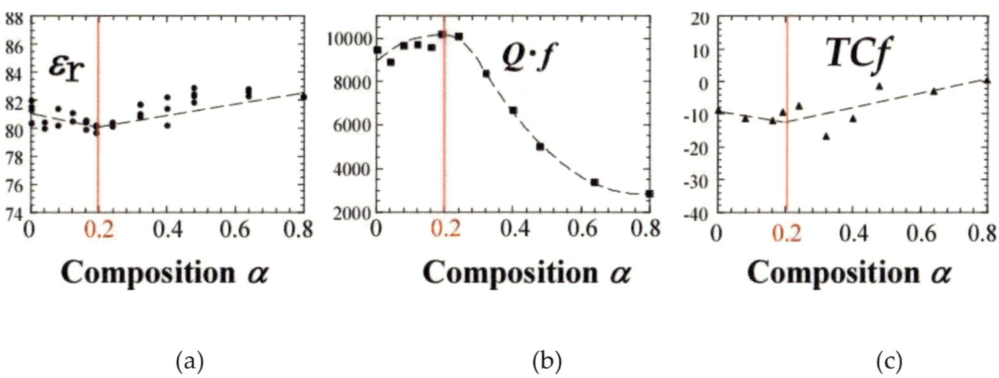

Figure 33. Microwave dielectric properties ε_r (a), $Q{\cdot}f$ (b) and *TCf* (c) of $(Ba_{1-\alpha}Sr_{\alpha})_{4.2}Sm_{9.2}Ti_{18}O_{54}$ solid solutions as a function of composition α.

3.2.2. Homologous compounds with perovskite layered structure

There are three kinds of homologous series composed of a perovskite layered structure. The perovskite layers have three different orientations such as (111) the plane series for $Ba_nLa_4Ti_{3+n}O_{12+3n}$, (100) the series for the Ruddlesden-Popper phase and (110) the series for $A_nB_nO_{3n+2}$ homologous compounds as shown in Tables 2(h) –(j). [87, 88]

- **(111) series for $Ba_nLa_4Ti_{3+n}O_{12+3n}$ homologous compounds**

The homologous compounds are also perovskite related compounds composed of a layered structure. The compounds are located in a R_2O_3-rich region, compared to pseudo-tungsten-bronze solid solutions in the BaO-RO_3-TiO_2 ternary phase diagram as shown in Fig. 22. The chemical formula is shown as $Ba_nLa_4Ti_{3+n}O_{12+3n}$, and there are four compounds at intervals n = 0, 1, 2, and 4 as shown in Fig. 34 [89–92]. As a compound with n = 4 is unstable below 1,450 °C, other compounds where n = 0, 1 and 2 are studied in this paper [93–97]. In particular, we mainly synthesized two compounds of n = 1 and 2. These compounds, n = 0: $La_4Ti_3O_{12}$ (2:3), n = 1: $BaLa_4Ti_4O_{15}$ (1:2:4), and n = 2: $Ba_2La_4Ti_5O_{18}$ (2:2:5), show hexagonal layered perovskite structures as shown in Fig. 35 [89]. Another *R* ion included in this homologous compound is the Nd ion, and the alkali earth ions Ca and Sr, substituted for Ba.

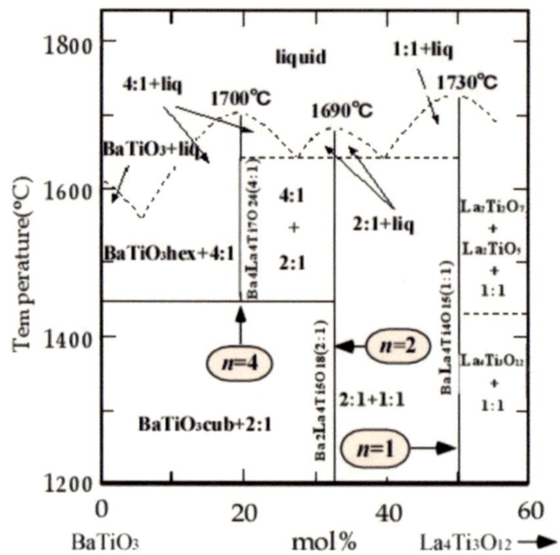

Figure 34. Binary phase diagram for $Ba_nLa_4Ti_{3+n}O_{12+3n}$ homologous compounds.

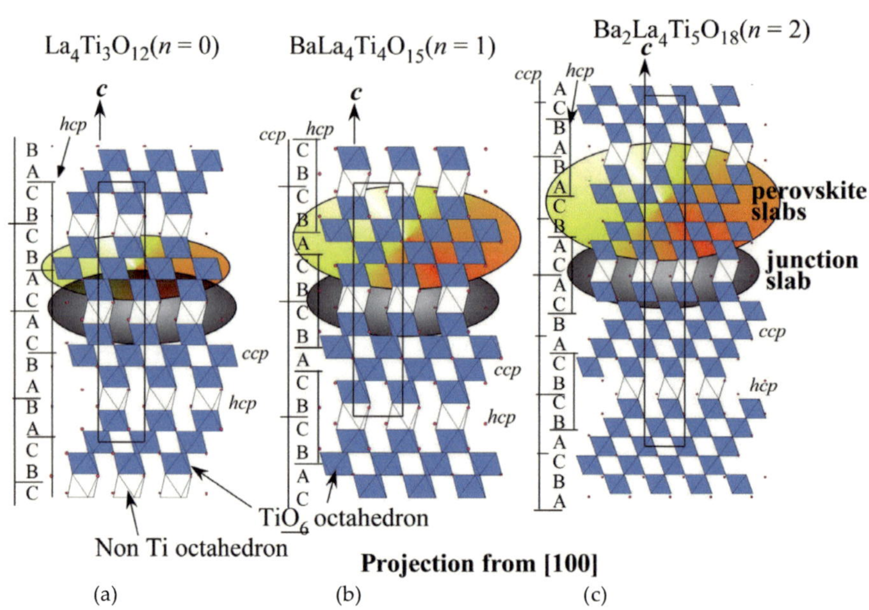

Figure 35. Crystal structure of $Ba_nLa_4Ti_{3+n}O_{12+3n}$ homologous compounds (a) $La_4Ti_3O_{12}$ (n = 0), (b) $BaLa_4Ti_4O_{15}$ (n = 1), and (c) $Ba_2La_4Ti_5O_{18}$ (n = 2).

http://dx.doi.org/10.5772/61718

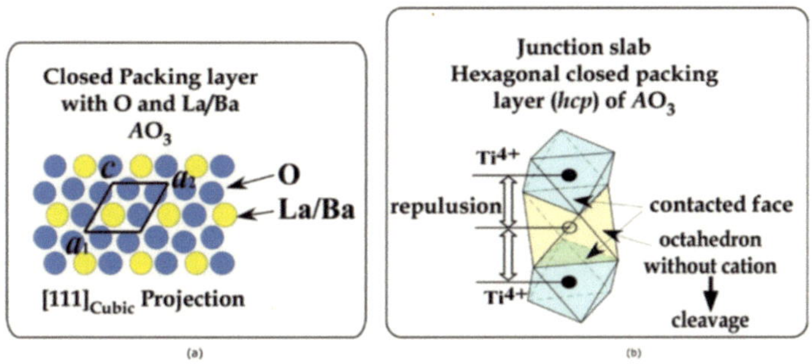

Figure 36. (a) Closed packing layer composed by oxygen and La/Ba atoms, which is one of specific features, (b) Junction slab composed by octahedron. Cleavage easily occurs at the center octahedron without cation.

The crystal structure of homologous $BaLa_4Ti_4O_{15}$ ($n = 1$) compounds is illustrated in Fig. 35(b). The crystal data are trigonal, $P\bar{3}c1$ (No. 165) and $Z = 2$. This compound has a sequence of five layers with La/BaO_3 closed packing as shown in Fig. 36 (a), like *hccch* of the $Ba_5Nb_4O_{15}$ type. Here, Ba is the cation located in the cuboctahedron of the perovskite structure, where *h* means hexagonal close packing (*hcp*) and *c* is cubic close packing (*ccp*). The perovskite slab with *ccp* is composed of four TiO_6 octahedral layers and the junction slab with *hcp* between perovskite slabs is composed of a three-octahedron string shared face with an empty octahedron in the center as shown in Fig. 36 (b).

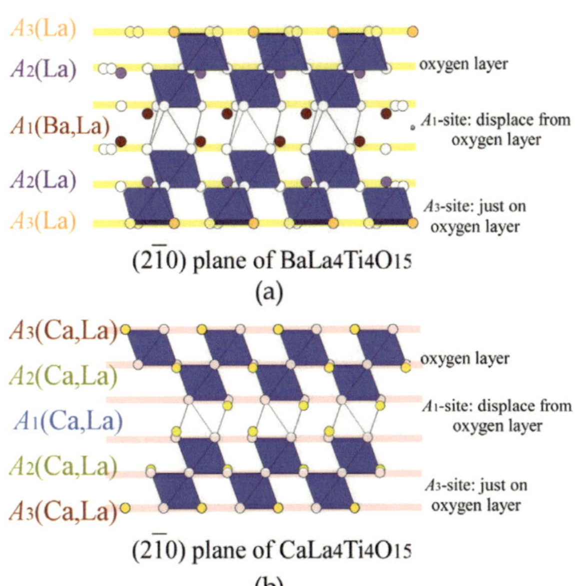

Figure 37. Alkaline earth elements in the A_1, A_2, and A_3-sites in the closed packing layer of $BaLa_4Ti_4O_{15}$ (a) and $CaLa_4Ti_4O_{15}$ (b) viewed along (2$\bar{1}$0). Ba ions with a large ionic radius are located only in A_1-sites of Ba analogues, while on the other hand, Ca ions of Ca analogues are distributed across all sites with La ions.

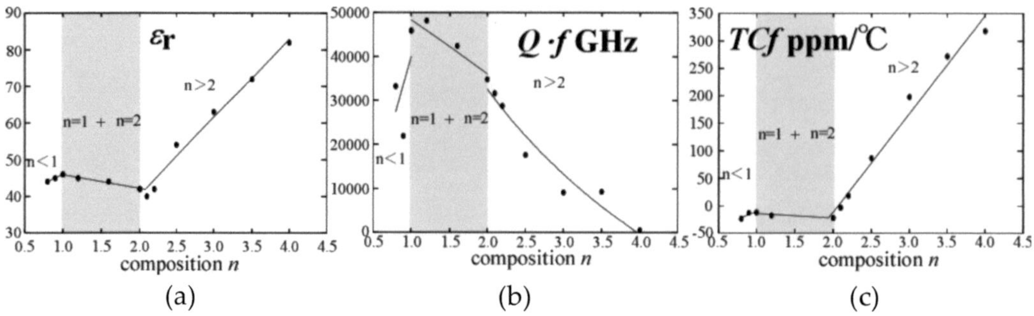

Figure 38. Microwave dielectric properties of $Ba_nLa_4Ti_{3+n}O_{12+3n}$ ceramics as a function of the composition *n*.

A	Sintering condition	D_r (%)	ε_r	*Q*•*f* (GHz)	*TCf* (ppm/°C)
Ba	1600°C 2 h	98.4	44.4	41,008	-26
Sr	1550°C 48 h	98.9	43.7	46,220	-8.4
Ca	1550°C 24 h	94.8	41.1	50,246	-25.5

Table 4. Relative density and microwave dielectric properties of $ALa_4Ti_4O_{15}$ (*A*=Ba, Sr and Ca).

The compound can include Sr and Ca ions substituted in place of Ba ions. The homologous $ALa_4Ti_4O_{15}$ (*A* = Ba, Sr, or Ca) compounds have a different ordering for *A*-site cations as shown in Fig. 37. In the case of Ba analogues, the ordering of Ba (r = 1.61 Å) and La ions (r = 1.36 Å) will occur as follows: Ba ions located in A_1-sites near the junction slab, and La ions in all A_1, A_2 and A_3-sites as shown in Fig. 37 (a) [96]. In the case of Sr and Ca analogues, the Sr ions (r = 1.44 Å) and the Ca ions (r = 1.34 Å) are all located in *A*-sites including La ions as shown in Fig. 37(b). As the space of A_1-sites is larger than those of A_2- and A_3-sites, Ba^{2+} ions with their large ionic radii predominantly occupy A_1-sites. On the other hand, as the ionic radii of Sr and Ca^{2+} are close to that of La^{3+} ions, the Sr and the Ca ions of the Sr and Ca analogues randomly occupy the *A*-sites [96].

The microwave dielectric properties of $Ba_nLa_4Ti_{3+n}O_{12+3n}$ are shown in Fig. 38 as functions of composition [93]. The sample with the composition n = 1 shows the best properties, such as the highest $Q{\cdot}f$ = 46,000 GHz, ε_r = 46, and TCf = −11 ppm/°C, which can be improved to near zero ppm/°C by means of the substitution of *Al* for La [94]. The microwave dielectric properties of the Ba, Sr, and Ca analogue ceramics are shown in Table 4 [97]. These samples showed excellent microwave dielectric properties for use in base stations of mobile phones, such as a value of ε_r greater than 40, a $Q{\cdot}f$ greater than 40,000 GHz and a *TCf* within ± 30 ppm/°C. The highest ε_r of 44.4 was observed in the case of the Ba analogue and the value decreased to 41.1 for Ca. The highest $Q{\cdot}f$ of 50,246 GHz was observed in the case of the Ca analogue, and the value decreased to 46,220 GHz for Sr and to 41,008 GHz for Ba. These values are much higher than those in an earlier report.

We would like to consider the reason for the large ε_r and the high $Q{\cdot}f$ based on the crystal structure. There are three characteristic points of the crystal structure: one is the size of the cation sites, another is the shift of the cation positions, and the third is the different divalent cation distributions. The large volume of cation sites results in the large ε_r. Divalent cation *A* with a large ionic radius, such as Ba, Ca and Sr, expands the lattice and brings an enlargement of cation sites. In particular, the expansion of the *B*-site volume affects Ti ion movement as a result of the rattling effect. In the case of the Ba analogue with the highest ε_r value, the volume is larger than that of the Ca analogue. The Ba ion with its large ionic radius of 1.61 Å is more effective than the Ca ion with $r = 1.34$ Å. In the second case of the shift of the cation positions, the ε_r of the Ba analogue with a large shift is larger than that of the Ca analogue. This shift might increase the movability of the La ion with a small ionic radius, so that the ε_r of the Ba analogue with a large shift is greater than that of the Ca analogue. The high $Q{\cdot}f$ value might come from the cation distribution and the volume of cation sites. In the case of the Ca analogue, as the shift of cations from the close packed layer of oxygen is smaller than it is for Ba analogue, then widely occupied *A*-sites might be distributed periodically with La ions to bring a high $Q{\cdot}f$. The ε_r values also depend on the ionic polarizations of Ba ($P = 6.4$), Sr ($P = 4.24$), and Ca ($P = 3.16$) [97]. These homologous compounds show characteristic near-zero ppm/°C values of the *TCf*. The *TCf* of the Sr analogue is near zero ppm/°C compared with that of the others, whose might come from the analogue of $SrTiO_3$ having a large positive *TCf* of 1,200 ppm/°C. Moreover, the *TCf* of the Ba analogue was improved to a near zero 1.3 ppm/°C, with a high ε_r of 44 and a $Q{\cdot}f$ of 47,000 GHz by substituting Al ions for La ions [97].

- **(100) series for Ruddlesden-Popper phase**

The Ruddlesden-Popper phase [98, 99] is shown with the chemical formula $A_{n+1}B_nO_{3n+1}$. Here, *A* is a cuboctahedral site and B is an octahedral site. These space groups are all *I4/mmm* (No. 139) with a center of symmetry *i*. The phase is composed of $SrTiO_3$ layers with $(100)_{cubic}$ plane as shown in Fig. 39(a). The layers are stacked, shifting each other by (1/2*a*, 1/2*b*). Among the layers, a SrO halite type structure is formed. The chemical formula is shown as $AO{\cdot}n(ABO_3)$. Here, $n = 1$: Sr_2TiO_4 [98] is K_2NiF_4 type stacking of one ABO_3 and three unit cells of perovskite; $n = 2$: $Sr_3Ti_2O_7$ is the stacking of two ABO_3 similar to $n = 1$, $n = 3$: $Sr_4Ti_3O_{10}$ is also the stacking of three ABO_3 with (100). $CaSmAlO_4$ ceramics with a K_2NiF_4 crystal structure are presented as microwave dielectrics [100]. The microwave dielectric properties are shown in Table 2 (i). Although the Auriviellius phase $A_{n+1}B_nO_{3n+3}$ [101, 102] and the Dion-Jacobson phase M^+ $[Ca_2Na_{n-3}Nb_nO_{3n+1}]^-$ exist in addition to the Ruddlesden-Popper phase in the (100) series [103,104], the microwave dielectric properties are not presented.

- **(110) series for $A_nB_nO_{3n+2}$ homologous compounds**

A (110) planar perovskite layered structure also creates a homologous series of $A_nB_nO_{3n+2}$. A typical example of this series is $Sr_2Ta_2O_7$ with $n = 4$ as shown in Fig. 39 (b) [105, 106]. The value of *n* is that of the octahedral layer in the single perovskite layer as seen in the figures. The layered structure was formed by the addition of oxygen atoms to non-bridged oxygen atoms of the BO_6 octahedron cut. The microwave dielectric properties are also shown in Table 2 (j).

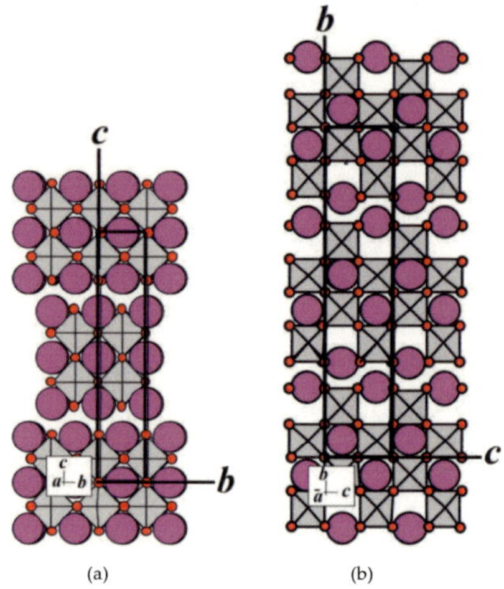

Figure 39. Crystal structure of layered perovskite: (a) (100) series: Ruddlesden-Popper phase of $Sr_3Ti_2O_7$ ($n = 2$), and (b) (110) series: $A_nB_nO_{3n+2}$ homologous compound of $Sr_2Ta_2O_7$ ($n = 4$).

- **Application of perovskite related compounds**

These perovskite related compounds are applied to microwave dielectrics. Dr. Okawa studied pseudo-tungsten-bronze solid solutions and (111) plane homologous series in his doctoral thesis [108], and these materials have been used in microwave applications. Microwave dielectrics based on $Ba_4(R_{1-y}Bi_y)_{9+1/3}Ti_{18}O_{54}$ (R: rare earth) solid solutions with Bi partially substituted for R were clarified with a high ε_r of > 80 [109,110]. They are used widely in the wireless communication systems of fire engines in Japan. An Al doped $BaLa_4Ti_4O_{15}$ homologous compound, which has the best microwave dielectric properties for use as a resonator was developed for use in the base stations of wireless communication systems [94]. It is currently used in base stations of the mobile communication systems of the Tokyo metro. In addition, at near $n = 0$ in $Ba_nLa_4Ti_{3+n}O_{12+3n}$ homologous compounds, a superior material was developed with properties of $\varepsilon_r = 42$, $Q{\cdot}f = 86{,}000$ GHz, and $TCf = -17$ ppm/°C [111, 112].

4. Flexibility of perovskite structure and microwave dielectric properties

As already described in many parts of this book, the structure of perovskite is flexible, producing many specific phases such as ferroelectrics and paraelectrics. This flexibility is due to the inclusion of many cations in the perovskite structure. There are three important features of the perovskite structure, detailed below :

- The first is the framework of octahedra connecting all their apexes with each other in three dimensions.

- The second is the closed packing layer of AO_3 instead of oxygen closed packing as shown in Fig. 36 (a).

- The third important feature is the large cation site, that is, cuboctahedron with basically 12 coordination.

The first produces two spaces for cations — octahedron and cuboctahedron — described as the third feature above, which many kinds of cations with different ionic radii and electric charges can occupy. This framework will be deformed and tilting. These features produce many kinds of ferroelectric and paraelectric properties. The closed packing of AO_3 discussed as the second feature of the perovskite structure, produces a high density, as heavy A ions are incorporated in the layer instead of oxygen. These high density and heavy materials are found in high pressure environments such as the deeper parts of the earth [113]. The cuboctahedron points are produced by connecting the AO_3 packing layer with B ions. This polyhedron includes special large cations such as Ba, Ca and Sr. If smaller sized ions such as Mg are present, the crystal structure formed will be that of ilmenite, similar to the Al_2O_3-type.

Furthermore, the flexibility of the perovskite structure is shown by the fact that different sized plural large cations can be also included in the deformed crystal structure. There are some examples as follows:

- Example 1: In the case of the order-disorder transition in complex perovskite, when the structure is ordered, two different sites appear under the changing crystal structure form from cubic to trigonal [38] as shown in Fig. 11.

- Example 2: As in the case of $SrTiO_3$ doped $LaAlO_3$, the structure creates a site partially occupied with Ti-ions under the space group change from $R\bar{3}c$ to $R\bar{3}$ [35, 36]. The crystal changes bring an improvement in $Q{\cdot}f$ due to a decrease in internal strain.

- Example 3: Although these examples do not change the framework of the perovskite structure, if the size difference becomes large enough, then the framework could be changed such as in the tungsten-bronze structure described above. In the case of tetragonal tungsten-bronze with a simple structure, the crystal structure change is illustrated in Fig. 40 [114]. When a perovskite block rotates slightly, the structure produces a pentagonal and a trigonal site from two cuboctahedra. Contrary, if the two different sized ions become the same size in the tungsten-bronze structure, then the crystal structure becomes a perovskite structure. In the case of pseudo-tungsten-bronze solid solutions, as the size difference between the Ba and La ions is small and causes the strain, the dielectric losses increase as described above (Fig. 28) [76].

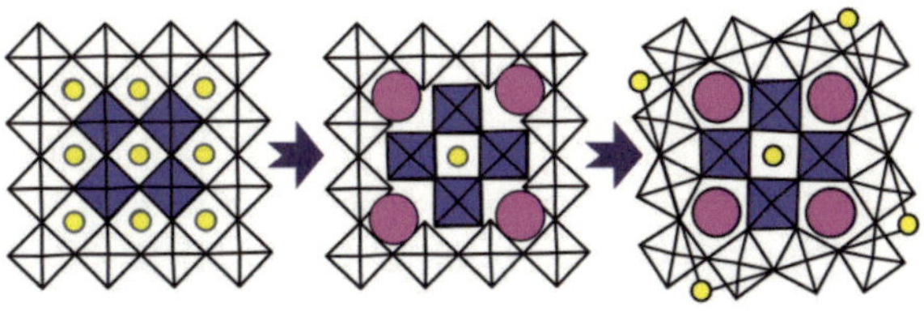

Figure 40. The structure of perovskite changes to a tungsten-bronze structure after the inclusion of two differently sized large ions, producing rhombic (A_1) and pentagonal (A_2) sites.

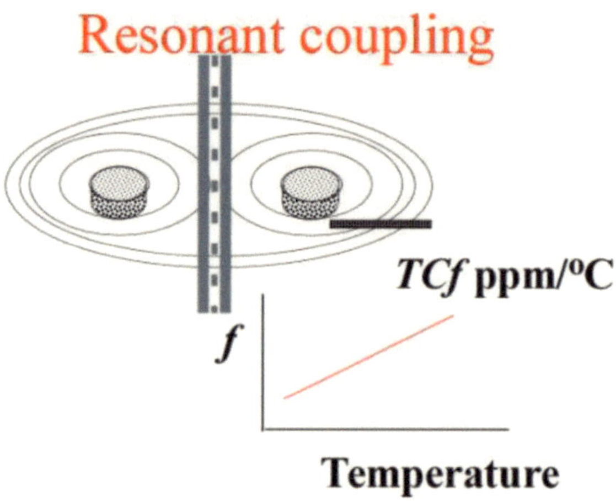

Figure 41. Resonant coupling for the temperature measurement of isolated places.

The intrinsic reasons for the ordering and symmetry effects on $Q{\cdot}f$ properties are that ordering reduces the internal stress and high symmetry reduces the formation of poles. Which effect is predominant ? As described above, in the absence of phase transition such as in pseudo-tungsten-bronze solid solutions, compositional ordering is predominant [74]. In the case of complex perovskite with an order-disorder transition, high symmetry is predominant rather than ordering, as described above [58, 65].

5. Functional advances in the next generation of microwave dielectrics

In this section, the future large scale application of microwave dielectrics will utilize some new and novel functions — based on microwave properties — as follows:

(1) electromagnetic resonance,

(2) electromagnetic wave shortening,

(3) electromagnetic wave delay,

(4) temperature variation of resonant frequency,

(5) electromagnetic wave absorption

(6) other functions such as transparency and refractive index.

As the content has been published in both "*A Handbook of Mutifunctional Ceramics*" [17] and a paper [115], please refer to those publications. Some important functions are presented below.

Fig. 41 shows a type of temperature sensor utilizing resonant coupling, which can measure the temperature on the opposite side of a wall without the need for an electric wire. The materials should have an extremely large *TCf* depending on the temperature. Fig. 42 shows a well-known

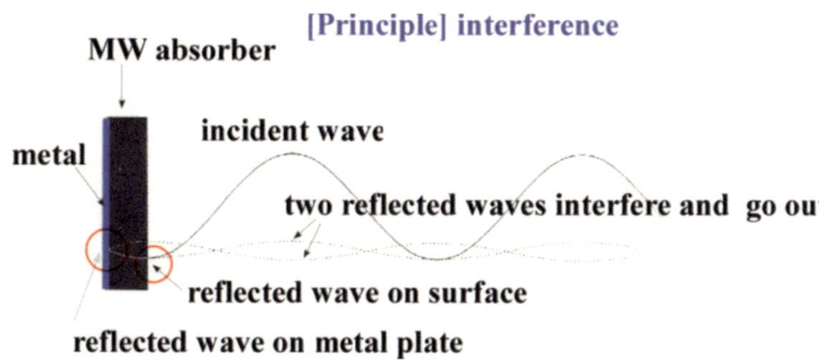

Figure 42. Principle of electromagnetic wave absorber.

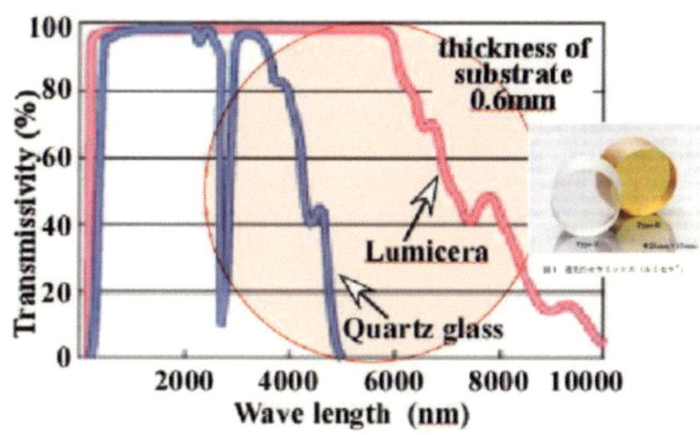

Figure 43. Transmissivity of Lumicera compared with quartz glass. Transparent region expanded to middle IR region.

principle of an electromagnetic wave absorber in a good design using wave interference, dielectric losses and wave retardation in the materials [116]. As in this example, new functions will be derived from the properties of the materials, and from physical principles. Fig. 43 shows the extreme transparency ceramics of Lumicera produced by Murata Manufacturing Co., Ltd. in Japan [117]. These materials are also microwave dielectrics, such as BMT — the 'king' of microwave dielectrics — as described above. The fabrication technology takes full advantage of our current understanding of the materials, such as controlled to cubic phase without birefringence.

6. Conclusion

Many kinds of microwave dielectrics with perovskite and related structure have been produced based on research into the relationship between the crystal structure and its properties. In this chapter, the following compounds related perovskite are introduced and discussed: simple perovskite, complex perovskite, pseudo-tungsten-bronze solid solutions and layered perovskite compounds such as $Ba_nLa_4Ti_{3+n}O_{12+3n}$ homologous with (111), Ruddlesden-Popper phase with (100) and $A_nB_nO_{3n+2}$ homologous with (110) oriented perovskite layers. Most of these compounds are paraelectrics with a center of symmetry *i*, and include many types of ion with different ionic radii and electric charges. They are also designed using stoichiometric techniques to develop superior properties. The superior microwave materials developed should be utilized in new and useful applications for the benefit of future generations.

Acknowledgements

I would like to thank Professors and graduate students of NIT, Meijo University and Hoseo University, and Doctors and researchers in the many companies which collaborated with NIT. A part of this work was supported by the following projects: (1) Support industries of Japan by Ministry of Economy, Trade and Industry (METI), Japan. (2) MEXT/JSPS KAKENHI Grant Number 25420721. (3) Adaptable & Seamless Technology Transfer Program (A-step) by MEXT, Japan

Author details

Hitoshi Ohsato[1,2*]

Address all correspondence to: ohsato.hitoshi@nitech.ac.jp

1 Nagoya Industrial Science Research Institute, Nagoya, Japan

2 Nagoya Institute of Technology, Nagoya, Japan

References

[1] Sebastian MT. Dielectric Materials for Wireless Communication. Amsterdam: Elsevier. 2008; ISBN-13:978-0-08-045330-9

[2] Sebastian MT, Ubic R, Jantunen H. Low-loss dielectric ceramic materials and their properties. International Materials Review. 2015;60:395-415.

[3] Ohsato H. High frequency dielectric ceramics. in: Adachi G editor. Materials Technology Hand Book for Rare-earth Elements. Tokyo: NTS Inc; 2008. p. 346–358. (Japanese).

[4] Ohsato H, Kagomiya I, Chae KW. Microwave dielectric ceramics with rare-earth elements (I). J Korean Phys Soc. 2012;61:971-979.

[5] Ohsato H, Kagomiya I, Kim JS. Microwave Dielectric Ceramics with Rare-Earth (II). Integrated Ferroelectrics. 2010;115:95-109.

[6] Wakino K. Recent Development of Dielectric Resonator Materials and Filters in Japan. Ferroelectrics. 1989;91:69-86.

[7] Tamura H. Progress of Ceramic Dielectric Materials for Microwave and Millimeter-wave Applications. MWE '99 Microwave Workshop Digest. 1999; p. 175-180.

[8] Kobayashi Y, Katoh M. Microwave Measurement of Dielectric Properties of Low-loss Materials by the Dielectric Resonator Method. IEEE Trans. on MTT. 1985;MTT-33:586-92.

[9] Ichinose N. High-frequency materials and their applications. In:New Ceramics & Electronic Ceramics. 1996;9(9):p. 1-50.

[10] Ohsato H, Ohhashi T, Kato H, Nishigaki S, Okuda T. Microwave Dielectric Properties and Structure of the $Ba_{6-3x}Sm_{8+2x}Ti_{18}O_{54}$ Solid Solutions. Jpn J Appl Phys. 1995;34:187-191.

[11] Ohsato H. High Frequency Dielectrics. In: Shiosaki T, editor. Development and Applications of Ferroelectric Materials. CMC; 2001. p. 135-147 (Japanese).

[12] Ohsato H. Development for Microwave dielectric ceramics. JCLS. 2003;55:1207-1218.

[13] Ohsato H. Research and Development of Microwave Dielectric Ceramics for Wireless Communications. J Ceram Soc Jpn. 2005;113:703-711.

[14] Ohsato H. New frontiers of microwave dielectrics with perovskite-type structure. Bulletin Japan Ceramics Society. 2008;43:597–609 (Japanese).

[15] Ohsato H. Research and Development for Electroceramics Based on Crystal Structure. In: Proceedings of the 4th Fulrath Memorial International Symposium on Advanced Ceramics; 4 April 2007; Tokyo; 2007. p. 1-6.

[16] Ohsato H. Design of Microwave Dielectrics Based on Crystallography. In: Akedo J, Chen XM, Tseng T, editors. Advances in Multifunctional Materials and Systems II, In: Ceramic Transactions. John Wiley & Sons; 2014;245:87-100.

[17] Ohsato H, Microwave dielectrics. In: Fukunaga O, Haneda H, Makishima A, editors. Handbook of Multifunctional Ceramics. Tokyo: NTS; 2011. P.152-166 (Japanese).

[18] Kajfez D, Guillon P, Dielectric resonators. 2nd ed. Atlanta: Noble Publishing Corporation; 1998.

[19] Ohsato H. Microwave Materials with High Q and Low Dielectric Constant for Wireless Communications. Mater Res Soc. Symp. Proc. 2005;833:55-62.

[20] Tsunooka T, Andou M, Higashida Y, Sugiura H, Ohsato H. Effects of TiO_2 on sinterability and dielectric properties of high-Q forsterite ceramics. J Eur Ceram Soc. 2003;23:2573-2578.

[21] Tsunooka T, Sugiyama H, Kakimoto K, Ohsato H, Ogawa H. Zero Temperature Coefficient τ_f and Sinterability of Forsterite Ceramics by Rutile Addition. J Ceram Soc Jpn Suppl. 2004;112:S1637-S1640.

[22] Guo Y, Ohsato H, Kakimoto K. Characterization and dielectric behavior of willemite and TiO_2-doped willemite ceramics at millimeter-wave frequency. J Eur Ceram Soc. 2006;26:1827-1830.

[23] Surendran KP, Santha N, Mohanan P, Sebastian M. Temperature stable low loss ceramic dielectrics in (1-x) $ZnAl_2O_{4-x}TiO_2$ system for microwave substrate applications. Eur Phys J B. 2004;41:301–306.

[24] Ohsato H, Kato K, Mizuta M, Nishigaki S, Okuda T. Microwave Dielectric Properties of the Ba_{6-3X} $(Sm_{1-Y}, R_Y)_{8+2X}Ti_{18}O_{54}$ (R = Nd and La) Solid Solutions with Zero Temperature Coefficient of the Resonant Frequency. Jpn J Appl Phys. 1995;34(9B):5413-5417.

[25] Ball CJ, Begg BD, Cookson DJ, Thorogood GJ, Vance ER. Structures in the System $CaTiO_3/SrTiO_3$. J Solid State Chem. 1998;139:283-247.

[26] Huang CH, Pana CL, Shim SJ. Liquid phase sintering of $MgTiO_3$-$CaTiO_3$ microwave dielectric ceramics. Mater Chem Phys. 2003;78:111-115.

[27] Yao GG, Liu P. Low temperature sintering and microwave dielectric properties of (1-x) $Mg_4Nb_2O_9$-$x$$CaTiO_3$ ceramics. Physica. 2010;B 405:547-551.

[28] Huang CH, Pana CL, Leeb WC. Microwave dielectric properties of mixtures of glass-forming oxides Zn–B–Si and dielectric ceramics $MgTiO_3$–$CaTiO_3$ for LTCC applications. J Alloys & Compounds. 2008;462:5-8.

[29] Cho SY, Kim IT, Hong KS. Microwave dielectric properties and applications of rare earth aluminates. J Mater Res. 1999;14:114-119.

[30] Cho SY, Kim CH, Kim DW, Hongs KS, Kim JH. Dielectric properties of $Ln(Mg_{1/2}Ti_{1/2})O_3$ as substrates for high-Tc superconductor thin films. J Mater Res. 1999;14:2484-2487.

[31] Geller S, Bala VB. Crystallographic studies of perovskite-like compounds. II. Rare earth aluminates. Acta Cryst. 1956;9:1019-1025.

[32] Inagaki Y, Suzuki S, Kagomiya I, Kakimoto K, Ohsato H, Sasaki K, Kuroda K, Shimada T. Crystal structure and microwave dielectric properties of $SrTiO_3$ doped $LaALO_3$ single crystal grown by FZ. J Euro Ceram Soc. 2007;27:2861-2864.

[33] Cho SY, Hong KS, Ko KH. Mixture-like behavior in the microwave dielectric properties of the (1−x)$LaAlO_3$–$x$$SrTiO_3$ system. Mater Res Bull. 1999;34:511–516.

[34] Shimada T, Kura K, Ohtsuki S. Dielectric properties and far infrared reflectivity of lanthanum aluminate–strontium titanate ceramics. J Euro Ceram Soc. 2006;26:2017–2021.

[35] Inagaki Y, Ishizawa N, Ohsato H, Kagomiya I, Kakimoto K, Shimada T. Structure of Sr and Ti codoped $LaAlO_3$ perovskite. In: XXI Congress of the International Union of Crystallography Congress and General Assembly. 2008; P11.11.42(C520).

[36] Ishizawa N, Inagaki Y, Kagomiya I, Kakimoto K, Ohsato H. Rhombohedral modification of Sr and Ti co-doped $LaAlO_3$. Photon Factory Activity Report 2007. 2008;25(Part B):186.

[37] Suvorov D, Valant M, Jancar B, Skapin SD. $CaTiO_3$-based ceramics: Microstructural development and dielectric-properties. Acta Chimi Slov. 2001;48:87-99.

[38] Galasso F, Pyle J. Ordering in compounds of the $A(B'_{0.33}Ta_{0.67})O_3$ type Inorg. Chem. 1963;2:482-484.

[39] Surendran KP, Sebastian MT, Mohanan P, Moreira RL, Dias A. Effect of Nonstoichiometry on the Structure and Microwave Dielectric Properties of $Ba(Mg_{0.33}Ta_{0.67})O_3$. Chem Mater. 2005;17:142-151.

[40] Kagata H, Kato J. Dielectric properties of Ca based complex perovskite at microwave frequencies. Jpn J Appl Phys. 1994;33:5463.

[41] Tamura H, Konoike T, Sakabe Y, Wakino K. Microwave Dielectric Properties of Ba-Nd-Ti-O System Doped with Metal Oxides. J Am Ceram Soc. 1984;67:C-59.

[42] Kageyama K. Research for synthesis and properties of low losses microwave dielectric ceramics [Dr. thesis]. Nagoya Institute of Technology; 1995;p. 59. (Japanese).

[43] Seabra MP, Avdeev M, Ferreira VM, Pullar RC, Alford NMN, Reaney IM. Structure–Property Relations in $xBaTiO_3$–$(1-x)La(Mg_{1/2}Ti_{1/2})O_3$ Solid Solutions. J Am Ceram Soc. 2004;87:584.

[44] Kageyama K. Microwave dielectric properties of CaO-Ga_2O_3-Ta_2O_5 ceramics. Ferroelectrics. 1990;109: 173-183.

[45] Wakino K, Minai K, Tamura H. Formation of Self-Organized Zirconium Titanate Nanotube Layers by Alloy Anodization. J Am Ceram Soc. 1984;67:278.

[46] Khalam LA, Sreemoolanathan H, Mohanan P, Sebastian MT. Preparation, characterization and microwave dielectric properties of $Ba(B'_{1/2}Nb_{1/2})O_3$ B' = La, Pr, Nd, Sm, Eu, Gd, Tb, Dy, Ho, Y, Yb and In ceramics. Mater Sc. Engn. 2004;B107:264.

[47] Silverman BD. Microwave Absorption in Cubic Strontium Titanate. Phys Rev. 1962;125:1921-1930.

[48] Wakino K, Murata M, Tamura H. Far Infrared Reflection Spectra of $Ba(Zn,Ta)O_3$-$BaZrO_3$ Dielectric Resonator Material. J Am Ceram Soc. 1986;69:34-37.

[49] Hiuga T, Matsumoto K. Ordering of $Ba(B_{1/3}B_{2/3})O_3$ Ceramics and Their Microwave Dielectric Properties. Jpn J Appl Phsy. 1989;S28-2:56-58.

[50] Kim ES,Yoon KH. Microwave Dielectric Properties of Complex Perovskite $Ba(Mg_{1/3}Ta_{2/3})O_3$. Ferroelect. 1992;133:1187.

[51] Lu CH, Tsai CC. Reaction kinetics, sintering characteristics, and ordering behavior of microwave. J Mater Res. 1996;11:1219-1227.

[52] Kawashima S, Nishida M, Ueda I, Ouchi H. $Ba(Zn_{1/3}Ta_{2/3})O_3$ Ceramics with Low Dielectric Loss at Microwave Frequencies. J Am Ceram Soc. 1983;66:421-423.

[53] Yokotani Y, Tsuruta T, Okuyama K, Kugimiya K. Low-Dielectric Loss Ceramics for Microwave Uses. National Technical Report. 1994;40:11-16. (Japanese).

[54] Matsumoto H, Tamura H, Wakino K. Ba(Mg, Ta)O_3-$BaSnO_3$ High-*Q* Dielectric Resonator. Jpn J Appl Phys. 1991;30:2347-2349.

[55] Koga E, Moriwake H. Effects of Superlattice Ordering and Ceramic Microstructure on the Microwave *Q* Factor of Complex Perovskite-Type Oxide $Ba(Zn_{1/3}Ta_{2/3})O_3$. J Ceram Soc Jpn. 2003;111:767-775 (Japanese).

[56] Koga E, Moriwake H, Kakimoto K, Ohsato H. Influence of Composition Deviation from Stoichiometric $Ba(Zn_{1/3}Ta_{2/3})O_3$ on Superlattice Ordering and Microwave Quality Factor *Q*. J Ceram Soc Jpn. 2005;113: 172-178 (Japanese).

[57] Koga E, Mori H, Kakimoto K, Ohsato H. Synthesis of Disordered $Ba(Zn_{1/3}Ta_{2/3})O_3$ by Spark Plasma Sintering and Its Microwave *Q* Factor. Jpn J Appl Phys. 2006;45(9B): 7484-7488.

[58] Koga E, Yamagishi Y, Moriwake H, Kakimoto K, Ohsato H. Order-disorder transition and its effect on Microwave quality factor *Q* in $Ba(Zn_{1/3}Nb_{2/3})O_3$ system. J Electroceram. 2006;17:375-379.

[59] Koga E, Yamagishi Y, Moriwake H, Kakimoto K, Ohsato H. Large *Q* factor variation within dense, highly ordered $Ba(Zn_{1/3}Ta_{2/3})O_3$ system. J Euro Ceram Soc. 2006;26:1961-1964.

[60] Kugimiya K. Crystallographic study on the *Q* of $Ba(Mg_{1/3}Ta_{2/3})O_3$ dielectrics. In:Abstract for Kansai branch academic meeting. 5 September 2003; Senri-Life Science. B-20. In: Abstract for Meeting of Microwave/Millimeterwave Dielectrics and Related Materials on the Ceramic Soc. 2004; Nagoya Institute of Technology:Nagoya Japan. (Japanese).

[61] Surendran KP, Sebastian MT, Mohanan P, Moreira RL, Dias A. Effect of Nonstoichiometry on the Structure and Microwave Dielectric Properties of $Ba(Mg_{0.33}Ta_{0.67})O_3$. Chem Mater. 2005;17:142-151.

[62] Koga E, Moriwake H, Kakimoto K, Ohsato H. Raman Spectroscopic Evaluation and Microwave Dielectric Property of Order/Disorder and Stoichiometric/Non-Stoichiometric $Ba(Zn_{1/3}Ta_{2/3})O_3$. Ferroelectrics. 2007;356:146–152.

[63] Izumi F, Ikeda T. A Rietveld-analysis program RIETAN-98 and its applications to zeolites. Mater Sci Forum. January 2000;321-324:198-203.

[64] Kolodiazhnyi T. Origin of extrinsic dielectric loss in 1:2 ordered, single-phase $BaMg_{1/3}Ta_{2/3}O_3$. J Euro Ceram Soc. 2014;34:1741-1753.

[65] Ohsato H, Koga E, Kagomiya I, Kakimoto K. Origin of High *Q* for Microwave Complex Perovskite. Key Eng Mat. 2010;421-422:pp77-80.

[66] Ohsato H, Koga E, Kagomiya I, Kakimoto K. Phase Relationship and Microwave Dielectric Properties in the Vicinity of $Ba(Zn_{1/3}Ta_{2/3})O_3$. Ceram Eng & Sci Proc. 2010;30:25-35.

[67] Ohsato H, Koga E, Kagomiya I, Kakimoto K. Dense Composition with High-*Q* on the Complex Perovskite Compounds. Ferroelectrics. 2009;387:28-35.

[68] Roth RS, Beach F, Antoro A, Davis K, Soubeyroux JL. Structural of the nonstoichiometric solid solutions $Ba_2RE_4[Ba_{x+}RE_{2/3-2/3x}]Ti_9O_{27}$ (RE = Nd, Sm). In:14 Int Congress Crystallog: Collected Abst; 1987; Perth: Australia; 07. 9-9.

[69] Ohsato H, Ohhashi T, Nishigaki S, Okuda T, Sumiya K, Suzuki S. Formation of solid solution of new tungsten bronze-type microwave dielectric compounds $Ba_{6-3x}R_{8+2x}Ti_{18}O_{54}$ (R = Nd and Sm, $0 \le x \le 1$). Jpn J Appl Phys. 1993;32:4323–4326.

[70] Matveeva, RG, Varforomeev MB, Il'yuschenko LS. Refinement of the composition crystal structure of $Ba_{3.75}Pr_{95-}Ti_{18}O_{54}$. Zh Neorg Khim. 1984;29:31–34 (Trans. Russ: J Inorg Chem. 1984;29:17-19.

[71] Ohsato H, Ohhashi T, Okuda T. Structure of $Ba_{6-3x}Sm_{8+2x}Ti_{18}O_{54}$ ($0 < x < 1$). In Ext. Abstr. AsCA '92 Conf; November 1992; Singapore: 14U-50.

[72] Ohsato H. Science of tungstenbronze-type like $Ba_{6-3x}R_{8+2x}Ti_{18}O_{54}$ (R= rare earth) microwave dielectric solid solutions. J Euro Ceram Soc. 2001;21:2703-2711.

[73] Ohsato H, Nishigaki S, Okuda T. Superlattice and dielectric properties of dielectric compounds. Jpn J Appl Phys. 1992;31(9B):3136–3138.

[74] Ohsato H, Imaeda M, Takagi Y, Komura A, Okuda T. Microwave quality factor improved by ordering of Ba and rare-earth on the tungstenbronze-type $Ba_{6-3x}R_{8+2x}Ti_{18}O_{54}$ (R = La, Nd and Sm) Solid Solutions. In Proceeding of the XIth IEEE International Symposium on Applications of Ferroelectrics. IEEE catalog number 98CH36245; 1998: p. 509-512.

[75] Ohsato H, Mizuta M, Ikoma T, Onogi Z, Nishigaki S, Okuda T. Microwave Dielectric Properties of Tungsten Bronze-Type $Ba_{6-3x}R_{8+2x}Ti_{18}O_{54}$ (R = La, Pr, Nd and Sm) Solid Solutions. J Ceram Soc Jpn. Int Edition. 1998;106-185:184-188.

[76] Ohsato H, Mizuta M, Okuda T. Crystal Structure and Microwave Dielectric Properties of Tungstenbronze-type $Ba_{6-3x}R_{8+2x}Ti_{18}O_{54}$ (R = La, Nd and Sm) Solid Solutions. In:

Morawiec H, Stroz D, editors. Applied Crystallography. World Scientific Publishing; 1998: p. 440-447.

[77] Ohsato H, Imaeda M, Komura A, Okuda T. Non-Linear Microwave Quality Factor Change Based on the Site Occupancy of Cations on the Tungstenbronze-type $Ba_{6-3x}R_{8+2x}Ti_{18}O_{54}$ (R = Rare Earth) Solid Solutions. In: K. M. Nair KM, Bhalla AS, editors. Dielectric Ceramic Materials. In: Ceramic Transaction. John Wiley & Sons; 1998;100:p41-50.

[78] Valant M, Sovorov D, Kolar D. X-Ray Investigations and Determination of Dielectric Properties of the Compound $Ba_{4.5}Gd_9Ti_{18}O_{54}$. Jpn J Appl Phys. 1996;35:144-150.

[79] Shannon RD. Dielectric polarizabilities of ions in oxides and fluorides. J Appl Phys. 1993;73:348.

[80] Ohsato H, Sugino J, Komura A, Nishigaki S, Okuda T. Microwave Dielectric Properties of $Ba_4(Nd2_{8/3-y}R_y)Ti_{18}O_{54}$ (R = Eu, Dy, Ho, Er and Yb) Solid Solutions. Jpn J Appl Phys. 1999;38(9B):5625-5628.

[81] Ohsato H. Crystallography and R&D for Material Science from Our Research: Electroceramics. Advanced Materials Research. 2006;11-12:95-100.

[82] Nagatomo T, Otagiri T, Suzuki M, Ohsato H. Microwave dielectric properties and crystal structure of the tungstenbronze-type like $(Ba_{1-\alpha}Sr_\alpha)_6(Nd_{1-\beta}Y_\beta)_8Ti_{18}O_{54}$ solid solutions. J Eur Ceram Soc. 2006;26:1895-1898.

[83] Suzuki M, Ohsato H, Kakimoto K, Nagatomo T, Otagiri T. Crystal structure and microwave dielectric properties of $(Ba_{1-\alpha}Sr_\alpha)_{6-3x}Sm_{8+2x}Ti_{18}O_{54}$ solid solutions. J Eur Ceram Soc. 2006;26:2035-2038.

[84] Mercurio D, Abou-Salama M, Mercurio PJ, Investigations of the Tungsten-bronze-Type $(Ba_{1-\alpha}Sr_\alpha)_{6-x}La_{8+2x/3}Ti_{18}O_{54}$ ($0 = x = 3$) Solid Solutions. J Eur Ceram Soc. 2001;21:2713-2716.

[85] Imaeda M, Ito K, Mizuta M, Ohsato H, Nishigaki S, Okuda T. Microwave dielectric properties of $Ba_{6-3x}Sm_{8+2x}Ti_{18}O_{54}$ solid solutions with Sr substituted for Ba. Jpn J Appl Phys. 1997;36(9B):6012-6015.

[86] Ohsato H, Suzuki M, Kakimoto K. Ionic Distribution and Microwave Dielectric Properties for Tungstenbronze-Type Like $Ba_{6-3x}R_{8+2x}Ti_{18}O_{54}$ (R = Sm, Nd and La) Solid Solutions. Ceramic Engineering and Science Proceedings. 2005;26(5):135-145.

[87] Ishizawa N. Layered Perovskite-Structural Classification. Ceramics. 1996;31:409-413 (Japanese).

[88] Magnéli A. Structures of the ReO_3-type with recurrent dislocations of atoms: `homologous series' of molybdenum and tungsten oxides. Acta Cryst. 1953;6:495–500. DOI: 10.1107/S0365110X53001381

[89] Trolliard G, Harre N, Mercurio D, Frit B. Cation-deficient perovskite-related (Ba, La)$_n$Ti$_{n-\delta}$ O$_{3n}$ ($n \geq 4\delta$) microphases in the $La_4Ti_3O_{12}$-$BaTiO_3$ system: An HRTEM approach. J Solid State Chem. 1999;145.678-693.

[90] Teneze N, Mercurio D, Trolliard G, Frit B. Cation-deficient perovskite-related compounds (Ba, La)$_n$Ti$_{n-1}$O$_{3n}$ (n = 4, 5, and 6): Rietveld refinement from neutron powder diffraction data. Mater Res Bull. 2000;35:1603–1614.

[91] Harre N, Mercurio D, Trolliard G, Frit B. Crystal structure of $BaLa_4Ti_4O_{15}$, member n=5 of the homologous series (Ba, La)$_n$Ti$_{n-1}$O$_{3n}$ of cation-deficient perovskite-related compounds. Mater Res Bull. 1998;33:1537-1548.

[92] Harre N, Mercurio D, Trolliard G, Frit B. Crystal structure of $BaLa_4Ti_5O_{15}$ member n= 6 of the homologous series (Ba,La)$_n$Ti$_{n-1}$O$_{3n}$ of cation deficient perovskite related compounds. J Solid State Inorg. Chem. 1998;35:77.

[93] Okawa T, Kiuchi K, Okabe H, Ohsato H. Microwave dielectric properties of $Ba_nLa_4Ti_{3+n}O_{12+3n}$ Homologous Series. Jpn J Appl Phys. 2001;40:5779-5782.

[94] Okawa T, Kiuchi K, Ohsato H. Microwave dielectric properties of $Ba_nLa_4Ti_{3+n}O_{12+3n}$ Homologous Compounds and Substitutetion of Trivalent Cations for La. Ferroelectrics. 2002;272:345-350.

[95] Ohsato H, Tohdo Y, Kakimoto K, Okabe H, Okawa T. Crystal structure and microwave dielectric properties of $BaLa_4Ti_{3+n}O_{12+3n}$ homologous compounds with high dielectric constant and high quality factor. Ceramic Engineering and science Proceedings. 2003;24:75-80.

[96] Tohdo Y, Kakimoto K, Ohsato H, Okawa T and Okabe H. Crystal Structure Analysis of Homologous Compounds ALa$_4$Ti$_4$O$_{15}$ (A=Ba, Sr and Ca) and their Microwave Dielectric Properties. Ceramic Engineering and Science Proceedings. 2005;26:147-153.

[97] Tohdo Y, Kakimoto K, Ohsato H, Yamada H, Okawa T. Microwave dielectric properties and crystal structure of homologous compounds ALa$_4$Ti$_4$O$_{15}$ (A = Ba, Sr and Ca) for base station applications. J Eur Ceram Soc. 2006;26:2039-2043.

[98] Ruddlesden SN, Popper P. New compounds of the K_2NIF_4 type. Acta Cryst. 1957;10:538–539. DOI: 10.1107/S0365110X57001929

[99] Ruddlesden SN, Popper P. The compound $Sr_3Ti_2O_7$ and its structure. Acta Cryst. 1958;11:54–55. DOI: 10.1107/S0365110X58000128

[100] Fan XC, Chen XM. Microstructure and Microwave Dielectric Properties of the $CaSmAlO_4$-Based Ceramics. J Am Ceram Soc. 2008;91:2917-22.

[101] Auriviellus B. Mixed bismuth oxides with layer lattices II. Structure of $Bi_4Ti_3O_{12}$. Arkiv Kemi. 1949;1:499.

[102] Galasso FS. Structure and Properties of Inorganic Solids. New York: Pergamon Press; 1970. 308 p.

[103] Dion M, Ganne M, Tournoux M. Nouvelles familles de phases $M^IM^{II}_2Nb_3O_{10}$ a feuillets "perovskites". Mater Res Bull. 1981;16:1429-1435.

[104] Jacobson AJ, Johnson JW, Lewandowski JT. Interlayer chemistry between thick transition-metal oxide layers: synthesis and intercalation reactions of $K[Ca_2Na_{n-3}Nb_nO_{3n+1}]$ (3.ltoreq. n.ltoreq. 7). Inorg Chem. 1985;24:3727-3729.

[105] Ishizawa N, Marumo F, Kawamura T, Kimura M. Compounds with perovskite-type slabs. II. The crystal structure of $Sr_2Ta_2O_7$. Acta Cryst.1976;B32:2564-2566.

[106] Ishizawa N, Marumo F, Iwai S. Compounds with perovskite-type slabs. IV. Ferroelectric phase transitions in $Sr_2(Ta_{1-x}Nb_x)_2O_7$ (x 0.12) and $Sr_2Ta_2O_7$. Acta Cryst. 1981;B37:26-31.

[107] Nanot PM, Queyroux F, Gilles JC, Chevalier R. Structure cristalline du composé $Nd_4Ca_2Ti_6O_{20}$, terme n= 6 de la série (Nd,Ca)n, Ti_nO_{3n+2}. Acta Cryst. 1976; B32: 1115-1120.

[108] Okawa T. Research for BaO-R_2O_3-TiO_2 (R = Rear Earth) microwave dielectric ceramics [Dr. thesis]. Nagoya Institute of Technology; 2003. (Japanese)

[109] Valant M, Suvorov D, Kolar D. Role of Bi_2O_3 in optimizing the dielectric properties of $Ba_{4.5}Nd_9Ti_{18}O_{54}$ based microwave ceramics. J Mat Res. 1996;11:928-931.

[110] Okawa T, Imaeda M, Ohsato H. Microwave Dielectric Properties of Bi-Added $Ba_4Nd_{9+1/3}Ti_{18}O_{54}$ Solid Solutions. Jpn J Appl Phys. 2000;39(9B):5645-5649.

[111] Yamada H, Okawa T, Tohdo Y, Ohsato H. Microwave dielectric properties of $Ba_xLa_4Ti_{3+x}O_{12+3x}$ (x = 0.0–1.0) ceramics. J Eur Ceram Soc. 2006;26:2059-2062.

[112] Ohsato H, Harada A, Okawa T, Okabe H. Microwave dielectric composite composition. US 7,046,258 B2, (2006).

[113] Murakami M, Hirose K, Kawamura K, Sata N, Ohishi Y. Post-perovskite phase transition in $MgSiO_3$. Science. 2004;304:855-858.

[114] Lundberg M, Sundberg M, Magneli A. The "Pentagonal Column" as a Building Unit in Crystal and Defect Structure of Some Groups of Transition Metal Compounds. J Solid State Chem. 1982:44:32-40.

[115] Ohsato H. Functional advances of microwave dielectrics for next generation. Ceramics International, 2012;38S:S141-S146.

[116] Tenue Publicity Group. Grain 29: TV ghost hunted/effect of natural resonating/electromagnetic-wave absorption. In: A guidebook of micro cosmos for ferrite/With ferrite. TDK Corporation; 2007: p. 86–87. (Japanese)

[117] Tanaka N. High refractive transparency ceramics. FC Report. 2003;21:90–91. (Japanese).

Chapter 10

ESR and Magnetization Studies of Bi-manganites

Rajender Singh and Ramesh Ade

Additional information is available at the end of the chapter

http://dx.doi.org/10.5772/61936

Abstract

The electron spin resonance (ESR) and magnetization (M) studies of Bi-manganites were undertaken to understand the coexistence of various magnetic phases and their effect on charge ordering as a function of composition and temperature. The studies on several compositions of Bi-manganites reveal that the electronic phase separation (PS) is an intrinsic phenomenon in this system.

Keywords: Bi-manganites, ESR, magnetization, electronic phase separation

1. Introduction

The perovskite structure is represented by general composition ABO_3. Manganites belong to this family with A as trivalent (La^{3+}, Pr^{3+}, Bi^{3+}, etc.) and/or divalent (Ba^{2+}, Ca^{2+}, Sr^{2+}, etc.) ion and B as Mn ion. Manganites have a complex phase diagram and display interesting properties like colossal magnetoresistance (CMR) and giant magnetoresistance (GMR) [1–7]. These properties are the result of strong coupling between the charge, spin, and orbital degrees of freedom. The strength of this coupling depends on hydrostatic pressure [8, 9], magnetic and electric fields [10, 11], grain size [12, 13], and disorder created due to substitution at A-/B-site. The trivalent rare-earth ions doping of different sizes into the perovskite structure causes disorder in the sample due to change in chemical pressure, leading to the evolution of coexistence of various magnetic phases viz: paramagnetic (PM), ferromagnetic (FM), canted-antiferromagnetic (C-AFM), and antiferromagnetic (AFM). The phenomenon of phase separation (PS) has been proposed to explain the properties of manganites in view of inhomogeneities arising due to doping of the material. The PS can be electronic or structural. The electronic PS occurs when cluster formation takes place at nanometric level. The structural PS, which is due to disorder, can induce formation of up to micrometer-size clusters, which assist in percolation leading to first-order transitions [14–16]. The Griffiths Phase (GP) concept [17] has been used to explain the coexistence of FM and AFM domains in different types of manganites.

$Bi_{1-x}Ca_xMnO_3$ (BCMO) and $Bi_{1-x}Sr_xMnO_3$ (BSMO) manganites are insulating over the entire range of compositions with various phases [18–20]. The charge ordering temperature (T_{CO}) is 300–500 K for these materials, which is high compared to rare-earth materials. The charge ordering (CO) in these materials does not depend on the one electron bandwidth mechanism like in other manganites as it is due to highly polarizable $6s^2$ lone pair of Bi^{3+}ions.

In this chapter, we will review the recent work reported on these systems. The effect of substitution of various elements including transition elements at Bi- and Mn-site of these systems is reviewed. The study on various systems is considered to elucidate the intrinsic nature of PS phenomenon in these systems. In order to understand the evolution of various magnetic phases as a function of dopants at different sites and grain size; the following series of samples synthesized by solid-state (SS) or sol–gel (SG) methods are considered.

1. $Bi_{1-x}Ca_xMnO_3$ ($x = 0.30$–0.90) (BCMO)
2. $Bi_{0.5-x}Pr_xCa_{0.5}MnO_3$ ($x = 0.0$–0.50) (BPCMO)
3. La-doped Bi-manganites [$Bi_{0.7-x}La_xCa_{0.3}MnO_3$ ($x = 0.07$–0.70) and $Bi_{0.30}La_{0.37}Ca_{0.33}MnO_3$]
4. $Bi_{0.5}Ca_{0.5}Mn_{0.95}TE_{0.05}O_3$ (TE = V, Cu, and Zn)
5. $Bi_{0.5}Ca_{0.5}Mn_{0.95}TM_{0.05}O_3$ (TM = Ni, Fe, Co, and Cr)
6. $Bi_{0.55}Ca_{0.45}MnO_3$ and $Bi_{0.5}Ca_{0.5}MnO_3$ nanoparticles

The properties of the samples were studied using the following characterization techniques.

a. X-ray diffraction (XRD)
b. Scanning electron microscopy (SEM)
c. Energy dispersive x-ray spectroscopy (EDX)
d. Electron spin resonance (ESR)
e. Magnetization by vibrating sample magnetometer.

2. Electron spin resonance

There are two experimental techniques, namely magnetic resonance and inelastic neutron scattering, which can be used to describe the spin dynamics. Since double-exchange (DE) interaction mechanism is an intrinsic dynamical process, the investigation of spin dynamics and their static properties is crucial to study the magnetic properties of manganites. Although neutron scattering can provide information about magnetically ordered phases, but it has limitations in probing the spin dynamics of PM phases. The electron spin/paramagnetic resonance (ESR/EPR) is a sensitive technique which can help to understand the spin structure and its dynamics in complex magnetic-ordered phases. This technique has been used by several scientists to study the spin dynamics of manganites. In the present work, the ESR technique is used to investigate various dynamical processes over a wide range of temperatures.

2.1. Origin of ESR in manganites

The Mn^{2+} ion ($S = 5/2$) is the most common ion detected by ESR. The Mn^{3+} ($3d^4$ with $S = 2$) exhibits a large zero-field splitting and strong spin-lattice relaxation and so unlikely to be ESR active [21]. The Mn^{4+} ions ($3d^3$ with $S = 3/2$) give an ESR signal usually at low temperatures. The manganese ions, Mn^{3+} and Mn^{4+}, are commonly seen in doped manganites. Theories suggesting that the two Mn^{3+} ions transforming into Mn^{2+} and Mn^{4+} ion pairs via thermally excited disproportionation were common [22–24]. It is generally accepted that manganites do not have, or have a very small ratio of, Mn^{2+} ions and hence do not contribute to the ESR signals exhibited by manganites. ESR signals obtained in La-based manganites have been attributed primarily to Mn^{4+} ions in an octahedral anion crystal electric field in ground state with a weak spin-lattice relaxation [25], which makes this ion ESR active even at higher temperatures [23]. Shengelaya et al. [25] attributes the ESR signal to Mn^{4+} in a system of three components such as Mn^{4+} ions, Mn^{3+} ions, and the lattice, and predicted an FM Curie–Weiss like behavior for the double-integrated line intensity, which is proportional to the static susceptibility arising from the FM couplings of the Mn^{4+} and Mn^{3+} magnetic subsystems [26]. But the ESR signals cannot be attributed to isolated Mn^{4+} ions. The origin of ESR signal observed in manganites has been assigned to some combinations or clusters of Mn^{3+}–Mn^{4+} ions coupled by a strong short-range FM–DE interaction [4, 27]. All the Mn ions are assumed to contribute to the ESR signal and the normalized double-integrated intensity (*DI*) is therefore proportional to the number of ESR centers and also a measure of ESR susceptibility (χ_{ESR}) [4, 28].

2.2. Theoretical background

The experimental ESR spectra (Fig. 1) was analyzed by Lorentzian expression given as

$$\frac{dP}{dH} = \frac{d}{dH} A \left(\frac{\Delta H}{4(H - H_0)^2 + \Delta H^2} + \frac{\Delta H}{4(H + H_0)^2 + \Delta H^2} \right) \tag{1}$$

where H_0 is the resonance field, ΔH is the linewidth, and A is a quantity proportional to the area under the curve which is related to the intensity of the signal [21, 29, 30]. The two terms in Eq. 1 describe the contribution from the clockwise and anticlockwise circularly polarized components of microwave radiation.

It was suggested [28] that when the ESR signals are intense near the critical point (Curie temperature, T_C), even for relatively small amount of samples, the ΔH depends strongly on the number of spins in the sample. Whenever large amount of samples are used, the sample size effects arise from overloading the cavity through magnetic losses [28, 31]. The ESR signals

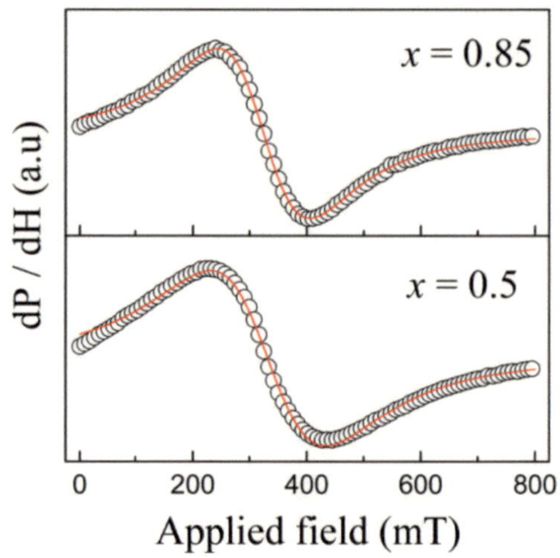

Figure 1. Lorentzian fit (solid red line) to the ESR spectra (open circles) of $Bi_{1-x}Ca_xMnO_3$ at 273 K [6].

remain Lorentzian as long as the magnetic losses are not large enough to drive the diode detector out of linearity [28]. When the sample is loaded in the cavity, at a fixed microwave frequency, ΔH can be given by the following expression

$$\Delta H^{obs} = \Delta H\left(1+b\right)^{1/2} \tag{2}$$

with

$$b = \frac{\left(4\pi/3\right)\eta Q_L \chi_{ESR}\omega}{\gamma \Delta H} \tag{3}$$

where χ_{ESR} is the static susceptibility corresponding to resonant species, η is the fitting factor, and Q_L is the loaded Q of the microwave cavity. Similarly, the changes in line intensity (I) also occur by size effects due to magnetic losses, and the expression for this can be given [28] as

$$I = \eta \chi_{ESR}\omega Q_L \left(1+b\right)^{-1/2} \tag{4}$$

These two equations (2) and (4) have certain limitations such as the value of b should be « 1. In addition to the sample dependent effects, frequency or applied field dependences of ΔH have also been reported [28, 32, 33]. Apart from the above-mentioned parameters, skin depth (δ) also affects the parameters I and ΔH of the ESR signal, and it is given by the expression

$$\delta = \left(\frac{\rho}{\mu_o \omega}\right)^{0.5} \tag{5}$$

http://dx.doi.org/10.5772/61936

where ρ is the resistivity of the sample measured at room temperature (RT), μ_o (= $4\pi \times 10^{-7}$ Vs/Am) is the permeability of free space, and ω (= $2\pi \times 9$ GHz) is the microwave frequency. The estimated values of δ are found to be in the range 0.2–1.5 mm, and it is larger than that of the size of samples [4]. In the present work, we have also taken care of cavity over loading problems. This ensures that the observed ESR signal intensity, ΔH, and resonance field (H_0) as a function of temperature are intrinsic in nature.

The study of magnetic properties of the samples with ESR technique is limited in the low temperatures due to its instrumental limitations. In order to support the ESR findings and to study the magnetic properties down to the temperature up to 5 K (from 350 K), the Quantum design Physical Property Measurement System (PPMS) was used. Magnetization as a function of temperature was measured in both zero field cooled (ZFC) and field cooled (FC) by applying a field of 500 Oe.

2.3. Studies on $Bi_{1-x}Ca_xMnO_3$ (x = 0.30–0.90)

Figure 2 depicts the Rietveld refinement of XRD pattern for the sample with x = 0.65. With the refinement, XRD data are well reproduced with *Pnma* space group orthorhombic structure of all the samples except the sample with x = 0.30. Using the refined parameters crystal structure is generated for all the samples, as shown in Fig. 3 for $Bi_{0.35}Ca_{0.65}MnO_3$ which is representative of all the samples. With increase in Ca content, decrease in unit cell volume is observed. This is ascribed to lower ionic radii of Ca^{2+} (1.12 Å) than that of Bi^{3+} (1.24 Å) ion.

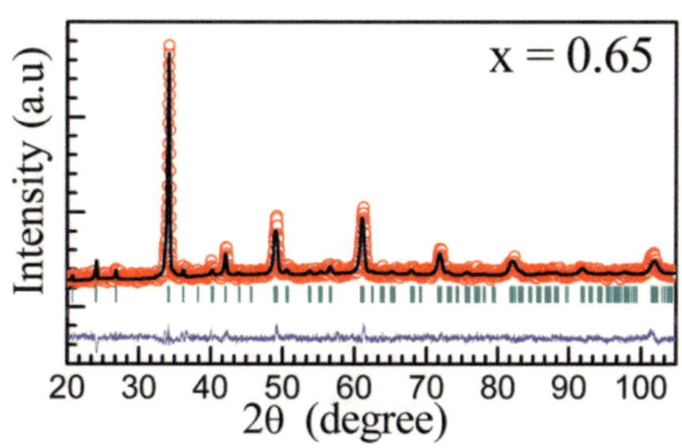

Figure 2. Rietveld refinement of the XRD data of $Bi_{1-x}Ca_xMnO_3$.

Figure 4 shows *DI* of ESR signal as a function of temperature. For samples with $0.40 \le x \le 0.80$, as the temperature decreases from high temperature (453K), *DI* increases and reaches its maximum, which is assigned as charge ordering temperature (T_{CO}) [5]. Below T_{CO}, *DI* decreases, evidencing the existence of AFM correlations and reaches a minimum and then it increases with further decrease in temperature for samples with $0.65 \le x \le 0.80$. For x = 0.30 sample, a weak T_{CO} is observed. For x = 0.85 and 0.90 samples, no T_{CO} is observed. For x = 0.85 and 0.90 samples, on cooling from high temperature, *DI* increases gradually. For $0.30 \le x \le 0.80$ samples,

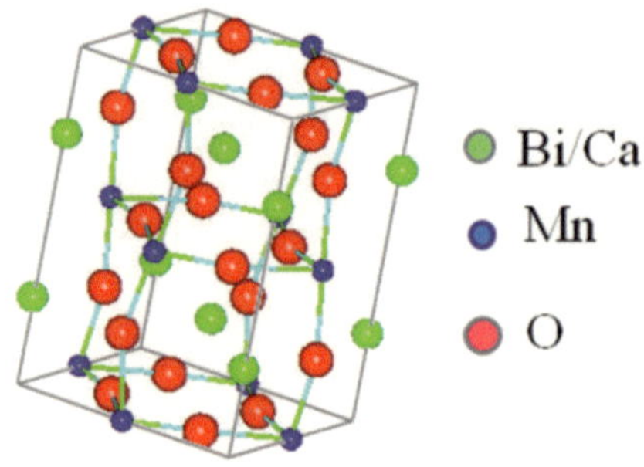

Figure 3. Crystal structure of $Bi_{0.35}Ca_{0.65}MnO_3$.

sharp decrease in *DI* below the temperature ~126–150 K is assigned as the temperature at which onset of long-range AFM ordering takes place, that is, Neel temperature (T_N).

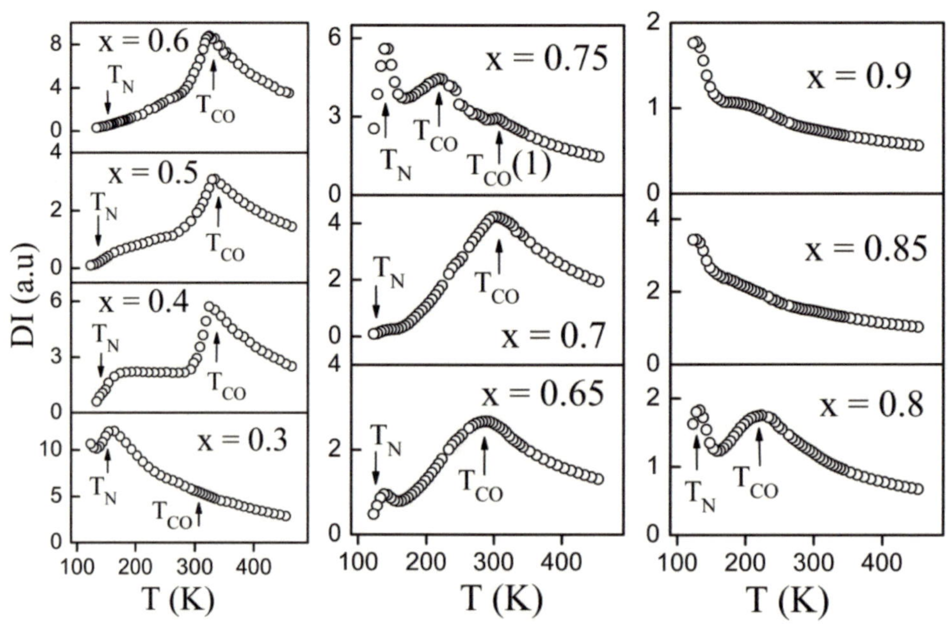

Figure 4. *DI* vs *T* plots of $Bi_{1-x}Ca_xMnO_3$ samples.

The 1/*DI* vs *T* plots for various samples are shown in Fig. 5. The linear fit in the high-temperature range is as per Curie–Weiss law [5]. The positive intercept on the x-axis in the temperature range $T > T_{CO}$ for samples with $0.30 \leq x \leq 0.80$, indicates the existence of FM–DE correlations. The charge ordering transition of $0.65 \leq x \leq 0.80$ samples is smeared compared to what is observed for samples with $x \leq 0.60$. The CO transition at 303 K is weak for sample with $x = 0.75$. The existence of Bi-rich CO phase, a chemical phase separation in the samples, with $x = 0.75$, 0.70, and 0.65 may be the reason for this. The upward and downward trend in 1/*DI* vs *T* plots

http://dx.doi.org/10.5772/61936

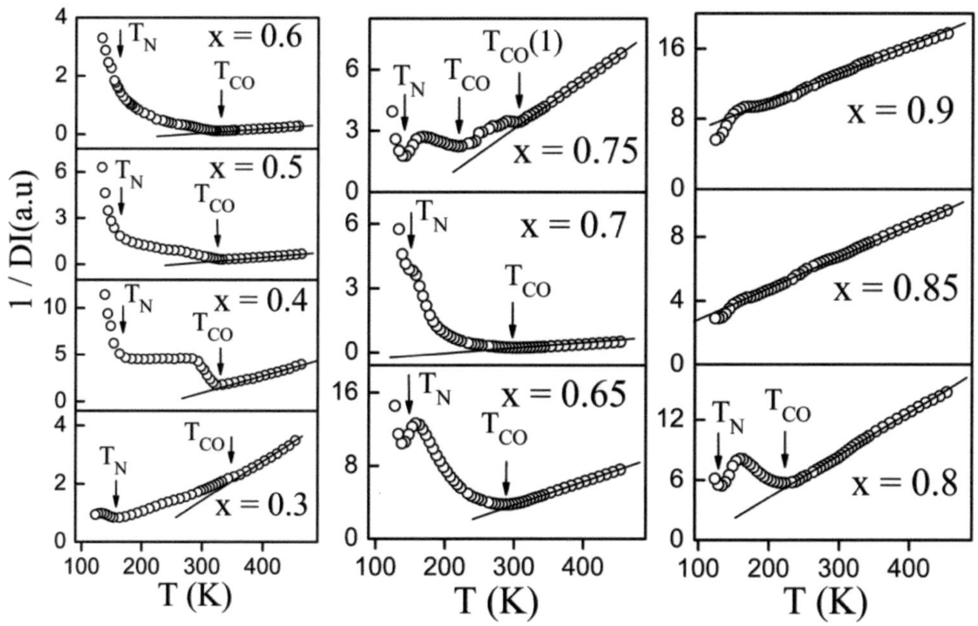

Figure 5. 1/*DI* vs *T* plots of $Bi_{1-x}Ca_xMnO_3$ samples. Ref. [5]

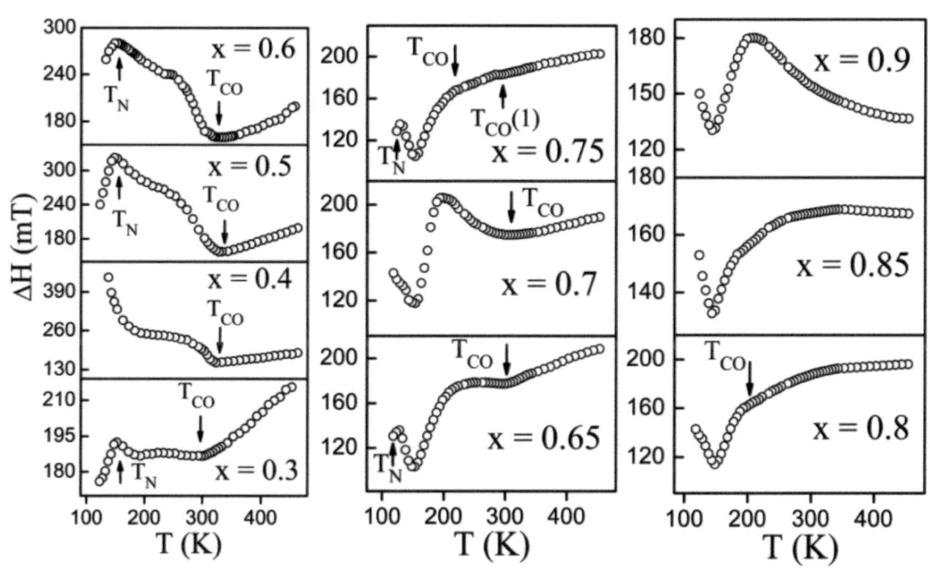

Figure 6. ΔH vs *T* plots of $Bi_{1-x}Ca_xMnO_3$ samples. Ref. [5].

below T_{CO} is an indication whether AFM or FM interactions dominate in the respective temperature range [5]. The minima in these plots indicate the onset of CO or AFM ordering. For samples with x = 0.65, 0.75, and 0.80, 1/*DI* vs *T* plots show a broad peak at ~160 K, which is attributed to C-type AFM correlations. The down turn behavior of $1/DI$ in the temperature range below 148 K for the samples with $x = 0.85$ and 0.90 can also be explained in view of short-range ordering of FM clusters in PM matrix as per Griffiths theory. The negative intercept on the x-axis for samples with $x = 0.85$ and 0.90 indicates dominant AFM-superexchange (SE) interactions above 150 K. Below 150 K, localized FM correlations dominate [4–6].

Figure 6 shows ΔH as a function of temperature. For samples with $x \geq 0.65$, ΔH shows prominent minima at ~150 K. The ESR signal persists only for a few degrees below 150 K for all the samples and then it disappears indicating the onset of long-range AFM state (T_N). ΔH increases linearly with increase in temperature in the temperature range $T \geq T_{CO}$. ΔH saturates with increase in temperature as T_{CO} decreases with increase in x. This effect is clearly seen for sample with $x = 0.8$. ΔH increases with decrease in temperature for the sample with $x = 0.90$, indicating the strengthening of AFM interactions with lowering of temperature.

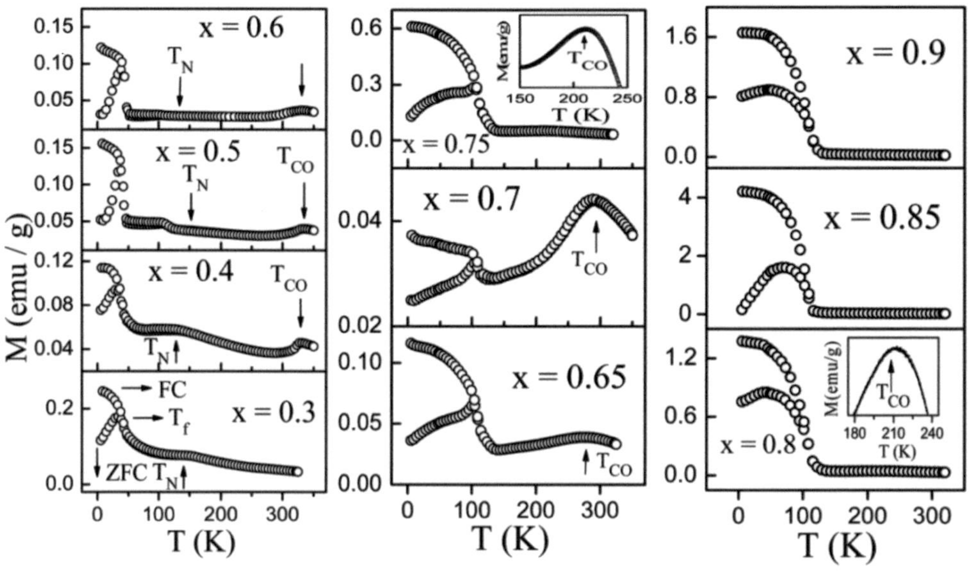

Figure 7. M vs T plots of $Bi_{1-x}Ca_xMnO_3$ samples.

Figure 7 shows magnetization vs temperature (M vs T) plots for various samples. As the temperature decreases from 350 K, both ZFC and FC curves show a peak at T_{CO} for samples with $0.30 \leq x \leq 0.80$. M decreases slightly on further decrease in temperature. The broad maximum at ~150 K observed for samples with $x \leq 0.50$ represents long-range AFM order [6]. The sharp rise in M for samples with $x \leq 0.60$ at ~50 K is related to the spin-glass (SG) or cluster-glass (CG) transitions which arises due to competition between FM interaction in $BiMnO_3$-type clusters and AFM coupling in clusters in which Mn^{3+} orbits are frozen in random orientations [6]. For these samples, M vs T plots in ZFC mode show a peak in M at spin freezing temperature (T_f). A small peak in M at ~120 K observed for sample with $x = 0.50$ is ascribed to field-induced spin canting in the AFM phase aided by the spontaneous canting moment [6]. The trend and the conclusions drawn from $1/\chi$ vs T (not shown) and $1/DI$ vs T plots are similar.

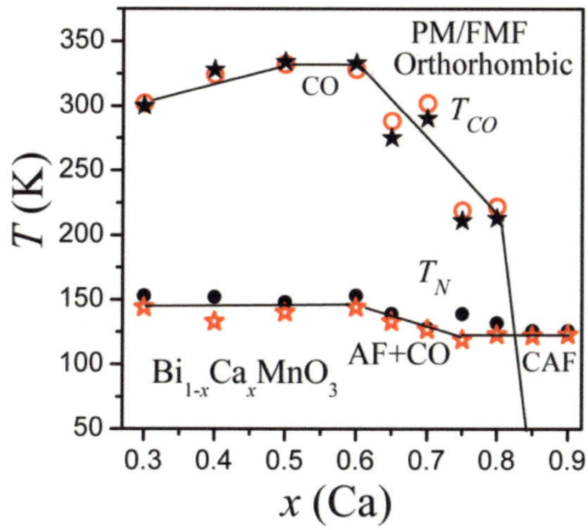

Figure 8. Phase diagram of BCMO system. Open circles (T_{CO}) and closed circles (T_N) are from ESR data and closed stars (T_{CO}) and open stars (T_N) from magnetization data [6]

Figure 8 shows the estimated values of T_{CO} and T_N from the ESR and magnetization data for $Bi_{1-x}Ca_xMnO_3$. The magnetic phase in the temperature range $T > T_{CO}$ is dominated by FM–DE interactions for samples with $0.30 \le x \le 0.80$. The PM–AFM transition at ~150 K coexists with PM to C-AFM transition at ~120 K for samples with $0.50 \le x \le 0.80$, whereas for samples with $x = 0.85$ and 0.90, only PM to C-AFM transition is observed.

2.4. Studies on $Bi_{0.5-x}Pr_xCa_{0.5}MnO_3$ ($0.0 \le x \le 0.50$) manganites

The XRD study of the $Bi_{0.5-x}Pr_xCa_{0.5}MnO_3$ ($0.0 \le x \le 0.50$) samples [34] confirms the single-phase orthorhombic nature with *Pnma* space group. A sharp decrease in the unit cell volume is observed as Pr content increases from $x = 0.0$ to 0.05. However, the unit cell volume increases steadily with further increase in Pr content from 0.05 to 0.50. The Bi^{3+} has effective ionic size of 1.170 or 1.24 Å depending on whether the $6s^2$ lone pair character is constrained or dominant, respectively [34]. The higher cell volume of undoped sample may be due to the strengthened dominant lone pair character of Bi^{3+} ion, whereas the sharp drop in the cell volume for sample with $x = 0.05$ indicates that the constrained effect of the $6s^2$ lone pair becomes strong due to Pr doping. The increase in cell volume for samples with $x > 0.05$ indicates that the constrained effect of the lone pair maintains the Bi^{3+} ionic size at 1.170 Å, that is, smaller than Pr^{3+} ionic size (1.179 Å).

The temperature dependence of ΔH of ESR signal is shown in Fig. 9. As temperature decreases from high temperature (453 K), ΔH decreases and reaches a minimum value at T_{CO}. At ~200 K, ΔH decreases sharply and reaches a minimum at 130, 150, and 160 K for samples with $0.05 \le x \le 0.15$, $0.20 \le x \le 0.40$, and $x = 0.50$, respectively. These minima indicate the existence of FM

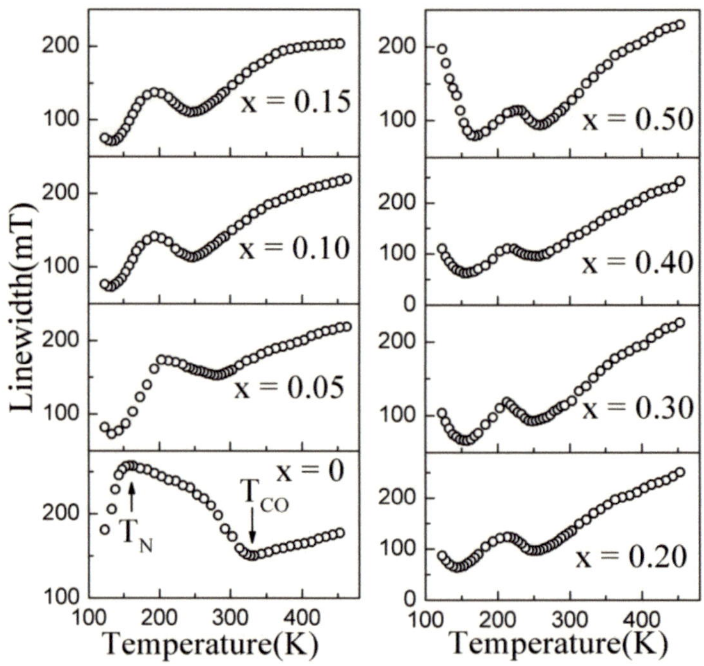

Figure 9. Linewidth versus Temperature plots of $Bi_{0.5-x}Pr_xCa_{0.5}MnO_3$ samples.

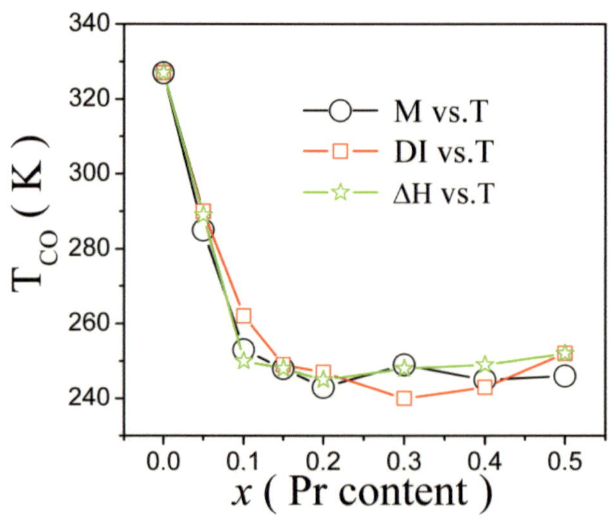

Figure 10. T_{CO} versus x of $Bi_{0.5-x}Pr_xCa_{0.5}MnO_3$ samples.

magnetic clusters embedded in the AFM matrix. Further, decrease in temperature leads to increase in ΔH due to the evolution of canted-AFM (C-AFM) phase.

The value of T_{CO}, as a function of Pr content (Fig. 10), estimated from ESR data matches quite well with those estimated from magnetization data analysis. T_{CO} decreases as x increases from 0.0 to 0.20 and becomes independent of doping content with further increase in x. For samples with $0.0 \leq x \leq 0.15$, T_{CO} decreases and Curie–Weiss temperature (θ_C) increases with increase in Pr doping.

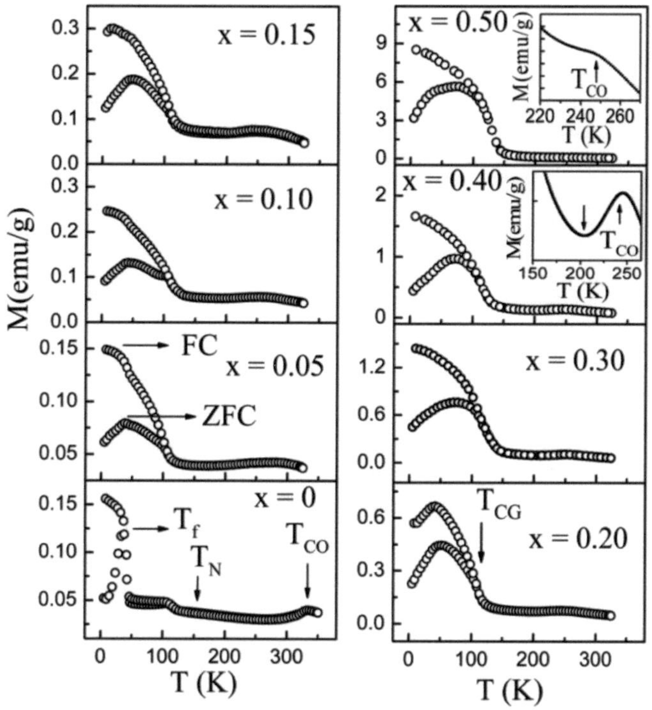

Figure 11. M versus T plots of $Bi_{0.5-x}Pr_xCa_{0.5}MnO_3$ samples sizes measured under the applied static field of 500 Oe.

The sharp rise in M (Fig. 11) value below 45 K for undoped sample and below 115 K for doped samples is observed, which is characteristic of FM-cluster glass (CG) behavior. The temperature T_f is called spin-freezing temperature. Below this temperature, M in ZFC mode decreases due to the competition between the evolving FM and AFM interactions, whereas M in FC mode continuously increases with decreasing temperature.

2.5. Studies on La-doped Bi-manganites

The base compound, $Bi_{0.7}Ca_{0.3}MnO_3$, has monoclinic structure, and the details are given in section 2.3, whereas the La-doped Bi-manganites, $Bi_{0.7-x}La_xCa_{0.3}MnO_3$ (x = 0.07, 0.14, 0.28, 0.35 and 0.7), have cubic structure. The unit cell volume increases with increasing La content.

Figure 12 shows the ESR spectra for x = 0.07 and 0.70 samples at some selected temperatures. For all samples, the ESR spectra above a particular temperature (~260–270 K) show a single resonance line at resonance field of 325 mT, which is characteristic of the existence of PM phase.

Below this temperature, FM correlations evolve in the PM matrix as indicated by the appearance of a shoulder (shown with '↓1') in the low-field region of the ESR signal. This shoulder can be assigned to a low-moment state. With further decrease in temperature, FM correlations grow at the expense of PM interactions, as a result of which a complete PM–FM transition occurs below ~180 K. This is indicated by the shift of the shoulder toward further lower-field region (shown with '↓2'). The data indicate development of a high-moment state as temperature decreases.

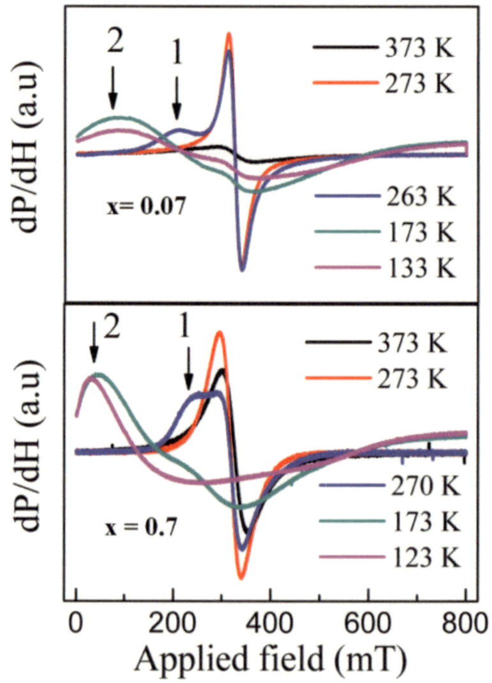

Figure 12. ESR spectra of $Bi_{0.7-x}La_xCa_{0.3}MnO_3$ (x = 0.07, 0.7).

Figure 13 shows M vs T plots of all the samples. M increases with increasing La content. The spin or cluster glass behavior of Bi- and La-containing samples is similar to that of $Bi_{1-x}Ca_xMnO_3$ ($x \leq 0.6$) system [6]. The samples show the coexistence of PM–FM phases over a wide temperature range. The estimated T_C, θ_C, and the temperature at which ΔH becomes minimum (T_{min}) increase with increase in La content.

The critical behavior of $Bi_{0.30}La_{0.37}Ca_{0.33}MnO_3$ (orthorhombic structure with *Pnma* space group) manganite at the critical point, where the system undergoes phase transition from PM to FM state, is investigated by using modified-Arrott plots (Fig. 14), Kouvel-Fisher method (Fig. 15), and critical isotherm analysis [35].

The sample shows second-order phase transition near the critical point [36]. The estimated critical exponents are close to as per the prediction by mean-field theory (MFT).

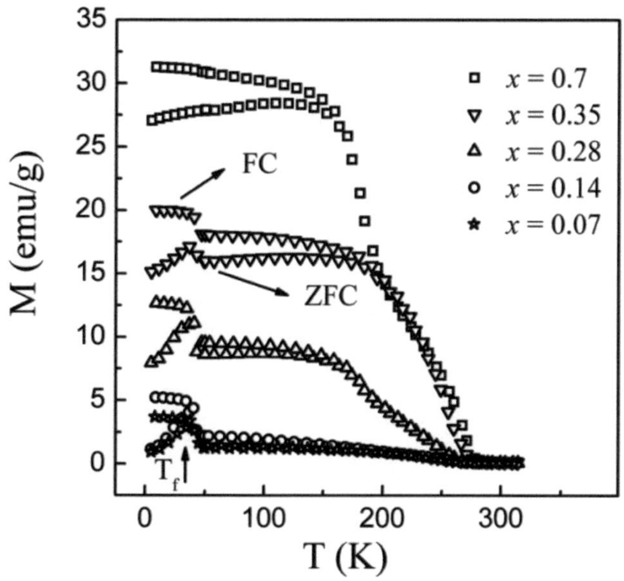

Figure 13. M versus T plots of $Bi_{0.7-x}La_xCa_{0.3}MnO_3$ samples measured under the applied static field of 500 Oe.

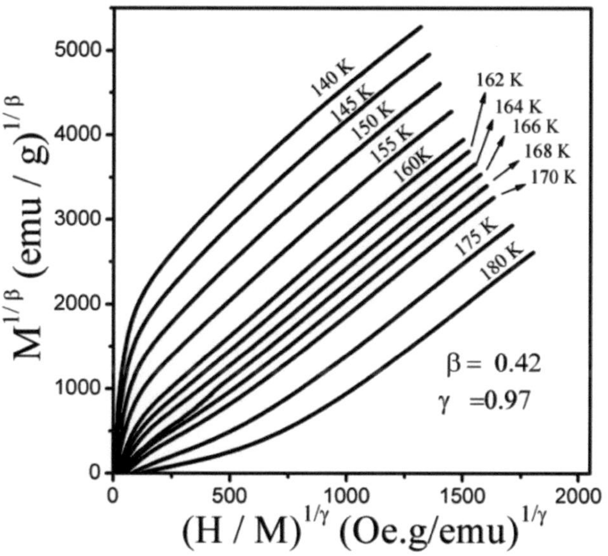

Figure 14. Modified Arrott plots of $Bi_{0.30}La_{0.37}Ca_{0.33}MnO_3$.

Using scaling equation,

$$M(H,\varepsilon) = \varepsilon^{\beta} f_{\pm}\left(H / \varepsilon^{\beta+\gamma}\right) \tag{6}$$

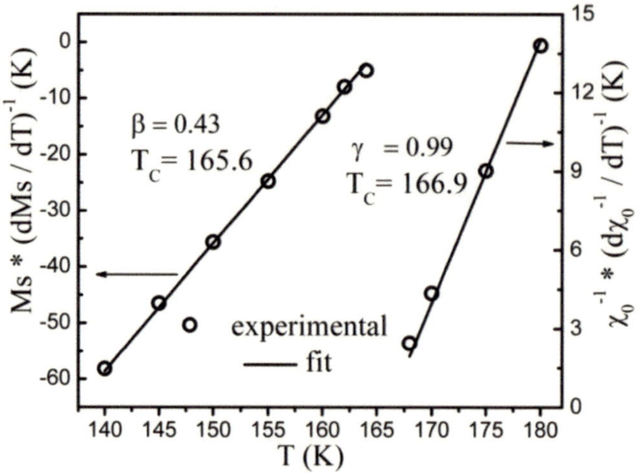

Figure 15. Kouvel-Fisher plots of $Bi_{0.30}La_{0.37}Ca_{0.33}MnO_3$.

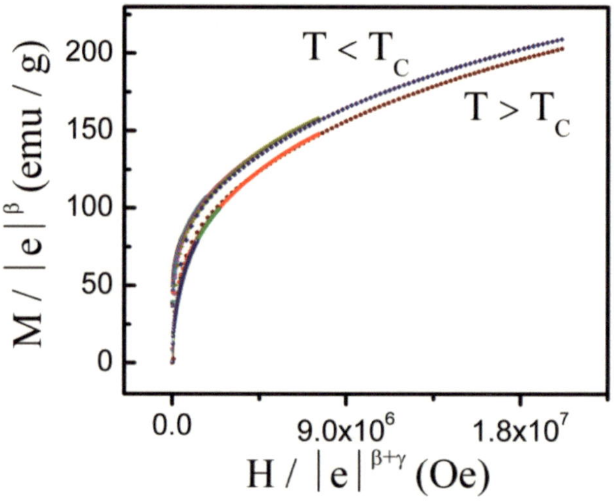

Figure 16. Scaling plots below and above T_C using the critical exponents determined from Kouvel–Fisher method of $Bi_{0.30}La_{0.37}Ca_{0.33}MnO_3$.

where f_+ for $T > T_C$ and f_- for $T < T_C$ are regular functions [35], we have plotted as shown in Fig. 16. There are two curves, one above the T_C and the other one below T_C, in agreement with the scaling theory. This indicates that the calculated critical exponents are reliable.

$Bi_{1-x}Ca_xMnO_3$ (BCMO) exhibits charge ordering property in the composition range $0.30 \le x \le 0.80$, which is related to first-order transition [3–6]. At critical point, $La_{1-x}Ca_xMnO_3$ (LCMO) manganite shows first-order transition in the composition range $0.30 \le x \le 0.40$ and second-order transition in the composition range $0.18 \le x \le 0.25$ and $0.40 \le x \le 0.48$ [37–39], whereas the present $Bi_{0.30}La_{0.37}Ca_{0.33}MnO_3$ sample also exhibits second-order transition which is similar to $La_{0.7}Sr_{0.3}MnO_3$ (LSMO) [40]. In the present system, due to La^{3+} doping, the average A-site ionic radius increases, as in the case of Sr^{2+}-doped La-manganite. The increased average A-site radius might be responsible for the observed second-order phase transition.

The ESR and magnetization studies on La-doped Bi-manganites reveal that the La-doping increases M as well as various transition temperatures describing the evolution of various magnetic phases. These changes are ascribed to the difference in ionic radii of Bi^{3+} and La^{3+} ions and the variations in $6s^2$ lone pair character of Bi^{3+} ions with La-doping.

2.6. Comparative studies on fixed doping level of transition element (TE = V, Cu and Zn) at Mn-site in $Bi_{0.5}Ca_{0.5}MnO_3$ system

XRD data analysis indicates that all the samples possess orthorhombic structure [41]. The difference between a and b lattice parameters decreases, and the c lattice parameter increases with doping. The change in the lattice parameters with doping confirms that the dopants are substituted at the Mn-site [41].

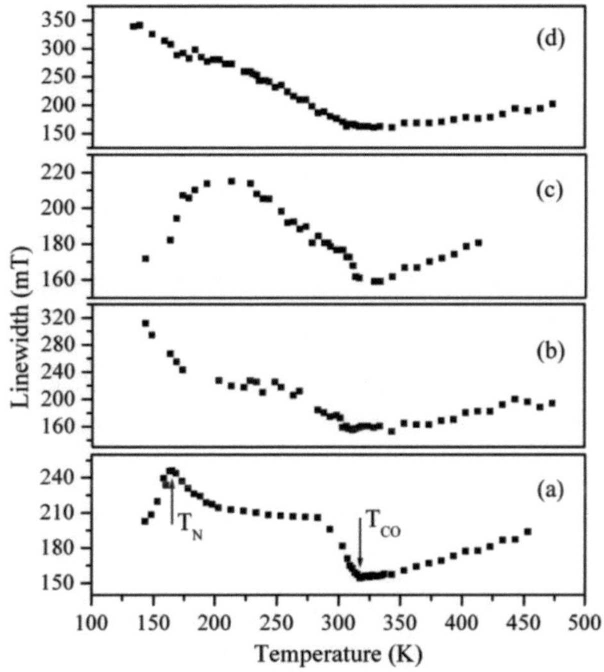

Figure 17. Linewidth versus temperature plots for $Bi_{0.5}Ca_{0.5}Mn_{0.95}TE_{0.05}O_3$ (a) undoped and doped samples with TE (b) V (c) Cu and (d) Zn [41]

ESR (Fig. 17) studies on $Bi_{0.5}Ca_{0.5}Mn_{0.95}TE_{0.05}O_3$ with TE = V, Cu, and Zn as a function of temperature and composition show a strong interplay between FM and AFM interactions. FM and AFM phases coexistence in the temperature range $T_{CO} > T > T_O$ for all the samples. In the temperature range $T > T_{CO}$, all samples show FM-dominated interactions in PM state.

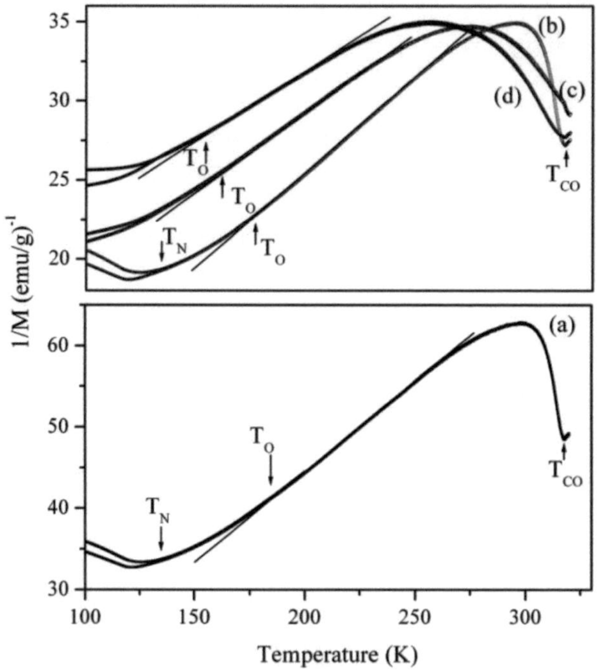

Figure 18. $1/M$ versus T plots for $Bi_{0.5}Ca_{0.5}Mn_{0.95}TE_{0.05}O_3$ (a) undoped and doped samples with TE (b) V (c) Cu (d) Zn [41]

The magnetization data in Fig. 18 supports the ESR observations. The estimated values of θ_C in the temperature range $T_{CO} > T > T_O$ (Fig. 18) are 5, –7, –93, and –172 K for undoped, V-, Cu-, and Zn-doped samples, respectively. The changes in magnetic properties of BCMO system with TE doping at Mn-site are due to the electronic nature of the dopants [41].

2.7. Comparative studies on fixed doping level of transition metals (TM = Ni, Fe, Co, and Cr) at Mn-site in $Bi_{0.5}Ca_{0.5}MnO_3$ system

All the samples have orthorhombic structure [42]. The Mn replacement by TM was confirmed by change in the lattice parameters with respect to lattice parameters of undoped one. This effect is weak for Co-doped sample compared to Fe-, Cr-, and Ni-doped samples. The peak widths of Cr-doped XRD pattern are essentially broader than for other TM-doped samples. It can be due to the non-homogeneous Cr-doped phase. It can also be due to smaller crystallite size in Cr-doped sample than other samples [42].

The ESR (Fig. 19) and magnetization (Fig. 20) data on the present BCMO and BCMTMO samples show that the FM correlations dominate in the temperature region $T > T_{CO}$. In the temperature range, $T_{CO} > T > T_O$, FM and AFM coexist for the undoped, Ni- and Fe-doped samples. The 3d TM doping melts the charge ordering and AFM ordering. The effect of Fe, Co, and Cr doping is same and stronger compared to Ni. This suggests that the dopants modify the interaction between Mn ions and possibly alter the band structure of the material instead of participating directly in the magnetic interaction mechanisms [42].

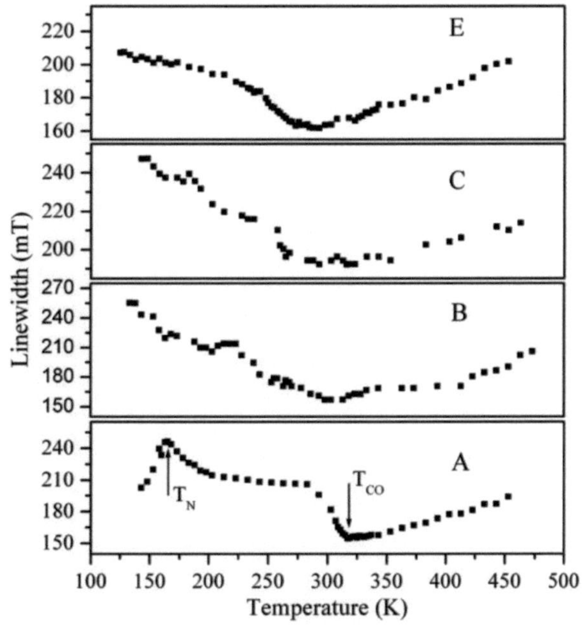

Figure 19. Linewidth versus Temperature plots for A-$Bi_{0.5}Ca_{0.5}MnO_3$ and $Bi_{0.5}Ca_{0.5}Mn_{0.95}TM_{0.05}O_3$ (TM = B-Ni, C-Fe and E-Cr)

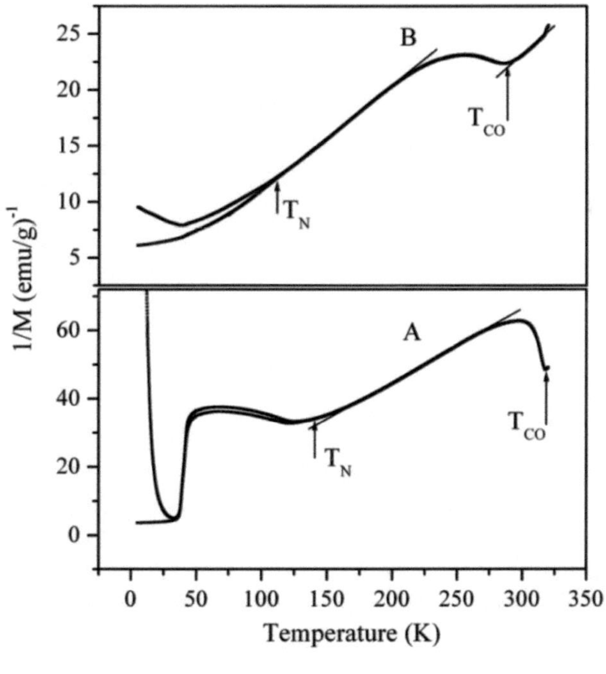

Figure 20. 1/*M* versus *T* plots for doped $Bi_{0.5}Ca_{0.5}Mn_{0.95}TM_{0.05}O_3$ (A-undoped, B-TM = Ni)

2.8. Studies on nanoparticles

The comparative ESR studies (Fig. 21) on bulk and nanometer-size (grain size 21 and 15 nm) samples of $Bi_{0.55}Ca_{0.45}MnO_3$ show that T_{CO} decreases from 313 to 306 K as sample grain size changes from bulk to 15 nm. The T_N value remains around 153 K for all the samples. In Fig. 21, the solid lines are as per Curie–Weiss law. Above $T_{CO,}$ the positive intercepts on the x-axis indicate the dominant FM correlations. In the temperature range, $T_{CO} > T > T_N$, the domains of FM and AFM spin correlations coexist [43]. The FM phase volume decreases, and the transitions become sharp with decrease in grain size [43].

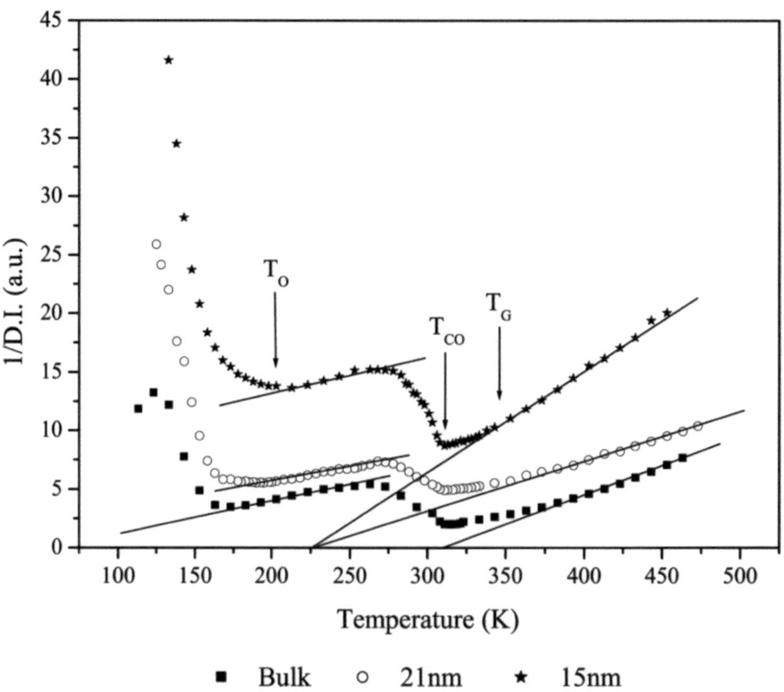

Figure 21. $1/DI$ versus T plots for the samples Bulk sample (filled squares), 21 nm nanoparticles (open circles) and 15 nm nanoparticles (filled stars).

The ESR data have been explained [43] in view of the Griffith's phase theory [9]. Griffith's theory suggests that above T_G large FM clusters form. Below T_G, smaller-sized clusters form, reducing the average size of the FM clusters. In Fig. 21, nonlinear decrease of $1/DI$ with decrease in temperature indicates existence of FM cluster in a PM matrix. In the temperature range, $T_{CO} > T > T_N$, orbital ordering sets in, bringing about long-range AFM ordering. In this temperature range, the Mn spins do not completely undergo OO and hence there is a coexistence of FM and AFM phases.

The effect of grain size on the properties of $Bi_{0.5}Ca_{0.5}MnO_3$ manganite samples synthesized by SG method is also studied. The samples with grain size (27, 450, and 1080 nm) were obtained

http://dx.doi.org/10.5772/61936

by appropriate heat treatment schedule. The magnetic behavior of samples with grain size 450 and 1,080 nm is similar to that of bulk sample described in (section 2.3), whereas for sample with GS 27 nm, the long-range charge ordering and AFM ordering transitions are suppressed (Fig. 22). Magnetization results (Fig. 23) also support the ESR findings.

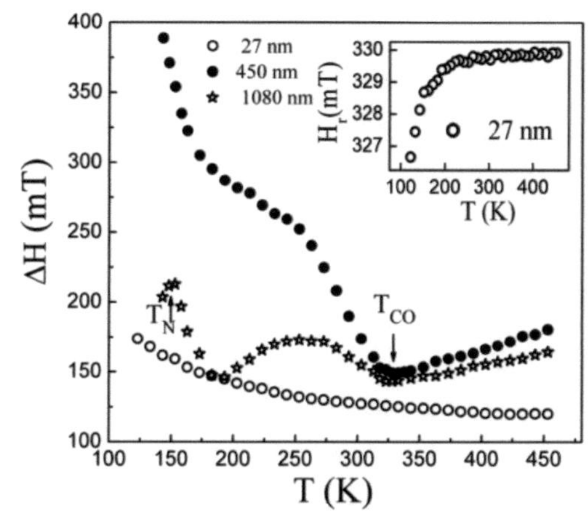

Figure 22. ΔH versus T plots of $Bi_{0.5}Ca_{0.5}MnO_3$ sample with different grain sizes. Inset shows H_r versus T plot for 27 nm sample.

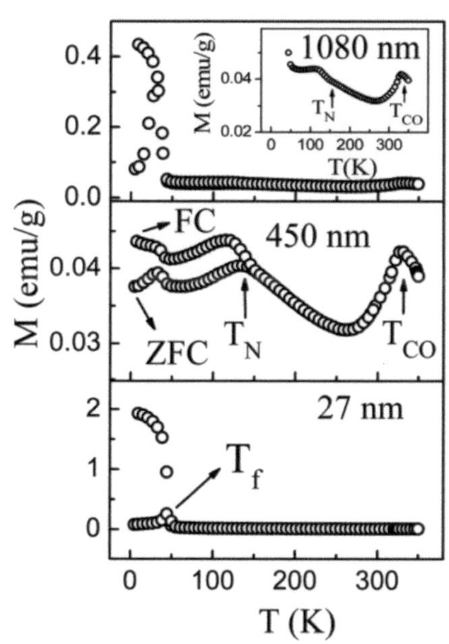

Figure 23. M versus T plots of $Bi_{0.5}Ca_{0.5}MnO_3$ sample with different grain sizes measured under the applied field of 500 Oe.

For these samples, M in FC mode increases with decreasing temperature. However, the ESR and magnetization results reveal that even for sample with GS 27 nm the AFM correlations still exist in the form of short-range order. The shift in FM–CG transition and T_f toward higher temperature and the sharp rise in M below 50 K for sample with GS 27 nm are observed. The evolving different magnetic correlations with decrease in GS are ascribed to increase in surface-to-volume ratio of grains.

3. Conclusions

The ESR and M studies of Bi-manganites of various compositions synthesized by SS or SG methods are reviewed. The effect of substitution of divalent (Ca^{2+}) and trivalent (Pr^{3+} and La^{3+}) cations at Bi-site and partial replacement of Mn by transition elements on the structure and magnetic properties of these materials is reviewed. The effect of grain size on the evolution of various magnetic interactions is also described. The study on various systems is useful in understanding the evolution of various magnetic phases as a function of composition and temperature. The studies on several compositions of Bi-manganites reveal that electronic phase separation is an intrinsic phenomenon in this system.

Acknowledgements

Ramesh Ade would like to thank the award of UGC-(JRF+SRF) from CSIR India.

Author details

Rajender Singh* and Ramesh Ade

*Address all correspondence to: rssp@uohyd.ernet.in

School of Physics, University of Hyderabad, Central University, Hyderabad, Telangana, India

References

[1] Jonker G H and Van Santen J H. Ferromagnetic compounds of manganese with perovskite structure. Physica, 1950; XVI: 337–349.

[2] Wollan E O and Koehler W C. Neutron diffraction study of the magnetic properties of the series of perovskite-type compounds [(1–x)La, xCa]MnO_3. Physical Review, 1955; 100: 545–563.

[3] Kurian J and Singh R. Electron spin resonance study of $Bi_{(1-x)}Ca_xMnO_3$, Journal of Applied Physics, 2008; 103: 07F707(1)–07F707(3).

[4] Kurian J and Singh R. Electron spin resonance and resistivity studies of charge-ordered $Bi_{1-x}Ca_xMnO_3$. Journal of Physics D: Applied Physics, 2008; 41: 215006–215015.

[5] Ramesh A and Singh R. Electron spin resonance studies of $Bi_{1-x}Ca_xMnO_3$ ($x \geq 0.65$). IEEE Transactions on Magnetics, 2012; 48: 4562–4565.

[6] Ramesh A and Singh R. Electron spin resonance and magnetization studies of $Bi_{1-x}Ca_xMnO_3$. Physica B, 2014; 448: 273–276.

[7] Uehara M, Mori S, Chen C H and Cheong S-W. Percolative phase separation underlies colossal magnetoresistance in mixed-valent manganites. Nature (London), 1999; 399: 560–563.

[8] Cui C and Tyson T A. Pressure effects on charge, spin, and metal-insulator transitions in the narrow bandwidth manganite $Pr_{1-x}Ca_xMnO_3$. Physical Review B, 2004; 70: 094409(1)–094409(8).

[9] Wang X, Cui Q, Pan Y, Gao W, Zhang J and Zou G. High pressure effects on the Jahn–Teller distortion in perovskite $La_{0.5-x}Bi_xCa_{0.5}MnO_3$. Journal of Alloys and Compounds, 2001; 321: 72–75

[10] Tomioka Y, Asamitsu A, Kuwahara H, Moritomo Y and Tokura Y. Magnetic-field-induced metal-insulator phenomena in $Pr_{1-x}Ca_xMnO_3$ with controlled charge-ordering instability. Physical Review B, 1995; 53: R1689–R1692.

[11] Yang C P, Chen S S, Guo D H, Wang H, Rao G H, Morchshakov V and Barner K. Electrical-field-induced order transitions in Sr-doped manganite perovskites. Journal of Alloys and Compounds, 2009; 467: 54–60.

[12] Zhang T and Dressel M. Grain-size effects on the charge ordering and exchange bias in $Pr_{0.5}Ca_{0.5}MnO_3$: The role of spin configuration. Physical Review B, 2009; 80: 014435(1)–014435(9).

[13] Rozenberg E, Shames A I and Auslender M. Nanometer sized effects on magnetic ordering in La–Ca manganites, probed by magnetic resonance. Nanoscience and Nanotechnology Letters, 2011; 3: 531–540.

[14] Dagotto E. Nanoscale phase separation and colossal magnetoresistance, Springer Series in Solid State Physics 136 (Springer, Berlin), 2003.

[15] Sarma D D, Topwal D and Manju U. Direct observation of large electronic domains with memory effect in doped manganites. Physical Review Letters, 2004; 93: 097202(1)–097202(4).

[16] Moreo A, Mayr M, Feiguin A, Yunoki S and Dagotto E. Giant cluster coexistence in doped manganites and other compounds. Physical Review Letters, 2000; 84: 5568–5571.

[17] Griffiths R B. Nonanalytic behavior above the critical point in a random Ising ferromagnet. Physical Review Letters, 1969; 23: 17–19.

[18] Woo H, Tyson T A, Croft M, Cheong S W and Woicik J C. Correlations between the magnetic and structural properties of Ca-doped $BiMnO_3$. Physical Review B, 2001; 63: 134412(1)–134412(12).

[19] Bokov V A, Grigoryan N A and Bryzhina M F. X-Ray diffraction and magnetic studies of solid solutions $Bi_{1-x}Ca_xMnO_3$. Physica Status Solidi, 1967; 20: 745–754.

[20] Bao W, Axe J D, Chen C H and Cheong S W. Impact of charge ordering on magnetic correlations in perovskite (Bi, Ca) MnO_3. Physical Review Letters, 1997; 78: 543–546.

[21] Banwell C N and McCash E M. Fundamental of molecular spectroscopy (Tata McGraw-Hill, 1997).

[22] Stevenson J W, Nasrallah M M, Anderson H U and Sparlin D M. Defect structure of $Y_{1-y}Ca_yMnO_3$ and $La_{1-y}Ca_yMnO_3$. Journal of Solid State Chemistry, 1993; 102: 175–184.

[23] Muller K A. Electron Paramagnetic Resonance of Manganese IV in $SrTiO_3$. Physical Review Letters, 1959; 2: 341.

[24] Muller K A, Berlinger W, Blazey K W and Albers J. Electron paramagnetic resonance of Mn^{4+} in $BaTiO_3$. Solid State Communications, 1987; 61: 21–25.

[25] Shengelaya A, Zhao G M, Keller H and Muller K A. EPR Evidence of Jahn-Teller polaron formation in $La_{1-x}Ca_xMnO_{3+y}$. Physical Review Letters, 1996; 77: 5296–5299.

[26] Gulley J E and Jacarrino V. Impure exchange-coupled paramagnets; Electron-Paramagnetic-Resonance Studies. Physical Review B, 1972; 6: 58–73.

[27] Oseroff B, Torikachvili M, Singley J and Ali S. Evidence for collective spin dynamics above the ordering temperature in $La_{1-x}Ca_xMnO_{3+\delta}$. Physical Review B, 1996; 53: 6521–6525.

[28] Causa M T, Tovar M, Caneiro A, Prado F, Ibanez G, Ramos C A, Butera A, Alascio B, Obradors X, Pinol S, Tokura Y and Oseroff S B. High-temperature spin dynamics in CMR manganites: ESR and magnetization. Physical Review B, 1998; 58: 3233–3239.

[29] Kumar A, Ghosh N, Joshi J P, Bhat H L and Bhat S V. Electron paramagnetic resonance studies of the insulating ferromagnetic manganite $Nd_{0.8}Pb_{0.2}MnO_3$ above the transition temperature. Solid State Communications, 2002; 123: 379–382.

[30] Anuradha K N, Rao S S and Bhat S V. EPR Evidence for premonitory charge-ordering fluctuations in hydrothermally grown $Pr_{0.57}Ca_{0.41}Ba_{0.02}MnO_3$ Nanowires. Applied Magnetic Resonance, 2009; 36: 347–356.

[31] Seehra M S. New method for measuring the static magnetic susceptibility by paramagnetic resonance. Review of Scientific Instruments, 1968; 39: 1044–1047.

[32] Ramos C A, Causa M T, Tovar M, Obradors X and Pinol S. Magnetization and FMR in $La_{0.7}Pb_{0.3}MnO_3$ single crystals. Journal of Magnetism and Magnetic Materials, 1998; 177–181: 867–868.

[33] Kurian J and Singh R. Electron spin resonance and resistivity studies of charge-ordered $Bi_{1-x}Sr_xMnO_3$. Journal of Alloys and Compounds, 2011; 509: 5127–5136.

[34] Ramesh A and Singh R. Effect of Pr doping on the properties of $Bi_{0.5-x}Pr_xCa_{0.5}MnO_3$ ($0 \leq x \leq 0.50$) manganites. Ceramics International, 2015; 41: 4759–4767.

[35] Oumezzine M, Pena O, Kallel S and Zemni S. Critical phenomena and estimation of the spontaneous magnetization through magnetic entropy change in $La_{0.67}Ba_{0.33}Mn_{0.98}Ti_{0.02}O_3$. Solid State Sciences, 2011; 13: 1829–1834.

[36] Banerjee S K. On a generalized approach to first and second order magnetic transitions, Physical Letters, 1964; 12: 16–17.

[37] Kim D, Revaz B, Zink B L, Hellman F, Rhyne J J and Mitchell J F. Tricritical point and the doping dependence of the order of the ferromagnetic phase transition of $La_{1-x}Ca_xMnO_3$. Physical Review Letters, 2002; 89: 227202(1)–227202(4).

[38] Panopoulos N, Koumoulis D, Diamantopoulos G, Belesi M, Fardis M, Pissas M, and Papavassiliou G. Spin order and lattice frustration in optimally doped manganites: A high-temperature NMR study. Physical Review B, 2010; 82: 235102(1)–235102(9).

[39] Adams C P, Lynn J W, Smolyaninova V N, Biswas A, Greene R L, Ratcliff II W, Cheong S-W, Mukovskii Y M and Shulyatev D A. First-order nature of the ferromagnetic phase transition in (La–Ca)MnO_3 near optimal doping. Physical Review B, 2004; 70: 134414(1)–134414(12).

[40] Ghosh K, Lobb C J, Greene R L, Karabashev S G, Shulyatev D A, Arsenov A A and Mukovskii Y. Critical phenomena in the double-exchange ferromagnet $La_{0.7}Sr_{0.3}MnO_3$. Physical Review Letters, 1998; 81: 4740–4743.

[41] Vijayan D, Kurian J and Singh R. ESR studies on $Bi_{0.5}Ca_{0.5}Mn_{0.95}TE_{0.05}O_3$ (TE = V, Cu, and Zn). IEEE Transaction on Magnetics, 2011; 47: 2701–2704.

[42] Vijayan D, Kurian J and Singh R. Electron spin resonance and magnetization studies on $Bi_{0.5}Ca_{0.5}Mn_{0.95}TM_{0.05}O_3$ (TM = Cr, Fe, Co and Ni). Journal of Applied Physics, 2012; 111: 07D716 (1)–07D716 (3).

[43] Kurian J and Singh R. Electron spin resonance studies of $Bi_{0.55}Ca_{0.45}MnO_3$ nanoparticles. Journal of Applied Physics, 2009; 105: 07D718(1)–07D718(3).

Chapter 11

Charge Carrier Dynamics in Organometal Halide Perovskite Probed by Time-Resolved Electrical Measurements

Carlito S. Ponseca Jr.

Additional information is available at the end of the chapter

http://dx.doi.org/10.5772/61631

Abstract

This chapter presents the fate of the charge carriers from the moment of its photogeneration in the perovskite to injection and transport into electrodes. Time-resolved electrical measurement techniques, terahertz (THz) spectroscopy and microwave (MW) conductivity, are primarily used to deconvolute ultrafast processes and to directly access behavior of charged species from the ps to μs timescales. Transient absorption and photoluminescence spectroscopy were also utilized to gain insight on carrier population dynamics and radiatively recombining charges. Photogenerated charged species were converted into highly mobile charges ($\mu_e = 12.5\ cm^2V^{-1}s^{-1}$ and $\mu_h = 7.5\ cm^2V^{-1}s^{-1}$) almost instantaneously (< 0.2 ps), while the remaining loosely bounded excitons dissociate into mobile charges after 2-3 ps. This high mobility is maintained for at least 1 ns as obtained by THz spectroscopy, while its lifetime is at least few tens of μs as measured by the MW conductivity technique. Lowering the temperature increases carrier mobilities with $T^{-1.6}$.Dependence and a 75 meV barrier energy is required for temperature-activated recombination. Finally, injection of hole from $MAPbI_3$ to Spiro-OMeTAD was found to be ultrafast and the state and population of dark holes dictate its recombination.

Keywords: THz spectroscopy, Time-resolved Microwave Conductivity (TRMC), photoconductivity, mobility

1. Introduction

The qualities of highly efficient solar cell material are its ability to absorb light with the widest spectral range possible, high light-to-charge conversion ratio, and transport of these charges to electrodes with least losses. This seems to be the case in organometal halide perovskite (OMHP)-based solar cell as its overall power conversion efficiencies (PCEs) have risen from a meagre 4% [1] to 20.1% [2] to date. Such increase has not been seen in any other solar cell technology since the conception of light harvesting. In fact, the PCE of dye-sensitized solar cell, where the OMHP was first used as a dye substitute, remained modest [2]. In recent years, spectroscopic studies on these materials have started to trickle the much-needed fundamental investigations. Most of its well-known characteristics include electron-hole diffusion lengths longer than one micrometer [3], high mobility, and very slow recombination [4]. Despite this, there is a long list of unexplained early timescale processes, which is fundamental to understating its solar cell function. For instance, details whether molecular excitons or to highly mobile charges are the initial photoproduct and how is this related to the exceptionally long diffusion lengths, remains unanswered. It is also unclear to what extent generation and recombination of mobile charges are affected by temperature. The use of metal oxide electrodes such as TiO_2 and Al_2O_3, as electron acceptor and isolating nanoparticles (NP), respectively, and their role on its electronic properties has not yet been understood. Many groups have also recently attempted to use organic electrodes, e.g., PCBM and Spiro-OMeTAD and were able to get decent PCE. Questions on its charge transfer mechanism, timescale, and details of injection are yet to yield convincing answers.

In this chapter, an ensemble of spectroscopic techniques, primarily time-resolved terahertz spectroscopy (TRTS) and time-resolved microwave conductivity (TRMC) complemented by photoluminescence and transient absorption, was used to monitor the creation of charged species induced by photoexcitation at the sub-ps timescale and probe its fate up to a hundred of microsecond. Both techniques have been used in an assortment of solar cell technologies including organic [5,6], dye-[7], and quantum dot- sensitized [8], and inorganic [9] systems. Neat methylammonium lead iodide ($MAPbI_3$) was used to study the intrinsic transport properties of perovskite material both at room and low temperatures. To determine how electron accepting metal and the role of NP in general, $MAPbI_3$ were introduced to TiO_2 and Al_2O_3, respectively. Charge transfer mechanism and the corresponding recombination dynamics when attached to organic electrodes, PCBM and Spiro-OMeTAD were also explored. Note that detailed discussion of different material contacts, their heterogeneity, and their morphologies' influence on charge carrier dynamics are presented in Ref. [10].

On light excitation, changes in conductivity from ground state (σ) to photo-induced or transient state ($\Delta\sigma$) are measured in the solar cell material. This conductivity when normalized with charge density yields mobility (μ) per charge carrier. This photoconductivity is measured at high frequencies, i.e., at the THz regime (0.1–1.5 THz) and at the GHz regime (12 GHz). As such, these high-frequency waves are propagating through free space and are able to interrogate the sample without any electrodes. The TRTS and TRMC signal size can be expressed as product of two quantities, quantum yield, and electron and hole mobilities ($\mu_e + \mu_h$). This product is calculated according to:

$$\Delta\sigma = \varphi \times \left(\mu_e + \mu_h\right) = \frac{\Delta\sigma L}{eI_0 F_A}$$

where $\Delta\sigma$ is the measured change in photoconductivity, L is the thickness, e is the elementary charge, I_0 is the number of photons per unit area per pulse, and F_A is the fraction of absorbed light. A rise in the photoconductivity kinetics signals either creation of charged species, since only these photoproducts can induce changes in conductivity or an increase in mobility. On the other hand, decay in the kinetics means disappearance of these species, through recombination, injection to lower mobility acceptor molecules, and/or decrease in mobility.

Section 2 of this chapter presents the early timescale generation mechanism of charge carrier in neat perovskite in order to understand the intrinsic property of this material. It also discusses the influence of NPs, TiO_2, and Al_2O_3 on the mobility of the photo-induced carriers. In Section 3, a discussion on the fate of the charges when temperature is lowered will be presented and relate these results to the origin of the very long recombination time. Lastly, in Section 4, the mechanism of electron and hole injection to organic electrodes, PCBM and Spiro-OMeTAD, is reported and the role of dark carriers in the recombination dynamics is discussed.

2. Probing the charge carrier dynamics of intrinsic $MAPbI3_3$

The early time THz photoconductivity kinetics of neat $MAPbI_3$, $MAPbI_3/Al_2O_3$ and $MAPbI_3/TiO_2$ is shown in Fig. 1a. For both $MAPbI_3$ and $MAPbI_3/Al_2O_3$, a two-step increase is observed. An instrument-limited rise with amplitude of about 70% is followed by 2-3 ps increase of about 30%. Tightly bound molecular excitons are, by definition, neutrally charged and hence cannot affect the conductivity of the material. This means that even if these excitons are generated, their contribution to the THz signal is nil. Therefore, the sub-ps increase in the photoconductivity kinetics comes from generation of charged species. It was reported that the binding energy of exciton in these materials may vary between 4 meV [11] to 50 meV [12]. The heterogeneity of the binding energy in a material could cause photogenerated species to dissociate at different rates. As such, the slower rise in the kinetics could be due to separation of loosely bounded charge pair becoming independent charges. Initially, charges bounded by Coulombic force require an energy that could separate them; in this case it seems enough that thermal energy, kT, dissociates them and only takes few ps, 2-3 ps rise in the THz kinetics of neat $MAPbI_3$ and $MAPbI_3/Al_2O_3$. Unlike the previous two samples, for $MAPbI_3/TiO_2$ the THz transient rises faster and reaches the maximum signal with a single-step instrument-limited rise time. It was reported before that light absorbers attached to metal oxide of high electron affinity, ultrafast electron injection can be expected. For example, RuN3 dye attached to TiO_2 [7] and in CdSe quantum dot attached to ZnO [8] shows ultrafast injection. Therefore, the ultrafast limited rise in THz kinetics of $MAPbI_3/TiO_2$ is due to sub-ps injection of electrons from the OMHP to TiO_2. Favorable alignment of energy band levels between the perovskite and metal oxide aides the separation of any remaining loosely bound electron–hole pair favoring ultrafast injection. This is the reason why there is no second-step rise seen in the THz kinetics. Shown in Fig. 1b is the THz kinetics per photon absorbed per pulse. It shows that the mobility of $MAPbI_3/TiO_2$ is 7.5 $cm^2V^{-1}s^{-1}$, about 3-4 smaller lower than the mobility in neat $MAPbI_3$ and the $MAPbI_3/Al_2O_3$, which is 20 $cm^2V^{-1}s^{-1}$. The difference in the mobility also supports the proposal that electron injection occurs in $MAPbI_3/TiO_2$. From the total mobility of neat $MAPbI_3$, electrons disappear due to their injection to TiO_2. Because of this, only the holes left in the perovskite are seen by the instrument, resulting in lower mobility.

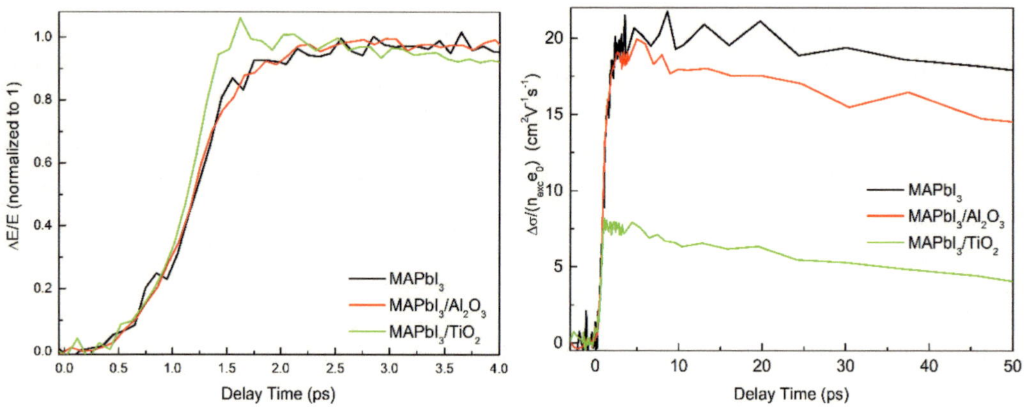

Figure 1. THz photoconductivity kinetics of neat $MAPbI_3$, $MAPbI_3/Al_2O_3$, and $MAPbI_3/TiO_2$. a. Normalized to 1 (λ_{pump} = 400 nm, I_{exc} = 1.7 x 10^{13} ph/cm^2 per pulse), b. Normalized with $n_{exc}e$. (Reprinted with permission from [57], Copyright 2014, American Chemical Society)

To further confirm electron injection, optical transient absorption (TA) was used to probe the arrival of electrons to TiO_2 after photoexciting the perovskite layer. The normalized TA kinetics of neat $MAPbI_3$ and $MAPbI_3/Al_2O_3$ are shown in Fig. 2a. It has a response-limited rise (negative), and then an approximately 2 ps of further decrease, very similar to the two-component increase in THz photoconductivity kinetics, showing that indeed there are charges that are not created right away. The negative signal, probed 970 nm is a part of the spectral region where there is no ground state absorption, and therefore indicative of stimulated emission. This is consistent with the ground state photoluminescence (PL) emission spectra (Fig. 2b) showing intense emission from these two samples. On the other hand, an ultrafast rise with only one component with positive sign, which means absorption, is obtained in the kinetics of $MAPbI_3$/ TiO_2. This is in agreement with the timescale of arrival of electrons in TiO_2 [10,13] and very similar to the THz kinetics (Fig. 1a). Despite the fact that there is still some emission, the strong quenching of photoluminescence in $MAPbI_3/TiO_2$ is an additional proof that there is injection of electrons. From the SEM image of the three samples (Fig. 3), $MAPbI_3/TiO_2$ shows no indication of domains of $MAPbI_3$ bigger than 500 nm, meaning that the mesoporous network of $MAPbI_3$ is formed within TiO_2 NPs analogous to that reported in Ref. [13]. Despite this, reduced domains of $MAPbI_3$ may still be formed brought about by voids due TiO_2 NPs. However, this could be small, as shown by the strong quenching of emission and the fast rise in THz photoconductivity kinetics.

As shown in Fig. 1b, the first 40 ps of the THz kinectics for the three samples manifests a slow decay. It is usually assumed that at the earliest timescale, quantum yield is 1. At later timescale, the THz photoconductivity becomes a product of charge concentration and mobility since photogenerated charges are either starting to recombine or lose their mobility. In such case, it is not clear to conclude, based only on the photoconducivity kinetics, if the decay is related to depopulation of charges or lowering mobility with time or both. Transient absorption is the appropriate technique to measure the charge population as a function of time. Shown in Fig.

http://dx.doi.org/10.5772/61631

4a are kinetic traces of neat $MAPbI_3$, obtained through TA and TRTS at similar pump levels (~10^{13} ph/cm^2 per pulse). Within the signal-to-noise, the decay of the two plots is identical. This implies that the decay in the THz kinetics should be coming from the depopulation of charge carriers since the TA kinetics has the same decay. Consequently, this means that mobility of charge carrier remains the same at least up to 1 ns, otherwise a faster decay in the THz kinetics would be observed. To determine the reason for the depopulation of charges, the excitation density dependence of THz kinetics was obtained. At the lowest pump excitation (Fig. 4b, 2.0 × 10^{12} ph/cm^2 per pulse), the kinetics remained flat until 200 ps, where mobility obtained is 25 cm^{2}V^{-1}s^{-1}. These highly mobile charges measured in THz for OMHP material are within the same order reported by Wehrenfennig, et al. [4]. At highest intensity, the decay is fastest and mobility lowest. These are strong indications that the decay is due to non-geminate recombination of charges, similar to those reported in bulk heterojunction solar cell [5,6].

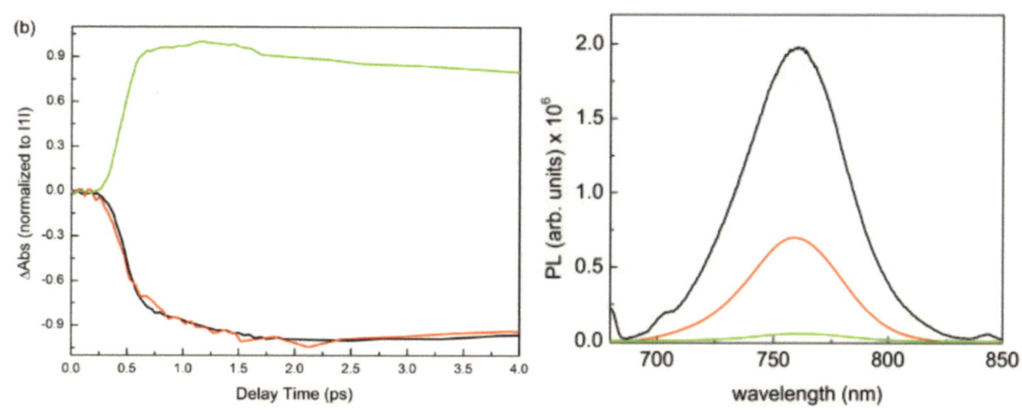

Figure 2. a. Transient absorption kinetics (λ_{pump} = 603 nm, λ_{probe} = 970 nm, I_{exc} = 6.0 x 10^{14} ph/cm^2 per pulse) b. Photoluminescence spectra (λ_{pump} = 550 nm) of the three samples. (Reprinted with permission from [57], Copyright 2014, American Chemical Society)

With the very high mobility of charges, it is interesting to know the nature of charge carrier transport. The THz photoconductivity spectra is a very useful tool in determining whether charges are behaving like free electron gas or in a confined, hopping-like manner. For this reason, the THz spectra of the three samples were measured after photoexcitation, i.e., at 10 ps as shown in Fig. 5. Within the signal-to-noise, the amplitude and also the shape of THz photoconductivity spectra are identical for neat $MAPbI_3$ and $MAPbI_3/Al_2O_3$. However, the spectral shape of $MAPbI_3/TiO_2$ is qualitatively different as well as its signal size is lower by four times. The real part of the conductivity is positive while the imaginary part is negative. From these observations several conclusions can be drawn. First, the THz photoconductivity kinetics as well as the spectra (Figs. 1a and 1b), and TA kinetics of neat $MAPbI_3$ and $MAPbI_3/Al_2O_3$ (Fig. 2a) show that Al_2O_3 NPs do not change the dynamical properties and mobility of charge carriers in perovskite, at least on the timescale probed by the THz measurements (1 ns). Secondly, the favorable band energy alignment between TiO_2 NPs and perovskite causes ultrafast injection, as shown in the reduction of signal size in the THz photoconductivity kinetics (Fig.1b) and spectra (Fig. 5). As consequence of injection to TiO_2, the mobility of electrons becomes the mobility of TiO_2 (0.1 $cm^2V^{-1}s^{-1}$), which leads to unbalanced transport of charges. Third, sign of real and imaginary parts of conductivity is a signature of confined motion of charges [6], which is rather counterintuitive considering the very high mobility obtained here. One could hypothesize that at the spatial vicinity of photogeneration, maybe tens of nm, the morphology of the material is very much favorable for fast motion of the charges. However, at larger spatial scale, charges may start encountering scattering centers, e.g., traps, which can make its motion more restricted. The THz spectra (Fig. 5), which is normalized, decay due to non-geminate recombination as well as with excitation density, at several delay times (100–950 ps), are identical, supporting the interpretation that mobility is constant for 1 ns but with confined mode of transport.

Figure 3. Scanning electron microscope image of a. neat $MAPbI_3$, b. $MAPbI_3/Al_2O_3$ and c. $MAPbI_3/TiO_2$. (Reprinted with permission from [57], Copyright 2014, American Chemical Society)

Differences in the mobilities of electrons and holes, typical with bulk heterojunction organic solar cells, which are sometimes several orders of magnitude, lowers the overall PCE due to built-in electric field brought about by space-charge-limited photocurrents [14]. This is because one of the charges has already reached the electrode while the other is still traversing the active layer of the solar cell. Therefore, it is important to assess not only the total mobility of the materials but to estimate both mobility of electrons and holes. Admittedly, THz measurement on one material alone would not provide this information, but rather by comparing it with known electron or hole mobilities this could be answered. The reported intrinsic electron mobility of porous TiO_2 films as measured by THz is <<1 $cm^2V^{-1}s^{-1}$ [15]. As shown in Fig. 1b, the mobility obtained for $MAPbI_3/TiO_2$ is 7.5 $cm^2V^{-1}s^{-1}$. From the discussion above, it was established that there is an ultrafast injection of electrons from perovskite to TiO_2, meaning that it should only be the mobility of holes that are left in the perovskite material that is seen by the instrument. As a consequence, the measured THz mobility of ~20 $cm^2V^{-1}s^{-1}$ for $MAPbI_3$ and $MAPbI_3/Al_2O_3$, where both electrons and holes are in the perovskite material and therefore contributing to the THz signal, 12.5 $cm^2V^{-1}s^{-1}$ should be coming from electrons since 7.5 $cm^2V^{-1}s^{-1}$ is from the holes. The resulting ratio of electron and hole mobilities in the perovskite is therefore about two, in agreement with the recent theoretical calculations of the relative effective masses of electrons and holes [16]. This also justifies the balanced long

http://dx.doi.org/10.5772/61631

diffusion lengths reported by the group of Stranks et al. [3, 17, 45]. The almost balanced electron and hole mobility is a unique key information rationalizing high PCE in OMHP-only or OMHP/Al_2O_3 solar cells.

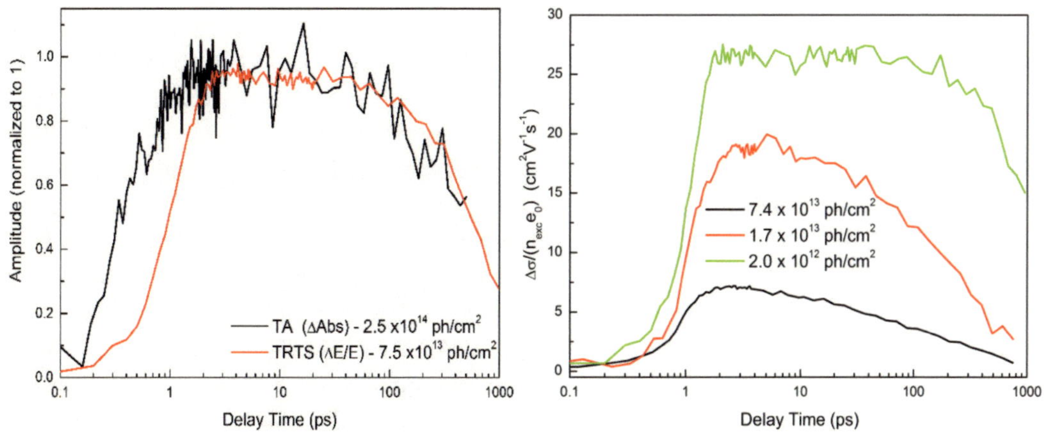

Figure 4. a. Comparison TA and TRTS kinetics for neat $MAPbI_3$ showing similar decay rates up to 1 ns. b. Intensity dependence THz kinetics of $MAPbI_3/Al_2O_3$. (Reprinted with permission from [57], Copyright 2014, American Chemical Society)

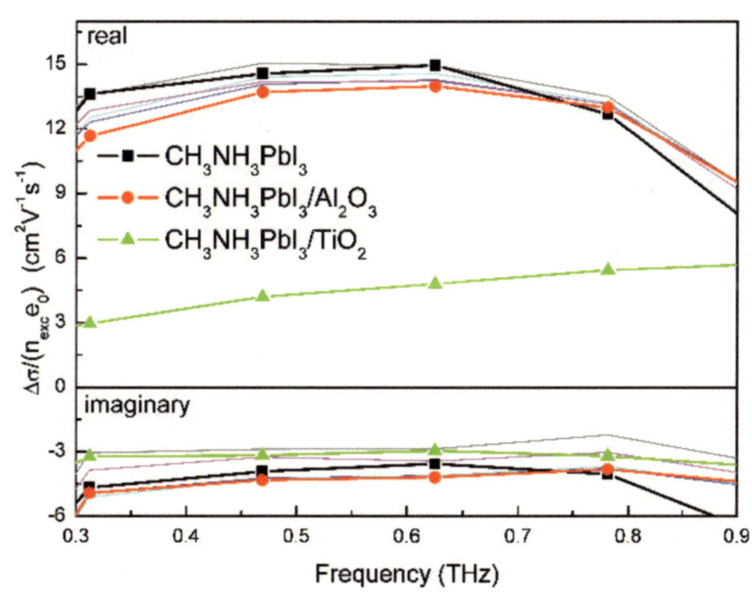

Figure 5. Photoconductivity spectra of $MAPbI_3/Al_2O_3$ at pump-probe delay of 10 ps normalized with $n_{exc}e$ (line with symbols). Solid lines are spectra at 100 ps (blue), 200 ps (cyan), 600 ps (magenta), and 950 ps (gray). (Reprinted with permission from [57], Copyright 2014, American Chemical Society)

Besides the ultrafast generation of highly mobile charges and balanced transport, another very important characteristic required for highly efficient solar cell is the timescale of recombination. It is desirable that charge carriers do not meet since every charge pair that recombines means one charge pair is not extracted. By using the TRMC, the photoconductivity measured by TRTS can be extended from tens of ns to a hundred microseconds. In addition, due to the superior stability of MW emitter, a significant increase in signal-to-noise is achieved allowing the use of excitation density two orders lower than TRTS. As a result, excitation-dependent second-order recombination is minimized, giving more precise dynamical information on recombination of charges. Figure. 6a is the plot of TRMC photoconductivity kinetics of the three samples measured at 5.9×10^9 ph/cm^2 per pulse for 1 μs. For $MAPbI_3$and $9MAPbI_3/Al_2O_3$, the mobility is 3 cm^{2}V^{-1}s^{-1}, while for $MAPbI_3/TiO_2$ it is 1 cm^{2}V^{-1}s^{-1}. The intrinsic mobility of electron in TiO_2 as measured by TRMC has been previously reported as <<0.1 cm^{2}V^{-1}s^{-1} [18,19]. As ultrafast injection in this material has been discussed above, the 1 cm^{2}V^{-1}s^{-1} of mobility in $MAPbI_3/TiO_2$ should be coming from holes only. Ergo, the mobility of electrons in $MAPbI_3$ and $MAPbI_3/Al_2O_3$ is 2 cm^{2}V^{-1}s^{-1}, which is consistent with the analysis of TRTS data. Such very high mobility is unique to this material especially in this later timescale since organic solar cell P3HT:PCBM, for example, has 0.045 cm^{2}V^{-1}s^{-1} only [18]. The signal-to-noise seems worse in Fig. 6a than in Fig. 1b but one should note that there is almost three orders' difference in their excitation intensity. At this very low fluence, the TRMC photoconductivity kinetics of the three materials is rather flat, up to 1 μs. This means that neither the charge population nor the mobility is decaying in this very long timescale. A much longer time window was used to determine the onset of recombination. Figures 6b and 6c are TRMC photoconductivity kinetics of $MAPbI_3/Al_2O_3$ and $MAPbI_3/TiO_2$ obtained from fluences of 5.9×10^9–6×10^{11} ph/cm^2 per pulse, which is 100 times lower than that used in TRTS and at timescale up to 100 μs. The decay is faster at higher intensities and slower with increasing lifetime and amplitude as fluence is lowered, a signature that charges are recombining non-geminately only [20,21]. This also means that there is no first-order recombination even at very low excitation, which suggests charges are diffusing rapidly away from its locus of generation aided by their very high mobility. TRMC signal is similar to photoconductivity measured in TRTS, which means that obtained response is from both charge concentration and mobility. Therefore, a single transient decay trace cannot give information whether it is charge carrier recombination or carrier relaxation that is observed. Nevertheless, it is clear in Fig. 6a that the TRMC kinetics is flat for 1-μs time, which means both its charge population and mobility do not change and any recombination or relaxation of mobility must all be considerably slower than 1 μs. Under ambient sunlight conditions, onset of charge recombination extends to tens of μs, which is the most important conclusion from these results.

Dynamical understanding of charge carriers of the perovskite samples can now be painted from the timescale of sub-ps to a hundred of μs at a wide range of excitation fluences. At excitation densities of 10^{13}-10^{14} ph/cm^2 per pulse (Figs. 1, 2, 4,), second-order non-geminate recombination directly leads to decay of photoconductivity signal. At low excitation fluence, the TRTS kinetics at lowest intensity remained flat up to 1 ns, which means that the carrier

http://dx.doi.org/10.5772/61631

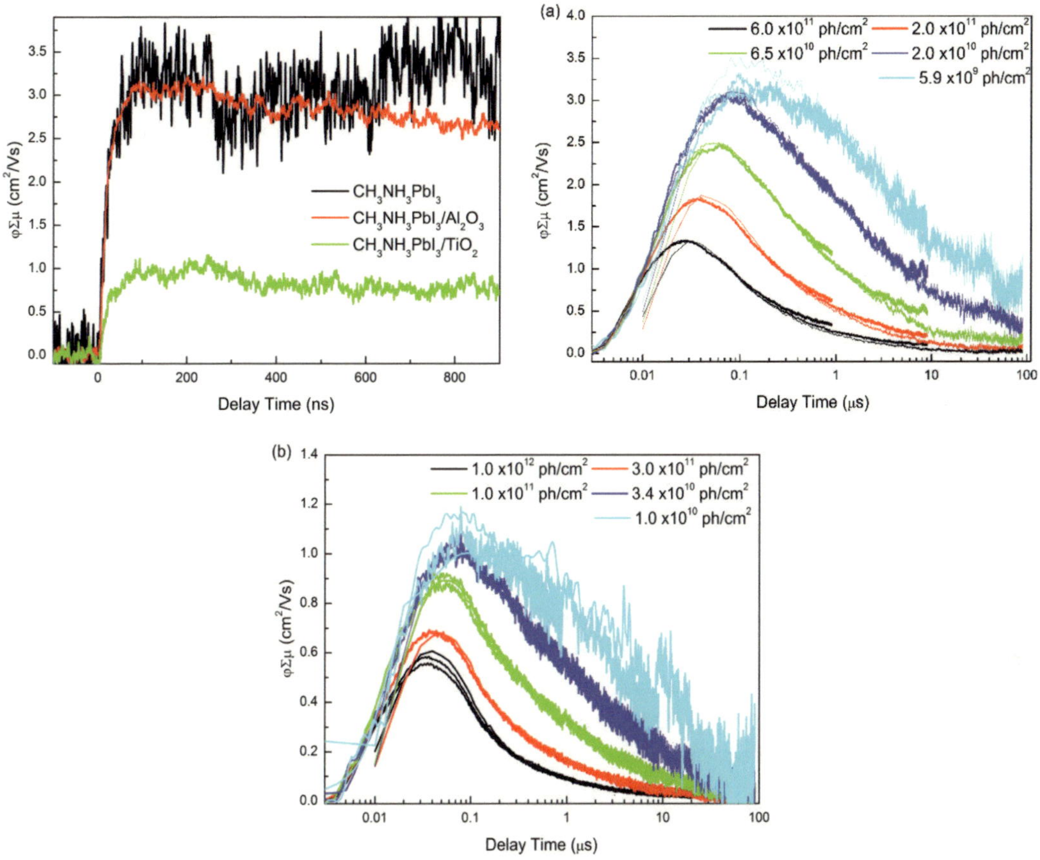

Figure 6. Time-resolved microwave conductivity kinetics of a. $MAPbI_3$, $MAPbI_3/Al_2O_3$, and $MAPbI_3/TiO_2$ measured at 5.9×10^9 ph/cm² per pulse for 1 μs and b. at different excitation densities measured up to 100 μs. (Reprinted with permission from [57], Copyright 2014, American Chemical Society)

mobility and its population occurs on a timescale longer than 1 μs, as shown by TRMC traces. Moreover, this means that charges can move over large distances allowing them to be extracted at the electrodes.

As a summary of this section, the combination of several time-resolved spectroscopy techniques, i.e., optical transient absorption, TRTS, and TRMC, characterization of the charge carrier dynamics in neat $MAPbI_3$ and $MAPbI_3$ attached with TiO_2 and Al_2O_3 was presented. Thin film of perovskite and that attached to Al_2O_3 have properties of an ideal solar cell device: Electron and holes are formed in sub-ps timescale with very high mobilities. Its estimated values suggest that they are almost balanced and carriers do not recombine until after tens of microseconds. These characteristics almost guarantee efficient charge collection, a very desirable property of solar cell device. Electron injection to TiO_2 lowers electron mobility resulting to unbalanced charge transport that could lead to built-in electric field. Engineering of electrodes such that balanced transport is still achieved is one possible way of improving its overall power conversion efficiency.

3. Influence of temperature on the charge dynamics

One of the key issues that need to be addressed in studying this class of materials is its dependence on temperature. This is due to the fact that until now there is a disagreement on how the charges are actually generated. As discussed in the previous section, at room temperature the heterogeneity in the exciton binding energy of the material may lead to different rates of charge dissociation. In such case, lowering the temperature should give an indication if it is indeed by excitonic means that charges are created or is it like bulk crystalline silicon that has band-to-band transition. It has been argued that on optical absorption of a photon with energy exceeding the bandgap, an electron is promoted conduction band leaving a hole in the valence band, thereby forming a correlated charge species or exciton. In the presence of, for example, TiO_2 NP electrons into the conduction band is injected thereby freeing exciton. Similar mechanism is found in hole transfer to Spiro-OMeTAD [23-25]. This is in line with the relatively high binding energies (E_B) estimated to be in the range of 19 and 45 meV for $MAPbI_3$ [26,27]. However, by replacing the TiO_2 with insulating Al_2O_3, where no injection is observed (see discussion above) and therefore exciton remained bound, similar high PCE has been achieved by Snaith et al. In fact, it was shown that it works efficiently well in a flat p-i-n configuration, where perovskite active layer serves as both light absorber, charge generation site and transporter [13,28,29]. Moreover, as presented in Section 1, both the TRTS and TRMC obtained very high mobility in the absence of electron or hole transporting electrodes, i.e., for neat $MAPbI_3$ and $MAPbI_3/Al_2O_3$. Due to these results, questions arise on the excitonic characteristics of the initial photoproduct. Crystallographic data show two-phase transitions, orthorhombic to tetragonal at about 160 K and tetragonal to cubic transition at 330 K [30-32]. By using PL and TRMC, exciton dissociation, charge carrier generation, and recombination is explored from 80 K to 300 K to elucidate the role of binding energy in determining the ratio between bound electron hole pairs and mobile charges. PL probes the emission from bound electron hole pairs, while TRMC probes unbound charges.

Shown in Fig. 7 is the plot of the peak emission intensity of $MAPbI_3$ as a function of temperature. A decrease in temperature results in a more intense emission which is a strong indication that generation of charges is thermally activated. At room temperature, the maximum number of mobile charges generated is reached where emission is at minimum. Using equation 1 of Ref. [33], an activation energy (E_B) of 32 ± 5 meV is obtained. It should be noted though that this E_B is specific to this particular sample and could vary depending on the preparation conditions as well as according to the technique used to probe it. In comparison with organic solar cell (E_B=0.3 eV-0.4 eV) [34], the binding energy obtained here is at least ten times lower and very similar to that of silicon, 15 meV. If one assumes that photoexcitation results only in emission or in charge carriers, then the resulting fit in Fig. 7 represents the charge carrier yield as a function of temperature as shown by the black trace.

Figure 8 are plots of TRMC kinetics for $MAPbI_3/Al_2O_3$ normalized with excitation intensity varied over a factor of 50 and measured at 165 K, 240 K, and 300 K. On the one hand, the fastest decay is observed at highest excitation intensity of the traces at 300 K, implying second-order recombination. On the other hand, at low intensities the lifetime of the charge carriers exceeds

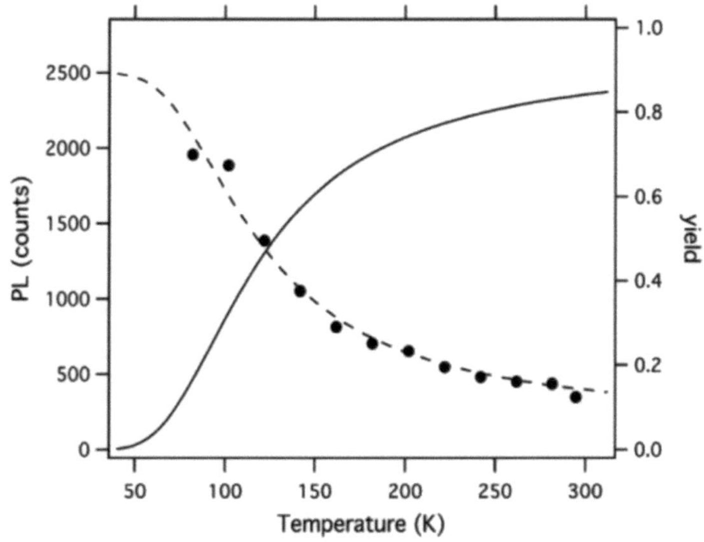

Figure 7. Temperature dependence of PL intensity of $MAPbI_3$ (black circles, λ_{exc} = 550 nm, λ_{dec} = 760 nm). Dashed line is exponential fit, while full line represents the yield of charges on assuming that photoexcitations lead either to radiative decay or to generation of charges. (Reprinted with permission from [33], Copyright 2014, American Chemical Society)

5 μs. Since the decay of TRMC kinetics is a lot slower than corresponding luminescence lifetimes (10 ns) [3,35] at similar intensities, it means that despite the radiative recombination in the ns timescale, there are long-lived charge carriers that survived and remained mobile. More importantly, the mobility of the charges is strongly dependent on the temperature, i.e., from 10.0 $cm^2V^{-1}s^{-1}$ at 165 K to 7.2 $cm^2V^{-1}s^{-1}$ at 240 K and 4.7 $cm^2V^{-1}s^{-1}$ at 300 K. If at these temperatures the changes in the photoconductivity are caused by the change in the yield of charges, then it should have been observed in the PL yield in Fig. 7. Hence, one can assign the changes as due to mobility only. The dependence of mobility on temperature is attributed to quieting of phonon at low temperature with $T^{-1.6}$ dependence, very similar to that of silicon [36]. An increase in mobility usually entails that the second-order recombination would be more efficient since there is higher probability of charges to meet. However, this is not the case in this material. With the increase of mobility, second-order recombination rate diminishes by a factor of 6, which is more than 100 times slower than the reported Langevin recombination, i.e., diffusion controlled recombination [4]. This is a strong indication that there is a barrier in recombination, which should be a temperature-activated process, i.e., meeting of electrons and holes does not necessarily result in recombination. Calculation results yielded a thermal activation over an energy barrier estimated to be ~75 meV [33]. There are two possible origins of this behavior currently reported in the literature: (1) due to the induced dipole moment brought about by the intermittent rotation of MA ions [30,37] and the (2) preferential spatial localization of charge carriers in different parts/materials of the perovskite unit cell. Density functional theory calculations have revealed that 6s- and 5p-orbitals of lead and iodine, respectively, consist of the maxima of the valence band, while 6p-orbitals of lead is where conduction band minima is mostly incorporated [38].

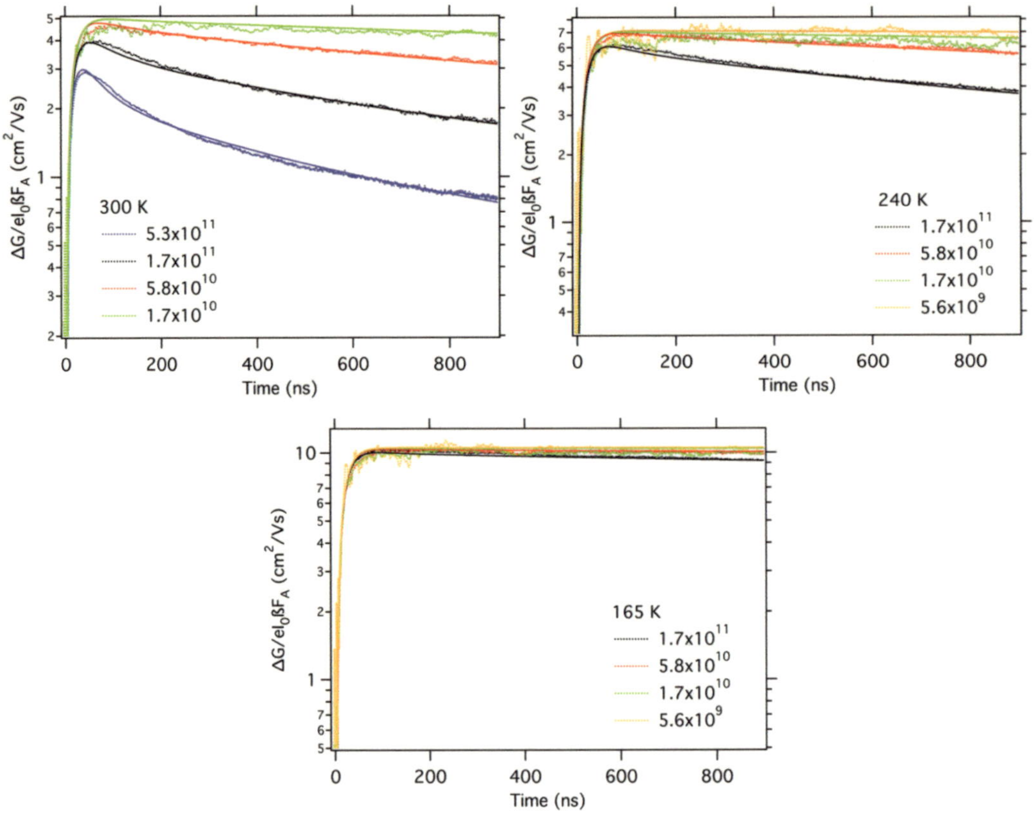

Figure 8. TRMC kinetics of $MAPbI_3/Al_2O_3$ at several excitation intensities measured at 165 K, 240 K, and 300 K (λ_{exc} = 410 nm). (Reprinted with permission from [33], Copyright 2014, American Chemical Society)

Temperature dependent behavior of MAPbI3 is presented in this section. Results show that this material has very similar characteristics as silicon, i.e. it has low exciton binding energy (32 meV)and its mobility has $T^{-1.6}$ dependence brought about by quieting of phonons. More importantly, charges carrier do not recombine when it they meet but instead require a thermal activation over an energy barrier of 75 meV, allowing the charges to be collected at the electrodes efficiently.

4. Role of dark carriers

In this section, the intrinsic property of $MAPbI_3$ is examined when organic electrodes, PCBM and Spiro-OMeTAD are deposited on top of the neat perovskite material. From these samples, the timescale of injection of charges and the dynamics of its recombination are analyzed. Plotted in Fig. 9 are the TRTS kinetics of neat $MAPbI_3$, $MAPbI_3$/PCBM, and $MAPbI_3$/Spiro-OMeTAD normalized with excitation density. At the earliest timescale, the mobility is found to be ~15 $cm^2V^{-1}s^{-1}$ for this particular neat $MAPbI_3$ sample. This mobility slightly decreased

http://dx.doi.org/10.5772/61631

after 1 ns due to second-order recombination, similar to that shown in Fig. 4b. In contrast, the decay of $MAPbI_3$/PCBM is faster, down to almost a third at the same timescale and initial mobility. Such decay is more clearly seen at 7 ns time window in Fig. 9b. For $MAPbI_3$/Spiro-OMeTAD, the initial mobility is 5 $cm^2V^{-1}s^{-1}$, i.e., three times smaller than the neat but remains flat up to 1 ns. At the later timescale, i.e., hundreds of ns to us (Fig. 10a), rapid decay is still seen in $MAPbI_3$/PCBM, while neat $MAPbI_3$ and $MAPbI_3$/Spiro-OMeTAD have very similar, if not identical, slower decay. In Fig.10b, the TRMC kinetics is normalized with excitation density and the corresponding mobility is plotted versus incident photon flux. For the neat $MAPbI_3$, mobility decreases on increasing fluences (>10^{11} ph/cm^2) due to second-order recombination. This behavior is also seen in the bilayer samples, but for $MAPbI_3$/PCBM, the threshold for this process is shifted to higher fluence (>10^{12} ph/cm^2).

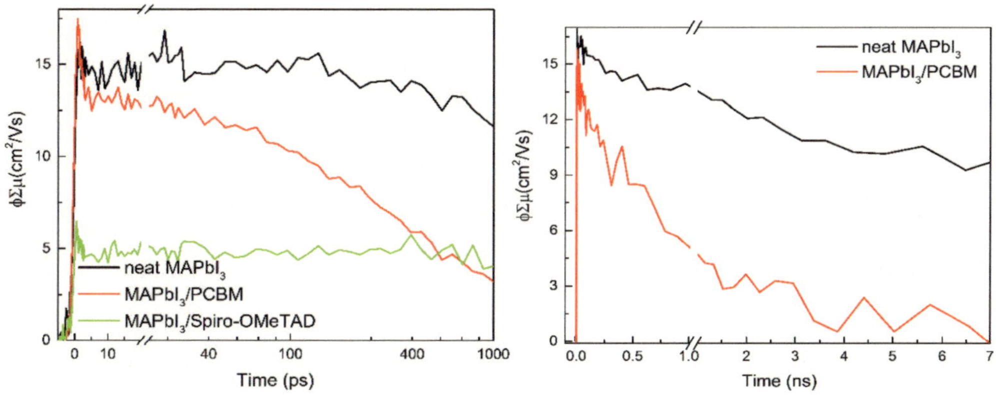

Figure 9. a. TRTS kinetics of neat $MAPbI_3$, $MAPbI_3$/PCBM, and $MAPbI_3$/Spiro-OMeTAD normalized to the excitation intensity (λ_{pump} = 590 nm). b. TRTS kinetics with 7 ns time window.

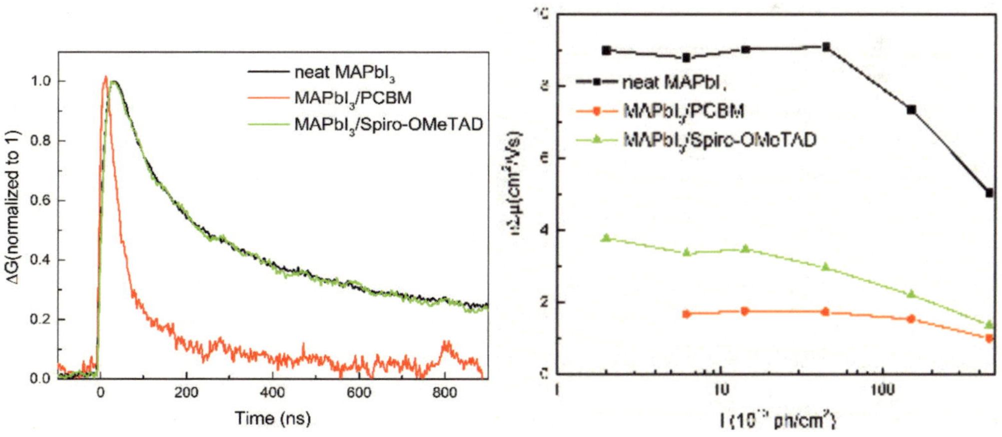

Figure 10. a. TRMC traces for the three samples recorded at an excitation intensity of 1.0 × 10^{11} photons/cm^2 per pulse (λ_{pump} = 600 nm) and normalized to unity. b. Mobility versus incident intensity for neat $MAPbI_3$ and bilayers.

It has been shown that Spiro-OMeTAD is a good hole transporting material (HTM) as it has been extensively used in solar cell devices. One of its favorable characteristics is its 0.5 eV valence band offset with respect to $MAPbI_3$ [23,39,40,41]. However, in order to extract decent PCE, additives like lithium bis(trifluoromethanesulfonyl) imide (LiTFSI) are necessary, which enables this HTM to substantially increase its very low intrinsic hole mobility, i.e., 10^{-8} S/cm [25,42]. Using TRTS, the mobility of $MAPbI_3$/Spiro-OMeTAD is found to be three times lower (5 $cm^2V^{-1}s^{-1}$) than in neat $MAPbI_3$ (15 $cm^2V^{-1}s^{-1}$) (see Fig. 9a). In agreement with this measurement, the mobility measured by TRMC has similar decrease, i.e., from 9 $cm^2V^{-1}s^{-1}$ in neat $MAPbI_3$ to 3.5 $cm^2V^{-1}s^{-1}$ in $MAPbI_3$/Spiro-OMeTAD (Fig. 10b). The decrease in the mobility can be interpreted as disappearance of either electrons or holes in the neat $MAPbI_3$ as a result of charge transfer. On the basis that Spiro-OMeTAD is an HTM, it should therefore be the holes that disappeared and were injected. The hole transfer is confirmed by both techniques and the timescale of injection as seen by TRTS is sub-ps. This suggests that the energy offset at the interface alone is sufficient to allow efficient sub-ps hole injection despite the fact that in this sample no additive was added. This finding consequently means that the mobility of 5 $cm^2V^{-1}s^{-1}$ in TRTS and 3.5 $cm^2V^{-1}s^{-1}$ in TRMC should have originated from electrons that are left in $MAPbI_3$. This process is schematically shown in Figure 11. From the initial 15 $cm^2V^{-1}s^{-1}$ in neat $MAPbI_3$ and 5 $cm^2V^{-1}s^{-1}$ in $MAPbI_3$/Spiro-OMeTAD (both from TRTS), this leads to a hole mobility of 10 $cm^2V^{-1}s^{-1}$ in the perovskite, implying that holes should have diffused at least 30 nm within 1 ps, which is consistent with a sub-ps injection time. Piatkowski et al. also recently reported that timescale of hole injection is 0.7 ps as measured by transient absorption spectroscopy [43].

It has been reported that that charge transfer over an interface typically leads to different decay kinetics as compared to dynamics of the carriers generated in a single semiconductor [44]. However, in the case of $MAPbI_3$/Spiro-OMeTAD, although hole injection is confirmed by TRTS, its decay in TRMC is similar to the neat $MAPbI_3$. In fact, even at different excitation intensity their decay is still very similar as shown in Fig. 12a. The identical TRMC kinetics imply that their decay pathways must be identical, if not very similar. It was previously reported that depending on the preparation conditions, perovskite could either be an n-doped or a p-doped semiconductor. For example, in the work of Leijtens et al., an n-type perovskite material was obtained when deposited to Al_2O_3 NPs [45]. In contrast, calculation of Shi et al. [46] suggests that p-type is usually obtained; while Kim et al. [47] reported that it can be controlled as n- or p-type depending on the defects. It is therefore not unrealistic, at least for a moment, to postulate that perovskite measured here has concentration of holes already in the dark (p_0), i.e., that $MAPbI_3$ is an unintentionally doped, p-type semiconductor. This would mean that once carriers are photogenerated and holes transferred to Spiro-OMeTAD, electrons left in the conduction band of $MAPbI_3$ will recombine with both, the dark holes in the valence band of $MAPbI_3$ and the photogenerated holes injected into Spiro-OMeTAD. So long as the concentration of photogenerated electrons is smaller than the total concentration of holes (dark and light-induced carriers), the electron hole recombination kinetics in $MAPbI_3$ should be barely dependent on whether there is hole injection or not. This scenario would then result in conductivity decay very similar to neat $MAPbI_3$. Since the obtained kinetics are identical, it can be surmised that the perovskite sample measured here is p-type. It should also be stressed

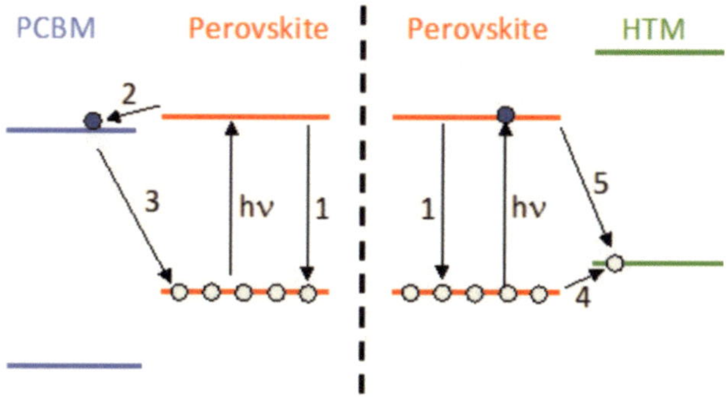

Figure 11. Schematic diagram of injection and recombination dynamics in from neat $MAPbI_3$ to PCBM and Spiro-OMeTAD.

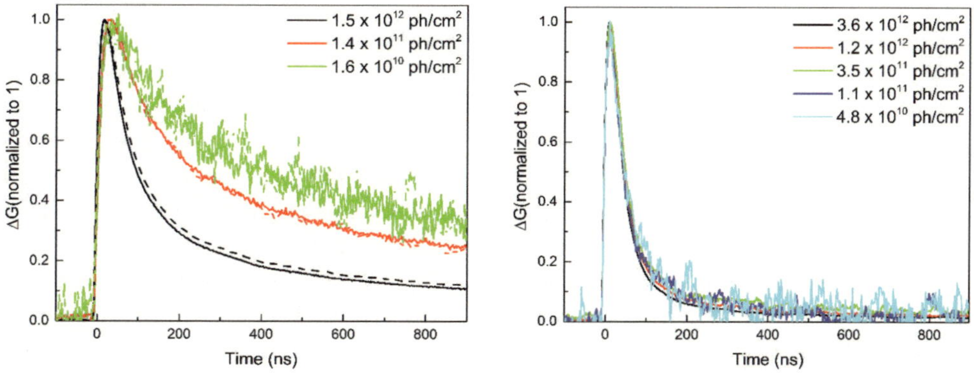

Figure 12. Excitation intensity dependence of TRMC kinetics of (a) $MAPbI_3$/Spiro-OMeTAD and (b) $MAPbI_3$/PCBM.

that both TRTS and TRMC probe only the change in conductivity due to optical excitation but not the dark conductivity.

For electron acceptors, PCBM has been extensively used for organic solar cells and lately has been also utilized for perovskite-based solar cells as well. Despite its small energy offset with respect to perovskite, the reported power conversion efficiency is over 10% [48,49]. This indicates that electrons are transferred from the perovskite to PCBM, which is schematically represented as process 2 in Fig. 11. The electron mobility in PCBM is quite small, i.e., 10^{-3} $cm^2V^{-1}s^{-1}$ [49-53], and therefore contributing insignificantly to the measured mobility in both TRTS and TRMC. Hence, the measured signal in Figs. 9 and 10b, only represent the mobile photogenerated charges in the perovskite. This implies that the decay measured in photoconductivity kinetics could represent not only charge recombination within the $MAPbI_3$ (process 1), electron injection from $MAPbI_3$ to PCBM (process 2), but the recombination at $MAPbI_3$/PCBM interface (process 3) as well. To disentangle the contributions of these processes, one can consider this qualitative consideration: that electron injection (process 2) is much faster than interfacial recombination between holes left in the perovskite and electrons injected to

PCBM (process 3). In this scenario, a clear nonexponential, slower decay is expected since mobile electrons in $MAPbI_3$ become immobile in PCBM. However, since there is no such plateau observed (Fig. 9b), it can be concluded that the rate constant for the interfacial electron hole recombination (process 3) is similar to, or exceeds, the electron injection rate, which is consistent with the fact that the difference in the energy gap in their conduction bands is small.

At the earliest timescale of the THz kinetics, $MAPbI_3$/PCBM bilayer and neat $MAPbI_3$ have the same mobility (~15 $cm^2V^{-1}s^{-1}$; Figs. 9a and 9b). This shows that mobile charges are rapidly formed (<1 ps) and that the photogenerated charges stay in the perovskite for at least a few ps. After 1 ns, the mobility is reduced to approximately a third, suggesting that charges disappear on this timescale. As discussed above, there are three different processes either consecutively or simultaneously occurring that may be responsible for this decay. The very small energy offset at the interface between $MAPbI_3$ and PCBM could retard electron injection to the ns timescale. Transient absorption spectroscopy as reported by Xing et al. estimated electron injection from perovskite to PCBM to be within several ns [17]. Therefore, electron injection into PCBM can be one of the processes leading to the THz decay. The second plausible origin of the THz decay is the inevitable recombination of electrons injected in the PCBM, which are pinned at the interface due to their low mobility, with dark and photogenerated holes in perovskite. Similar to $MAPbI_3$/Spiro-OMeTAD, concentration of dark holes here is expected to be at least the same order since the preparation method used is similar. Third, the excitation-dependent second-order recombination within the perovskite process could also manifest as decay of the conductivity signal. The THz kinetics of neat $MAPbI_3$ (Figs. 9 a and Figs. 9) show that at 2.1×10^{12} ph/cm^2 per pulse, this process occurs on the many ns timescale, significantly slower than the decay of the $MAPbI_3$/PCBM conductivity. Hence, this process only weakly contributes. The THz conductivity decay can therefore be assigned to convolution of electron injection to PCBM and electron–hole recombination at the perovskite/PCBM interface wherein both processes are occurring on a similar timescale.

The TRMC measurements of $MAPbI_3$/PCBM in Fig. 10a show extended time window of conductivity measurements and were obtained at lower excitation intensity than the TRTS measurements. Because of this, the second-order recombination could only occur on the microsecond timescale [33] and will only marginally influence the decay. Hence, the origin of the decay should only be from electron injection and interfacial recombination. The excitation-dependent mobility in Fig. 10b provides another indication of the timescale of electron injection and interfacial recombination. On the one hand, $MAPbI_3$/PCBM has mobility of 2 $cm^2V^{-1}s^{-1}$, representing mobile holes only since injection is apparent. On the other hand, $MAPbI_3$/Spiro-OMeTAD has hole mobility of 5.5 $cm^2V^{-1}s^{-1}$, i.e., 9 $cm^2V^{-1}s^{-1}$ from the neat minus 3.5 $cm^2V^{-1}s^{-1}$. There is almost a three-time difference between the hole mobilities measured from the two samples. Assuming 100% electron injection into PCBM after several ns, it means that two-thirds of the photogenerated mobile holes have disappeared, most probably through interfacial recombination, on the timescale similar to injection time. Admittedly, Wojciechowski et al. reported TRMC kinetics decay similar to that observed here. However, the decay was assigned to electron injection under the assumption that the difference electron and hole mobility is at least ten times [54]. This is in contrast with several theoretical papers on the electrons and holes'

http://dx.doi.org/10.5772/61631

effective masses stating that they do not differ more than a factor 2 [16,55,56], to experimental works showing balanced electron–hole diffusion lengths [3,17], and to previous THz and TRMC conductivity measurements [33,57]. Moreover, using the results presented here it is possible to understand the TRMC kinetics as ns electron injection into PCBM in convolution with interfacial recombination between immobile electrons in PCBM and photogenerated and dark holes in the perovskite. The resulting picture is an oppositely charged bilayer material wherein recombination dynamics is not influenced by the excitation density. In fact, as shown in Figure 12b, the TRMC kinetic decays of $MAPbI_3$/PCBM do not have any dependency on excitation. This is a strong indication of the first-order character of the recombination.

These results have far-reaching implications in understanding the fundamental photophysical processes in these materials and to the operation of perovskite solar cells. Utmost care should be taken in interpreting photophysical data as these are strongly influenced by the state and population of defects that control their doping, as reported by several papers [45-47]. Recombination at the $MAPbI_3$/PCBM interface represents a loss mechanism and therefore is detrimental in the operation of solar cells. Similar to strategy in optimizing the performance of perovskite devices by adding dopants to Spiro-OMeTAD in order to increase its conductivity, PCBM should also be doped. Moreover, developing new methods in reducing unintentional doping would certainly be beneficial. In summary of this section, it was found that hole transfer from $MAPbI_3$ into Spiro-OMeTAD occurs on a sub-ps timescale, while its recombination dynamics is identical to neat $MAPbI_3$ and controlled by a high concentration of dark holes. Electron injection in PCBM is slower, few ns, which is convoluted with the interfacial recombination between the electrons residing in PCBM and the photogenerated and dark holes in $MAPbI_3$. The positively charged majority carriers brought about by unintentional doping, dictate not only the recombination of photoexcited carriers in neat $MAPbI_3$ layer but also control the charge injection dynamics in bilayer samples. Finally, reduction of the hole concentration in the perovskite could help to retard the recombination yielding a higher overall power conversion efficiency.

5. Conclusion

Despite the advancement in the understanding of the device properties of perovskite solar cells, research on its fundamental electrical characteristics have remained scarce. This work has shown that using time-resolved THz spectroscopy and microwave conductivity measurements, complemented by transient absorption and photoluminescence spectroscopy, an in-depth understanding may be achieved. Among its nearly ideal solar cell characteristics are ultrafast charge generation (<0.2 ps), high mobility (μ_e = 12.5 $cm^2V^{-1}s^{-1}$ and μ_h = 7.5 $cm^2V^{-1}s^{-1}$) which remained constant up to at least 1 ns, charge lifetime of tens of μs, and recombination barrier energy of 75 meV. One of the challenges that need to be addressed by the solar cell community is to standardized protocol on growth/preparation methods such that the state and concentration of defects, thereby, dark carriers will be controlled as desired. As presented in this chapter, such dark carriers play a vital role in recombination dynamics and hence could spell success or failure of the device.

Acknowledgements

This work was supported by the Swedish Energy Agency (STEM), the Swedish Research Council, the Knut&Alice Wallenberg Foundation, the European Research Council (Advanced Investigator Grant to VS, 226136-VISCHEM), the Nanometer Consortium at Lund University (nmc@LU), and the Lund Laser Center. The time-resolved THz setup is partly developed by Pascher Instruments (www.pascherinstruments.com). The author would also like to express his gratitude to Prof. Tom J. Savenije of Delft University of Technology, The Netherlands, where all the TRMC measurements were done. Finally, this work is dedicated to Prof. Villy Sundström, a mentor, a colleague, and a friend.

Author details

Carlito S. Ponseca Jr.*

Address all correspondence to: carlito.ponseca@chemphys.lu.se

Division of Chemical Physics, Lund University, Lund, Sweden

References

[1] Kojima A, Teshima K, Shirai Y, Miyasaka T. Organometal Halide Perovskites as Visible-Light Sensitizers for Photovoltaic Cells. *J Am Chem Soc* 2009; 48: 6050–6051.

[2] Anon National Renewable Energy Labs (NREL) efficiency chart (2015); http://www.nrel.gov/ncpv/images/efficiency_chart.jpg Natl Renew Energy Labs Effic chart (2015); http//www.nrel.gov/ncpv/images/efficiency_chart.jpg

[3] Stranks SD, Eperon GE, Grancini G, et al. Electron-Hole Diffusion Lengths Exceeding 1 micrometer in an Organometal Trihalide Perovskite Absorber. *Science* 2013; 342:341–344.

[4] Wehrenfennig C, Eperon GE, Johnston MB, et al. High Charge Carrier Mobilities and Lifetimes in Organolead Trihalide Perovskites *Adv Mater* 2013; 26:1584–1589.

[5] Ponseca CS, Yartsev A, Wang E, Andersson MR, et al. Ultrafast Terahertz Photoconductivity of Bulk Heterojunction Materials Reveals High Carrier Mobility up to Nanosecond Time Scale. *J Am Chem Soc* 2012; 134:11836–11839.

[6] Ponseca CS, Němec H, Vukmirović N, et al. Electron and Hole Contributions to the Terahertz Photoconductivity of a Conjugated Polymer:Fullerene Blend Identified *J Phys Chem Lett* 2012; 3:2442–2446.

[7] Nemec H, Rochford J, Taratula O, et al. Influence of the Electron-Cation Interaction on Electron Mobility in Dye-Sensitized ZnO and TiO_2 Nanocrystals: A Study Using Ultrafast Terahertz Spectroscopy. *Phy Rev Lett* 2010; 104:197401.

[8] Žídek K, Zheng K, Ponseca CS, et al. Electron Transfer in Quantum-Dot-Sensitized ZnO Nanowires: Ultrafast Time-Resolved Absorption and Terahertz Study. *J Am Chem Soc* 2012; 134:12110–12117.

[9] Ponseca CS, Nemec H, Wallentin J, et al. Bulk-Like Transverse Electron Mobility in an Array of Heavily n-Doped InP Nanowires Probed by Terahertz Spectroscopy. *Phys Rev B* 2014; 90:085405.

[10] Rothenberger G, Fitzmaurice D, Gratzel M. *J Phys Chem* 1992; 96:5983-5987.

[11] Lin Q, Armin A, Nagiri RR, Burn PL, et al. Electro-Optics of Perovskite Solar Cells. *Nat Photonics* 2014; 9:106–112.

[12] Sum TCT, Mathews N. Advancements in Perovskite Solar cells: Photophysics behind the Photovoltaics. *Energy Environ Sci* 2014; 7:2518–2534.

[13] Lee MM, Teuscher J, Miyasaka T., et al. Efficient Hybrid Solar Cells Based on Meso-Superstructured Organometal Halide Perovskites. *Science* 2012; 338:643–645.

[14] Blom PWM, Mihailetchi VD, Koster LJA, et al. Device Physics of Polymer:Fullerene Bulk Heterojunction Solar Cells. *Adv Mater* 2007; 19:1551–1556.

[15] Hendry E, Koeberg M, O'Regan B. Local Field Effects on Electron Transport in Nano-structured TiO2 Revealed by Terahertz Spectroscopy. *Nano Lett* 2006; 6:755–759.

[16] Quarti C, Mosconi E, De Angelis F. Interplay of Orientational Order and Electronic Structure in Methylammonium Lead Iodide: Implications for Solar Cell Operation. *Chem Mater* 2014; 26:6557–6569.

[17] Xing G, Mathews N, Sun S, et al. Long-Range Balanced Electron- and Hole-Transport Lengths in Organic-Inorganic CH3NH3PbI3. *Science* 2013; 342:344–347.

[18] Murthy DHK, Gao K, Vermeulen MJW, et al. Mechanism of Mobile Charge Carrier Generation in Blends of Conjugated Polymers and Fullerenes: Significance of Charge Delocalization and Excess Free Energy. *J Phys Chem C* 2012; 116:9214–9220.

[19] Fravventura, MC, Deligiannis D, Schins JM, et al. What Limits Photoconductance in Anatase TiO2 Nanostructures? A Real and Imaginary Microwave Conductance Study. *J Phys Chem C* 2013; 117:8032–8040.

[20] De S, Pascher T, Maiti M, et al. Geminate Charge Recombination in Alternating Poly-fluorene Copolymer/Fullerene Blends. *J Am Chem Soc* 2007; 129:8466–8472.

[21] Pal SK, Kesti T, Maiti M, et al. Geminate Charge Recombination in Polymer/Fullerene Bulk Heterojunction Films and Implications for Solar Cell Function. *J Am Chem Soc* 2010; 132:12440–12451.

[22] Bi D, Yang L, Boschloo G, et al. Effect of Different Hole Transport Materials on Recombination in CH3NH3PbI3 Perovskite-Sensitized Mesoscopic Solar Cells. *J Phys Chem Lett* 2013; 4:1532–1536.

[23] Kim HS, Lee C-R, Im J.H, et al. Lead Iodide Perovskite Sensitized All-Solid-State Submicron Thin Film Mesoscopic Solar Cell with Efficiency Exceeding 9%. *Sci Rep* 2012; 2:591-595.

[24] Jeng, JY, Chiang, YF, Lee MH, et al. CH3NH3PbI3 Perovskite/Fullerene Planar- Heterojunction Hybrid Solar Cells. *Adv Mater* 2013; 25:3727–3732.

[25] Marchioro A, Teuscher J, Friedrich D, et al. Unravelling the Mechanism of Photoinduced Charge Transfer Processes in Lead Iodide Perovskite Solar Cells. *Nat Photonics* 2014; 8:250–255.

[26] Ishihara T. Optical-Properties of PbI-Based Perovskite Structures. *J Lumin* 1994; 60(1): 269–274.

[27] Sun S, Salim T, Mathews N, et al. The Origin of High Efficiency in Low-Temperature Solution-Processable Bilayer Organometal Halide Hybrid Solar Cells. *Energy Environ Sci* 2014; 7:399–407.

[28] Liu M, Johnston MB, Snaith HJ. Efficient Planar Heterojunction Perovskite Solar Cells by Vapour Deposition. *Nature* 2013; 6:1739–1743.

[29] Ball JM, Lee MM, Hey A, Snaith HJ. Low-Temperature Processed Meso-Superstructured to Thin-Film Perovskite Solar Cells. *Energy Environ Sci* 2013; 6:1739–1743.

[30] Wasylishen RE, Knop O, Macdonald JB. Cation Rotation in Methylammonium Lead Halides. *Solid State Commun* 1985; 56:581–582.

[31] Onodayamamuro N, Yamamuro O, Matsuo T, Suga HP-T. Phase-Relations of CH3NH3PbX3 (X = CL, BR, I) Crystals. *J Phys Chem Solids* 1992; 53:277–281.

[32] Onodayamamuro N, Matsuo T, Suga H. Dielectric Study of CH3NH3PbX3 (X = CL, BR, I). *J Phys Chem Solids* 1992; 53:935–939.

[33] Savenije TJ, Ponseca CS, Kunneman L, et al. Thermally Activated Exciton Dissociation and Recombination Control the Carrier Dynamics in Organometal Halide Perovskite *J Phys Chem Lett* 2014; 5:2189–94.

[34] Servaites JD, Ratner MA, Marks TJ. Organic Solar Cells: A New Look at Traditional Models. *Energy Environ Sci* 2011; 4:4410–4422.

[35] Stoumpos CC, Malliakas, CD, Kanatzidis MG. Semiconducting Tin and Lead Iodide Perovskites with Organic Cations: Phase Transitions, High Mobilities, and Near-Infrared Photoluminescent Properties. *Inorg Chem* 2013; 52:9019–9038.

[36] van Zeghbroeck B. Principles of Semiconductor Devices. University of Colorado: Boulder CO, 2006.

http://dx.doi.org/10.5772/61631

[37] Knop O, Wasylishen RE, White MA, et al. Alkylammonium Lead Halides 0.2. CH3NH3PbCl3, CH3NH3PbBr3, CH3NH3PbI3 Perovskites – Cuboctahedral Halide Cages with Isotropic Cation Reorientation. *Can J Chem: Rev Can Chim* 1990; 68:412–422.

[38] Even J, Pedesseau L, Jancu JM, Katan C. Importance of Spin–Orbit Coupling in Hybrid Organic/Inorganic Perovskites for Photovoltaic Applications. *J Phys Chem Lett* 2013; 4:2999–3005.

[39] Abrusci A Stranks SD, Docampo P, et al. High-Performance Perovskite-Polymer Hybrid Solar Cells via Electronic Coupling with Fullerene Monolayers. *Nano Lett* 2013; 13:3124–3128.

[40] Shen Q, Ogomi Y, Chang J, et al. Charge Transfer and Recombination at the Metal Oxide/CH3NH3PbClI2/spiro-OMeTAD Interfaces: Uncovering the Detailed Mechanism behind High Efficiency Solar Cells. *Phys Chem Chem Phys* 2014; 16:19984–19992.

[41] Edri E, Kirmayer S, Mukhopadhyay S, et al. Elucidating the Charge Carrier Separation and Working Mechanism of CH3NH3PbI3−xClx Perovskite Solar Cells. *Nat Commun* 2014; 5:3461–3465.

[42] Nguyen WH, Bailie, CD, Unger EL, McGehee MD. Enhancing the Hole-Conductivity of Spiro-OMeTAD without Oxygen or Lithium Salts by Using Spiro (TFSI) 2 in Perovskite and Dye-Sensitized Solar Cells. *J Am Chem Soc* 2014; 36:10996–11001.

[43] Piatkowski P, Cohen B, Javier Ramos F, et al. Direct Monitoring of Ultrafast Electron and Hole Dynamics in Perovskite Solar Cells. *Phys Chem Chem Phys* 2015; 17:14674–14684.

[44] Kroeze JE, Savenije TJ, Vermeulen MJW, Warman JM. Contactless Determination of the Photoconductivity Action Spectrum, Exciton Diffusion Length, and Charge Separation Efficiency in Polythiophene-Sensitized TiO2 Bilayer. *J Phys Chem B* 2003; 107:7696–7705.

[45] Leijtens T, Stranks SD, Eperon GE, et al. Electronic Properties of Meso-Superstructured and Planar Organometal Halide Perovskite Films: Charge Trapping, Photodoping, and Carrier Mobility. ACS Nano 2014; 8:7147–7155.

[46] Shi T, Yin W-J, Yan Y. Predictions for p-Type CH3NH3PbI3 Perovskites. *J Phys Chem C* 2014; 118:25350–25354.

[47] Kim J, Lee S-H, Lee J-H, Hong K-H. The Role of Intrinsic Defects in Methylammonium Lead Iodide Perovskite. *J Phys Chem Lett* 2014; 5:1312–1317.

[48] Malinkiewicz O, Yella A, Lee YH, et al. Perovskite Solar Cells Employing Organic Charge-Transport Layers. *Nat Photon* 2013; 8:128–132.

[49] Gao P, Grätzel M, Nazeeruddin MK, Organohalide Lead Perovskites for Photovoltaic Applications. *Energy Environ Sci* 2014; 7:2448–2463.

[50] Von Hauff E, Dyakonov V, Parisi J. Study of Field Effect Mobility in PCBM Films and P3HT:PCBM Blends. *Sol Energy Mater Sol Cells* 2005; 87:149–156.

[51] Mihailetchi VD, Xie H, De Boer B, et al. Charge Transport and Photocurrent Generation in Poly(3-hexylthiophene): Methanofullerene Bulk-Heterojunction Solar Cells. *Adv Funct Mater* 2006; 16:699–708.

[52] Warman JM, De Haas MP, Anthopoulos TD, De Leeuw DM. The Negative Effect of High-Temperature Annealing on Charge-Carrier Lifetimes in Microcrystalline PCBM. *Adv Mater* 2006; 18;2294–2298.

[53] Savenije TJ, Ferguson AJ, Kopidakis N, Rumbles G. Revealing the Dynamics of Charge Carriers in Polymer:Fullerene Blends Using Photoinduced Time-Resolved Microwave Conductivity. *J Phys Chem C* 2013; 117:24085–24103.

[54] Wojciechowski K, Stranks, SD, Abate A, et al. Heterojunction Modification for Highly Efficient Organic–Inorganic Perovskite Solar Cells. *ACS Nano* 2014; 8:12701–12709.

[55] Chang Y, Park C, Matsuishi K. First-Principles Study of the Structural and the Electronic Properties of the Lead-Halide-Based Inorganic-Organic Perovskites (CH3NH3)PbX3 and CsPbX3 (X = Cl, Br, I). *J Korean Phys Soc* 2004; 44: 889–893.

[56] Giorgi G, Fujisawa, JI, Segawa H, Yamashita K. Small Photocarrier Effective Masses Featuring Ambipolar Transport in Methylammonium Lead Iodide Perovskite: A Density Functional Analysis. *J Phys Chem Lett* 2013; 4:4213–4216.

[57] Ponseca, CS, Savenije, TJ, Abdellah M, et al. Organometal Halide Perovskite Solar Cell Materials Rationalized: Ultrafast Charge Generation, High and Microsecond-Long Balanced Mobilities, and Slow Recombination. J Am Chem Soc 2014; 136:5189–5192.

Chapter 12

Photoexcitations and Emission Processes in Organometal Trihalide Perovskites

Michele Cadelano, Michele Saba, Nicola Sestu, Valerio Sarritzu, Daniela Marongiu, Feipeng Chen, Roberto Piras, Francesco Quochi, Andrea Mura and Giovanni Bongiovanni

Additional information is available at the end of the chapter

http://dx.doi.org/10.5772/61282

Abstract

Organometal halide perovskites have recently attracted widespread attention among scientists, as they combine the advantages of low-cost processability with strong light absorption, band-gap tunability from the near-infrared to the visible region of the electromagnetic spectrum, efficient light emission and charge transport. Such combination of features is unique among solution-processed materials and makes perovskites appealing for several optoelectronic applications, in particular those related to energy sustainability, which could help the advent of a new generation of low-cost but efficient solar cells and large-area light-emitting devices.This chapter reports a critical review of the efforts that scientists have made until now to understand the photophysics of organometal halide perovskites. We address the ongoing debate on the nature of the photoexcited species, namely the role played by free carriers and excitons, the determination of the exciton binding energy as a measure of the Coulomb interaction strength in these materials, the competition between radiative and non-radiative processes, the role and density of charge carrier traps, and last but not least a critical analysis of those phenomena at the base of laser action, highlighting the most relevant results and possible solutions to issues that still remain open.

Keywords: Hybrid perovskites, excitons, free carriers, amplified spontaneous emission, laser

1. Introduction

Organometal halide perovskites are solution-processed semiconductors showing efficient charge transport, favorable emission properties, strong light absorption and optical gap tunability from the visible to the near-infrared spectrum [1–16]. Such interesting properties make these materials very appealing for the realization of solar cells and optical emitters [17–19]. Despite the widespread research activities on these novel semiconductors, some fundamental aspects of the photophysics underlying perovskites have been elusive, especially for what concerns the excited state dynamics [20–29]. For instance, due to the hybrid organic–inorganic nature of organometal perovskites, in principle it is not clear if the excited state properties are dominated by bound or unbound electron-hole states, that is excitons or free carriers. While initial reports assumed the presence of excitons, much like in organic materials, recent optical spectroscopy evidences have emerged that the majority of band-edge optical excitations at room temperature appears to be free carriers [6,22,30].

The architecture of a solar cell depends on this feature. Indeed, if the majority of photoexcited carriers happens to be bound electron-hole states, a heterojunction is necessary to split charges and then produce a current flow in a photovoltaic device. On the other hand, if photoexcitation results in the formation of free carriers, the solar cell structure is simplified, since charges are already separated and can be easily collected at the electrodes.

A lively debate has then started to understand the physical reasons that underpin the prevalence of free carriers over excitons. Clearly, within such debate, a precise and reliable determination of the exciton binding energy is of paramount importance. Reports of exciton binding energies have varied widely, from less than 5 meV to over 50 meV [22,30–37]; it has even been suggested that the exciton binding energy may be temperature-dependent, due to ionic screening effects [33,34,38].

Beyond the potential application in photovoltaics, organometal halide perovskites exhibit also excellent emission properties. In fact, amplified spontaneous emission (ASE) was also demonstrated for photoexcited carrier densities above a threshold value [17–19], representing the first step towards a laser device. Until now, many experiments probed optical amplification under impulsive excitation, meaning that the pumping pulse duration is much shorter than the typical lifetime of the excited states of the crystals [17–19,39–41]. In these kinds of experiments, ASE occurs within a sub-nanosecond time window, which is far from the continuous wave (cw) operation of a real laser device. Hence, it is not clear how long these materials can sustain optical amplification and what warming issues and parasitic processes are involved. Understanding these features is crucial in the perspective of the realization of a perovskite-based cw laser [19].

The aim of this chapter is to summarize the progress made in understanding the physics of excitons and free carriers in perovskites, both in the low excitation regime relevant for solar cells and in the high excitation regime needed for optical amplification. Experimental results obtained from optical spectroscopy measurements carried out in the Department of Physics of the University of Cagliari will be analyzed and critically compared with results published by other research groups.

Sections 2–5 of this chapter describe the absorption and the emission properties of pure and mixed methylammonium lead iodide perovskite films ($MAPbI_3$ and $MAPbI_{3-x}Cl_x$, respectively), showing that the prevailing excited species in these materials is a correlated electron-hole plasma at room temperature, from injected carrier densities typical of solar illumination to those typical to obtain light amplification [30].

An investigation about amplified spontaneous emission in organolead halide perovskites is provided in Sections 6–10, where both the temporal and the spectral photoluminescence (PL) signals have been studied under short- and long-pulsed excitation at high laser fluence, focusing on the physical parameters that allow or inhibit optical amplification in methylammonium lead iodide and bromide ($MAPbBr_3$) perovskite thin films.

2. Exciton binding energy in perovskite films

A large absorption coefficient across the visible spectrum and the consequent efficient solar light harvesting is one of the strong points of perovskites. Initial works rationalized the strong absorption close to the energy-gap and the nature of optical excitations close to the band-gap in terms of excitons, as in organic semiconductors; further investigations have recently converged to state that photoexcitation results in the generation of free carriers in a wide range of pumping intensities, even at ones comparable to solar illumination, as typical of inorganic semiconductors. Such finding is consistent with small values of the exciton binding energy, for which thermal energy is sufficient to ionize excitons in free electrons and holes.

A reliable framework usually adopted in solid state physics to estimate the exciton binding energy is Elliot's theory of Wannier excitons [42], which is valid for bulk semiconductors having exciton binding energies much smaller than the energy gap (Wannier excitons). Such theory accurately describes the optical transitions near the band-gap in inorganic semiconductors like GaAs and GaP [43,44] and models the shape of the optical absorption coefficient $\alpha(\hbar\omega)$ by the following equation:

$$\alpha(\hbar\omega) \propto \frac{\mu_{cv}^2}{\hbar\omega}\sum_j \left|\varphi_j(r=0)\right|^2 \delta\left(\hbar\omega - E_j\right) \propto \frac{\mu_{cv}^2}{\hbar\omega}\left[\sum_j \frac{4\pi\sqrt{E_b^3}}{j^3}\delta\left(\hbar\omega - E_j^b\right) + \frac{2\pi\sqrt{E_b}\,\theta\left(\hbar\omega - E_g\right)}{1 - e^{-2\pi\sqrt{\frac{E_b}{\hbar\omega - E_g}}}}\right] \quad (1)$$

The absorption coefficient depends on the weighted density of electron-hole pair states, with the weight provided by the probability for an electron and a hole to be at the same position, that is $|\varphi_j(r=0)|^2$, while φ_j represents the wave functions of bound and unbound states. μ_{cv} represents the transition dipole moment between conduction and valence bands, and $\hbar\omega$ is the photon energy involved in the transition. In the third expression of Eq. (1), the first term describes the transitions to bound exciton states E_j^b, while the second term takes into account the band-to-band transitions above the energy gap E_g. $\delta(x)$ and $\theta(x)$ are the Dirac-delta and the Heaviside step function, respectively, and E_b is the exciton binding energy.

Eq. (1) can be used to fit the measured absorption spectra of perovskite films, provided that it is convoluted with a bell-shaped function accounting for line broadening. In addition, the introduction of the joint valence-band energy-momentum dispersion leads to a better approximation between theory and experiment, as it overcomes the limits of the parabolic bands approximation. Upon these improvements, Eq. (1) can be modified as follows:

$$\alpha(\hbar\omega)=\frac{\alpha_{\mathrm{scal}}}{\hbar\omega}\left[\sum_j \frac{2E_{\mathrm{b}}}{j^3}\operatorname{sech}\left(\frac{\hbar\omega-E_j^{\mathrm{b}}}{\Gamma}\right)+\int_{E_{\mathrm{g}}}^{\infty}\operatorname{sech}\left(\frac{\hbar\omega-E}{\Gamma}\right)\frac{1}{1-e^{-2\pi\sqrt{\frac{E_{\mathrm{b}}}{E-E_{\mathrm{g}}}}}}\left[1+\frac{10\mu b}{\hbar^4}\left(E-E_{\mathrm{g}}\right)\right]dE\right] \tag{2}$$

The density of states has been developed in series at the first order for small values of b, which is a factor that takes into account the non-parabolic dispersion. The hyperbolic secant function of width Γ accounts for thermal and inhomogeneous broadening; α_{scal} is a constant that is adjusted in order to obtain the right scaling between the model and the experimental absorption spectrum; in the fitting procedure, we also allowed for absorbance offsets often found in experimental curves due to unbalanced reference or zero spectra in spectrophotometers.

In order to establish that the model can describe the absorption of $MAPbI_3$ perovskite films, regardless of the particular morphology and growth conditions of each sample, we fitted Eq. (2) with a least square method to several absorption spectra reported in literature by different research groups. The binding energy is obtained as a fitting parameter, and the fitting range has been extended as much as possible at low and high energy in order to minimize uncertainties connected to the calculation procedure. Published data were extracted from the pdf files through the CurveSnap software, freely available online.

First, we analyzed the absorption measurements at low temperature, where the broadening is less pronounced and therefore the fitting procedure is more sensitive to the value of the exciton binding energy. Figures 1a,b show a comparison between the absorption spectra reported by D'Innocenzo et al. and our experimental data, recorded at 160 and 170 K, respectively [22,30]; the relative contributions of excitons and free carriers to the absorption spectrum are also shown. While the broadening is different in the two cases, the values of the exciton binding energy are comparable and coincide within the uncertainty. Figures 1c–f show the same analysis carried out at room temperature, applied to absorption spectra reported in references [5,22,45]. Even at room temperature, the fitted values for E_{b} do not differ more than 20% from the average of 23 meV. Such value is intermediate between the binding energy typical of Frenkel excitons in organic semiconductors (>100 meV) and the one typical of Wannier excitons in inorganics (<10 meV) [46,47].

The overall picture emerging from the analysis of optical absorption data seems therefore converging towards a value of 23 meV for the exciton binding energy. These results are in contrast with what observed with magnetometry, as some recent publications have reported values of the exciton binding energy that depend on temperature, as a consequence of screening effects. Such works report exciton binding energy to be nearly 15 meV at low

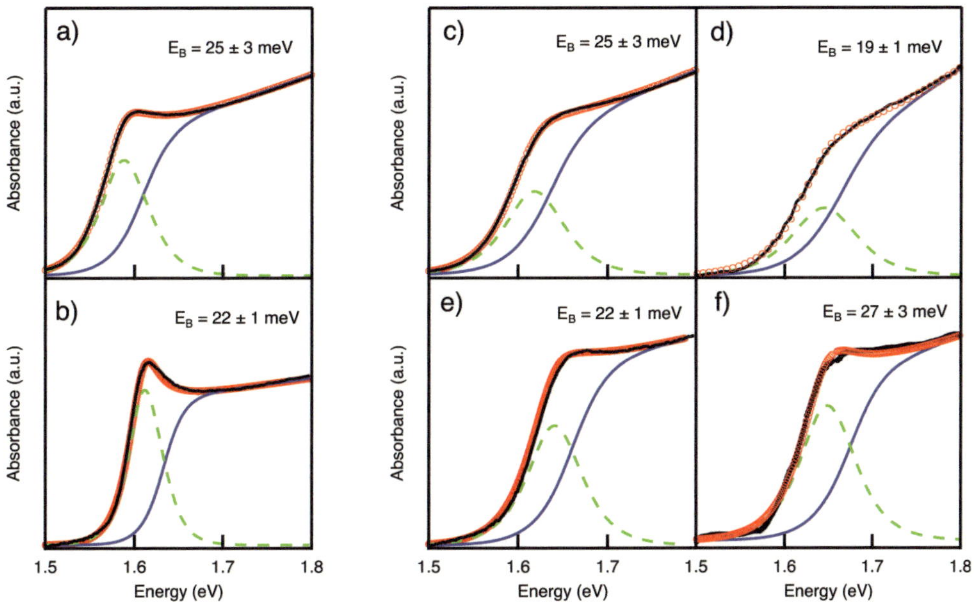

Figure 1. Absorption spectra in $MAPbI_3$ perovskite films. The red empty circles represent the theoretical fits to the experimental data (continuous black lines). The contributions to the absorption due to both excitonic and band-to-band transitions are modelled in the framework of the Elliott's theory of Wannier excitons. The dotted green lines are relative to excitonic transitions, while the continuous blue lines are relative to band-to-band contributions with the inclusion of Coulomb interactions between electrons and holes. (a,b) Absorption spectrum reported by Saba et al. at 170 K and by D'Innocenzo et al. at 160 K, respectively [22,30]. (c,d,e,f) Absorption spectra reported at room temperature by Saba et al., Sutherland et al., D'Innocenzo et al. and Stranks et al., respectively [5,22,30,45].

temperature and a few meV at 300 K [33,34,38,48], while previous works with similar techniques reported values as large as 50 meV [31,49]. Since the interpretation of magnetometry data requires several analysis steps, at the moment it is not clear what the origin of this discrepancy is.

3. Emission properties from free carriers

Linear absorption spectroscopy is extremely useful to obtain information about the optical response and quickly highlights benefits and disadvantages of the material in terms of light harvesting, which represents the main ingredient for photovoltaics applications. Further insight into the electronic properties is provided by the study of the excited state emission properties, as the techniques probe how optical excitations populate the available states. Particularly, photoluminescence under pulsed, femtosecond excitation is useful in determining relaxation rates and excited state dynamics.

Emission in $MAPbI_3$ and $MAPbI_{3-x}Cl_x$ perovskites occurs close to the excitonic resonance and the high energy side of the PL peak overlaps with the continuum of band-to-band transitions in a wide excitation range, as shown in Figure 2a. Short-pulse excitation experiments show that the PL signal rises instantaneously after the laser pulse arrival; the time decay of photoluminescence can be observed from Figures 2b,c. Two different samples are presented, $MAPbI_3$ and $MAPbI_{3-x}CL_x$ perovskites; absorption and emission properties are quite similar in both perovskites, since chlorine content has been determined to be a small fraction, not larger than 2%. However, the dynamics of the PL is different in the two samples, with much longer decay in the $MAPbI_{3-x}CL_x$ film; several reports in literature state that the role of Cl is to alter the perovskite crystallization dynamics and, as a result, $MAPbI_{3-x}CL_x$ perovskites have lower trap densities than what observed in $MAPbI_3$ [50,51], with a corresponding longer excited state lifetime. Inset in Figure 2a shows a clear difference between perovskites with and without chlorine in terms of PL lifetimes at low laser intensity, according to what reported in literature.

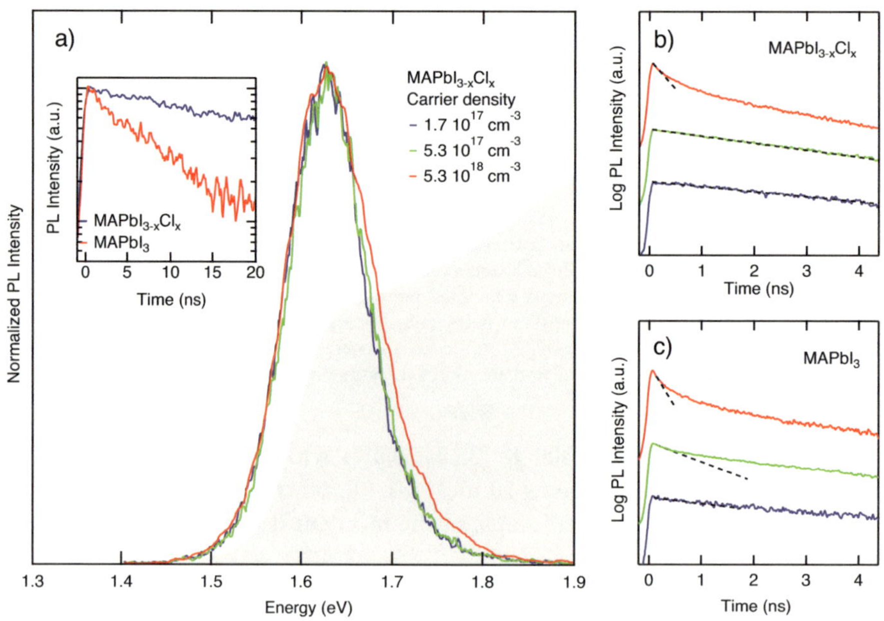

Figure 2. (a) Transient photoluminescence spectra for mixed $MAPbI_{3-x}Cl_x$ films at 300 K, integrated in 60 ps (temporal resolution of our streak camera), compared with the absorption spectrum at the same temperature (gray shadow). Photoluminescence was excited by 130-fs-long laser pulses with a repetition rate of 1 kHz and 3.18 eV photon energy. The legend reports the injected carrier density, calculated from laser pulse fluence and film absorbance. Inset: time resolved PL signal from mixed $MAPbI_{3-x}Cl_x$ (blue line) and pure $MAPbI_3$ (red line) films, relative to an excitation density of at $1.7x10^{17}cm^{-3}$. The different mean PL lifetime is attributed to different trap densities in the samples. (b,c) Transient PL signal as a function of the injected electron–hole pair density at the film surface in $MAPbI_{3-x}Cl_x$ (b) and $MAPbI_3$ films (c). The initial decays of the signal are fitted by an exponential function (black dotted lines). The injected carrier densities are relative to the PL spectra shown in (a).

Radiative recombination processes can be investigated studying the intensity of the photoluminescence signal immediately after the laser excitation (PL_0), before electronic states are depopulated by slow recombination mechanisms. The electron-hole pair density n_0 at the sample surface, injected from a short laser pulse, can be estimated multiplying the laser pulse fluence times the absorption coefficient of the sample at the excitation wavelength. The evolution of PL_0 with n_0 shows what type of carrier prevails as outcome of the photoexcitation.

Figure 3 shows PL_0 as a function of n_0 in $MAPbI_3$ and $MAPbI_{3-x}Cl_x$ perovskites. The PL intensity scales quadratically in the injected carrier densities from less than 10^{16} to 10^{18} cm^{-3}, which corresponds to laser pulse fluence values from less than 0.1 to about 10 μJ/cm². Such quadratic scaling is the signature of bimolecular recombination and is consistent with the radiative recombination from a gas of unbound electron-hole pairs [52]. On the other hand, isolated excitons should give rise to monomolecular recombination, with the PL_0 intensity linear in the exciting laser fluence. PL_0 tends to saturate above $2.5x10^{18}$ cm^{-3}, as a consequence of the band filling.

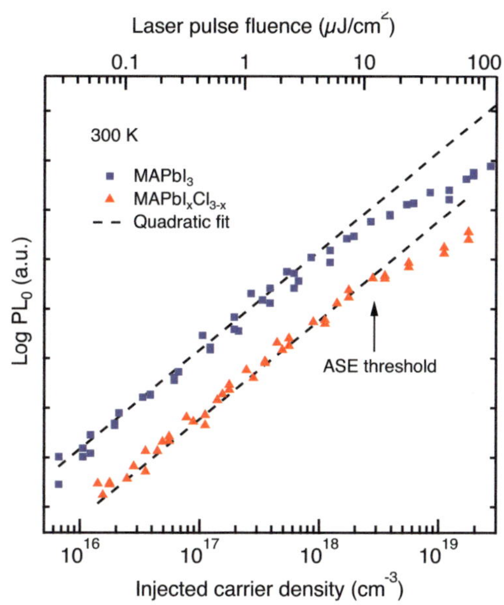

Figure 3. $MAPbI_3$ and $MAPbI_{3-x}Cl_x$ photoluminescence emission intensity estimated at t=0 (PL_0) as a function of injected electron-hole density (lower axis) and laser pulse fluence (upper axis). PL_0 values relative to laser pulse fluences higher than 100 μJ/cm² could not be extracted from experimental decays, since they were much faster than the temporal resolution of our instrument. The black-dashed lines represent the PL_0 quadratic dependence on n_0, and are reported as a guide for eyes.

Focusing only at time zero, all the recombination processes that occur later after laser excitation are neglected. On the other hand, they can be investigated studying the quantum yield (QY), defined by the ratio between the time-integral of the PL signal (TIPL) and the laser fluence ϕ.

Figure 4a shows the QY as a function of the laser fluence, which can be explained in terms of interplay between radiative and non-radiative recombination processes. The rise of the QY below $n_0 \sim 10^{17}$ cm^{-3} is due to the fact that carrier trapping is the fastest recombination process in such excitation regime, while radiative recombination becomes faster for growing carrier densities, so that a larger fraction of optical excitations recombine radiatively. The increase in QY reaches a maximum above $n_0 \sim 10^{17}$ cm^{-3}, as the radiative recombination rate overcomes the carrier trapping and becomes the fastest recombination channel; under such conditions, a further increase in the radiative recombination rate only makes recombination faster, but it does not significantly increase the fraction of carriers that recombine radiatively, which is already close to unity. For even higher excited densities, above $n_0 \sim 10^{17}$ cm^{-3}, non-radiative density-dependent Auger recombination, whose rate is cubic in the density, and therefore increases faster than the radiative rate, becomes dominant, causing the observed QY decrease [30]. The remarkable high values of the QY (30% for $MAPbI_3$ and 70% for $MAPbI_{3-x}Cl_x$) are reached for injected carrier densities close to the amplified spontaneous emission thresholds reported for these materials [18,19].

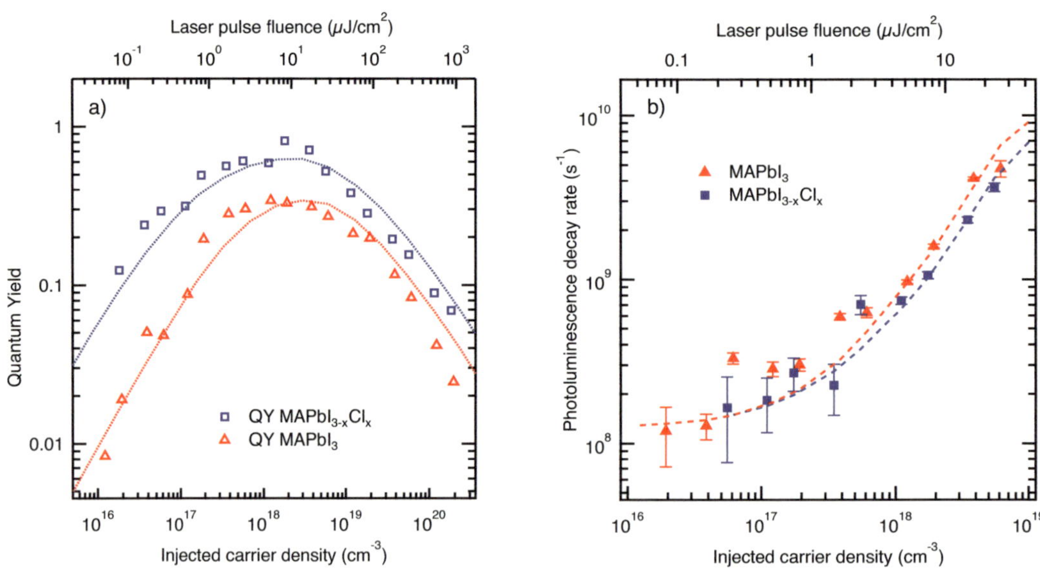

Figure 4. Photoluminescence quantum yield and photoluminescence decay rates for the two perovskite samples ($MAPbI_3$ red triangles, $MAPbI_{3x}Cl_x$ blue squares). (a) The emission quantum yields are calculated as TIPL/ ϕ, where ϕ is the laser pulse fluence. The injected carrier density is calculated multiplying the laser pulse photon fluence by the absorption coefficient of the films. As a reference, the laser pulse fluence directly measured in the experiments is also reported on the top axis. Initially, the QY grows with fluence for both films, as the bimolecular recombination becomes faster and a growing fraction of the injected excitations recombine radiatively, instead of being trapped. At excitations 4×10^{19} cm^{-3}, non-radiative Auger processes dominate. The absolute QY is scaled to match theoretical predictions. The maximum QY values are about 30% for $MAPbI_3$ and 70% for $MAPbI_{3-x}Cl_x$. The dotted lines represent predictions from a rate equation model accounting for the main relaxation channels for electrons and holes. Simulations take into account the exponential spatial profile of the electron-hole density created by laser pulses. The very good agreement between model and data indicates that the main photophysical processes are accounted for in the model. (b) Decay rates for the transient PL signal plotted as a function of the injected carrier density on bottom axis and the laser pulse fluence on top axis. The rates, defined as $k_{PL} = \left[\frac{dPL}{dt} \cdot \frac{1}{PL}\right]_{t=0}$, are extracted from the data in Figures 2b,c. Such rates represent the initial decay and should not be mistaken for the average photoluminescence decay rate obtained by fitting the entire decay with an exponential function. The error bars represent the standard deviation from a least square fit to an exponential decay and are reported only when they exceed the size of the marker. The rates are very similar

for the two samples (although the average exponential decay rates are significantly different) and grow together for growing injected carrier densities. Such rates measure the intrinsic density-dependent bimolecular and Auger recombination processes. The dotted lines represent the results of the rate equation used to model the experimental data: the agreement with the experiment is satisfactory also in this case.

Also the analysis of the initial PL decay rate at time zero, extracted right after laser pulse excitation, confirms our interpretation of the hierarchy of the recombination rates. Figure 4b shows that the initial PL decay rate increases with n_0, as a consequence of the activation of density-dependent recombination mechanisms. Differently from the average PL decay rate, calculated at low excitation intensity, which is different in various samples (inset in Figure 2a), the initial PL decay rate is quite similar among them. This fact reflects that while the mean PL decay time is sensitive to extrinsic effects like traps and defects, which are significant in all solution-processed semiconductors, the initial PL decay rate is dictated only by intrinsic nonlinear processes [30]. For growing excitation densities, the initial recombination rates increase in both perovskite samples, as radiative and Auger recombination become faster and faster.

4. Recombination rates

Fundamental parameters of semiconductors can be estimated theoretically by analyzing absorption and photoluminescence experimental results. As well as the absorption coefficient and the emission wavelength, bimolecular and Auger recombination rates represent two constants that characterize the material and can be used as figures of merit for some optoelectronic applications. Values of the bimolecular recombination constant from 10^{-11} to 10^{-9} cm^3 s^{-1} have been reported in literature for $MAPbI_3$ and $MAPbI_{3-x}Cl_x$ perovskites [20,30,53]. One would expect that for such an intrinsic process similar values are to be obtained in all samples. Such wide range of reported values can be attributed to the different methods used to calculate it. It is common to extract the bimolecular constant from the PL decay curves, fitting them to a multi-power decay. Such method, however, does not provide reliable estimates whenever applied to noisy curves with several competing processes that are simultaneously relevant; in the particular case of perovskites, a monomolecular decay from trapping, a bimolecular one from radiative recombination and trimolecular one for Auger need to be accounted simultaneously, all convoluted with the temporal resolution of the experimental apparatus. On the other hand, extracting the bimolecular constant from the absorption coefficient is a widely used method in solid state physics that makes use of the symmetry between absorption and emission processes and is completely independent on non-radiative recombination processes. In order to extract the bimolecular coefficient, one has to fit the absorption spectrum with the Kubo-Martin-Schwinger relation, as detailed in references [54,55]. Our estimate of the bimolecular constant in $MAPbI_3$ perovskites, extracted from absorption, is 2.6×10^{-10} cm^3 s^{-1}, a value included between the two extrema reported in literature.

Differently from the bimolecular constant, the Auger recombination rate could significantly change among samples, as it increases with the defect density of a crystal, which can depend on the synthesis method and the surface morphology [56,57]. Recent works report Auger recombination constants from 10^{-29} to $10^{-28}cm^6$ s^{-1} for pure and mixed methylammonium lead iodide perovskites. Our estimates of $2x10^{-28}$ cm^6 s^{-1} for $MAPbI_{3-x}Cl_x$ and $4x10^{-28}$ cm^6 s^{-1} for $MAPbI_3$ perovskites are consistent with other published results [20,30]. Auger recombination rate is higher in pure iodide perovskites than in those with chlorine, according to the less crystallinity in samples synthesized without Cl.

5. Steady-state photoluminescence

The experimental results obtained under impulsive regime have provided numerical estimates of fundamental parameters in perovskites and a clearer view of the excited state dynamics. Nevertheless, such kinds of experiments do not investigate what occurs when samples are continuously excited and there is interplay between absorption and relaxation processes, as happens in steady-state operation. In perspective of the realization of perovskite-based working devices, it is important to have a deep understanding of these processes, as real devices work under continuous operation. A reliable instrument of investigation of the steady-state properties is represented by experiments carried out exciting samples under cw pumping or with pulses much longer in time than the PL lifetime of the excited states.

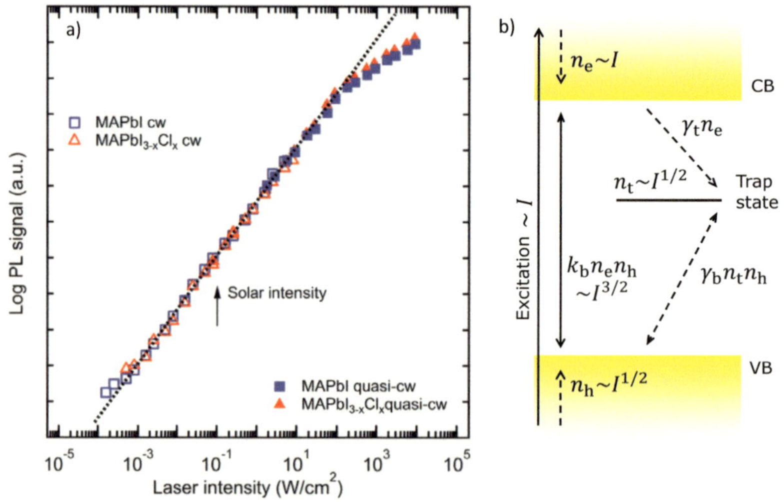

Figure 5. Steady-state photoluminescence. (a) Photoluminescence signal as a function of the intensity of the exciting laser. The empty markers represent the measurements obtained with a cw Nd:Yag 532 nm laser; the filled markers are instead measured in quasi-cw conditions, exciting the samples with 300-ns-long laser pulses from a Q-switched 527 nm Nd:Ylf laser; the pulse duration is much longer than all the relevant relaxation rates, so that steady-state conditions are expected to be achieved during laser excitation. The PL signal grows as $I^{3/2}$ for a wide range of excitations. Investigations were extended from laser intensities much lower than the solar one to intensities large enough to generate population inversion and optical gain. The unusual 3/2 power law is attributed to intra-gap trap states with only electrons or only holes, but not both of them. The dashed black line shows the PL dependence in laser intensity as a result of the trap model, under the assumption that electrons are the trapped species. (b) shows a sketch of the relaxation of optical excitations under steady-state conditions (VB is the valence band and CB the conduction band).

Figure 5a shows our experimental results carried out under cw excitation, spanning light intensities from 10^{-4} to 10^{4} W/cm^2. Below 100 W/cm^2, the PL signal follows a 3/2 power law in the laser intensity for five decades excitation intensity. Such 3/2 power dependence implies that the radiative recombination is not the only process that governs the electron-hole dynamics under this regime; otherwise a linear dependence of the PL in the laser intensity would be expected, if radiative processes would be dominant. The particular 3/2 power law could be explained accounting the role of intra-gap states that act only on one kind of carrier, meaning traps only for electrons or only for holes (Figure 5b). Above 10^{2} W/cm^2, the deviation from the 3/2 power law to a linear dependence is attributed to the increase of radiative recombination with respect to carrier trapping [30]. Trap density is estimated to be of the order of 10^{16}-10^{17}cm^{-3}, in agreement with what reported by other works [6]. Despite such large trap density, carrier trapping does not prejudice electronic properties, as PL lifetimes exceed several nanoseconds from intensities smaller than solar illumination to those typical to obtain light amplification. Hence, the resulting carrier mobility is sufficiently high to justify the remarkable transport properties and the efficient charge collection in perovskite-based solar cells reported in literature.

6. Organolead halide perovskites as active media in laser devices

The strong absorption at the band-edge, in addition to the ambipolar transport with high carrier mobility and gap-tunability, makes organometal halide perovskites very promising not only for the realization of solar cells but also as optical gain media in laser devices, as highlighted by many reports showing amplified spontaneous emission under intense photoexcitation [17–19,39,45,58]. Deschler et al. and Xing et al. reported optically pumped lasing and amplified spontaneous emission from perovskites arranged in cavity resonators and in cavity-free configuration, respectively [17,18]. Recently, Zhu et al. reported threshold densities of the order of 10^{16} cm^{-3} from single-crystal lead halide perovskite nanowires, which is a value about two orders of magnitude smaller than what was reported by the previous works [39]. In the light of these results, it is clear that organometal halide perovskites have a potential application as active media in laser devices. However, demonstrations reported up to date report ASE under impulsive excitation, a condition well away from real laser devices, which work under continuous operation. Probing emission properties with ns-long-pulse excitation simulates a regime more similar to what happens in a real device, and provides further information about the potentialities and the shortcomings of perovskites as optical gain media [19]. The first report about nanosecond excitation is the one published by Sutherland et al., which shows lasing from $MAPbI_3$ perovskite thin layers deposited on spherical glass resonators, exciting samples with 2-ns-long pulses [45]. Nevertheless, this regime is still far from a real cw excitation and other experiments are required to have a deeper understanding of emission properties in perovskites [19]. Results obtained in different excitation regimes, from the impulsive to the quasi-cw, and improvements of our experiments with respect to other works are reported in the following sections and compared with the most recent researches.

7. Lasing and ASE under impulsive excitation

ASE in $MAPbI_3$ and $MAPbBr_3$ perovskite thin films manifests itself through a sharp peak that appears under short-pulse excitation in the low energy side of the PL spectrum of both films, when the injected carrier density reaches a threshold value. Figure 6 shows emission spectra at different excitation intensities from methylammonium iodide and bromide perovskite thin films at room temperature.

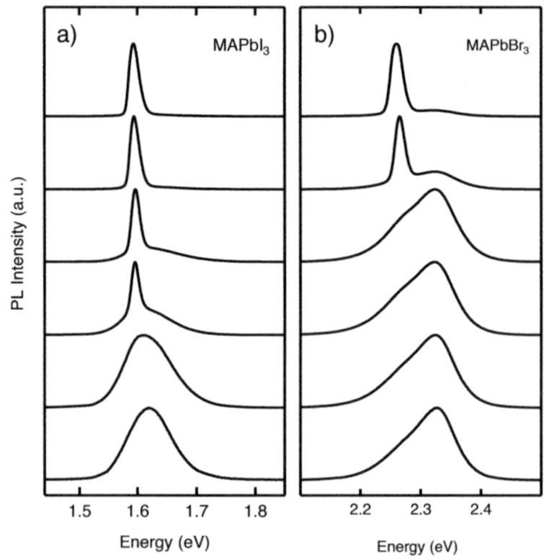

Figure 6. Time integrated photoluminescence spectra of $MAPbI_3$ (a) and $MAPbBr_3$ (b) perovskites at 300 K. Emission was excited by 130-fs-long laser pulses with a repetition rate of 1 kHz and 3.18 eV photon energy. Spontaneous and amplified spontaneous emissions were detected by a CCD camera. From top to bottom: laser fluence = 90 μ J/cm^2, 70 μ J/cm^2, 40 μ J/cm^2, 30 μ J/cm^2, 20 μ J/cm^2, 10 μ J/cm^2.

ASE threshold densities are calculated by averaging the laser fluence times the absorption coefficient at the excitation wavelength over the film thickness. Even if samples realized with the same procedure are excited with the same laser source, they show slightly different values of the ASE threshold, but for most samples the threshold is in the interval $2-5 \times 10^{18}$ cm^{-3} [19]. Such values are comparable to the ones reported by Xing et al. (1.7×10^{18} cm^{-3} in $MAPbI_3$ films), who investigated $MAPbX_3$ (where X = Cl, Br, I) thin films under short-pulse excitation [18]. The dispersion of the ASE thresholds among the same set of samples can be attributed to their surface morphology, which is the main responsible for optical losses. Instead, the large difference in ASE values reported in various publications can be associated to the sample architecture, in addition to the issues originated from the surface morphology. In the experiments reported by Deschler et al., $MAPbI_{3-x}Cl_x$ perovskite layers were placed between dielectric and evaporated gold top mirrors [17]. The same perovskite layers show ASE in cavity-free configuration, but at fluences above 1 mJ/cm^2, about two orders of magnitude higher than what observed in both our results and the ones reported by Xing et al. [18]. The remarkable low lasing thresholds reported by Zhu et al. (1.5 x10^{16} cm^{-3}) are attributed to both the high crystallinity of nanowires and their morphology, as the linear shape of nanowires acts as waveguide [39].

Even though such investigations are extremely useful in perspective of the realization of a perovskite-based laser, they are not sufficient to state if these materials could be effectively employed as optical gain media. Actually, experiments carried out under impulsive regime do not take into account any interplay between laser excitation rates and relaxation processes that occur in real cw lasers, as the carrier injection is ultrafast and the emission occurs immediately after, when pumping has already stopped. Real lasers work under cw operation, a more complicated condition than what happens under impulsive regime, as excitation and emission occur at the same time and optical amplification is observed until the population inversion condition is kept. Under cw excitation, issues connected to warming could considerably affect the light emission properties, especially at high-injected carrier densities (above 10^{18} cm^{-3}), where non-radiative Auger recombination begins to compete with optical emission [19,30].

8. Amplified spontaneous emission under ns excitation

Investigation of emission properties under excitation pulses having duration comparable or longer than the PL lifetime supplies the limits of the ultrafast pumping. Under 4-ns long-pulse excitation, ASE is observable at room temperature in both $MAPbI_3$ and $MAPbBr_3$ thin films at threshold densities of the order of 10 - 15 kW/cm^2 (~40−60 μ J/cm^2) [19]. Such values are comparable with what reported by Sutherland et al. (65 ± 8 μ J/cm^2), who excited perovskite layers of ~75 nm in thickness with pulses of 2 ns [45]. At this excitation intensity, the PL lifetime of our samples, measured under femtosecond laser excitation, is about 1 ns, which is a value lower than the 4-ns laser pulse.

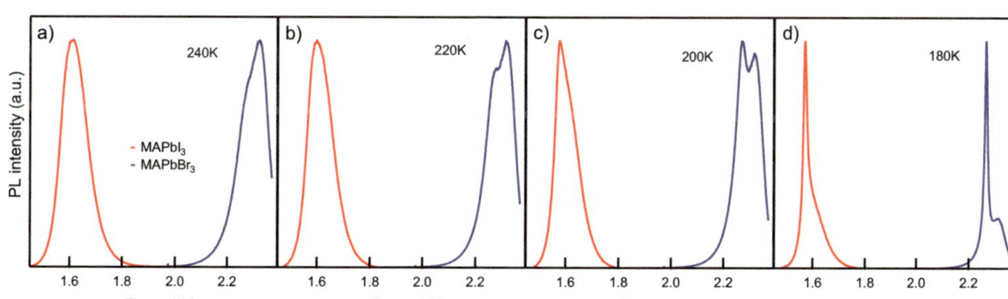

Figure 7. Quasi-steady-state stimulated emission in trihalide perovskite films. Photoluminescence was excited at different temperatures by 300-ns-long laser pulses with a repetition rate of 6 Hz and 2.35 eV photon energy. (a–d) Time-integrated photoluminescence spectra of $MAPbI_3$ (red lines) and $MAPbBr_3$ (blue lines) perovskites detected by a CCD camera; ASE occurs at cryogenic temperatures (180 K), but disappears around 220 K.

Despite using excitation pulses of 4 ns could be considered a quasi-cw excitation regime, such pumping is not sufficient to guarantee if perovskites can act as active media in a real device.

Using 300-ns-long pulses as laser source, no ASE is observed at room temperature in any perovskite film. A very small hint of amplification is observed at 220 K and a clear ASE peak appears at lower temperatures, as can be observed from Figure 7. Such findings suggest that some processes connected to temperature clearly affect the ASE threshold density under cw excitation. ASE inhibition is likely due to the higher amount of energy deposited by the long pulses, which results in more warming than what obtained under short-pulse excitation [19].

9. Optical thermometry

The plasma temperature can be extracted from PL spectra by fitting the high energy tail of the emission peak to a Ae^{-E/k_BT_P} Boltzmann function, where A is an arbitrary multiplication factor, E the photon energy, k_B the Boltzmann constant and T_P the plasma temperature [59].

Inset in Figure 8 illustrates the fitting procedure and shows the plasma temperature as a function of the laser fluence. As one may expect, the plasma temperature increases for increasing laser fluence. The main plasma heating sources under photoexcitation are due to the excess energy of the laser photons with respect to the energy gap and to non-radiative processes from both carrier trapping (at low intensities) and Auger recombination (at high intensities) [19]. Sample excitation with different laser wavelengths, from energies higher than the energy gap to about resonant with it, shows direct effects of excess energy on plasma temperature. Experimental results, reported in Figure 8, show that both contributions are significant, as larger excess energy causes larger warming, but some warming occurs even for quasi-resonant excitation, when Auger is the main source of heating. The role of Auger recombination is also evident at high densities by the rapid increase of the plasma temperature with the excitation intensity.

More significant warming effects can be observed using 300-ns-long excitation pulses and measuring the time-resolved PL evolution at different lattice temperatures T_L and, for each T_L, at different excitation intensities. Here, the thermometry is applied to each of the spectra in the time-resolved spectrogram and therefore yields information about the temporal dynamics of the plasma temperature. At low excitation intensities, the time-resolved PL signal follows instantaneously the laser pulse, while for densities comparable and higher than ASE threshold, a clear temporal reshape of the PL signal is observed. Such process is evident at high laser intensity even when ASE is not noticeable from PL spectra as shown in Figure 9c,d. The temperature dynamics helps providing a physical explanation, showing that the temporal reshape is accompanied by an increase in the plasma temperature.

Both the temporal reshaping and the observation of ASE only at low lattice temperature under long-pulse excitation can be interpreted in terms of decrease of the radiative efficiency with increasing laser intensity. Such process is a consequence of the plasma warming generated during the intense pulse [19]. An interesting point is that every time ASE is observed, it stops when the plasma temperature overcomes about 370 K, independently from the initial temperature of the lattice (Figures 9e,f).

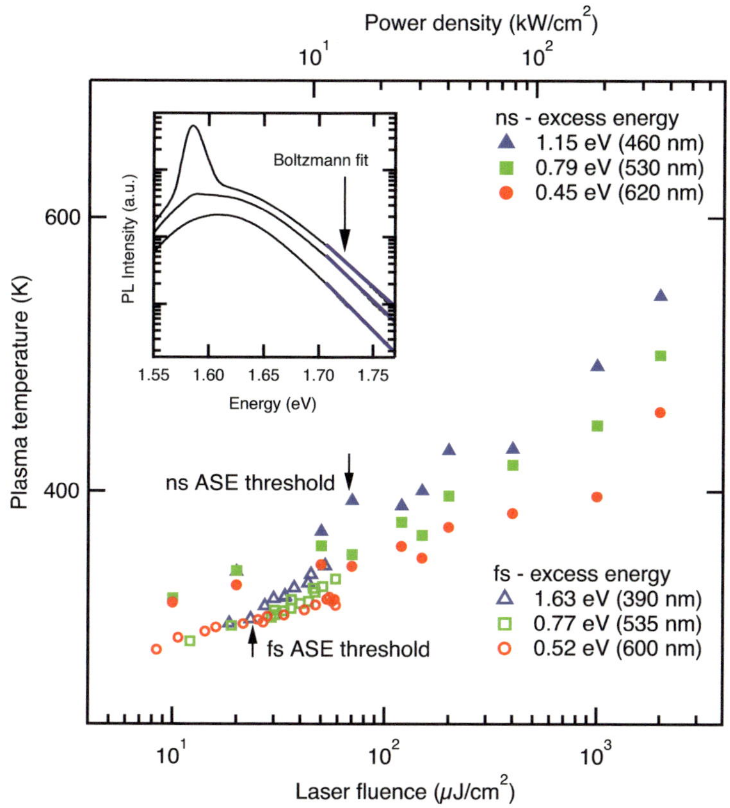

Figure 8. Plasma temperatures of a $MAPbI_3$ film as a function of the laser fluence. The empty markers represent the measurements obtained exciting the sample at a lattice temperature of 300 K with 130-fs-long laser pulses at repetition rate of 1 kHz and different photon energies; the filled markers represent the opposite regime, with 5-ns-long laser pulses at a repetition rate of 10 Hz and different photon energies. The arrows represent the ASE thresholds for the two excitation regimes. The corresponding average excitation power density during the ns laser pulses is also reported on the top axis as a reference; such axis does not apply to measurements under fs excitation. The excess energy was calculated as the difference between the excitation photon energy and the energy gap (1.55 eV); the wavelength indicated in brackets is the actual central wavelength of the excitation laser. Inset: plasma temperatures are extracted fitting the high energy tail of photoluminescence spectra with a $Ae^{-E/k_B T_P}$ exponential function, representing a Boltzmann thermal distribution, E being the photon energy, T_P the temperature, k_B the Boltzmann constant and A an arbitrary multiplication factor. The blue lines represent the fitting functions.

As a confirmation, one can measure the ASE threshold densities carried out under short-pulse excitation as a function of the lattice temperature, from 300 K to 160 K. Experimental results, reported in Figure 10, show that ASE threshold fluence follows a quadratic dependence in the lattice temperature in both $MAPbI_3$ and $MAPbBr_3$ perovskites [19]. Since absorption does not change significantly from room temperature to 160 K, such dependence cannot be ascribed to a variation of the injected electron-hole pairs density with temperature, but to the influence of temperature on the radiative recombination rate. Indeed, it is known that the radiative rate of an electron-hole plasma is inversely proportional to the plasma temperature [52].

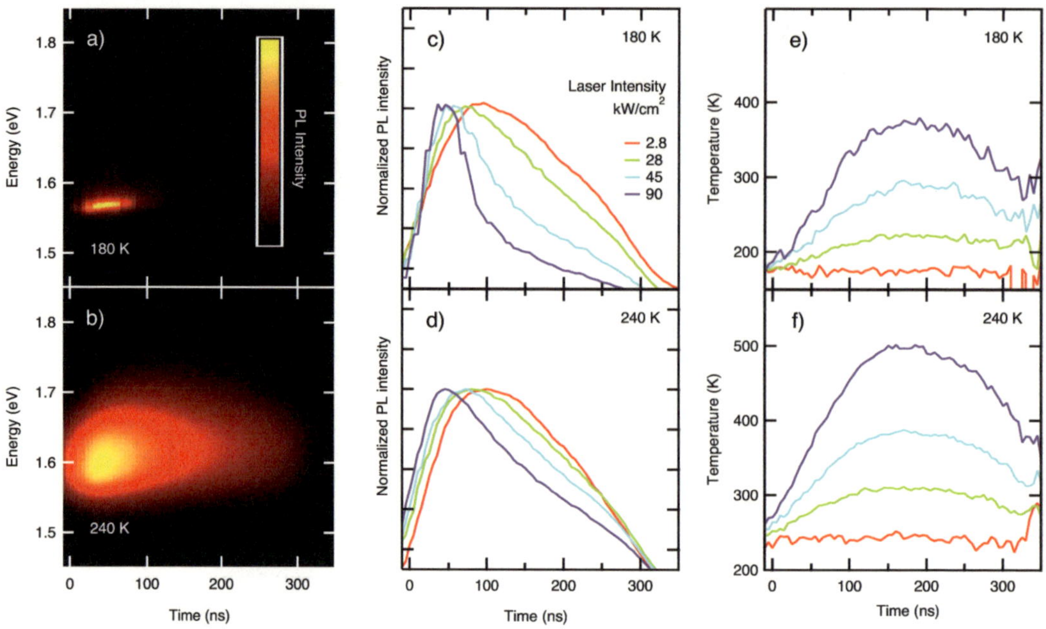

Figure 9. Photoluminescence was excited at different lattice temperatures by 300-ns-long laser pulses with a repetition rate of 6 Hz and 2.35 eV photon energy. (a,b) Time-resolved PL spectrograms of a $MAPbI_3$ film, acquired with a gated intensified camera at different lattice temperatures. (c,d) Corresponding PL temporal profiles, which demonstrate a narrowing of the emission profile with respect to the exciting laser pulse duration; we attributed the effect to stimulated emission and Auger recombination. (e,f) Plasma temperature extracted from $MAPbI_3$ photoluminescence spectrograms in panels a,b, at different lattice temperatures, with the fitting procedure reported in the inset of Figure 8. Significant warming occurs during the excitation pulse and limits the duration of ASE.

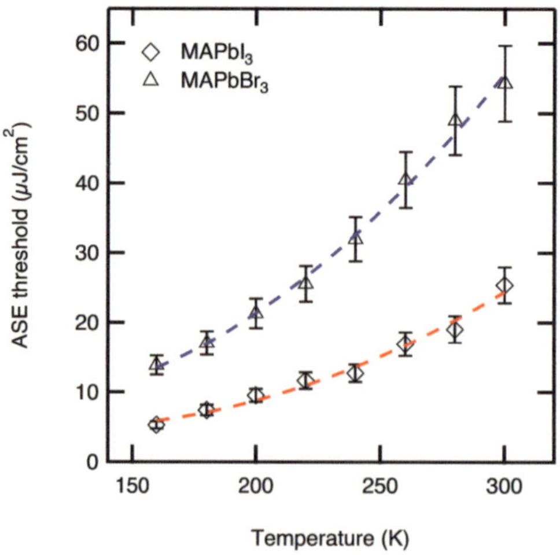

Figure 10. ASE threshold fluence as a function of the lattice temperature in $MAPbI_3$ and $MAPbBr_3$ perovskites, under short-pulse excitation. The dashed lines are fits of measured values to a quadratic dependence of the ASE threshold on temperature.

10. Comparison with nitride semiconductors

Table 1 compares the values of parameters relevant to optical amplification obtained for perovskites to those known for nitride semiconductors at room temperature. Such comparison is very instructive in perspective of the realization of a perovskite-based cw laser. In fact, the advancement of nitride-based lasers has overtaken the problems concerning warming with success and may serve as a useful guide to improve perovskite device performances.

	n_{thr} (cm^{-3})	τ_{thr}(ns)	γ_A(cm^6 s^{-1})	R_{th}(K kW^{-1} cm^2)
Perovskites [18,19,39,60]	10^{16}-10^{18}	1-3	4×10^{-28}	2-4
Nitrides [61]	10^{18}-10^{19}	2-5	5×10^{-30}	0.5

Table 1. Comparison between perovskites and nitrides ASE threshold density n_{thr}, lifetime at ASE threshold τ_{thr}, Auger recombination constant γ_A and thermal resistance R_{th}, respectively. Values are relative to room temperature.

It can be observed that τ_{thr}, that is, the carrier lifetime at ASE threshold density, is slightly longer in nitrides than in perovskites, thanks to the significantly lower value of the Auger recombination coefficient γ_A, which has a value almost two orders of magnitude smaller than in perovskites. Auger recombination is the major responsible for the efficiency reduction in nitride-based LEDs at high power and the value of γ_A is higher in the presence of disorder due to defects [56,57]. It is therefore important to reduce the Auger recombination coefficient in perovskites, for example, acting in order to decrease both the trap density and the ASE threshold, for example, improving the crystallinity and the surface morphology quality. Another crucial parameter is the thermal resistance γ_A, which is about four to eight times higher in perovskites than in nitrides. Furthermore, the substrate on which perovskites are deposited could seriously contribute to enhance the value of the thermal resistance, thus increasing the ASE threshold, particularly under long-pulse excitation. Such arguments make the outlook for perovskite-based cw lasers very promising, provided that a great effort should be made in terms of reduction of ASE threshold, Auger recombination and thermal resistance.

11. Conclusions

This chapter has reviewed the most recent progresses concerning the investigation of the photophysical properties in organolead trihalide perovskites. The absorption spectrum in $MAPbI_3$ perovskites shows the influence of excitonic states at the band-edge, even if there is evidence that the majority of photogenerated carriers results in free electrons and holes, in a wide excitation range at room temperature. Based on a large body of optical absorption data, the exciton binding energy is estimated to be 23±4 meV by applying the Elliot's theory of Wannier excitons to published absorption spectra; the estimated value stays constant from cryogenic to room temperature. This value is in contrast with what evidenced by magnetometry measurements, which provide temperature-dependent values of the exciton binding energy. Additional investigations should be addressed to understand the origin of the discrepancy between what observed by optical spectroscopy and magnetometry.

Trap density in perovskites is significant, like in most solution-processed semiconductors, and its effects emerge particularly under continuous excitation. Although trapping has a negative impact in optoelectronic devices, its capture cross-section happens to be low, as attested by the typical PL lifetime in perovskites exceeding several nanoseconds. The resulting values of carrier mobility can exceed 10 $cm^2 V^{-1} s^{-1}$ and justify the efficient charge collection reported for perovskite-based PV devices.

Emission properties can also play an essential role in the development of new devices based on hybrid perovskites. Many reports provide evidence of amplified spontaneous emission from perovskite thin films under short-pulse excitation, with ASE threshold densities comparable to those of the best state-of-art organic crystals. Under cw pumping, warming suppresses light amplification, in the same way it happens in nitride semiconductors. Similarities between these latter and hybrid perovskites could suggest a way to improve light emission performances in perovskites, in perspective of the realization of a real perovskite-laser.

Author details

Michele Cadelano, Michele Saba, Nicola Sestu, Valerio Sarritzu, Daniela Marongiu, Feipeng Chen, Roberto Piras, Francesco Quochi, Andrea Mura and Giovanni Bongiovanni*

*Address all correspondence to: giovanni.bongiovanni@dsf.unica.it

Dipartimento di Fisica, Università degli Studi di Cagliari, Monserrato, Italy

References

[1] Kojima A, Teshima K, Shirai Y, Miyasaka T. Organometal halide perovskites as visible-light sensitizers for photovoltaic cells. J Am ChemSoc 2009;131(17):6050.

[2] Im J, Lee C, Lee J, Park S, Park N. 6.5% efficient perovskite quantum-dot-sensitized solar cell. Nanoscale 2011;3(10):4088.

[3] Kim HS, Lee CR, Im JH, Lee KB, Moehl T, Marchioro A,Moon SJ, Humphry-Baker R, Yum JH, Moser JE, Grätzel M, Park NG.Lead iodide perovskite sensitized all-solid-state submicron thin film mesoscopic solar cell with efficiency exceeding 9%. Sci Rep 2012;2:591.

[4] Burschka J, Pellet N, Moon S, Humphry-Baker R, Gao P, Nazeeruddin MK, Grätzel M. Sequential deposition as a route to high-performance perovskite-sensitized solar cells. Nature 2013;499(7458):316.

[5] Stranks SD, Eperon GE, Grancini G, Menelaou C, Alcocer MJP, Leijtens T, Herz LM, Petrozza A, Snaith HJ. Electron-hole diffusion lengths exceeding 1 micrometer in an organometal trihalide perovskite absorber. Science 2013;342(6156):341.

[6] Xing G, Mathews N, Sun S, Lim SS, Lam YM, Grätzel M, Mhaisalkar S, Sum TC. Long-range balanced electron- and hole-transport lengths in organic-inorganic CH3NH3PbI3. Science 2013;342(6156):344.

[7] Snaith HJ. Perovskites: the emergence of a new era for low-cost, high-efficiency solar cells. J PhysChem Lett 2013;4(21):3623.

[8] Malinkiewicz O, Yella A, Lee YH, Espallargas GM, Graetzel M, Nazeeruddin MK, Bolink HJ. Perovskite solar cells employing organic charge-transport layers. Nat Photonics 2013;8(2):128.

[9] Park N. Organometal perovskite light absorbers toward a 20% efficiency low-cost solid-state mesoscopic solar cell. J PhysChem Lett 2013;4(15):2423.

[10] Noh JH, Im SH, Heo JH, Mandal TN, Seok Il S. Chemical management for colorful, efficient, and stable inorganic–organic hybrid nanostructured solar cells. Nano Lett 2013;13(4):1764.

[11] Docampo P, Ball JM, Darwich M, Eperon GE, Snaith HJ. Efficient organometal trihalide perovskite planar-heterojunction solar cells on flexible polymer substrates. Nat Commun2013;4:2761.

[12] Gao P, Grätzel M, Nazeeruddin MK. Organohalide lead perovskites for photovoltaic applications. Energy Environ Sci 2014;7(8):2448.

[13] Ogomi Y, Morita A, Tsukamoto S, Saitho T, Fujikawa N, Shen Q, Toyoda T, Yoshino K, Pandey SS, Ma T, Hayase S. CH3NH3SnxPb(1-x)I3perovskite solar cells covering up to 1060 nm. J PhysChem Lett 2014;5(6):1004.

[14] Eperon GE, Stranks SD, Menelaou C, Johnston MB, Herz LM, Snaith HJ. Formamidinium lead trihalide: a broadly tunable perovskite for efficient planar heterojunction solar cells. Energy Environ Sci 2014;7(3):982.

[15] Noel NK, Stranks SD, Abate A, Wehrenfennig C, Guarnera S, Haghighirad A, Sadhanala A, Eperon GE, Pathak SK, Johnston MB, Petrozza A, Herz LM, Snaith HJ. Lead-free organic-inorganic tin halide perovskites for photovoltaic applications. Energy Environ Sci 2014;7(9):3061.

[16] Lee MM, Teuscher J, Miyasaka T, Murakami TN, Snaith HJ. Efficient hybrid solar cells based on meso-superstructuredorganometal halide perovskites. Science 2012;338(6107):643.

[17] Deschler F, Price M, Pathak S, Klintberg LE, Jarausch D, Higler R, Hu◎ttner S, Leijtens T, Stranks SD, Snaith HJ, Atatu◎re M, Phillips RT, Friend RH. High photolumi-

nescence efficiency and optically pumped lasing in solution-processed mixed halide perovskite semiconductors. J PhysChem Lett 2014;5(8):1421.

[18] Xing G, Mathews N, Lim SS, Yantara N, Liu X, Sabba D, Grätzel M, Mhaisalkar S, Sum TC. Low-temperature solution-processed wide wavelength tunableperovskites for lasing. Nat Mater. 2014;13(5):476.

[19] Cadelano M, Sarritzu V, Sestu N, Marongiu D, Chen F, Piras R, Corpino R, Carbonaro CM, Quochi F, Saba M, Mura A, Bongiovanni G. Can trihalide lead perovskites support continuous wave lasing? Adv Optic Mater 2015. DOI: 10.1002/adom.201500229.

[20] Wehrenfennig C, Eperon GE, Johnston MB, Snaith HJ, Herz LM. High charge carrier mobilities and lifetimes in organolead trihalide perovskites. Adv Mater 2013;26(10):1584.

[21] Marchioro A, Teuscher J, Friedrich D, Kunst M, van de Krol R, Moehl T, Grätzel M, Moser JE. Unravelling the mechanism of photoinduced charge transfer processes in lead iodide perovskite solar cells. Nat Photonics 2014;8(3):250.

[22] D'Innocenzo V, Grancini G, Alcocer MJP, Kandada ARS, Stranks SD, Lee MM, Lanzani G, Snaith HJ, Petrozza A. Excitons versus free charges in organo-lead tri-halide perovskites. Nat Commun2014;5:3586.

[23] Roiati V, Colella S, Lerario G, De Marco L, Rizzo A, Listorti A, Gigli G. Investigating charge dynamics in halide perovskite-sensitized mesostructured solar cells. Energy Environ Sci 2014;7(6):1889.

[24] De Wolf S, Holovsky J, Moon S, Löper P, Niesen B, Ledinsky M, Haug FJ, Yum JH, Ballif C. Organometallic halide perovskites: sharp optical absorption edge and its relation to photovoltaic performance. J PhysChem Lett 2014;5(6):1035.

[25] Zhao Y, Nardes AM, Zhu K. Solid-State Mesostructured perovskite CH3NH3PbI3solar cells: charge transport, recombination, and diffusion length. J PhysChem Lett 2014;5(3):490.

[26] Giorgi G, Fujisawa J, Segawa H, Yamashita K. Small photocarriereffective masses featuring ambipolar transport in methylammonium lead iodide perovskite: adensity functional analysis. J PhysChem Lett 2013;4(24):4213.

[27] Filippetti A, Mattoni A. Hybrid perovskites for photovoltaics: insights from first principles. Phys Rev B 2014;89(12).

[28] Mosconi E, Amat A, Nazeeruddin MK, Grätzel M, De Angelis F. First-principles modeling of mixed halide organometal perovskites for photovoltaic applications. J PhysChem C 2013;117(27):13902.

[29] Kim J, Ho PKH, Murphy CE, Friend RH. Phase separation in polyfluorene-based conjugated polymer blends: lateral and vertical analysis of blend spin-cast thin films. Macromolecules 2004;37(8):2861.

[30] Saba M, Cadelano M, Marongiu D, Chen F, Sarritzu V, Sestu N, Figus C, Aresti M, Piras R, Lehmann AG, Cannas C, Musinu A, Quochi F, Mura A, Bongiovanni G. Correlated electron-hole plasma in organometal perovskites. NatCommun2014;5:5049.

[31] Hirasawa M, Ishihara T, Goto T, Uchida K, Miura N. Magnetoabsorption of the lowest exciton in perovskite-type compound (CH3NH3)PbI3. Physica B 1994;201:427.

[32] Tanaka K, Takahashi T, Ban T, Kondo T, Uchida K, Miura N. Comparative study on the excitons in lead-halide-based perovskite- type crystals CH3NH3PbBr3 CH3NH3PbI3. Solid State Commun 2003;127(9-10):619.

[33] Even J, Pedesseau L, Katan C. Analysis of multivalley and multibandgapabsorption and enhancement of free carriers related to exciton screening in hybrid perovskites. J PhysChem C 2014;118(22):11566.

[34] Yamada Y, Nakamura T, Endo M, Wakamiya A, Kanemitsu Y. Photoelectronicresponses in solution-processed perovskite CH3NH3PbI3solar cells studied by photoluminescence and photoabsorptionspectroscopy. IEEE J Photovoltaics 2015;5(1):401.

[35] Lin Q, Armin A, Nagiri RCR, Burn PL, Meredith P. Electro-optics of perovskite solar cells. Nat Photonics 2014;9(2):106.

[36] Sun S, Salim T, Mathews N, Duchamp M, Boothroyd C, Xing G, Sum TC, Lam YM. The origin of high efficiency in low-temperature solution-processable bilayer organometal halide hybrid solar cells. Energy Environ Sci 2014;7(1):399.

[37] Sestu N, Cadelano M, Sarritzu V, Chen F, Marongiu D, Piras R, Mainas M, Quochi F, Saba M, Mura A, Bongiovanni G. Absorption F-Sum Rule for the Exciton Binding Energy in Methylammonium Lead Halide Perovskites. J PhysChem Lett2015;6:4566.

[38] Eames C, Frost JM, Barnes PRF, O'Regan BC, Walsh A, Islam MS. Ionic transport in hybrid lead iodide perovskite solar cells. Nat Commun 2015;6(7497).

[39] Zhu H, Fu Y, Meng F, Wu X, Gong Z, Ding Q, Gustafsson MV, Trinh MT, Jin S, Zhy XY. Lead halide perovskite nanowire lasers with low lasing thresholds and high quality factors. Nat Mater 2015;14(6):636.

[40] Tian J, Gao R, Zhang Q, Zhang S, Li Y, Lan J, Qu X, Cao G. Enhanced performance of CdS/CdSe quantum dot cosensitizedsolar cells via homogeneous distribution of quantum dots in TiO2film. J PhysChem C 2012;116(35):18655.

[41] Dhanker R, Brigeman AN, Larsen AV, Stewart RJ, Asbury JB, Giebink NC. Random lasing in organo-lead halide perovskite microcrystal networks. ApplPhys Lett 2014;105(15):151112.

[42] Elliot RJ. Intensity of Optical Absorption by Excitons. Phys Rev 1957;108(6):1384.

[43] Sturge MD. Optical absorption of gallium arsenide between 0.6 and 2.75 eV. Phys Rev 1963;129(2835).

[44] Sell DD, Lawaetz P. New analysis of direct exciton transitions: application to GaP. Phys Rev Lett 1971;26(6):331.

[45] Sutherland BR, Hoogland S, Adachi MM, Wong CTO, Sargent EH. Conformal organohalideperovskites enable lasing on spherical resonators. ACS Nano 2014;8(10): 10947.

[46] Shah J. Ultrafast Spectroscopy of Semiconductors and Semiconductor Nanostructures. Springer; 1999.

[47] Lanzani G. The Photophysics behind Photovoltaics and Photonics. Wiley; 2012.

[48] Miyata A, Mitioglu A, Plochocka P, Portugall O, Wang JT, Stranks SD, Snaith HJ, Nicholas RJ. Direct measurement of the exciton binding energy and effective masses for charge carriers in organic-inorganic tri-halide perovskites. Nat Phys 2015;11(7): 582.

[49] Kitazawa N, Watanabe Y, Nakamura Y. Optical properties of CH3NH3PbX3 (X= halogen) and their mixed-halide crystals. J Mater Sci 2002;37(17):3585.

[50] Buin A, Pietsch P, Xu J, Voznyy O, Ip AH, Comin R, Sargent EH. Materials processing routes to trap-free halide perovskites. Nano Lett 2014;14(11):6281.

[51] Yu H, Wang F, Xie F, Li W, Chen J, Zhao N. The role of chlorine in the formation process of «CH3NH3PbI3-xClx» perovskite. AdvFunct Mater 2014;24:7102.

[52] Bongiovanni G, Staehli JL. Density dependence of electron-hole plasma lifetime in semiconductor quantum wells. Phys Rev B 1992;46(15):9861.

[53] Manser JS, Kamat PV. Band filling with free charge carriers in organometal halide perovskites. Nat Photonics 2014;8(9):737.

[54] Kubo R. Statistical-mechanical theory of irreversible processes. I. J PhysSocJpn. 1957;12(6):570.

[55] Martin PC, Schwinger J. Theory of Many-Particle Systems. I. Phys Rev 1959;115(6): 1342.

[56] Smart G. New clues to LEDs' efficiency droop. Phys Today 2013;66(7):12.

[57] Kioupakis E, Rinke P, Delaney KT, Van de Walle CG. Indirect auger recombination as a cause of efficiency droop in nitride light-emitting diodes. ApplPhys Lett 2011;98(16):161107.

[58] Xing J, Liu XF, Zhang Q, Ha ST, Yuan YW, Shen C, Sum TC, Xiong Q. Vaporphase synthesis of organometal halide perovskite nanowires for tunableroom-temperature nanolasers. Nano Lett 2015;15(7):4571.

[59] Leo K, Ruhle W, Ploog KH. Hot-carrier energy-loss rates in GaAs/AlxGa1-xAs quantum wells. Phys Rev B 1988;38(3):1947.

[60] Pisoni A, Jaćimović J, Barišić OS, Spina M, Gaál R, Forró L, Horvàth E. Ultra-low thermal conductivity in organic-inorganic hybrid perovskite CH3NH3PbI3. J Phys-Chem Lett 2014;5(14):2488.

[61] Nakamura S, Senoh M, Nagahama S, Iwasa N, Yamada T, Matsushita T, Kiyoku H, Sugimoto Y, Kozaki T, Umemoto H, Sano M, Chocho K. Continuous-wave operation of InGaN/GaN/AlGaN-based laser diodes grown on GaN substrates. ApplPhys Lett 1998;72(16):2014.

Section 3

Perovskite Materials: Applications

Chapter 13

Optical Absorption, Charge Separation and Recombination Dynamics in Pb and Sn/Pb Cocktail Perovskite Solar Cells and Their Relationships to the Photovoltaic Properties

Shen Qing, Ogomi Yuhei, Toyoda Taro, Yoshino Kenji and Hayase Shuzi

Additional information is available at the end of the chapter

http://dx.doi.org/10.5772/62101

Abstract

Due to the unique characteristics such as simple low-temperature preparation method and high efficiency with a record of over 20%, organometal trihalide perovskite ($CH_3NH_3PbI_3$)-based solid-state hybrid solar cells have attracted an increasing interest since 2012 when it was reported. During the last several years, some of the fundamental photophysical properties of perovskite related to the high photovoltaic performance have been investigated. Optical absorption, charge separation and recombination are very important factors influencing the perovskite solar cell performance. In this chapter, our recent results of optical absorption, charge separation (electron and hole injection) and charge recombination dynamics at each interface in perovskite solar cells, and their relationships to photovoltaic properties will be introduced. Our results suggest that charge recombination is a key factor in improving the performance of the perovskite solar cells.

Keywords: Urbach energy, charge separation, charge recombination, optical absorption, Sn/Pb cocktail perovskite, perovskite

1. Introduction

Organolead halide perovskites in the format of AMX3 (A=organic molecule, e.g., CH_3NH_3(MA), B=Pb, X=Cl, Br, and I) can be easily crystallized from solution at relatively low temperature (i.e., ≤100 °C), which makes it possible to use them as light absorbing materials in different kinds of solar cells. Following the recently reported certified high power conversion efficiency (PCE) of over 20%, [1-12] the interest in organolead halide perovskite-based organic-inorganic hybrid solid-state solar cells has increased dramatically over the past several years. The higher PCEs of organolead halide perovskites (especially $MAPbI_3$) result from their unique properties that are key for achieving high photovoltaic performance, which are (1) a direct band gap and a high optical absorption coefficient; [13, 14] (2) large dielectric coefficient leading to smaller exciton binding energy; [15] (3) long photoexcited carrier lifetimes (>100 ns) and long diffusion lengths (100 – 1000 nm or even longer); [16, 17] (4) no deep state defects and very small Urbach energy. [18]

It is reasonable to expect that further improvements in photovoltaic performance can be achieved by increasing the light harvesting up to NIR wavelengths of 1000 nm, since $MAPbI_3$ perovskite only absorbs light at wavelengths below 800 nm constrained by its optical band gap of 1.5 eV. In addition, practically, Pb-free organometal halide perovskites are preferred due to the potential toxicology issue of Pb. Replacing Pb with Sn or mixing Pb and Sn in organometal halide perovskites can result in increased light harvesting in the NIR region up to 1000 nm [19, 20] and, at the same time, reduce the toxicity issue related to Pb. Several research groups have reported, very recently, Sn-based or Pb/Sn cocktail $MASn_xPb_{1-x}I_3$ ($0 \leqq x \leqq 1$) perovskite solar cells. [21-24] However, it is found that the PCE of Sn/Pb cocktail $MASn_xPb_{1-x}I_3$ perovskite solar cells is far inferior to that of $MAPbI_3$ perovskite solar cells.

In order to improve the photovoltaic performance of Pb and Sn/Pb cocktail perovskite solar cells, it is critical to gain a thorough understanding of the optical absorption properties, the photoexcited carrier lifetimes, as well as the charge separation and recombination dynamics at each interface. In this chapter, we will focus on recent studies of the optical absorption, photoexcited carrier lifetime, the charge separation and charge recombination dynamics at each interface in perovskite solar cells, including Pb-based and Sn/Pb cocktail perovskite solar cells. The relationships of each of these physical properties to the photovoltaic performance of the solar cells will be discussed and the methodologies for improving the photovoltaic performance of perovskite solar cells will be proposed.

2. Experimental

2.1. Sample preparation

Samples of Pb-based perovskite hybrid solar cells were prepared by the following method. [25] F-doped SnO_2 layered glass (FTO glass, Nippon Sheet Glass Co. Ltd) was patterned using Zn powder and 6 N HCl aqueous solution. Titanium diisopropoxide bis(acetylacetonate) solution in ethanol was sprayed onto this patterned FTO glass at 300 °C to prepare compact TiO_2 layers. Porous TiO_2 layers were fabricated by spin-coating TiO_2 pastes of different nanoparticle sizes (18 nm: PST-18NR or 30 nm: PST-30NRD, JGC Catalysts and Chemicals Ltd.) in ethanol (TiO_2 paste : ethanol = 1:2.5 weight ratio for PST-18NR or TiO_2 paste : ethanol = 2:7 weight ratio for PST-30NRD), followed by heating the substrates at 550 °C for 30 min. For some TA measurements, glass, instead of FTO was used as the substrate and a porous Y_2O_3 layer was fabricated on the glass substrates. Next, CH_3NH_3I and $PbCl_2$ were mixed with a 3:1 molar ratio for preparing a 40 % solution of perovskite in N,N-dimethylformamide and the mixture was spin-coated on the TiO_2 and Y_2O_3 porous substrates. After heating at 100 °C for 45 minutes, the substrates were spin-coated with a mixture of 55 mM of tert-butylpyridine, 9 mM of lithium bis(trifluoromethylsyfonyl)imide salt, and 68 mM of *spiro*-OMeTAD. Finally, Ag and Au electrodes were fabricated by vacuum deposition for the photovoltaic measurements. The photovoltaic performance was evaluated using an AM1.5G 100 mW/cm^2 irradiance solar simulator (CEP-2000, Bunkoukeiki Inc) with a 0.4 cm x 0.4 cm mask.

The Sn/Pb cocktail perovskite samples were prepared using the following method. [23] PbI_2 (Purity: 99.999 %), SnI_2, and regioregular poly(3-hexylthiophene-2,5-diyl) (P3HT) were purchased from Sigma Aldrich and used as received. F-doped SnO_2 coated glass ((FTO glass), Nippon Sheet Glass Co. Ltd) was patterned using Zn and a 6 N HCl aqueous solution. A compact TiO_2 layer was produced by spraying titanium diisopropoxide bis(acetylacetonate) solution in ethanol onto this patterned glass at 300 °C. The substrate was then dipped in a 40 mM solution of $TiCl_4$ in water for 30 min. Then the substrate was baked at 500 °C for 20 min. Next, a porous TiO_2 layer was produced by spin-coating with a TiO_2 paste (PST-18NR, JGC Catalysts and Chemicals Ltd.) in ethanol (TiO_2 paste : ethanol = 2:7 weight ratio). The substrate was then heated at 550 °C for 30 min. For certain TA measurements, glass substrates other than FTO ones, with a porous Al_2O_3 layer deposited on them, were used. The substrates were then spin-coated with a mixture of SnI_2, PbI_2, and CH_3NH_3I (0.5 : 0.5 : 1.0 (molar ratio)) in dimethylformamide (DMF) (40 wt %), and they were baked at 70 °C for 30 min. Next, P3HT in chlorobenzene solution (15 mg/ml) was spin-coated on the prepared perovskite layer and the substrate was put in nitrogen at ambient temperature for 1 h. For conducting TA measurements, samples of Sn/Pb cocktail perovskite deposited on both Al_2O_3 and TiO_2 substrates with and without P3HT as a hole transport material (HTM) were employed. Ag and Au electrodes were finally deposited by vacuum deposition for characterizing the photovoltaic performance, which was evaluated using an AM 1.5G 100 mW/cm^2 irradiance solar simulator (CEP-2000SRR, Bunkoukeiki Inc) with a 0.4 cm x 0.4 cm mask.

2.2. Characterization methods

2.2.1. Optical absorption measurements

A gas-microphone photoacoustic (PA) technique [26] was used to study the optical absorption properties of the samples. The light source was a 300 W xenon arc lamp. By passing the light through a monochromator, a monochromatic light beam was obtained. This beam was modulated with a mechanical chopper and focused onto the surface of the sample placed in a sealed PA cell filled with N_2 gas. The measurements of PA spectrum were carried out in the wavelength range of 500–1200 nm with a modulation frequency of 33 Hz at room temperature. The PA signal was measured by first passing the microphone output through a preamplifier and then to a lock-in amplifier. The PA spectra were normalized with the PA spectrum obtained from a carbon black sheet.

2.2.2. Transient Absorption (TA) Measurements

Two different TA setups were used: the charge separation (electron injection and hole injection) dynamics was characterized by a femtosecond TA technique (fs-TA) [27, 28] and charge recombination dynamics was characterized by a nanosecond TA technique (ns-TA) [28-31]. In the fs-TA setup for characterizing the charge separation, [27, 28, 31] the laser source used was a titanium/sapphire laser (CPA-2010, Clark-MXR Inc.) with a wavelength of 775 nm, a repetition rate of 1 kHz, and a pulse width of 150 fs. The light was separated into two parts. One part was used as a probe pulse. The other part was used to pump an optical parametric amplifier (OPA) (a TOAPS from Quantronix) to generate light pulses with a wavelength tunable from 290 nm to 3 μm. This was used as a pump light to excite the sample. In this study, a pump light wavelength of 470 nm and a probe beam wavelength of 775 nm were used. Deschler and co-workers [32] reported very recently that the TA response of ground state bleaching (GSB) for $CH_3NH_3PbClI_2$ on glass peaked at 1.65 eV and was spectrally unchanged up to times beyond 200 ns. Therefore, the probe light wavelength of 775 nm (i.e., the photon energy of 1.64 eV) used in this study is appropriate to monitor the GSB of $CH_3NH_3PbClI_2$ on Y_2O_3 and TiO_2 substrates with or without *spiro*-OMeTAD, and thus electron transfer and hole transfer processes at each interface can be investigated systematically. On the other hand, in the case of Sn/Pb cocktail perovskite, the probe light of 775 nm was used to detect the photo-excited carrier absorption.

In the ns-TA setup for characterizing the charge recombination [28-31], the pump light source was an OPO (Surelite II – 10FP) output excited by a Nd:YAG nanosecond pulse laser (Panther, Continuum, Electro-Optics Inc.), with a pulse width of 5 ns and a repetition rate of 0.5 Hz. A pulse light with a wavelength of 470 nm was used as the pump light to excite the sample. The probe light was produced from a fiber coupled CW semiconductor. Three different probe wavelengths, of 785 nm, 658 nm and 1310 nm, were used. Specifically, the probe beam of 785 nm was employed to measure the TA responses of GSB for $MAPbClI_2$ on Y_2O_3 substrates, i.e., the recombination of electrons and holes in $MAPbClI_2$.[32] The probe beam of 658 nm was used to measure the trapped electrons in TiO_2 [33] based on the research of Yoshihara and co-workers, which was used to investigate charge recombination between the electrons in TiO_2 and the holes in the perovskite. The probe beam of 1310 nm was used to monitor the holes in *spiro*-OMeTAD [34] and thus measure the charge recombination between holes in *spiro*-OMeTAD and electrons in TiO_2 and/or in perovskite. For all measurements, the pump and probe lights were irradiated from the glass side and the TA measurements were carried out in a nitrogen atmosphere.

3. Optical Absorption Study: Bandgap and Urbach Energy of Pb and Sn/Pb cocktail perovskite

Figure 1 shows the optical absorption spectra of the Pb ($MAPbI_2$) and Sn/Pb cocktail ($MASn_{0.5}Pb_{0.5}I_3$) perovskite samples on a porous TiO_2 substrate measured using the PA technique (we refer to this as the PA spectrum in the following) at room temperature. From the position of the shoulder in each PA spectrum, [35] the bandgap energies of $MAPbI_2$ and $MASn_{0.5}Pb_{0.5}I_3$ are determined to be 1.52 and 1.21 eV, respectively, which are almost the same as those given in our earlier reports. [23, 25] We find that, below the shoulder, the trend of the absorption coefficient is exponential. The slope of this exponential tail, known as the Urbach

http://dx.doi.org/10.5772/62101

tail, corresponds to the absorption tail states, and is usually quantitatively expressed by the Urbach Energy E_U. [18, 31, 36] Investigation of these exponential tails can offer important information on the band structure, disorder, defects, impurities, and electron-phonon interactions in semiconductor materials. The numerical relationship between the optical absorption coefficient α in the exponential tail and the photon energy (hv) is shown in eq. (1), through which the values of E_U of the samples can be calculated. [37, 38]

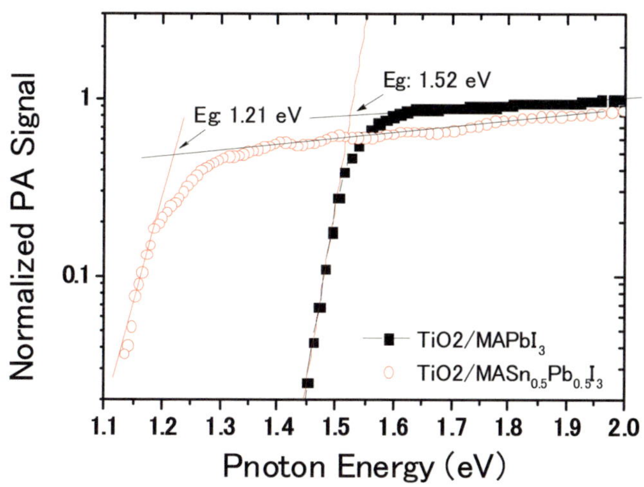

Figure 1. Optical absorption spectra of MAPbI3 and Sn/Pb cocktail $MASn_{0.5}Pb_{0.5}I_3$ perovskite measured using a photoacoustic (PA) technique. The bandgap energies were determined to be 1.52 eV and 1.21 eV, respectively. The Urbach energies were determined to be 22 meV and 34 meV, respectively.

$$\alpha = \alpha_0 \exp\left(\frac{hv - hv_0}{E_u}\right) \tag{1}$$

where h is Planck's constant, and α_0, v_0, E_u are fitting parameters. The values of E_u for Pb and Sn/Pb cocktail perovskites were determined to be 22 meV and 34 meV, respectively. The value of E_u for Sn/Pb cocktail perovskite is larger than that for the Pb perovskite. It can be assumed that the value of E_u is a reflection of the disorder and/or defects in the semiconductor crystal. [18, 36-38] Thus, the width of the exponential tail increases with increasing density of defects. Therefore, the larger E_u determined for Sn/Pb cocktail perovskite (i.e., 34 meV) implies that there could be higher defect states in the sample compared to $MAPbI_3$ (i.e., 22 meV). This will lead to faster recombination of photoexcited carriers in the Sn/Pb cocktail $MASn_{0.5}Pb_{0.5}I_3$ perovskite and lower values of V_{oc} and FF, which we will discuss in detail below together with the TA measurement results.

4. Dynamics of Photoexcited Carrier Recombination and Charge Transfer in Pb-based Perovskite Solar Cells and their Relationships to the Photovoltaic Properties [25, 39, 40]

To perform systematic investigations on the charge separation and recombination dynamics in perovskite solar cells, TA measurements were conducted for $MAPbI_3$ on either Y_2O_3 or TiO_2 substrates, with and without a *spiro*-OMeTAD layer. For perovskite deposited on Y_2O_3 substrate, the lifetime of photoexcited charge carriers in $MAPbI_3$ and the charge separation dynamics at the $MAPbI_3$/*spiro*-OMeTAD interface were studied by the fs-TA technique. The pump light wavelength used to excite $MAPbI_3$ only was 470 nm, and the probe light wavelength used was 775 nm, which is just at the optical absorption edge of $MAPbI_3$. Figure 2 shows the normalized TA response of $MAPbI_3/Y_2O_3$ for a time scale of up to 3 ns with a lower pump light intensity (0.9 μJ/cm²) with and without the *spiro*-OMeTAD layer. We can observe a bleaching signal with a very slow decay in the figure. Because electron injection from the perovskite to Y_2O_3 cannot occur due to the large band gap of Y_2O_3, it is reasonable to attribute the slow decay to the slow recombination process of photoexcited charge carriers in $MAPbI_3$. Two processes were found in the TA decay process and the TA signal obtained can be fitted to the biexponential function shown below very well:

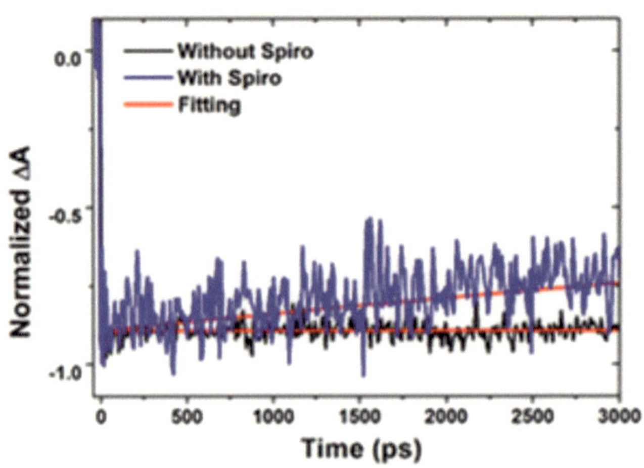

Figure 2. Normalized TA responses of $MAPbClI_2/Y_2O_3$ with and without *spiro*-OMeTAD as a hole transport material (HTM). The red solid lines represent the fitting results with eq. (2). (reference: 25)

$$Y = A_1 e^{-t/t_1} + A_2 e^{-t/t_2} \tag{2}$$

where t_1 and t_2 are two time constants, A_1 and A_2 are the contributions from the corresponding components. The time constants t_1 and t_2 of the two charge recombination processes were calculated to be 34 ps ($A_1/(A_1+A_2)$: 9%) and much larger than 10 ns ($A_2/(A_1+A_2)$: 91%), respectively. The faster decay process could be attributed to nonradiative recombination of electrons and holes through defects or trap states in $MAPbI_3$ or at the $MAPbI_3/Y_2O_3$ interface. The relative contribution of the faster decay is less than 10%, which suggests that the defect or trap state density in $MAPbI_3$ or at the interface is very small, which is consistent with a thermally stimulated current measurement. [39, 40] The slower decay process could be attributed to the recombination of free electrons and holes in $MAPbI_3$. This result suggests that the lifetime of the photoexcited charge carriers was very long, which was confirmed to be as much as the order of microseconds by the ns-TA measurement results shown in Figure 3. The TA decay was fitted with eq. (2), and two decay processes were found with lifetimes of 3.7±0.1 μs (70%) and 60±1 μs (30%), respectively. Combining the fs-TA and ns-TA results for $MAPbI_3/Y_2O_3$, the lifetime of the photoexcited charge carriers in $MAPbI_3$ can be mostly considered to be as long as the order of microseconds.

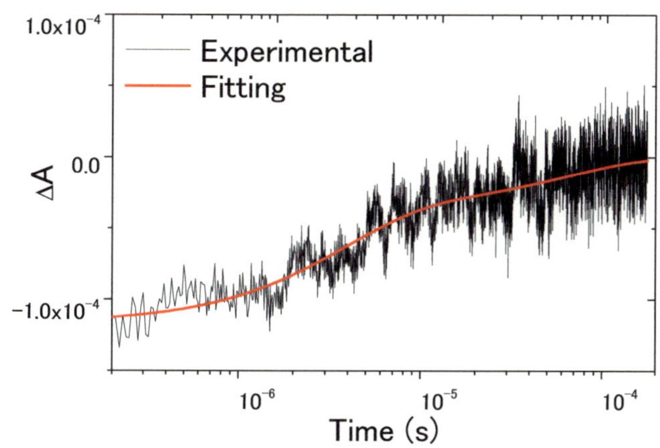

Figure 3. TA responses of $MAPbI_3/Y_2O_3$ without *spiro*-OMeTAD for a time scale of 200 μs measured with a pump light wavelength of 470 nm and a probe light wavelength of 785 nm. The red solid lines represent the fitting results with eq. (2). (reference: 25)

The TA response of $MAPbI_3/Y_2O_3$ with *spiro*-OMeTAD was fitted to eq. (2), and only one exponential decay with a time constant of 16±2 ns was found. Compared with the long lifetime of the TA decay in $Y_2O_3/MAPbI_3$, the fast decay process in $Y_2O_3/MAPbI_3$/*spiro*-OMeTAD can be considered to originate from the photoexcited hole injection from $MAPbI_3$ to *spiro*-OMeTAD. This result indicates that charge separation occurred at the interface between $MAPbI_3$ and *spiro*-OMeTAD. [25]

Figure 4 shows the TA response of $MAPbI_3/Y_2O_3$ with a layer of *spiro*-OMeTAD measured at a probe wavelength of 1310 nm. This result clearly indicates that charge separation at the $MAPbI_3$/*spiro*-OMeTAD interface truly occurred, which is entirely consistent with the fs-TA results shown in Figure 2. The decrease of the TA signal in the sample of $MAPbI_3/Y_2O_3$ with a layer of *spiro*-OMeTAD originated from the recombination process between the electrons in $MAPbI_3$ and the holes in *spiro*-OMeTAD, which can be fitted very well with eq. (2). Only one exponential decay process was observed, with a time constant determined to be 0.37±0.07 μs, which corresponds to the recombination time [25]. Figure 5 shows the photoexcited charge carrier (electrons and holes) dynamics in $MAPbI_3$ deposited on a Y_2O_3 substrate with *spiro*-OMeTAD as a HTM.

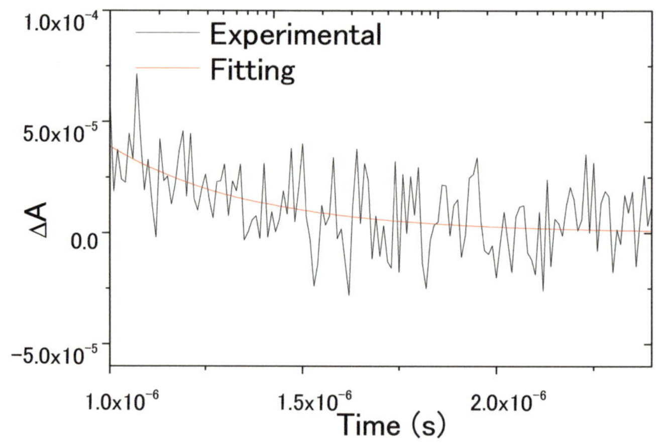

Figure 4. TA responses of $MAPbI_3/Y_2O_3$ with *spiro*-OMeTAD measured with a pump light wavelength of 470 nm and a probe light wavelength of 1310 nm. The red solid lines represent the fitting results with eq. (2). (reference: 25)

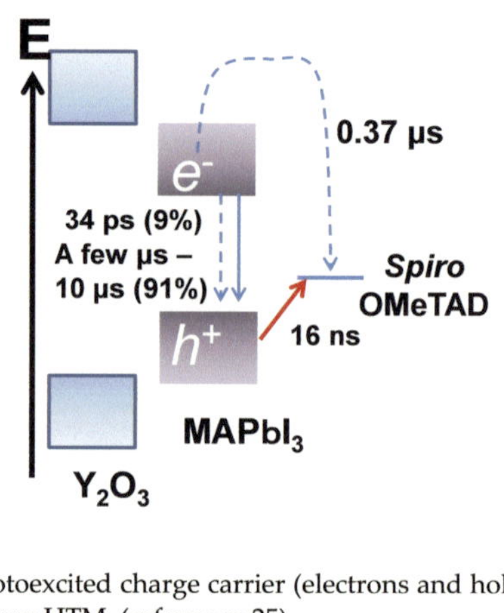

Figure 5. Schematic illustration of photoexcited charge carrier (electrons and holes) dynamics in $MAPbI_3$ deposited on a Y_2O_3 substrate with *spiro*-OMeTAD as a HTM. (reference: 25)

The photoexcited electron injection and the recombination dynamics were measured using the TA techniques for $MAPbI_3$ deposited on TiO_2 substrates. Figure 6(a) shows the normalized TA responses of $MAPbI_3/TiO_2$ for 400 ps measured with different pump light intensities. The decay processes in the normalized TA responses overlapped very well with each other when the pump light intensity was lower. However, when the pump light intensity became larger than 3.75 μJ/cm^2, a faster decay process, resulting from Auger recombination, appeared in the TA response. [25] To determine the electron injection dynamics from $MAPbI_3$ to TiO_2, the TA

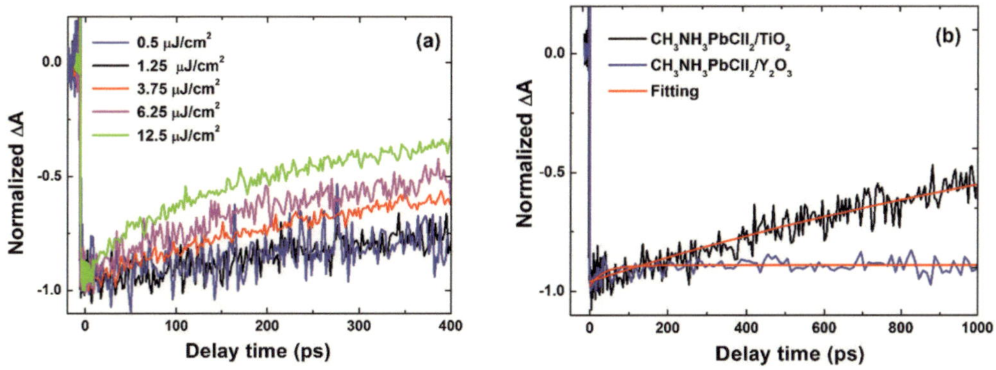

Figure 6. (a) Dependence of the normalized TA responses of $CH_3NH_3PbClI_2/TiO_2$ on pump light intensity and (b) theoretical fitting results of the TA response (pump light intensity: 1.25 μJ/cm²) with a biexponential function (eq. (2)). For comparison, the TA response of $CH_3NH_3PbClI_2/Y_2O_3$ is also shown in (b). The pump light wavelength is 470 nm and the probe light wavelength is 775 nm. (reference: 25)

response of $MAPbI_3/TiO_2$ measured under a lower pump light intensity (0.9 μJ/cm²) was used to make a comparison with that of $MAPbI_3/Y_2O_3$ as shown in Figure 6(b). The TA response of $MAPbI_3/TiO_2$ decayed much faster than that of $MAPbI_3/Y_2O_3$ and can be fitted to a single exponential decay function very well, with a time constant determined to be 1.8±0.1 ns, which is approximately 2 - 3 orders smaller than the photoexcited charge carrier lifetimes (~μs) of $MAPbI_3$ as shown above. Then the electron injection time t_{ET} and electron injection rate k_{ET} were calculated to be about 1.8 ns and 5.5×10^8 s^{-1}, respectively. [25] Thus, the electron injection efficiency η_{Einj} was calculated to be about 100% for all electrons with lifetimes longer than 100 ns. It is supposed here that the other 9% of charge carriers that recombined with a lifetime of 34 ps were not present in the $MAPbI_3$ prepared on TiO_2, because no faster decay component was observed in the TA responses of $MAPbI_3/TiO_2$. [25]

Next, the recombination dynamics in $MAPbI_3/TiO_2$ without the layer of *spiro*-OMeTAD, i.e., the recombination process between the electrons in TiO_2 and the holes in $MAPbI_3$ was measured using ns-TA. Figure 7 shows the TA response of $MAPbI_3/TiO_2$ measured with a pump light wavelength of 470 nm and a probe light wavelength of 658 nm. [25] According to the TA signal in nanocrystalline TiO_2 films reported by Yoshikawa [33] and our earlier studies, [18-21] it is reasonable to think that the TA response in $MAPbI_3/TiO_2$ observed at 658 nm corresponds to the electrons in TiO_2 injected from $MAPbI_3$. [25] Then the recombination time between the electrons in TiO_2 and the holes in perovskite was determined to be 0.14 μs by fitting Figure 7 to eq. (2).

Then, the recombination dynamics in $MAPbI_3/TiO_2$ with *spiro*-OMeTAD as a hole transport material, i.e., the recombination between the electrons in TiO_2 and the holes in *spiro*-OMeTAD, were measured using ns-TA. Figure 8 shows the TA responses of $MAPbI_3/TiO_2$ without and with *spiro*-OMeTAD, which were measured with a pump light wavelength of 470 nm and a probe light wavelength of 1310 nm, which was used to monitor the relaxation dynamics of holes in *spiro*-OMeTAD. From fig. 8 we can see that no TA signal was observed for the

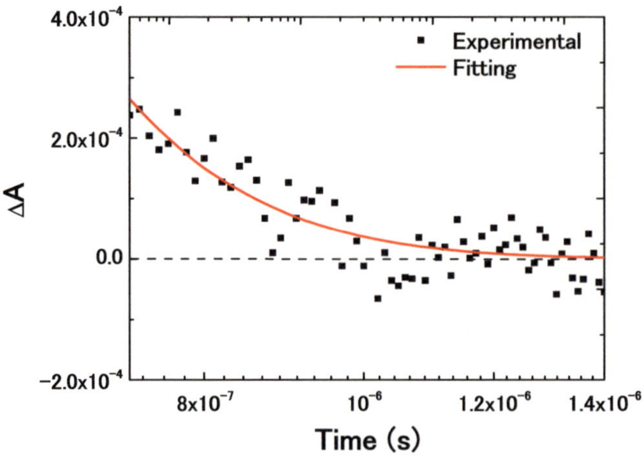

Figure 7. TA response of $MAPbI_3/TiO_2$ without a HTM measured with a pump light wavelength of 470 nm and a probe light wavelength of 658 nm. The red solid line represents the fitting result with a single exponential decay function with a time constant of 0.14 μs. (reference: 25)

sample without *spiro*-OMeTAD. However, for the sample with *spiro*-OMeTAD, an absorption signal can be clearly observed, which confirms that the TA signal probed at 1310 nm originated from the holes in *spiro*-OMeTAD. The TA decay in Figure 8 can be fitted very well with a single exponential decay function and the time constant was determined to be 60 ±0.5 μs. This is a typical result for cells with 18 nm sized TiO_2 nanoparticles. It is important to understand how the carrier dynamics relate to the photovoltaic properties of the cells. Several cells were measured and the energy conversion efficiency was typically 5-7%. These cells are called cell A.

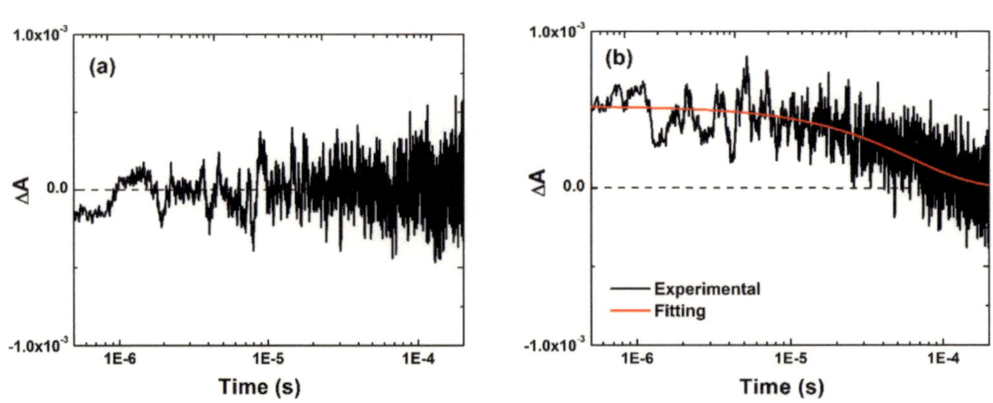

Figure 8. TA responses of $MAPbI_3/TiO_2$ without (a) and with (b) *spiro*-OMeTAD as a HTM measured with a pump light wavelength of 470 nm and a probe light wavelength of 1310 nm. The red solid line in (b) represents the fitting result with a single exponential decay function with a time constant of 60 μs. (reference: 25)

Then another kind of cell with 30 nm sized TiO_2 nanoparticles were also studied, which we call cell B. The energy conversion efficiency of cell B was typically 8-10%. It was found that the recombination dynamics in the Perovskite solar cells has a great influence on the IPCE, Jsc and the energy conversion efficiency. [25] Two kinds of TiO_2/$MAPbI_3$/*spiro*-OMeTAD solar cells showing different IPCE spectra and photovoltaic performance were studied, and the relationships between the charge separation and recombination dynamics and the photovoltaic properties were investigated. [25] Figure 9 shows the IPCE spectra and current-voltage (I-V) curves of the two kinds of cell. For cells A and B, the TiO_2 mesoporous layers were made of TiO_2 nanoparticles with sizes of 18 nm and 30 nm, respectively. The short circuit current J_{sc}, open circuit voltage V_{oc}, fill factor *FF* and energy conversion efficiency η of cell A are 11.81 mA/cm^2, 0.79 V, 0.71 and 6.59%, respectively. The IPCE value at 470 nm (the pump light wavelength used in the TA measurements) is 58%. On the other hand, for cell B, J_{sc}, V_{oc}, *FF* and η are 17.60 mA/cm^2, 0.78 V, 0.69 and 9.54%, respectively. The IPCE value of cell B at 470 nm is about 85%. To investigate the correlation between the photoexcited charge carrier dynamics and the photovoltaic properties, especially the IPCE, the charge separation and recombination dynamics of cell B were also characterized [25]. Figure 10 compares the charge carrier dynamics of the two cells A and B, it can be seen that the recombination time of electrons in TiO_2 and holes in the perovskite was almost identical (i.e., 0.13-0.14 μs), and the electron injection time was a little faster (0.7 ns) in cell B. However, the recombination time of electrons in TiO_2 and holes in *spiro*-OMeTAD in cell B became as long as 600 μs, which is ten times longer than that in cell A (60 μs). It is important to see how these changed separation and recombination dynamics influence the IPCE and J_{sc} of the two kinds of cell. It is well known that IPCE is proportional to the charge separation efficiency η_{Csep} and the charge collection efficiency η_{Ccol}, where η_{Csep} is the product of the electron injection efficiency η_{Einj} and the hole injection efficiency η_{Hinj}. η_{Einj}, η_{Hinj} and η_{Ccol} can be calculated using the photoexcited carrier lifetimes, electron and hole injection times, and recombination times as shown in Figures 5 and 10. As discussed in detail in reference 25, the charge separation efficiency of the two kinds of cell were found to be almost the same, about 90%. However, the charge collection efficiency of cell B was found to be almost 100%, which is much larger than that of cell A, only 70%.

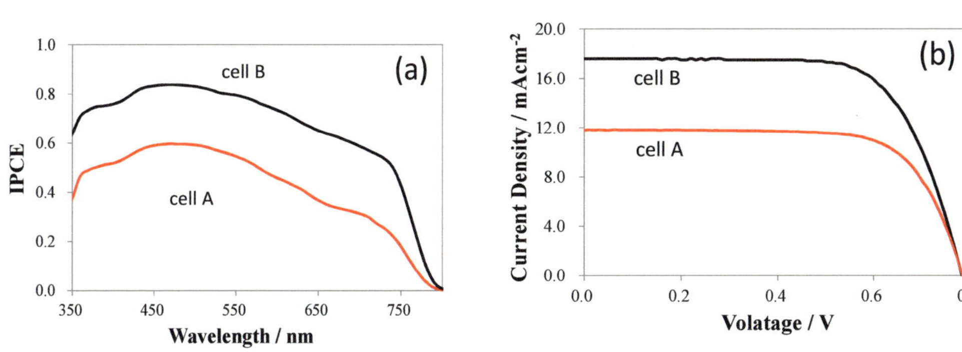

Figure 9. Incident photon to current conversion efficiency (IPCE) spectra of the solar cells A (red) and B (black) (a) and current voltage (I-V) curves of TiO_2/ $MAPbI_3$/*spiro*-OMeTAD solar cells A (red) and B (black) (b). (reference: 25)

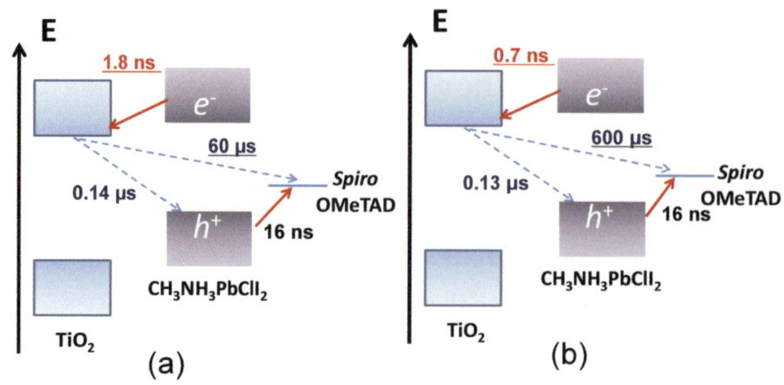

Figure 10. Schematic illustration of photoexcited electron injection and recombination dynamics in two TiO_2/$MAPbI_3$ / *spiro*-OMeTAD solar cells A (a) and B (b). (reference: 25)

5. Dynamics of Photoexcited Carrier Recombination and Charge Transfer in Sn/Pb Cocktail Perovskite Solar Cells and their Relationships to the Photovoltaic Properties [31]

To find the mechanism responsible for the low PCE, especially the low V_{oc} and *FF* in Sn/Pb cocktail perovskite solar cells, which has been reported [23], both the fs-TA and ns-TA techniques were used to investigate the charge separation and recombination dynamics in these solar cells systematically. [31] Sn/Pb cocktail perovskites deposited on either Al_2O_3 or TiO_2 films, with and without P3HT, were measured by the TA techniques using probe beams with different wavelengths. Similar to the Pb-based perovskite solar cells discussed above, firstly, photoexcited carrier lifetimes, hole injection to P3HT and recombination at the Sn/Pb cocktail perovskite /P3HT interface were clarified as shown in Figure 11. [31] The photoexcited carrier lifetimes are 4.2±0.4 ps ($A_1/(A_1+A_2+y_0)$: 55%), 650±57 ps ($A_2/(A_1+A_2+y_0)$: 25%), and much longer than 3 ns ($y_0/(A_1+A_2+y_0)$: 20%), respectively. Therefore, the recombination dynamics and lifetimes of photoexcited carriers in Sn/Pb cocktail perovskite are much faster and shorter than those in Pb-based perovskite, for which case they are much greater than 100 ns for over 90% of the photoexcited carriers as mentioned above. It is considered that the faster decay processes are due to some nonradiative recombination of electrons and holes through defects or trap states in the Sn/Pb cocktail perovskite, which is consistent with the measured larger Urbach energy in the Sn/Pb cocktail perovskite (i.e., 34 meV). It is larger than that in Pb-based perovskite (i.e., 22 mev). On the other hand, the charge separation at the interface between Sn/Pb cocktail perovskite and P3HT occurred in approximately 1 ps and interfacial charge recombination between the Sn/Pb cocktail perovskite (on an Al_2O_3 substrate) and P3HT occurred in approximately 16 ps. [31]

Secondly, the photoexcited charge separation and recombination dynamics of Sn/Pb cocktail perovskite deposited on a TiO_2 substrate with P3HT as a hole transport material were evaluated, which is shown in Figure 12. The charge separation at the perovskite/TiO_2 interface occurs in as fast as 1 ps. It is worth noting that the recombination time at the Sn/Pb cocktail perovskite/

http://dx.doi.org/10.5772/62101

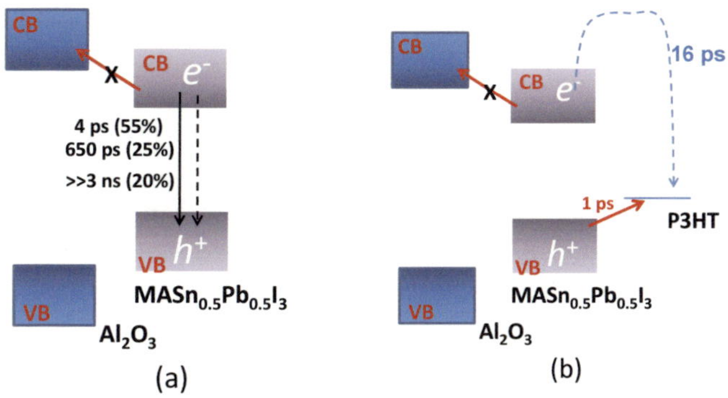

Figure 11. Schematic illustration of photoexcited charge carrier (electrons and holes) dynamics in Sn/Pb cocktail $MASn_{0.5}Pb_{0.5}I_3$ perovskite deposited on an Al_2O_3 substrate without (a) and with (b) P3HT as a hole transport material. (reference: 31)

TiO_2 interface is 880 μs, which is about two to three orders of magnitude slower compared to that occurring at the $MAPbI_3/TiO_2$ interface as shown in Figure 10. On the other hand, the recombination time at the TiO_2/P3HT interface is 190 μs, which is much faster compared to that without P3HT. This indicates that the recombination of electrons in the TiO_2 becomes faster when P3HT is used, which originates from the recombination at the TiO_2/P3HT interface. This recombination is one of the main reasons for the lower V_{oc} and *FF* of the Sn/Pb cocktail solar cells. [23] Pinhole-free Sn/Pb cocktail perovskite needs to be prepared in order to reduce direct interaction between the TiO_2 and P3HT, thus suppressing potential recombination. [31]

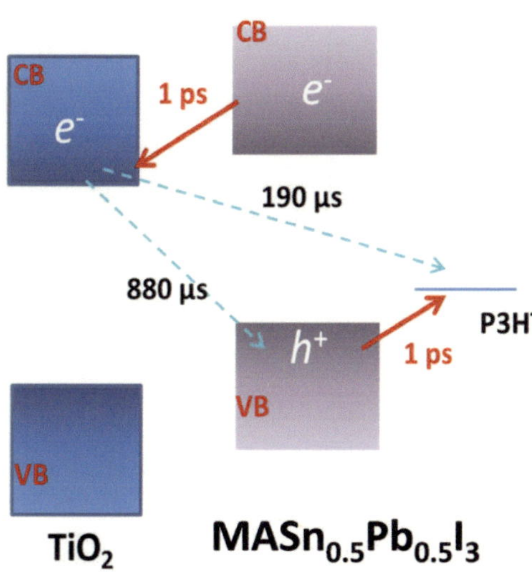

Figure 12. Schematic illustration of photoexcited charge separation and recombination dynamics in Sn/Pb cocktail $MASn_{0.5}Pb_{0.5}I_3$ perovskite solar cells with TiO_2 as an electron transport material (ETM) and P3HT as a HTM. (reference: 31)

It is important to understand how the optical absorption, charge separation and recombination dynamics relate to the incident photon to current conversion efficiency (IPCE) spectrum and the current-voltage (I-V) characteristics of the solar cell. Figure 13 shows a typical IPCE spectrum of a Sn/Pb cocktail perovskite solar cell. Based on the photoexcited carrier relaxation dynamics of Sn/Pb perovskite on an Al_2O_3 substrate, and the charge transfer dynamics at TiO_2/perovskite/P3HT interfaces, which are summarized in Table 1, it is possible to calculate the charge separation efficiency. [31] First, it is determined that the hole injection efficiency η_{Hinj} is almost 100% and the electron injection efficiency η_{Einj} is about 86%. Thus, the charge separation efficiency η_{Csep} ($\eta_{Csep}=\eta_{Hinj}\eta_{Einj}$) is 86%. The 14% loss in η_{Csep} results from the fast recombination component, with a lifetime of 4 ps, of photoexcited carriers in the Sn/Pb cocktail perovskite, which may originate from the nonradiative recombination owing to the defects in the sample. Therefore, in order to achieve 100% charge separation efficiency, the crystalline quality of the Sn/Pb cocktail perovskite has to be improved. The optical absorption is estimated to be about 95% in this case, due to the fact that the perovskite material is much more absorbing over a broader range up to 1000 nm, which could result in complete absorption (such as at 470 nm) in thin films such as 500 nm. [23] As shown in Figure 13, the IPCE value at 470 nm, corresponding to the excitation wavelength used for the TA measurements, is 68%. So the charge collection efficiency η_{Ccol} is estimated to be 83%. [31] The 17% loss in η_{Ccol} mainly originates from the charge recombination occurring at the TiO_2/P3HT interface. In order to improve *IPCE* and *J*sc for the solar cell, direct contact between P3HT and TiO_2 should be avoided or at least suppressed by preparing a pinhole-free Sn/Pb cocktail perovskite layer on TiO_2 or by inserting a barrier/blocking layer at the interface. [31]

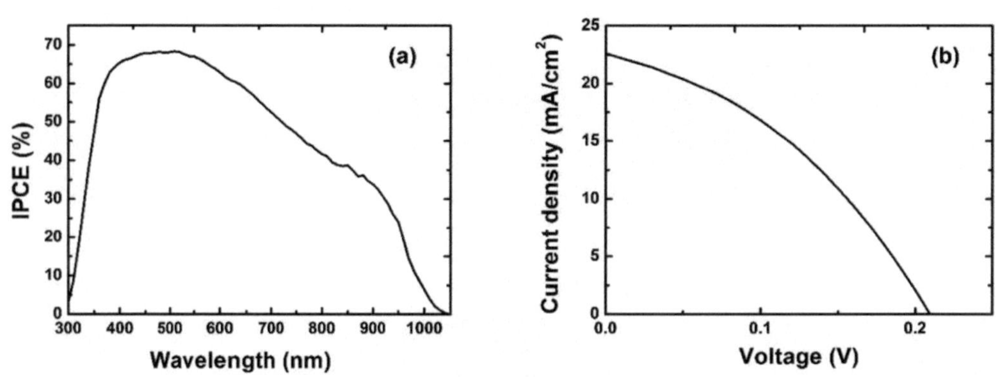

Figure 13. A typical incident photon to current conversion efficiency (IPCE) spectrum (a) and current-voltage (I-V) curve (b) of a Sn/Pb cocktail $MASn_{0.5}Pb_{0.5}I_3$ perovskite solar cell.(reference: 31)

As shown in Figure 13(b), J_{sc}, V_{oc}, *FF* and PCE of the Sn/Pb cocktail perovskite solar cells are 22.61 mA/cm^2, 0.21 V, 0.37 and 1.77%, respectively. It is obvious that the low PCE is mainly due to the low V_{oc} and FF. J_{sc} could be improved to as high as 30 mA/cm^2 if the charge separation efficiency can be increased to 100% by reducing the fast nonradiative recombination rate and if the charge collection efficiency can be increased to more than 95% through suppression of

Samples	Al_2O_3/ $MASn_{0.5}Pb_{0.5}I_3$	Al_2O_3/ $MASn_{0.5}Pb_{0.5}I_3$/ P3HT	TiO_2/ $MASn_{0.5}Pb_{0.5}I_3$	TiO_2/ $MASn_{0.5}Pb_{0.5}I_3$/ P3HT
Charge separation	N/A	Hole injection	Electron injection	Electron/hole injections
		1 ps	1 ps	1 ps/1 ps
Bulk recombination (photoexcited carrier lifetimes)	4 ps (55%) 650 ps (25%) >>3 ns (20%	N/A	N/A	N/A
Interfacial recombination	N/A	16 ps	880 μs	190 μs

Table 1. Photoexcited carrier lifetimes, hole injection and electron injection dynamics as well as charge recombination at each interface in $MASn_{0.5}Pb_{0.5}I_3$ perovskite solar cells. [31]

the recombination occurring at the TiO_2/P3HT interface. For V_{oc}, the bandgap-voltage offset $E_G - qV_{oc}$ is calculated to be as large as 1 eV, where E_G is the bandgap of 1.21 eV and q is the elementary charge. This bandgap-voltage offset is quite similar to that of amorphous silicon, and is thought to be closely related to the larger Urbach energy as shown in Figure 14. The recombination resistance from the I-V curve was estimated to be 27 Ω cm², which could also result in a lower V_{oc}. The smaller recombination resistance estimated is mostly due to the faster recombination of electrons and holes at the TiO_2/P3HT interface. Thus, the lower V_{oc} in the Sn/Pb cocktail perovskite solar cells is due to two reasons: (1) the larger Urbach energy; (2) the smaller recombination resistance. The lower *FF* is also due to reason (2). Therefore, these findings imply that the photovoltaic performance of the Sn/Pb cocktail perovskite solar cells can be further improved by decreasing the Urbach energy, i.e., reducing the defects in the perovskite, and by reducing the recombination occurring at the TiO_2/P3HT interface through appropriate interface engineering such as passivation or inserting a barrier/blocking layer. [31]

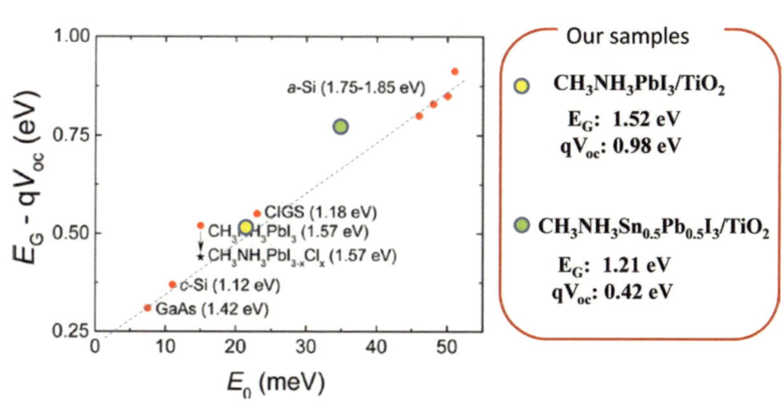

Figure 14. The bandgap-voltage offset ($E_G - qV_{oc}$) versus Urbach energy for typical photovoltaic absorber materials (from reference 18) and for the Pb and Sn/Pb cocktail perovskites used in our study at room temperature.

6. Conclusions

In summary, by conducting a systematic study on the optical absorption, photoexcited charge carrier lifetimes, and charge separation and charge recombination dynamics, we have explored the ways through which the photovoltaic performance of Pb and Sn/Pb cocktail perovskite solar cells can be improved.

For $MAPbI_3$ solar cells, we find that the great differences in the IPCE and J_{sc} of the two kinds of cell result from the great difference in the charge recombination between the electrons in TiO_2 and holes in *spiro*-OMeTAD. These results indicate that the key for improving the IPCE and J_{sc} of the perovskite-based solar cells is charge collection efficiency, instead of charge separation efficiency. Thus suppressing the recombination through proper interfacial engineering for perovskite-based solid hybrid solar cells is crucial for achieving a high efficiency.

For the Sn/Pb cocktail perovskite solar cells, it is determined that the bandgap is 1.21 eV and light harvesting can be extended to a wider wavelength over 1000 nm. The Urbach energy is calculated to be 34 meV, which is more than twice that of the $MAPbI_3$ perovskite. Three recombination processes for the photoexcited carriers were found with lifetimes being 4 ps (55%), 650 ps (25%) and one much larger than 3 ns (20%). The larger Urbach energy and the faster recombination suggest that there are defects in the prepared Sn/Pb cocktail perovskite. These results are significantly different from those of the $MAPbI_3$ perovskite, in which case the photocarrier lifetimes are larger than 100 ns and almost no defects were observed. Moreover, the charge separation and charge recombination dynamics were explored. We find that both the electron injection into TiO_2 and hole transfer to the P3HT layer occurred on a timescale of 1 ps. It was surprising to find that the charge recombination lifetime at the Sn/Pb cocktail perovskite/TiO_2 interface was as long as 880 μs, which is 2-3 orders of magnitude greater than that occurring at the $MAPbI_3/TiO_2$ interface. However, the charge recombination lifetime became shorter, i.e., 190 μs, when P3HT was employed as a HTM. This implies that the charges can be more effectively collected without the HTM. On the basis of the above results, we find that the loss in the charge separation efficiency originates from fast photoexcited carrier recombination with a lifetime of 4 ps, and the loss in charge collection efficiency was due to the charge recombination occurring at the TiO_2/P3HT interface, thus leading to lower IPCE values. Also, the low V_{oc} and FF were found to result from the large Urbach energy and the recombination occurring at the TiO_2/P3HT interface. These findings indicate that the photovoltaic performance of Sn/Pb cocktail perovskite solar cells can be further improved by reducing the defects in the material and the recombination occurring at the TiO_2/HTM interface through proper interfacial engineering.

Author details

Shen Qing[1*], Ogomi Yuhei[1], Toyoda Taro[1], Yoshino Kenji[3] and Hayase Shuzi[2*]

*Address all correspondence to: shen@pc.uec.ac.jp; hayase @life.kyutech.ac.jp

http://dx.doi.org/10.5772/62101

1 Department of Engineering Science, Faculty of Informatics and Engineering, The University of Electro-Communications, Chofu, Tokyo, Japan

2 Graduate School of Life Science and Systems Engineering, Kyushu Institute of Technology, Wakamatsu, Kitakyushu, Japan

3 Department of Electrical and Electronic Engineering, Miyazaki University, Kibanadai-nishi, Miyazaki, Japan

This work was supported by the CREST program, Japan Science and Technology Agency (JST).

References

[1] http://www.nrel.gov/ncpv/.

[2] H. S. Jung and N.-G. Park, *Small*, 2014, 11, 10-25.

[3] N. J. Jeon, J. H. Noh, Y. C. Kim, W. S. Yang, S. Ryu and S. I. Seok, *Nat. Mater.*, 2014, 13, 897-903.

[4] H. Zhou, Q. Chen, G. Li, S. Luo, T.-b. Song, H.-S. Duan, Z. Hong, J. You, Y. Liu and Y. Yang, *Science*, 2014, 345, 542-546.

[5] M. A. Green, A. Ho-Baillie and H. J. Snaith, *Nat. Photon*, 2014, 8, 506-514.

[6] M. Z. Liu, M. B. Johnston and H. J. Snaith, *Nature*, 2013, 501, 395-398.

[7] G. E. Eperon, V. M. Burlakov, A. Goriely and H. J. Snaith, *ACS Nano*, 2014, 8, 591-598.

[8] J. Burschka, N. Pellet, S. J. Moon, R. Humphry-Baker, P. Gao, M. K. Nazeeruddin and M. Gratzel, *Nature*, 2013, 499, 316-319.

[9] H. S. Kim, C. R. Lee, J. H. Im, K. B. Lee, T. Moehl, A. Marchioro, S. J. Moon, R. Humphry-Baker, J. H. Yum, J. E. Moser, M. Gratzel and N. G. Park, *Sci. Rep.*, 2012, 2, 591.

[10] M. M. Lee, J. Teuscher, T. Miyasaka, T. N. Murakami and H. J. Snaith, *Science*, 2012, 338, 643-647.

[11] J. H. Im, J. Chung, S. J. Kim and N. G. Park, *Nanoscale Res. Lett.*, 2012, 7, 353.

[12] N. G. Park, *J. Phys. Chem. Lett.*, 2013, 4, 2423-2429.

[13] G. Hodes, *Science*, 2013, 342, 317-318.

[14] H. J. Snaith, *J. Phys. Chem. Lett.*, 2013, 4, 3623-3630.

[15] S. Sun, T. Salim, N. Mathews, M. Duchamp, C. Boothroyd, G. Xing, T. C. Sum and Y. M. Lam, *Energy Environ. Sci.*, 2014, 7, 399-407.

[16] S. D. Stranks, G. E. Eperon, G. Grancini, C. Menelaou, M. J. P. Alcocer, T. Leijtens, L. M. Herz, A. Petrozza and H. J. Snaith, *Science*, 2013, 342, 341-344.

[17] G. C. Xing, N. Mathews, S. Y. Sun, S. S. Lim, Y. M. Lam, M. Gratzel, S. Mhaisalkar and T. C. Sum, *Science*, 2013, 342, 344-347.

[18] S. De Wolf, J. Holovsky, S.-J. Moon, P. Löper, B. Niesen, M. Ledinsky, F.-J. Haug, J.-H. Yum and C. Ballif, *J. Phys. Chem. Lett.*, 2014, 5, 1035-1039.

[19] C. C. Stoumpos, C. D. Malliakas and M. G. Kanatzidis, *Inorg. Chem.*, 2013, 52, 9019-9038.

[20] Y. Takahashi, R. Obara, Z.-Z. Lin, Y. Takahashi, T. Naito, T. Inabe, S. Ishibashi and K. Terakura, *Dalton Trans.*, 2011, 40, 5563-5568.

[21] F. Hao, C. C. Stoumpos, D. H. Cao, R. P. H. Chang and M. G. Kanatzidis, *Nat. Photon*, 2014, 8, 489-494.

[22] N. K. Noel, S. D. Stranks, A. Abate, C. Wehrenfennig, S. Guarnera, A. Haghighirad, A. Sadhanala, G. E. Eperon, S. K. Pathak, M. B. Johnston, a. petrozza, L. Herz and H. Snaith, *Energy Environ. Sci.*, 2014, 7, 3061-3068.

[23] Y. Ogomi, A. Morita, S. Tsukamoto, T. Saitho, N. Fujikawa, Q. Shen, T. Toyoda, K. Yoshino, S. S. Pandey, T. Ma and S. Hayase, *J. Phys. Chem. Lett.*, 2014, 5, 1004-1011.

[24] F. Hao, C. C. Stoumpos, R. P. H. Chang and M. G. Kanatzidis, *J. Am. Chem. Soc.*, 2014, 136, 8094-8099.

[25] Q. Shen, Y. Ogomi, J. Chang, S. Tsukamoto, K. Kukihara, O. Takuya, N. Osada, K. Yoshino, K. Katayama, T. Toyoda and S. Hayase, *Phys. Chem. Chem. Phys.*, 2014, 16, 19984-19992.

[26] A. Rosencwaig and A. Gersho, *J. Appl. Phys.*, 1976, 47, 64-69

[27] Q. Shen, Y. Ogomi, B.-w. Park, T. Inoue, S. S. Pandey, A. Miyamoto, S. Fujita, K. Katayama, T. Toyoda and S. Hayase, *Phys. Chem. Chem. Phys.*, 2012, 14, 4605-4613.

[28] Q. Shen, Y. Ogomi, S. K. Das, S. S. Pandey, K. Yoshino, K. Katayama, H. Momose, T. Toyoda and S. Hayase, *Phys. Chem. Chem. Phys.*, 2013, 15, 14370-14376.

[29] N. Maeda, H. Hata, N. Osada, Q. Shen, T. Toyoda, S. Kuwahara and K. Katayama, *Phys. Chem. Chem. Phys.*, 2013, 15, 11006-11013.

[30] N. Osada, T. Oshima, S. Kuwahara, T. Toyoda, Q. Shen and K. Katayama, *Phys. Chem. Chem. Phys.*, 2014, 16, 5774-5778.

[31] Q. Shen, Y. Ogomi, J. Chang, T. Toyoda, K. Fujiwara, K. Yoshino, K. Sato, K. Yamazaki, M. Akimoto, Y. Kuga, K. Katayama, and S. Hayase, *J. Mater. Chem. A*, 2015, 3, 9308-9316.

[32] F. Deschler, M. Price, S. Pathak, L. E. Klintberg, D.-D. Jarausch, R. Higler, S. Hüttner, T. Leijtens, S. D. Stranks, H. J. Snaith, M. Atatüre, R. T. Phillips and R. H. Friend, *The Journal of Physical Chemistry Letters*, 2014, 5, 1421-1426.

[33] T. Yoshihara, R. Katoh, A. Furube, Y. Tamaki, M. Murai, K. Hara, S. Murata, H. Arakawa and M. Tachiya, *The Journal of Physical Chemistry B*, 2004, 108, 3817-3823.

[34] R. Plass, S. Pelet, J. Krueger, M. Grätzel and U. Bach, *The Journal of Physical Chemistry B*, 2002, 106, 7578-7580.

[35] J. J. Prías-Barragán, L. Tirado-Mejía, H. Ariza-Calderón, L. Baños, J. J. Perez-Bueno and M. E. Rodríguez, *J. Cryst. Growth*, 2006, 286, 279-283

[36] F. Urbach, *Phys. Rev.*, 1953, 92, 1324.

[37] K. T. H., *Phys. Rev.*, 1966, 144, 582.

[38] K. S. and N. W., *Phys. Rev. B*, 1999, 59, 12940.

[39] Y. Ogomi, A. Morita, S. Tsukamoto, T. Saitho, Q. Shen, T. Toyoda, K. Yoshino, S. S. Pandey, T.L. Ma, and S. Hayase, *J. Phys. Chem. C*, 2014, 118, 16651–16659.

[40] Y. Ogomi, K. Kukihara, Q. Shen, T. Toyoda, K. Yoshino, S. Pandey, H. Momose, and S. Hayase, *Chemphyschem*, 2014.

Chapter 14

Optical, Excitonic, and Electronic Properties of $CH_3NH_3PbI_3$ Thin Films and Their Application in Photovoltaics

Sheng Hsiung Chang, Hsin-Ming Cheng, Sheng-Hui Chen and Kuen-Feng Lin

Additional information is available at the end of the chapter

http://dx.doi.org/10.5772/61278

Abstract

In the past two years, the highest power conversion efficiency of perovskite absorber (PA)–based photovoltaics has been 20.2%. The PA can be fabricated on flat substrates (for example, ZnO, TiO_2, and PEDOT:PSS) using solution processes, which have a low-cost advantage in terms of industry production. In this report, the recent advances of PA-based photovoltaics will be mentioned. Then, the optoelectronic properties of PA, material fabrication, and photovoltaic performance will be discussed. On the other hand, we used scanning electron microscopy, two-dimensional X-ray diffractometer, and photoluminescence spectroscopy to investigate the fundamental properties of $CH_3NH_3PbI_3$ thin films fabricated with and without toluene washing treatment, which provides an assessment of the development potential of PA-based photovoltaics.

Keywords: Perovskite photovoltaic, Exciton, Nanoplasmonic structure, 2D X-ray diffraction

1. Introduction

The perovskite structure is named after the structure of calcium carbonate ($CaTiO_3$) compound and its molecular formula is ABX_3. Perovskite oxide materials ($SrTiO_3$ and $LaAlO_3$) [1] and nonoxide perovskite materials ($MgCNi_3$) [2] have high-temperature superconductivity properties, which indicate that perovskite structures have excellent electrical properties. In the past two years, the most popular light-absorbing materials for photovoltaic cells have been organic–inorganic lead halide perovskite absorbers ($CH_3NH_3PbI_3$), in which absorption bandgap is approximately 1.57 eV [3] and exciton binding energy is less than 50 meV [4], hence they have excellent power conversion efficiency (PCE). The molecular structure of the earliest application of perovskite materials in an optoelectronic device ($(C_6H_5C_2H_4NH_3)_2PbI_4$) was in 1994, in Kyushu University, Japan, where Tsutsui's research focused on the fluorescence characteristics of materials via electrical excitation [5]. Additionally, the emission wavelength of the luminous element was 520 nm at a low temperature, the full-width at half-maximum (FWHM) of fluorescence was about 10 nm, and the luminance of the perovskite-based device was 10,000 cd/m^2 with an injection current density of 2 A/cm^2. The brightness of perovskite-based light-emitting diodes was equivalent to four to five times of the organic light-emitting diode. In April 2009, Miyasaka's research team from the University of Tokyo published the world's first article relating to perovskite-sensitized solar cells. Their study showed that $CH_3NH_3PbI_3$ and $CH_3NNH_3PbBr_3$ could convert sunlight into electrical energy as light-absorbing materials through dye-sensitized solar cell structures, and the obtained PCE was 3.81% [6]. Although the study was published in an important academic journal, it did not receive much attention because of poor efficiency. In 2012, Gratzel from the Swiss Federal Institute of Technology, Lausanne (EPFL), optimized the device parameters of the dye-sensitized solar cells to enhance the PCE to 7.28% [7] under 0.1 sunlight intensity. In the same year, Snaith led his research team from the University of Oxford in England to publish their research on organic–inorganic lead halide perovskite photovoltaics using solution process in a scientific journal. They used an insulator Al_2O_3 nanoparticle film to replace the TiO_2 nanoparticle film for the first time, and named the structure meso-superstructured solar cell (MSSC). Additionally, the PCE reached a record of 10.9% [8]. Since then, scientists have launched a series of battles for higher PCE. In July 2013, Bruschka and Pellet adopted a two-step solution process method to produce $CH_3NH_3PbI_3$ on the porous TiO_2 substrate and the efficiency obtained was 12.9% [9]. Also, in September 2013, Snaith produced high-quality perovskite light-absorbing materials using a vapor deposition method and obtained a PCE of 5.4% [10]. In addition, in July 2014, Professor Wu from the National Central University used $PC_{71}BM$ thin film as an electron transport layer and improved the PCE up to 16.3% [11]. In August 2014, Yang's team optimized the energy level of each layer of the solar cells and increased the PCE to 19.3% [12]. Gratzel and Park also optimized the proportion of CH_3NH_3I and PbI_2 in the two-step solution process method and achieved an average PCE of up to 16.3% (±0.35%) [13].

This chapter will describe the optoelectronic properties, molecular structure, and fabrication method of perovskite materials, and calculate the effects of the nanoplasmonic structure on exciton generation and the dissociation of perovskite materials. In addition, it will use scanning electron microscope (SEM), two-dimensional X-ray diffractometer (2D-XRD), UV-vis spectrometer and photoluminescence to explore the morphology, structure, and electronic and excitonic properties of $CH_3NH_3PbI_3$ thin films formed by a solution process with and without a nonpolar solvent washing treatment [14], and will discuss the development direction and potential of the perovskite photovoltaics.

2. The optical and electronic properties of perovskite absorbers

The perovskite structure is shown in Figure 1. According to the first-principles calculation, each CH_3NH_3 and Pb provides one and two electrons to I_3, and the structure of CH_3NH_3 and

Pb-I can maintain electric neutrality due to van der Waals force, so CH_3NH_3 cations hardly affect the characteristics of Pb in the conduction band and "I" determines the properties of the valence band [15]. Although the CH_3NH_3 cation has no significant effect on the electronic structure, the orientation of the CH_3NH_3 cation can influence the dielectric constant of $CH_3NH_3PbI_3$ materials. As shown in Table 1, the exciton binding energy can be calculated by $E_b = e^2 /[4\pi\varepsilon_0\varepsilon_d r]$, where, r is the Bohr radius (2.8 nm) [16] and ε_d is the dielectric constant [17] of $CH_3NH_3PbI_3$. Exciton binding energy is as low as possible to be more favorable for exciton self-dissociation at room temperature and exciton dissociation at the interface to produce photocurrent, so the orientation of CH_3NH_3 cation is very essential to excitonic properties. The most direct physical parameter for device efficiency is the exciton diffusion length. Snaith estimated that the exciton diffusion length could be over 1 μm, through measurement of the exciton's lifetime. The long diffusion length in poly-crystalline $CH_3NH_3PbI_3$ thin films has been directly observed using a spatial and temporal imaging system [18]. Therefore, photons can be effectively converted into photocurrent after being absorbed. Experimental results showed that the exciton lifetime of the perovskite materials could exceed 100 ns [19].

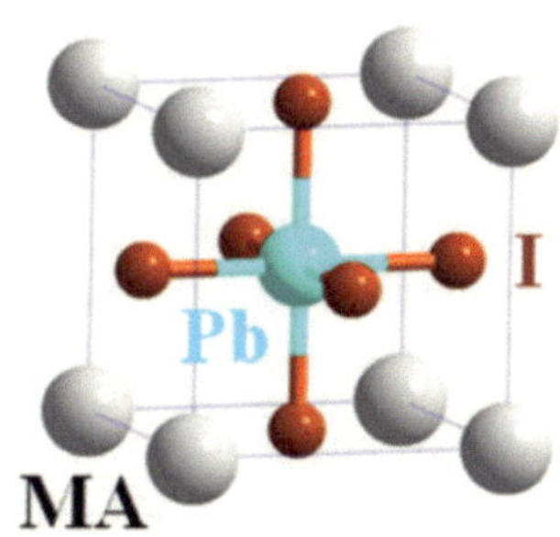

Figure 1. Perovskite structure. MA is $[CH_3NH_3]^+$ [15].

CH_3NH_3 Orientation	$\varepsilon_d (x, y, z)$	$E_b (x, y, z) (meV)$
(001)	22.39, 27.65, 17.97	36.7, 29.7, 45.7
(011)	17.95, 23.56, 22.67	45.8, 34.9, 36.2
(111)	36.52, 37,28, 24.94	22.5, 22.0, 32.9

Table 1. Anisotropic dielectric constants and exciton binding energies of $CH_3NH_3PbI_3$ crystals.

From the perspective of materials design, "I" can be replaced by Br or Cl, and hence the absorption bandgap of $CH_3NH_3PbI_3$, $CH_3NH_3PbBr_3$, and $CH_3NH_3PbCl_3$ are 1.6, 1.95, and 2.46 eV [20], respectively, so the absorption bandgap will increase with the decrease of halogens' atomic number. For application, it can be used with different ratios of halogen to achieve colorful commercial applications. However, the greater the absorption bandgap, the lower the amount of sunlight that can be absorbed. Also, the toxicity of heavy metal Pb is a key factor hindering the development of perovskite solar cells; thus, one of the directions for future development is to use nontoxic Sn to replace Pb. As the conduction band of $CH_3NH_3SnI_3$ is 0.24 eV lower than that of $CH_3NH_3PbI_3$, $CH_3NH_3SnI_3$ has a lower absorption bandgap at about 1.3 eV [21]. So far, the maximum PCE of $CH_3NH_3SnI_3$-based photovoltaics is about 5.2%, so there is still a lot of room for improvement in the material and device production.

For the part of the organic molecules, based on the theoretical calculation, Walsh proposed in October 2013, that the absorption bandgap of NH_4PbI_3 could decrease to 1.20 eV by replacing CH_3NH_3 with NH_4, which was expected to absorb more sunlight [17]. That following year, Walsh pointed out that the carrier recombination rate [22] in the perovskite absorbers was strongly associated with the strength of electric dipole moment of the organic cation, and he suggested that the use of larger electric dipole moments of CF_3NH_3 to replace CH_3NH_3 could decrease the carrier recombination rate in the perovskite absorbers. In January 2014, Snaith used $HC(NH_2)_2$ to replace CH_3NH_3 and formed $HC(NH_2)_2PbI_3$ to obtain a lower absorption bandgap at about 1.48 eV, and the PCE of $HC(NH_2)_2PbI_3$-based photovoltaics reached 14.2% [23]. In August 2014, Park produced $HC(NH_2)_2PbI_3$ on the substrate of mesoporous TiO_2/FTO and obtained device PCE of up to 16% [24]. Although the exciton diffusion length in perovskite materials could reach to micrometers, the thickness of light absorption materials in perovskite-based photovoltaics shall be controlled at around 300 to 400 nm to obtain a better PCE: the optimal short-circuit current density (J_{SC}) is about 22 mA/cm^2, depending on the thickness and bandgap of light absorption materials; the maximum open circuit voltage (V_{OC}) is about 1.1 V depending on device structure; and the optimal fill factor (FF) is about 75%, which is lower than the FF of gallium arsenide photovoltaics (FF = 86%) [25]. Recently, a high PCE of 20.2% was achieved using $HC(NH_2)_2PbI_3$ ($FAPbI_3$) as the light absorber in photovoltaics [26].

The FF of photovoltaics indicates the recombination status—the carriers will recombine inside and on the interface of light absorption materials. When the light absorption materials are thicker, the carrier recombination mainly occurs inside the light absorption materials. The decrease of thickness of the light absorption material will reduce the amount of absorbed sunlight, resulting in a decrease in short-circuit current density. For the same light absorption material, the appropriate thickness can be calculated by the product of optimal J_{sc} and FF. At present, there are theoretical calculations for optimizing the thickness of perovskite light absorption materials [27], but there are still no experimental results on the relationship between the changes in thickness of perovskite absorbers and the PCE of photovoltaics.

3. Exciton and carrier in perovskite materials

The energy is obtained by absorbing the photon, and thus the electron is excited from the ground state shifting to the excited state. Such an excited state can be regarded as an electron and a hole that is bound together by electrostatic attraction, and such an electron–hole pair is called an exciton.

Figure 2 shows when the electron in the ground state is excited to the excited state in perovskite materials, such an electron in the excited state will be called a hot electron. The hot electron can interact with the crystal lattice to generate coherent and incoherent collisions with two different paths back to the ground state and the conduction band. Raman scattering signal is generated by the coherent collision caused by hot electrons and lattice vibration. However, as the frequency of the hot electron vibration is much higher than that of the lattice vibration, Raman scattering signal strength is very weak. On the other hand, after the incoherent collision produced by the interaction between hot electrons and lattice vibration, the electron which is relaxed in the conduction band is called a cold electron. At this time, the electrons in the conduction band and the holes in the valence band will form excitons. For $CH_3NH_3PbI_3$ absorbers, the lifetime of hot electrons is about 1 ps [28], and the exciton lifetime associated with the material structure is approximately 10–100 ns.

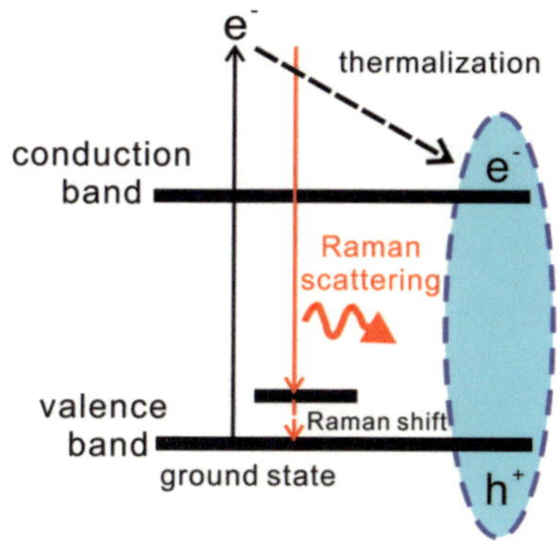

Figure 2. Energy relaxation pathways of photoexcited electrons.

Exciton binding energy is a very important parameter for the light absorption materials of photovoltaics. When the exciton in the heterogeneous interface of different materials is dissociated by the built-in electric field and becomes a free electron and hole after overcoming the binding energy. The exciton binding energy for organic materials is about 150 to 1000 meV, so with the exciton dissociation at the region of charge transfer radius, the potential difference of p–n interface must be larger than the exciton binding energy. Taking P3HT/PCBM as an example, the exciton binding energy of P3HT is about 300 meV, and LUMO potential difference of P3HT-PCBM interface is about 800 meV (3.9–3.1 eV); therefore, the exciton in the P3HT-PCBM interface can be smoothly dissociated with approximately 1.2 ps exciton dissociation time [29]. The exciton binding energy of $CH_3NH_3PbI_3$ is smaller than 50 meV but higher than the thermal energy of room temperature (K_{BT} = 25.6 meV, T = 25°C). Therefore, about half of the excitons will self-dissociate to become free carriers at room temperature [30], and the other half of the excitons will have to diffuse to the interface to dissociate and become free carriers (electrons and holes).

The electrons and holes formed by the self-dissociation of excitons at room temperature must not recombine, and each has to transport to the electrodes to form a photocurrent. In March 2014, Walsh, based on theoretical calculations, predicted that the transporting paths of electrons and holes inside $CH_3NH_3PbI_3$ are different (see Figure 3) [22], and estimated that the carrier recombination rate in $CH_3NH_3PbI_3$ was very low. The arrow direction shown in Figure

3 is the orientation of the CH_3NH_3 cation (electric dipole). Electrons move along the lowest potential (blue arrow) while the holes flow along the maximum potential (red arrow). Therefore, the electric dipole of the organic cation would affect the recombination probability of electrons and holes. Also Walsh pointed out that the CF_3NH_3 cations with larger electric dipole moments should be able to effectively reduce the carrier recombination rate in perovskite absorbers.

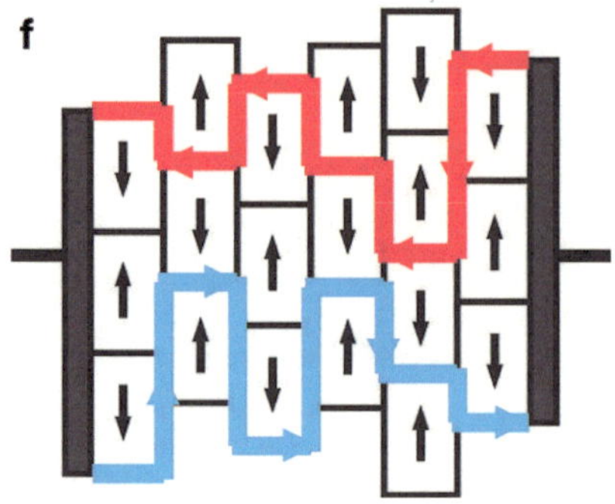

Figure 3. The propagation pathways of electrons and holes in $CH_3NH_3PbI_3$ [22].

4. Device architecture and manufacturing process of perovskite-based photovoltaics

Perovskite-based photovoltaics with more than 17% PCE can be grouped into three structures: the dye-sensitized solar cell structures, regular-type organic photovoltaic structures, and inverted-type organic photovoltaic structures. The dye-sensitized solar cell structures, based on the process order, includes FTO-conductive glass substrate/TiO_2 compact layer/mesoporous TiO_2/$HC(NH_2)_2PbI_3$ perovskite absorber/PTAA hole transporting layer/Au and its currently published maximum PCE is 20.2% [26]. For regular-type organic photovoltaic structures, its process sequence is ITO-conductive glass/PEDOT:PSS/$CH_3NH_3PbI_3$ perovskite absorber/PCBM electron transport layer/Al and its currently published maximum PCE is 17.8% [29]. For inverted-type organic photovoltaic structures, its process sequence is ITO-conductive glass/PEIE/Y:TiO_2/$CH_3NH_3PbI_{3-x}Cl_x$ perovskite absorber/Spiro-OMeTAD hole transport layer/Au, and its currently published maximum PCE is 19.3% [12]. Currently, the manufacturing process of a regular-type organic photovoltaic structure is the easiest.

Much research has been conducted on the energy conversion of perovskite-based photovoltaics in the last two years, mainly because perovskite absorbers can be produced through simple solution process methods. The main fabrication methods include one-step and two-step solution processes. The one-step method is first adopted so that the precursor of perovskite materials is dissolved in dimethyl formamide (DMF), followed by soaking of the perovskite materials on the substrate. However, the biggest drawback of this method is the poor coating for the film, which can easily result in short-circuits due to the contact between the upper and lower electrodes. The use of an appropriate upper electrode can reduce the incidence of short-circuits [32]. As shown in Figure 4, using p-type P3HT can fill the pores of $CH_3NH_3PbI_{3-x}Cl_x$ thin films. The addition of a small amount of 1,8-diiodooctane (DIO) inside the perovskite precursors [33] can effectively reduce the generation of voids. In addition, the perovskite film with high coverage can also be obtained on the rough FTO substrate. Thus, the surface properties of the substrate have an essential effect on the continuity of perovskite thin film.

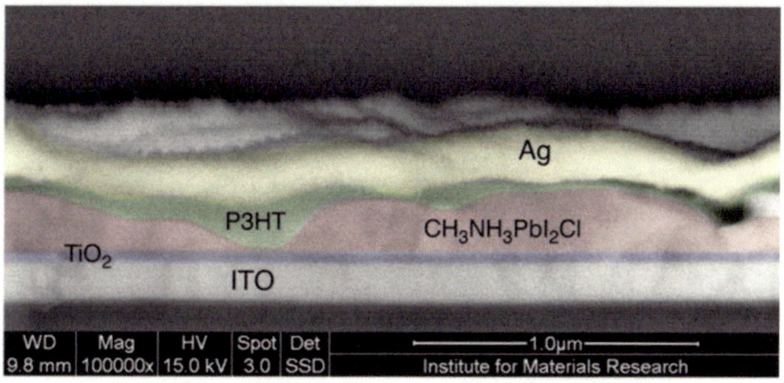

Figure 4. Cross-sectional view of device under a scanning electron microscope [30].

In July 2014, Spiccia and Seok spontaneously used organic solvent (chlorobenzene or toluene) [34,35] to inject perovskite precursor thin films during the spin coating process, which can greatly enhance the film continuity of perovskite thin films. In September 2014, Cheng sprayed argon gas positively into the substrate to improve the solvent evaporation speed in the spin coating process of perovskite precursors and achieved up to 16.97% PCE. Figure 5 shows that perovskite thin films with argon gas processing have good film continuity and crystallinity; on the contrary, the surface of perovskite thin films without gas treatment would appear as reticular defect structures [36]. In November 2014, Jen used different solvents (toluene, chlorobenzene, and dichlorobenzene) to fabricate perovskite thin films with high continuity on top of PEDOT:PSS/ITO/glass substrate, and obtained a high PCE of 13.7% [37]. Moreover, in the manufacturing process, when the heating temperature decreases from 100°C down to 70°C, the photovoltaics still have excellent PCE of 12%. Therefore, such low-temperature deposition manufacturing techniques can be integrally applied to soft substrates.

Bruschka and Pellet from EPFL were the first to adopt a two-step coating method [9] to produce perovskite materials in the porous TiO_2 substrates. They first used a spin coating process for even distribution of PbI_2 on the mesoporous TiO_2 thin films, and then soaked the substrate into the IPA solvent with solved CH_3NH_3I. Hence, the PbI_2 on the TiO_2 surface would be reacted to $CH_3NH_3PbI_3$. Using a long-term interdiffusion process [38], the two-step solution process method can also produce high continuity of perovskite thin films on a flat PEDOT:PSS/ITO/glass substrate, which is shown in Figure 6.

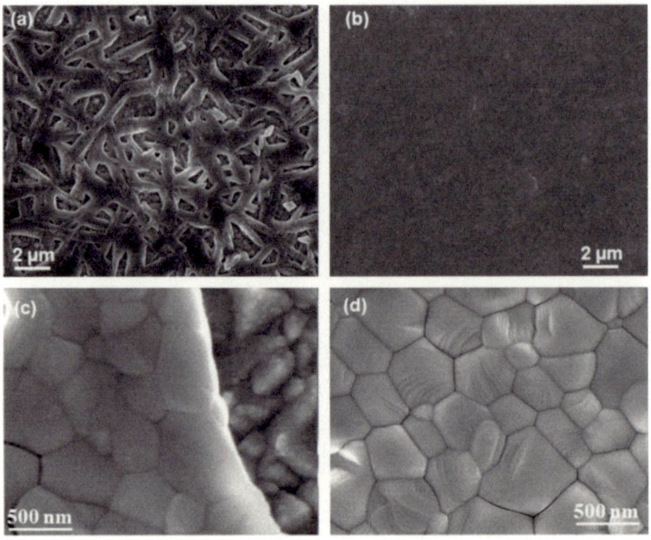

Figure 5. Scanning electron microscopy image of the $CH_3NH_3PbI_3$ surface: (a) and (c) are fabricated without argon gas; (b) and (d) are fabricated with argon gas [36].

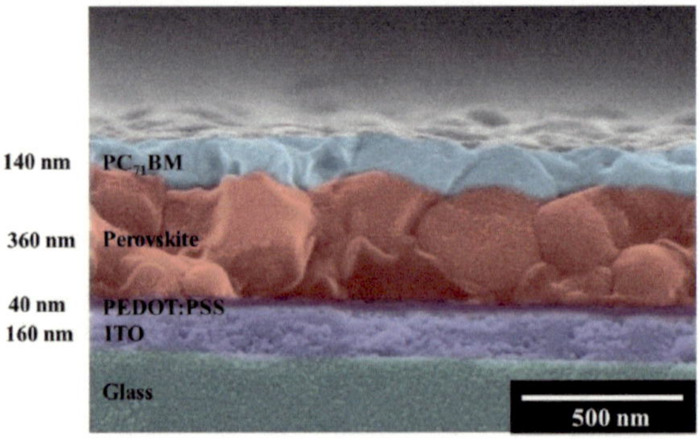

Figure 6. Cross-sectional view of multilayer film under a scanning electron microscope [11].

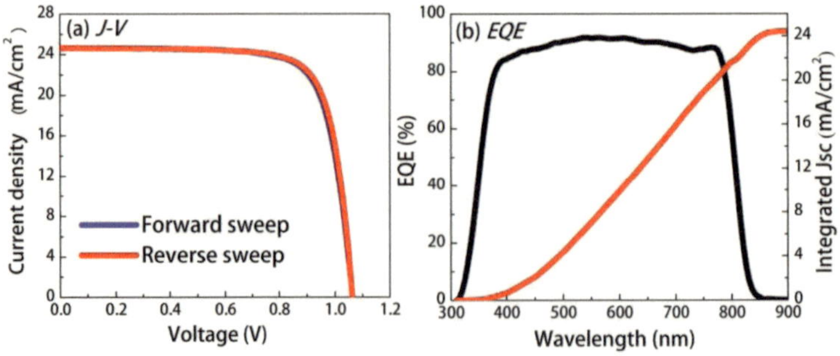

Figure 7. The J–V curves and external quantum efficiency (EQE) spectrum of $CF_3NH_3PbI_3$–based photovoltaics [26].

http://dx.doi.org/10.5772/61278

At present, among the perovskite thin film production methods with three matching photovoltaic structures and two fabrication methods, the two-step solution process method can obtain the highest PCE of 20.2% but has the longest manufacturing process, and its photovoltaic performances are shown in Figure 7. In addition, the one-step solution process method with argon gas treatment, although with a slightly lower PCE of 16.97%, can save the most time and is suitable for making large area high-quality perovskite thin films.

5. Exciton properties in the interface between metal particles and perovskite absorbers

In order to be able to simultaneously improve the J_{SC} and FF of perovskite-based photovoltaics, we calculate the exciton properties generated by the nanoplasmonic structure embedded in perovskite absorbers to analyze the effects of nanoplasmonic structures on the properties of perovskite-based photovoltaics.

The essential parameters for electromagnetic simulations are optical parameters (refractive index and absorption coefficient), using a transfer-matrix method with Lorentz model to describe the dielectric constant of the perovskite absorber. Figure 8 shows the transmittance spectrum of perovskite/substrate, which are used for calculating the refractive index and extinction coefficient of perovskite absorber, as shown in Figure 9. Moreover, we buried a nanoplasmonic structure inside the perovskite thin film (see Figure 10), and selected appropriate structure parameters (the period = 100 nm and the metal ellipsoid length = 150 nm). Figure 11 shows the absorption spectra of nanoplasmonic structures embedded in perovskite absorber with different ellipsoid gap. When the distance between the ellipsoid is 30 nm (the short axis of the ellipsoid is 70 nm), it has a better absorption enhancement effect; at the same time, when the thickness of the perovskite absorber is decreased to 300 nm, it is expected to have a better FF.

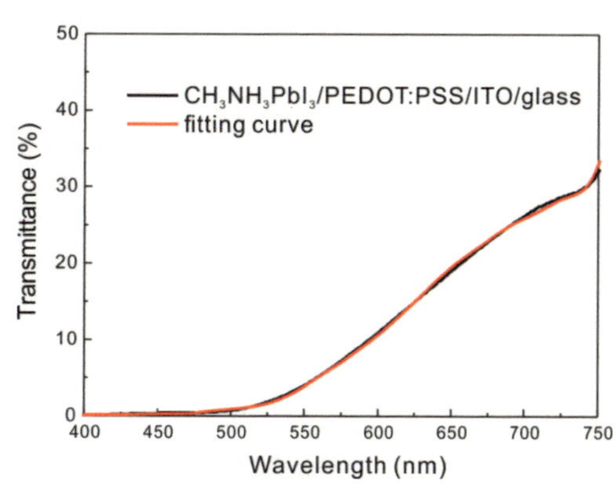

Figure 8. Transmittance spectra of $CH_3NH_3PbI_3$/substrate [39].

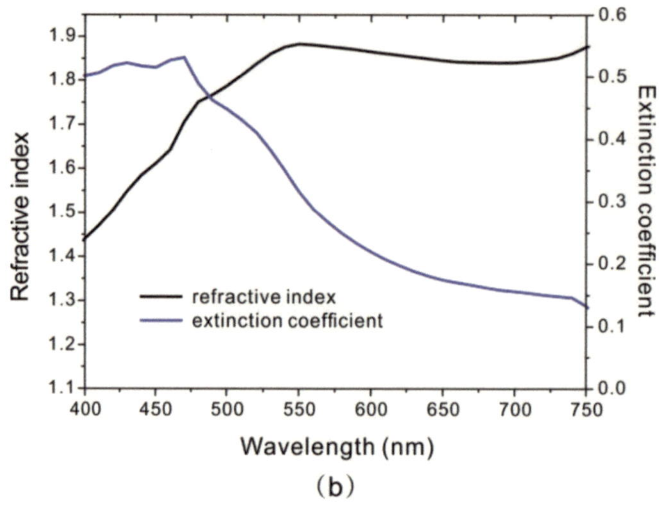

Figure 9. Refractive index and extinction coefficient of $CH_3NH_3PbI_3$ thin film [39].

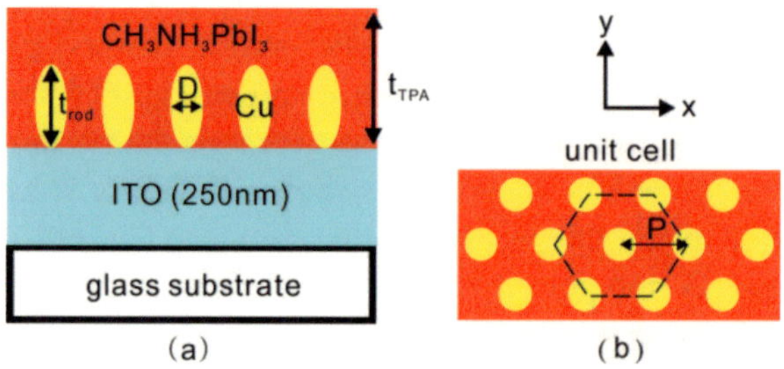

Figure 10. Side view and top view of nanoplasmonic structure embedded in the perovskite absorber [39].

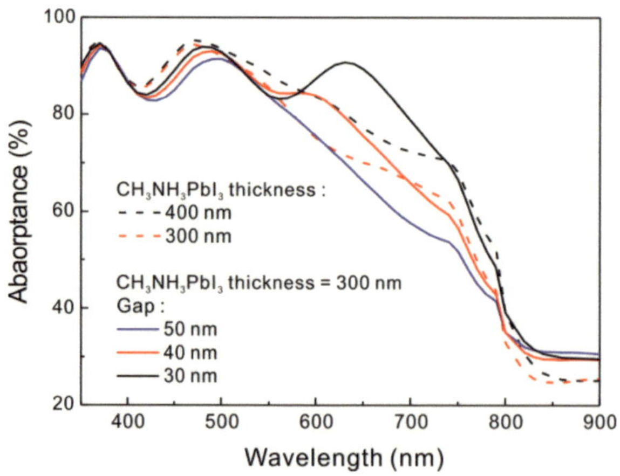

Figure 11. Absorption spectra of nanoplasmonic structures embedded in perovskite absorber with different structure parameters [39].

http://dx.doi.org/10.5772/61278

The nanoplasmonic structures enhanced absorptions have currently been realized in the organic solar cells [40]. However, the effect of the spatial distribution of the exciton around nanoplasmonic structures is seldom discussed. Figure 12 shows the spatial distribution of electric field (exciton). When the sunlight transmits from the ITO/glass substrate to the nanoplasmonic structure, the light field is localized at the interface between the $CH_3NH_3PbI_3$ and the Cu ellipsoid, thus most of the excitons are generated at the interface, and the exciton can be dissociated at the interface between $CH_3NH_3PbI_3$ (valance band = –5.4 eV) and Cu (Fermi energy level = –4.94 eV). After the exciton dissociation, the hole can propagate along the Cu ellipsoid to the ITO electrode, while the electron can propagate along the $CH_3NH_3PbI_3$, which will help to inhibit the carrier recombination inside the $CH_3NH_3PbI_3$. Thus, the introduction of the nanoplasmonic structure can improve the FF of $CH_3NH_3PbI_3$-based photovoltaics while enhancing the light absorption.

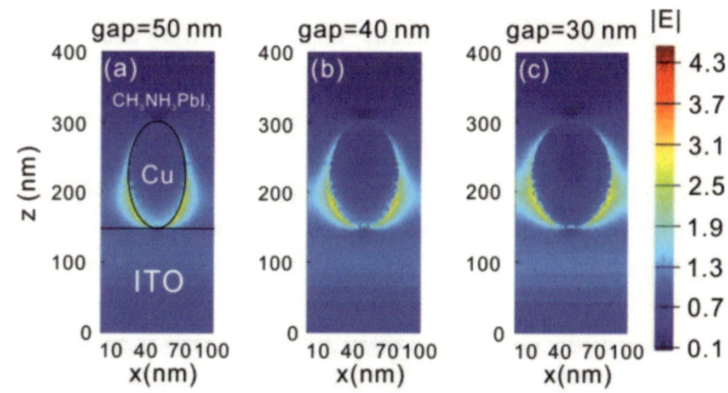

Figure 12. The distribution of the electric field (exciton) with different ellipsoid gap [39].

6. Future development direction

The PCE of GaAs single-junction photovoltaics is 28.9%, which is the closest to the Shockley–Queisser (SQ) limit of 33.7%. According to SQ theory, if the photon energy is larger than the energy gap (E_g) of light absorption materials, the energy of the incident photon will be converted from the electron to the excited state. The electron in the excited state (hot electron) and lattice will have incoherent collision and then relax to the conduction band to form exciton with the hole in the valence band, and such excitons must be dissociated to generate photocurrent. Also, assuming that the remaining energy of the absorbed photon (E_P = 2.26 eV) after thermal relaxation is equal to the energy gap of the material (E_g = 1.43 eV), so if a photon can produce a free electron, the voltage loss of this photoelectric conversion process is 0.83 V (E_P/e-E_g/e). Additionally, we can also use a simple formula to evaluate the SQ limit:

$$PCE = A \times FF \times (E_g / E_P), \tag{1}$$

where *PCE* is theoretical PCE, *A* is the amount of sunlight which can be absorbed by the material, and *FF* is fill factor. Taking GaAs and $CH_3NH_3PbI_3$ as examples, their SQ limits are 32.7% (0.6 × 0.86 × 1.43/2.26) and 26.6% (0.5 × 0.75 × 1.6/2.26), respectively, which indicates that the solar cells with $CH_3NH_3PbI_3$ as the main material still have space to improve.

In the past ten years, low production costs have been an advantage for organic photovoltaics and dye-sensitized solar cells, but the PCE is always less than 15%. Except that the absorption bandgap of materials cannot be effectively extended to the near-infrared region, the interaction between the light and the material is also the reason for the constrained efficiency. Taking P3HT-based photovoltaics for example, they are constrained by the exciton binding energy (~300 meV) and exciton diffusion length (~10 nm), and they must use the p–n interface with larger potential difference ($\Delta V_{P3HT\text{-}ICBA} = 0.64$ V) and blended thin film structure (P3HT:ICBA) in order to effectively convert excitons into the photocurrent. If we use formula (1) to estimate the theoretical PCE of P3HT:ICBA, E_g has to be replaced with eV_{OC} to obtain a PCE of 8.16% (0.32 × 0.75 × 0.87/2.26), which is much closer to the experimental value (7.4%) [41]. In addition, the absorption bandgap and the exciton binding energy seem to limit the development of organic photovoltaics. Table 2 indicates different photovoltaic performances and the characteristics of light absorption materials, and exciton binding energy of organic P3HT is much larger than inorganic materials (E_b < 20 meV). Also noteworthy is the Raman shift of organic materials (1000 to 1700 cm^{-1}), which is larger than that of inorganic materials (<600 cm^{-1}), even much larger than Raman scattering of $CH_3NH_3PbI_3$ (Pb-I stretching: 91 cm^{-1}). Also, the phonon energy of organic materials is much larger than that of inorganic materials. Because a large part of the energy is converted into thermal energy after the photons are absorbed by organic materials, it thus restricts the development of organic photovoltaics and dye-sensitized solar cells.

Absorber	V_{OC} (V)	PCE (%)	E_g (eV)	E_b (meV)	Raman shift (cm^{-1})
GaAs	1.12	28.9	1.43	4	LO 295
Si	0.71	25.0	1.11	18	520
InP	0.88	22.1	1.34	5	LO 348
CIGS	0.72	19.8	1.2	12	Cu-S 472
CdTe	0.86	19.6	1.5	10.6	LO 162
$CH_3NH_3PbI_3$	1.13	19.3	1.6	50	LO 91
Prophyrin dye	0.91	12.3	1.79	330	1000–1700
P3HT	0.75	7.4	1.9	300	1000–1700

Table 2. Different photovoltaic performances and the characteristics of light absorption materials

According to an estimation of the SQ limit, there is still space for the development of perovskite-based photovoltaics. From the perspective of the manufacturing process, most of the research over the past two years has focused on how to enhance the continuity of the

$CH_3NH_3PbI_3$ film to avoid contact between the upper and lower electrodes, which can result in a short-circuit situation. At first, using perovskite materials with a mixture of Cl can obtain better film continuity. As the mixed amount of Cl is very small, the contribution of Cl is to decrease the crystallinity of $CH_3NH_3PbI_3$ and reduce the agglomeration of particles due to the rapid crystallization to further enhance the continuity of the films [42]. Moreover, the different surface roughness of substrate (ITO and FTO) can influence the continuously of $CH_3NH_3PbI_3$ films, $CH_3NH_3PbI_3$ fabricated in a more rough substrate will have a better film continuity, which is interpreted as a rough surface that is unfavorable for $CH_3NH_3PbI_3$ crystallization. Furthermore, organic solvent or inert gas is injected into $CH_3NH_3PbI_3$ precursors during the spin coating process, which can obtain $CH_3NH_3PbI_3$ thin films with good continuity and excellent PCE of 17.8%. In order to verify the mechanism of this method, we used a SEM, a 2D-XRD and fluorescence spectroscope to analyze $CH_3NH_3PbI_3$ thin films fabricated with and without toluene washing treatment [14].

Figure 13 shows the morphologies of $CH_3NH_3PbI_3$ thin films fabricated with and without toluene treatment. The $CH_3NH_3PbI_3$ thin film without toluene treatment appeared as a flower-like structure on rough surfaces, indicating that $CH_3NH_3PbI_3$ may have specific crystallization orientation. Additionally, the $CH_3NH_3PbI_3$ thin film with toluene treatment appeared to be a more even and continuous film.

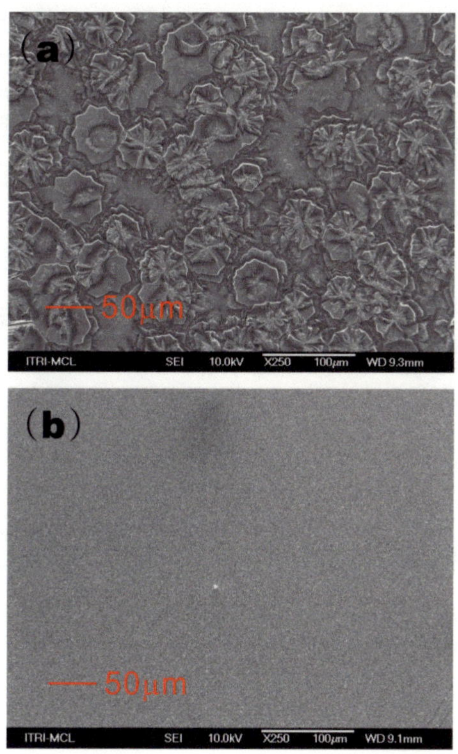

Figure 13. The SEM images of $CH_3NH_3PbI_3$ thin films; (a) without toluene treatment; (b) with toluene treatment [14].

In order to understand the crystallization orientation of $CH_3NH_3PbI_3$ materials, we use 2D-XRD as a detection tool. Figure 14 shows 2D-XRD images with and without toluene treatment. For the $CH_3NH_3PbI_3$ thin film without toluene treatment, the preferred orientation of the crystallization tendency is at (112), while the $CH_3NH_3PbI_3$ thin film with toluene treatment has no specific crystallization orientation, which indicates that $CH_3NH_3PbI_3$ precursors with the toluene treatment will interfere with the crystallization of $CH_3NH_3PbI_3$ and thus obtain more smooth and continuous $CH_3NH_3PbI_3$ thin films. Analysis of the FWHM of X-ray diffraction peak can also obtain the crystal domain size of $CH_3NH_3PbI_3$ thin films with and without toluene treatment at 17.6 nm and 29.6 nm, respectively, indicating that the toluene treatment indeed inhibited the crystallization of $CH_3NH_3PbI_3$.

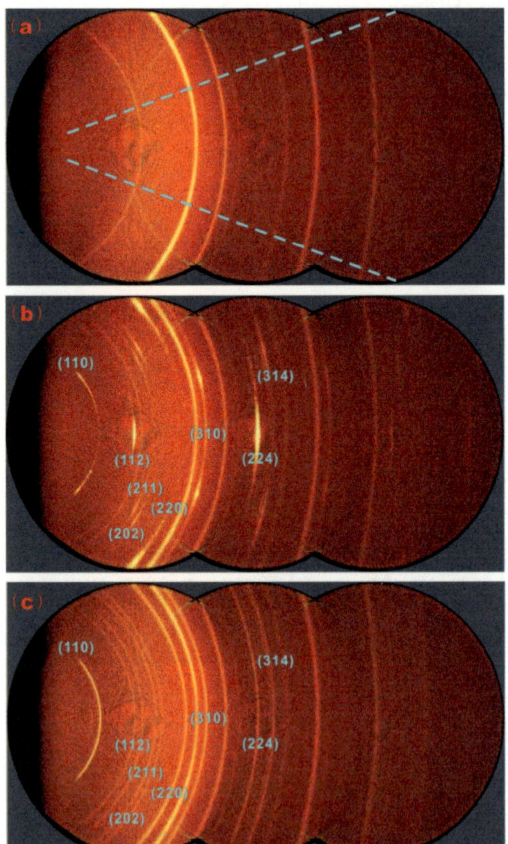

Figure 14. Two-dimensional X-ray diffraction patterns: (a) PEDOT:PSS/ITO/glass; (b) $CH_3NH_3PbI_3$/PEDOT:PSS/ITO/glass without toluene treatment; (c) $CH_3NH_3PbI_3$/PEDOT:PSS/ITO/glass with toluene treatment [14].

Figure 15 displays the absorbance spectra of perovskite absorbers on PEDOT:PSS/ITO/glass substrate fabricated with and without the toluene treatment. The perovskite films have two absorption peaks at ~480 nm and ~750 nm, which indicate the formation of $CH_3NH_3PbI_3$. When neglecting the reflectance, the absorbance can be treated as absorption. The absorption coefficient can be obtained by $\alpha = A/D$, where A is the absorbance, and D (~260 nm) is the film thickness measured by an α step. The absorption coefficient of the $CH_3NH_3PbI_3$ film fabricated with (without) the toluene treatment is 6.02×10^4 cm^{-1} (4.63×10^4 cm^{-1}) at a wavelength of 479 nm.

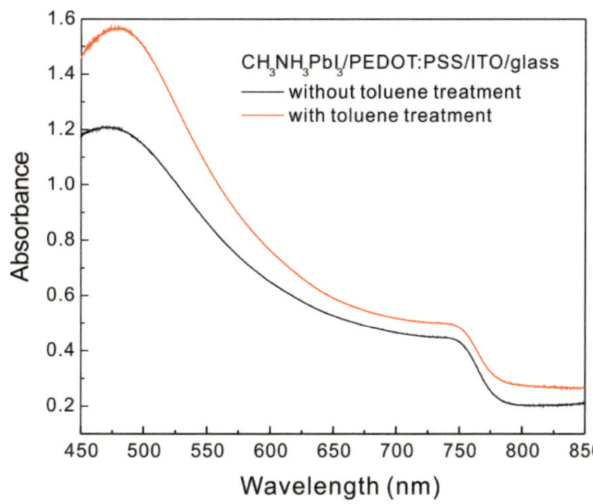

Figure 15. Absorbance spectra of $CH_3NH_3PbI_3$/PEDOT:PSS/ITO/glass without and with toluene treatment [14].

In order to understand the exciton characteristics of the CH_3NH_3PbI thin film with toluene treatment, we measured the fluorescence spectra of the sample, as shown in Figure 16. The $CH_3NH_3PbI_3$ thin film with toluene treatment had weaker fluorescence, showing that the excitons in the $CH_3NH_3PbI_3$ thin film with toluene treatment had shorter lifetimes. The experimental data showed that the toluene treatment can improve the continuity of $CH_3NH_3PbI_3$ thin films but inhibit crystallinity, leading to a reduction in the exciton's lifetime. In addition, we also used this method to fabricate photovoltaics with structures starting with the substrate sequentially to glass/ITO/PEDOT:PSS/$CH_3NH_3PbI_3$/PCBM/Ag, and the obtained highest PCE was 11.7% (V_{OC} = 0.92 V, J_{SC} = 18.2 mA/cm^2, FF = 0.7) [14], which proves that toluene-treated CH_3NH_3PbI can be used as a light absorption material for excellent PCE.

Although the $CH_3NH_3PbI_3$ absorbers with toluene treatment have good continuity, and the PCE can easily exceed 10%, the morphologies of these perovskite thin films and exciton properties showed that the substrate in the absence of specific crystallization orientation has to sacrifice the crystallinity of perovskite thin films to achieve the good continuity, which could produce high-efficiency photovoltaics through solution process methods. Therefore, we believe that it is necessary to find substrates with crystallinity or specific structure for the production of perovskite-based photovoltaics in order to manufacture perovskite absorbers with both high continuity and crystallization by solution process methods to be able to improve the V_{OC} and FF of the photovoltaics.

Since the hot electron lifetime of $CH_3NH_3PbI_3$ is about 1 ps, high conductive nanostructures can be used to improve the efficiency of hot electron injection from $CH_3NH_3PbI_3$ to electron

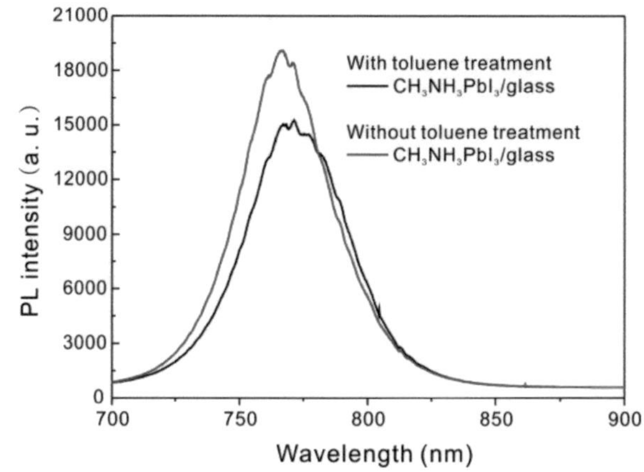

Figure 16. The photoluminescence spectra of $CH_3NH_3PbI_3$ thin films [14].

acceptors. The issue of hot electron injection and multiple exciton generation has been widely discussed in studies of quantum dot–sensitized solar cells. In theoretical estimation, the limit of the PCE of the photovoltaics with hot electron injection is 42% [43], which is higher than the SQ limit of 33.7%. Therefore, we predicted that if we use ZnO with nanorod structures to collect hot electron injection, it should be able to improve the V_{OC} of perovskite-based photovoltaics. Because the hot electrons inject into ZnO at a higher energy potential, and if the conductivity of ZnO remains high enough that the hot electrons are able to maintain its energy potential transporting to the electrode, then higher V_{OC} should be achieved.

Moreover, in the aging test of perovskite solar cells, Park conducted tests measuring the stability of PCE under different humidity storage conditions [44]. Under 55% relative humidity, the PCE of $CH_3NH_3PbI_3$-based photovoltaics decreased from 11% to 3.5% after 20 days of storage. In addition, if using $CH_3NH_3PbI_{2.4}Br_{0.6}$ as the light absorption material under the same storage conditions (55% relative humidity), the PCE could be maintained at 9.5% for 20 days. Although "I" of the $CH_3NH_3PbI_3$ was replaced by Br, which could improve the stability of perovskite-based photovoltaics, an increase of the Br substituted amount would also increase the absorption bandgap of the perovskite materials, which would result in the decrease of PCE from 11% to 9.5%, at the expense of light absorption. The results for aging tests of long-term exposure to sunlight have not yet been published in internationally important journals. As a result, it is still doubtful whether perovskite photovoltaics have long-term stability, which shall be further tested under actual conditions for the properties of the high PCE of perovskite-based photovoltaics.

7. Conclusion

The power conversion efficiency (PCE) of perovskite-based photovoltaics have exceeded 20%, and the efficiency in different types of substrates by solution process methods can reach over

http://dx.doi.org/10.5772/61278

17%, indicating that there is very high possibility for commercializing perovskite-based photovoltaics. However, the light absorption materials contain Pb, which is the biggest obstacle for commercialization of perovskite-based photovoltaics. The absorption bandgap of $CH_3NH_3SnI_3$ with Pb replaced with Sn is about 1.3 eV, which is a promising light absorption material. So far, the highest PCE of $CH_3NH_3SnI_3$-based photovoltaics is about 5.2% far away from the Shockley–Queisser limit of 33.7%. Therefore, research on $CH_3NH_3SnI_3$-based photovoltaics have been slowed due to the benefit issues, and thus it is essential to conduct more basic research to further evaluate the limits of perovskite-based photovoltaics for the arrival of earlier commercialization process.

Acknowledgements

This work was supported by the Material and Chemical Research Laboratories, Industrial Technology Research Center and the National Science Council under Grant NSC 101-2731-M-008-002-MY3.

Author details

Sheng Hsiung Chang[1*], Hsin-Ming Cheng[2], Sheng-Hui Chen[3] and Kuen-Feng Lin[1,3]

*Address all correspondence to: shchang@ncu.edu.tw

1 Research Center for New Generation Photovoltaics, National Central University, Taoyuan, Taiwan, ROC

2 Material and Chemical Research Laboratories, Industrial Technology Research Center, Hsinchu, Taiwan, ROC

3 Department of Optics and Photonics, National Central University, Taoyuan, Taiwan, ROC

References

[1] J. G. Bednorz and K. A. Muller. Perovskite-type oxides—The new approach to high-Tc superconductivity. *Reviews of Modern Physics* 60 (1988) 585–600.

[2] T. He, Q. Huang, A. P. Ramirez, Y. Wang, K. A. Regan, N. Rogado, M. A. Hayward, M. K. Haas, J. S. Slusky, K. Inumara, H. W. Zandbergen, N. P. Ong, and R. J. Cava. Superconductivity in the non-oxide perovskite $MgCNi_3$. *Nature* 411 (2001) 54–56.

[3] G. Xing, N. Mathews, S. Sun, S. S. Lim, Y. M. Lam, M. Gratzel, S. Mhaisalkar, and T. C. Sum. Long-range balanced electron- and hole-transport lengths in organic–inorganic $CH_3NH_3PbI_3$. *Science* 342 (2013) 344–347.

[4] K. Tanaka, T. Takahashi, T. Ban, T. Kondo, K. Uchida, and N. Miura. *Solid State Communications* 127 (2003) 619–623.

[5] M. Era, S. Morimoto, T. Tsutsui, and S. Saito. Organic–inorganic heterostructure electroluminescent device using a layered perovskite semiconductor (C6H5C2H4NH3)2PbI4. *Applied Physics Letters* 65 (1994) 676–678.

[6] A. Kojima, K. Teshima, Y. Shirai, and T. Miyasaka. Organometal halide perovskite as visible-light sensitizers for photovoltaic cells. *Journal of the American Chemical Society* 131 (2009) 6050–6051.

[7] L. Etgar, P. Gao, Z. Xue, Q. Peng, A. K. Chandiran, B. Liu, Md. K. Nazeeruddin, and M. Gratzel. Mesocopic $CH_3NH_3PbI_3/TiO_2$ heterojunction solar cells. *Journal of The American Chemical Society* 134 (2012) 17396–17399.

[8] Michael M. Lee, Joël Teuscher, Tsutomu Miyasaka, Takurou N. Murakami, Henry J. Snaith. Efficient hybrid solar cells based on meso-superstructured organometal halide perovskites. *Science* 338 (2014) 643–647.

[9] J. Burschka, N. Pellet, S-J. Moon, R. Humphry-Baker, P. Gao, M. K. Nazeeruddin, and M. Gratzel. Sequential deposition as a route to high-performance perovskite-sensitized solar cells. *Nature* 499 (2013) 316–319.

[10] M. Liu, M. B. Johnston, and H. J. Snaith. Efficient planar heterojunction perovskite solar cells by vapour deposition. *Nature* 501 (2013) 395–398.

[11] C-H. Chiang, Z-L. Tseng, and C-G. Wu. Planar heterojunction perovskite/$PC_{71}BM$ solar cells with enhanced open-circuit voltage via a (2/1)-step spin-coating process. *Journal of Materials Chemistry A* 2 (2014) 15897–15903.

[12] H. Zhou, Q. Chen, G. Li, S. Luo, T-B. Song, H-S. Duan, Z. Hong, J. You, Y. Liu, and Y. Yang. Interface engineering of highly efficient perovskite solar cells. *Science* 345 (2014) 542–546.

[13] J-H. Im, I-H. Jang, N. Pellet, M. Gratzel, and N-G. Park. Growth of $CH_3NH_3PbI_3$ cuboids with controlled sized for high-efficiency perovskite solar cells. *Nature Technology* 9 (2014) 927–932.

[14] K-F. Lin, S-H. Chang, K-H. Wang, H-M. Cheng, K. Y. Chiu, K-M. Lee, S-H. Cheng, and C-G. Wu. Unraveling the high performance of tri-iodide perovskite absorber based photovoltaics with a non-polar solvent washing treatment. *Solar Energy Materials and Solar Cells*, accepted for publication.

[15] W-J. Yin, T. Shi, and Y. Yan. Unusual defect physics in $CH_3NH_3PbI_3$ perovskite solar cell absorber. *Applied Physics Letters* 104 (2014) 063903.

[16] M. Hirasawa, T. Ishihara, T. Goto, K. Uchida, and N. Miura. Magnetoabsorption of the lowest exciton in perovskite-type compound $(CH_3CH_3)PbI_3$. *Physica B: Condensed Matter* 201 (1994) 427–430.

[17] F. Brivio, A. B. Walker, and A. Walsh. Structural and electronic properties of hybrid perovskites for high-efficiency thin-film photovoltaics from first-principles. *APL Materials* 1 (2013) 042111.

[18] Z. Guo, J. S. Manser, Y. Wan, P. V. Kamat, and L. Huang. Spatial and temporal imaging of long-range charge transport in perovskite thin films by ultrafast microscopy. *Nature Communications* 6 (2015) 7471.

[19] S. D. Stranks, G. E. Eperon, G. Grancini, C. Menelaou, M. J. P. Alcocer, T. Leijtens, L. M. Herz, A. Petrozza, and H. J. Snaith. Electron–hole diffusion lengths exceeding 1 micrometer in an organometal trihalide perovskite absorber. *Science* 342 (2013) 341–344.

[20] K. T. Butler, J. M. Forst, and A. Walsh. Band alignment of the hybrid halide perovskites $CH_3NH_3PbCl_3$, $CH_3NH_3PbBr_3$, and $CH_3NH_3PbI_3$. *Materials Horizons* accepted 28 Oct. 2014 (DOI: 10.1039/C4MH00174E).

[21] F. Hao, C. C. Stoumpos, D. H. Cao, R. P. H. Chang, and M. G. Kanatzidis. Lead-free solid-state organic–inorganic halide perovskite solar cells. *Nature Photonics* 8 (2014) 489–494.

[22] J. M. Frost, K. T. Butler, F. Brivio, C. H. Hendon, M. van Schilfgaarde, and A. Walsh. Atomistic origins of high-performance in hybrid halide perovskite solar cells. *Nano Letters* 14 (2014) 2584–2590.

[23] G. E. Eperon, S. D. Stranks, C. Menelaou, M. B. Johnston, L. M. Herz, and H. J. Snaith. Formamidinium lead trihalide: a broadly tunable perovskite for efficient planar heterojunction solar cells. 7 (2014) 982–988.

[24] J-W. Lee, D-J. Seol, A-N. Cho, and N-G. Park. High-efficiency perovskite solar cells based on the black polymorph of $HC(NH_2)_2PbI_3$. *Advanced Materials* 26 (2014) 4991–4998.

[25] M. A. Green, K. Emery, Y. Hishikawa, W. Warta, and E. D. Dunlop. Solar cell efficiency table (version 44). *Progress in Photovoltaics: Research and Applications* 22 (2014) 701–710.

[26] W. S. Yang, J. H. Noh, N. J. Jeon, Y. C. Kim, S. Ryu, J. Seo, and S. I. Seok. High-performance photovoltaic perovskite layers fabricated through intromolecular exchange. *Science* (DOI: 10.1126/science.aaa9272).

[27] F. Liu, J. Zhu, J. Wei, Y. Li, M. Lv, S. Yang, B. Zhang, J. Yao, and S. Dai. Numerical simulation: toward the design of high-efficiency planar perovskite solar cells. *Applied Physics Letters* 104 (2014) 25358.

[28] H-Y. Hsu, C-Y. Wang, A. Fathi, J-W. Shiu, C-C. Chung, P-S. Shen, T-F. Guo, P. Chen, Y-P. Lee, and E. W-G. Diau. Femtosecond excitonic relaxation dynamics of perovskite on mesoporous films of Al_2O_3 and NiO nanoparticles. *Angewandte Chemie* 53 (2014) 9339–9342.

[29] S. H. Chang, C-H. Chiang, H-M. Cheng, C-Y. Tai, and C-G. Wu. Broadband charge transfer dynamics in P3HT:PCBM blended film. *Optics Express* 38 (2013) 5342–5345.

[30] T. C. Sum and N. Mathews. Advancements in perovskite solar cells: photophysics behind the photovoltaics. *Energy Environmental Science* 7 (2014) 2518–2534.

[31] W. Nie, H. Tsai, R. Asadpour, J-C. Blancon, A. J. Neukirch, G. Gupta, J. J. Crochet, M. Chhowalla, S. Tretiak, M. A. Alam, H-L. Wang, and A. D. Mohite. High-efficiency solution-processed perovskite solar cells with millimeter-scale grains. *Science* 347 (2015), 522–525.

[32] B. Conings, L. Baeten, C. D. Cobbelaere, J. D'Haen, J. Manca, and H-G. Boyen. Perovskite-based hybrid solar cells exceeding 10% efficiency with high reproducibility using a thin film sandwich approach. *Advance Materials* 26 (2014) 2041–2046.

[33] P-W. Liang, C-Y. Laio, C-C. Chueh, F. Zuo, S. T. Williams, X-K. Xin, J. Lin, and A. K-Y. Jen. Additive enhanced crystallization of solution-processed perovskite for highly efficiency planar-heterojunction solar cells. *Advanced Materials* 26 (2014) 3748–3754.

[34] M. Xiao, F. Huang, W. Huang, Y. Dkhissi, Y. Zhu, J. Etheridge, A. Gray-Weale, U. Bach, Y-B. Cheng, and L. Spiccia. A fast deposition-crystallization procedure for highly efficient lead iodide perovskite thin-film solar cells. *Angewandte Chemie* 126 (2014) 10056–10061.

[35] N. J. Jeon, J. H. Noh, Y. C. Kim, W. S. Yang, S. Ryu, and S. I. Seok. Solvent engineering for high-performance inorganic–organic hybrid perovskite solar cells. *Nature Materials* 13 (2014) 897–903.

[36] F. Huang, Y. Dkhissi W. Huang, M. Xiao, I. Benesperi, S. Rubanov, Y. Zhu, X. Lin, L. Jiang, Y. Zhou, A. Gray-Weale, J. Etheridge, C. R. McNeill, R. A. Caruso, U. Bach, L. Spiccia, and Y-B. Cheng. Gas-assisted preparation of lead iodide perovskite films consisting of a monolayer of single crystalline grains for high efficiency planar solar cells. *Nano Energy* 10 (2014) 10–18.

[37] J. W. Jung, S. T. Williams, and A. K-Y. Jen. Low-temperature processed high-performance flexible perovskite solar cells via rationally optimized solvent washing treatments. *RSC Advances* 4 (2014) 62971–62977.

[38] C. Bi, Y. Shao, Y. Yuan, Z. Xiao, C. Wang, Y. Gao, and J. Huang. Understanding the formation and evolution of interdiffusion grown organolead halide perovskite thin films by thermal annealing. *Journal of Chemistry A* 2 (2014) 18508–18514.

[39] S. H. Chang, K-F. Lin, C-H. Chiang, S-H. Chen, and C-G. Wu. Plasmonic structure enhanced exciton generation at the interface between the perovskite absorber and copper nanoparticles. *Scientific World Journal* 2014 (2014) 128414.

[40] S. H. Chang. Modeling and design of Ag, Au, and Cu nanoplasmonic structures for enhancing the absorption of P3HT:PCBM-based photovoltaics. *IEEE Photonics Journal* 5 (2013) 4800509.

[41] X. Guo, C. Cui, M. Zhang, L. Huo, Y. Huang, J. Hou, and Y. Li. High efficiency polymer solar cells based on poly(3-hexylthiophene)/indene-C70 bisadduct with solvent additive. *Energy Environmental Science* 5 (2012) 7943–7949.

[42] S. T. Williams, F. Zuo, C-C. Chueh, C-Y. Liao, P-W. Liang, and A. K-Y. Jen, Role of chloride in the morphological evolution of organo-lead halide perovskite thin films. *ACS Nano* 8 (2014) 10640–10654.

[43] M. C. Beard, A. G. Midgett, M. C. Hanna, J. M. Luther, B. K. Hughes, and A. J. Nozik. Comparing multiple exciton generation in quantum dots to impact ionization in bulk semiconductors: implications for enhancement of solar energy conversion. *Nano Letters* 10 (2010) 3019–3027.

[44] N-G. Park. Perovskite solar cells: an emerging photovoltaic technology, *Materials Today* 18 (2015) 65–72.

Chapter 15

Numerical Simulations on Perovskite Photovoltaic Devices

Bernabé Marí Soucase, Inmaculada Guaita Pradas and Krishna R. Adhikari

Additional information is available at the end of the chapter

http://dx.doi.org/10.5772/61751

Abstract

Organometal halide perovskites have recently attracted tremendous attention due to their potential for photovoltaic applications, and they are also considered as promising materials in light emitting devices. In particular, in the last years promising photovoltaic devices with efficiencies above 20% have already been prepared using organometal halide perovskites as absorbent materials.

A planar heterojunction perovskite-based solar cell is made of three main layers sandwiched between the two conducting electrodes. The standard design for a planar heterojunction perovskite-based solar cell is: Back electrode/ Hole Transport Material (HTM)/ Perovskite absorber / Electron Transport Material (ETM) / Transparent electrode. For planar heterojunction-based solar cells, recent efforts have revealed that increasing conductivity of the hole transport materials by doping and optimizing charge collection by adjusting the absorber thickness could bring a positive impact on the efficiency. Electron transporting materials are also a crucial component in perovskite-based solar cells. The effect of different electron transporting materials in the final behaviour of the PV device can also be numerically simulated. Several PV parameters such as thicknesses of the absorber, HTM and ETM, respectively can be optimized by simulation methods and subsequently implemented by experimentalists. The hole mobility and acceptor concentration of the HTM, interface trap density and work-function of back contact metal have shown significant influence on the device performance. Even with these strong merits, enhancement of hole mobility and conductivity of HTM, stability of perovskite and TiO_2 and replacement of toxic lead are still crucial. Through suitable processing/synthesizing of the perovskite absorbers, best engineering the selective contact, and increasing conductivity of HTM and ETM will boost the stability as well as performance of the device.

This chapter presents a review of the evolution of perovskite materials from their discovery to their present significance as the main constituent of a new class of photovoltaic devices. We also evaluate the use of numerical simulation methods for determining the optimal configuration of perovskite-based solar cells and analyzing their optoelectronic behavior. The outcome of a simulation study on organometal halide perovskite focusing on the role of the different components of the solar cell using Solar Cell Capacitance Simulator as a simulation tool are discussed. A photoconversion efficiency of 22.7%, V_{OC} =

1.12 V, J_{SC} = 24.66 mA/cm^2 and FF = 82.1% has been found after optimizing the different parameters involved. These results still are slightly higher than the experimental ones but the confluence between both are expected in the short time.

Keywords: Organometal halide perovskites, $CH_3N_3PbI_3$, Hole Transport Materials, Electron Transport Materials, photovoltaic solar cell, photocurrent, photoconversion efficiency

---

1. Introduction

The world is highly mechanized, competitive, depends almost upon machinery, life style is being improving day by day consequently energy consumption rate is increasing unexpectedly. To fulfill the current energy demand, the world depends, to date, upon fossil fuel based (coal, natural gas, and petroleum products) energy resource. At the mean time, its reservoirs are limited and getting depleted gradually, cost is highly influencing by socio-economic and political situations, and its use creates an adverse effect to the environment as well as to the human health. Thus, everyone should think its best alternative before it comes to an end. So, it is the high time to look at and develop the unexhausted, environment friendly green/clean energy and associated technologies. As a source of energy solar energy is free, unlimited, and available everywhere in this world. The best option, hereby, might be solar energy and solar cell technology which is the main purpose of this paper and discussed briefly below. However, to date the relatively low efficiency to cost ratio of photovoltaic solar cells limiting its use. Thus, increasing efficiency and decreasing cost of solar cell is the major challenge to the researchers, academicians and industrialists.

1.1. Solar cells

In every step and activity in our life, such as: cooking a food, lighting a home and a street, opening a college and a hospital, running a vehicle and a factory etc we need energy. Thus energy is extremely crucial factor for better quality of life for one and all, employment creation and industrial as well as economic development in a country. Global population, their living style/standard, industries and hence the energy demand is increasing day by day where world energy generation capacity is nearly 18 TW [1] and otherwise almost remaining the same. The main resource of energy, worldwide, is fossil fuel based energy. Actually it is untenable and also contributes substantially to climate change and global warning. World nuclear energy report 2014 highlighted the fact that "the nuclear share in the world's power generation declined steadily from a historic peak of 17.6% in 1996 to 10.8% in 2013". In the world, 31 countries with 388 reactors are currently generating 333 GW. 67 reactors are under construction since July 2014 with a total capacity of 64 GW [2]. Nuclear energy data indicates that it would not meet the current energy demand and it is quite hazardous to the mankind too. The renewable resources; hydroelectricity, geothermal, wind, bio-fuels are limited. On the other hand solar energy is inexhaustible, accessible, nonhazardous and environment friendly too which can directly be changed into heat and electricity. In facts feasible capacity of generation of electricity from the sun light is 1000 times higher than the current world energy demand [1]. Moreover, production of photo-electricity on one day is sufficient for one year by using even a less efficient photovoltaic solar cell. So, it would be the best option to resolve the world future energy demand and crisis.

Solar cell is usually called photovoltaic which literary means the conversion of light or photon into electricity. Edmund Becquerel discovered photovoltaic effect in 1839 by illuminating silver chloride in an acidic solution and measuring the current whereas this effect was observed in 1876 by W. G. Adams and R. E. Day for the first time in a solid: selenium [3, 4].

In 1883, Charles Fritts demonstrated thin film selenium cell and he had realized enormous potential of photovoltaic devices [3]. Albert Einstein discovered the photoelectric effect in1905, and awarded by the Nobel Prize in Physics 1921 for this theoretical work [5]. The next significant step forward came 34 years later, after Fritts, where the work was carried out with the copper-cuprous oxide by Grondahl in 1927 [6] and he described the outcome of the work as development of a rectifier and a photovoltaic cell.

Almost at the same time, silicon had drawn the attention of researchers as it was emerging for use as point contact rectifier. Russel S. Ohl, Bell laboratory engineer, developed a "diode" in 1939 by doping one side of silicon with electron donor and other side with acceptor material. After one year he had observed an electric voltage across the ends of the so formed "p-n junction diode" when light shone it. It was the first invention of the photovoltaic device based on silicon semiconductor thus; Ohl got a patent for his work and this effect. This fact revealed after their paper published in 1952 [7]. In the same laboratory in 1954 D. M. Chapin, C. S. Fuller, and G. L. Pearson announced the first "modern photovoltaic silicon cell" with the efficiency of 6% [8]. Vanguard I was the first satellite to use solar power launched by USA in 1958. After that Explorer III, Vanguard II and Sputnik-3 were launched with PV-powered system on board.

In order to design a solar cell and realize the role of different components we must understand basic steps or processes they are undertaken in the solar cell layers. Following are the processes that undergo to convert electromagnetic energy into electrical energy in a photovoltaic [9] where this phenomenon is known as photovoltaic action:

- absorption of incident light by an absorber i.e., interaction between the photon and an absorber to generate either electron-hole pairs or excitons so, the absorber jumps to the excited state i.e., an electron from valence band to conduction band,

- the separation of charge carrier of opposite types and to move in opposite directions i.e., toward contact the electrodes (anode and cathode),

- the separate extraction of these carriers to an external circuit via contacts where energetic electron leaves an electrode, does something useful at an electrical load, hereby loose energy, and finally returns back to initial position through another electrode to complete one round trip,

- at the end of the round trip, the electron recombine with a hole so that the absorber returns back to the ground state and ready to absorb another photon for next trip.

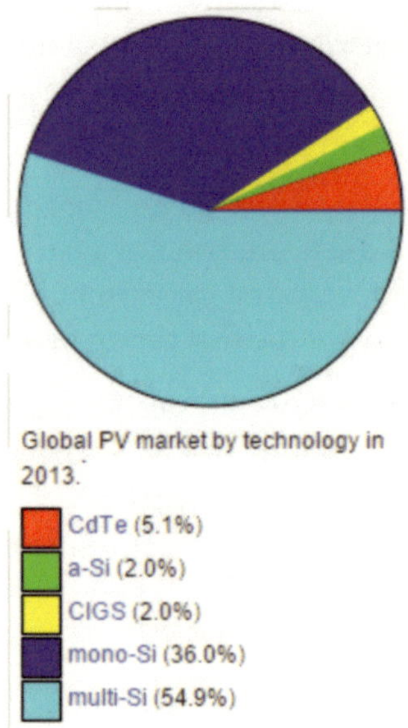

Source: [10, 11]

Figure 1. Global PV market by Technology

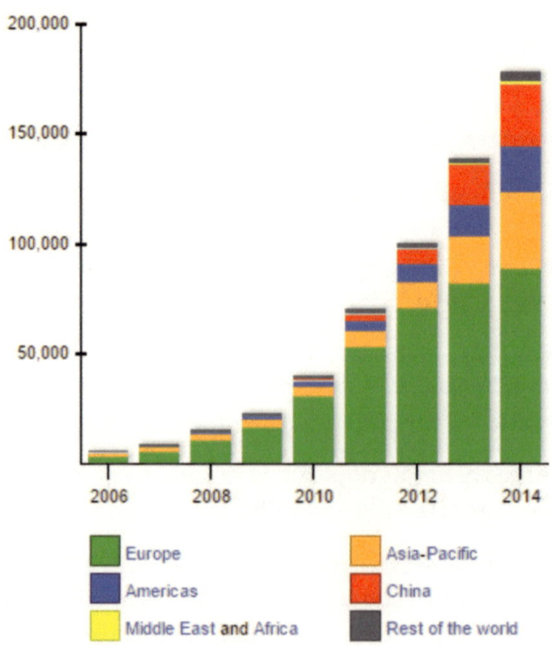

Source: [10, 11]

Figure 2. Worldwide growth of Photovoltaics

http://dx.doi.org/10.5772/61751

Since that time different solar cell technologies have been developed and achieved different efficiencies. Now NREL divided the solar cell technologies into 5 categories they are: Crystalline silicon cells, single junction GaAs, multijunction cells, thin film technologies and emerging PV. Gradual development of solar cells and their best efficiency is illustrated in Figure 3 [12]. Leonid A. Kosyachenko categorized the solar cell technologies as so called first, second and third generation photovoltaics. Based on Kosyachenko classification, all types of silicon wafer based and GaAs solar cells are represented as the first generation photovoltaics. Amorphous silicon (a-Si) and non silicon based thin film such as CIGS, CdTe based solar cells are so-called second generation photovoltaics, relatively younger dye-sensitized solar cells (DSSCs), organic and quantum dots solar cells are the third generation photovoltaics [1], whereas hybrid solar cells are considered as fourth generation solar cells. There is always more space in third and fourth generation photovoltaic technology to identify and develop further new materials and solar cell devices to achieve a low cost and high efficiency.

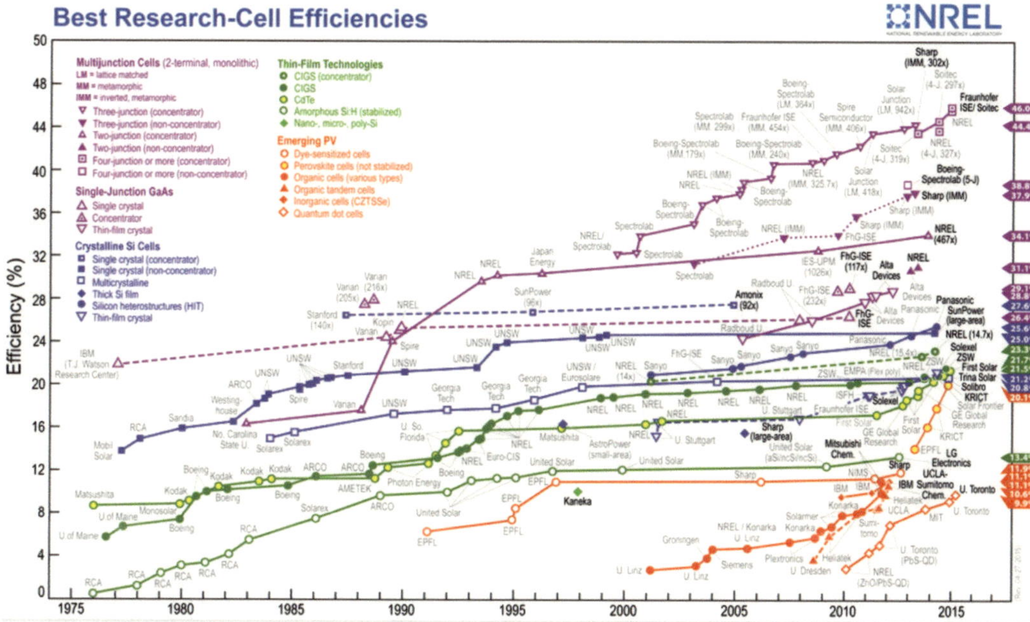

Figure 3. Best research cell efficiencies [12]

Silicon is indirect band gap (BG) material even though its BG is 1.1eV thus, silicon based solar cell consumes a huge material and energy, as a consequence, a low productivity, high labor cost and hence a very low efficiency to cost ratio even for mass production with feasible efficiency of 16-17% and 13-15% in case of single- and polycrystalline materials respectively [13]. Even with such demerits, since the date of first comercial manufacture of PV solar cells around in 60's decade of 20th century to date the silicon based solar cells have been extremly dominating the photovoltaic technology. However, in the last decade new generation of materials and devices (eg; dye sensitized, cadmium telluride and copper indium gallium selenide based solar cells) has spread the photovoltaics region.

It is illustrated by Figure 1. Figure 2, 3 and 4, inspiring gestures, show the worldwide orientation towards the solar cell technologies; hopefully it would be the first in near future. Organic sensitizers in dye sensitized solar cells (DSSCs) have low absorption coefficients and narrow absorption bands which limit the photovoltaic performance [15]. Liquid organic electrolyte in DSSCs is volatile and can easily be degraded by ultravoilet radiation, so stability is the another major issue in this technology [1].

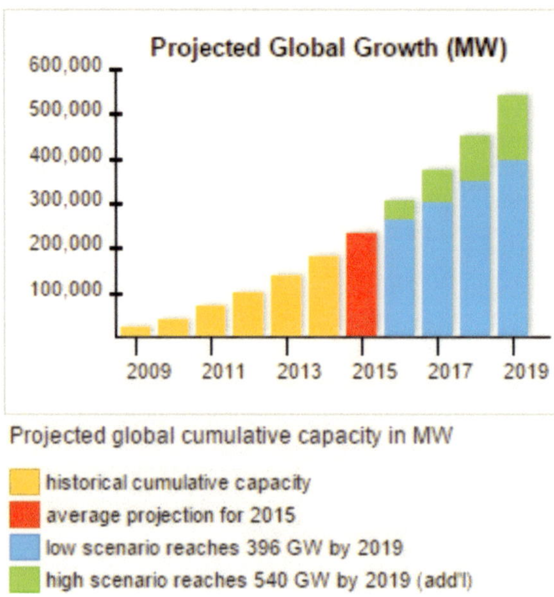

Figure 4. Projected global growth of PV. [14] (Source: Growth of photovoltaics)

Cadmium, tellurium, indium and galium are rare elements and almost they are expensive too furthermore, Cd and Te are toxic as well as scattered elements [1, 16]. These facts about existing materials always look for the low cost, easy processability, low consumption of material, reproducible, environment friendly, non toxic and easy abundant absorbing material for the solar cell industries.

1.2. Perovskites

The main concern in designing a photovoltaic solar cell is to maximize the efficiency to cost ratio i.e., to reduce the total cost, increase the efficiency and life time of PV module [17]. There are various factors they affect the efficiency of a solar cell. The first and foremost factor is the percentage of electromagnetic energy reaching the absorber which the solar cell converts into electricity. Secondly, the types of materials since different materials have different absorption coefficients and band gap and hence different maximum theoretical efficiency [18]. Third factor is the structure of the material or semiconductor used in the cell. Higher performance of solar cell is expected and usually observed in perfect crystalline structure with suitable dopants. Fourth is thickness of the absorbing material. Too thin and too thick; both are not suitable for good photovoltaic action. In the former one the thickness may not sufficient for diffusion of charge carriers whereas latter case increases the cost and reduces efficiency. Fifth factor is the amount of light reaching the absorbing material i.e., reflectance, transmittance and absorbance of the material. Sixth factor affecting the efficiency is the temperature since different materials have different response to temperature higher than room temperature.

1.2.1. Structure of perovskites

In 1839, Gustav Rose discovered the mineral calcium titanate ($CaTiO_3$) in Ural Mountain, Russia and gave the name Perovskite to honor the Russian mineralogist L. A. Perovski [19]. Any material that resembles the crystal structure of mineral $CaTiO_3$ is termed as perovskite structure or simply perovskite. It is generally represented by the formula ABX_3 where A is a big-sized cation (either organic or inorganic), B is divalent small metallic cation (Cu^{2+}, Mg^{2+}, Ge^{2+}, Sn^{2+}, Pb^{2+}, Eu^{2+}, Yb^{2+}, etc), and X is halide ion (Cl^-, Br^- and I^-) which binds to both cations [20]. The perovskites can be divided into two main categories: alkali halide perovskite and organo-metal halide perovskite.

Perovskites possess different astonishing optoelectronic behaviors which make the perovskite as promising candidate of photovoltaic absorber. Ferroelectric behavior, discovered half a century ago, is one of the characteristics among them.

In 1979, Salau reported the potassium lead iodide as a direct band gap material with the value 1.4 to 2.2 eV, which suits the solar spectrum [21, 22]. Capability of halide perovskite to convert light to electricity was discovered in 1990s and fabricated for LED.

α, β, γ, & δ are four possible phases of perovskite where α is high temperature phase T > 327 K and has cubic structure (eg $CsSnI_3$). This structure allows only one formula unit per unit cell, so $CH_3NH_3^+$ cannot obtain cubic structure. For a temperature T < 327 K perovskite changes from α to β phase usually found in tetragonal structure with lattice parameters a = 8.855 Å and c = 12.659 Å where exact values depend on molecular orientation [23].

Generally, ABX_3 has cubic structure and α phase where B has 6 nearest neighbor X ions (octahedral) and A has twelve fold coordination sites as shown in Figure 5.

One crucial parameter to maintain the cubic structure of perovskite is the tolerance factor, $t = (R_A+R_X)/\sqrt{2}\,(R_B+R_X)$. It should be close to one to obtain in the cubic structure where, R is radius of the ions and suffixes A, B and X are as defined above. Hendon et al. mentioned that stable perovskite is found in the range 0.7< T < 1 [24], which guides cation A must be larger than B in this regard, CH_3NH_3 ion is one of the best options. Smaller t could be lead to lower symmetry tetragonal β phase or orthorhombic γ phase. For $MAPbI_3$ perovskite, transition from α to β to γ occurs at 327 and 160^0 K respectively. The transition of perovskite is depending on the tilting and rotation of the BX_6 polyhedra in the lattice. Whereas the fourth phase δ is non perovskite phase [23, 24, 25, 26, 27].

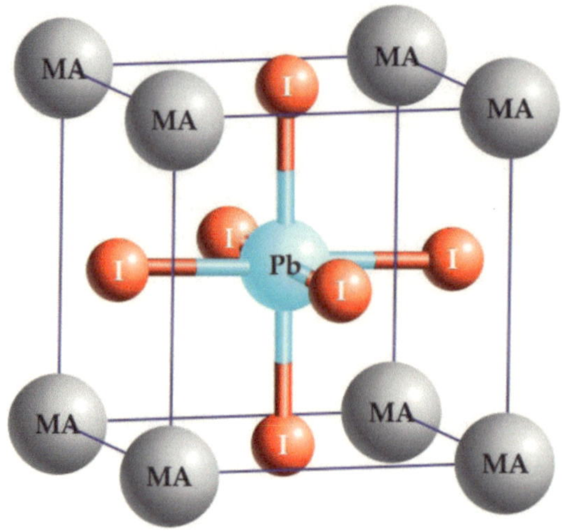

Source: Perovskite http://www.surrey.ac.uk/chemistry/perovskite/what/history/index.htm

Figure 5. Cubic structure of $MAPbI_3$ perovskite

1.2.2. Electronic Behavior

Pb has an occupied 6s orbital, which is below the top of valence bands of the perovskite. This lone pair of s electrons often gives rise to unusual behaviors in perovskite [28]. Unlike GaAs and CdTe, in the first principle study carried out by Walsh et al., 2011 density of states (DOS) and partial charge density plots of halides perovskite showed the coupling between Pb s and I p (anti-bonding state) contribute to VBM but CBM is derived almost from the Pb p state [28, 29]. And perovskite gets ionic and covalent, dual nature in electronic structures. Thus organic part/ion does not play a direct role to determine electronic behaviors but takes part in stabilizing perovskite structure and changing the lattice constants.

Experimental/research works and DFT-PBE calculations show the different values of band gap for perovskites. Band gap of perovskite depends on synthesizing process and the size of organic/inorganic cation, metallic ion and very less in halide ion. Band gap of $CH_3NH_3PbI_3$, $CH_3NH_3PbBr_3$, $CH_3NH_3PbCl_3$, $CH_3NH_3Pb_{3-x}Cl_x$ are 1.49-1.61, 1.95 eV, 2.46 eV and 1.59 eV respectively [20, 30] whereas mostly used band gap for $CH_3NH_3PbI_3$ is 1.5 eV [27].

1.2.3. Ambipolar conductivity

The electronic structure of $MAPbI_3$ perovskite is different compared to conventional semiconductor. A cation Pb p orbital has a much higher energy level than anion p orbital as in p-s semiconductor and hence CBM of $MAPbI_3$ is more dispersive. At the meantime VBM is also dispersive due to strong s-p coupling. Based upon the formula $m^* = h^2 [\frac{\partial^2 \varepsilon(k)}{\partial^2 k}]^{-1}$ [31] effective mass of electron is balanced by that of hole in $MAPbI_3$ perovskites which finally results into an ambipolar charge transport behavior in perovskite based solar cells.

1.2.4. Optical properties

Absorber is the key player in photovoltaic solar cells to achieve good performance. In this context, optical absorption strength and range are the crucial factor of the materials. The edge transition for $MAPbI_3$ perovskite comes from mixed s-p coupling and Pb p orbital so that transition probability from Pb s to Pb p is high [27]. Moreover, perovskite is direct band gap material and hence it has high optical absorption strength and wider range to absorb sufficient solar energy to achieve high value of power conversion efficiency. Figure 6 shows the absorbance spectra for three $MAPbX_3$ perovskites with different halide components; X=I, Br and Cl. Notice that the band gap changes with the selection of halide whereas $MAPBI_3$ has the best band gap for photovoltaic applications.

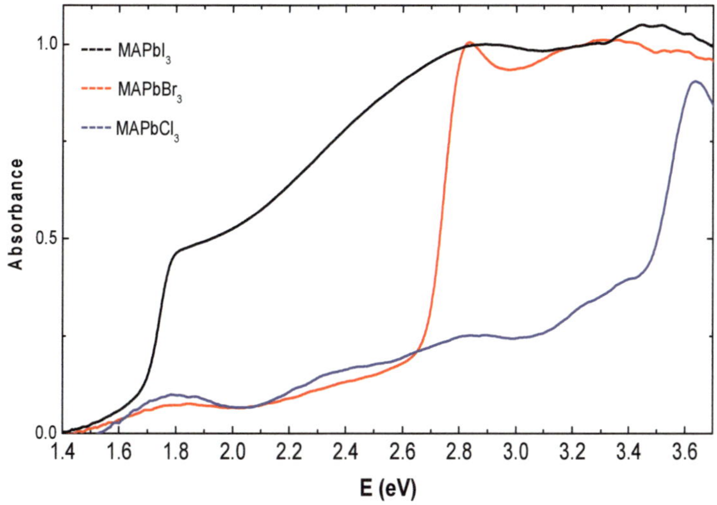

Figure 6. Absorbance for different organometal halide perovskites $MAPbI_3$, $MAPbBr_3$ and $MAPbCl_3$.

The performance of the photovoltaic solar cells mainly depends upon absorption value, and thickness of the absorption layer. Plot in Figure 7 shows the variation of performance of the solar cell with thickness of the perovskite. Usually performance of solar cell increases with increase in thickness of the perovskite. Furthermore, the plot shows that thin layer of $MAPbI_3$ layer yields high fill factor (FF).

1.2.5. Defects in perovskites: intrinsic/point defect

Efficiency of a solar cell is highly affected by defects such as point/intrinsic and grain boundaries, crystalline structure and amount of doping of an absorbing material when processed via low-cost method. Electron and hole diffusion length and open circuit voltage (V_{OC}) of a solar cell is greatly influenced by the point (Schottky and Frenkel) defects [27]. The defect densities of perovskite depend on the formation energy and hence chemical potential, it corresponds to precursors, partial pressure and temperature. The defects with low formation

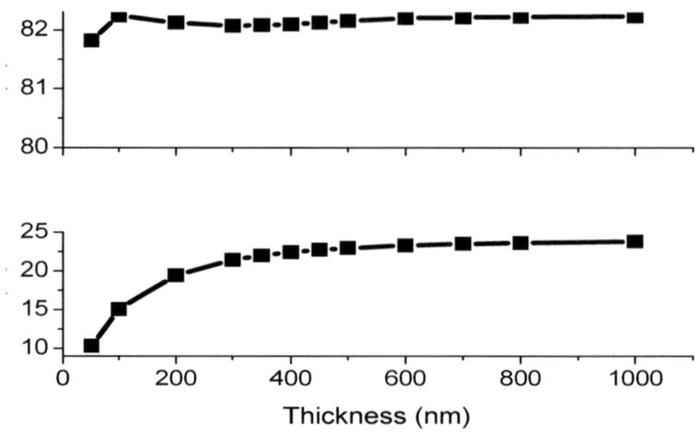

Figure 7. Variation of FF and PCE with thickness of perovskite

energy create only shallow level which results the long electron-hole diffusion length and high V_{OC}. On the other hand defects with deep levels have high formation energy it results into unpleasant effect on electron-hole diffusion length and V_{OC}. Experimental and simulation results of the long electron-hole diffusion length and high V_{OC} conforms the unusual shallow defect levels [27].

In CIGS and CdTe, p-type doping is easier in equilibrium and n-type doping is rather difficult due to self-compensation. But in $MAPbI_3$ the formation energy of methylammonium (MA) interstitial defect (donor like) and iodine vacancy defect (acceptor like) have similar values that make both; p-type and n-type doping possible. Unlike other researchers, Agiorgousis et al. and Baumann et al. suggested and pointed out the strong covalent bonds and deep level defects by using first principles calculations [32, 33]. Despite that the presence of defects and traps in perovskite and particulars of their impact are so far under discussion and investigation.

1.2.6. Progress in perovskite solar cells

In 2009, Miyasaka et al. opened up first perovskite solar cell based on mesoporous TiO_2 photo-anode and observed the power conversion efficiency (PCE) 3.81% and 3.13 % for $MAPbI_3$ and $MAPbBr_3$ respectively along with their poor cell stability [15].

In 2012, Kanatzidis and his coworkers synthesized alkali metal perovskite (floride doped $CsSnI_3$) as p-type Hole Transporting Material (HTM) in dye sensitized solar cells where 10.2% PCE was reported. At the mean time, Michael Gratzel & coworkers with N. G. Park used $MAPbI_3$ as light absorber with SpiroMETAD on mesoporous TiO_2 where measured efficiency was 9.7%. Snaith in collaboration with Miyasaka boosted PCE and V_{OC} to 10.9 % and 1.13 V respectively by replacing the n type mesoporous titanium oxide by an inert Al_2O_3 scaffold [34] in the consequence of faster diffusion of electron through the perovskite.

Liu and Snaith et al. in 2013 further boosted the efficiency to 15.4% via architecturing the vapor deposition heterojunction solar cell without electron conduction scaffold [35]. Since that time to about end of 2014, in reference [36 - 40] reported 15.6%, 15.9%, 16.7%, 19.3% and 20.1% PCE respectively. Astonishingly it is a great improvement in the efficiency for perovskite based solar cells, as shown in Figure 3, also makes the perovskite as a premising candidate for immediate future PV solar cells. The main aim of photovoltaic design is to optimize efficiency to cost ratio. In this context, among the perovskite family, $MAPbI_3$ is one who proved itself as a best and champaign material due to its favorable opto-electric behavior, long lifetime, low temperature solution processability, ferroelectricity and hence a superb photovoltaic performance. There are lots of space to further improve the efficiency and deep insight into excellent optoelectronic behavior, thermodynamic stability of the perovskite absorber and the formation mechanism of the dominant intrinsic defects.

2. Numerical Simulations

Simulation is a crucial technique to realize deep insight into the physical operation, viability of proposed physical explanation and effect of physical changes on performance of the solar cell devices. There are various simulation models (SCAPS, AMPS, SCAP, etc) for solar cells simulation. SCAPS (Solar Cell Capacitance Simulator) is one-dimensional simulation program with seven semiconductor input layers developed by a goup of solar cell researcher at the department of Electronics and Information System, University of Gent, Belgium [41]. It is impractical as well as wastage of time and money to design a solar cell without simulation works. It minimizes not only the risk, time and money rather analyzes layers properties and role to optimize the solar cell performance. In order to simulation a device all the basic input parameters should be well defined so that it behaves as a real counterpart. The perovskite-based solar cells have employed a similar structure with inorganic semiconductor solar cells, such as CIGS, and Wannier-type exciton in the perovskite is found. Thus SCAPS like 1D simulator can be employed to simulate the perovskite based solar cells [42].

The main features of the latest version of SCAPS, to address the basic parameters, are as follows [41]:

- up to 7 semiconductor input layers can be defined as shown in Figure 8 and 9,
- almost all parameters can be graded: band gap (Eg), electron affinity (χ), dielectric constant (ε), valence and conduction band density of states (N_V *and* N_C), charge carriers mobility (μ_n *and* μ_p), accetor and donor dopant concentration (N_A, N_D), thermal velocity of charge carriers (V_{thn} *and* V_{thp}), working temperature (T) and absorption constant (α m^{-1}) all traps (defects) N_t are to be defined,
- recombination mechanisms: band-to-band (direct), Auger, SRH-type,
- defect levels: in bulk or at interface; their charge state and recombination is accounted for,
- defect levels, charge type: no charge (idealization), monovalent (single donor, acceptor), divalent (double donor, double acceptor, amphoteric), multivalent (user defined),

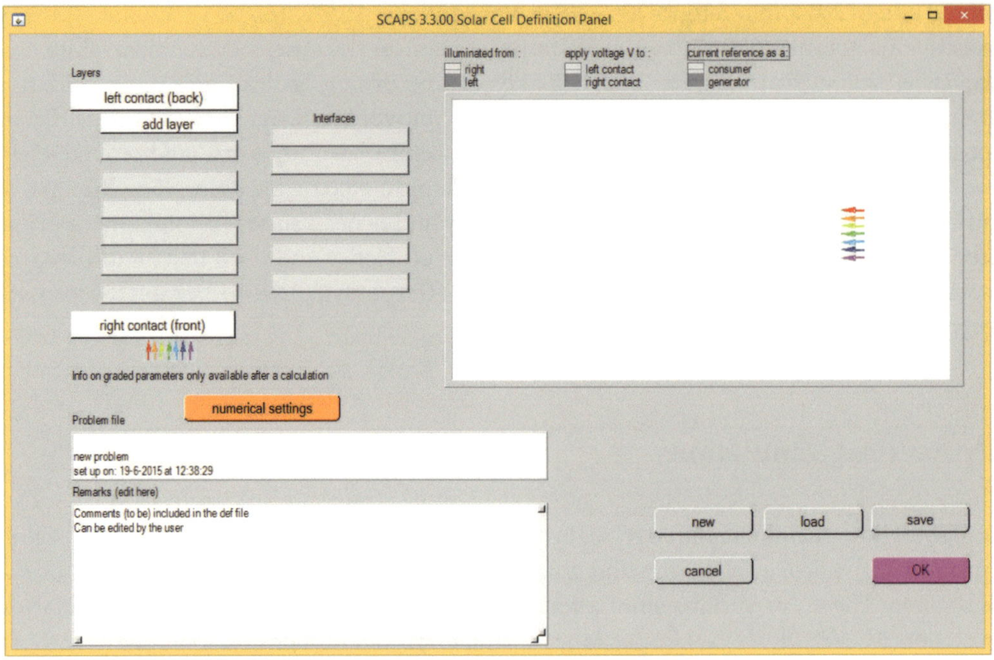

Figure 8. SCAPS-Solar cell definition panel.

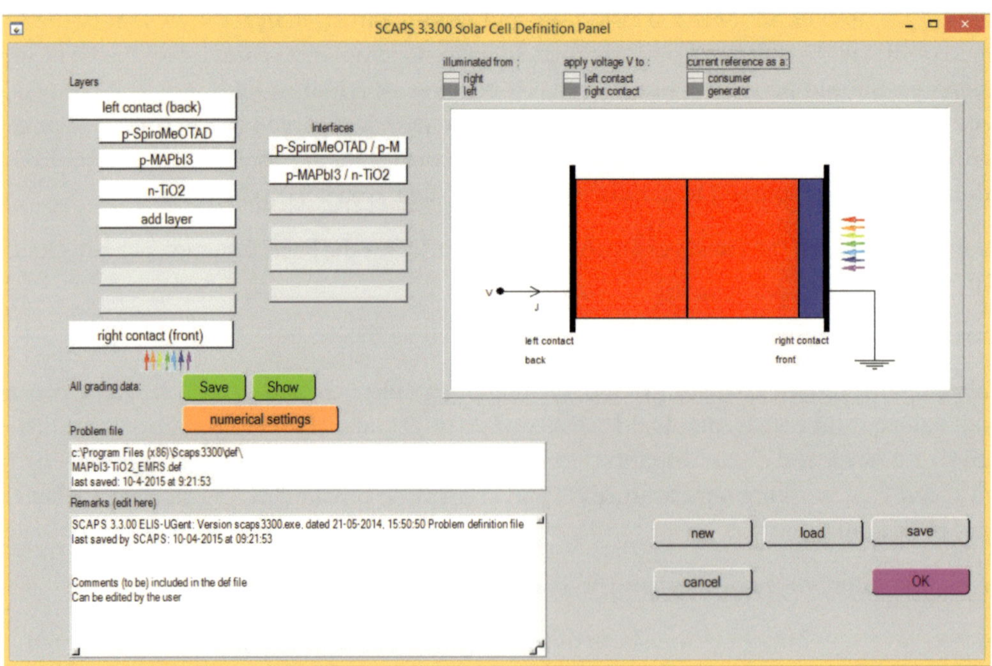

Figure 9. SCAPS-Solar cell definition panel with simulation

- defect levels, energetic distributions: single level, uniform, Gauss, tail, or combinations are available as shown in Figure 10,
- defect levels, optical property: direct excitation with light possible (impurity photovoltaic effect, IPV),

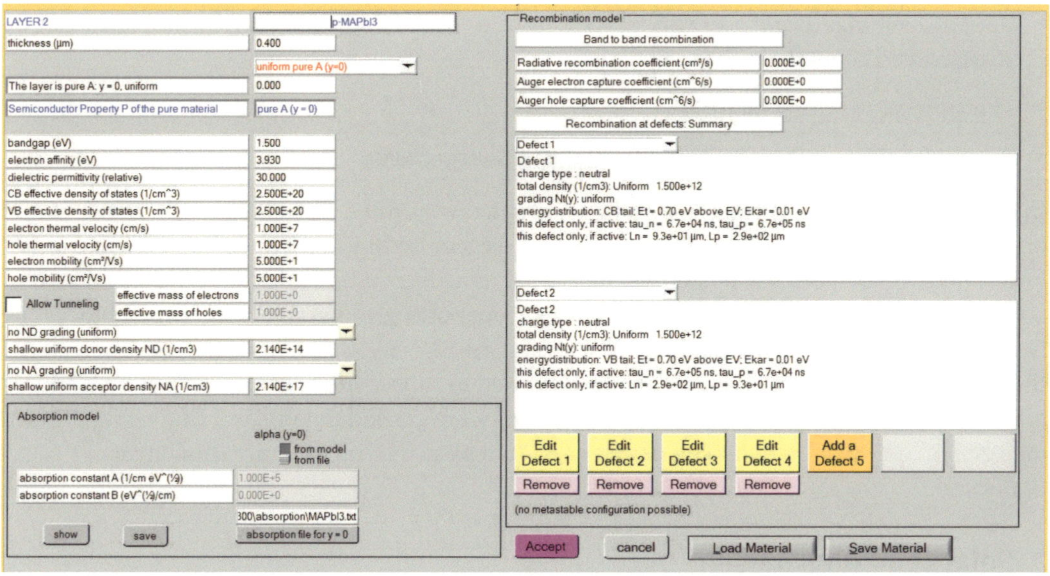

Figure 10. SCAPS-Solar cell simulations with sample data and defects.

- meta-stable defects: transitions between acceptor and donor configurations for known meta-stable defects in CIGS: the V_Se and the In_Cu defect; also custom set meta-stable transitions implemented,
- contacts: work function or flat-band; optical property (reflection of transmission filter) filter,
- tunneling: intra-band tunneling (within a conduction band or within a valence band); tunneling to and from interface states,
- generation: either from internal calculation or from user supplied $G(x)$ file,
- illumination: a variety of standard and other spectra included (AM0, AM1.5D, AM1.5G, AM1.5G edition2, monochromatic, white, etc),
- illumination: from either the p-side or the n-side; spectrum cut-off and attenuation,
- working point for calculations: voltage, frequency, temperature can be chosen by user,
- the program calculates energy bands, concentrations and currents at a given working point, J-V characteristics, ac characteristics (C and G as function of V and/or f), spectral response (also with bias light or voltage),

- batch calculations possible; presentation of results and settings as a function of batch parameters,
- loading and saving of all settings; startup of SCAPS in a personalized configuration; a script language including a free user function,
- very intuitive user interface,
- a script language facility to run SCAPS from a "script file"; all internal variables can be accessed and plotted via the script,
- a built-in curve fitting facility,
- a panel for the interpretation of admittance measurements.

In this study, perovskite based solar cell is simulated with three input layers; where two layers are p-type one is Hole Transporting Material and Electron Blocked Layer (say HTM), another low doped organic lead halide perovskite is used as active/absobing layer. N-type transparent conducting oxide can be used as the Electron Transporting Material (ETM) which blocks the holes. Thicknesses of HTM/$MAPbI_3$/ETM layers are 400/450/90 nm, respectively, if not stated there. Furthermore, front and back contact are used to collect and transport the charge carriers effectively to the external circuit. The simulations were carried out under 1 Sun using differnt material to optimize the solar cell devices and to realize the role of individual layer.

3. Solar cells physics

The basic physical principles [9, 41, 43] of a solar cell device are given below:

3.1. Absorbance and absorption coefficient

When an absorbing material of thickness d is being illuminated by an electromagnetic energy (usually light) of wavelength λ and intensity I (λ), some of light is reflected (R), some is absorbed (A) and remaining is transmitted (T) through the material. Now absorbance [A_{abs} (λ)] and absorption coefficient [α (λ)] can be defined as

$$A_{abs}(\lambda) = -\log_{10}[\frac{T(\lambda)}{I(\lambda) - R(\lambda)}] \tag{1}$$

$$\alpha(\lambda) = \frac{A_{abs}(\lambda)}{d} \tag{2}$$

3.2. Density of States

The valence band (N_V) and conduction band (N_C) effective densities of states are temperature dependent material properties which are given by formulae

http://dx.doi.org/10.5772/61751

$$N_V = \frac{(m_p^*)^{\frac{3}{2}}\sqrt{2}}{\pi^2\hbar^3}\sqrt{(E - E_V} \tag{3}$$

$$N_C = \frac{(m_n^*)^{\frac{3}{2}}\sqrt{2}}{\pi^2\hbar^3}\sqrt{(E - E_C} \tag{4}$$

where m_p^* and m_n^* are effective mass of hole and electron, E, E_V and E_C are energy levels at steady state, valence band edge and conduction band edge respectively.

3.3. Carrier concentration

In thermal equilibrium, number of free electrons and holes charge densities; n and p respectively are given by

$$n = N_C \exp(\frac{E_F - E_C)}{kT_n} = N_C \exp(\frac{V_n}{kT_n}) \tag{5}$$

$$p = N_v \exp(\frac{E_V - E_F)}{kT_p} = N_C \exp(\frac{V_p}{kT_p}) \tag{6}$$

Where E_F if Fermi level and V is built in potential, k Boltzmann constant and T is spatially varying electron and hole temperature.

3.4. Poisson's equation

Poisson's equation defines the electric field (E) that is modified as a result of the current flowing and the charge in the delocalized states, traps and recombination centers in the device as given below

$$\frac{\partial}{\partial x}(\varepsilon \frac{\partial \Psi}{\partial x}) = -q[p - n + N_D^+ - N_A^- + \frac{\rho_{def}}{q}] \tag{7}$$

where, ψ is electrostatic potential, ε is dielectric constant and q is electronic charge.

The first two terms in right side are free charge carriers per volume, third and fourth are ionized donor and acceptor-like dopants i.e,. localized states and ϱ_{def} is defect charge density.

3.5. Formula for p-n junctions diode

Solar cell behaves as a p-n junction diode. The voltage and current in ideal diode are given by formulae as below:

$$V_0 = \frac{kT}{q} \ln(\frac{N_A N_D}{n_i^2}) \tag{8}$$

where n_i is intrinsic carriers concentrations.

$$I = I_0[\exp(\frac{qT}{kT}) - 1] \tag{9}$$

where I_0 is saturation current.

3.6. Solar Cell Formulae

$$V_{OC} = \frac{nkT}{q} \ln(\frac{I_{SC}}{I_0} + 1) \tag{10}$$

$$J = J_0[\exp(\frac{qT}{nkT}) - 1] - J_{SC} \tag{11}$$

where, V_{OC}, J_{SC} and J_0 are open circuit voltage, short circuit current and saturation current density respectively.

3.7. Electron and hole Diffusion lengths

Collection length, a distance over which light absorption caused excitaions can be transported, is limited by diffusion, drift, or combination of the two. The electron and hole diffusion length are given by

$$L_n^{Diff} = [D_n \tau_n]^{\frac{1}{2}} \tag{12}$$

elecron diffusion length.

$$L_p^{Diff} = [D_p \tau_p]^{\frac{1}{2}} \tag{13}$$

hole diffusion length.

where $D_n = kT_n\mu_n$ is electron diffusion coefficient, $D_p = kT_p\mu_p$ hole diffusion coefficient, τ s life time and μs mobility of electron and hole.

3.8. Transport equation

Mobility of free electrons, holes, variation of carrier concentrations and hence quasi Fermi level with position result electron and hole currents. Thus, in thermal equilibrium

$$J_n = -\frac{\mu_n n}{q}\frac{\partial E_{Fn}}{\partial x} \tag{14}$$

$$J_p = +\frac{\mu_p p}{q}\frac{\partial E_{Fp}}{\partial x} \tag{15}$$

where μ_n and μ_p are electron and hole mobilities, and n and p are free electrons and holes density, E_{Fn} and E_{Fp} are electron and hole quasi-Fermi level respectively.

3.9. Continuity equations

In the absence of illumination and recombinations, in steady state, the free electron current density at any point in a solar cell device must be equal to that at some other piont. Same is valid for holes too. In a device generation and recombinations both are undertaken. Thus, the conservation of free electrons and free holes in the device is expressed as continuty equations

$$\frac{\partial n}{\partial t} = -\frac{\partial J_n}{\partial x} + G - \Re_n \tag{16}$$

$$\frac{\partial p}{\partial t} = -\frac{\partial J_p}{\partial x} + G - \Re_p \tag{17}$$

where Gs are photo generation rates and $\Re$ s are recombination rates in the device.

3.10. Recombination

Shockley-Read-Hall (SRH), Radiative and Auger recombinations are the possible recombination mechanisms in solar cells, they are given as

$$\Re^{L} = \frac{v\sigma_n\sigma_p N_T[np - n_i^2]}{\sigma_p[p + p_1] + \sigma_n[n + n_1]} \tag{18}$$

SRH recombination

$$\Re^{R} = [\frac{g_{th}^{R}}{n_i^2}](pn - n_i^2) \tag{19}$$

Radiative recombination

$$\Re_n^{A} = [\frac{n - n_0}{\tau_n^{A}}] \tag{20}$$

$$\Re_p^{A} = [\frac{p - p_0}{\tau_p^{A}}] \tag{21}$$

(Eqs. 20 and 21 are Auger recombinations)

where σs are capture cross-sections for electrons and holes, v electron thermal velocity, N_T number of gap states per volume, n_i intrinsic number density, g_{th}^{R} gives the number of electrons in conduction band and holes in valence band generated per unit time per unit volume and τ_n^{A} & τ_p^{A} are electron and hole lifetimes.

3.11. Power conversion efficiency

The fraction of incident power converted into usefull electricity by the solar cells is termed as Power conversion efficiency (η). It is given as below:

$$\eta = \frac{P_{out}}{P_{in}} = \frac{FF \times V_{OC}J_{SC}}{P_{in}} \tag{22}$$

$$\text{Fill Factor } FF = \frac{V_{mp}J_{mp}}{V_{OC}J_{SC}} \tag{23}$$

where V_{mp} and J_{mp} are voltage and current at maximum power point respectively.

3.12. Work function of contacts

In most of the cases work function of the front and back contacts are well defined. Otherwise, in flat band case the work function of contact (s) is calculated as follows:

$$\Phi_m = \chi + k_B T \ln(\frac{N_C}{N_D - N_A}) \quad (24)$$

when contact is n-type

$$\Phi_m = \chi + E_g - k_B T \ln(\frac{N_C}{N_A - N_D}) \quad (25)$$

when contact is p-type

$$\Phi_m = \chi + k_B T \ln(\frac{N_C}{n_i}) \quad (26)$$

when contact is intrinsic.

where Φ_m is workfunction of metal/material, χ is electron affinity, k_B is Boltzmann constant, T is operating temperature, N_C is conduction band effective density of states, and N_A and N_D are shallow acceptor and donor dopant concentrations respectively.

4. Results and Discussions

Figure 11 is the absorptance plots where the absorption coefficient, α (1/m) of titanium oxide is derived from reference [44], Zinc oxide is from SCAPS abs. file and that of $MAPbI_3$ is based on [45, 46]. The absorption coefficient for $MAPbI_3$ perovskite displayed in the Figure is very similar to that measured for perovskite synthesized in UPV laboratory, Valencia, Spain. The energy band diagram for the solar cell used in the simulation is shown in Figure 12.

From Figure 11, it is clear that the cutoff wavelength of perovskite lays around 825 nm so, the band gap ≈ 1.5 eV, which is the closer value required for ultimate theoretical limit according to W. Shockley and H. J. Queisser [18]. The band gap of perovskite depends on size of organic/inorganic (larger cation), metal (small cation), and halide (anion) components chosen during the processing and preparation. Increase in size of the cations and anion changes the covalent character between the cation (s) and anion and size of ionic radius which ultimately yields blue shift and red-shift respectively of the band gap [22, 47]. The band gap of the absorbing material is a crucial parameter for photovoltaic action as the absorber is the key material in a solar cell device [9]. Thus, suitable combination of the cationic components and anionic component in perovskite results into an extremely suitable absorber to yield the solar cells of better performance.

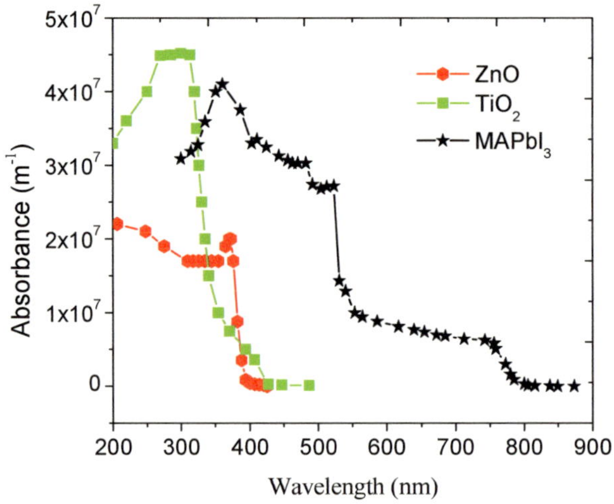

Figure 11. Absorbance of ZnO, TiO_2 and $MAPbI_3$

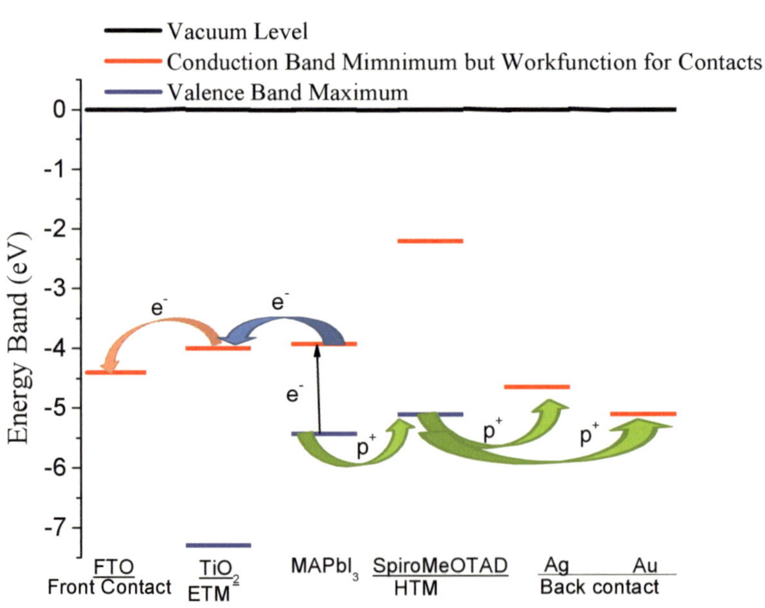

Figure 12. Energy band allignment

The **absorber layer, charge transporting layers**, front and back contacts, defects, interface and surface property are the major components and properties they have effect on the performance of the solar cell systems. **Operating temperature** of the solar cells plays a vital role in the performance of the PV devices. Considering these facts the simulation works are undertaken to find out role of individual layer and their associated properties which are briefly discussed below.

4.1. Role of absorber

As absorber is the key component in the solar cell devices the deep knowledge and understanding about this matter is crucial to design and fabricate a solar cell. Here our study has focused on realization of role of absorber to make more practical and optimization of the performance of the PV devices. For compensation (N_A/N_D) ratio and Gaussian energy distribution we have chosen two ETMs; TiO_2 and ZnO. Simulation with different aspects of the absorbing material is discussed below.

4.1.1. Variation of thickness of the absorber

4.1.1.1. Without considering interface trap density of states

Performance of the cell depends mainly on optoelectronic characteristics and thichness of the absorber layer. Thus simulations are carried out to examine the device parameters with thickness of the absorber from 50 nm to 700 nm under 1 Sun (AM1.5G) illumination. First of all simulations were carried out without considering interface trap density of states neither shallow minority carrier concentration. But inputs value of band tail density of states, and Gaussian acceptor/donor states of $MAPbI_3$ were 10×10^{14} $eV^{-1}cm^{-3}$ and 10×10^{14} cm^{-3} respectively. The short circuit current and PCE both are found to be increased sharply with increase in thickness up to 500 nm as shown in Figure 13. After this, increment is very slow and reaches to almost optimal efficiency 25.22%, V_{OC} 1.2 V, J_{SC} 25.49 mA/cm^2 and FF 82.56% at 700 nm which is closed to the detailed bance limit [18]. At the 700 nm thickness MAPbI3 absorbs almost incident photons to create the electron-hole pairs and the photo generated almost carriers are separated and transproted to the HTM and ETM by the built in field with minimum recombination thus, it can be consider as the length of optimal photovoltaic action.

4.1.1.2. Considering interface trap density of states

Dark current and current under illumination in PV solar cells are illustrated in Figure 14. In dark, solar cell behaves as a large flat diode and produces a very small current due to the minority charge carriers in the device structure which resembles to J-V characteristics of a pn junction diode. Device without trap defect is not possible so, to make more practical solar cell we introduce the total trap density = 1.0×10^9 cm^{-2} in the interfaces and added shallow level donor and acceptor density N_D = 10% of N_A and N_A = 10% of N_D in HTM and ETM respectively. Simulation results shown in Figures 7 is quite interesting and suggested to more rational. State of affairs here is somewhat different than the former case. It reveals that: i) the device parameters are found to be decreasing with introduction of interface traps due to change in shunt

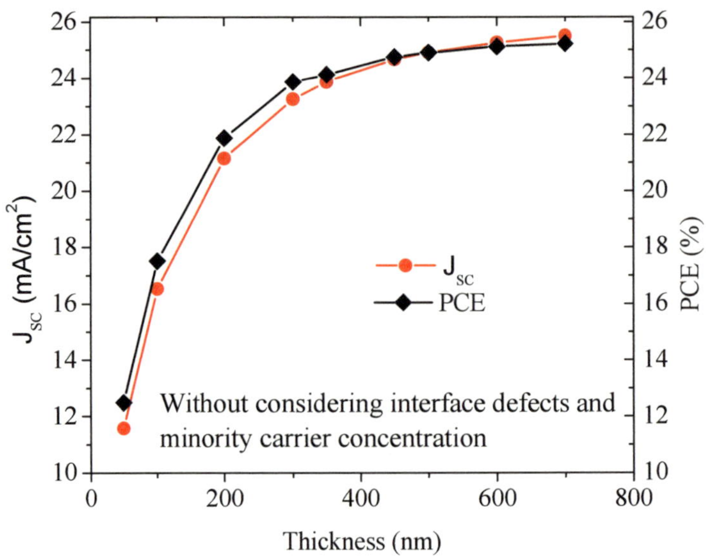

Figure 13. J_{SC} and PCE vs thickness

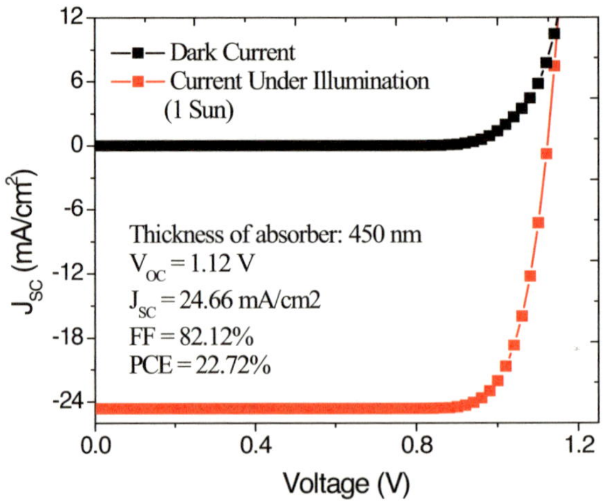

Figure 14. J-V curve as a function of illuminations

resistance, ii) parameters are increasing more or less steeply up to 350 nm and very slowly beyound 450 nm, iii) fill factor is observed maximum value of 82.24% for 100 nmthickness where other parameters also have significant values. The results show that 450 nm thichness is also sufficient for good photovoltaic action. Furthermore, 700 nm is sufficient for optimal photovoltaic performance where V_{OC}, J_{SC}, FF and PCE are observed 1.12 V, 25.49 mA/cm^2, 82.29% and 23.48% respectively. Beyond this thickness there might increase in resistance that

http://dx.doi.org/10.5772/61751

results in insignificant increment in PCE. The rate of the photogeneration of charge carriers depends on the amount of light reaching, absorption range and thichness of the absorber, here $MAPbI_3$ perovskite since it is the key component of the solar cell. Figure 15 is the quantum efficiency curves as a function of wavelength of incident light for different thickness of the absorber also verifies the above mentioned upshot. Besides, our results for an absorber thickness of 400 nm is similar but expected to improve to move towards the practicality than reported by [45].

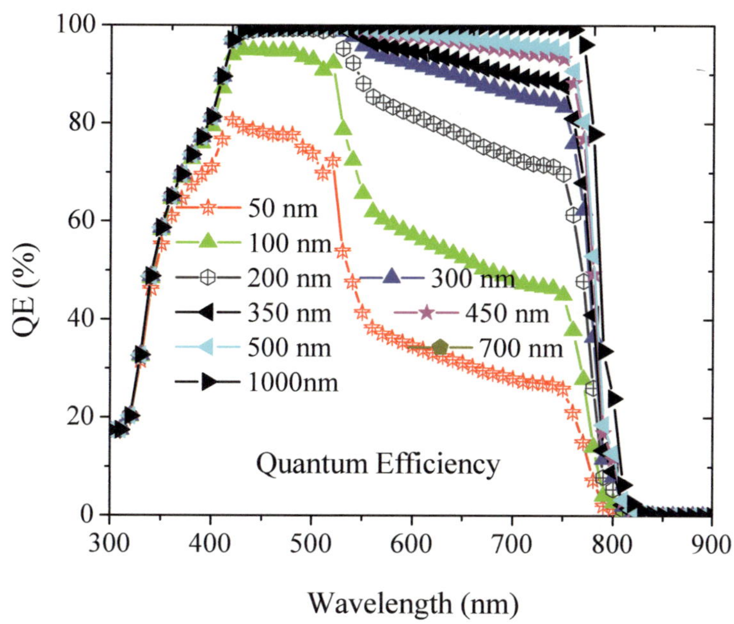

Figure 15. Variation of quantum efficiency with thickness of absorber.

4.1.2. Band tail density of states of perovskite

The input parameters for band tail density has varied from 1.0×10^{12} $eV^{-1}.cm^{-3}$ to 1.0×10^{17} $eV^{-1}.cm^{-3}$. J-V curves as a function of band tail defect is shown in Figure 16, which shows the defects in the absorber has greatly influenced the V_{OC}, and finally to the performance of the device. Infact, band tail density of states are the intra-band gap states which may act as recombination centers for charge carriers and are also called defects in layer (s) of the solar cell. Simulation studies have envisaged that similar nature is not observed for HTM and ETM as reported by [45] too, since photogeneration of the charge carriers is taken place in the absorber say active layer. The Figure 16 tells that defects do not significantly influence the J_{SC} since amount of incident photon and thickness are remaing unchanged. Relatively higher value of V_{OC} is observed which is similar to [45, 48, 49]. It is also observed that V_{OC} and J_{SC} are almost same from 1.0×10^{12} to 1.0×10^{14} eV^{-1} cm^{-3}. These are the consequences of the larger absorption rang, high crystalinity, low pinhole, shallow level defect densities and a prolonged electron-hole lifetime in perovskite which are characteritics of materials and also created unintentionally/purposefully during the deposition/coating/processing the layer (s). Similar result is reported by L. M. Herz et. al. [49].

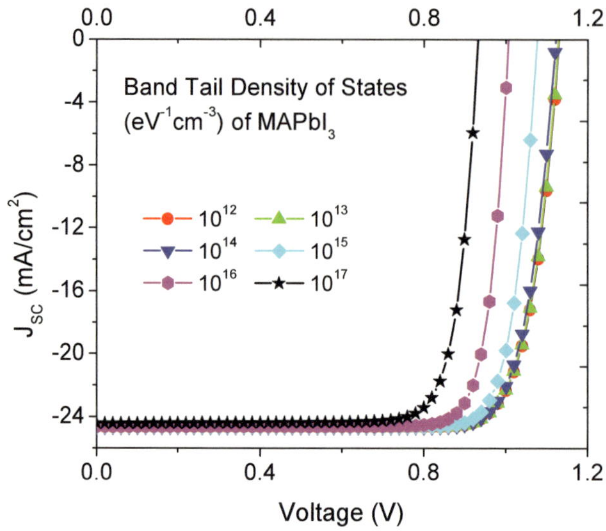

Figure 16. JV curve as a function of band tail density of states

4.1.3. Role of interface trap density

Research process always endeavor towards the perfect by realizing and optimizing the overall behaviours of the material. So, simulations were carried out with setting the interface trap densities from 1.0×10^8 cm^{-2} to 1.0×10^{14} cm^{-2} and studied the results. Figure 17 shows that V_{OC} and PCE are decreasing with increase in interface trap density upto 1.0×10^{13} cm^{-2} and remaining almost same beyond this. Figure 18 shows that QE is considerably small when trap density is above 1.0×10^{11} cm^{-2}. The plots show the crucial role of interface trap density in efficiency of the solar cells since interface traps at high level are also the recombination centers and hence change in shunt resistance. Optimizing doping of the HTM and ETM, and formation of homogeneous, smooth, and flat surface will significantly reduce the interface trap density.

4.1.4. Compensation ratio (N_A/N_D) of the absorber

For this study, simulation works have been carried out with two different ETMs; TiO_2 and ZnO. Figure 19 shows the effect of compensation ratio in $MAPbI_3$ perovskite. In this case, simulation works have been carried out with variation of compensation ratio from 0 to 20%. Increase in compensation ratio of perovskite, an absorber, and results into slightly decreased in the performance of the PV devices in both cases suggested to increase in recombination/loss within it. The similar effect which has been observed in both ETMs increases the aptness of substituting TiO_2 by ZnO.

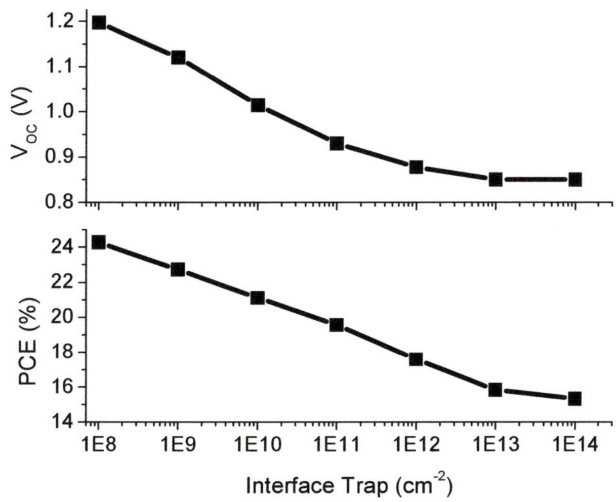

Figure 17. V_{OC} and PC vs interface traps

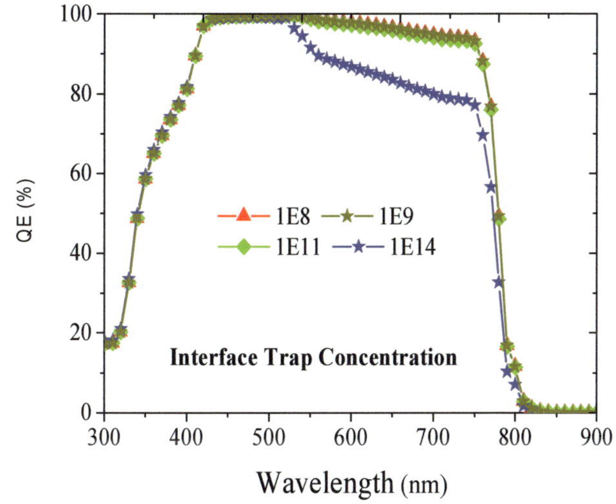

Figure 18. QE as a function of interface traps

4.1.5. Gaussian Energy Distribution in Perovskite

Figures 20 and 21 are the JV curves as the function of the Gaussian energy distribution (defect) of MAPbI3 perovskite. Simulations have been carried out with varying the defect values from 10^{12} to 10^{19} cm^{-3}. It is found that V_{OC} and hence PCE of PV devices have been decreasing with increase in Gaussian defect for both ETMs. Beyond 10^{17} cm^{-3} i.e., at high defect levels performance of the device goes significantly decreasing due to more recombination, loss and change in resistance.

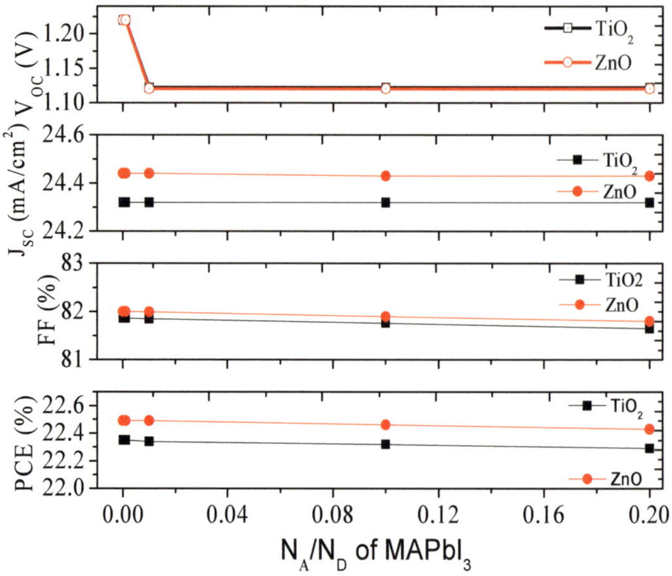

Figure 19. PVC Parameters vs Compensation ratio (N_A/N_D) of $MAPbI_3$

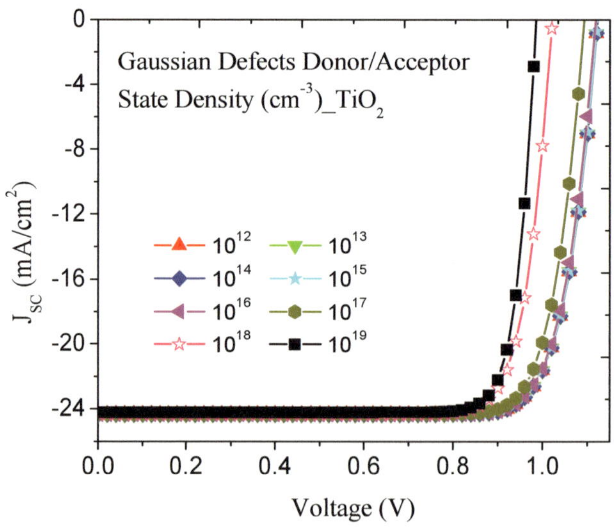

Figure 20. Gaussian Defects of Perovskite with TiO_2

4.2. Role of Electron Transporting Materials

Two usual conducting oxides; titanium oxide and zinc oxide have been used to analyze the role of electron transporting layer in the heterojunction thin film solar cells. The band gap and

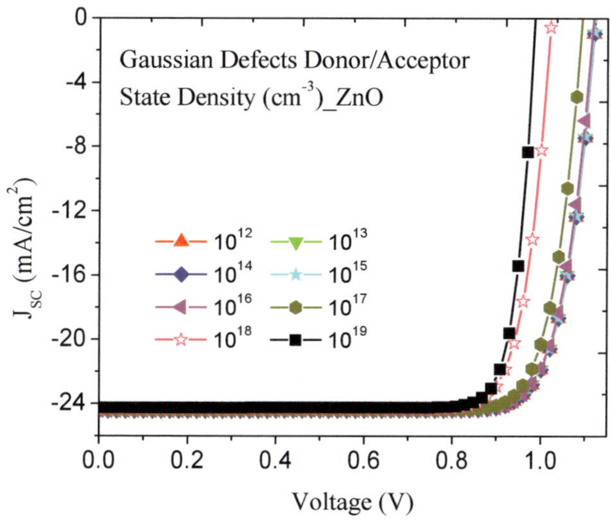

Figure 21. Gaussian Defects of Perovskite with ZnO

electron affinity of the materials should be such that it conducts the free electrons and creates barrier to the holes. Here simulation works are carried out with varying layer properties of two transparent conducting oxides as ETMs. They are briefly discussed below.

4.2.1. Thickness of the ETMs

Figure 22 is the plot of solar cell parameters; V_{OC}, J_{SC}, FF and PCE versus thickness of the ETMs; TiO_2 and ZnO. In both cases V_{OC}, J_{SC} and PCE are gradually decreasing due to fractional absorption of incident light by the ETMs layer, the bulk recombination and surface recombination at the interface and change in series resistance. Thickness of ETMs has been varied from 50 nm to 450 nm to make the practical devices. Observation showed that TiO_2 is more sensitive than that of ZnO due to its high absorption coefficient and reflectance and less transmittance than ZnO. It is obvious that the increase in thickness of the ETM lessen the performance of the solar cells due to increase in partial absorption of photons and resistance of the device.

Moreover, we have carried out simulation works with practically viable thickness of TiO_2 - ETM as shown in Figure 23 which is the plot of FF and PCE as a function of thickness of TiO_2. FF and PCE are gradually decreasing due to fractional absorption of incident light by the TiO_2 layer, the bulk recombination and surface recombination at the interface [22]. But fill factor in this case is found to be slightly increasing with increase in thickness suggested the higher conductivity of the TiO_2 than $MAPbI_3$ and SpiroMeOTAD and partial absorption of the light.

4.2.2. Band tail density of states of perovskite

Figure 24 and 25 are the J-V curves as a function of band tail defect varies from 10^{12} to 10^{17} $eV^{-1}.cm^{-3}$. The plots show that the defects in the absorber has influenced the V_{OC}, and finally to the performance of the device due to the change in shunt resistance. Band tail density of states

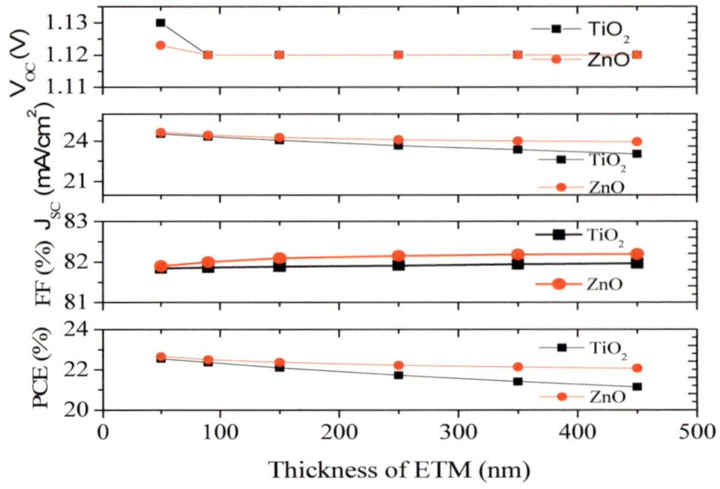

Figure 22. Variation of PV cells parameters with thickness of ETMs

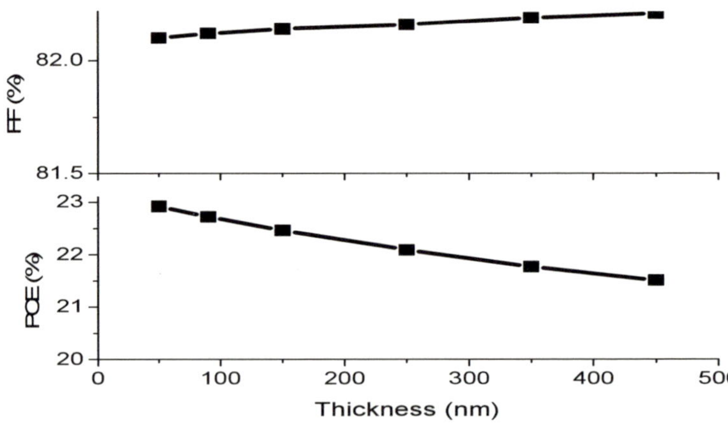

Figure 23. FF and PCE vs thickness of TiO_2.

are particularly intra-band gap recombination centers are also called defects. Both the Figures tell that defects do not significantly influence the J_{SC} since amount of incident photon and thickness of system are remaing unchanged. Relatively higher value of V_{OC} is observed as [48, 49]. It is also observed that V_{OC} and J_{SC} are almost same from $1.0{\times}10^{12}$ to $1.0{\times}10^{14}$ eV^{-1} cm^{-3}.

These are the consequences of the larger absorption rang, high crystalinity, low pinhole, shallow level defect densities and a prolonged electron-hole lifetime in perovskite which are characteritics of materials and also created unporposefully/purposefully during the coating/processing/fabrication of the layer (s). This is similar result as reported by [49]. Furthermore, TiO_2 has found relatively less sensitive than in ZnO due to small electron mobility and higher doping concentration than ZnO.

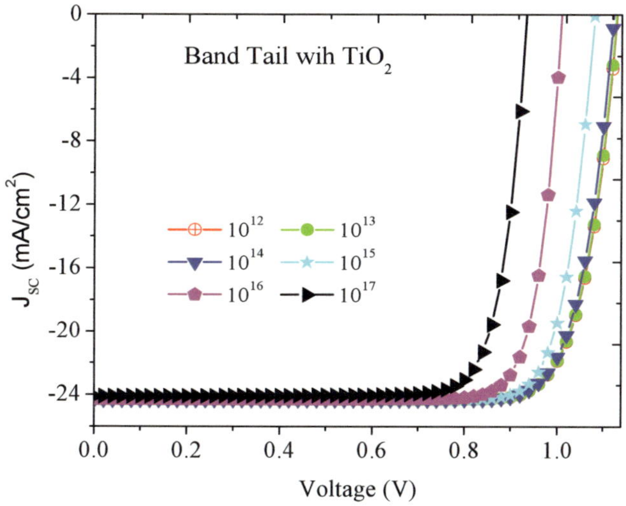

Figure 24. Band Tail of Perovskite with TiO_2 as ETM

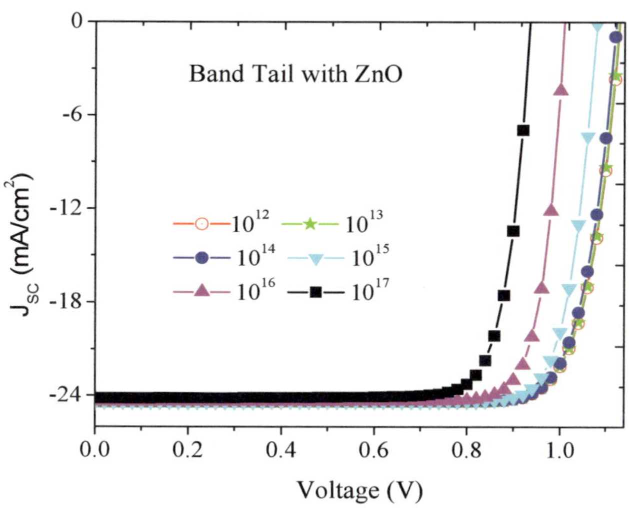

Figure 25. Band Tail of Perovskite with ZnO as ETM

4.2.3. Interface trap density of states

Figures 26 and 27 are the JV curves as a function of interface trap density varied from 1.0×10^8 to 1.0×10^{15} cm^{-2}. It is quite clear from the plots that V_{OC} and PCE are decreasing with increase in interface trap density upto 1.0×10^{14} cm^{-2}. It shows that interface trap increases the recombination centers and hence change in shunt resistance. There is decrease in J_{SC} with the interface trap beyond 1.0×10^{12} cm^{-2} suggested to increase in series resistance. Relatively more sensitiveness of ZnO than TiO_2 towards interface trap indicates the relatively small defect and small electron mobility of TiO_2. Optimizing doping of the ETM, and formation of homogeneous, smooth, and flat surface will significantly reduce the interface trap density and hence increase the performance.

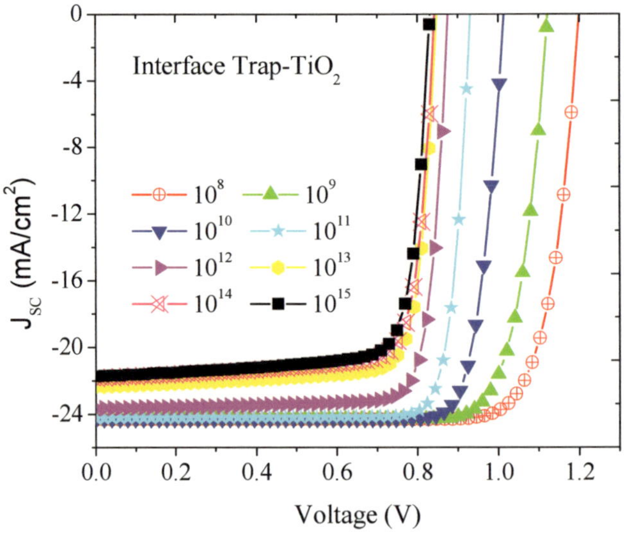

Figure 26. Interface Trap of Perovskite with TiO_2

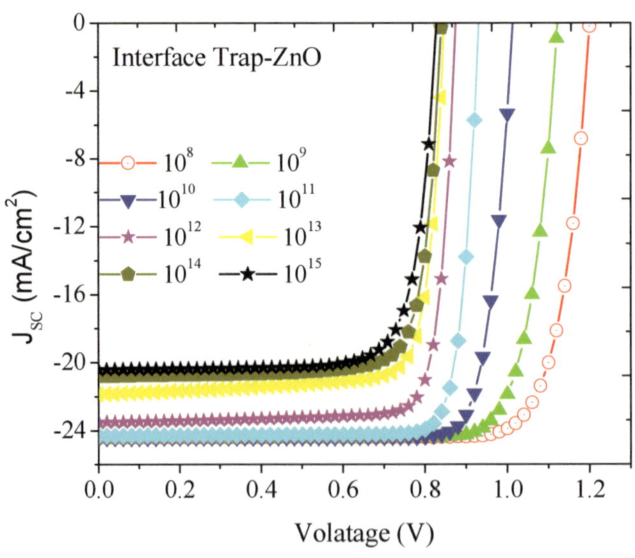

Figure 27. Interface Trap of Perovskite with ZnO

4.2.4. Role of Dopant Concentrations of ETMs.

The plot of PV cell parameters as a function of dopant concentration (N_D) of both ETMs TiO_2 and ZnO is shown in Figure 28. The study has been carried out from 10^{15} to 10^{21} cm^{-3} values. Both ETMs have exhibited the similar performance here too. V_{oc}, FF and PCE have been increasing with increase in dopant concentration of TiO_2 and ZnO up to around 10^{18} cm^{-3} due to the increase in conductivity of ETM. Although dopant concentrations have been increasing beyond this value PV parameters remain unchanged due to Moss-Burstein effect [50].Variation of dopant concentration of TiO_2 is found to be little bit more sensitive than that of ZnO up to around 10^{18} cm^{-3} due to smaller value of electron mobility and hence small conductivity of TiO_2.

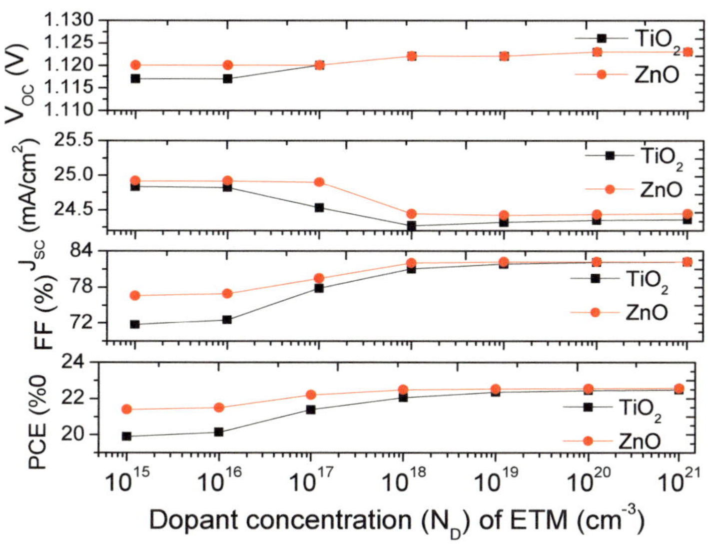

Figure 28. PVC Parameters vs Dopant concentrations (N_D) of TiO_2 and ZnO

4.2.5. Interface Trap Electron/Hole Capture Cross-Section.

Figure 29 shows the effect of interface trap electron/hole capture cross-section for both interfaces on performance of the device where the value varies. The electron/hole capture cross-sections have been varied from $10^{-14}/10^{-15}$ and $10^{-15}/10^{-14}$ to $10^{-20}/10^{-21}$ and $10^{-21}/10^{-20}$ cm^2 for first and second interfaces respectively. Plots show that V_{OC} and PCE both are decreasing with increase in cross-section of TiO_2 and ZnO due to increase in recombination/loss and change in shunt resistance. ZnO is observed more sensitive than that of TiO_2 toward the higher values of interface trap electron/hole capture cross-sections. It is suggested that ZnO has such more defects than TiO_2 and hence less efficient in real practice.

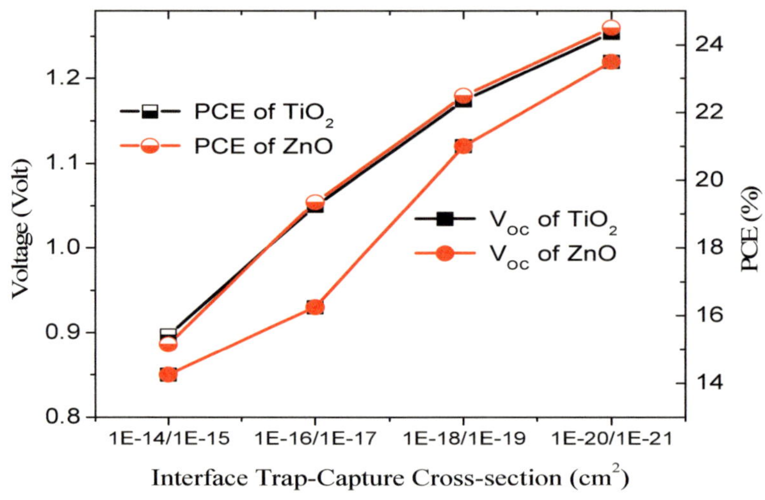

Figure 29. V_{OC} and PCE vs Interface Trap Electron/Hole Capture Cross-section.

4.3. Role of HTM

4.3.1. Role of thickness of the HTM

Figures 30 is the plot of FF and PCE versus practically viable thickness of the

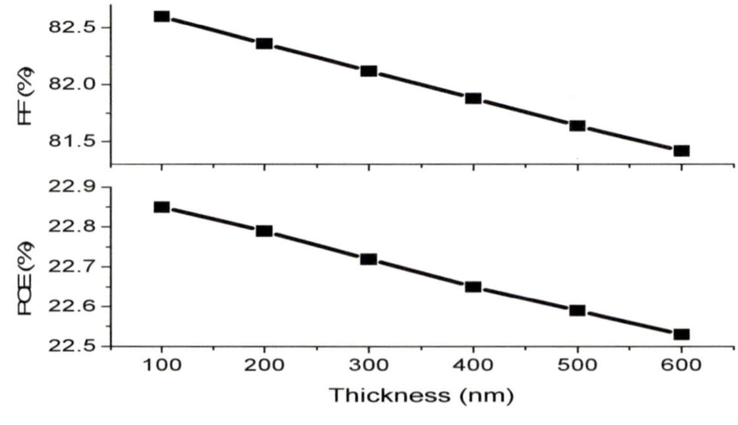

Figure 30. Variation of FF and PCE with thickness of HTM.

HTM (Spiro-MeOTAD). In the case of HTM, fill factor and PCE both are decreasing with increase in thickness suggested the increase in recombination and resistance. Thus, a superb junction diode like contact between the absorbar and HTM is necessary to increase the fill

http://dx.doi.org/10.5772/61751

factor and hence to improve the efficiency of the device. It is obvious that the decrease in thickness of the HTM improves the performance of the solar cells due to decrease in recombination and resistance of the device.

4.3.2. Role of hole mobility and acceptor concentration of the HTM

Figures 31 and 32 show the effect of hole mobolity and acceptor contrantration of HTM respectively on the device performance. The Spiro-MeOTAD has relatively low hole mobility value. Note that different researcher [34, 45, 51] have reported different values of hole mobility of SpiroMeOTAD but $2.0{\times}10^{-4}$ cm^2/V s has been used in this study. Due to small value of hole mobility and acceptor concentration low value of FF and hence the low PCE are observed. The value goes on increasing with increase in hole mobility and acceptor concentration. Properties such as hole mobility and dopant concentration (N_A), are responsible for resistance/conductance of the HTM, here Spiro-MeOTAD, they highly influence the performance of the device. Although Spiro-MeOTAD has merits either enhencement of conductivity and dopant concentration of this material or replacement of it by suitable HTM is highly appreciable for better performance of the device.

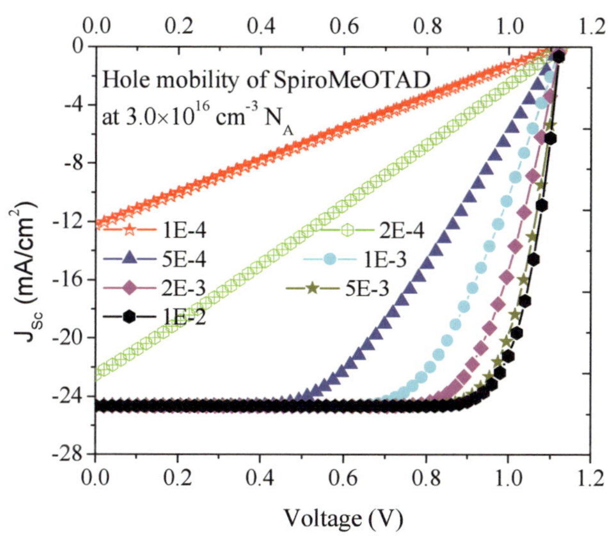

Figure 31. J-V curve as a function of hole mobolity.

4.4. Role of front and back contacts

4.4.1. Role of workfunction of the back contact metal

In this case, simulations were carried out at thicknesses 400/400/90 nm of the device layers so as to compare the outcome with references [45, 46]. Figure 33 is the plot of fill factor and PCE versus workfunction of the counter electrode. It is found that fill factor and PCE both are

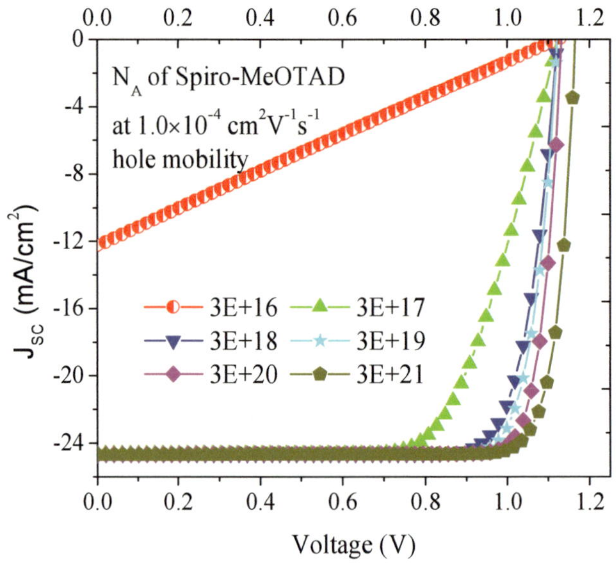

Figure 32. J-V curve as a function of dopant concentrations (N_A).

decreasing with decrease in workfunction of the back contact metal. Back contact is the counter electrode connected to the HTM to collects the holes or to enter the almost relaxed electrons into the device from the external circuit. Figure 12 shows the energy band alignment between different layers of the device.

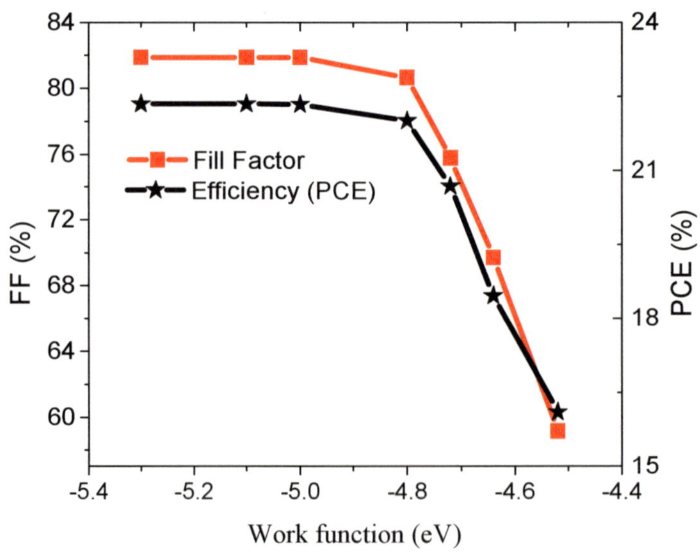

Figure 33. FF and PCE vs workfunction of back contact.

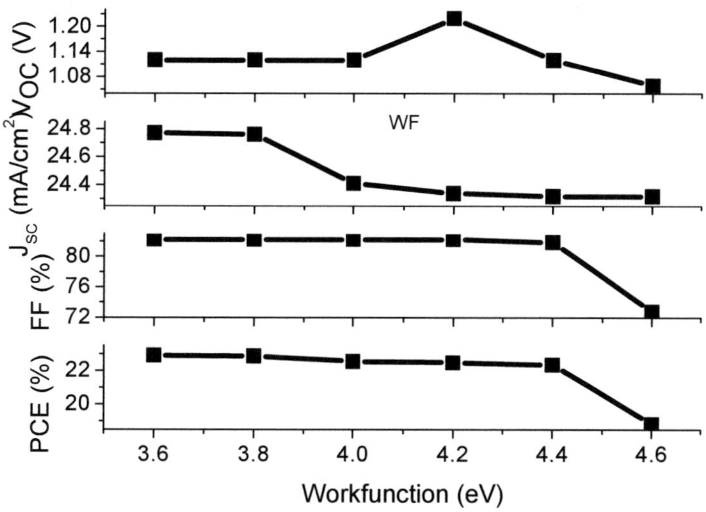

Figure 34. PV all parameters vs workfunction of front contact.

An ohmic contact between them is necessary to transport the holes efficiently to the back contact. As workfunction of back contact is 5.0 eV or above than the HTM i.e., workfunction of the back contact is nearly equal to or slightly greater than that of HTM then holes transport effectively and there would be barrier for electrons. Below 5.0 eV a dipole oriented positive in the metal and negative in the HTM results an electrostatic barrier to the holes or Schottky barrier [9] and hence difficultly arise in holes transportation at contact. This indicates the requirement of gold (workfunction = 5.1 eV) like higher workfunction material to develop ohmic contact to conduct the hole to the electrode [44, 45, 47]. Moreover, it is quite clear from the Figure 33 that, at workfunction 4.64 eV i.e., of Ag [52], PCE is decreased to about 18 % with V_{OC} = 1.12 V,J_{SC} = 24.32 mA/cm^2 and FF = 69 %. This result has close agreement with the experimentally determined value 15. 4% PCE, 1.07 V V_{OC}, 21.5 mA/cm^2 J_{SC} and 67% FF for $MAPbI_3$ perovskite based planner solar cell with the Ag as counter electrode [46]. Thus, our result claims the realism in performance of device than reported by [45]. Consequently, it always opens the pathway to replace the Au-counter electrode if someone is focused on the cost. Besides, there are other options too with favorable high value workfunction material to replace the counter electrode [53].

4.4.2. Role of workfunction of the front contact

For the sake of convenient and to make real solar cells simulations were carried out at thicknesses 400/400/90 nm of the device layers. Figure 34 is the the PV all parameters versus workfunction of the electrode/front contact. It is found that fill factor and PCE both are increasing with decrease in workfunction of the front contact material. Front contact is the electrode connected to the ETM to collects the energetic electrons to go round the external circuit to do some useful work over there and at the mean time it allows the light to reach the absorber. Figure 12 shows the energy band alignment between different layers of the device. An ohmic contact between them is necessary to transport the electrons efficiently to the front contact. As workfunction of front contact is slightly smaller than or nearly equal to the workfunction of the ETM, there is ohmic contact for the electrons and barriers for the holes. Thus, a comparatively low workfunction and transparent material is suitable to use as front contact material.

4.5. Role of temperature

Usually in simulation studies and fabrications 300K is chosen as the operating temperature where it is regarding as the room temperature. In actual practice the operating temperature varies with latitude, altitude, day of the year and time of the day of the place concerned. Moreover, performance of a solar cell is affected by the operating temperature as well. In this context we have varied the temperature from 277K to 410K to realize actual behavior of the perovskite based solar cells.

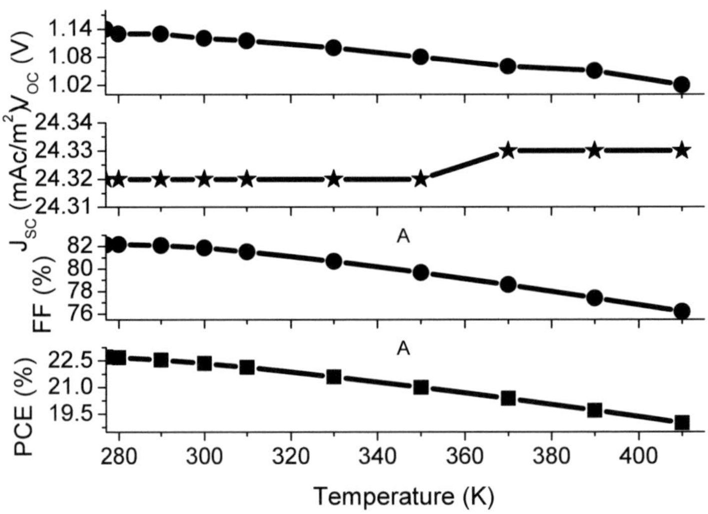

Figure 35. Variation of PV parameters with temperature (K)

Figure 35 shows the variation of PV parameters with variation of operating temperature of the solar cells where performance of the solar cells is influenced by operating temperature. At temperature lower than room temperature the solar cells have exhibited higher performance and at room temperature all PV parameters have moderate value. The temperatures above room temperature have detrimental to the overall performance. At higher temperature the carrier concentrations, mobility of the charge carries, resistance and band gap of the materials would be greatly affected that would ultimately alter PV parameters. Figure 36, JV curve as a function of temperature, conforms the variation of V_{OC}, FF and hence resistance in the solar cell devices.

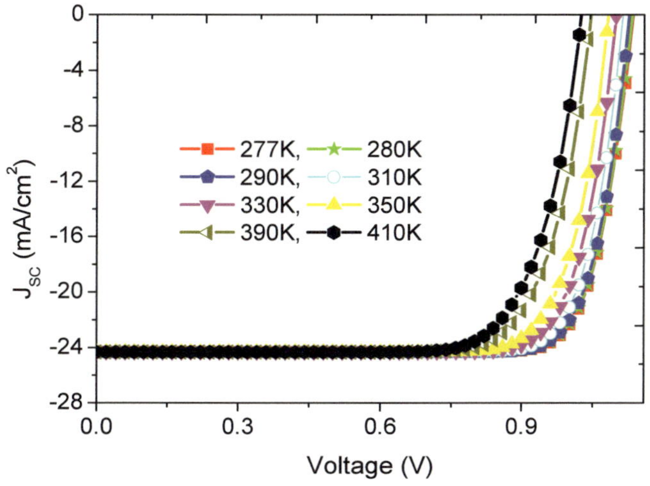

Figure 36. JV curve as a function of temperature

4.6. Concentrator solar cells and multi-Sun

Figure 37 is the solar irradiance at air mass zero (AM0, i.e., extraterrestrial irradiance) and air mass 1.5 global (AM1.5G, terrestrial irradiance). When sunlight enters the Earth's atmosphere then there is absorption, transmission, reflection, and scattering the light by ozone, oxygen, water in different phases, carbon dioxide and atmospheric contents. As a consequence the extraterrestrial radiation greatly modifies into global (direct plus diffuse) radiation on the Earth's surface. Extraterrestrial radiation is concerned mainly to the PV devices in space where as global is significant to the terrestrial activities including PV solar cell devices.

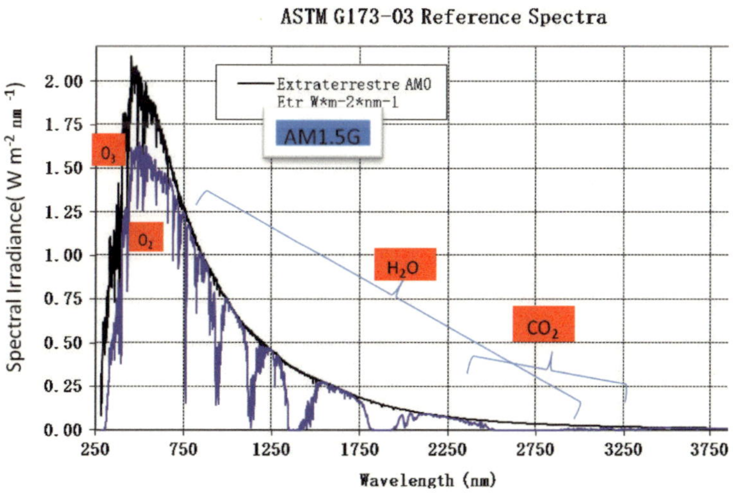

Figure 37. Solar irradiance vs wavelength for different air mass.

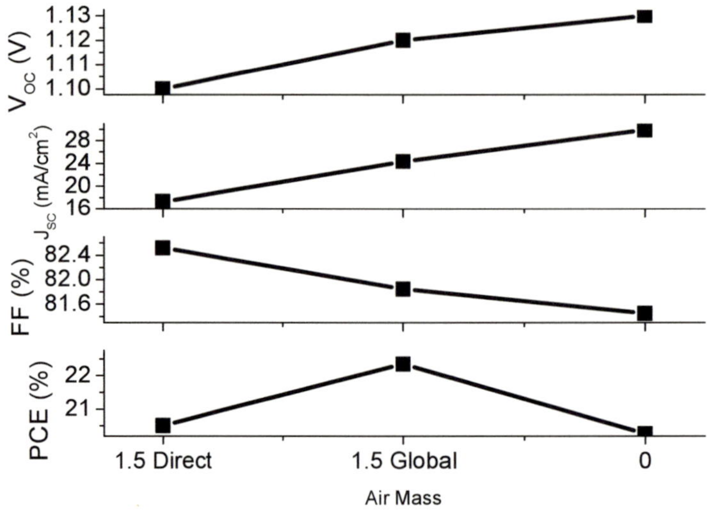

Figure 38. PV parameters vs Air Mass.

Figure 38 shows the variation of all PV parameters with solar spectrum in different air mass for perovskite as absorber. Efficiency is maximum at AM1.5G. Fill factor is went regularly decreasing from AM1.5D to AM0. Intense light would be changed the temperature, carrier concentrations, band gap, mobility of the charge carriers and resistance they would ultimately change the performance of the PV devices.

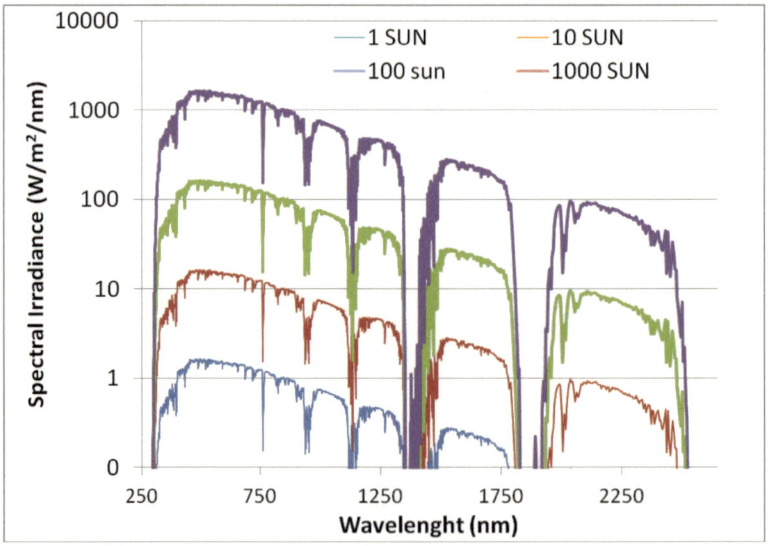

Figure 39. Solar spectrums at different Suns.

Variation of solar irradiance for different 'Sun' is illustrated in figure 39. An array of mirrors or lenses have been employed in order to focus onto a small sized absorber which is thereby subjected to intense sunlight for the generation of energy in one form to another [54] this is so called 'multi-sun' which enhances efficiency and reduce the size of the solar cell devices and the PV device is called concentrator solar cells.

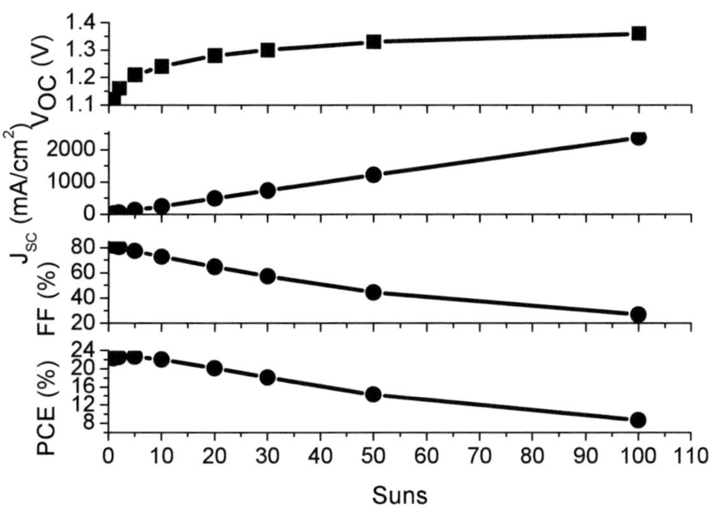

Figure 40. PV parameters vs Suns.

Figure 40 shows the effect of 'multi-sun' on the performance of the solar cell devices. Observations found that this is effective from 2 'Sun' to around 30 'Suns' and maximum values are found at around 5 Suns for the perovskite solar cells. After 30 'Suns' they produce detrimental effect to the performance which would be due to change in carrier concentrations, mobility of the charge carries, resistance and band gap of the materials.

5. Conclusions

In this paper we have discussed the outcome of a simulation study on organometal halide perovskite focusing on the role of the different layers of the solar cell using SCAPS as a simulation tool. An interestingly high performance; PCE 22.72%, V_{OC} = 1.12 V, J_{SC} = 24.66 mA/cm^2, and FF = 82.12% have been observed for 450 nm thickness of absorber. Increase in thickness of the absorber has increased the PCE. In the case of HTM and ETM performance has been decreasing with increase in thickness. The higher values of J_{SC}, V_{OC}, FF, and quantum efficiency plots have proved the low pinholes and shallow defect density, outstanding absorption range and strength. The hole mobility and acceptor/donor concentration of the HTM and ETM, Band tail of perovskite, interface trap density and workfunction of back and front contacts (electrode and counter electrode) have shown significant influence on the device performance. Even with these strong merits of HTM and ETM, enhancement of hole mobility and conductivity of HTM and ETM, stability of perovskite and TiO_2 and replacement of toxic lead are still crucial. Through suitable processing/synthesizing of the perovskite absorbers, best engineering the selective contact, and increasing conductivity of HTM and ETM will boost the stability as well as performance of the device.

This paper has also been focused on the study of two Electron Transporting Materials on organometal halide perovskite based solar cells. An interestingly high PCE, in both cases, have been found for 400 nm thickness of absorber. ZnO as ETM has been found slightly more efficient than TiO_2 by using similar baseline parameters. In the case of increase in thickness of the ETMs PCE have found significant influence but the effect in TiO_2 is more prominent than in ZnO due to low hole mobility and somewhat more absorption as well as less transmission of light through TiO_2. Furthermore, dopant concentrations of ETMs and compensation ratio (N_D/N_A) of perovskite have also changed the performance of the devices due to increase in conductivity and recombination/loss respectively. This study has pointed out two things - first, ZnO is found slightly more defective than TiO_2 as practical solar cell with ZnO is less efficient. Although simulations have shown ZnO to be more efficient than TiO_2 in real practice it is less efficient since ZnO possessed relatively more band tail and interface traps. Second, ZnO is a good alternative of TiO_2 for highly efficient solar cells to reduce the cost and enhance the electron mobility.

Change in temperature and illumination have affected the performance of the solar cells due to change in carrier concentrations, mobility of the charge carries, resistance and band gap of the materials.

Acknowledgements

This work has been supported by the Spanish Government through Ministerio de Economía y Competitividad (ENE2013-46624-C4-4-R) and Generalitat valenciana (Prometeus 2014/044)

Author details

Bernabé Marí Soucase[1*], Inmaculada Guaita Pradas[2] and Krishna R. Adhikari[3]

*Address all correspondence to: bmari@fis.upv.es;adhikarikrishna@wrc.edu.np

1 Departament de Física Aplicada-IDF, Universitat Politècnica de València, València, Spain

2 Departament d'Economia i Ciències Socials, Universitat Politècnica de València, València, Spain

3 Tribhuvan University, Kathmandu, Nepal

References

[1] Kosyachenko LA. Thin-Film Photovoltaics as a Mainstream of Solar Power Engineering. Solar Cells - Thin-Film Technologies, INTECH; 2011.1-39 p. ISBN: 978-953-307-570-9.

[2] Mycle Schneider, Antony Froggatt et al. World Nuclear Industry Status Report. 2014.

[3] Green MA. Photovoltaics: coming of age, Conference: Photovoltaic Specialists Conference, Conference Record of the Twenty First IEEE 1990.

[4] PV Lighthouse: http://www.pvlighthouse 2015/06/11

[5] Nobel Prize in Physics 1921: Albert Einstein: (http://nobelprize.org/nobel_prizes/physics/ laureates/1921/), Nobel Prize Official Page.

[6] Grondahl LO. The copper-cuprous oxide rectifier and phoelectric cell. Rev. Modern Phys. 1933;5:141.

[7] Kingsbury EF and Ohl RS. Photo-electric properties of ionically bombarded silicon. Bell Syth. J. 1952; 31: 802-815.

[8] Chapin M, Fuller CS, and Pearson GL. A New Silicon p-n Junction Photocell for Converting Solar Radiation into Electrical Power. Journal of Applied Physics. 1954; 25: 676–677.

[9] Fonash SJ. Solar Cell Device Physics, 2nd ed. Academic Press; 2012.

[10] http://www.solarpowereurope.org/fileadmin/user_ upload/ documents/Publications/Global_Market_Outlook_2015_-2019_lr_v23.pdf.

[11] Growth of photovoltaics https://en.wikipedia.org/wiki/Growth_of_photovoltaics

[12] NREL, 4514 (2014) http://www.nrel.gov/ncpv/images/efficiency_chart.jpg

[13] Szlufcik J, Sivoththaman S, Nijs JF, Mertens RP, and Overstraeten RV. Low cost industrial manufacture of crystalline silicon solar cells. In: Markvart T and Castaner L, editors. Practical Handbook of Photovoltaics: Fundamentals and Applications, Elsevier, New York; 2003. p.155-184.

[14] Global Market Outllok 2015-2019. Global_Market_Outlook_2015_-2019_lr_v23.pdf

[15] Miyasaka T, Kojima A, Teshima K, and Shirai Y. Organometal halide perovskite as visible- light sensitizer for photovoltaic cells. Jour. of Americ. Chem. Soc. 2009; 131: 6050-6051.

[16] Kammen DM, Wadia C, and Alivisatos AP. Materials availibility expands the opportunity of large scale photovoltaics deployment. *Environment Science Technology*. 2009; 43: 2072-2077.

[17] Bottaro D and Moscowitz J. Current photovoltaic technology: current progress and future prospects. 1977. MIT Energy laboratory Reprot- MIT-EL 77-041W.

[18] Shockley W and Queisser HJ. Detailed balance limit of efficiency of pn junction solar cells. Journal of Applied Physics. 1961; 31: 510-519.

[19] chemistry/perovskite/what/history/index; http://www.chemistry/perovskite/what/history/ index.htm

[20] Luo S and Daoud WA. Recent progress in organic-inorganic halide perovskite solar cells: mechanism and material design. J. Mater. Chem. A. 2015. doi:10.1039/c4ta04953e

[21] Shalau AM. Fundamental absopion edge in lead (II) iodode:Potassium iodide alloys. Sol. Energy Mater. 1980; 2 (3): 327-332.

[22] Gao P, Gratzel M, and Nazeeruddin MK. Organohalide lead perovskites for photovoltaic applications. Energy and Environmental Science. 2014; 7: 2448-2463.

[23] Bretschneider SA, Weickert J, Dorman JA, and Schmidt-Mende L. Research update: Physical and electrical characteristics of lead halide perovskites for solar cells applications. APL Materials. 2014; 2: 040701 1-9.

[24] Frost JM, Butler KT, Brivio F, Hendon CH, Schilfgaarde MV and Walsh A. Atomistic origin of high performance in hybrid halide perovskite solarcells. Nano Lett. 2014; 14: 2484-2590.

[25] Baikie T, Fang YN, Kadro JM, Schreyer M, Wei FX, Mhaisalkar SG, Gratzel M and White TJ. Synthesis and crystal chemistry of the hybrid perovskite $(CH_3NH_3)PbI_3$ for solid-state sensitised solar cell applications. J. Mater. Chem. A. 2013; 1: 5628–5641.

[26] Stoumpos C, Malliakas CD and Kanatzidis MG. Semiconducting Tin and Lead Iodide Perovskites with Organic Cations: Phase Transitions, High Mobilities, and Near-Infrared Photoluminescent Properties. Inorg. Chem. 2013; 52: 9019–9038.

[27] Yin WJ, Yang JH, Kang J, Yan Y and Wei SH. Halide perovskite materials for solar cells: a theretical review. J. Mater. Chem. A. 2014.

[28] Walsh A, Watson GW. The origin of the stereochemically active Pb (II) lone pair: DFT calculations on PbO and PbS. Journal of Solid State Chemistry. 2005; 178: 1422–1428;

[29] Walsh A, Payne DJ, Egdell RG and Watson GW. Stereochemistry of post-transition metal oxides: revision of the classical lone pair model. Chem.Soc. Rev. 2011; 40: 4455–446.

[30] Keith T. Butler, Jarvist M. Frost and Aron Walsh,. Band alignment of the hybrid halide provskites $CH_3NH_3PbCl_3$, $CH_3NH_3PbBr_3$, and $CH_3NH_3PbI_3$. Materials Horizons. 2014; 2: 228-231.

[31] Giorgi G, Fujisawa JI, Segawa H and Yamashita K. Small Photocarrier Effective Masses Featuring Ambipolar Transport in Methylammonium Lead Iodide Perovskite: A Density Functional Analysis. J. Phys. Chem. Lett. 2013; 4: 4213–4216.

[32] Agiorgousis ML, Sun YY, Zeng H and Zhang S. Strong covalency-induced recombination centers in perovskite solar cell material $CH_3NH_3PbI_3$. Jour. Americ. Chem. Soc. 2014; 136: 14570-14577.

[33] Baumann SV, Rieder P, Heiber MC, Tvingstedt K, and Dyakonov V. Identification of traps states in perovskite solar cells. Journ. Phys. Chem. Lett. 2015; 6: 2350-2354.

[34] Lee MM, Teuscher J, Miyasaka T, Murakami TN, and Snaith HJ. Efficient hubrid solar cells based on meso-superstructured organometal halide perovskites. Science. 2012; 338: 643-647.

[35] Liu M, Johnston MB and Snaith HJ. Efficient planar heterojunction perovskite solar cell by vapor deposition. Nature. 2013;501:395-398.

[36] Wang JT, Ball JM, Barea EM, Abate A, Alexander-Webber JA, Huang J, Saliba M, Mora-Sero I, Bisquert J, Snaith HJ and Nicholas RJ. Low-temperature processed electron collection layers of graphene/TiO_2 nanocomposites in thin film perovskite solar cells. Nano Letters. 2014;14:724-730.

[37] Wojciechowski K, Saliba M, Leijtens T, Abate A and Snaith HJ. Sub-150^0 C processed meso-superstructered perovskite solar cells with enhanced efficiency. Energy Environ. Sci. 2014;7:1142-1147.

[38] Jeon NJ, Lee HG, Kim YC, Seo J, Noh JH, Lee J and Seok SI. O-Methoxy Substituents in Spiro-OMeTAD for Efficient Inorganic–Organic Hybrid Perovskite Solar Cells. J. Am. Chem. Soc. 2014;136:7837–7840.

[39] Zhou H, Chen Q, Li G, Luo S, Song T, Duan HS, Hong Z, You J, Liu Y, Yang Y. Interface engineering of highly efficient perovskite solar cells. Science. 2014;345: 542-546.

[40] Green MA, Emery K, Hishikawa Y, Warta W and Dunlop ED. Solar cell efficiency tables (Version 45). Prog. Photovolt: res. Appl. 2015;23:1-9.

[41] Niemegeers A, Burgelman M, Decock K. SCAPS Manual. University of Gent. 2014.

[42] Minemoto T, Murata M. Device modeling of perovskite solar cells based on structural similarity with thin film inorganic semiconductor solar cells. Journal of Applied Physics. 2014;116: 5, 054505.

[43] PV EDUCATION.ORG, Retrieved from http://www.PVEDUCATION.ORG

[44] Deng Q, Han X, Gao T, Shao G. Remarkable optical red shift and extremly high optical absorption coefficient of V-Ga co-doped TiO_2. Journal of Appilid Physics. 2012;112:013523 (1-8).

[45] Liu F, Zhu J, Wei J, Li Y, Lv M, Yang S, Zhang B, Yao J, and Dai S. Numerical simulation: Toward the design of high efficiency planar perovskite solar cells. Applied Physics Letters. 2014;104: 253508.

[46] Wolf SD, Holobsky J, Moon SJ, Loper P, Niesen B, Ledinsky M, Haug FJ, and Ballif C. Organometallic halide perovskites: Sharp optical absorption edge and its relation to photovoltaic performance. The Journal of Physical Chemistry Letters. 2014; 5:1035-1039.

[47] Xiang HJ, Gong XG, Lang L, Yang JH, and Liu HR. First principles study on the electric and optical properties of cubic ABX_3 halide perovskite. Physics Letters A. 2014;378:290-293.

[48] Park NG, Kim HS, and Im HS. Organohalide perovskite: New horizons in solar cell research. Journal of Physical Chemistry C. 2014;118:5615-5625.

[49] Laura M. Hertz, Christian Wehrenfennig, Giles E. Eperon, Michael B. Johston and Henry J. Snaith. (2014). High charge carrier mobilities and lifetimes in organolead trihalide perovskite. *Advanced Materials*. 26, 1584-1589.

[50] Trukhanov VA, Bruevich VV and Paraschuk DY. Effect of doping on performance of organic solar cells. Physics Review B. 2011;84: 205318.

[51] Wang MK, Gratzel C, Moon SJ, Humphry-Baker R, Rossier-Iten N, Zakeeruddin SM, and Gratzel M. Surface design in solid state dye sensitized solar cells: Effect of Zwitterionic co-adsorbants on photovoltaic performance. Advanced fuctional Materials. 2009;19:2163-2172.

[52] Kittel C. Introduction to Solid State Physics. John Wiley and Sons, Inc. (2005)

[53] Yamben SD, Haldar A, Liao KS, Dillon EP, Barron AR, and Curran SA. Optimization of organic solar cells with thin film Au as anode. Solar Energy Materials and Solar Cells. 2011;95:2424-24.

[54] Middy Gerald W. Solar insolation and concentration by coupled fiber optic, United States Patents. 1984.

Chapter 16

Tantalate-based Perovskite for Solar Energy Applications

Yiguo Su, Junyu Lang, Chunfang Du and Xiaojing Wang

Additional information is available at the end of the chapter

http://dx.doi.org/10.5772/61390

Abstract

To realize a sustainable society in the near future, the development of clean, renewable, cheap and sustainable resources and the remediation of environmental pollutions using solar energy as the driving force would be important. During the past few decades, plenty of efforts have been focused on this area to develop solar light active materials to meet the increased energy and environmental crisis. Owning to the unique perovskite-type structure, tantalate-based semiconductors with unable chemical composition show high activities toward the conversion of solar radiation into chemical energy. Moreover, various engineering strategies, including crystal structure engineering, electronic structure engineering, surface/interface engineering, co-catalyst engineering and so on, have been developed in order to modulate the charge separation and transfer efficiency, optical absorption, band gap position and photochemical and photophysical stability, which would open a realm of new possibilities for exploring novel materials for solar energy applications.

Keywords: Tantalate, photocatalysis, solar energy, perovskite

1. Introduction

In view of global energy crisis and environmental pollution, the search for renewable and clean energy resources and the development of eco-friendly systems for environmental remediation have received great attention. Solar energy is the prime renewable source of energy for every life on the earth. The amount of solar energy that strikes the Earth yearly in the form of sunlight is approximately ten thousand times the total energy that is consumed on this planet [1]. However, sunlight is diffuse and intermittent, which impedes its collecting and storage that play critical roles in the full exploitation of its potentials. As one of the most promising solutions for storing and converting solar energy, semiconductor photocatalysis has attracted much attention, since it provides an environmental benign strategy for splitting water into hydrogen and oxygen, reducing carbon dioxide into useful chemicals and fossil fuels, and completely eliminating all kinds of contaminants under the sunlight illumination under ambient conditions [2–4]. Generally, the fundamental principles of semiconductor photocatalysis have been extensively reported in previous works [5]. A photocatalytic reaction is initiated on the basis of the formation of photogenerated charges (such as electrons and holes) after capture of sunlight by a semiconductor. Consequently, electrons can transit from valence band (VB) to conduction band (CB) leaving behind holes in the VB. If the photogenerated electron–hole pairs' separation is maintained, the photogenerated carriers can move to the semiconductor surfaces to react with the adsorbed small molecules (dioxygen and water), generating the redox foundations of active species which lead to water splitting and/or destruction of organic compounds [6]. It is also noted that the photogenerated electrons in the CB can also recombine with the photogenerated holes in the VB to dissipate the input energy in the form of heat or radiated light (see Figure 1). From the perspective of efficient utilization of solar energy, the recombination between the photogenerated electrons and holes is not desired, which limits the efficiency of a semiconductor photocatalyst. For better photocatalytic performance, the photogenerated electrons and holes must be separated effectively, and charges have to be transferred rapidly at the mean time across the photocatalysis to astrict the recombination.

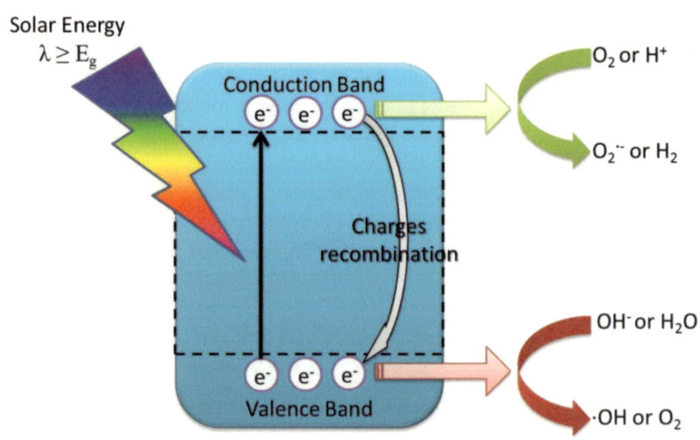

Figure 1. Schematic illustration of the principle of photocatalysis.

To date, several semiconductors, including TiO_2, ZnO, SnO_2, $BiVO_4$ and so forth, have been extensively investigated [7–10]. Among them, tantalate-based semiconductors with perovskite-type structure have certainly verified to be one of the brilliant photocatalysts for producing hydrogen from water and the oxidative disintegration of much organic containment [11]. For instance, $NaTaO_3$ with a perovskite-type structure showed a quantum yield of 56% under ultraviolet light irradiation after lanthanum doping and NiO co-catalyst loading [12]. Nevertheless, because of their broad band gap, most tantalate-based semiconductors can only react under ultraviolet or near-ultraviolet radiation, which reduces the utilization of ~43% of the solar spectrum To efficiently utilize the sunlight in visible region, the design of visible-light response tantalate-based catalysts is current demanded. Up to now, numerous methodologies have been developed to prepare different visible-light-driven tantalate-based photocatalysts, including doping strategy, heterojunction, facet control and so on [13,14].

This chapter emphasizes certain topical works that concentrate on tantalate-based photocatalysts for solar energy application. The aim is to display that the rational design, fabrication and modifications of tantalate-based semiconductors have tremendous effects onto their final photocatalytic activity, simultaneously, providing some stimulating perspectives on the future applications.

2. Alkali tantalate-based perovskite semiconductors

2.1. Synthetic strategies of alkali tantalates

Alkali tantalate-based perovskite semiconductors (likewise $LiTaO_3$, $NaTaO_3$ and $KTaO_3$) have a general formula of ABO_3 and have drawn a lot of attention due to the peculiar superconductivity, photocatalytic property, electrochemical reduction and electromagnetic features. There are two kinds of totally different cationic sites in a perovskite photocatalysis, in which A-site is coordinated by twelve O^{2-}, and is usually occupied by relative bigger cations (Li, Na and K). The B-site is taken up by smaller cations (Ta) with a coordination of six, as illustrated in Figure 2. The bond angles of Ta–O–Ta are 143° for $LiTaO_3$, 163° for $NaTaO_3$ and 180° for $KTaO_3$, respectively [15]. Wiegel and coworkers have reported the relationship between crystal structures and energy delocalization for alkali tantalates. When the Ta–O–Ta bond angle is close to 180°, the migration of excitation energy can be accelerated and the band gap decreases [16]. Thereby, the delocalization of excited energy of $LiTaO_3$, $NaTaO_3$ and $KTaO_3$ increases in turn. This result suggests that $KTaO_3$ may be predicted to be with the best photocatalytic activity.

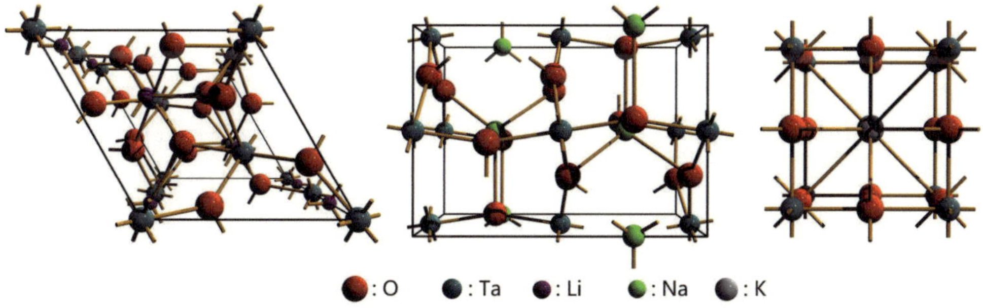

Figure 2. Crystal structure of $LiTaO_3$, $NaTaO_3$ and $KTaO_3$, respectively.

Alkali tantalate with different sizes, morphologies and compositions can be prepared via traditional solid-state method, solvothermal, sol–gel, molten salt and other methods. Basically, the traditional solid-state method is quite often used to prepare alkali tantalates, which includes the high temperature processing of the combination of alkali salts and tantalum pentaoxide. Kudo and coworkers successfully prepared $ATaO_3$ (A = Li, Na and K) materials with high crystallinity via solid-state method. It is found that all alkali tantalate showed superior photocatalytic activity toward stoichiometric water splitting under ultraviolet condition [17]. The high photocatalytic activity is chiefly depending on the high CB level consisting of Ta 5d orbitals [15]. Among them, $KTaO_3$ is the most photocatalytic active, which may be ascribed to the fact that $KTaO_3$ can absorb the most of photons and possesses the least distorted perovskite structure, being consistent with the above-mentioned discussion. The evolution rate of H_2 and O_2 was determined to be 29 and 13 μmolh^{-1}, respectively. To improve more photocatalytic activity of $ATaO_3$, a modified solid-state method was adopted by adding extra amount of alkali to compensate the loss [18]. When preparing the alkali tantalates with the existence of excess alkali, the photocatalytic activity of $LiTaO_3$, $NaTaO_3$ and $KTaO_3$ materials were improved ten to hundred times. $LiTaO_3$ is the naked alkali tantalate photocatalyst which showed the highest activity. This is because $LiTaO_3$ possesses higher conduction band levels than that of $NaTaO_3$ and $KTaO_3$, which may predict higher transfer rate of excited energy and the subsequent higher phtocatalytic activity. This type of phenomena was likewise observed for $CaTa_2O_6$, $SrTa_2O_6$ and $BaTa_2O_6$ photocatalysts with similar crystal structures [19].

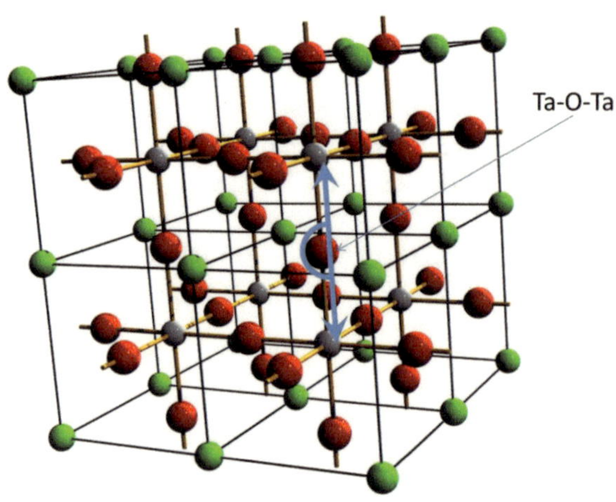

Figure 3. Ta–O–Ta bond angle in a 2 × 2 × 2 supercell of monoclinic phase NaTaO3. Red ball, green ball and grey ball represent O, Na and Ta atoms, respectively.

One should note that the synthetic strategy also has great influence on the structural features as well as photocatalytic activity. For instance, sol–gel method was also used to prepare $NaTaO_3$ nanoparticles. By using $CH_3COONa\ 3H_2O$ and $TaCl_5$ as the raw materials and citric acid as the complexing agent, $NaTaO_3$ nanoparticles with monoclinic phase that shows an indirect band gap, high densities of states near the band edges and a Ta–O–Ta bond angle close to 180° (Figure 3) are obtained. This result is quite different to $NaTaO_3$ that was synthesized via solid-state method, which formed the orthorhombic phase that has a direct band gap and a Ta–O–Ta bond angle of 163°. It is found that monoclinic $NaTaO_3$ has lots of effective states available for the photogenerated charge pairs. Meanwhile, the larger surface area and the advantageous features in the electronic and crystalline structures for the monoclinic $NaTaO_3$ have resulted in a remarkably higher photocatalytic activity for the sol–gel synthesized $NaTaO_3$ than that for the solid-state derived orthorhombic $NaTaO_3$ [20]. Besides sol–gel method, the molten-salt approach is also adopted to prepare alkali tantalate materials [21,22]. By a convenient molten-salt process, a series of $NaTaO_3$ and $KTaO_3$ efficient photocatalysts is successfully synthesized, which are highly crystallized single crystal nanocubes (about 100 nm large). Doping tetravalent Zr^{4+} and Hf^{4+} in $NaTaO_3$ and $KTaO_3$ efficiently increases the activity and stability of catalyst at the same time, although the energy levels have no change. Moreover, Zr^{4+} and Hf^{4+} doping can also led to particle size reduction and nearly monodispersed feature of $NaTaO_3$ and $KTaO_3$ nanoparticles. In the absence of co-catalyst, the photocatalytic activity can reach 4.65 and 2.31 $mmolh^{-1}$ toward H_2 and O_2 production, respectively [22]. A novel kind of strontium-doped $NaTaO_3$ mesocrystals was also prepared by a common molten-salt way. The obtained three-dimensional architectures showed high crystallinity, preferred orientation growth and high surface area. The ability for hydrogen generation of photocatalyst achieves 27.5 and 4.89 $mmolh^{-1}$ for methanol aqueous solution and pure water splitting under ultra-violet light irradiation [23]. However, either solid-state method or molten-salt approach often leads to ultra-low surface areas of alkali tantalates, which limits the photocatalytic activity. Hydrothermal synthesis is advantageous for regular nucleation of nanocrystals with well-defined particles, morphologies, crystallinity and surface areas [24]. For instance, nano-sized Ta_2O_5 and $NaTaO_3$, $KTaO_3$ and $RbTaO_3$ cubes are prepared by a facile hydrothermal method [25]. It is observed that pH influences much in the process of tantalum compound nanoparticles preparation. The obtained morphologies ranging from agglomerated particles in acidic medium over sticks at neutral pH value to cubes in elementary media can be achieved, which are similar to titanates [26]. A microwave-assisted hydrothermal technique was reported using Ta_2O_5 and NaOH as starting materials under quite mild conditions with short reaction time. The BET surface area of $NaTaO_3$ nanoparticles prepared by microwave-assisted hydrothermal method is about 1.5 times than that prepared by conventional hydrothermal method [27]. After loading NiO as co-catalyst, this photocatalyst showed photocatalytic activity for overall water splitting more than two times greater than those prepared by conventional hydrothermal process [28]. As an outstanding example, $NaTaO_3$ nanoparticles through hydrothermal treatment highly improved the photocatalytic activity by a factor of 8 toward water splitting in comparison with the photocatalysts obtained by traditional solid-state method, which is attributed to their smaller particle size, larger surface area and higher crystallinity [29,30].

2.2. Electronic structure engineering

2.2.1. Doping strategies

Introducing foreign elements, including metal ions or non-metal ions, into semiconductor host matrix is one of the most effective methods to modulate the electronic structure of the host semiconductor and produce enhanced photocatalytic performance. Owing to a big difference in radius of A- and B-site ions in alkali tantalates, the dopants can selectively permeate into the A or B sites, which determine chemical composition, surface features, electronic structure and their photocatalytic properties. To date, studies on the alkali tantalates derived by doping strategy are thoroughly investigated. La-doped $NaTaO_3$ is the most active photocatalyst in photocatalytically splitting water area [12]. In this case, the catalytic activity of $NaTaO_3$ is extremely modulated by doping with La^{3+}. For example, the crystallinity growths and a surface stair structure with nanometer-scale features are constructed, which improve the separation efficiency of the photogenerated electron–hole pairs and the photocatalytically splitting water activity. The surface step structure is also formed in alkaline earth metal ion doped $NaTaO_3$, which showed improvement of photocatalytic water splitting properties [31]. Bi^{3+} doped $NaTaO_3$ nanoparticles are prepared under different initial stoichiometric ratio by traditional solid-state reaction, which showed visible light absorption and tunable photocatalytic activity. Controlling the original molar ratio of the reactants, the intrusion of bismuth at sodium site and tantalum site in $NaTaO_3$ can be well-modulated and the optimum performance can be easily changed. Occupancy of Bi atom at Na site of $NaTaO_3$ is not contributing to increase visible light absorption while occupancy of Bi at Ta site or at both Na and Ta site induces visible light absorption and the subsequent methyl blue degradation under visible light [32,33]. La, Cr codoping $NaTaO_3$ system is also developed by spray pyrolysis from aqueous and polymeric precursor solution. The hydrogen evolution rate of La, Cr codoped $NaTaO_3$ was enhanced 5.6 times to 1467.5 μmol $g^{-1}h^{-1}$, and the induction period was shortened to 33%, compared to the identical values achieved by the Cr-doped $NaTaO_3$ photocatalyst prepared from aqueous precursor solution [34]. Besides metal ion doping, several non-metal ions are also incorporated into the host matrix of alkali tantalates for improved visible light absorption and photocatalytic performance [35,36]. A plane-wave-based density functional theory calculation is conducted to predict the doping effects on the variations of the band structure of non-metal ions doped $NaTaO_3$. There were studies about nitrogen, sulfur, carbon and phosphorus monodoping and nitrogen–nitrogen, carbon–sulfur, phosphorus–phosphorus and nitrogen–phosphorus codoping. Nitrogen and sulfur monodoping can improve the valence band edge to higher and keep the ability to split water into H_2 and O_2 remain unchanged, as is shown in Figure 4. Double hole-mediated codoping can decrease the band gap dramatically. Nitrogen–nitrogen, carbon–sulfur and nitrogen–phosphorus codoping could narrow band gap to 2.19, 1.70 and 1.34 eV, respectively, which could absorb visible light.

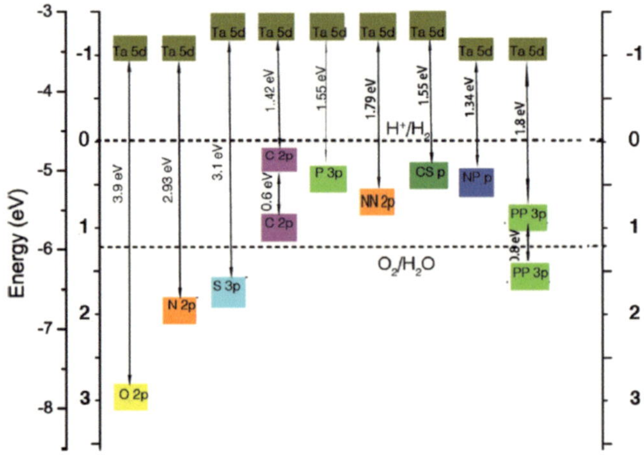

Figure 4. Band alignment of non-metal ions doped $NaTaO_3$. The position of the valence band edge of pure $NaTaO_3$ is adopted from experiment [37].

2.2.2. Defect chemistry engineering

Defect chemistry plays an important role in modulating the electronic structure, charge carrier conductivity and photocatalytic performance [38]. Defect chemistry often shows different impacts on the photocatalytic efficiency for most of the semiconductors. Previous literature on $NaTaO_3$ indicated that the accretion of the extra quantity of Na in the synthesis of $NaTaO_3$ blocked construction of sodium ion defects in $NaTaO_3$ crystals, leading to the extreme enhancement of photocatalytic activity [18]. Basically, the native defects, such as oxygen vacancies and sodium ion defects, are often observed in $NaTaO_3$. Oba and coworkers investigated the formation energies and electronic structure of lattice vacancies, antisite defects and lanthanum impurities in $NaTaO_3$ using first-principles calculations based on density-functional theory [39]. Under oxygen-poor environments, oxygen vacancy as a double donor is a main defect. In La-$NaTaO_3$, the replacement of La at Ta site is similar to make up as a shallow acceptor under oxygen-rich environments whereas the replacement of La at Na site forms as a double donor under oxygen-poor environments. The location predilection of lanthanum leads to self-compensation in heavily doped cases, which have great impact on the change in carrier concentration and photocatalytic activity [12]. Defective center not only alters the carrier concentration but also induces visible light absorption. In Eu^{3+} doped $NaTaO_3$, a nonstoichiometric Na/Ta molar ratio led to site-selective occupation of Eu^{3+} dopant ions, which resulted in a monotonous lattice expansion and local symmetry distortion [11]. The site-selective occupation of Eu^{3+} gave rise to certain types of defective centers due to the charge difference between Eu^{3+} ions and Na^+ and/or Ta^{5+} ions, which is crucial to the modification of absorption in visible region and photocatalytic activity.

2.3. Surface/interface engineering

2.3.1. Heterojunction of nano-/microarchitectures

The constructions of heterojunction by combining a semiconductor with other semiconductors have attracted much research attention because of their perfect effectiveness in the separation of the photogenerated charge carriers and boosting the photocatalytic activity. In the past few years, a lot of significant findings have been described on the heterojunction of nano-/microarchitectures. Nano-Cu_2O/$NaTaO_3$ composite for the degradation of organic pollutants have also been successfully developed [13]. Nano-Cu_2O/$NaTaO_3$ composite exhibits highly enhanced photocatalytic activity in comparison to their individual counterpart. Furthermore, C_3N_4/$NaTaO_3$ and C_3N_4/$KTaO_3$ composite photocatalysts were also developed [40,41]. Loading of C_3N_4 is a good strategy to achieve the visible light photocatalytic activity (Figure 5). Photogenerated electron jumped from the VB to CB of C_3N_4 could unswervingly insert into the conduction band of $NaTaO_3$ or $KTaO_3$, making C_3N_4/$NaTaO_3$ and C_3N_4/$KTaO_3$ as visible light-driven photocatalyst. Both of the composites showed superior photocatalytic activity toward Rhodamine B degradation under visible light irradiation, being close to commercial P25. Yin and coworkers reported the preparation of novel C–$NaTaO_3$–Cl–TiO_2 composites via a facile solvothermal method. When C–$NaTaO_3$ is joined with Cl–TiO_2 to construct a core shell configuration, the visible light-induced degradation activity toward NO_x of the catalysts under visible light irradiation could be highly improved because of the suppression of the recombination of photogenerated charge carriers [42]. Zaleska et al. prepared a series of novel binary and ternary composite photocatalysts based on the combination of $KTaO_3$, CdS and MoS_2 semiconductors via hydro/solvothermal precursor route. They found that the highest photocatalytic activity toward phenol degradation under both UV-Vis and visible light irradiation and superior stability in toluene removal was observed for ternary hybrid obtained by calcination of $KTaO_3$, CdS and MoS_2 powders at the 10: 5: 1 molar ratio [43].

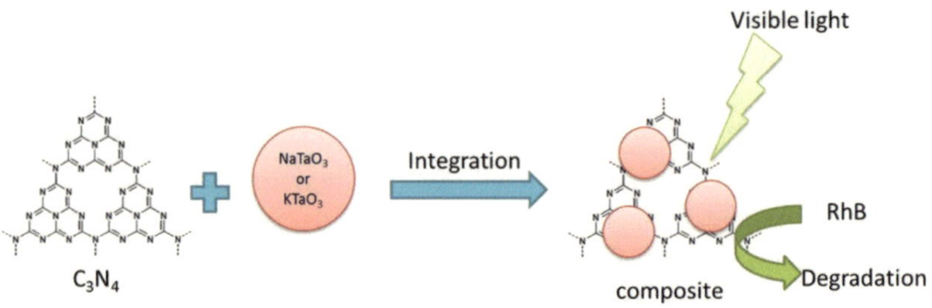

Figure 5. Schematic illustration of the photocatalytic degradation process of RhB by visible light-irradiated C_3N_4/$NaTaO_3$ or C_3N_4/$KTaO_3$.

2.3.2. Mesoporous structures construction

As one of the most important factors, surface area also imposes a big effect on the photocatalytic activity of the semiconductors. The majority of photocatalytic reactions occur at semiconductor surfaces, and therefore the photocatalytic activities of semiconductor oxides are usually greatly improved by the increase in surface area [44]. To further improve the surface area, nanocrystalline $NaTaO_3$ thin films with ordered three-dimensional mesoporous and nanostick-like constructions were successfully produced by PIB-*b*-PEO polymer-based sol–gel method. $NaTaO_3$ prepared at 650°C exhibits a BET surface area of about 270 m^2cm^{-3}, which is much larger than the ever reported values [45]. These nanocrsytalline mesoporous $NaTaO_3$ samples show both enhanced ultraviolet light photocatalytic activity and can keep steady performance. A confined space synthesis process was also used for preparing colloidal array of $NaTaO_3$ by using three-dimensional mesoporous carbon as the hard template. This method brings about the creation of a colloidal collection of mesoporous $NaTaO_3$ particles (20 nm). After NiO loading, the mesoporous $NaTaO_3$ nanoparticles showed photocatalytic activity for overall water splitting more than three times as high as non-structured bulk $NaTaO_3$ particles [46]. A carbon modified $NaTaO_3$ mesocrystal nanoparticle was also successfully synthesized by a one-pot solvothermal method by employing $TaCl_5$, NaOH and glucose as the starting materials and distilled H_2O/ethylene glycol mixed solution as a reaction solvent. The as-synthesized mesocrystal nanoparticles exhibited a high specific surface area of 90.8 m^2g^{-1} with large amounts of well-dispersed mesopores in the particles. The carbon-modified $NaTaO_3$ mesocrystal demonstrated excellent efficiency for continuous NO gas destruction under visible light irradiation, which is considerably superior to those of the unmodified $NaTaO_3$ specimen and commercial Degussa P25, owning to large specific surface area, high crystallinity and visible light absorption [47].

2.4. Co-catalyst engineering

2.4.1. Noble metal co-catalyst engineering

As well documented in previous literatures, co-catalyst introduces two positive factors into the photocatalyst, including promotion on the separation of photogenerated charge carriers and construction of active sites for reduction and/or oxidation reaction. Several noble metals have been commonly used as co-catalysts for photocatalytic applications. For example, water splitting activity of $NaTaO_3$:La was improved when Au was loaded either by photodeposition method or by impregnation method. Moreover, Au/$NaTaO_3$:La prepared by impregnation method exhibits much higher and more stable photocatalytic activity toward water splitting due to the fact that O_2 reduction on photodeposited Au co-catalyst was more efficient than that of impregnated Au co-catalyst [48]. Besides Au nanoparticles, Pt is also frequently used as co-catalyst for increasing the photocatalytic activity of alkali tantalates. With the deposition of Pt nanoparticles as co-catalyst, rare earth (including Y, La, Ce and Yb) doped $NaTaO_3$ exhibits a clear improvement of the hydrogen evolution, which is due to the fact that Pt nanoparticles act as electron scavengers reducing the photogenerated charge carrier recombination rate and facilitating the electron move to metal sites from the CB of $NaTaO_3$, being as the catalytic center for hydrogen generation [49]. Moreover, Pd nanoparticles are also used as a co-catalyst for H_2 production from water containing electron donor species. Su et al. prepared novel Pd/NiO core/shell nanoparticles as co-catalyst, which are placed on the surface of La doped $NaTaO_3$ photocatalyst. It is noted that Pd nanoparticles are more effective for H_2 generation from water containing methanol, while Pd/NiO core/shell nanoparticles exhibit a higher H_2 generation by splitting pure water. The presence of NiO not only provides hydrogen evolution sites and suppresses the reverse reactions on Pd-based catalysts but also improves the stability of the Pd nanoparticles on the La doped $NaTaO_3$ surfaces [50]. In another case, when RuO_2 (1 wt.%) was introduced as co-catalyst, the ability for H_2 generation of $NaTaO_3$ prepared by an innovative solvo-combustion reaction was improved significantly, reaching around 50 mmol of H_2 after 5 h, which is the best of other reports in literature [51].

2.4.2. Earth abundant elements co-catalyst engineering

Due to too much scarcity and expense of noble metal co-catalyst to apply for wider scope solar energy applications, the development of high-efficiency and low-cost noble-metal-free co-catalysts is acutely necessary. Lately, co-catalysts composed of earth abundant elements have been explored extensively to replace noble metal co-catalysts for solar energy applications [52]. NiO is a p-type semiconductor with a band gap energy ranging within 3.5–4.0 eV, which is widely used as the co-catalyst of tantalates-based semiconductors for enhancing photocatalytic activity [53]. In the case of NiO/$NaTaO_3$:La photocatalyst with high photocatalytic reactivity, NiO acts as co-catalyst loading as ultrafine NiO particles, which possesses characteristic absorption bands at 580 and 690 nm, The ultrafine NiO particles were highly active for hydrogen evolution as well as Pt of an excellent co-catalyst [12]. A detailed study on the structural features of NiO nanoparticles indicated that the interdiffusion of Na^+ and Ni^{2+} cations created a solid–solution transition zone on the outer sphere of $NaTaO_3$. The high photocatalytic activity resulting from a low NiO loading suggests that the interdiffusion of cations heavily doped the p-type NiO and n-type $NaTaO_3$, reducing the depletion widths and facilitating charge transfers through the interface barrier [54]. Besides NiO, Ni metallic nanoclusters were also used as co-catalyst. For instance, a series of nickel-loaded $La_xNa_{1-x}TaO_3$ photocatalysts was synthesized by a hydrogen peroxide-water based solvent method. Systematical investigation indicated that the activity of hydrogen generation from pure water is in sequence: Ni/NiO > NiO > Ni, whereas the activity sequence with respect to aqueous methanol is: Ni > Ni/NiO > NiO. Ni metallic nanoclusters exhibit the most active sites and facilitate the formation of hydrogen from aqueous methanol. In the case of Ni/NiO core/shell structure, Ni metallic nanoclusters induce the migration of photogenerated electrons from the bulk to catalyst surface, while NiO acts as H_2 evolution site and prevents water formation from H_2 and O_2 [55].

2.4.3. Molecular co-catalyst engineering

Molecular co-catalyst engineering have received much research attention in recent years. In a molecular/semiconductor hybrid system, the noble-metal-free molecular complex as co-catalyst can not only facilitate the charge separation but also help us to understand the mechanisms of hydrogen evolution and carbon dioxide reduction at molecular level [56]. Although the study on molecular sensitized alkali tantalates is limited, an excellent research has been done by Hong and coworkers. In this case, by using a molecular co-catalyst $[Mo_3S_4]^{4+}$, the photocatalytic activity of $NaTaO_3$ was significantly improved. The hydrogen production rate is about 28 times higher than pure $NaTaO_3$ because $[Mo_3S_4]^{4+}$ clusters can provide a large number of effective active sites for hydrogen evolution and the matching of the conduction band of $NaTaO_3$ and the reduction potential of $[Mo_3S_4]^{4+}$ also acts as one of the major determinants for the enhancement of the photocatalytic activity [57].

3. Alkaline earth and transition metal tantalate-based perovskite semiconductors

3.1. Synthetic methodologies of alkaline earth and transition metal tantalates

Solid-state reaction method and hydrothermal method are used routinely to synthesize alkaline earth and transition metal tantalates. Almost all the alkaline earth and transition metal tantalates can be obtained by high-temperature solid-state method using Ta_2O_5 and other salts as starting materials. For instance, $Sr_2Ta_2O_7$ [58], $Sr_{0.8}Bi_{2.2}Ta_2O_9$ [59], $Bi_2SrTa_2O_9$ [60] $H_{1.81}Sr_{0.81}Bi_{0.19}Ta_2O_7$ [61], $Ba(Zn_{1/3}Ta_{2/3})O_3$ [62], $Ba(Mg_{1/3}Ta_{2/3})O_3$ [63], $Ba_4Ta_2O_9$ [64] and $Ba_5Ta_4O_{15}$ [65] have been synthesized successfully by this method, which show prospects in many application including photocatalytic semiconductor, solar cells and electronic device. The high-temperature treatment of traditional solid-state reaction will increase the size of particles and thus decrease the surface area. $Sr_2Ta_2O_7$ photocatalysts of layered perovskite structures gotten from the solid state reaction had better activity which is mainly because of their more negative conduction band. $H_2ATa_2O_7$ (A = Sr or $La_{2/3}$) [66] and $LiCa_2Ta_3O_{10}$ [67] were reported to be obtained by similar way with extra alkali, which can supply the loss at high temperature to suppress defects formation. This makes the crystal structure grow well and has better catalytic efficiency than others synthesized with a theoretical ratio in most cases. This improved solid-state reaction method would efficiently inhibit the recombination of photo-carrier to enhance the photocatalytic activity. A new polymerizable complex technique is one of the preparation methods of alkaline earth tantalates, which has a relative moderate condition. This method includes the provision of Ta-base compound and then come into being the sticky sol–gel, after the treatment at 600–700 °C. Comparing with solid-state method, the tantalate-based photocatalysts synthesized by a polymerizable complex way often have greater crystallinity and better crystal size, which will lead to remarkably increase the photocatalytic efficiency [68]. Comparing with solid-state method, hydrothermal method has been widely used in synthesizing perovskite tantalates with very lower reaction temperature. Lots of alkaline earth tantalate could be prepared by the hydrothermal method exhibiting higher activity. In 2006, Zhu and coworkers synthesize monomolecular-layer $Ba_5Ta_4O_{15}$ nanosheets by hydrothermal method [65], which show enhanced activity ten times better than that of solid-state method-derived $Ba_5Ta_4O_{15}$ particles in photodegradation reactions of Rhodamine B solution. Perovskite $Ca_2Ta_2O_7$ has also been synthesized by hydrothermal process in aqueous KOH solution at 373 K for 120 h, which shows photocatalytic water splitting activity under UV-light irradiation [69]. Moreover, sol–gel route, as a common way to prepare the nanomaterials, also can be used for preparation of some perovskite tantalates. One typical case is that the ferroelectric $SrBi_2Ta_2O_9$ [70] and $SrBi_2Ta_2O_9$ nanowires [71] were synthesized using ethylene glycol as solvent, which showed greater dielectric and ferroelectric properties than the ceramics prepared by the solid-state reactions owning to a denser and more homogeneous microstructure with a better distribution of grain orientations. Sol–gel method is also used to prepare metastable phase like $Sr_{0.5}TaO_3$ nanosheets [72] with photocatalytic activities of water splitting under ultraviolet light irradiation. Several transition metal tantalates with perovskite structure can also be synthesized by these methods, including $AgTaO_3$ [73], $LaTaO_4$ [74], $H_2La_{2/3}Ta_2O_7$ [75] and so forth.

3.2. Crystal structure engineering

For ideal perovskite alkaline earth and transition metal tantalates, the cubic-symmetry structure has the Ta atom in 6-fold coordination, surrounded by an octahedron of oxygen atoms, and the alkaline earth or transition metal cation in 12-fold cuboctahedral coordination (Figure 6). The relative ion size requirements for stability of the cubic structure are quite stringent, so slight buckling and distortion can produce several lower-symmetry distorted versions, in which the coordination numbers of cations are reduced. [77] But in fact, almost all perovskite alkaline earth and transition metal tantalates have compound perovskite structures covering two different cations at Ta site in TaO_6 octahedra or at cation site in 12-fold cuboctahedral coordination. This led to the chance of alternatives between ordered and disordered. Crystal structure is a very important factor manipulating the band gaps of perovskite tantalates containing the following aspects: (1) the bond angle of tantalum and oxygen ions of octahedra units; (2) the interlayer spacing of the perovskite; (3) the interaction between perovskite layers; (4) the polarization ability of cations at the interlayer toward the oxygen ions of octahedra facing the interlayer. Some of alkaline earth and transition metal tantalates with layered perovskite exhibited outstanding photo catalytic activity, including $Ba_5Ta_4O_{15}$, $H_{1.81}Sr_{0.81}Bi_{0.19}Ta_2O_7$, $SrBi_2Ta_2O_9$, $LaTaO_4$, $H_2La_{2/3}Ta_2O_7$ and $Sr_{0.5}TaO_3$ reported by many groups [61,65,70,72,74,75]. This kind of perovskite composites are promising materials with multiple elements, perovskite framework and layer-like structures, which can be classified into three category structures by the different interlayer structure. On the other hand, it has been reported that some perovskite alkaline earth tantalates with double-perovskite structure also show photocatalytic activity under ultraviolet light irradiation [79]. And some transition metal tantalates have simple cubic perovskite-type structure like $AgTaO_3$, $Ba_3ZnTa_2O_9$, Sr_2GaTaO_6 [80] and $NaCaTiTaO_6$, $NaCaTiNbO_6$, $NaSrTiTaO_6$ and $NaSrTiNbO_6$ [81].

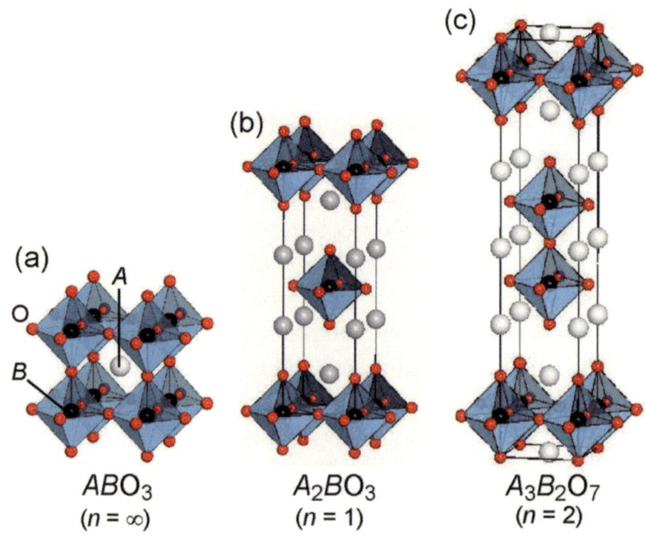

Figure 6. General structure of perovskite and layered perovskite [76].

3.3. Metal/non-metal doping strategies for band gap engineering

Introducing external ions into crystal structure has been generally approved as a positive way to improve the visible-light photocatalytic activity of semiconductors with larger band gaps. For nitrogen-doped layered oxide $Sr_5Ta_4O_{15-x}N_x$, the extension of the visible light absorption has been ascribed to the substitution of nitrogen for oxygen atoms as well as the formation of Ta–N bonds. The N 2p states mixed with pre-existing O 2p states shift the valence band maximum upward and result in wide visible light absorption [82]. A slight N dopant led to hinder the recombination of photo-generated charge pairs. N-doped $Ba_5Ta_4O_{15}$ also displays a brilliant photocatalytic activity under solar condition. The doping resulted in a significant narrowing of the band gap from 4.06 eV to ca. 1.76 eV, indicating that it can use more visible light [83]. Furthermore, Sun et al. investigated affection of band gap doping with several metal and non-metal by DFT calculation [84]. It is found that, in most perovskite cases, the valence band levels were shifted upwards, in which the maximum contribution to valence band maximum comes from the p orbitals of the dopant anions, which shift. On the other hand, the dopant cations shift the CB level downwards because the CBM is chiefly governed by the d orbitals of foreign cations. This conclusion was applicable to perovskite structure tantalates system (Figure 7) directly by Liu and coworkers [85].

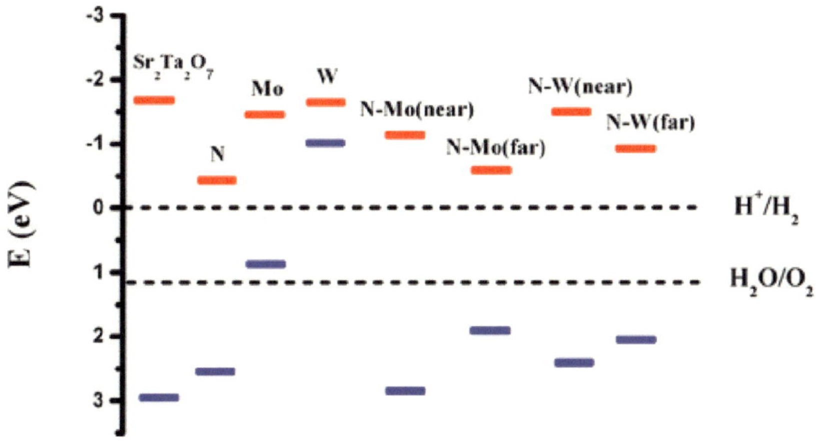

Figure 7. The electronic band edge positions with respect to the water reduction and oxidation potential levels for the pure and doped $Sr_2Ta_2O_7$ systems [85].

3.4. Multi-component heterojunction

Multi-component semiconductor combination tactic shows effectivity to improve photocatalytic activity by separation of the photo-generated charge carriers with a formation of a heterojunction structure. Heterojunction structure is the interface that is located at two areas of different crystalline semiconductors. This kind of material has to consider the following points, including near crystal structure, similar interatomic spacing and close coefficient of thermal expansion. Otherwise, they should have discrepant band gap values, which is exact contrary to a homojunction. It is benefited to regulate the electronic energy bands. To promote the redox ability and photocatalytic activity, composite photocatalysts involving two or more components were extensively studied. One type of such composites is usually constructed by coupling semiconductors with larger band gap for the purpose of the higher redox ability. A charming work is the $Ba_5Ta_4O_{15}/Ba_3Ta_5O_{15}$ composite reported by Roland Marschall et al., which synthesized through the sol–gel method showed brilliant activities in OH radical generation and photocatalytic hydrogen production [86]. The outstanding activity is expected to come from enhanced charge carrier separation. In 2011, Wang and coworkers present Pt-loaded graphene-$Sr_2Ta_2O_{7-x}N_x$ (Figure 8) with enlarged visible light absorption region and enhanced photocatalytic hydrogen generation [87].

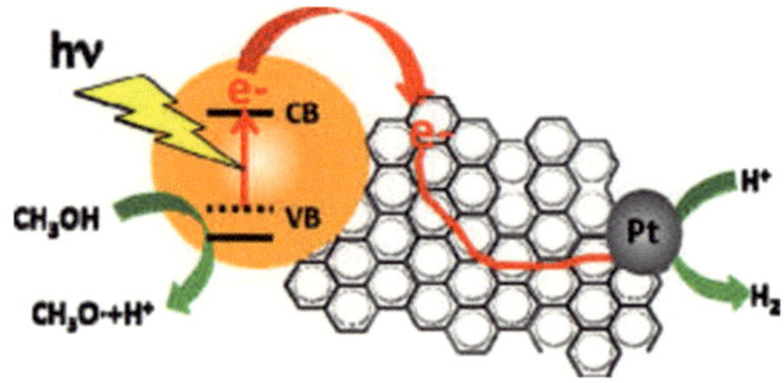

Figure 8. Schematic diagram for Pt-loaded graphene-$Sr_2Ta_2O_{7-x}N_x$ photocatalyst under simulated solar light irradiation [87].

3.5. Co-catalyst surface modification

Transition metals and their oxides are usually used as practical co-catalysts for photocatalysis. The role of the co-catalysts attached on the surface of the semiconductor material is particularly significant. It increases the overall photocatalytic activity by helping to separate charge pairs, which can work for both bulk and surface electron/hole pathway. The chemical reaction that took place at surface is promoted by the co-catalysts. Various metals and oxides loaded on the surface of semiconductor show different effects. In most photocatalytic water splitting systems, several metals like Au and Pt can accelerate the rate of reduction of hydrogen observably [88, 89]. On the other hand, some oxides like NiO, NiO_x and RuO_2 can promote the rates of both hydrogen and oxygen production [90–92]. Among them, NiO_x exhibited highest activity in photocatalytic process [78]. As a hydrogen evolution site, the co-catalyst has to extract the photogenerated electrons from the CB of host materials. Thus, the conduction band level of co-catalyst should be below that of photocatalyst. In addition, photocatalytic water splitting is sensitive to the deposition methods of co-catalysts. Kudo et al. reported that photocatalysts show diversity in photocatalytic water splitting with different deposition methods [59]. Moreover, transition-metal sulfides like MS (M = Ni, Co, Cu) have also been developed as co-catalysts to improve the photocatalytic activity. These sulfides have the same effects with other co-catalysts in reaction process [93].

4. Summary and outlook

Tantalate-based perovskite semiconductors are well known for their wide spread applications in photocatalysis, ionic conductors, luminescence host materials and ferroelectric ceramics. Drawbacks of wide band gap and low charge separation efficiency inhibit the further development of tantalate-based perovskite semiconductors as superior photocatalysts. The combination of various strategies, such as doping, heterojunction and co-catalyst engineering, induces a thrilling beginning for exploring visible light active and highly efficient photocatalysts for solar energy applications. However, the studies on tantalate-based perovskite semiconductors are currently unsystematic. Meanwhile, the as-mentioned strategies and the derived photocatalytic systems with high efficiency and stability still need to be further developed.

Acknowledgements

This work is financially supported by the National Natural Science Foundation of China (Grants 21267014, 21367018, 21563021), the Project of Scientific and Technological Innovation Team of Inner Mongolia University (12110614), Fund of Key Laboratory of Optoelectronic Materials Chemistry and Physics, Chinese Academy of Sciences (2008DP173016-1410).

Author details

Yiguo Su, Junyu Lang, Chunfang Du and Xiaojing Wang*

*Address all correspondence to: wang_xiao_jing@hotmail.com

College of Chemistry and Chemical Engineering, Inner Mongolia University, Hohhot, Inner Mongolia, P. R. China

References

[1] Chen HM, Chen CK, Liu RS, Zhang L, Zhang J, Wilkinson DP. Nano-architecture and material designs for water splitting photoelectrodes. Chem Soc Rev 2012;41:5654–71. DOI: 10.1039/C2CS35019J

[2] Martindale BCM, Hutton GAM, Caputo CA, Reisner E. Solar hydrogen production using carbon quantum dots and a molecular nickel catalyst. J Am Chem Soc 2015;137:6018–25. DOI: 10.1021/jacs.5b01650

[3] Tu W, Zhou Y, Zou Z. Photocatalytic conversion of CO_2 into renewable hydrocarbon fuels: state-of-art accomplishment, challenges, and prospects. Adv Mater 2014;26:4607–26. DOI: 10.1002/adma.201400087

[4] Lin Y, Li D, Hu J, Xiao G, Wang J, Li W, Fu X. Highly efficient photocatalytic degradation of organic pollutants by PANI-modified TiO_2 composite. J Phys Chem C 2012;116:5764–72. DOI: 10.1021/jp211222w

[5] Cargnello M, Diroll BT. Tailoring photocatalytic nanostructures for sustainable hydrogen production. Nanoscale 2014;6:97–105. DOI: 10.1039/C3NR05383K

[6] Li L, Liu J, Su Y, Li G, Chen X, Qiu X, Yan T. Surface doping for photocatalytic purposes: relations between particle size, surface modifications, and photoactivity of SnO_2:Zn^{2+} nanocrystals. Nanotechnology 2009;20:155706–9. DOI: http://dx.doi.org/10.1088/0957-4484/20/15/155706

[7] Lin H, Li L, Zhao M, Huang X, Chen X, Li G, Yu R. Synthesis of high-quality brookite TiO_2 single-crystalline nanosheets with specific facets exposed: tuning catalysts from inert to highly reactive. J Am Chem Soc 2012;134:8328–31. DOI: 10.1021/ja3014049

[8] Schumann J, Eichelbaum M, Lunkenbein T, Thomas N, Galván MCÁ, Schlögl R, Behrens M. Promoting strong metal support interaction: doping ZnO for enhanced activity of Cu/ZnO:M (M = Al, Ga, Mg) catalysts. ACS catalysis. 2015;5:3260–70. DOI: 10.1021/acscatal.5b00188

[9] Long J, Xue W, Xie X, Gu Q, Zhou Y, Chi Y, Chen W, Ding Z, Wang X. Sn^{2+} dopant induced visible-light activity of SnO_2 nanoparticles for H_2 production. Catal Commun 2011;16:215–9. DOI: 10.1016/j.catcom.2011.10.002

[10] Thalluri SM, Hussain M, Saracco G, Barber J, Russo N. Green-synthesized $BiVO_4$ oriented along {040} facets for visible-light-driven ethylene degradation. Indust Eng Chem Res 2014;53:2640–6. DOI: 10.1021/ie403999g

[11] Su Y, Peng L, Guo J, Huang S, Lv L, Wang X. Tunable optical and photocatalytic performance promoted by nonstoichiometric control and site-selective codoping of trivalent ions in $NaTaO_3$. J Phys Chem C 2014;118:10728–39. DOI: 10.1021/jp412236u

[12] Kato H, Asakura K, Kudo A. Highly efficient water splitting into H_2 and O_2 over lanthanum-doped $NaTaO_3$ photocatalysts with high crystallinity and surface nanostructure. J Am Chem Soc 2003;125:3082–9. DOI: 10.1021/ja027751g

[13] Su Y, Lang J, Cao N, Wang T, Zhu B, Wang X. Morphological reconstruction and photocatalytic enhancement of $NaTaO_3$ nanocrystals via Cu_2O loading. J Nanoparticle Res 2015;17:63–9. DOI: 10.1007/s11051-015-2877-9

[14] Wang B, Kanhere PD, Chen Z, Nisar J, Pathak B, Ahuja R. Anion-doped $NaTaO_3$ for visible light photocatalysis. J Phys Chem C 2013;117:22518–24. DOI: 10.1021/jp407025r

[15] Kato H, Kudo A. Photocatalytic water splitting into H_2 and O_2 over various tantalate photocatalysts. Catal Today 2003;78:561–9. DOI: 10.1016/S0920-5861(02)00355-3

[16] Wiegel M, Emond MHJ, Stobbe ER, Blasse G. Luminescent of alkali tantalates and niobates. J Phys Chem Solid 1994;55:773–8. DOI: 10.1016/0022-3697(94)90030-2

[17] Kato H, Kudo A. New tantalate photocatalysts for water decomposition into H_2 and O_2. Chem Phys Lett 1998;295:487–92. DOI: 10.1016/S0009-2614(98)01001-X

[18] Kato H, Kudo A. Water splitting into H_2 and O_2 on alkali tantalate photocatalysts $ATaO_3$ (A = Li, Na, and K). J Phys Chem B 2001;105:4285–92. DOI: 10.1021/jp004386b

[19] Kato H, Kudo A. Photocatalytic decomposition of pure water into H_2 and O_2 over $SrTa_2O_6$ prepared by a flux method. Chem Lett 1999;28:1207–8. DOI: 10.1246/cl.1999.1207

[20] Hu CC, Teng H. Influence of structural features on the photocatalytic activity of $NaTaO_3$ powders from different synthesis methods. Appl Catal A: Gen 2007;331:44–50. DOI: 10.1016/j.apcata.2007.07.024

[21] Lee S, Teshima K, Mizuno Y, Yubuta K, Shishido T, Endo M, Oishi S. Growth of well-developed sodium tantalate crystals from a sodium chloride flux. Cryst Eng Comm 2010;12:2871–7. DOI: 10.1039/b921092j

[22] Sun J, Chen G, Li Y, Jin R, Wang Q, Pei J. Novel (Na, K)TaO_3 single crystal nanocubes: Molten salt synthesis, invariable energy level doping and excellent photocatalytic performance. Energy Environ Sci 2011;4:4052–60. DOI: 10.1039/c1ee01259b

[23] Sun J, Chen G, Pei J, Jin R, Wang Q, Guang X. A simple approach to strontium sodium tantalite mesocrystals with ultra-high photocatalytic properties for water splitting. J Mater Chem 2012;22:5609–14. DOI: 10.1039/c2jm16387j

[24] Su Y, Li L, Li G. Synthesis and optimum luminescence of $CaWO_4$-based red phosphors with codoping of Eu^{3+} and Na^+. Chem Mater 2008;20:6060–7. DOI: 10.1021/cm8014435

[25] Gömpel D, Tahir MN, Panthöfer M, Mugnaioli E, Brandscheid R, Kolb U, Tremel W. Facile hydrothermal synthesis of crystalline Ta_2O_5 nanorods, $MTaO_3$ (M = H, Na, K, Rb) nanoparticles and their photocatalytic behaviour. J Mater Chem A 2014;2:8033–40. DOI: 10.1039/c4ta00183d

[26] Koll D, Andrusenko I, Mugnaioli E, Birkel A, Panthöfer M, Kolb U, Tremel W. Snapshots of the formation of $NaTi_3O_6(OH)$ $2H_2O$ nanowires: a time-resolved XRD/HRTEM study. Zeitschrift für anorganische und allgemeine Chemie. 2013;639:2521–6. DOI: 10.1002/zaac.201300411

[27] Yin S. Microwave-assisted hydrothermal synthesis of nitrogen doped titania nanoparticles. Funct Mater Lett 2008;01:173 DOI: 10.1142/S1793604708000319

[28] Shi J, Liu G, Wang N, Li C. Microwave-assisted hydrothermal synthesis of perovskite $NaTaO_3$ nanocrystals and their photocatalytic properties. J Mater Chem 2012;22:18808–13. DOI: 10.1039/c2jm33470d

[29] Lee Y, Watanabe T, Takata T, Hara M, Yoshimura M, Domen K. Hydrothermal synthesis of fine $NaTaO_3$ powder as a highly efficient photocatalyst for overall water splitting. Bull Chem Soc Jpn 2007;80:423–8. DOI: 10.1246/bcsj.80.423

[30] Liu JW, Chen G, Li ZH, Zhang ZG. Hydrothermal synthesis and photocatalytic properties of $ATaO_3$ and $ANbO_3$ (A = Na and K). Int J Hydrogen Energy 2007;32:2269–72. DOI: 10.1016/j.ijhydene.2006.10.005

[31] Iwase A, Kato H, Kudo A. The effect of alkaline earth metal ion dopants on photocatalytic water splitting by $NaTaO_3$ powder. Chem Sus Chem 2009;2:873–7. DOI: 10.1002/cssc.200900160

[32] Kanhere PD, Zheng J, Chen Z. Site specific optical and photocatalytic properties of bi-doped $NaTaO_3$. J Phys Chem C 2011;115:11846–53. DOI: 10.1021/jp2003936

[33] Kanhere P, Tang Y, Zheng J, Chen Z. Synthesis, photophysical properties, and photocatalytic applications of Bi doped $NaTaO_3$ and Bi doped $Na_2Ta_2O_6$ nanoparticles. J Phys Chem Solid 2013;74:1708–13. DOI: 10.1016/j.jpcs.2013.06.013

[34] Kang HW, Lim SN, Park SB, Park AA. H_2 evolution under visible light irradiation on La and Cr co-doped $NaTaO_3$ prepared by spray pyrolysis from polymeric precursor. Int J Hydrogen Energy 2013;38:6323-6334. DOI: http://dx.doi.org/10.1016/j.ijhydene.2013.03.048

[35] Li FF, Liu DR, Gao GM, Xue B, Jiang YS. Improved visible-light photocatalytic activity of $NaTaO_3$ with perovskite-like structure via sulfur anion doping. Appl Catal B Environ 2015;166–167:104–11. DOI: 10.1016/j.apcatb.2014.10.049

[36] Liu DR, Jiang YS, Gao GM. Photocatalytic degradation of an azo dye using N-doped $NaTaO_3$ synthesized by one-step hydrothermal process. Chemosphere. 2011;83:1546–52. DOI: 10.1016/j.chemosphere.2011.01.033

[37] Wang B, Kanhere PD, Chen Z, Nisar J, Pathak B. Anion-doped $NaTaO_3$ for visible light photocatalysis. J Phys Chem C 2013;117:22518–24. DOI: 10.1021/jp407025

[38] Chen S, Wang LW. Intrinsic defects and electronic conductivity of TaON: First-principles insights. Appl Phys Lett 2011;99:222103–3. DOI: 10.1063/1.3664346

[39] Choi M, Oba F, Tanaka I. First-principles study of native defects and lanthanum impurities in $NaTaO_3$. Phys Rev B 2008;78:014115–8. DOI: 10.1103/PhysRevB.78.014115

[40] Yong Z, Ren J, Hu H, Li P, Ouyang S, Xu H, Wang D. Synthesis, characterization, and photocatalytic of g-C_3N_4/$KTaO_3$ composites under visible light irradiation. J Nanomater 2015;2015:821986–7. DOI: 10.1155/2015/821986

[41] Kumar S, Kumar B, Surendar T, Shanker V. g-C_3N_4/$NaTaO_3$ organic-inorganic hybrid nanocomposite: high-performance and recyclable visible light driven photocatalyst. Mater Res Bull 2014;49:310–8. DOI: 10.1016/j.materresbull.2013.09.013

[42] Wu X, Yin S, Dong Q, Sato T. Preparation and visible light induced photocatalytic activity of C-$NaTaO_3$ and C-$NaTaO_3$-Cl-TiO_2 composite. Phys Chem Chem Phys 2013;15:20633–40. DOI: 10.1039/c3cp53437e

[43] Bajorowicz B, Cybula A, Winiarski MJ, Klimczuk T, Zaleska A. Surface properties and photocatalytic activity of $KTaO_3$, CdS, MoS_2 semiconductors and their binary and ternary semiconductor composites. Molecules 2014;19:15339–60. DOI: 10.3390/molecules190915339

[44] Su Y, Hou L, Du C, Peng L, Guan K, Wang X. Rapid synthesis of Zn^{2+} doped $SnWO_4$ nanowires with the aim of exploring doping effects on highly enhanced visible photocatalytic activities. RSC Adv 2012;2:6266–73. DOI: 10.1039/c2ra20401k

[45] Reitz C, Brezesinski K, Haetge J, Perlinch J, Brezesinski T. Nanocrystalline $NaTaO_3$ thin film materials with ordered 3D mesoporous and nanopillar-like structures through PIB-*b*-PEO polymer templating: towards high-performance UV-light photocatalysts. RSC Adv 2012;2:5130–3. DOI: 10.1039/c2ra20203d

[46] Yokoi T, Sakuma J, Maeda K, Domen K, Tatsumi T, Kondo JN. Preparation of a colloidal array of $NaTaO_3$ nanoparticles via a confined space synthesis route and its photocatalytic application. Phys Chem Chem Phys 2011;13:2563–70. DOI: 10.1039/c0cp02141e

[47] Wu X, Yin S, Liu B, Kobayashi M, Kakihana M, Sato T. A carbon modified $NaTaO_3$ mesocrystal nanoparticle with excellent efficiency of visible light induced photocatalysis. J Mater Chem A 2014;2:20832–40. DOI: 10.1039/c4ta04132a

[48] Iwase A, Kato H, Kudo A. The effect of Au cocatalyst loaded on La-doped $NaTaO_3$ on photocatalytic water splitting and O_2 photoreduction. Appl Catal B: Environ 2013;136–137:89–93. DOI: 10.1016/j.apcatb.2013.02.006

[49] Jana P, Montero CM, Pizarro P, Coronado JM, Serrano DP, O'Shea VADLP. Photocatalytic hydrogen production in the water/methanol system using Pt/RE:$NaTaO_3$ (RE = Y, La, Ce, Yb) catalysts. Int J Hydrogen Energy 2014;39:5283–90. DOI: 10.1016/j.ijhydene.2013.12.182

[50] Husin H, Su WN, Pan CJ, Liu JY, Rick J, Yang SC, Chuang WT. Pd/NiO core/shell nanoparticles on $La_{0.02}Na_{0.98}TaO_3$ catalyst for hydrogen evolution from water and aqueous methanol solution. Int J Hydrogen Energy 2013;38:13529–40. DOI: http://dx.doi.org/10.1016/j.ijhydene.2013.07.116

[51] Gómez-solis C, Ruiz-Gómez MA, Torres-Martínez LM, Juárez-Ramírez I, Sánchez-Martínez D. Facile solvo-combustion synthesis of crystalline $NaTaO_3$ and its photocatalytic performance for hydrogen production. Fuel 2014;130:221–7. DOI: 10.1016/j.fuel.2014.04.019

[52] Ran J, Zhang J, Yu J, Jaroniec M, Qiao SZ. Earth-abundant cocatalysts for semiconductor-based photocatalytic water splitting. Chem Soc Rev 2014;43:7787–812. DOI: 10.1039/c3cs60425j

[53] Tsai CW, Chen HM, Liu RS, Asakura K, Chan TS. Ni@NiO core-shell structure-modified nitrogen-doped $InTaO_4$ for solar-driven highly efficient CO_2 reduction to methanol. J Phys Chem C 2011;115:10180–6. DOI: 10.1021/jp2020534

[54] Hu CC, Teng H. Structural features of p-type semiconducting NiO as a co-catalyst for photocatalytic water splitting. J Catal 2010;272:1–8. DOI: 10.1016/j.jcat.2010.03.020

[55] Husin H, Su WN, Chen HM, Pan CJ, Chang SH, Rick J, Chuang WT, Sheu HS, Hwang BJ. Photocatalytic hydrogen production on nickel-loaded $La_xNa_{1-x}TaO_3$ prepared by hydrogen peroxide-water based process. Green Chem 2011;13:1745–54. DOI: 10.1039/c1gc15070g

[56] Wen F, Li C. Hybrid artificial photosynthetic systems comprising semiconductors as light harvesters and biomimetic complexes as molecular cocatalysts. Account Chem Res 2013;46:2355–64. DOI: 10.1021/ar300224u

[57] Seo SW, Park S, Jeong HY, Kim SH, Sim U, Lee CW, Nam KT, Hong KS. Enhanced performance of $NaTaO_3$ using molecular co-catalyst $[Mo_3S_4]^{4+}$ for water splitting into H_2 and O_2. Chem Commun 2012;48:10452–4. DOI: 10.1039/c2cc36216c

[58] Kudo A, Kato H, Nakagawa S. Water splitting into H_2 and O_2 on new $Sr_2M_2O_7$ (M = Nb and Ta) photocatalysts with layered perovskite structures: factors affecting the photocatalytic activity. J Phys Chem B 2000;104:571–5. DOI: 10.1021/jp9919056

[59] Sugandha, Jha AK. Effect of the sintering temperature on nanocrystalline non-stoichiometric $Sr_{0.8}Bi_{2.2}Ta_2O_9$ ferroelectric ceramics. Ferroelectrics 2014;459:160–71. DOI: 10.1080/00150193.2013.849518

[60] Peng Z, Xing X, Chen X. Preparation and structure of a new layered organic-inorganic hybrid between the protonated form of a perovskite $Bi_2SrTa_2O_9$ and tetraphenylporphyrin by the intercalation behavior. J Alloys Compounds 2006;425:323–8. DOI: 10.1016/j.jallcom.2006.01.051

[61] Li Y, Chen G, Zhou C, Li Z. Photocatalytic water splitting over a protonated layered perovskite tantalate $H_{1.81}Sr_{0.81}Bi_{0.19}Ta_2O_7$. Catal Lett 2008;123:80–3. DOI: 10.1007/s10562-008-9397-5

[62] Nedelcu L, Toacsan MI, Banciu MG, Loachim A. Microwave properties of $Ba(Zn_{1/3}Ta_{2/3})O_3$ dielectric resonators. J Alloys Compounds 2011;509:477–81. DOI: 10.1016/j.jallcom.2010.09.069

[63] Park TK, Kim NK. Phase development in $Ba(Mg_{1/3}Ta_{2/3})O_3$ via conventional and B-site precursor routes. Ceramics Int 2008;34:1955–8. DOI: 10.1016/j.ceramint.2007.07.013

[64] Ling CD, Avdeev M, Kharton VV, Yaremchenko AA, Macquart RB, Hoelzel M. Structures, phase transitions, hydration, and ionic conductivity of $Ba_4Ta_2O_9$. Chem Mater 2010;22:532–540. DOI: 10.1021/cm903170t

[65] Xu TG, Zhang C, Shao X, Wu K, Zhu YF. Monomolecular-layer $Ba_5Ta_4O_{15}$ nanosheets: Synthesis and investigation of photocatalytic properties. Adv Funct Mater 2006;16:1599–607)DOI: 10.1002/adfm.200500849

[66] Shimizu K, Itoh S, Hatamachi T, Kitayama Y, Kodama. Pillaring of Ruddlesden–Popper perovskite tantalates, $H_2ATa_2O_7$ (A = Sr or $La_{2/3}$), with n-alkylamines and oxide nanoparticles. J Mater Chem 2006;16:773–9. DOI: 10.1039/B514066H

[67] Ikeue K, Mitsuyama T, Arayama K, Tsutsumi A, Machida M. Effect of heat treatment on local structure and photocatalytic water splitting activity of Ni-loaded $LiCa_2Ta_3O_{10}$. J Ceramic Soc Jpn 2009;117:1161–5. DOI: doi.org/10.2109/jcersj2.117.1161

[68] Yoshioka K, Petrykin V, Kakihana M, Kato H, Kudo A. The relationship between photocatalytic activity and crystal structure in strontium tantalates. J Catal2005;232:102–7. DOI: 10.1016/j.jcat.2005.02.021

[69] Ikeda S, Fubuki M, Takahara YK, Matsumura M. Photocatalytic activity of hydrothermally synthesized tantalate pyrochlores for overall water splitting. Appl Catal A: Gen 2006;300:186–90. DOI: 10.1016/j.apcata.2005.11.007

[70] Babooram K, Ye ZG. New soft chemical routes to ferroelectric $SrBi_2Ta_2O_9$. Chem Mater 2006;18:532–40. DOI: 10.1021/cm051866c

[71] Wang W, Ke H, Rao JC, Feng M, Zhou Y. Sol-Gel synthesis of $SrBi_2Ta_2O_9$ nanowires. J Alloys Compound 2010;504:367–70. DOI: 10.1016/j.jallcom.2010.05.118

[72] Inaba K, Suzuki S, Noguchi Y, Miyayama M, Toda K, Sato M. Metastable $Sr_{0.5}TaO_3$ perovskite oxides prepared by nanosheet processing. Euro J Inorg Chem 2008;2008(35):5471–5. DOI: 10.1002/ejic.200800776

[73] Li M, Zhang J, Dang W, Cushing SK, Guo D, Wu N, Yin P. Photocatalytic hydrogen generation enhanced by band gap narrowing and improved charge carrier mobility in $AgTaO_3$ by compensated co-doping. Phys Chem Chem Phys 2013;15:16220–6. DOI: 10.1039/C3CP51902C

[74] Nyman M, Rodriguez MA, Rohwer LES, Martin JE, Waller M, Osterloh FE. Unique $LaTaO_4$ polymorph for multiple energy applications. Chem Mater 2009;21:4731–7. DOI: 10.1021/cm9020645

[75] Shimizu K, Itoh S, Hatamachi T, Kodama T, Sato M, Toda K. Photocatalytic water splitting on Ni-intercalated Ruddlesden-Popper tantalate $H_2La_{2/3}Ta_2O_7$. Chem Mater 2005;17:5161–6. DOI: 10.1021/cm050982c

[76] Tsujimoto Y, Yamaura K, Muromachi ET. Oxyfluoride chemistry of layered perovskite compounds. Appl Sci 2012;2:206–19. DOI: 10.3390/app2010206

[77] Bao NZ, and Arunava G. Inorganic spintronic materials. Encyclo Inorg Bioinorg Chem 2004;03:01–15.

[78] Miseki Y, Kato H, Kudo A. Water splitting into H_2 and O_2 over niobate and titanate photocatalysts with (111) plane-type layered perovskite structure. Energy Environ Sci 2009;2:306–14. DOI: 10.1039/B818922F

[79] Bharti C, Sinha TP. Synthesis, structure and dielectric properties of a rare earth double perovskite oxide Ba_2CeTaO_6. Mater Res Bull 2011;46:1431–6. DOI: 10.1016/j.materresbull.2011.05.007

[80] Ibberson RM, Moussa SM, Rosseinsky MJ, Fitch AN, Iddles D, Price T. In situ neutron and X-ray powder diffraction study of cation ordering and domain growth in the dielectric ceramic $Ba_3ZnTa_2O_9$-Sr_2GaTaO_6. J Am Ceramic Soc 2006;89:1827–33. DOI: 10.1111/j.1551-2916.2006.01047.x

[81] Rajendran DN, Nair KR, Rao PP, Sibi KS, Koshy P, Vaidyan VK. New perovskite type oxides: $NaATiMO_6$ (A=Ca or Sr; M=Nb or Ta) and their electrical properties. Mater Lett 2008;62:623–8. DOI: 10.1016/j.matlet.2007.06.022

[82] Chen S, Yang J, Ding C, Li R, Jin S, Wang D, Han H, Zhang F, Li C. Nitrogen-doped layered oxide $Sr_5Ta_4O_{15-x}N_x$ for water reduction and oxidation under visible light irradiation. J Mater Chem A 2013;1:5651–9. DOI: 10.1039/C3TA10446J

[83] Mukherji A, Sun C, Smith SC, Lu GQ, Wang L. Photocatalytic hydrogen production from water using N-doped $Ba_5Ta_4O_{15}$ under solar irradiation. J Phys Chem 2011;115:15674–8. DOI: 10.1021/jp202783t

[84] Guo Z, Sa B, Pathak B, Zhou J, Ahuja R, Sun Z. Band gap engineering in huge-gap semiconductor $SrZrO_3$ for visible-light photocatalysis. Int J Hydrogen Energy 2014;39:2042–8. DOI: 10.1016/j.ijhydene.2013.11.055

[85] Liu P, Nisar J, Ahuja R, Pathak B. Layered perovskite $Sr_2Ta_2O_7$ for visible light photocatalysis: a first principles study. J Phys Chem C 2013;117:5043–50. DOI: 10.1021/jp310945e

[86] Marschall R, Soldat J, Wark M. Enhanced photocatalytic hydrogen generation from barium tantalate composites. Photochem Photobiol Sci 2013;12:671–7. DOI: 10.1039/C2PP25200G

[87] Mukherji A, Seger B, Lu GQ, Wang L. Nitrogen doped $Sr_2Ta_2O_7$ coupled with graphene sheets as photocatalysts for increased photocatalytic hydrogen production. ACS Nano 2011;5:3483–92. DOI: 10.1021/nn102469e

[88] Iwase A, Kato H, Kudo A. Nanosized Au particles as an efficient cocatalyst for photocatalytic overall water splitting. Catal Lett 2006;108:7–10. DOI: 10.1007/s10562-006-0030-1

[89] Hagiwara H, Inoue T, Kaneko K, Ishihara T. Charge-transfer mechanism in Pt/$KTa(Zr)O_3$ photocatalysts modified with porphyrinoids for water splitting. Chem – Euro J 2009;15:12862–70. DOI: 10.1002/chem.200901772

[90] Sreethawong T, Ngamsinlapasathian S, Suzuki Y, Yoshikawa S. Nanocrystalline mesoporous Ta_2O_5-based photocatalysts prepared by surfactant-assisted templating sol-gel process for photocatalytic H_2 evolution. J Molecul Catal A – Chem 2005;235:1–11. DOI: 10.1016/j.molcata.2005.03.021

[91] Chen X, Chen W, Gao H, Yang Y, Shangguan W. In situ photodeposition of NiO_x on CdS for hydrogen production under visible light: Enhanced activity by controlling solution environment. Appl Catal B: Environ 2014;152–153:68–72. DOI: 10.1016/j.apcatb.2014.01.022

[92] Jorge AB, Martin DJ, Dhanoa MTS, Rahman AS, Makwana N, Tang J, Sella A, Corà F, Firth S, Darr JA, McMillan PF. H_2 and O_2 evolution from water half-splitting reactions by graphitic carbon nitride materials. J Phys Chem C 2013;117:7178–85. DOI: 10.1021/jp4009338

[93] Wang J, Li B, Chen J, Li N, Zheng J, Zhao J, Zhu Z. Enhanced photocatalytic H_2-production activity of $Cd_xZn_{1-x}S$ nanocrystals by surface loading MS (M = Ni, Co, Cu) species. Appl Surface Sci 2012;259:118–23. DOI: 10.1016/j.apsusc.2012.07.003

Energy Production, Decontamination, and Hydrogenation Reactions over Perovskite-Type Oxide Catalyst

Gina Pecchi, Nestor Escalona, I. Tyrone Ghampson and Ruddy Morales

Additional information is available at the end of the chapter

http://dx.doi.org/10.5772/61522

Abstract

Heterogeneous catalysis is an important tool in industrial processes because of the recoverability of the catalysts. Transition metal perovskites-type oxides, with the general formula ABO_3, offer attractive alternative to noble metal catalysts due to their high activity, high thermal stability, and low cost. Moreover, their physicochemical properties can be tailored to create a family of catalysts by varying the compositions of A and B. Indeed, the partial substitution at the A- and/or B-site with another metal cation stabilizes unusual oxidation states of the B cation with the simultaneous formation of structural defects. In particular, lanthanum-based perovskites have been used extensively and can be grouped into: (i) perovskites with oxygen vacancies as catalysts for oxidation reactions and (ii) perovskites as precursors to prepare nanosized catalysts for hydrogenation reactions. This chapter focuses on the use of pure and doped lanthanum perovskites as active and selective heterogeneous catalysts for catalytic energy production reaction (DME combustion), decontamination reactions (methane, acetyl acetate, toluene, n-hexane, and soot combustion), and hydrogenation reactions (guaiacol, glycerol, and xylose hydrogenation).

Keywords: Lanthanum, substituted, oxidation, hydrogenation, energy

1. Introduction

Perovskite-type oxides have been extensively studied since the pioneering work of Voorhoeve et al. [1] in the early 1970s on their potential as a cheaper alternative to noble metals as an automobile exhaust catalyst. Since then, the literature has been inundated with studies expanding their applications, particularly involving their use in combustion. Perovskites have been demonstrated to be effective in several oxidation reactions. For environmental processes, the main objective is to achieve high conversions with almost no by-products or pollutants. Recently, there has been a particular emphasis on developing catalysts capable of either transforming pollutants into less harmful products or valorizing the by-products.

The tendency of perovskites to form unique structures is responsible for their reactivity. In general, perovskite-type oxides are represented by the general formula ABO_3, wherein the larger A-site metals, usually rare earth metals, have a dodecahedral coordination and the smaller B-site elements, first-row transition metals, have a sixfold coordination. The ability to tailor the properties of perovskites due to the different combinations of elements employed and the high degree of flexibility in their composition allow for their use in a wide range of applications.

In heterogeneous catalysis, the creation of oxygen vacancies, generally obtained by substitution in the A- and B-site, accounts for their use in catalyzing chemically and environmentally relevant reactions, such as oxidation reactions. For pure perovskites-type oxides (ABO_3), the combustion activity is dependent mainly on the B component oxides. In our previous work, we have mainly utilized Mn, Fe, Co, and Ni for the B-site of the perovskite-oxides due to their high catalytic activities; on the other hand, we have exclusively used La cation for the A-site position.

In addition to the research conducted into perovskite-based oxidation reactions, advances in preparation methods have led to perovskites being used as precursors to prepare nanosized catalysts for hydrogenation reactions. This is achieved by reduction or reduction–oxidation cycles of perovskites-type oxides under controlled conditions to leave the metal in the B-site in a high degree of dispersion on a matrix of the Ln_2O_3 oxide. Studies have demonstrated that during the reduction–oxidation cycles of $LnCoO_3$ (Ln = La, Pr, Nd, Sm, Gd) perovskites, the stability of the perovskite is affected by the nature of the Ln, with the largest lanthanide ion in the series, La, forming the most stable perovskite structure.

This chapter focuses on the use of lanthanum perovskites as active and selective catalysts in heterogeneous reactions for catalytic energy production, decontamination, and hydrogenation reactions.

2. Energy production

2.1. Fischer-Tropsch reaction

The Fischer-Tropsch synthesis (FTS) is the most important catalytic process for the synthesis of gasoline and/or diesel from syngas (a mixture of CO and H_2 formed by methane reforming) [2]. Efforts are being made worldwide to shift from conventional feedstock such as coal due to their negative environmental impact, and biomass appears to be an attractive alternative as a feedstock for the production of hydrocarbons. The gasification of biomass stream produces a gas mixture called "biosyngas" consisting mainly of CO_2, CO, CH_4, H_2, and N_2. The H_2/CO ratio in the resulting biosyngas is close to 1 and can be adjusted using the water–gas shift reaction. The advantages of hydrocarbons produced from FTS using a biosyngas feed include being near CO_2-neutral and being free of sulfur and nitrogen compounds. However, there is a dearth of systematic studies of catalytic systems in FTS using biosyngas [3-5]. Perovskites have been shown to be highly active in FTS using a H_2/CO mixture, and its activity was related to the properties of the perovskite precursor: stability, crystalline structure, and degree of reduction [6, 7]. These results paved the way for the evaluation of perovskites for FTS of biosyngas. Hence, this study was focused on FTS over $LaFe_{1-x}Co_xO_3$ perovskites using biosyngas as the feed.

2.2. B-site substitution

The FTS reaction using a simulated gas mixture similar to biosyngas over $LaFe_{1-x}Co_xO_3$ perovskites presented a wide distribution of liquid hydrocarbons (C_6 to C_{18+}) and a high production of CH_4 attributed to Co metal segregation on the surface. The distribution of the liquid products was related to the average size of iron particles [8].

A series of $LaFe_{1-x}Co_xO_3$ perovskites ($0.0 \leq x \leq 1.0$) were prepared using the amorphous citrate precursor method and characterized by N_2 physisorption, X-ray diffraction (XRD), Fourier-transform IR spectroscopy (FTIR), O_2-Temperature programmed desorption (O_2-TPD), Thermal-programmed reduction (TPR), and Scanning electron microscopy (SEM). FTS was carried out in a stainless steel fixed-bed reactor at 300°C and 1 MPa. The BET surface areas of the samples (13 m^2g^{-1}) were as expected for these kinds of structures. The reduction behaviors of the catalysts (TPR_1-TPR_2) are shown in Fig. 1. This analysis provides information on the thermal stability of the catalyst (segregation of Co particles) under reducing conditions, deduced from the reducibility of the oxide precursor. The procedure involved: (i) a first reduction step under H_2/Ar flow up to 700°C (TPR_1); (ii) cooling under Ar flow to room temperature; (iii) an oxidation step under flowing O_2/He mixture up to 700°C; and (iv) a second reduction treatment (TPR_2) as in step 1. Figure 1 shows that the largest differences between TPR_1 and TPR_2 were observed for pure $LaFeO_3$ and $LaCoO_3$ perovskites, while the substituted perovskites showed similar profiles. Pure $LaFeO_3$ is essentially irreducible. On the other hand, the profile for pure $LaCoO_3$ revealed two well-defined peaks: a peak at 350°C is indicative of the formation of an oxygen-deficient perovskite structure, while the second peak above 500°C corresponds to the complete reduction of the perovskite to form Co^0. The substitution of Fe^{3+} by Co^{3+} affected the initial reduction stage, showing only one reduction peak, probably by passing an intermediate reduction state. The subsequent reduction of Co^{3+}/Co^{2+} to Co^0 occurred at a higher temperature, then decreased in intensity with increasing Fe, and subsequently disappeared when the level of substitution reached 50%. The slight difference in the TPR_2 profiles in comparison with TPR_1 is indicative of the reversibility of the redox cycle. The small and broad peaks observed at 410°C for $LaFeO_3$ in the TPR_2 profile are associated with Fe^{3+} segregation during the first reduction step, since this perovskite is a Co-free sample and La is not reducible under the experimental conditions. Thus, during TPR_1, metallic iron is expelled from the perovskite structure at 620°C. The quantification of the first reduction peak at temperatures below 500°C is representative of the extent of Co insertion into the perovskite structure: our data revealed a near linear increase in hydrogen consumption with x, an indication of the progressive insertion of Co into the perovskite structure. The XRD patterns of the calcined perovskites displayed two crystal systems, orthorhombic for $x<0.5$ and rhombohedral for $x\geq 0.5$. On the other hand, XRD analyses of reduced perovskite showed no significant changes after H_2 reduction, indicative of the high thermal stability of these $LaFe_{1-x}Co_xO_3$ perovskite solid solutions. The diffraction lines broadened upon substitution of Fe^{3+} by Co^{3+}, which suggests a decrease in the crystallinity. The highest stability under hydrogen was obtained for the perovskite with x = 0.5, while the thermal stability of x = 0.1 and 0.3 perovskites were similar to that of the pure $LaCoO_3$ and x = 0.4 perovskites, respectively. The observed behavior of the x = 0.2 perovskite indicates a loss in the perovskite structure, suggesting segregation of the B cation. FTIR spectra indicated that for the $x \leq 0.5$ perovskites, Co^{3+} ions were incorporated into the orthorhombic $LaFeO_3$ perovskite structure and that a new rhombohedral structure appeared for x = 0.5. The pure $LaFeO_3$ tolerated Co insertion with no structural changes up to values of $x \leq 0.3$. For x = 0.4 and x = 0.5, the presence of a mixture of orthorhombic and rhombohedral structures were detected. The catalytic activity indicated a high intrinsic CO conversion for the substitution degree of x = 0.1, 0.2, and 1.0, in line with their orthorhombic structures and the presence of segregated Co metal. The x = 0.2 perovskite displayed the highest CO intrinsic conversion attributed to the presence of highly dispersed, segregated Co species. The absence of segregated Co species in the x = 0.3 and 0.4 perovskites explains their lower activities. In fact, these perovskites possessed higher thermal stability under reducing atmospheres due to the presence of a mixture of orthorhombic and rhombohedral crystalline structures, which implies that they did not segregate Co species. In relation to selectivity, an increase in x led to a decrease in CH_4 selectivity, while a decrease in x favored selectivity to longer-chain length condensable hydrocarbons: this was attributed to changes in the average particle size of segregated Co species. The smaller particles of the x = 0.2 and 0.1 perovskites favored the formation of longer-chain hydrocarbons. Pure $LaCoO_3$ perovskite displayed the highest extent of poorly dispersed segregated Co, favoring the formation of short-chain hydrocarbons (C_8-C_9). Thus, the catalytic activity of $LaFe_{1-x}Co_xO_3$ perovskites in FTS can be explained in terms of the crystalline structure, segregated Co content, and particle size.

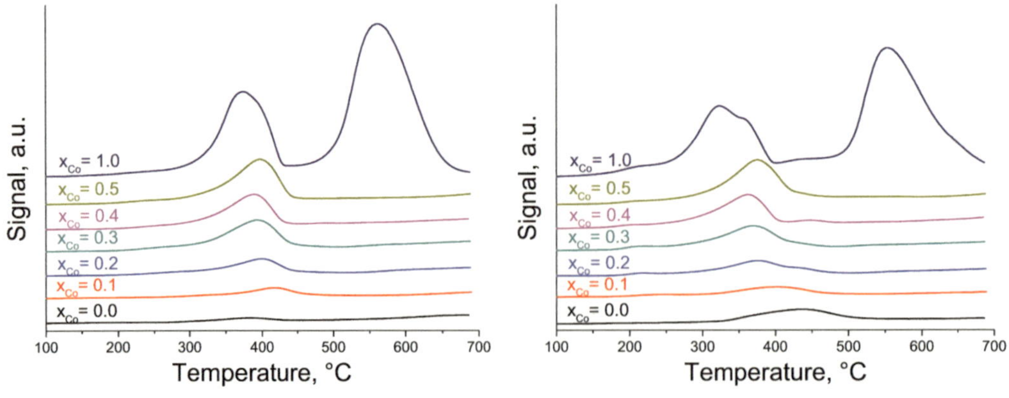

Figure 1. Temperature-programmed reduction profiles of $LaFe_{1-x}Co_xO_3$ perovskites: left: TPR_1 and right: TPR_2

3. Dimethyl Ether (DME) combustion

The easier transportation, lower emission of polluting particles, and higher thermal efficiency of dimethyl ether (DME) make it a useful, low-cost replacement for liquefied petroleum gas (LPG) and diesel fuel. Recently, the catalytic combustion of DME has been studied on Ce-doped manganese oxide octahedral molecular sieves (OMS-2) [9]. Mixed oxides with perovskite-type structures, in particular $LaMnO_{3.15}$, have the potential to be highly active catalysts for the deep oxidation of DME. This catalyst has been previously used for the catalytic oxidation of toluene [10], carbon monoxide [11], and NH_3 [12].

3.1. B-site substitution

The electronic properties and performance of $La(Mn_{0.5}B_{0.5})O_{3.15}$ (B = Cr, Fe, Co, Ni, Cu) perovskites in the combustion of dimethyl ether (DME) revealed that an increase in the catalytic performance, expressed as intrinsic activity ($\mu molm^{-2}h^{-1}$), is contingent on B-site substitution in $LaMnO_3$ and a decrease in the required calcination temperature to obtain the perovskite structure. The similar tolerance factor (t) of the substituted perovskites, a measure of the geometric effect, plays a secondary role on the surface and catalytic performance. Consequently, the electronic effect explains the relationship between catalytic performance, chemisorbed oxygen species, and reducibility of the substituted perovskites. XRD results indicated that calcined pure $LaMnO_{3.15}$ and Cr-, Co-, and Fe-substituted perovskites presented well-defined rhombohedral structure, whereas Cu- and Ni-doped perovskites displayed La_2CuO_4 and La_2NiO_4 phases. The largest surface area corresponded to the $LaMnO_{3.15}$ perovskite with 11 m^2g^{-1}, and subsequently decreased upon B-substitution. The TPR profiles obtained for the pure and oxygen-rich $LaMnO_{3.15}$ perovskite presented a concentration of 32% Mn^{4+} ions after H_2 reduction. An observed reduction peak at 350°C with a shoulder at 420°C corresponds to the reduction of Mn^{4+} to Mn^{3+}, followed by the destruction of the perovskite structure (Mn^{3+} to Mn^{2+}) at 800°C [13-15]. Upon B- substitution, the two reductions steps of pure $LaMnO_{3.15}$ perovskite were maintained when Co and Ni were used, while only one reduction step was detected for the Cr- and Fe-substituted perovskite. This result is closely related to the thermal behavior of nonreducible perovskites such as $LaCrO_3$ and $LaFeO_3$. Thus, Fe- and Cr-substituted perovskites showed lower reducibility compared to the parent $LaMnO_{3.15}$ but higher reducibility in comparison with the nonreducible $LaCrO_3$ and $LaFeO_3$ perovskites. The observed increase in the reducibility of the Co- and Ni-substituted perovskites is similar to the effect induced by La substitution of divalent alkaline-earth cations. The high stabilities of Co^{2+} and Ni^{2+} helped stabilize the Mn^{4+} ions, with corresponding increase in the reduction peaks. On the other hand, the TPR results for the Cu-substituted perovskite showed a single-step complete reduction of the Cu ions to the metallic state, similar to the reduction of CuO but at lower temperatures. The O_2-TPD experiments were used to evaluate the redox properties of the catalysts. The data revealed two different behaviors, consistent with the different patterns observed from TPR results: for the Cr- and Fe-substituted perovskites, a behavior similar to the parent $LaMnO_{3.15}$ perovskite was observed, with the presence of α- and β-oxygen species beginning at ~600°C. Conversely, the absence of β-oxygen at temperatures below 700°C for the Co-, Ni-, and Cu-substituted perovskites is related to the lattice oxygen-deficient structures. XPS analyses showed that the $La3d_{5/2}$ spectra presented peaks with BE at 834.3 eV and 838.1 eV that correspond to La^{3+} in the perovskite. The small difference between the BEs of Mn^{3+} at 641.3 eV and Mn^{4+} at 642.4 eV makes it difficult to distinguish them. Regarding the Cr $2p_{3/2}$ spectra, two peaks with BEs centered at 576.6 and 579.4 eV were detected: the peaks were assigned to surface Cr^{3+} and Cr^{6+} species, respectively. The main peak of the Fe $2p_{3/2}$ spectra was located at 710.8 eV, and is attributed to surface Fe^{3+} species. For the Co $2p_{3/2}$ spectra, the two peaks detected at 780.6 and 786.9 eV are associated with Co^{3+} and Co^{2+} species, respectively. Because the Ni $2p_{3/2}$ signal coincides with the La $3d_{3/2}$ signal, the BE of Ni $2p_{1/2}$ was measured instead. The displacement of the Ni BE to 873.2 eV suggests the presence of surface Ni^{2+} species. Considering the low BET surface area of the substituted perovskites, the surface composition was similar to the nominal composition in the bulk, and the surface analysis may be extrapolated to the bulk. The BE of O 1s peak less than 532 eV suggests the absence of surface oxygen vacancies. Therefore, the substitution of Mn^{3+} with a first-row transition metal in the (III) oxidation state preserved the oxygen-rich perovskite structure for the $LaMn_{0.5}Cr_{0.5}O_{3.15}$ and $LaMn_{0.5}Fe_{0.5}O_{3.15}$ perovskites. On the other hand, the substitution by Co^{2+}, Ni^{2+}, and Cu^{2+} generated a stoichiometric cubic perovskite structure. These differences in behavior are in agreement with interpretations deduced from TPR and O_2-TPD results. The catalytic activity of the B-site substituted perovskites were tested in the combustion of DME with excess oxygen in a flow reactor. The light-off curves allow the calculation of the reaction rate of the oxidation reaction at low conversion levels (<10%) at 180°C. The reaction rate was correlated with the H_2 consumption calculated from TPR profiles and the α-oxygen derived from O_2-TPD (both parameters normalized by the specific surface area), and shown in Fig. 2. It can be inferred from this correlation that the high stability of (III) oxidation state of Cr^{3+} and Fe^{3+}cations did not increase the redox character of the Mn^{4+}/Mn^{3+} pair; on the other hand, the higher redox properties and catalytic performances of the $LaMn_{0.5}Co_{0.5}O_3$ and $LaMn_{0.5}Ni_{0.5}O_3$ catalysts stem from the high stability of (II) oxidation state of Co^{2+} and Ni^{2+}, leading to a progressive decrease of Mn^{3+} species, while stabilizing the Mn^{4+} species in the network.

4. Decontamination reactions

4.1. VOCs abatement

Metal oxides with a perovskite structure have been consistently proposed during the last two decades as alternative catalysts for the deep oxidation of hydrocarbons. It is well-known that perovskite-like mixed oxide substitution of the trivalent A-site metal ion with a bivalent or tetravalent metal cation (A′) is accompanied by a modification of the oxidation state of the B-site metal cation, thus modifying its catalytic activity. Moreover, modification of the oxidation state of the B-site metal cation by insertion of A′ may be accompanied by the formation of structural defects, thus leading to nonstoichiometry. This usually means the creation of oxygen defects in the Co-containing perovskites, while in the Mn-perovskites it results in excess of oxygen [13]. Fe-based [12] and Ni-based perovskites [14] were shown to exhibit an intermediate behavior. Good catalytic performances in combustion reactions were displayed by La- or

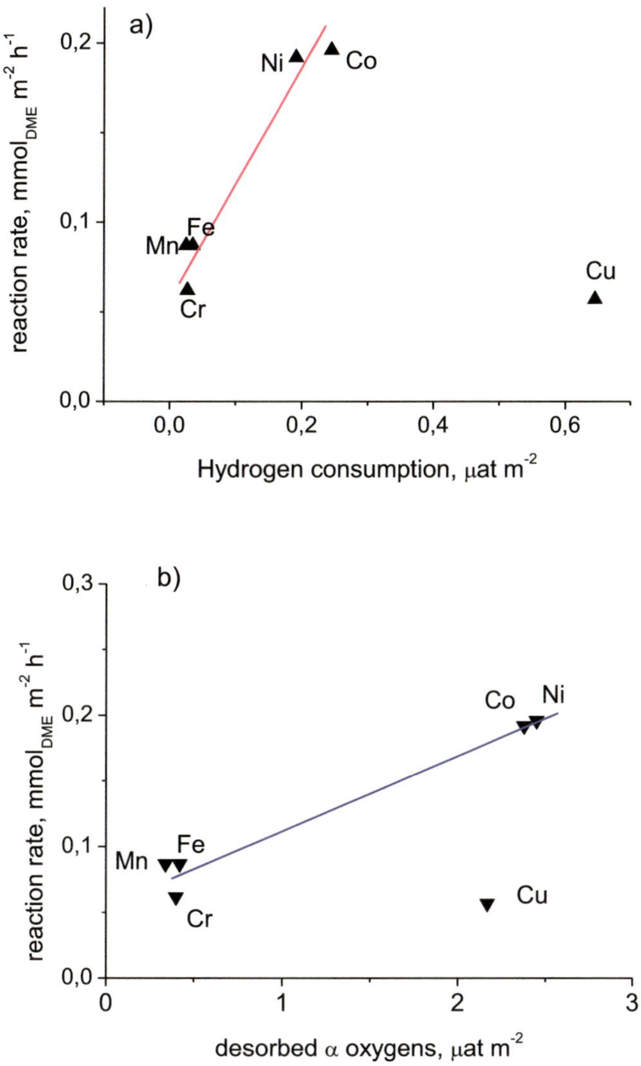

Figure 2. Intrinsic activity as a function of quantity of hydrogen consumed (above) and quantity of α-oxygen desorbed

La_xSr_{1-x}-based perovskites containing Co, Fe, or Mn as B cation [15, 16]. A major drawback for their application, however, is their low surface area, and hence significant efforts are underway to increase the specific surface area of conventional $A_{1-x}A'_xBO_{3-\delta}$ perovskites through various preparation methods [17, 18].

The insertion of calcium ions into the $LaFeO_3$ perovskite lattice results in a solid with structural and electronic defects possessing highly active sites for redox reactions. A continuous increase in number of oxygen vacancies in $LaFeO_3$ perovskites upon A-site substitution of La^{3+} by Ca^{2+} was previously reported [19]. In this work, $La_xCa_{1-x}FeO_3$ (x = 0.0, 0.1, 0.2, 0.3, 0.4) perovskites prepared by precursor citrate method were characterized and evaluated in catalytic combustion reaction. The effect of La substitution by calcium on the physical, chemical, and catalytic properties for the total combustion of acetyl acetate and methane was studied. Surface area values ranging from 19 to 39 m^2g^{-1} were obtained as the calcium content of the perovskites varied from 0.0 to 0.4. XRD patterns indicated that for pure $LaFeO_3$ an orthorhombic perovskite structure was observed; on the other hand, a shift toward larger 2θ angles appeared as La^{3+} was partially substituted by Ca^{2+}, evidence of lattice distortion. Even though the ionic radii are very similar (La^{3+} = 0.136 nm; Ca^{2+} = 0.134 nm), the substitution of La^{3+} with a cation of lower valence like Ca^{2+} entails either the oxidation of the cation at the B-site (i.e., $Fe^{3+} \rightarrow Fe^{4+}$) or the formation of oxygen vacancies in order to retain electroneutrality. Since the ionic radius of Fe^{4+} (0.0585 nm) is smaller than that of Fe^{3+} (0.0645 nm), the distortion can be explained by the presence of Fe^{4+} in the perovskite structure, in agreement with previous results reported for other iron-based perovskites. The shift of the 2θ angles as a function of the degree of calcium substitution is illustrated in Fig. 3: a linear shift of the 2θ values is observed with increasing Ca content, a clear indication of the insertion of calcium in the studied range.

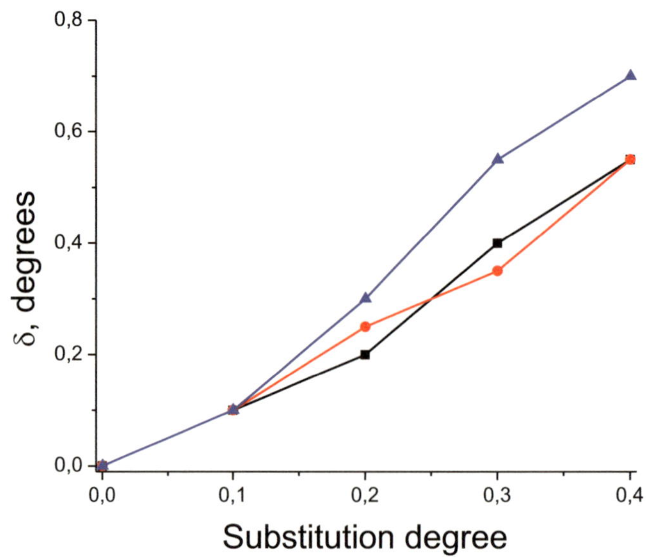

Figure 3. Shift of the peak position as a function of substitution degree 2θ (●) 40°; (■) 46°; (▲) 58°

IR spectra of the perovskites show a band at 865 cm^{-1}, attributed to O-C-O bending in the carbonates. This peak appeared as a shoulder for x = 0.1 and increased linearly when x ≥ 0.2, without showing any change in the wavelength. This band is ascribed to the presence of calcium and corresponds to the out-of-plane bending of the carbonate ions supporting the incorporation of calcium in the perovskite structure. O_2-TPD profiles showed clear increase of the desorbed oxygen as the degree of Ca substitution increased. In addition, an appreciable desorption peak was observed at temperatures above 500°C for the perovskite with the highest degree of substitution. This desorbed oxygen corresponds to the so-called β oxygen, which is associated with lattice oxygen or with oxygen species occupying the inner vacancies created

http://dx.doi.org/10.5772/61522

by substitution of Ca by La. The plot of the amount of O_2 desorbed from 150°C up to 700°C, which involves α-oxygen species, as a function of Ca substitution showed a continuous increase. Desorption of large amount of lattice oxygen in the bulk from the samples indicates a large substitution vacancy and/or defects. TPR profiles for the pure $LaFeO_3$ showed no reduction peaks, indicative of the nonreducibility of this perovskite. However, the TPR profile for the substituted perovskites showed a pronounced peak at lower temperatures that shifted to higher temperatures and increased in intensity as the degree of Ca substitution increased. The replacement of part of La^{3+} by Ca^{2+} led to a well-defined H_2 consumption peak due to the higher amount of Fe^{4+} ions generated to compensate for the unbalanced charge, consistent with the deduction from XRD results. In situ XRD measurements also revealed no detectable differences between diffraction patterns of pre- and post-TPR measurements, suggesting that the integrity of the structure was retained after hydrogen treatment, although the nature of the phases present in the sample changed. The post-TPR XRD pattern of the pure $LaFeO_3$ perovskite was essentially the same as that obtained for the fresh perovskite: the only difference was a minimal increase in the intensity of the peaks and small shifts toward lower angles, indicating increased crystallinity with a great expansion of the cell parameters. The catalytic activity in the oxidation of methane and acetyl acetate in a flow reactor using an excess amount of oxygen showed two contrasting behaviors: a near independency of the ignition temperature with Ca substitution for the methane reaction; and a decrease in ignition temperature with Ca substitution for the acetyl acetate reaction. These results clearly demonstrate the differences between a suprafacial and intrafacial reaction mechanism. The presence of substituted Ca ions enhances the catalytic activity of suprafacial reactions (e.g., acetyl acetate combustion), which typically occurs at lower temperatures. Conversely, the activity of intrafacial reactions (e.g., methane combustion), which requires a much higher temperature, is only slightly modified by Ca^{2+} insertion into the perovskite.

4.2. A-site substitution

A different behavior was detected upon the substitution of La^{2+} by Ca^{2+} in $LaNiO_3$ and $LaFeO_3$ perovskite-type oxides. In the Fe series ($La_{1-x}Ca_xFeO_3$), the perovskite structure is maintained within the studied range of substitution. In contrast, a completely different behavior was observed for the Ni series ($La_{1-x}Ca_xNiO_3$), whereby a Brownmillerite-type structure ($La_2Ni_2O_5$) with a large degree of segregated phase as highly dispersed mixed oxides was observed [20]. There was no monotonic dependence of the S_{BET} values on the extent of La substitution. The XRD patterns shown in Fig. 4 revealed the presence of orthorhombic structure for the Fe series; however, for the pure $LaNiO_3$ counterpart, the peak at $2\theta = 43.3°$ indicated the presence of segregated NiO. For the Ni series, the most prominent diffraction line at $2\theta = 32.7°$ changed from a singlet for x = 0.0 to a doublet for both x = 0.2 and 0.4, as shown in the inset of Fig. 4. As x increased, the diffraction line at $2\theta = 32.7°$ corresponding to $LaNiO_3$ structure decreased in intensity and, consequently, a new diffraction line at $2\theta = 32.4°$ corresponding to $La_2Ni_2O_5$ brownmillerite-type structure appeared. Additionally, NiO and CaO were also present as segregated phases for the x = 0.2 and 0.4 solids. Thus, it is difficult to substitute La^{3+} with Ca^{2+} in $LaNiO_3$ even to a limited extent. TPR profiles showed increase in the extent of reduction in the Fe series, explained by the higher amount of Fe^{4+} ions generated to compensate for the unbalanced charge. For the Ni series, a large amount of H_2 consumption indicated a total reduction of Ni^{3+} ions in $LaNiO_3$ to yield the $La_2Ni_2O_5$ phase, followed by an H_2 consumption peak at 500°C, ascribed to Ni° deposited on La oxide. The evolution of oxygen during TPD measurements showed the expected presence of physisorbed species at low temperatures in the Fe series, in addition to α-oxygen, which appeared as a shoulder for the pure $LaFeO_3$. There was a noticeable increase in the intensity of the peak attributed to the so-called β-oxygen upon Ca^{2+} substitution. Since this peak is associated with the lattice oxygen or with oxygen species occupying inner vacancies, their evolution is a measurement of the replacement of La^{3+} by Ca^{2+} in the perovskite structure. For the Ni series, the desorption profiles for the pure $LaNiO_3$ and the substituted samples showed a large increase in the desorption peak detected at 350°C, indicative of either the presence of segregated metal oxide phases in a high degree of dispersion or crystalline oxides (such as $La_2Ni_2O_5$), as detected by XRD. With regard to the XPS spectra of the Fe series, the BE of the most intense peaks of doublets for $La3d_{5/2}$ at 834.8 eV and $Fe2p_{3/2}$ at 710.3 eV were almost constant. For the Ni counterpart, on the other hand, the overlap of the $La3d_{3/2}$ and $Ni2p_{3/2}$ peaks may mask the accurate measurement of the intensity of the Ni BE; thus, the $Ni2p_{1/2}$ was measured instead. The least intense $Ni2p_{1/2}$ peak at 873.4 eV was almost constant, and the detected satellite line at 881.2 eV is indicative of the presence of Ni^{2+} ions on the surface. The Ca $2p_{3/2}$ doublet has two peaks at BE of 346.2 and 346.8 eV, corresponding to surface Ca-O and carbonated surface Ca^{2+} ions, respectively. O1s peak was deconvoluted into three components: at BE of 529.2 eV assigned to surface O^{2-} species, at 531.0 eV which arises from lattice [La-O-M] bonds and at 532.3 eV due to hydroxyl/carbonate groups. In the Fe series, x = 0.2 showed the largest extent of Ca incorporated into the perovskite structure at 346.2 eV, while a clear trend was not observed for the Ni series. The (Fe/La+Ca) and (Ni/La+Ca) surface ratios were higher than the nominal compositions: i.e., the solid surface became enriched with Ni and Fe. Since the carbonation degree is related to the extent of segregated phases, the (CO_3^{2-}/La+Ca) surface ratio for the Fe and Ni series can provide an insight into the surface structure of the solid. A decrease in the carbonation degree upon Ca substitution was obtained for the Ca series; conversely, a larger increase with increasing x was detected for the Ni series. This finding is in agreement with the previous characterization results, and demonstrates that upon Ca substitution, Ni^{3+} ions are more difficult to stabilize than Fe^{3+} ions in the perovskite lattice. The perovskites were tested in toluene combustion (4000 ppmv) in a flow reactor under an excess of oxygen using a GHSV of 47000 h^{-1}. For the Fe series, the substituted perovskites displayed higher catalytic activity than the pure $LaFeO_3$ perovskite and the maximum catalytic activity was obtained for the perovskite with x = 0.2. The observed trend in activity indicated no simple dependence on the metal ion oxidation state, and appeared to be due to the effect of the pair Fe^{4+}/Fe^{3+} on the vacancy ordering, in that a disordered distribution of oxygen vacancies led to an increase in the catalytic activity. The lower activation energy displayed by $La_{0.8}Ca_{0.2}FeO_3$ can be attributed to cationic vacancies compensating the amount of Fe^{4+} involved in the reaction. Conversely, for the Ni series, substitution resulted in solids with lower activity than the pure $LaNiO_3$ perovskite due to the loss of the perovskite structure and the corresponding change in the oxidation state of Ni^{3+} to Ni^{2+}.

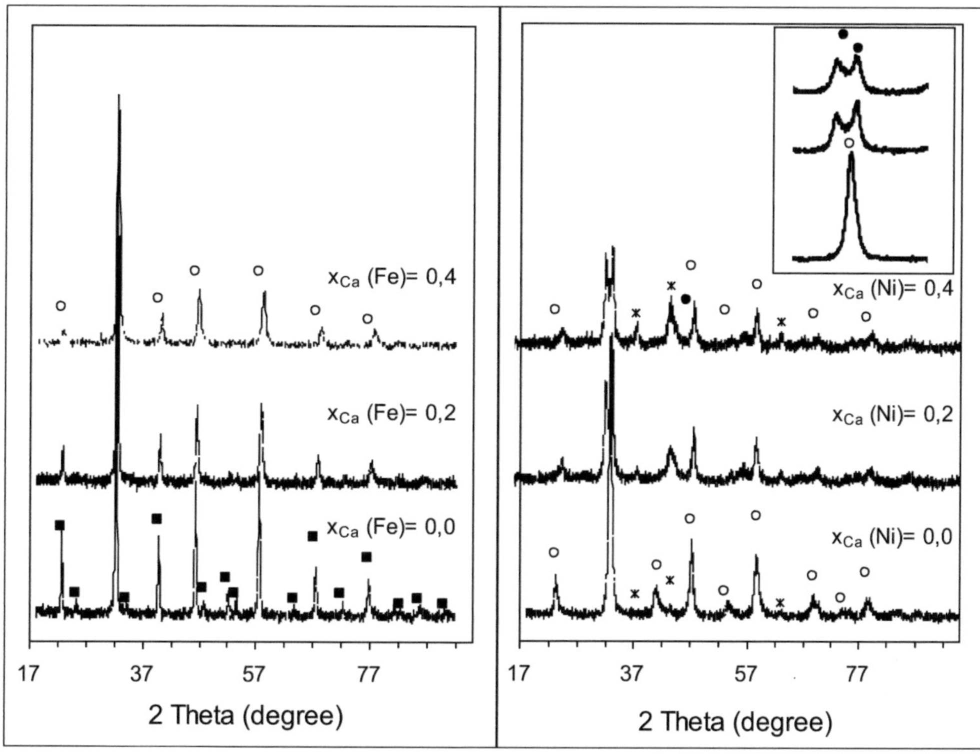

Figure 4. X-ray diffraction patterns of (a) $La_{1-x}Ca_xFeO_3$: (■) orthorhombic; (○) cubic $LaFeO_3$; (b) $La_{1-x}Ca_xNiO_3$; (○)$LaNiO_3$; (■) NiO; (●) $La_2Ni_2O_5$

4.3. B-site substitution

The catalytic activity expressed as intrinsic activity (mol m^{-2} h^{-1}) for the total combustion of acetyl acetate and ethanol over substituted $LaFe_{1-y}Ni_yO_3$ perovskites revealed that the simultaneous presence of Ni^0 and Ni^{2+} is required to achieve optimum performance for these perovskites-type oxides [21]. The largest BET surface area of 20 m^2g^{-1} was obtained for the unadulterated $LaFeO_3$ perovskite, and then decreased progressively upon Ni substitution. XRD patterns indicated an orthorhombic phase for $LaFeO_3$ and a rhombohedral phase for $LaNiO_3$. Upon substitution of Fe^{3+} ions by Ni^{3+} ions, a shift toward larger 2θ angles was observed. The shift of the peak reached a maximum for a substitution degree y = 0.1, and then remained constant for higher substitutions. These findings suggest complete insertion of nickel for substitution y = 0.1, whereas a simple mixed NiO_x phase became segregated for substitutions y> 0.1. FTIR spectra indicated that upon Ni^{3+} substitution the band at 560 cm^{-1} shifted toward higher wavenumbers and increased in intensity compared to that for pure $LaFeO_3$. The dependence of these two parameters on the substitution degree runs in parallel with the shift observed in the main diffraction lines of the XRD patterns, confirming that Ni^{3+} ions were incorporated within the perovskite structure. The TPR_1-TPR_2 cycles of the substituted perovskites did not show significant differences between them, indicating that their structures were restored after intermediate oxidation. No reduction peaks were observed for the nonreducible $LaFeO_3$, while two reduction peaks were observed for $LaNiO_3$, attributed to reduction of Ni^{3+} ions of $LaNiO_3$ to yield $La_2Ni_2O_5$. The second reduction peak at 500°C led to Ni^0. Only one less-intense peak was observed for the substituted perovskite with almost no change in intensity (in comparison with TPR_2), indicative of the high stability of $LaFe_{1-y}Ni_yO_3$ (y = 0.1, 0.2, 0.3). The evolution of oxygen during TPD experiments indicated the presence of α-oxygen in the pure $LaNiO_3$ and the two highest substituted perovskites. The intensity of the desorption peaks increased as the degree of substitution increased. There was no evidence of the presence of α-oxygen on substituted perovskite with y = 0.1, and this was related to the absence of segregated metal oxide phases in this sample. The BEs of the XPS spectra for O1s was deconvoluted into three components: 529.2 eV attributed to surface O^{2-} species; 531.0 eV, which arose from lattice [La-O-M] bonds; and 532.3 eV, which was due to hydroxyl/carbonate groups. The intensity of the two peaks at 529.2 eV and 531.0 eV changed almost linearly in a perfect opposite way, which was related to the incorporation of Ni into the structure. The increase in intensity of the lattice oxygen component in the sample with increase in Ni substitution was related to the increasing catalytic activity in combustion reactions. The surface atomic ratios Fe/La, Ni/La, and CO_3^{2-}/La was plotted as a function of Ni substitution degree and shown in Fig. 5: upon Ni substitution, the surface of the substituted perovskites became Ni- and Fe-enriched. The Fe/La surface ratio for the nonsubstituted $LaFeO_3$ is consistent with the slight La-enrichment often reported for La-containing perovskites, whereas the larger Ni/La surface ratio for $LaNiO_3$ and also for the substituted $LaFe_{1-y}NiO_3$ samples indicated that Ni^{3+} ions are more difficult to be stabilized than Fe^{3+} ions in the perovskite lattice and results in surface segregation of a separate NiO phase. The extent of carbonation was higher for $LaFeO_3$ due to basic character of the slight La-enrichment observed in this perovskite. The catalytic activity tested in the combustion of acetyl acetate and ethanol in a flow reactor under an excess of oxygen showed differences in catalytic behavior depending on the nature of the substrate to be combusted. Ethanol reacted at lower temperatures and exhibited a higher reaction rate. The catalytic activity of $LaFeO_3$ increased upon nickel substitution but declined for $LaNiO_3$. The approach undertaken to modulate oxygen adsorption and release properties by inserting Ni^{3+} ions into the $LaFe_{1-y}Ni_yO_3$ structure while retaining a separate NiO phase is crucial to enhance the performance of substituted perovskites in the combustion of ethanol and acetyl acetate.

4.4. B-site substitution

For substituted $LaMn_{1-y}Co_yO_3$ perovskite-type oxides, a crystal phase transformation occurred for y values above 0.5 and the general trend obtained was: for y ≤0.5, Co ions were inserted into the manganite structure; for y >0.5, the Mn ions were inserted into the cobaltite structure. The enhancement of the ferromagnetic properties and the thermal stability against reduction for y = 0.5 was attributed to the optimized Co^{2+}-Mn^{4+} interactions [22, 23].

The manganite had the highest surface area (38 m^2g^{-1}), and that upon Co substitution, a progressive decrease in surface area was observed until reaching the cobaltite surface value of 16 m^2g^{-1}. XRD analysis showed the existence of cubic and rhombohedral perovskites phases

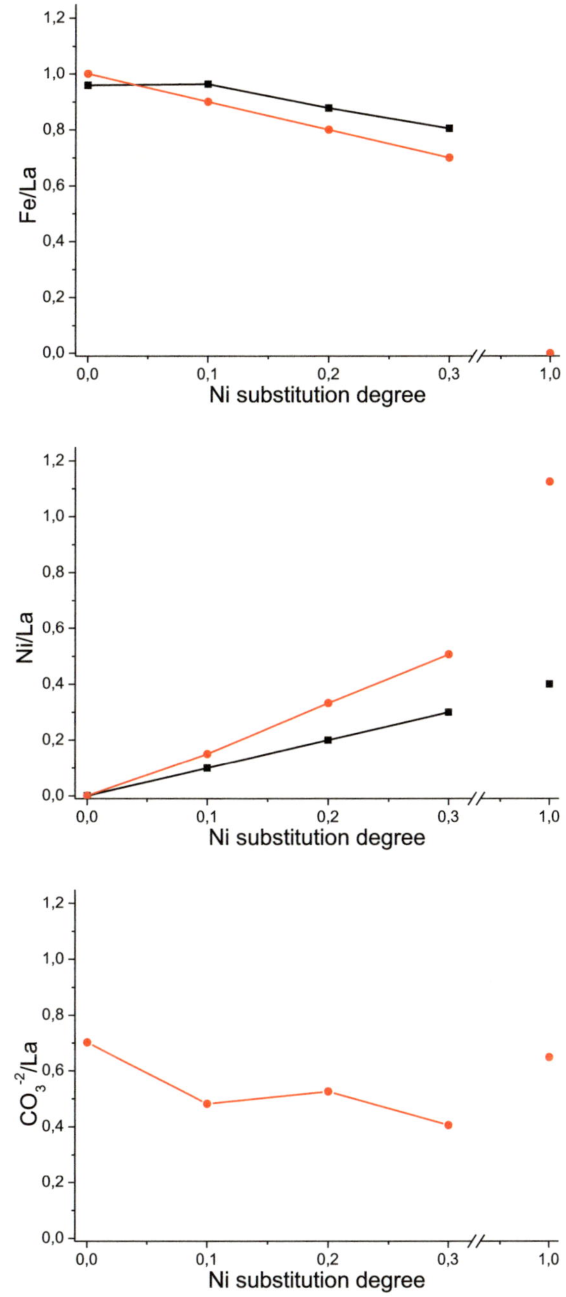

Figure 5. XPS ratios for the $LaFe_{1-y}Ni_yO_3$ perovskite: (■) atomic surface ratio, (●) nominal (a) Fe/La, (b) Ni/La, (c) CO_3^{2-}/La

for pure $LaMnO_3$ and $LaCoO_3$, respectively. The general trend obtained for the $LaMn_{1-y}Co_yO_3$ perovskites were as follows: (i) the XRD patterns were quite similar to the cobaltite's rhombohedral structure for the y>0.5 samples; (ii) the structure of the perovskite was modified to form a new orthorhombic structure for y<0.5; (iii) and for the case of y=0.5, the pattern was practically equal to the one obtained for pure $LaMnO_3$. The vibration frequencies in the infrared region allowed us to conclude that the insertion of Mn into the cobaltite resulted in notable distortions of the highly symmetrical rhombohedral structure, and that manganite can withstand Co insertion without large structural changes up to values of $y \leq 0.3$. With regard to TPR profiles, $LaCoO_3$ and $LaMnO_3$ displayed two reduction peaks, ascribed to the formation of oxygen-deficient intermediate and the reduction of the perovskite structure at higher temperatures. The substituted perovskites exhibited nearly identical behavior with temperature displacements, which was attributed to the previously discussed distortions. The magnetic properties were investigated in the paramagnetic and magnetically ordered regimes: in both cases the data revealed two well-identified regions separated by the equimolar composition y = 0.5. For samples with y > 0.5 (Mn substitutes Co in the cobaltite $LaCoO_3$) the Curie-Weiss temperature sharply decreased, reaching negative values at y = 1. This nonmonotonous variation was also observed in the ordering temperature T_c, which reached a maximum value for y = 0.5. The magnetic moment of the paramagnetic regime showed a linear decrease during the first half of the cobalt substitution, which was interpreted as the progressive transformation of Mn^{3+} into Mn^{4+} ions. This decrease became more pronounced above y = 0.5 since Mn ions were all converted to Mn^{4+}, and cobalt was being introduced as Co^{3+}. The overall behavior observed through the compositional variation of the Curie-Weiss temperature is explained by the varying character of the magnetic interactions between Co and Mn. During the first half of the series ($y \leq 0.5$), strong Co^{2+}-Mn^{4+} double-exchange ferromagnetic interactions were triggered when Co^{2+} was introduced into the lattice, while antiferromagnetism was augmented when Co was introduced as Co^{3+} (y>0.50). To investigate the ordered regime, the thermal dependence (ZFC/FC cycles) and the field dependence (M loops) of the magnetization was carried out. Figure 6 compares the (ZFC/FC cycles) regions separated by the equimolar composition y = 0.5. A strong ferromagnetic behavior was observed for the first half of the series since the magnetization M_{FC} reached a characteristic plateau and the ordering temperature increased toward a maximum value of 235 K. For y compositions above 0.5, the magnetization decreased by one or two orders of magnitude, meaning that the ferromagnetic Co^{2+}-Mn^{4+} interactions were weaker, being progressively replaced by the less effective Co^{2+}-Co^{3+} interactions. Thus, magnetic iso-field studies (ZFC-FC) revealed that, for $y \leq 0.50$, the system presented an antiferromagnetic canted-like ordering of the Mn/Co sublattice, in which the presence of divalent Co ion created Mn^{3+}-Mn^{4+} pairs that interact ferromagnetically through the oxygen orbital. This was confirmed by the magnetization loops in which the magnetic moment increased when substituting Mn for Co. The catalytic activity in total acetyl acetate combustion in a flow reactor under an excess of oxygen (60000 $cm^3g^{-1}h^{-1}$) indicated that lower ignition temperature (i.e., the higher activity) was displayed by manganite, while higher ignition temperature was exhibited by the cobaltite. The substituted perovskite displayed ignition temperatures between these two values, indicating that changes in y values modified the active sites required for the oxidation of organic compound. The larger intrinsic activity (molm^{-2}h-1) of the substituted perovskites with y= 0.5 was attributed to the magnetic order of Mn^{4+}-Co^{2+}. This ordered state confirmed the presence of two regimes, separated by the equimolar composition $LaMn_{0.5}Co_{0.5}O_3$: strong ferromagnetic interactions for $y_{Co} \leq 0.5$ and a superposition of antiferromagnetism and ferromagnetism for $y_{Co} > 0.5$.

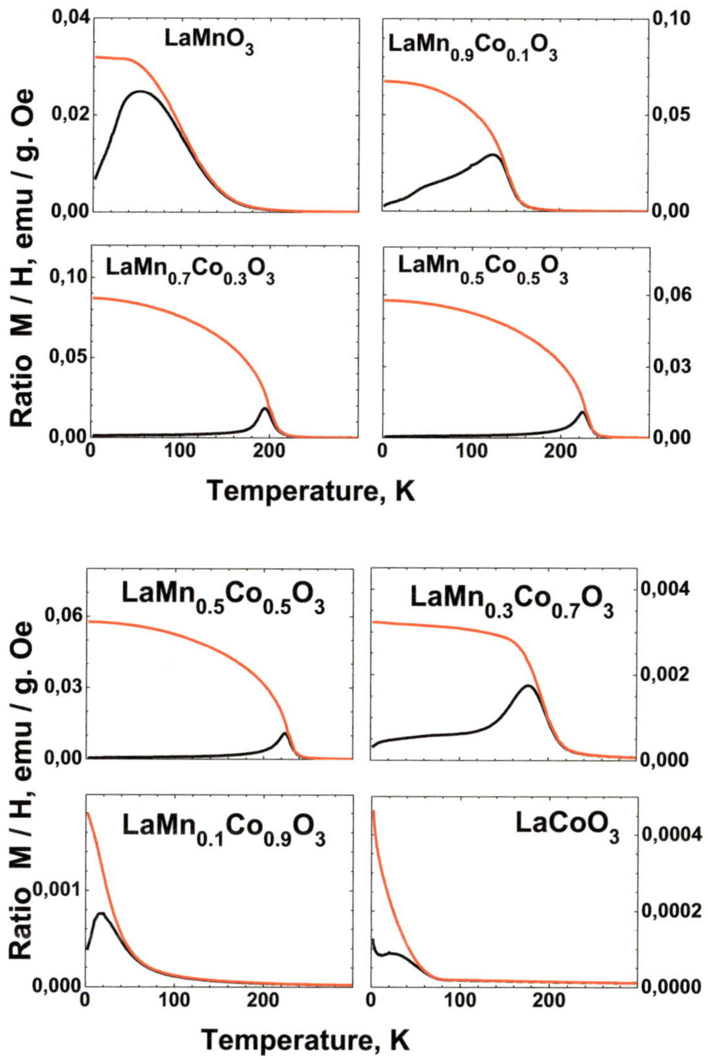

Figure 6. Magnetization cycles for $LaMn_{1-y}Co_yO_3$ measured under 0.025 T: above: y = 0.0, 0.1, 0.3, 0.5; below: y = 0.5, 0.7, 0.9, 1.0

4.5. A- and B-site substitution

Partial substitution of La^{3+} by Ag^+ in $LaMn_{0.9}Co_{0.1}O_3$ substantially increased the catalytic activity in the removal of n-hexane in a flow reactor under an excess of oxygen, attributed to the interaction between O- and the low coordinated metal site. Silver segregated phases further increased the catalytic activity [24].

The BET surface areas of the perovskites varied only slightly with substitution degree. The XRD profiles revealed that the substitution of small amounts of La^{3+} by Ag^{+} cations in a Co-substituted manganite ($LaMn_{0.9}Co_{0.1}O_3$) did not alter its rhombohedral crystallographic structure, and that it only decreased their crystallinity. Segregated silver oxide (Ag_2O) phase was also detected upon increasing the Ag content. TPR profiles indicated that upon introduction of Ag, two features were seen: the first reduction peak shifted toward lower temperatures and became wider upon increasing Ag content; the second reduction peak also started at a lower temperature but became more intense upon increasing Ag content. Since the first reduction step involves the formation of oxygen-deficient perovskite, and the shoulder includes the reduction of individual segregated oxides, quantification of this peak can be taken as a measurement of the extent of Ag insertion in the perovskite structure. The increase in H_2 consumption in the first reduction step upon Ag-substitution is indicative of the appearance of a higher proportion of micro crystals from segregated silver, MnO_2 phases, and/or surface carbonates species. The shift of the peak toward lower temperatures indicates a growth in crystal size. These results are in line with the XRD patterns, which showed diminished crystallinity as silver increased. The FTIR spectra showed an increase in the asymmetric and symmetric O-C-O stretching vibration modes of carboxylate structures with the substitution degree associated with the presence of segregated phases in the form of carbonates in the Ag-substituted perovskites. The XPS spectra of La $3d_{5/2}$ with BE at 834.8 eV and Co $2p_{3/2}$ with BE at 780.6 eV demonstrated the presence of La^{3+} and Co^{3+} species on the surface, whereas manganese appeared at slightly higher BE than is expected for Mn^{3+} ions, suggestive that the surface manganese was present as Mn^{4+}. Differences in the surface O-species upon silver substitution was observed. For the $x = 0.0$ sample, the two surface oxygen species were the ones expected for a nonsubstituted perovskite. On the other hand, upon Ag substitution, the lower oxidation state of Ag^{+} inserted into the network of the $LaMn_{0.9}Co_{0.1}O_3$ perovskite induced the appearance of O^{2-} species at BE higher than 532 eV known as oxygen vacancies, which is necessary to maintain the electroneutrality of the solid. A relative proportion of this kind of oxygen species of 18% was exhibited by the perovskite with an Ag substitution of $x = 0.1$, which suggests that the maximum content of Ag with almost no segregated phases was at $x = 0.1$, and that this segregated phase disappeared at larger quantity of Ag. The EDAX and surface Mn/La and Co/La atomic ratios were much higher than the nominal compositions, and the differences were even larger upon increasing Ag substitution. These results indicate that the surface of the silver-substituted perovskites became Mn- and Ag-enriched, in line with deductions from oxygen-TPD profiles, which showed that Ag-substituted perovskites displayed a new desorption peak associated with the redox cycle of the B cation. This result further indicated that in these perovskites the Mn^{4+}/Mn^{3+} redox cycle was almost constant, in agreement with XPS results. Therefore, the charge compensation due to the substitution of La^{3+} by Ag^{+} occurred mainly via a nonstoichiometric decrease rather than through a change in the Mn^{4+}/Mn^{3+}ratio. The catalytic activity of the perovskites was evaluated in the total combustion of n-hexane in a flow reactor under an excess of oxygen and shown in Fig. 7. The typical S-shaped curves represent the total conversion of n-hexane oxidation as a function of the reaction temperature up to complete combustion. There was an appreciable increase in the catalytic activity, expressed in terms of the total conversion, when Ag was substituted into the perovskite. For Ag substitution of x = 0.1, a noticeable shift of the curve toward lower temperatures is indicative of a higher catalytic activity of the perovskite upon a 10% Ag substitution of La. The changes in the activity of the $La_{1-x}Ag_xMn_{0.9}Co_{0.1}O_3$ systems are mainly dependent on their crystalline structure and surface composition. The highest activity displayed by the x = 0.3 sample can be explained in terms of the large amount of segregated silver on the surface. Sample x = 0.2 was only slightly less active, although the silver surface segregation was much lower than in its x = 0.3 counterpart. This could be due to the less intense diffraction lines of the rhombohedral phase displayed by the x = 0.3 perovskite. For the sample x = 0.1, in which the rhombohedral structure and almost no silver segregation existed, the relation between catalytic activity and the nonstoichiometric oxygen ratio suggests an intrafacial reaction mechanism.

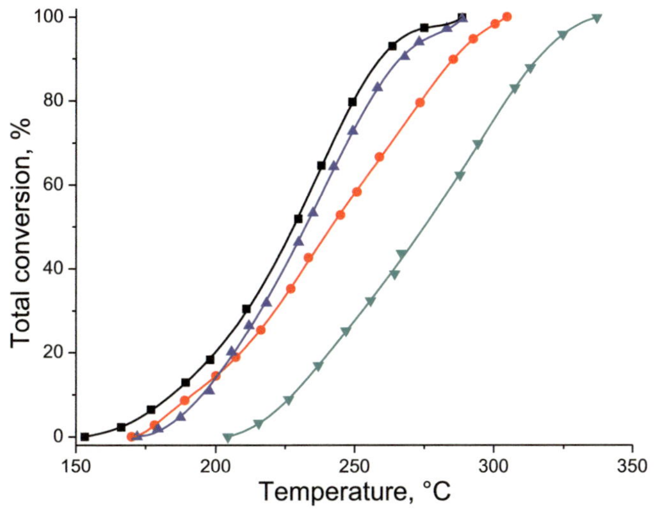

Figure 7. Stationary-state conversion over $La_{1-x}Ag_xMn_{x0.90}Co_{0.1}O_3$ perovskite for (▼) x = 0.0, (●) x = 0.1, (▲) x = 0.2, and (■) x = 0.3.

5. Soot combustion

Diesel engines play a vital role in modern society because they are widely used for heavy-duty transportation and power generation. However, they are one of the largest contributors to environmental pollution because they emit particulate matter and nitrogen oxide. Particulate matter consists mostly of carbonaceous soot, minor amounts of volatile organic fraction from unburned fuel, as well as inorganic compounds such as ash and sulfur compounds. These materials are detrimental to human health, plants physiology, the environment, and building materials [25, 26]. Thus, stringent emission standards for soot have been set by most countries, leading to continuous efforts in emission control technology. Diesel particulate filter (DPF) is the most effective and widely employed approach to reduce soot in exhaust gases. However, one drawback for using the filter is that it requires regeneration through continuous combustion to remove accumulated soot. Recently, perovskite-type oxides have received considerable attention as catalytic systems for soot combustion [27-29]. Pure and doped lanthanum chromites, manganites, and cobaltites have been tested with mixed results in the catalytic combustion of soot [30]. Cerium has been usually reported as a good promoter in perovskite lattice. Partial substitution of La by Ce or Sr on cobalt-based perovskite led to an increase in catalytic oxidation [31]. When perovskite-type oxides were reduced, their low thermal behavior under reducing atmosphere allowed the formation of highly dispersed metal nanoparticles [22], which consequently modified the oxygen storage capacity of the catalyst. The substitution of the lanthanum cation La^{3+} by an oxygen storage cation such as cerium, with the capacity of the redox Ce^{3+}/Ce^{4+} pair, will have a direct effect in the reducibility of the perovskite structure and the dispersion degree of the pure or mixed La_2O_3 and CeO_2 oxides.

5.1. A-site substitution

The catalytic activity of soot combustion over oxidized and partially reduced substituted $La_{1-x}Ce_xCoO_3$ perovskites is enhanced by two factors: (1) the increase in CeO_2 content, closely related to the increase in the material's capacity for oxygen transfer to the carbonaceous surface; and (2) the formation of larger CeO_2 crystals in the higher-Ce-content samples after H_2 reduction, responsible for the improvement of the exchange of oxygen and number of contact points between catalyst and soot [32].

The BET specific surface areas of the calcined solids ranged from 10 to 19 m^2g^{-1}, and increased upon incorporation of Ce. After reduction, S_{BET} decreased and ranged from 7 to 16 m^2g^{-1}. XRD patterns of the calcined perovskites indicated that Ce was introduced into the lattice of the parent $LaCoO_3$ solid at lower degree of substitution (i.e., $x \leq 0.5$), while for larger Ce content ($x \geq 0.7$) formation of CeO_2 phase was detected. Figure 8 showed the XRD of the (a) calcined and the (b) reduced solids. Larger structural changes after H_2 treatment at 700 ºC were observed for the samples with lower Ce content ($x \leq 0.3$), while calcined and reduced samples with higher Ce content ($x = 0.7$ and 0.9) showed nearly identical XRD patterns. For the samples with lower Ce content ($x = 0.1, 0.3, 0.5$), the most intense diffraction peak of the perovskite-type structure ($2\theta = 33°$) fully disappeared after reduction treatment, and the reduced solids showed the presence of La_2O_3, La_2CoO_4, CeO_2 as segregated phases as well as $Co°$, which appeared as two broad peaks at $2\theta = 44.3°$ and 51.5°. TPR analysis showed that the reduction of pure $LaCoO_3$ solid occurred in two steps: at 365°C forming an intermediate oxygen-deficient perovskite structure ($LaCoO_{3-\delta}$); and, at 530°C which corresponds to the complete reduction of the perovskite, leading to the formation of La_2O_3, $La(OH)_3$, H_2O, and Coº. With the addition of Ce, there was a noticeable shift of the peaks to lower temperatures, which implies an increase in the reducibility of the catalysts. The evolution of oxygen during TPD analysis showed a beneficial effect on the presence of α-oxygen with increasing Ce content. The importance of these surface oxygen species is their close relation with the oxygen vacancies in the perovskite structure that can activate molecular oxygen in soot. The catalytic activity determined by nonisothermal thermogravimetric analysis indicated that the activity of calcined catalysts increased with Ce content. $La_{0.1}Ce_{0.9}CoO_3$, the most active catalyst, was the sample with the highest Ce/La mass ratio and possessed CeO_2 and Co_3O_4 crystalline phases instead of perovskite-type structure (from their XRD patterns). These results suggest that CeO_2 played the main role in soot combustion. The higher activity of the higher Ce-substituted catalysts indicated the availability of reactive oxygen species on the surface of the catalyst, attributed to the weaker binding of oxygen. The enhanced activity of the two catalysts with larger Ce content (x = 0.7 and 0.9) pointed out the importance of structural defects and oxygen mobility by lattice oxygen via La^{3+}-doped CeO_2 in the CeO_2/soot interface for the catalytic oxidation of soot. In order to evaluate the effect of perovskite-type structure on soot combustion, the destruction of the perovskite structure was carried out in hydrogen at 700°C. After reduction, an increase in the catalytic activity was observed in varying degrees: only a slight increase of the activity for the samples with low CeO_2 content (x≤0.5), while the activity for the catalysts with higher amount of CeO_2 (x = 0.7, 0.9) increased sharply. These results suggest that the activity of these materials for soot combustion is improved by increasing the CeO_2 content, closely related to the higher material capacity for oxygen transfer from catalyst to carbonaceous surface and the formation of larger CeO_2 crystals in the higher-Ce-content samples after reduction.

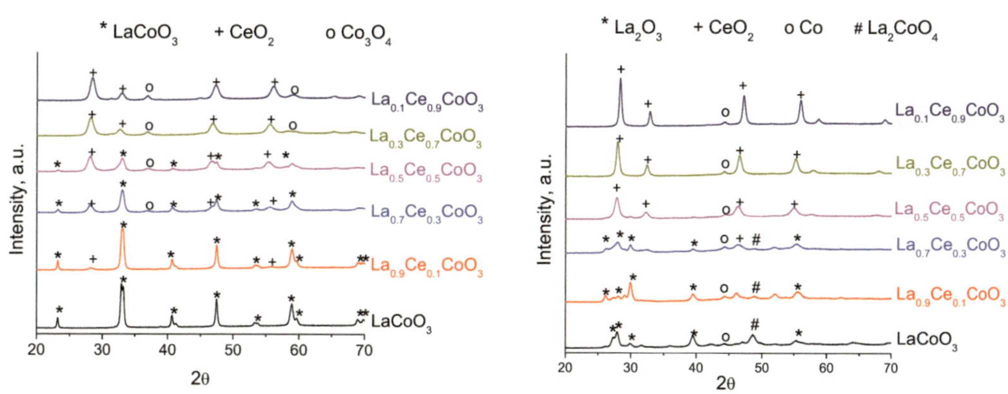

Figure 8. XRD profiles for $La_{1-x}Ce_xCoO_3$ catalysts: (left) calcined; (right) reduced

5.2. A- and B-site substitution

The increase in the catalytic performance in soot combustion by Ag^+ substitution in $LaMn_{0.9}Co_{0.1}O_3$ perovskite structure occurred due to increase in surface Ag species and superoxide ions content. These species have been found to be responsible for the higher capacity for oxygen transfer from the catalyst to the carbonaceous material in the solids with higher-Ag-content [33].

No significant differences between BET specific areas upon Ag substitution were found. XRD profiles revealed that the crystallinity of the rhombohedral structure of the Co-substituted manganite ($LaMn_{0.9}Co_{0.1}O_3$) decreased upon Ag substitution. For the higher Ag-substituted perovskites, the intensities of three diffraction lines- indexed to a segregated Ag_2O phase- increased with increasing Ag substitution: a mean crystal size of 7 nm was estimated by the Debye Scherrer equation at 2θ = 37.96°. The main differences observed in relation to the TPR profiles of Ag-substituted perovskites with respect to Ag-free perovskite were as follows: the first reduction peak became more intense with increasing x, and the two reduction peaks become closer and shift toward lower temperatures. The increase in H_2 consumption upon Ag-substitution is indicative of the appearance of a higher proportion of Ag_2O segregated crystallites and/or other segregated species. The shift of the peaks toward lower temperatures coincided with the increase in crystallite size; furthermore, it indicates that the redox properties of the solids increased with Ag content. These results were consistent with XRD patterns, which showed a drop in crystallinity of the structure of the perovskites upon increasing silver content, indicating that the Ag^+ inserted into the network made the formation of the perovskite structure more difficult. The BEs from XPS spectra revealed the presence of carbonates strongly attached to the surface, as expected for La-containing perovskites. The core-level spectra indicated the presence of La^{3+} and Co^{3+} species, whereas Mn appeared at slightly higher BE than that expected for Mn^{3+} ions, suggesting that surface Mn was present as Mn^{4+}. Other notable observations were: (i) the $Ag3d_{5/2}$ peaks with BE at 368.2 and 369.3 eV indicated the presence of surface Ag^0 and Ag^+ species, respectively; (ii) the O1s peak was attributed to surface lattice oxygen species (O_{latt}^{2-}) at 529.8 eV, hydroxyl and/or carbonate groups at 531.0–531.4 eV, and loosely adsorbed superperoxide species (O_{2ads}^-) at 532.4–532.6 eV. The surface Ag/La+Ag atomic ratio of the silver-substituted perovskites indicated the enrichment of Ag upon increasing substitution; in addition, the nearly unchanged surface O_{2ads}^-/O_{latt}^{2-} atomic ratio of the Ag-free and the Ag-containing perovskites means that the charge compensation due to the substitution of La^{3+} by Ag^+ occurred via nonstoichiometric decrease rather than through a change in the Mn^{4+}/Mn^{3+} ratio. Since identical trends were obtained for the Ag^+/Ag and the O_{2ads}^-/O_{latt}^{2-} surface ratios for the Ag-containing perovskites, it was proposed that these superoxides ions acted as the most important active species, facilitating oxygen transportation for soot oxidation. In addition, due to the fact that surface Ag and superoxides' ions formation increased with increasing Ag content, surface Ag_2O was proposed as an active phase in the reaction. The evolution O_2-TPD experiments revealed a continuous increase in total α-oxygen species upon Ag doping, supporting the hypothesis of the formation of Ag^+ sites in the lattice of the perovskite. The catalytic activity for soot combustion was related to the temperature corresponding to the maximum from the DTG curve shown in Fig. 9. Higher values for T_m mean lower catalytic activity. T_m value for the uncatalyzed CB combustion was found to be 650°C, while much lower T_m values were observed for all the catalyzed reactions, indicative of the catalytic nature of the materials for soot combustion. Catalytic activity was observed to increase with the silver content, with the best catalyst being the one with Ag content of x = 0.3 (burned off soot at 370°C). This important catalytic effect was attributed to the presence of oxygen vacancies due to the substitution of La^{3+} by Ag^+ in the perovskite lattice. These vacancies enhanced the availability and reactivity of oxygen species on the catalytic surface, because of the weaker oxygen binding. The results show that structural defects and oxygen mobility are important factors that control the catalytic activity of these types of crystalline structures. The presence of surface Ag_2O phase improved the O_2 exchange and number of contact points between catalyst and soot, with no significant effect on BET surface area and crystal size. These results contrast the proposed mechanism for soot combustion on La-containing perovskites, in which the oxidation of soot particles is attributed to the surface oxygen species, formed on the perovskite structure.

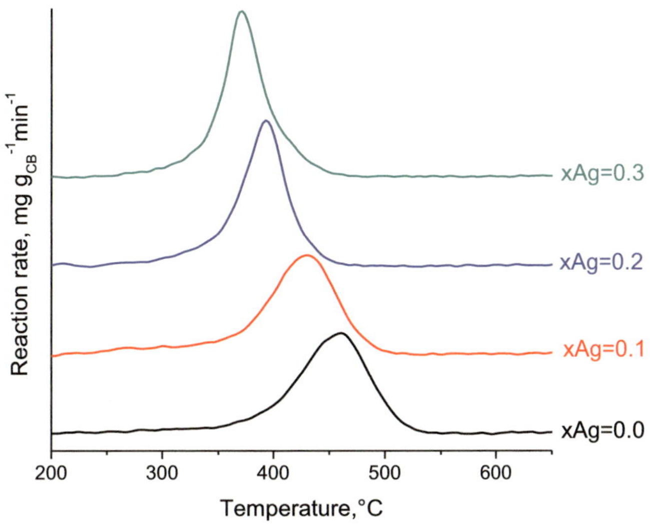

Figure 9. Thermogravimetric assays for the catalytic combustion of CB

6. Hydrogenation reactions

For these kinds of heterogeneous catalytic reactions, perovskite-type oxides are not used directly as catalysts but rather as precursors to obtain metal nanoparticles catalysts. Therefore, prior to the reaction, perovskites are activated *ex situ* under H_2 at 500°C.

7. Guaiacol conversion

Catalytic studies focusing on hydrodeoxygenation (HDO) of guaiacol, the most representative phenolic compound in bio-oil, has provided insights into catalyst optimization for lignin valorization technologies. Different active phases, including sulfides [34-36] and metals [37-39], have been investigated as catalysts for the HDO of guaiacol. Nickel-based catalysts have recently received extensive attention for the HDO of lignin derived compounds, and have consistently been proven to be highly active.

8. A-site substitution

The preparations of Ni catalysts from the reduction of $La_{1-x}Ce_xNiO_3$ perovskite-type oxides were effective in catalyzing the conversion of guaiacol to cyclohexanol, an important feedstock for the chemical industry. The most active catalyst was the solid with Ce content x = 0.5 and 0.7, which was ascribed to the highest thermal stability under reductive atmosphere of the CeO_2-solid solution partially substituted with La [40].

The BET specific surface areas of the calcined and reduced solids showed a monotonic dependence of the surface area on Ce content with a slight decrease after reduction treatment. The main difference in the XRD patterns of the as-calcined and as-reduced solids after Ce substitution was the absence of the perovskite structure and the appearance of segregated phases in the latter. The most crystalline perovskites corresponded to the undoped $LaNiO_3$ material and the lowest Ce-substituted perovskite ($La_{0.9}Ce_{0.1}NiO_3$). Upon increasing the Ce content (x≥0.5), there were neither diffraction peaks characteristic of the perovskite structure nor single or mixed La oxides; however, new crystalline phases corresponding to CeO_2 and NiO appeared. The XRD patterns of the Ni-reduced solids indicated a complete loss of the $LaNiO_3$ perovskite structure and the formation of Ni^0 for the material with Ce content x = 0.0 and 0.1. For higher Ce contents (i.e., x≥0.5), there were no observable changes in the crystalline structure after H_2 treatment, and a shift of the main CeO_2 peak toward higher 2θ angles upon Ce doping can be taken as a clear indication of the modification of the crystalline structures, suggestive of the insertion of La^{3+} into the CeO_2 structure. These results indicate the formation of the perovskite structure with low thermal stability under H_2 for the samples with low Ce content x≤0.3, and the formation of a solid solution of CeO_2-La_2O_3 with higher thermal stability for samples with higher degree of Ce substitution x≥ 0.5. The Ni-reduced catalyst with Ce content x = 0.5 displayed the lowest crystallinity and the highest Ni^0 dispersion. The TPR_1-TPR_2 cycles demonstrated clear differences in the H_2 consumption peaks between TPR cycles, an indication of the structural changes under H_2 treatment. The TPR_1 profiles showed two different behaviors attributed to the reduction of Ni^{2+} to Ni^0: two reduction peaks for the samples with Ce contents x≤ 0.3 and only one reduction peak for those with Ce contents x≥0.5. The amount of H_2 consumed during TPR_1 and TPR_2 cycles is plotted as a function of Ce content and shown in Fig. 10. The TPR_1 profile showed larger H_2 consumption for the perovskites with Ce content x≤0.3 and almost constant value for those with Ce content x≥0.5. Considering that the theoretical hydrogen uptake for sample reduction assuming that NiO is stoichiometrically reduced to Ni° is 4.1 mmol g^{-1}, the larger obtained values suggest that other oxides entities are also being reduced. Since XRD results did not show any structural changes under reduction atmosphere, it is possible to calculate the degree of reduction of the perovskite structure considering only the reduction of Ni^{3+} to Ni^0. The obtained reduction degree of 24% fits quite well with previously reported results corresponding to the total reduction of Ni present in the original perovskites. The TEM micrographs of all of the Ni-reduced catalysts showed a narrower particle size distribution for metallic Ni between 5 and 25 nm. The H_2 chemisorption capacity assuming a stoichiometry of $H/Ni_s = 1$ was smaller than the expected one considering the particle size obtained from TEM and XRD, indicating that Ni particles were partially inserted and not available for H_2 uptake. In the conversion of guaiacol over Ni-reduced catalysts, cyclohexanol and methoxycyclohexanol were principally obtained via the hydrogenation and demethylation routes. In other words, the Ni-reduced catalysts displayed a high hydrogenation activity and a low activity toward hydrogenolysis of C-OH bond. The products' distribution calculated both at 20% conversion of guaiacol and at 4 h of reaction time showed that all the Ni-reduced catalysts displayed similar trends in product distribution. The changes in Ni particle size and the presence of segregated phase influenced by the degree of Ce substitution did not affect the selectivity. This result suggests that the simultaneous presence of highly dispersed nickel partially inserted into a CeO_2-solid solution is a prerequisite for improved catalytic performance for the conversion of guaiacol over these catalytic systems.

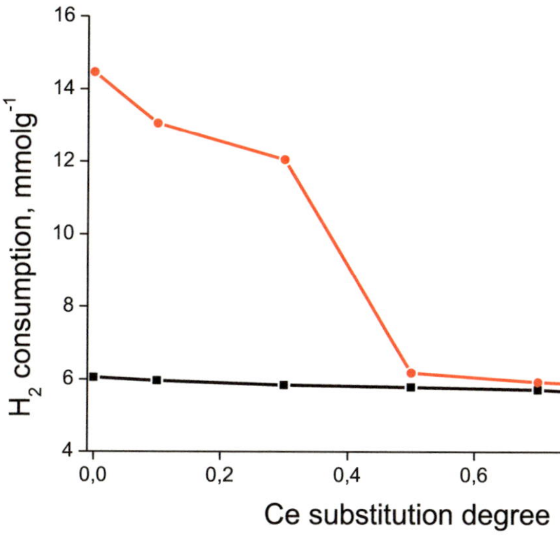

Figure 10. Plot of hydrogen consumption of the $La_{1-x}Ce_xNiO_3$ vs. nominal Ce content (●) TPR_1; (■) TPR_2

9. Glycerol reforming

The use of environmentally friendly and renewable feedstock for clean energy production such as hydrogen has been extensively studied, with particular focus on the catalytic green reforming process of glycerol [41]. Even though the reaction pathway is a complex process, as a result of the occurrence of wide variety of reactions, the type of catalysts used play a key role in determining the product distribution. For supported metal catalysts as well as Ni-based catalysts, high basic character of the catalysts has consistently been observed to lead to higher activity and resistance toward deactivation [42, 43].

9.1. A-site substitution

Mixed oxides with the general formula $La_{1-x}Ce_xNiO_3$ used as precursors for supported Ni catalysts have led to active catalysts for the production of hydrogen-rich gas stream. Substitution of 50% of La by Ce provided more deactivation-resistant catalyst, allowing only minor amount of deactivation, associated with the formation of CeO_2-La_2O_3 solid solutions, which acted as an oxygen buffer and, hence, facilitated the removal of carbon deposits on the catalysts surface [44].

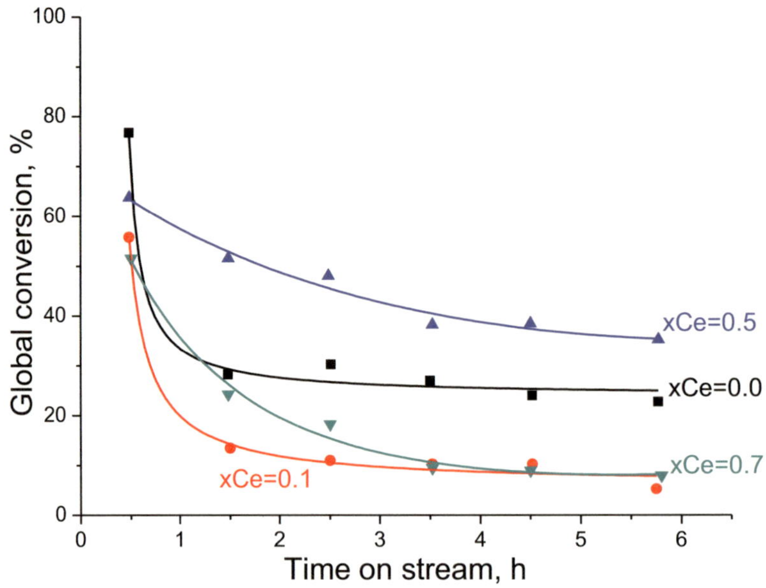

Figure 11. Evolution of global glycerol conversion as a function of time on stream for glycerol steam reforming at 500 °C over all mixed oxide-derived catalysts

The $La_{1-x}Ce_xNiO_3$ perovskites used as precursors for supported Ni catalysts were previously fully characterized and used as catalysts in glycerol steam reforming. Figure 11 illustrates the global performance of the catalysts for glycerol reforming. Regardless of the features of the catalysts or degree of La substitution by Ce, a drop in global conversion versus time on stream was observed. However, the sample formed from the mixed oxide with an equivalent content of La^{3+} and Ce^{3+} (x=0.5) was much more stable and promising, which may be associated with the formation of a CeO_2-La_2O_3 solid solution as detected by XRD analysis. The catalysts with Ce substitution other than 50% led to a dramatic decrease in glycerol conversion, revealing a strong deactivation process: indeed x = 0.1 and x = 0.7 samples presented conversion within 5 and 20% right after the first hour. To investigate the causes of deactivation, Fe-SEM analysis revealed that the deactivation of unsubstituted $LaNiO_3$ perovskite-derived catalysts (x = 0.0) was due to the formation of plentiful amounts of filamentous carbon. Even though, the formation of the carbon nanostructures did not occur equally over all Ce-containing catalysts due to varying degrees of Ce substitution, considerable amounts of carbon filaments and similar FE-SEM images were obtained even for unsubstituted $LaNiO_3$ perovskite-derived catalyst (x=0.0). Conversely, after a much more detailed and meticulous analysis, only few filaments could be found on the x = 0.5 solid. Thus, it can be argued that although deactivation occurred irrespective of catalysts composition, the sample with 50% La substituted by Ce was more resistance to deactivation compared to other loadings. TEM micrographs allowed the identification of carbon nanotubular structures with an open end, where it was suggested that Ni nanoparticles were detached. From this analysis, it was suggested that Ni nanoparticles were intimately involved in the formation of those carbon filaments. TPO/TGA analyses combined with mass spectrometry were complementarily performed for the used catalysts.

The evolution of CO_2 (m/z=44) observed in the analysis originated from the gasification of carbon deposits. The x = 0.5 sample possessed the lowest amount of the more reactive carbon deposits. Thus, the minimization of the deactivation process in glycerol reforming requires an appropriate Ce content, which can lead to the formation of CeO_2-La_2O_3 solid solutions.

10. Xylose hydrogenation

Liquid phase hydrogenation of aldoses is an important process in the synthesis of mono- and disaccharides, which have long been used as natural sweetening agents. Xylitol, an excellent sweet molecule, can also be commercialized because its sweetening capacity exceeds that of sugar by 20-25%, in addition to having no insulin requirements [45]. Several well-known hydrogenation catalysts such as noble metals or nickel-based [46] catalysts have been used for xylose hydrogenation [47]. Due to their lower price, the main advantage of Ni catalysts is their utilization in a typical slurry batch reactor, leading to good activity and selectivity. Nevertheless, the major drawback of Ni catalysts (which has been reported for Raney Ni) is their high deactivation rate as a result of poisoning of the active sites and metal leaching.

10.1. A-site substitution

A series of active and selective partially Ni-substituted $La_{1-x}Ce_xAl_{0.18}Ni_{0.82}O_3$ perovskite catalysts with a constant Ni loading of 20 wt.% were tested in the hydrogenation of xylose to xylitol to gain insight into the importance of the perovskite structure in preventing leaching.

The specific surface areas obtained showed that the solids with lower Ce substitution degree displayed lower values between 9 and 15 m^2g^{-1}, while higher values above 30 m^2g^{-1} were obtained for solids with higher Ce substitution degrees. The XRD patterns of the reduced solids shown in Figure 12 indicated that the Ce-free solid presented perovskite structure while upon Ce substitution segregated phases appeared. The x = 0.1 solid showed both the perovskite phase and the presence of mixed oxides. For larger Ce substitution degree the perovskite structure was not present and only peaks corresponding to Ce-La oxide solid solution were found. TPR results show only two reduction peaks corresponding to reduction of Ni species. TEM micrographs revealed that Ni was dispersed on the oxide support without forming nanoparticles. Catalytic activity results indicated that all the catalysts were active in the hydrogenation of xylose to xylitol. The catalysts with lower Ce substitution degree (x=0.0, 0.1) displayed higher catalytic activity, demonstrating the importance of crystalline perovskite structure in this reaction. Xylitol selectivity between 40 and 60% was obtained at a conversion of 30%. Hydrogenolysis products were also detected. Atomic absorption spectroscopy (AAS) analysis of the catalyst before and after reaction indicated that for the unsubstituted and lower Ce substituted solids (x = 0.0, 0.1), a constant Ni content close to the nominal loading was obtained, indicative of the high stability under aqueous medium. On the other hand, the solid with the higher Ce substitution degree (x = 0.5, 0.7) showed loss of Ni by leaching. The results from this study demonstrated clear dependence on the reactivity and stability of the catalyst on the presence of crystalline perovskite, modulated by the degree of Ce substitution.

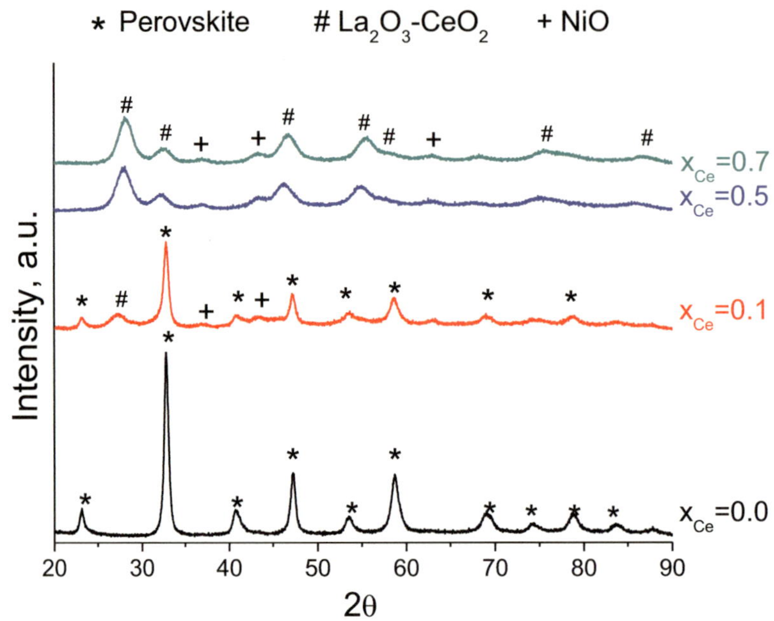

Figure 12. XRD pattern of Ni-substituted $La_{1-x}Ce_xAl_{0.18}Ni_{0.82}O_3$ catalyst

11. Conclusion

Perovskite-type oxides with La in the A-site and a d-transition metal in the B-site position are the most widely used materials for catalytic purposes. The activity of La perovskites can be noticeably improved by partial replacement of the A-site and/or B-site metal. The different combinations of partial substitution of A-site by an adequate A′ metal with similar ionic radii but lower oxidation state modifies the oxidation state of the B-site metal cation and/or oxygen vacancies formation. This substitution also affects the thermal stability, crystalline structure, and activity of the perovskites. Two clear applications can be identified:

- Perovskites with oxygen vacancies to be used as catalysts in oxidation reactions
- Perovskites to be used as precursors to prepare nanosized catalysts to be used in hydrogenation reactions

Due to their tunable properties, perovskites show different crystallization structures and high flexibility in their composition, and, hence, it is important to understand how different perovskites behave under different reaction conditions and what would be the most effective metal combination for a given catalytic process.

Acknowledgements

The authors thank CONICYT Fondecyt No 1130005 and Red Doctoral REDOC, and the MINEDUC project UCO1202.

Author details

Gina Pecchi[1*], Nestor Escalona[2,3], I. Tyrone Ghampson[2] and Ruddy Morales[1]

*Address all correspondence to: gpecchi@udec.cl

1 Dpto Físico Química, Facultad Ciencias Químicas, Universidad de Concepción, Concepción, Chile

2 Dpto Ingeniería Química y Bioprocesos, Escuela de Ingeniería, Pontificia Universidad Católica de Chile, Macul, Santiago, Chile

3 Facultad de Química, Pontificia Universidad Católica de Chile, Santiago, Chile

References

[1] Voorhoeve RJH, Remeika JP, Trimble LE, Cooper AS, Disalvo FJ, Gallagher PK. Perovskite-like La1−xKxMnO3 and related compounds: Solid state chemistry and the catalysis of the reduction of NO by CO and H2. *J Solid State Chem*. 1975 8//;14(4): 395-406.

[2] John R. Anderson MB. M. *Dry*: Springer-Verlag Berlin Heidelberg; 1981.

[3] Escalona N, Medina C, García R, Reyes P. Fischer Tropsch reaction from a mixture similar to biosyngas. Influence of promoters on surface and catalytic properties of Co/SiO2 catalysts. *Catalysis Today*. 2009 5/15/;143(1–2):76-9.

[4] Medina C, García R, Reyes P, Fierro JLG, Escalona N. Fischer Tropsch synthesis from a simulated biosyngas feed over Co(x)/SiO2 catalysts: Effect of Co-loading. *Appl Catalys A: General*. 2010 1/31/;373(1–2):71-5.

[5] Ubilla P, Garcia R, Fierro JLG, Escalona N. Hydrocarbons synthesis from a simulated biosyngas feed over FE/SIO2, catalysts. *J Chilean Chem Soc*. 2010;55:35-8.

[6] Bedel L, Roger AC, Estournes C, Kiennemann A. Co0 from partial reduction of La(Co,Fe)O3 perovskites for Fischer–Tropsch synthesis. *Catalysis Today*. 2003 10/15/;85(2–4):207-18.

[7] Bedel L, Roger AC, Rehspringer JL, Kiennemann A. Structure-controlled La-Co-Fe perovskite precursors for higher C2-C4 olefins selectivity in Fischer-Tropsch synthe-

sis. In: Xinhe B, Yide X, editors. *Studies in Surface Science and Catalysis*. Volume 147: Elsevier; 2004. p. 319-24.

[8] Escalona N, Fuentealba S, Pecchi G. Fischer–Tropsch synthesis over LaFe1−xCoxO3 perovskites from a simulated biosyngas feed. *Appl Catalysis A: General*. 2010 6/15/;381(1–2):253-60.

[9] Yu L, Diao G, Ye F, Sun M, Zhou J, Li Y, et al. Promoting Effect of Ce in Ce/OMS-2 Catalyst for Catalytic Combustion of Dimethyl Ether. *Catal Lett*. 2011 2011/01/01;141(1):111-9. English.

[10] Liang JJ, Weng H-S. Catalytic properties of lanthanum strontium transition metal oxides (La1-xSrxBO3; B = manganese, iron, cobalt, nickel) for toluene oxidation. *Indus Engin Chem Res*. 1993 1993/11/01;32(11):2563-72.

[11] Teraoka Y, Nii H, Kagawa S, Jansson K, Nygren M. Influence of the simultaneous substitution of Cu and Ru in the perovskite-type (La,Sr)MO3 (M Al, Mn, Fe, Co) on the catalytic activity for CO oxidation and CO–NO reactions. *Appl Catalys A: General*. 2000 3/13/;194–195:35-41.

[12] Wu Y, Yu T, Dou B-s, Wang C-x, Xie X-f, Yu Z-l, et al. A comparative study on perovskite-type mixed oxide catalysts A′xA1 − xBO3 − λ (A′ = Ca, Sr, A = La, B = Mn, Fe, Co) for NH3 oxidation. *J Catalys*. 1989 11//;120(1):88-107.

[13] Tejuca LG, Fierro JLG, Tascón JMD. Structure and Reactivity of Perovskite-Type Oxides. In: D.D. Eley HP, Paul BW, editors. *Advances in Catalysis*. Volume 36: Academic Press; 1989. p. 237-328.

[14] Kharton VV, Viskup AP, Naumovich EN, Tikhonovich VN. Oxygen permeability of LaFe1−xNixO3−δ solid solutions. *Mater Res Bull*. 1999 10/8/;34(8):1311-7.

[15] Rao CNR, Gopalakrishnan J. *New Directions in Solid State Chemistry*. Cambridge University Press; 1997.

[16] Yu Z, Gao L, Yuan S, Wu Y. Solid defect structure and catalytic activity of perovskite-type catalysts La1-SrNiO3-[small lambda] and La1-1.333ThNiO3-[small lambda]. *J Chem Soc Faraday Trans*. 1992;88(21):3245-9.

[17] Popa M, Frantti J, Kakihana M. Lanthanum ferrite LaFeO3+d nanopowders obtained by the polymerizable complex method. *Solid State Ionics*. 2002 12/2/;154–155:437-45.

[18] Sumathi R, Johnson K, Viswanathan B, Varadarajan TK. Selective oxidation and dehydrogenation of benzyl alcohol on ABB′O3 (A=Ba, B=Pb, Ce, Ti and B′=Bi, Cu, Sb)-type perovskite oxides-temperature programmed reduction studies. *Appl Catalysis A: General*. 1998 8/24/;172(1):15-22.

[19] Pecchi G, Reyes P, Zamora R, Campos C, Cadús LE, Barbero BP. Effect of the preparation method on the catalytic activity of La1−xCaxFeO3 perovskite-type oxides. *Catalysis Today*. 2008 4//;133–135:420-7.

[20] Pecchi G, Jiliberto MG, Delgado EJ, Cadús LE, Fierro JLG. Effect of B-site cation on the catalytic activity of La1−xCaxBO3 (B = Fe, Ni) perovskite-type oxides for toluene combustion. *J Chem Technol Biotechnol*. 2011;86(8):1067-73.

[21] Pecchi G, Reyes P, Zamora R, Cadús LE, Fierro JLG. Surface properties and performance for VOCs combustion of LaFe1−yNiyO3 perovskite oxides. *J Solid State Chem*. 2008 4//;181(4):905-12.

[22] Pecchi G, Campos C, Peña O. Thermal stability against reduction of LaMn1−yCoyO3 perovskites. *Mater Res Bull*. 2009 4/2/;44(4):846-53.

[23] Pecchi G, Campos C, Peña O, Cadus LE. Structural, magnetic and catalytic properties of perovskite-type mixed oxides LaMn1−yCoyO3 (y = 0.0, 0.1, 0.3, 0.5, 0.7, 0.9, 1.0). *J Mol Catalys A: Chemical*. 2008 3/3/;282(1–2):158-66.

[24] Pecchi G, Campos CM, Jiliberto MG, Delgado EJ, Fierro JLG. Effect of additive Ag on the physicochemical and catalytic properties of LaMn0.9Co0.1O3.5 perovskite. *Appl Catalys A: General*. 2009 12/15/;371(1–2):78-84.

[25] Li Q, Meng M, Dai F, Zha Y, Xie Y, Hu T, et al. Multifunctional hydrotalcite-derived K/MnMgAlO catalysts used for soot combustion, NOx storage and simultaneous soot–NOx removal. *Chem Engin J*. 2012 3/1/;184:106-12.

[26] Li Z, Meng M, Dai F, Hu T, Xie Y, Zhang J. Performance of K and Ni substituted La1−xKxCo1−yNiyO3−δ perovskite catalysts used for soot combustion, NOx storage and simultaneous NOx-soot removal. *Fuel*. 2012 3//;93:606-10.

[27] Pecchi G, Cabrera B, Buljan A, Delgado EJ, Gordon AL, Jimenez R. Catalytic oxidation of soot over alkaline niobates. *J Alloys Compounds*. 2013 2/25/;551:255-61.

[28] Pecchi G, Cabrera B, Delgado EJ, García X, Jimenez R. Activity of KNbO3 as catalyst for soot combustion: Effect of the preparation method. *Appl Catalys A: General*. 2013 2/26/;453:341-8.

[29] Taniguchi K, Okinaka N, Akiyama T. Preparation and characterization of La1−xKxFeO3 (x = 0–1) by self-propagating high-temperature synthesis for use as soot combustion catalyst. *J Alloys Compounds*. 2011 3/10/;509(10):4084-8.

[30] Fino D, Russo N, Saracco G, Specchia V. The role of suprafacial oxygen in some perovskites for the catalytic combustion of soot. *J Catalys*. 2003 7/25/;217(2):367-75.

[31] Zhang R, Luo N, Chen B, Kaliaguine S. Soot Combustion over Lanthanum Cobaltites and related oxides for diesel exhaust treatment. *Energy Fuels*. 2010 2010/07/15;24(7):3719-26.

[32] Pecchi G, Morales R, Delgado EJ, Jimenez R, Fraga M. Catalytic combustion of soot onCe-doped Lanthanum Cobaltites. *J Chilean Chem Soc*. 2014;59:2725-30.

[33] Pecchi G, Dinamarca R, Campos CM, Garcia X, Jimenez R, Fierro JLG. Soot oxidation on silver-substituted LaMn0.9Co0.1O3 perovskites. *Indus Engin Chem Res*. 2014 2014/06/18;53(24):10090-6.

[34] Bui VN, Laurenti D, Afanasiev P, Geantet C. Hydrodeoxygenation of guaiacol with CoMo catalysts. Part I: Promoting effect of cobalt on HDO selectivity and activity. *Appl Catalys B: Environmental*. 2011 1/14/;101(3–4):239-45.

[35] Laurent E, Delmon B. Study of the hydrodeoxygenation of carbonyl, carylic and guaiacyl groups over sulfided CoMo/γ-Al2O3 and NiMo/γ-Al2O3 catalysts: I. Catalytic reaction schemes. *Appl Catalys A: General*. 1994 2/17/;109(1):77-96.

[36] Sepúlveda C, Escalona N, García R, Laurenti D, Vrinat M. Hydrodeoxygenation and hydrodesulfurization co-processing over ReS2 supported catalysts. *Catalysis Today*. 2012 11/15/;195(1):101-5.

[37] Gutierrez A, Kaila RK, Honkela ML, Slioor R, Krause AOI. Hydrodeoxygenation of guaiacol on noble metal catalysts. *Catalysis Today*. 2009 10/15/;147(3–4):239-46.

[38] Lee CR, Yoon JS, Suh Y-W, Choi J-W, Ha J-M, Suh DJ, et al. Catalytic roles of metals and supports on hydrodeoxygenation of lignin monomer guaiacol. *Catalys Commun*. 2012 1/5/;17:54-8.

[39] Nimmanwudipong T, Runnebaum RC, Block DE, Gates BC. Catalytic conversion of guaiacol catalyzed by platinum supported on alumina: Reaction network including hydrodeoxygenation reactions. *Energy Fuels*. 2011 2011/08/18;25(8):3417-27.

[40] Escalona N, Aranzaez W, Leiva K, Martínez N, Pecchi G. Ni nanoparticles prepared from Ce substituted LaNiO3 for the guaiacol conversion. *Appl Catalys A: General*. 2014 7/5/;481:1-10.

[41] Simonetti DA, Kunkes EL, Dumesic JA. Gas-phase conversion of glycerol to synthesis gas over carbon-supported platinum and platinum–rhenium catalysts. *J Catalys*. 2007 4/25/;247(2):298-306.

[42] Dieuzeide ML, Jobbagy M, Amadeo N. Glycerol steam reforming over Ni/γ-Al2O3 catalysts, modified with Mg(II). Effect of Mg (II) content. *Catalys Today*. 2013 9/15/;213:50-7.

[43] Iriondo A, Güemez MB, Barrio VL, Cambra JF, Arias PL, Sánchez-Sánchez MC, et al. Glycerol conversion into H2 by steam reforming over Ni and PtNi catalysts supported on MgO modified γ-Al2O3. In: E.M. Gaigneaux MDSHPAJJAM, Ruiz P, editors. *Studies in Surface Science and Catalysis*. Volume 175: Elsevier; 2010. p. 449-52.

[44] Franchini CA, Aranzaez W, Duarte de Farias AM, Pecchi G, Fraga MA. Ce-substituted LaNiO3 mixed oxides as catalyst precursors for glycerol steam reforming. *Appl Catalys B: Environmental*. 2014 4/5/;147:193-202.

[45] Mikkola J-P, Sjöholm R, Salmi T, Mäki-Arvela P. Xylose hydrogenation: kinetic and NMR studies of the reaction mechanisms. *Catalysis Today*. 1999 1/27/;48(1–4):73-81.

[46] Wisniak J, Hershkowitz M, Leibowitz R, Stein S. Hydrogenation of xylose to xylitol. *Product R&D*. 1974 1974/03/01;13(1):75-9.

[47] Wisniak J, Hershkowitz M, Stein S. Hydrogenation of xylose over platinum group catalysts. Product R&D. 1974 1974/12/01;13(4):232-6.

Chapter 18

Improvement of Catalytic Performance of Perovskites by Partial Substitution of Cations and Supporting on High Surface Area Materials

Fabio Souza Toniolo and Martin Schmal

Additional information is available at the end of the chapter

http://dx.doi.org/10.5772/61279

Abstract

In this chapter, we present two relevant strategies to improve the activity and selectivity of perovskite-mixed oxides ABO_3 in heterogeneous catalytic reactions such as the oxidation of hydrocarbons, soot combustion and CO selective oxidation, for which the surface sites and lattice oxygen species play important roles for the chemical transformations. Besides, we focus on synthesis of higher alcohols, partial oxidation of methane, oxidative reforming of diesel and dry reforming of methane for which the perovskite is a precursor that leads to a dispersed metal active phase over an oxide matrix. But which strategies are we talking about? First, the partial substitution of cations A and B by different elements, which change atomic distances, causes unit cell distortions, stabilizes multiple oxidation states or induces cationic or anionic vacancies within the lattice. And all these new features perturb the solid reactivity by changing the reaction mechanism on the catalyst surface. Thus, appropriate cations substitutions may lead to better catalysts. The second strategy comprises supporting the perovskite, which usually presents low surface area, on high surface area materials to maximize the exposed surface sites.

Keywords: Catalysis, Partial substitution, Vacancy, Monolith

1. Introduction

Perovskite-type oxides with ABO_3 structure have attracted significant interest in many areas of solid-state chemistry, including catalysis. Several ABO_3 formulations have demonstrated success, especially in total oxidation and partial oxidation of hydrocarbons, carbon monoxide oxidation, alkenes hydrogenation, alkanes hydrogenolysis, alcohols synthesis, dry reforming and water gas shift reaction. However, these materials present some limitations for broader application in catalysis such as low activity and stability in certain experimental conditions, low metal dispersion when the perovskite structure collapses to form a supported-type catalyst B/AO_x or even low surface area resulting from high calcination temperature.

To overcome these limitations aiming to improve the catalytic activity, selectivity and stability of perovskite-type oxides, one can design these materials by substituting partially and properly the cations A and B or even supporting the perovskite on high surface area materials (porous oxides, monolith-type structure). By substituting the original A and B cations, one can control the extent of substitution and decide for an appropriate cation that will bring significant structural changes, such as lattice distortions, stabilization of multiple oxidation states or generation of cationic and anionic vacancies, all having as a direct consequence the change in catalytic activity. Upon spreading the perovskite on a support, one can choose the best matrix to accommodate the oxide particles and expose the largest amount of active sites in order to improve the catalytic performance.

In this sense, we discuss in the next pages some recent advances that apply perovskite-type oxides for reactions of interest in catalysis and how the above strategies impact on the performance of the catalysts. All visited literature was critically analyzed and here we present well-structured discussions and hypothesis based on extensive experiments and precise characterization techniques.

2. Partial substitution of cations as a strategy to enhance catalytic performance

Perovskite-type oxides ABO_3 can be properly modified by the partial substitution of atoms at A and/or B-sites, creating isostructural $A_{1-x}A_xB_{1-y}B_yO_3$ crystals (or $AA'BB'O_3$ for simplicity), which may stabilize unusual oxidation states of the B component, induce structural distortions and create cationic or anionic vacancies. The nature and extent of the dopant that substitutes A and B elements impact on the physicochemical properties and catalytic activity of the material [1].

From the catalysis point of view, a solid containing surface and lattice defects presents more structural and electronic instability, which is compensated by the interaction with molecules from gas phase through chemical reactions that break and form bonds. Defects are welcome in catalysis, and the partial substitution of cations within the perovskite lattice tends to enhance the defects density of the crystal structure.

Perovskite oxides may also be considered as catalytic precursors for generating highly active metals dispersed onto a matrix, B/AO_x, that can be obtained *in situ* by reducing the precursor ABO_3 [2–4]. The partial substitution of cations can increase the dispersion of B in reducing treatments and avoid metal sintering during reaction conditions. In this sense, high metal dispersion from perovskite oxides emerges as an interesting alternative to conventional metal supported catalysts.

For instance, several lanthanum-containing perovskites have been considered promising for different applications such as soot combustion in automotive exhaust treatments, oxidation of volatile organic compounds (VOCs) in air pollution control and reforming or partial oxidation of hydrocarbons for hydrogen production [5]. Though perovskites have a remarkable activity and some desired features such as low-cost synthesis and high thermal stability, their industrial application is still missing due to drawbacks related to the low surface area and low density of structural defects, which both can be improved by selecting an appropriate synthesis method to (i) insert different cations into the crystal lattice and (ii) spread the oxide phase over a porous support.

2.1. Perovskites for oxidation reactions

An interesting study about the effect of partial substitution of single-site (A or B) and dual-site (A and B) in the $LaCoO_3$ perovskite on the activity for toluene oxidation and methane oxidation was presented in ref. [5]. This investigation focused onto the effect of single- and dual-site modifications, while most of the researches explain only single-site substitution. The oxidation reactions on perovskite-type oxides usually take place through different mechanisms, either involving weakly chemisorbed oxygen (i.e., α-O in the toluene oxidation, for instance) denoted as suprafacial mechanism, or comprising surface lattice oxygen species (i.e., α'-O in the methane oxidation) denoted as intrafacial mechanism. The latter is directly related to the mobility and amount of oxygen species within the crystal lattice. The following catalysts were prepared by continuous flow hydrothermal synthesis method [5]: $LaCoO_3$ (ABO_3), $LaCaCoO_3$ ($AA'BO_3$), $LaCoMgO_3$ ($ABB'O_3$), and $LaCaCoMgO_3$ ($AA'BB'O_3$). After heat treatments at 700, 800 and 900°C, these materials presented mostly the perovskite phase, but some traces of La_2O_3, Co_3O_4 and MgO (depending on the catalyst) and the specific surface area was lower than 30 m^2/g. Authors evidenced the calcium substitution by verifying distortions in the angle α of the rhombohedral structure of $LaCaCoO_3$, since a significant contraction of the unit cell should not be expected due to the similar ionic radii of Ca^{2+} (0.100 nm) and La^{3+} (0.103 nm). On the other hand, magnesium substitution led to more severe angular distortions and unit cell expansion, because Mg^{2+} has ionic radius (0.072 nm) larger than Co^{3+} (0.063 nm), as well as the oxidation state of Co^{3+} changed to Co^{4+} to guarantee the structural electroneutrality, as revealed by X-ray photoelectron spectroscopy (XPS).

By applying those materials in the oxidation of 500 ppm toluene (in the absence and presence of water 3 vol.%), which is believed to take place through the suprafacial mechanism, the activity sequence was $LaCaCoMgO_3$ > $LaCoMgO_3$> $LaCoO_3$> $LaCaCoO_3$. According to XPS measurements [5], the insertion of Ca^{2+} into the perovskite lattice slightly increased the amount of oxygen species chemisorbed on the surface (O^- and $O_2{}^-$, denoted as O_{ads}) relating to lattice oxygen (O_{lat}). However, Mg^{2+} insertion had a more relevant impact and practically doubled O_{ads}/O_{lat} ratio, suggesting that Mg^{2+} (with valence lower than Co^{3+} and La^{3+}) generated more oxygen vacancies than Ca^{2+} to ensure charge compensation. As a consequence, the surface vacancies were occupied by oxygen species from gas phase. According to the authors, these findings agreed with the temperature-programmed desorption of oxygen (O_2-TPD) which showed that Mg^{2+}-substituted materials released larger amount of weakly chemisorbed oxygen

(α-O), which were adsorbed on surface oxygen vacancies. Since toluene oxidation makes use of surface oxygen (suprafacial mechanism), we should expect $LaCaCoMgO_3$ and $LaCoMgO_3$ (which contain this type of oxygen) to present higher activity.

On the other hand, Ca^{2+} insertion decreased the catalyst reducibility (shifting reduction peaks to higher temperature in H_2-TPR experiments), which explained higher activity of $LaCoO_3$ instead of $LaCaCoO_3$, since the non-substituted perovskite would release surface lattice oxygen easier to oxidize toluene than $LaCaCoO_3$. Furthermore, the authors estimated the activation energy for the different catalysts in the temperature range of 140–280°C and found out that the dual-insertion of Ca^{2+} and Mg^{2+} into the $LaCoO_3$ perovskite induced the activation energy to decrease from 143 kJ/mol ($LaCoO_3$) to 34 kJ/mol ($LaCaCoMgO_3$), which is comparable or even smaller than some noble metals applied in this suprafacial oxidation process (31–37 kJ/mol on Au/$LaCoO_3$ [6]; 86 kJ/mol on Pt/Al_2O_3 [7].

For the oxidation of 1 vol.% CH_4 (in the absence and presence of water 3 vol.%), which is believed to occur by an intrafacial mechanism, ref. [5] found the sequence of the catalytic performance $LaCoMgO_3 > LaCaCoMgO_3 > LaCoO_3 > LaCaCoO_3$ to agree with the sequence of surface lattice oxygen (α'-O) released in O_2-TPD measurements, i.e., 19.9 > 13.6 > 6.2 > 2.7 μmol O_2/g, confirming the important role of surface lattice oxygen in this type of reaction. Calcium substitution had a negative impact on the generation of surface lattice oxygen (decreasing α'-O which releases during O_2-TPD). Differently, magnesium substitution increased considerably surface lattice oxygen species, which participates in methane oxidation reactions (intrafacial mechanism). The estimated activation energy within the temperature range 390–450°C reduced from 152 kJ/mol on $LaCoO_3$ to 80 kJ/mol on $LaCoMgO_3$, which is similar to noble metal-based catalysts ($BaZrMO_3$, M = Rh, Pd, Pt, Mn, Ni, Co, activation energy = 72–106 kJ/mol) [8].

Thus, the dual-site substitution of Ca^{2+} and Mg^{2+} was beneficial for suprafacial oxidation processes by increasing oxygen species chemisorbed on the surface oxygen vacancies, but had a less beneficial effect on intrafacial oxidation processes, for which only the single substitution of Co^{3+} by Mg^{2+} is highly indicated to increase the amount of surface lattice oxygen.

The work of Liu *et al.* [9] brings insights on the role of the partial substitution of La^{3+} by Ce^{4+} into $La_{1-x}Ce_xMnO_3$ ($0 < x < 0.1$) which is a very active perovskite for the oxidation of volatile organic compounds (VOCs). The enhancement in activity is usually explained in terms of oxygen excess in the lattice, cationic vacancies, structural defects and the presence of multiple Mn oxidation states (Mn^{3+}/Mn^{4+}). Inserting cerium into the perovskite structure is a great strategy due to the Ce^{3+}/Ce^{4+} redox behavior, which along with Mn^{3+}/Mn^{4+} may increase oxygen transfer within the lattice. These researchers found that the activity of $La_{1-x}Ce_xMnO_3$ perovskites (all presenting specific area approximately 25 m^2/g) in the oxidation of 1,000 ppm benzene at the temperature range of 100–450°C correlated directly with the extent of cerium substitution (x value), *i.e.*, the higher the cerium content, the higher the benzene conversion according to the sequence $x = 0 < 0.025 \approx 0.05 < 0.075 < 0.1$. But how to explain the promoter effect of cerium on the catalytic oxidation of benzene?

Firstly, authors confirmed cerium insertion into the perovskite lattice (using X-ray diffraction measurements, XRD) by verifying absence of segregated CeO_2 phase, presence of lattice distortions and the decrease of $La_{1-x}Ce_xMnO_3$ crystalline domain size as a function of cerium content. That should be expected since the higher Ce^{4+} coordination with their surrounding oxygen atoms (within the same crystal plan) than trivalent La^{3+} tends to inhibit crystal growth, resulting in smaller crystal size.

Then, XPS measurements revealed the existence of Ce^{3+} and Ce^{4+} in the perovskites with higher cerium content ($x = 0.05; 0.075; 0.10$), as well as an increasing Mn^{4+}/Mn^{3+} ratio and decreasing O_{ads}/O_{lat} ratio as a function of cerium substitution, apparently unexpected. Ce^{4+} substitution should induce Mn to have lower oxidation states, or at least decrease Mn^{4+}/Mn^{3+} ratio as a charge compensation mechanism. Similarly, in the presence of Ce^{4+} and Mn^{4+}, the O_{ads}/O_{lat} ratio should increase, representing more oxygen species adsorbed on the surface to compensate the positive cationic charges, but the opposite occurred. According to [9], these findings indicated that cerium substitution created non-stoichiometric Ce-Mn-O compounds on the perovskite surface, and plenty of oxygen was accommodated into the lattice decreasing O_{ads}/O_{lat} ratio as the cerium content increased. Releasable oxygen from lattice justifies the higher activity of cerium-substituted perovskites in the benzene oxidation, which is believed to involve oxygen species supplied by the reducible oxide structure. The H_2-TPR results supported this hypothesis by showing that cerium substitution increased the reducibility, specially at the temperature range of 300–580°C, suggesting that Ce^{4+} increased the amount of oxygen species within the lattice (to reach structural neutrality) and also their diffusion from bulk to the surface.

The reactivity of substituted lanthanum cobaltite perovskites in the carbon oxidation (a solid–solid–gas catalytic reaction) has also been studied [10]. Changes in the $LaCoO_3$ perovskite structure and the existence of gradient in the cobalt valence from bulk to surface were evaluated as a function of partial substitution of La^{3+} by Sr^{2+} and temperature reaction. $LaCoO_3$ and $La_{0.5}Sr_{0.5}CoO_3$ were prepared by spray pyrolysis method (specific surface area < 14 m^2/g), and then ground with carbon at catalyst:carbon = 9:1 ratio (w/w) to give a tight contact. O_2-TPD measurements revealed that the maximum oxidation rate of carbon occurred at 650°C in the absence of catalyst, but $LaCoO_3$ perovskite reduced this temperature to 540°C, and then Sr^{2+} substitution ($La_{0.5}Sr_{0.5}CoO_3$) led to the lowest temperature 510°C. CO and CO_2 were the byproducts in the carbon oxidation. The direct oxidation (absence of catalyst) led to similar amounts of CO and CO_2 (50%/50%), but $LaCoO_3$ and $LaSrCoO_3$ were more selective to CO_2, 89% and 94%, respectively. To explain the improvement in the activity and selectivity, the authors investigated the cations oxidation state and composition of the catalytic systems by using EELS (electron energy loss spectroscopy, which is a bulk sensitive analysis) and XPS (surface sensitive), before, at an intermediate temperature, and after the carbon oxidation process. The conclusions are illustrated in Figure 1, which shows a catalyst–carbon particle model for the catalytic phenomenon.

The partial substitution of La^{3+} by Sr^{2+} created plenty of oxygen vacancies within the perovskite lattice (as a charge compensator mechanism), which also caused the cobalt oxidation state to decrease from Co^{3+} (in $LaCoO_3$) to values between +2.5 and +2.7 (in $La_{0.5}Sr_{0.5}CoO_3$). Authors [10] estimated that the as-prepared Sr-substituted perovskite ($La_{0.5}Sr_{0.5}CoO_3$) presented 50%Co^{3+}

and 50%Co^{2+} (Figure 1a), which was consistent with the unit cell expansion due to the higher ionic radius of Co^{2+} (0.074 nm) related to Co^{3+} (0.063 nm). Surprisingly, the tight contact between carbon and perovskite at room temperature immediately decreased Co^{3+}/Co^{2+} ratio in the bulk, as well as enhanced Co^{2+} (60–70%) on the surface (Figure 1b). This partial reduction of cobalt in the bulk and especially on the catalyst surface suggests oxygen mobility from bulk to surface and a remarkable surface reactivity for carbon oxidation, even at room temperature, which agreed with the increase of carbonate species on the surface, as seen by XPS.

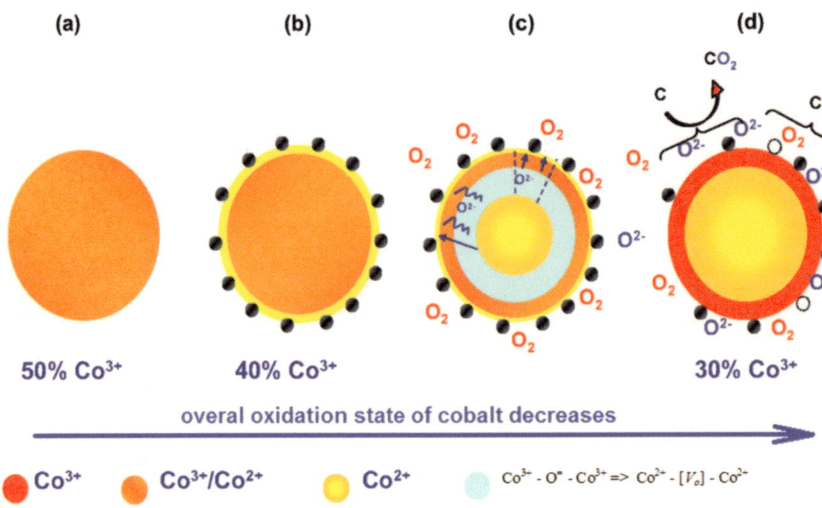

Figure 1. Scheme of cobalt oxidation states distribution within the perovskite $La_{0.5}Sr_{0.5}CoO_3$ as a consequence of the interaction and reaction with carbon particulates: (a) perovskite as prepared (without carbon); (b) after grinding carbon and catalyst at room temperature; (c) heating treatment at 400°C in the presence of 0.5 Torr O_2; (d) final state of the perovskite after removal of carbon by prolonged thermal oxidation. The expressed percentages of cobalt were deduced from EELS calculations. Reprinted from reference [10], copyright (2008) with permission from Elsevier Inc.

As depicted in Figure 1c, by increasing the temperature reaction in the presence of oxygen, the carbon oxidation activity enhanced because O^{2-} species migrated from bulk to surface promoting surface carbonates elimination (as proved by XPS). For this reason, the overall oxidation state of cobalt decreased by the end of reaction and the temperature for the maximum oxidation rate of carbon reduced by 140°C (from 650°C in the absence of catalyst to 510°C on $LaSrCoO_3$). In the presence of oxidative atmosphere, carbon particulates on the catalyst surface were oxidized by O_2 from gas phase and also by lattice O^{2-} anions leading to the consumption of lattice oxygen, which decreased the overall oxidation state of cobalt. After complete carbon oxidation, the perovskite surface presented abundance of Co^{3+} (90–95%) while bulk showed oxygen deficit (Co^{2+}-rich) as shown in Figure 1d, due to the diffusion of oxygen species through the structural vacancies towards surface or re-establishment of the surface lattice with oxygen from gas phase. The overall reaction (1) represents the interaction between the perovskite surface and carbon particles, summarizing the phenomenon in Figure 1 ($[V_o]$ means an anionic vacancy):

$$\left(Co^{3+} - O^{=} - Co^{3+}\right) + C \rightarrow \left(Co^{2+} - \left[V_o\right] - Co^{2+}\right) + CO_x \quad (1)$$

Therefore, Sr^{2+} substitution into the perovskite lattice created oxygen vacancies, increasing cobalt reducibility and enhancing the mobility of lattice oxygen species, which were an additional source of oxygen for the carbon oxidation towards CO_2. But the partial substitution of A and/or B sites into the ABO_3 mixed oxide may lead to consequences other than the generation of anionic vacancies and stabilization of unusual oxidation states, such as the stabilization of the crystal lattice avoiding the segregation of undesired phases (which worsen the catalytic performance), as reported in the reference [11] for $LaCoO_3$ substituted by K^+ and Mg^{2+} applied for the soot oxidation.

$La_{1-x}K_xCo_{1-y}Mg_yO_3$ perovskites ($x = 0 – 0.4$; e $y = 0 – 0.2$) prepared by citric acid complexation [11] showed an increasing segregation of Co_3O_4 as La^{3+} was substituted by K^+ (no Mg substitution then). The phase segregation can be explained because of the larger ionic radius of K^+ (0.133 nm) than La^{3+} (0.103 nm), which caused severe lattice distortions and unstable occupancy of the lanthanum site by K^+ leading to cobalt to segregate as Co_3O_4. However, the increasing Mg^{2+} substitution into the perovskite containing the higher amount of K^+ (*i.e.*, $x = 0.4$: $La_{0.6}K_{0.4}Co_{1-y}Mg_yO_3$) led to a gradual diminishment of segregated Co_3O_4, since the larger ionic radius of Mg^{2+} (0.072 nm) than Co^{3+} (0.063 nm) gave more stability for the crystal lattice avoiding cobalt segregation. This hypothesis was confirmed by XRD, FTIR, TEM and EDS results which evidenced Co_3O_4 phase decreasing as a function of Mg substitution. By replacing both cations K^+ and Mg^{2+} into the perovskite, Co^{4+} was detected by XPS, as a charge compensator mechanism. According to ref. [11], the activity of the perovskites in the oxidation of carbon (using synthetic air 20 vol.% O_2/N_2 and tight catalyst/carbon contact) was established as a function of the temperature for the maximum oxidation rate (T_m). The non-catalyzed reaction (absence of catalyst) presented T_m = 590°C, but the perovskite $LaCoO_3$ led to T_m = 448°C, which gradually decreased to T_m = 370°C with increasing substitution of La^{3+} by K^+ within the perovskite. By inserting Mg^{2+} into the B site of $La_{0.6}K_{0.4}Co_{1-y}Mg_yO_3$, the structure stability increased and cobalt segregation decreased (ensuring the existence of anionic vacancies useful for oxygen mobility) and as a consequence, the lowest T_m = 359°C was reached on $La_{0.6}K_{0.4}Co_{0.9}Mg_{0.1}O_3$. Therefore, the improvement in the carbon oxidation followed exactly the increasing amount of K^+ up to $x = 0.4$ (into A site) and increasing amount of Mg^{2+} up to $y = 0.1$ (into B site), which is very consistent with the increasing reducibility of the perovskites (and redox properties) as a function of the partial substitution of K^+ and Mg^{2+} (as evidenced by H_2-TPR measurements). In particular, Mg^{2+} stabilized the crystal lattice and reduced cobalt segregation. That ensured the integrity of the perovskite oxide structure and the mobility of lattice oxygen towards surface which enhanced carbon particulate oxidation.

2.2. Perovskites for CO selective oxidation

Carbon monoxide selective oxidation reaction (CO-SELOX) is an interesting and economic approach to remove CO from H_2-rich gas streams, but a selective catalyst should avoid H_2 consumption since CO and H_2 oxidation reactions compete each other. Pt-supported catalysts are typically used because CO adsorbs more strongly on Pt surface than H_2. Other classes of CO selective oxidation catalysts comprise non-noble metals such as Ag, Au and also oxides, e.g., CuO-CeO_2, CeO_2, Cu-LaO_2-CeO_2 as well as cobalt-based oxides. Literature reports bulk and supported cobalt oxides applied for SELOX, but rarely cobalt-based perovskites. The main constraint for cobalt oxides application is their low stability in H_2-rich stream and consequent metallic cobalt formation on the catalyst surface which catalyzes undesirable reactions such as H_2 oxidation and methanation. For instance, CoO_x/ZrO_2 showed high CO conversion at 175°C, but above this temperature H_2 oxidation took place, and methanation reaction occurred above 250°C [12].

For this reason, a suitable catalyst for CO selective oxidation should not only be able to keep the oxide structure during the process but also allow oxygen mobility through the crystal lattice in order to enhance the CO oxidation rates towards CO_2 on the surface active sites. Cobalt-containing perovskites seem to satisfy all these requirements.

Magalhães *et al.* [13] applied Ce-substituted $LaCoO_3$ perovskites in the CO-SELOX reaction in H_2-rich feedstream (1% CO, 1% O_2, 60% H_2, He balance, vol.%) and found out the new structures containing cerium increased the resistance to undesirable reactions and did not damage the high selectivity to CO_2 presented by the non-substituted perovskite. They obtained single phase $La_{1-x}Ce_xCoO_3$ perovskites (x = 0; 0.05 and 0.10) with low surface area (< 15 m^2/g) and verified, by Rietveld refinements, that the Ce^{4+} substitution into La^{3+} sites decreased both cell parameters of the rhombohedral unit cell and crystalline domain sizes, since the ionic radius of Ce^{4+} is smaller than La^{3+}, exactly as verified by ref. [9] for $LaCeMnO_3$. Figure 2 illustrates the effect of cerium insertion into the perovskite on the CO conversion to CO_2 and H_2O byproducts during temperature-programmed reactions: region 1 evidenced the selective CO oxidation in which CO and O_2 were exclusively converted to CO_2, and one can note that the presence of cerium affected the reaction kinetics shifting CO conversion to higher temperatures. In region 2 the undesired H_2 oxidation gained importance leading to water formation, and finally methanation reaction occurred at higher temperatures at region 3 (leading to CH_4 and H_2O formation). Though cerium substitution shifted the range of temperature for CO-SELOX reaction up to 240°C, all the undesired reactions also shifted to higher temperature.

Authors [13] affirmed that Ce^{4+} changed some Co^{3+} to Co^{2+} to keep the charge neutrality within the structure, consequently decreasing the amount of active Co^{3+} sites on the surface and decreasing the activity for CO conversion (for this reason, $La_{0.90}Ce_{0.10}CoO_3$ showed CO conversions at higher temperature than $LaCoO_3$ in Figure 2). However, the formation of cationic vacancies into A and B sites as a second mechanism to keep the charge neutrality would stabilize the entity Co^{3+}/O_2 on the surface, ensuring high CO_2 selectivity for cerium-substituted perovskites. At temperatures higher than 350°C, Co^{n+} species coexist with metallic Co^0 which leads to undesirable methanation reaction. This hypothesis was supported by H_2-TPR experiments that showed perovskite reduction above 300°C (i.e., $T_{reduction}$ > 300°C). These results clearly evidenced that Co^{3+} as part of a perovskite structure is more resistant to reduction when compared to typical supported cobalt catalysts, such as CoO_x/CeO_2 ($T_{reduction}$ > 275°C) [14], CoO_x/ZrO_2 ($T_{reduction}$ > 250°C) [12] and $Co_3O_4/CeZrO_2$ ($T_{reduction}$ > 200°C) [15].

http://dx.doi.org/10.5772/61279

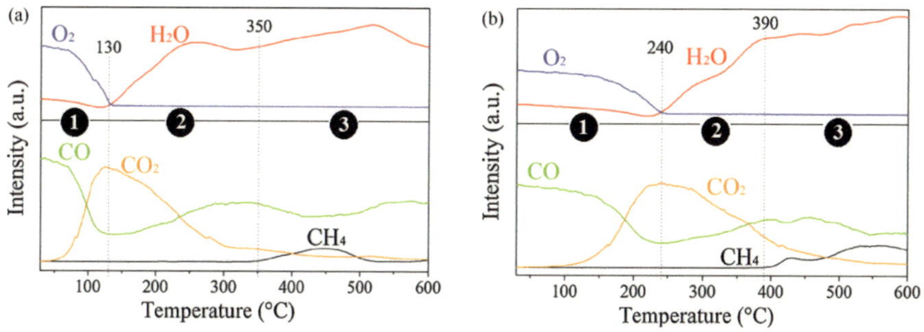

Figure 2. Effect of temperature and cerium substitution on compounds distribution in the CO-SELOX reaction for (a) $LaCoO_3$ and (b) $La_{0.90}Ce_{0.10}CoO_3$ (1% CO, 1% O_2, 60% H_2, He balance, vol.%). Reprinted from reference [13], copyright (2010) with permission from Elsevier B.V.

Schmal *et al.* [16] investigated the effect of strontium insertion into $La_{1-x}Sr_xCoO_3$ ($x = 0$; 0.05) on the catalytic performance in the CO-SELOX reaction. Differently from cerium, Sr^{2+} as a cationic dopant is expected to increase cobalt oxidation state and/or create oxygen vacancies within the crystal lattice, which could enhance oxygen mobility and provide lattice oxygen for CO oxidation on the surface. These researchers found out evidences for complete Sr^{2+} substitution into the rhombohedral crystalline structure of the perovskite (by means of XRD and Raman spectroscopy), which led to distortions and expansion of the unit cell volume, since the ionic radius of Sr^{2+} (0.132 nm) is larger than La^{3+}. XPS indicated that Sr^{2+} did not modify the cobalt oxidation state (Co^{3+}) in $La_{0.95}Sr_{0.05}CoO_3$, suggesting that anionic vacancies have been created within the lattice as a charge compensator. This hypothesis was supported by the enhancement in both activity to CO conversion and selectivity to CO_2 after Sr^{2+} substitution: the temperature to convert 50% of CO decreased from 170°C ($LaCoO_3$) to 155°C ($La_{0.95}Sr_{0.05}CoO_3$), and selectivity to CO_2 (at 29% isoconversion of CO) increased slightly from 87% to 91%. Interestingly $La_{0.95}Sr_{0.05}CoO_3$ was more stable than $LaCoO_3$ on stream at 170°C, presenting constant CO conversion 75% for 48 h, while the non-substituted perovskite showed decreasing CO conversion from ca. 75% to 45% at the same conditions.

How to explain that? The dopant Sr^{2+} caused distortions in the lattice due to its size and oxidation state, giving rise to anionic (oxygen) vacancies, which initially neutralized the structure charge but then enhanced the oxygen transfer through the lattice, i.e., from bulk to surface during reaction (indeed, the full redox process also comprises the re-establishment of lattice oxygen from oxygen coming from gas phase). As a consequence, this oxygen supplying from the lattice enhanced the CO oxidation on the catalyst surface.

A short communication [17] highlighted the influence of cerium substitution in $La_{1-x}Ce_xNiO_3$ ($x = 0$; 0.05; 0.1) on the catalysts properties and performance in the CO-SELOX reaction (2.5% CO, 5% O_2, 33% H_2 and N_2 balance, vol.%). The surface area of the catalysts with rhombohedral structure did not change with cerium addition, which usually takes place only at Ce content above $x > 0.4$ simultaneously to the appearance of segregated CeO_x phases [18]. CO conversion was higher for the cerium-doped perovskites than for $LaNiO_3$ at the entire range of 150–230°C, and undesirable H_2 oxidation took place only above 230°C (H_2 conversion < 12%). The authors discussed the promoter effect of cerium in terms of (i) very active surface Ce^{3+}/Ce^{4+} redox cations that directly adsorbs and activates O_2, (ii) stabilization of reduced Ni^{n+} species ($n < 3$), not discarding (iii) the enhanced lattice oxygen mobility from bulk to surface supplying oxygen for the CO oxidation.

Literature reports that CO oxidation over perovskites proceeds through a suprafacial mechanism involving weakly chemisorbed oxygen. The following mechanistic proposal has been suggested by Tascon *et al.* [19] to explain CO oxidation on $LaCoO_3$: molecular oxygen chemisorbs on Co^{2+} cations as an $O_2^{\,-}$ anion, dissociating to form atomic oxygen (O^-) on the cobalt sites. Simultaneously, CO chemisorbs on surface oxide ions yielding a labile species that interacts with adsorbed atomic oxygen, producing carbonates (the rate-determining step) which decompose towards CO_2 and oxygen.

$$O_{2(g)} \rightarrow O_2^{-}{}_{(ad)} \rightarrow 2O^{-}{}_{(ad)} \tag{2}$$

$$CO_{(g)} \rightarrow CO_{(ad)} \tag{3}$$

$$CO_{(ad)} + 2O^{-}{}_{(ad)} \rightarrow CO_3^{2-}{}_{(ad)} \text{ rate-limiting step} \tag{4}$$

$$CO_3^{2-}{}_{(ad)} \rightarrow CO_{2(ad)} + O^{2-}{}_{(ad)} \rightarrow CO_{2(g)} + O^{2-}{}_{(ad)} \tag{5}$$

Though the mentioned mechanism (equations (2)–(5)) does not clarify about the role of the lattice oxygen within the elementary steps, we may infer that those mobile oxygen species, specially the surface lattice oxygen, participate more easily in the redox process and contribute in some extent into the elementary step (4).

2.3. Perovskite as a catalytic precursor

Considering only geometric factors, the Goldschmidt tolerance factor (t) for perovskites indicates that lanthanum, which is the largest lanthanide ion in the series, leads to the most stable perovskite structure [20]. However, the thermal stability of ABO_3 perovskites strongly depends on cations at both positions A and B. Particularly, the thermal stability in a reducing atmosphere such as H_2, during reduction treatments or redox cycles, has been the focus of research because the metal B can reach high degree of dispersion over an AO_x matrix when Figure 2 these treatments. Considering the importance of dispersed metals in heterogeneous catalysis, the reduction or reduction–oxidation of perovskites under controlled conditions and atmosphere may offer a promising methodology for the preparation of highly active and dispersed catalysts [20]. In this sense, we can face a perovskite-type oxide as a catalytic precursor for a dispersed metal catalyst.

Copper-substituted materials have been widely investigated in the literature, not only in perovskite oxides, since copper may modify the carbon formation mechanism avoiding coke deposition. Tien-Thao *et al.* [21] verified that replacing cobalt by copper into $LaCo_{1-x}Cu_xO_3$ perovskites ($0 < x < 0.6$; specific surface area 10–60 m^2/g) increased the reducibility and caused strong cobalt–copper interaction that enhanced metallic dispersion of cobalt and prevented metal sintering. The main catalytic consequence of Cu^{2+} substitution was the notable selectivity to higher alcohols in the syngas conversion ($H_2/CO = 2$ diluted to 20 vol.% in He, 1000 psi, 275°C) by using the perovskites partially reduced in H_2. According to these authors, X-ray diffraction of the as-prepared materials proved the copper substitution due to the clear structure distortions, which affected the redox properties and stability of these mixed oxides. Temperature-programmed desorption of O_2 showed large amount of oxygen releasing from Cu^{2+}-substituted perovskites, reflecting a high concentration of surface oxygen vacancies (Co^{3+} substituted by Cu^{2+} creates anionic vacancies for charge compensation). The partially reduced $LaCo_{1-x}Cu_xO_3$ perovskites were tested in the syngas conversion and yielded a homologous series of linear hydrocarbons (from methane to undecane) and a set of linear primary alcohols with the chain growth probability values ranging from 0.34 to 0.42.

The authors of ref. [21] expressed they did not expect to have higher alcohols on the cobalt-based perovskites, since cobalt is indeed known as a good Fischer-Tropsch catalyst (forming higher hydrocarbons) but not active for higher alcohols. However, the basic properties of the amorphous matrix La_2O_3 catalyzed or promoted in some extent the synthesis of these alcohols (Fischer-Tropsch metal catalysts can yield alcohols when promoted with a base). And at the reaction conditions, dispersed metallic cobalt and copper species were expected to coexist over intermediate oxides ($La_2Co_{0.75}Cu_{0.25}O_{4\pm\delta}$ and La_2O_3) as evidenced by XRD for samples partially reduced in H_2. Furthermore, Tien-Thao *et al.* [21] emphasized the selectivity to higher alcohols showed to be dependent on the content of intra-lattice copper in the perovskite. The selectivity to alcohols (36.5–49.5%) and productivity (a maximum at 49.6 mg/(g_{cat}.h) for $LaCo_{0.7}Cu_{0.3}O_3$) indicated that by increasing copper content, the selectivity to alcohol (C_2 $^+$OH) also increased simultaneously with a decrease in the selectivity to methane and C_2 $^+$ hydrocarbon. Those authors stated that a uniform distribution of the metallic cobalt–copper on the catalyst surface, as well as a high metal dispersion, seem to be the key for the synthesis of higher alcohols, but they did not prove that copper increased the cobalt dispersion. However, this kind of evidence was discussed in ref. [22] in the reduction of cobalt-based perovskites substituted with copper.

$LaCoO_3$ and $LaCo_{0.8}Cu_{0.2}O_3$ perovskite oxides with low surface area (< 10 m^2/g) were obtained by the polymerizable complex route, which is based on the polyesterification between citric acid and ethylene glycol in solution containing the soluble precursor anions. Then, the authors of ref. [22] confirmed the partial substitution of copper due to the increase in cell parameters caused by the larger ionic radius of the hexacoordinated Cu^{2+} and also due to the absence of segregated copper phases. By applying these perovskites in the partial oxidation of methane $CH_4/O_2/He = 2/1/37$ and 5/1/64, they verified a dynamic structural transformation in which the perovskite collapsed towards lanthanum-based matrix and metallic cobalt and copper species. The perovskites showed comparable performance up to 650°C, but remarkable differences in between 600 and 900°C in which $LaCoO_3$ favored CH_4 dissociation leaving carbon on the surface, while the copper-substituted perovskite was more efficient to produce syngas and suppress carbon deposition (thermogravimetry showed up to 18.5 times less carbon on $LaCo_{0.8}Cu_{0.2}O_3$ than on $LaCoO_3$). In this study, addition of copper increased the reducibility of the perovskite, decreased the temperature to obtain syngas and inhibited carbon deposition, and authors attributed that to a higher cobalt dispersion caused by copper. And how did they prove that?

According to their results, the presence of copper doubled cobalt dispersion and metal specific area, S_m, ($S_m = 6.2\ m^2/g_{Co}$ and $12.1\ m^2/g_{Co}$ for $LaCoO_3$ and $LaCo_{0.8}Cu_{0.2}O_3$, respectively) since the amount of H_2 able to chemisorb on the surface doubled on the sample with Cu as seen in Figure 3 during H_2 chemisorption measurements. However, H_2 does not chemisorb on copper at such conditions [23], and therefore this finding suggests a higher cobalt dispersion promoted by Cu due to a strong cobalt–copper interaction as proposed by Tien-Thao *et al.* [21] for similar materials. Therefore, copper may be a very important dopant for perovskites applied in reactions with hydrocarbons. Differently from catalytic systems involving transition metals (iron, nickel or cobalt), copper does not tend to catalyze Fisher-Tropsch reactions or processes involving carbonium ion chemistry due to its low activity for breaking C–O bonds or forming C–C bonds, which avoids both wax formation in CO/H_2 reactions and coke formation from hydrocarbons [24].

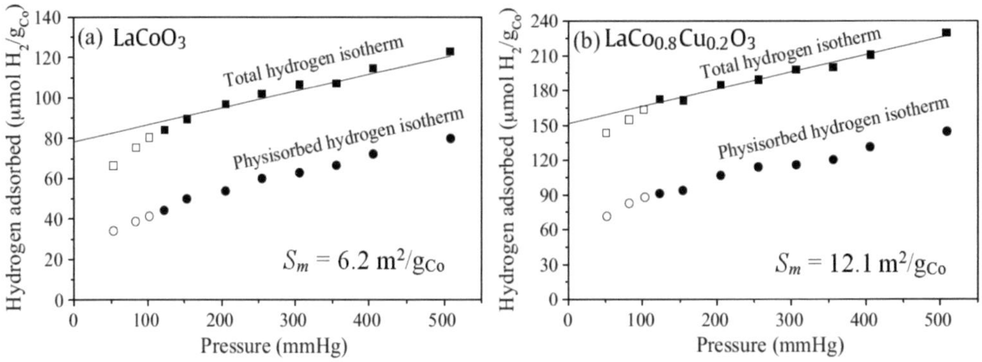

Figure 3. Hydrogen adsorption on (a) $LaCoO_3$ and (b) $LaCo_{0.8}Cu_{0.2}O_3$ after complete reduction of the perovskite structure. Symbols: (■,□) total H_2 isotherm; (●,○) reversible H_2 isotherm. In the literature, total H_2 uptake is reported to be used for quantifying chemisorbed H_2 over cobalt, since the process is activated at 150°C [4, 25–27]. Extrapolation of the straight-line portion of the total adsorption isotherm to zero pressure gives the chemisorbed H_2 over cobalt, and full symbols represent the data used to calculate dispersion and metal specific area. Approximately double H_2-uptake can be observed for copper-substituted perovskite. Reprinted from reference [22], copyright (2012) with permission from Elsevier B.V.

Reforming and partial oxidation of hydrocarbons aiming syngas production still find the deactivation as the main challenge to be overcome. Noble metals such as Rh, Ru, Pd, Pt and Ir are very active and more stable against coke deposition and metal sintering than Ni, but the latter is highly available and presents a lower price, which makes Ni more appropriate for industrial applications. By considering perovskite-type oxides ABO_3, nickel can occupy the B position and also be partially substituted by another metal in order to obtain a structure more stable, resistant to carbon deposition and more active ($ABB'O_3$). As discussed previously, the treatment under reducing atmosphere may cause a collapse of the oxide structure, leading the B and B' metals to be dispersed on the surface of AO_x. The dopant B' may contribute to increase dispersion of B, but B' can also be an active phase.

Ruthenium-substituted $LaCoO_3$ perovskites were investigated as catalytic precursors in the oxidative reforming of diesel for hydrogen production [28] and authors concluded that Co substitution by Ru (i) increased the material reducibility, (ii) led to smaller particle size of La_2O_3 and metallic Co^0 phases and (iii) increased the amount of Ru atoms on the catalyst surface, affecting directly the catalytic behavior: the greater Co^0 + Ru^0 exposition and higher extension of the $La_2O_2CO_3$ phase derived from the perovskite structure, the higher the activity and stability of the catalysts.

The effect of replacing nickel by zinc in $LaNi_{1-x}Zn_xO_3$ perovskites ($0 < x < 1$) on the methane dry reforming ($CH_4/CO_2/He$ = 1/1/1, 750°C) was studied in ref. [29], showing that, for $x \leq 0.4$, the partial substituted oxides were highly active towards syngas production. Particularly, $LaNi_{0.8}Zn_{0.2}O_3$ presented the highest CH_4 and CO_2 conversions and also the highest resistance to carbon deposition among all the catalysts. Characterization before and after catalytic evaluation showed decomposition of the rhombohedral structure of the perovskite towards metallic Ni^0 and hexagonal La_2O_3 phases (for high Zn content, ZnO and La_2NiO_4 phases also emerged). Authors found out that the partial substitution of Ni by Zn in $LaNiO_3$ increased the reduction temperature of the perovskite, suggesting a more stable structure under reaction conditions which could avoid Ni sintering. For this reason $LaNi_{1-x}Zn_xO_3$ samples ($x \leq 0.4$) presented activity and stability improvements in the methane dry reforming, opposite to the lower activity shown by $LaNiO_3$ and Zn high-content perovskites (the latter containing insufficient active nickel phase). Other investigators claim that $LaNiO_3$-derived catalysts form oxycarbonate species such as $La_2O_2CO_3$ during methane dry reforming, due to CO_2 adsorption on La_2O_3, and these species electronically stabilize nickel particles inhibiting metal sintering [30].

Therefore, the above discussion comprises the application of perovskites for important catalytic reactions and points out the efforts made to better comprehend the role of the active sites on perovskite-type oxides. Investigations focus on elucidating why the *substituting cations* enhance the activity and stability of this type of oxide structure or how they modify the characteristics of the supported metal catalysts obtained from the precursor perovskite under reducing treatments.

More examples are found in details in the literature: [31] studied $LaSrCoO_3$ perovskites and concluded that Sr substitution could increase the amount of chemisorbed oxygen species over the perovskite, improving the catalytic performance in the toluene oxidation. Refs. [32] and [33] showed that by substituting La^{3+} by Ca^{2+} into $LaFeO_3$ lattice, iron valence changed from Fe^{3+} to Fe^{4+}, improving the catalytic performance in oxidation reactions. Similar substitution of La^{3+} by Ca^{2+} into $LaCoO_3$ increased the surface oxygen vacancy density, yielding higher catalytic activities in the propane oxidation [34]. A series of B site substitutions over $LaCoO_3$

perovskite showed that Mn^{2+}, Fe^{2+}, Ni^{2+}, Cu^{2+} dopants could enhance CO oxidation [35]. All these interesting researches illustrate the relevance of perovskites as catalytic precursors or as active oxides for direct application in heterogeneous catalysis. Efforts on *in situ* studies are a strong tendency and will bring important insights about the physicochemical properties and dynamic transformations of the perovskites under relevant conditions in catalysis.

3. Supporting perovskites on high surface materials as a strategy to enhance catalytic performance

Bulk perovskites prepared via conventional procedures exhibit rather surface area lower than 30 m^2/g [36], which strongly limits the application of these materials as catalysts. The $LaCoO_3$ reported in [37] showed high CO oxidation activity, but authors observed a decrease in the catalytic activity for samples synthesized at high temperatures, which was ascribed to the abrupt loss of surface area. An alternative to overcome this drawback is to support perovskites on traditional porous solid matrices such as Al_2O_3 and SiO_2 in order to spread the perovskite particles and increase the exposed active sites.

$LaCoO_3$ perovskite (low surface area < 10 m^2/g) has been supported on alumina, x%$LaCoO_3/\gamma$-Al_2O_3 (x = 10, 20, 40 wt.%) by using a physical mixture and thermal treatment [38], leading to materials with higher surface area: 146, 131 and 96 m^2/g, respectively. As the perovskite load increased on alumina (164 m^2/g), the overall surface area and mesopore volume decreased due to blocking of support pores. The 40%$LaCoO_3/\gamma$-Al_2O_3 catalyst presented the best performance in the CO-SELOX reaction showing the lowest temperature for 50%CO conversion $T_{50\%CO}$ = 168°C, while bulk $LaCoO_3$ had intermediate activity with $T_{50\%CO}$ = 240°C. The higher effective exposition of the perovskite active phase over the alumina support explains the catalytic performance. The authors estimated the metal cobalt surface from H_2-chemisorption experiments as an indirect measurement of the oxide perovskite dispersion on the alumina surface (since the active catalyst in the CO-SELOX reaction is indeed the oxide perovskite and not the reduced structure, a H_2-reduction pretreatment was exclusively performed only before H_2-chemisorption measurements, and not before the catalytic tests). The supported 40%$LaCoO_3/\gamma$-Al_2O_3 perovskite had ca. twice as many metallic area than $LaCoO_3$ (13.5 and 6.2 m^2/g_{Co}, respectively), which suggests that the particles of oxide perovskite over the alumina-support also presented approximately twice exposed surface area than the bulk perovskite. For this reason, 40%$LaCoO_3/\gamma$-Al_2O_3 showed higher CO conversion. Besides, the catalyst with higher load of perovskite showed the highest selectivity to CO_2 (>75%) at the temperature range of 100–170°C, above that the H_2 oxidation side reaction also took place decreasing CO_2 selectivity for all the catalysts. By comparing the performance of 40%$LaCoO_3/\gamma$-Al_2O_3 with noble metal-based catalysts, this supported perovskite showed higher selectivity to CO_2 (i.e., 75% at $T_{50\%CO}$ = 168°C) than 1wt% Pt/Al_2O_3 (46%) evaluated in the CO-SELOX at similar conditions of space velocity and isoconversion [39]. The physical mixture between support and perovskite and thermal treatment as described in ref. [38], allowed a tight contact between those phases, but not a uniform spreading of the perovskite over the alumina as revealed by electron microscopy and energy-dispersive X-ray spectroscopy (EDS). In this sense, new methodologies should be investigated in order to improve the formation of smaller perovskite particles and their uniform distribution on the support. Simple methods such as precipitation, coprecipitation and wet impregnation should be considered.

Once again, these findings indicate that supported perovskites may be a very good alternative for CO-SELOX reaction. Other investigations also took advantage of supporting perovskite oxides on Al_2O_3-La_2O_3 [40], $Ce_{1-x}Zr_xO_2$ [41], ZrO_2 [42] and MCM-41 molecular sieve [43] to enhance the catalytic performance by increasing the number of exposed perovskite active sites.

Also attempts to support perovskites onto a monolithic structure have been reported, onto a metallic [44–46] or ceramic one [35, 46–48]. The ceramic monoliths are usually made of synthetic cordierite, which is wash-coated with γ-Al_2O_3 to increase the contact area between gas phase and solid surface. Monolith, by itself, presents several advantages over conventional catalysts, such as good mass transfer, easier product separation, good thermal and mechanical properties, easy scale up and low pressure drop, making this structure particularly suitable for high gas space velocities. Compared to conventional powdered perovskites used in fixed-bed reactors, monolith-supported perovskites have some unquestionable advantages: the pressure drop remains very low for high-cell density monoliths, and the desired extent of conversion is obtained with a smaller amount of perovskite.

Brackmann *et al.* [47] coated the cordierite substrate with γ-Al_2O_3 by repeating six times the wash-coating procedure and obtained 10 wt.% alumina well dispersed on the cordierite. The alumina had high adherence, since negligible mass loss was verified after the alumina/cordierite pieces underwent ultrasonic bath for 30 min. Then, $LaCoO_3$ synthesized by Pechini route was deposited on?γ-Al_2O_3/cordierite by successive dip coatings, in which 11 immersion cycles led to 10 wt.% mass increment on the support. This technique was efficient for depositing $LaCoO_3$ perovskite on γ-Al_2O_3/cordierite by avoiding formation of other oxides and by promoting a good adherence of the perovskite phase, since only 3.5% mass loss was verified after pieces underwent ultrasonic bath for 30 min. The pH of the perovskite suspension in the presence of γ-Al_2O_3/cordierite was properly controlled to increase electrostatic repulsion among particles and avoid agglomeration. This procedure tends to promote a uniform distribution of the perovskite on the γ-Al_2O_3/cordierite. Figure 4 shows pieces of the cordierite structure coated with γ-Al_2O_3 and $LaCoO_3$/γ-Al_2O_3 calcined at 500 and 700°C, respectively.

$LaCoO_3$/γ-Al_2O_3/cordierite was active for syngas production in the partial oxidation of methane (800°C; $CH_4/O_2 = 4/1$; 1 atm; $W/F = 6.67 \times 10^{-5}$ g_{cat}.min.cm^{-3}) yielding the following byproducts distribution: 42, 54, 3.5% to H_2, CO, CO_2, respectively, at 36% CO conversion for 30 h. The H_2/CO ratio lower than unit was ascribed to the reverse water–gas shift and dry reforming of methane reactions taking place at those conditions [47]. The high stability was discussed in terms of (i) basic properties of La_2O_3 (which originated from the perovskite collapse during activation in H_2), (ii) stable and dispersed cobalt particles onto La_2O_3/Al_2O_3/cordierite, (iii) dispersion of metal cobalt on the top of carbon nanotubes (which grow up at reaction conditions). The monolith structure was pointed out as a promising support to increase the exposition of the perovskite phase, showing excellent adherence properties for γ-Al_2O_3 and $LaCoO_3$.

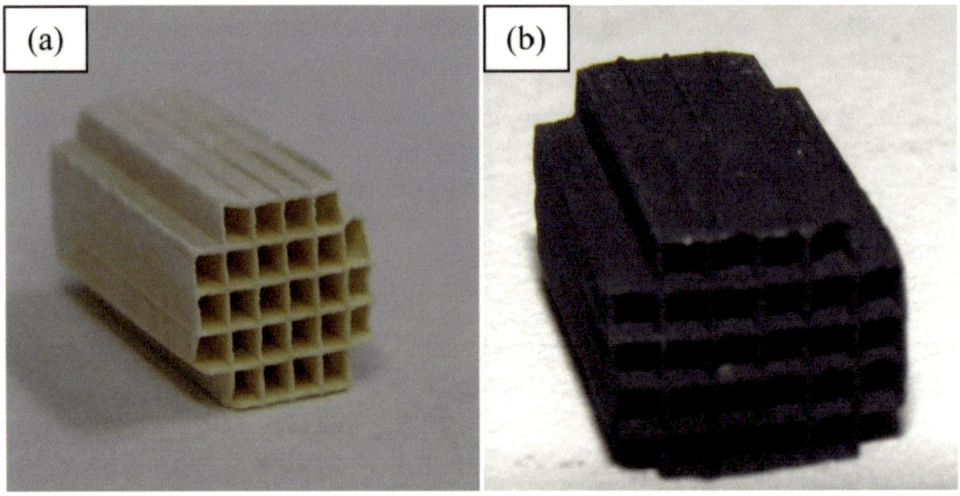

Figure 4. Pieces of honeycomb-type cordierite structure approximately 1.5 cm length, 0.8 cm wide, 0.65 cm high, cell density of 400 cell/in^2. (a) γ-Al_2O_3/cordierite calcined at 500°C and (b) $LaCoO_3$/γ-Al_2O_3/cordierite calcined at 700°C. Wash-coating and dip coating procedures were efficient to load alumina and perovskite phase onto the honeycomb-type cordierite. The perovskite $LaCoO_3$ was previously prepared by Pechini route. Reprinted from reference [47], copyright (2014) with permission from Hydrogen Energy Publications LLC and Elsevier Ltd.

An important progress on the synthesis of supported perovskites on monolith structure has been reported by ref. [48]. Initially, three different catalysts based on typical $LaCoO_3$ perovskite were obtained as (i) powder $LaCoO_3$ prepared by citrate method, (ii) $LaCoO_3$/monolith prepared by traditional wash-coating and (iii) $LaCoO_3$/monolith directly synthesized by a modified citrate route, all calcined at 650°C/6 h and then tested in the oxidation of CO and hydrocarbons (environmental applications) as a model reaction to validate the direct synthesis, which showed to be simple, sustainable, fast and reproducible. Results confirmed the expected formation of crystalline $LaCoO_3$ for the powdered sample (low surface area 9.8 m^2/g) and also over the cordierite (which presented 21 m^2/g) for both methods (ii) and (iii). By supporting the perovskite on the cordierite using the traditional wash-coating, a minimum of eight cycles were mandatory to achieve a uniform coating, since only afterwards the XPS revealed the cordierite species to disappear from the surface spectra. SEM images also confirmed that eight cycles were necessary to completely cover the cordierite and form a grain-type morphology. Activity measurements indicated that more than eight cycles (in the wash-coating procedure) did not lead to further significant increase of catalytic activity. Therefore, we can learn from wash-coating method that a significant number of cycles is required as well as lots of labor and time in order to ensure reproducibility and a uniform coating (which is a requisite for high activity).

The direct synthesis of the perovskite on the bare monolith (no γ-Alumina supported in advance) was carried out in a single step by immersing the monolith pieces in a solution containing La^{3+} and Co^{2+} precursors, the piece was then calcined [48]. XPS measurements proved the interaction between the cordierite and the deposited perovskite, and SEM images indicated a layer of porous perovskite on the cordierite, and this layer can be modulated by using a precursor solution with appropriate concentration and by controlling the immersion time. The directly synthesized $LaCoO_3$/monolith proved to be as effective as the best wash-coated catalyst (obtained after 8 cycles) in the oxidation of CO, C_3H_6 and C_3H_8 (environmental applications) leading to similar performances. Thus, we can learn that the direct method is simpler, faster, reproducible and does not deal with organic solvents, making it very attractive for industrial application through a sustainable process.

4. Conclusion

This chapter presented two strategies to overcome some limitations of perovskite-type oxides in catalysis: (1) the partial substitution of cations, which stabilizes unusual oxidation states of the metal components and creates anionic or cationic vacancies within the perovskite lattice. This is interesting for oxidation reactions, since vacancies can increase oxygen mobility from lattice to surface, increasing the catalytic activity of oxidation or improving the selectivity to products of interest. If the perovskite is a precursor, the partial substitution of cations can increase the reducibility and metal dispersion of the catalyst. The strategy (2) comprises to support perovskites on porous materials like a monolith or a conventional high surface area material to increase the number of exposed perovskite active sites.

Nomenclature

CO-SELOX: carbon monoxide selective oxidation reaction

EDS: energy-dispersive X-ray spectroscopy

EELS: electron energy loss spectroscopy

FTIR: Fourier transform infrared spectroscopy

H_2-TPR: temperature-programmed reduction (by H_2)

O_2-TPD: temperature-programmed desorption of oxygen

SEM: scanning electron microscopy

TEM: transmission electron microscopy

VOCs: volatile organic compounds

XPS: X-ray photoelectron spectroscopy

XRD: X-ray diffraction

Author details

Fabio Souza Toniolo* and Martin Schmal*

*Address all correspondence to: toniolo@peq.coppe.ufrj.br; schmal@peq.coppe.ufrj.br

Chemical Engineering Program – COPPE, Federal University of Rio de Janeiro, Rio de Janeiro, RJ, Brazil

References

[1] Ferri D, Forni L. Methane combustion on some perovskite-like mixed oxides. Appl Catal B 1998;16:119–26. DOI:10.1016/S0926-3373(97)00065-9

[2] Provendier H, Petit C, Estournès C, Libs S, Kiennemann A. Stabilisation of active nickel catalysts in partial oxidation of methane to synthesis gas by iron addition. Appl Catal A 1999;180:163–73. DOI:10.1016/S0926-860X(98)00343-3

[3] Goldwasser MR, Rivas ME, Lugo ML, Pietri E, Pérez-Zurita J, Cubeiro ML, Griboval-Constant A, Leclercq G. Combined methane reforming in presence of CO_2 and O_2 over $LaFe_{1-x}Co_xO_3$ mixed-oxide perovskites as catalysts precursors. Catal Today 2005;107–108:106–13. DOI:10.1016/j.cattod.2005.07.073

[4] Echchahed B, Kaliaguine S, Alamdari H. Well Dispersed Co^0 by Reduction of $LaCoO_3$ Perovskite. Int J Chem Reactor Eng 2006;4:A29. DOI:10.2202/1542-6580.1332

[5] Zhang J, Tan D, Meng Q, Weng X, Wu Z. Structural modification of $LaCoO_3$ perovskite for oxidation reactions: the synergistic effect of Ca^{2+} and Mg^{2+} co-substitution on phase formation and catalytic performance. Appl Catal B 2015;172–173:18–26. DOI: 10.1016/j.apcatb.2015.02.006

[6] Li X, Dai H, Deng J, Liu Y, et al. Au/3DOM $LaCoO_3$: high-performance catalysts for the oxidation of carbon monoxide and toluene. Chem Eng J 2013;228:965–75. DOI: 10.1016/j.cej.2013.05.070

[7] Łojewska J, Kołodziej A, Dynarowicz-Łątka P,Wesełucha-Birczyńska A. Engineering and chemical aspects of the preparation of microstructured cobalt catalyst for VOC combustion. Catal Today 2005;101:81–91. DOI:10.1016/j.cattod.2005.01.005

[8] Gallucci K, Villa P, Groppi G, Usberti N, Marra G. Catalytic combustion of methane on $BaZr_{(1-x)}Me_xO_3$ perovskites synthesised by a modified citrate method. Catal Today 2012;197:236–42. DOI:10.1016/j.cattod.2012.08.034

[9] Liu G, Li J, Yang K, Tang W, et al. Effects of cerium incorporation on the catalytic oxidation of benzene over flame-made perovskite $La_{1-x}Ce_xMnO_3$ catalysts. Particuology 2015;19:60–8. DOI:10.1016/j.partic.2014.07.001

[10] Hueso JL, Caballero A, Ocaña M, González-Elipe AR. Reactivity of lanthanum substituted cobaltites toward carbon particles. J Catal 2008;257:334–44. DOI:10.1016/j.jcat.2008.05.012

[11] Fang S, Wang L, Sun Z, Feng N. Catalytic removal of diesel soot particulates over K and Mg substituted $La_{1-x}K_xCo_{1-y}Mg_yO_3$ perovskite oxides. Catal Commun 2014;49:15–9. DOI:10.1016/j.catcom.2014.01.029

[12] Yung MM, Zhao Z, Woods MP, Ozkan US. Preferential oxidation of carbon monoxide on CoO_x/ZrO_2. J Molecul Catal A 2008;279:1–9. DOI:10.1016/j.molcata.2007.09.026

[13] Magalhães RNSH, Toniolo FS, da Silva VT, Schmal M. Selective CO oxidation reaction (SELOX) over cerium-doped $LaCoO_3$ perovskite catalysts. Appl Catal 2010;388:216–24. DOI:10.1016/j.apcata.2010.08.052

[14] Woods MP, Gawade P, Tan B, Ozkan US. Preferential oxidation of carbon monoxide on Co/CeO_2 nanoparticles. Appl Catal B 2010;97:28–35. DOI:10.1016/j.apcatb.2010.03.015

[15] Zhao Z, Lin X, Jin R, Dai Y, Wang G. High catalytic activity in CO PROX reaction of low cobalt-oxide loading catalysts supported on nano-particulate CeO_2–ZrO_2 oxides. Catal Commun 2011;12:1448–51. DOI:10.1016/j.catcom.2011.05.031

[16] Schmal M, Perez CAC, Magalhães RNSH. Synthesis and characterization of perovskite-type oxides $La_{1-x}M_xCoO_3$ (M = Ce, Sr) for the selective CO oxidation (SELOX). Top Catal 2014;57:1103–11. DOI:10.1007/s11244-014-0275-7

[17] Maluf SS, Assaf EM. CO preferential oxidation (CO-PROX) on $La_{1-x}Ce_xNiO_3$ perovskites. Catal Commun 2011;12:703–6. DOI:10.1016/j.catcom.2010.12.022

[18] Crespin M, Levitz P, Gatineau L. Reduced forms of $LaNiO_3$ perovskite. Part 1 – evidence for new phases: $La_2Ni_2O_5$ and $LaNiO_2$. J Chem Soc, Faraday Transac 2 1983;79:1181–94. DOI: 10.1039/F29837901181

[19] Tascón JMD, Fierro JLG, Tejuca LG. Kinetics and mechanism of CO oxidation on $LaCoO_3$. Zeitschrift für Physikalische Chemie 1981;124:249–57. DOI:10.1524/zpch.1981.124.2.249 (http://goo.gl/gk6nlc)

[20] Peña MA, Fierro JLG. Chemical structures and performance of perovskite oxides. Chem Rev 2001;101:1981–2017. DOI: 10.1021/cr980129f

[21] Tien-Thao N, Alamdari H, Zahedi-Niaki MH, Kaliaguine S. $LaCo_{1-x}Cu_xO_{3?\delta}$ perovskite catalysts for higher alcohol synthesis. Appl Catal A 2006;311:204–12. DOI: 10.1016/j.apcata.2006.06.029

[22] Toniolo FS, Magalhães RNSH, Perez CAC, Schmal M. Structural investigation of $LaCoO_3$ and $LaCoCuO_3$ perovskite-type oxides and the effect of Cu on coke deposition in the partial oxidation of methane. Appl Catal B 2012;117 –118:156–65. DOI:10.1016/j.apcatb.2012.01.009

[23] Figueiredo JL, Ribeiro FR. Catálise Heterogênea. 1st ed. Fundação Calouste Gulbenkian. Lisbon/Portugal; 1987. 347 p. (pp. 166–168).

[24] Twigg MV, Spencer MS. Deactivation of copper metal catalysts for methanol decomposition, methanol steam reforming and methanol synthesis. Top Catal 2003;22:191–203. DOI:10.1023/A:1023567718303

[25] Reuel RC, Bartholomew CH. The stoichiometries of H_2 and CO adsorptions on cobalt: effects of support and preparation. J Catal 1984;85:63–77. DOI: 10.1016/0021-9517(84)90110-6

[26] Xiong J, Borg Ø, Blekkan E A, Holmen A. Hydrogen chemisorption on rhenium-promoted γ-alumina supported cobalt catalysts. Catal Commun 2008;9:2327–30. DOI: 10.1016/j.catcom.2008.05.017

[27] Silva RRCM, Schmal M, Frety R, Dalmon JA. Effect of the support on the fischer–tropsch synthesis with Co/Nb_2O_5 catalysts. J Chem Soc, Faraday Transac 1993;89:3975–80. DOI: 10.1039/FT9938903975

[28] Mota N, Navarro RM, Alvarez-Galvan MC, Al-Zahrani SM, Fierro JLG. Hydrogen production by reforming of diesel fuel over catalysts derived from $LaCo_{1-x}Ru_xO_3$ perovskites: effect of the partial substitution of Co by Ru (x = 0.01-0.1). J Power Sources 2011;196:9087–95. DOI:10.1016/j.jpowsour.2010.11.143

[29] Moradi GR, Rahmanzadeh M, Khosravian F. The effects of partial substitution of Ni by Zn in $LaNiO_3$ perovskite catalyst for methane dry reforming. J CO_2 Utilization 2014;6:7–11. DOI:10.1016/j.jcou.2014.02.001

[30] Su YJ, Pan KL, Chang MB. Modifying perovskite-type oxide catalyst $LaNiO_3$ with Ce for carbon dioxide reforming of methane. Int J Hydrogen Energy 2014;39:4917–25. DOI:10.1016/j.ijhydene.2014.01.077

[31] Pereñíguez R, Hueso JL, Gaillard F, Holgado JP, Caballero A. Study of oxygen reactivity in $La_{1-x}Sr_xCoO_{3?\delta}$ perovskites for total oxidation of toluene. Catal Lett 2012;142:408–16. DOI:10.1007/s10562-012-0799-z

[32] Barbero BP, Gamboa JA, Cadús LE. Synthesis and characterisation of $La_{1-x}Ca_xFeO_3$ perovskite-type oxide catalysts for total oxidation of volatile organic compounds. Appl Catal B 2006;65:21–30. DOI:10.1016/j.apcatb.2005.11.018

[33] Pecchi G, Jiliberto MG, Delgado EJ, Cadús LE, Fierro JLG. Effect of B-site cation on the catalytic activity of $La_{1-x}Ca_xBO_3$ (B = Fe, Ni) perovskite-type oxides for toluene combustion. J Chem Technol Biotechnol 2011;86:1067–73. DOI:10.1002/jctb.2611

[34] Merino NA, Barbero BP, Grange P, Cadús L E. $La_{1-x}Ca_xCoO_3$ perovskite-type oxides: preparation, characterisation, stability, and catalytic potentiality for the total oxidation of propane. J Catal 2005;231:232–44. DOI:10.1016/j.jcat.2005.01.003

[35] Yan X, Huang Q, Li B, Xu X, et al. Catalytic performance of $LaCo_{0.5}M_{0.5}O_3$ (M = Mn, Cr, Fe, Ni, Cu) perovskite-type oxides and $LaCo_{0.5}Mn_{0.5}O_3$ supported on cordierite for CO oxidation. J Indust Eng Chem 2013;19:561–65. DOI:10.1016/j.jiec.2012.09.026

[36] Tejuca LG, Fierro JLG, Tascón JMD. Structure and reactivity of perovskite-type oxides. Adv Catal 1989;36:237–328. DOI:10.1016/S0360-0564(08)60019-X

[37] Taguchi H, Yamada S, Nagao M, Ichikawa Y, Tabata K. Surface characterization of $LaCoO_3$ synthesized using citric acid. Mater Res Bull 2002;37:69–76. DOI:10.1016/S0025-5408(01)00799-1

[38] Chagas CA, Toniolo FS, Magalhães RNSH, Schmal S. Alumina-supported $LaCoO_3$ perovskite for selective CO oxidation (SELOX). Int J Hydrogen Energy 2012;37:5022–31. DOI:10.1016/j.ijhydene.2011.12.052

[39] Manasilp A, Gulari E. Selective CO oxidation over Pt/alumina catalysts for fuel cell applications. Appl Catal B 2002;37:17–25. DOI:10.1016/S0926-3373(01)00319-8

[40] Liotta LF, Di Carlo G, Longo A, Pantaleo G, et al. Structural and morphological properties of Co–La catalysts supported on alumina/lanthana for hydrocarbon oxidation. J Non-Crystall Solids 2004;345–346:620–3. DOI:10.1016/j.jnoncrysol.2004.08.109

[41] Alifanti M, Florea M, Parvulescu VI. Ceria-based oxides as supports for $LaCoO_3$ perovskite; catalysts for total oxidation of VOC. Appl Catal B 2007;70:400–5. DOI: 10.1016/j.apcatb.2005.10.037

[42] Villoria JA, Alvarez-Galvan MC, Navarro RM, Briceno Y, et al. Zirconia-supported $LaCoO_3$ catalysts for hydrogen production by oxidative reforming of diesel: optimization of preparation conditions. Catal Today 2008;138:135–40. DOI:10.1016/j.cattod.2008.06.016

[43] Makshina EV, Nesterenko NS, Siffert S, Zhilinskaya EA, et al. Methanol oxidation on LaCo mixed oxide supported onto MCM-41 molecular sieve. Catal Today 2008;131:427–30. DOI:10.1016/j.cattod.2007.10.088

[44] Kucharczyk B, Tylus W. Metallic monolith supported $LaMnO_3$ perovskite-based catalysts in methane combustion. Catal Lett 2007;115:122–32. DOI: 10.1007/s10562-007-9076-y

[45] Kucharczyk B, Tylus W. Effect of Pd additive on the activity of monolithic $LaMnO_3$-based catalysts for methane combustion. Catal Today 2008;137:318–23. DOI:10.1016/j.cattod.2007.11.049

[46] Arendt E, Maione A, Klisinska A, Sanz O, Montes M, et al. Structuration of $LaMnO_3$ perovskite catalysts on ceramic and metallic monoliths: physico-chemical characterisation and catalytic activity in methane combustion. Appl Catal A 2008;339:1–14. DOI:10.1016/j.apcata.2008.01.016

[47] Brackmann R, Perez CAC, Schmal M. LaCoO3 perovskite on ceramic monoliths: pre and post reaction analyzes of the partial oxidation of methane. Int J Hydrogen Energy 2014;39:13991–4007. DOI:10.1016/j.ijhydene.2014.07.027

[48] Guiotto M, Pacella M, Perin G, Iovino A, et al. Washcoating vs. direct synthesis of $LaCoO_3$ on monoliths for environmental applications. Appl Catal A 2015;499:146–57. DOI:10.1016/j.apcata.2015.04.013

Chapter 19

Copper-based Perovskite Design and Its Performance in CO_2 Hydrogenation to Methanol

Feng Li, Haijuan Zhan, Ning Zhao and Fukui Xiao

Additional information is available at the end of the chapter

http://dx.doi.org/10.5772/61520

Abstract

Three series of perovskite-type catalysts, i.e., La–M–Mn–Cu–O (M = Mg, Y, Zn, Ce), La–M–Cu–Zn–O (M = Ce, Mg, Zr, Y), and La–Mn–Zn–Cu–O, were designed and applied in CO_2 hydrogenation to methanol. The materials were characterized by XRD, N_2-adsorption, N_2O-adsorption, ICP-OES, XPS, and TPD techniques. Perovskite structures were observed and the "metal on oxide" could be realized via reduction. Upon the introduction of the fourth elements, more structure defects, smaller particles, higher Cu dispersion, larger amount of hydrogen desorption at low temperature, and more amount of basic sites were obtained. The selectivity for methanol and the TOF values were higher for the catalysts derived from perovskite-type precursors. The catalytic performance was related to $Cu^{\alpha+}$ and/or Cu^0 species, low-temperature H_2 adsorption on the unit, and the weak basic sites.

Keywords: Perovskite, Copper, CO_2, Methanol, Hydrogenation

1. Introduction

Perovskite-type oxides have received significant attention because of their important electric, magnetic, ferromagnetic, pyroelectric, and piezoelectric properties [1,2]. Recently, much attention has been paid to perovskite-type oxides as catalysts due to their high activity and thermal stability. For a typical ABO_3 perovskite, A-site is a larger rare earth and/or alkaline earth cation and B-site is a smaller transition metal cation. In such structure, the A-site keeps the structure and the B-site provides the catalytic activity site. B-site cations could be reduced to well-dispersed metallic species supported on the A-site cations oxide, which leads to ideal catalyst precursors for many reactions that involve metal as active sites [3,4]. Besides, perovskite-type A_2BO_4 mixed oxides with the K_2NiF_4 structure, consisting of alternating layers of

ABO_3 perovskite and AO rock salt, have also been studied [5], which exhibit variable oxygen stoichiometry. The replacement of A-site and/or B-site cations by other metal cations often results in the formation of crystal microstrain and adjustable activity [6].

CO_2 is the main greenhouse gas, and various strategies have been implemented to reduce its concentration [7-10]. An important CO_2 utilization is the hydrogenation to methanol, which is considered as the most valuable product since it can be used as solvent, alternative fuel, and raw material, and it can be converted to olefins, aromatics, or gasoline derived from traditional petrochemical processes [11,12].

The synthesis of methanol over Cu/ZnO-type catalysts has been studied for many years. However, several important problems still remain open, such as the working oxidation of copper and the reaction mechanism [13-15]. In addition, the low activity and stability of catalysts, which are partly attributed to Cu sintering accelerated by the presence of the by-product, water vapor, create major barriers for practical application [16]. It is found that the catalysts with higher Cu dispersion, easier reduction property, and better adsorption properties for relative gases could achieve better catalytic performance for methanol synthesis [16].

Few work on the application of Cu-based perovskite-type oxides for CO_2 hydrogenation has been investigated. In the present work, three series of perovskite-type-based catalysts were prepared and tested for CO_2 hydrogenation to methanol, and the relationship between physical–chemical property and catalytic performance was discussed.

2. Catalyst preparation

The perovskite-type oxides were prepared by sol-gel method using citric acid as complexing agent. The precursor salts were $La(NO_3)_3$. nH_2O; $Mn(NO_3)_2$, 50% solution; $Cu(NO_3)_2$. $3H_2O$; $Mg(NO_3)_2$. $6H_2O$; $Y(NO_3)_3$. $6H_2O$; $Zn(NO_3)_2$. $6H_2O$; $Ce(NO_3)_3$. $6H_2O$; $ZrO(NO_3)_2.2H_2O$. Adequate amounts of the precursor salts along with citric acid were dissolved in deionized water at a molar ratio of 2:1 (metal cations: citric acid). The solution was heated to 353 K to remove the water, and then the temperature was increased to 423 K until ignition. The resulting powder was finally calcined under air at 673 K for 2 h and then at 1073 K for 4 h. The three series of catalysts were: (1) For doped La–M-Mn–Cu–O based (M= Mg, Y, Zn, Ce) perovskite materials, the ratio for La, M, Mn, Cu is 0.8: 0.2: 0.5: 0.5. The La–Mn–Cu–O catalyst and Mg, Y, Zn, Ce doping catalysts were then denoted as P, Mg–P, Y–P, Zn–P, and Ce–P, respectively. (2) A series of La–M-Cu–Zn–O (M= Ce, Mg, Zr, Y) based perovskite-type catalysts, i.e., $LaCu_{0.7}Zn_{0.3}O_x$, $La_{0.8}Ce_{0.2}Cu_{0.7}Zn_{0.3}O_x$, $La_{0.8}Mg_{0.2}Cu_{0.7}Zn_{0.3}O_x$, $La_{0.8}Zr_{0.2}Cu_{0.7}Zn_{0.3}O_x$ and $La_{0.8}Y_{0.2}Cu_{0.7}Zn_{0.3}O_x$ samples were prepared, of which the subscripts were the nominal composition. The catalysts were then denoted as LCZ-173, LCCZ-8273, LMCZ-8273, LZCZ-8273 and LYCZ-8273, respectively. (3) The $LaZn_{0.4}Cu_{0.6}O_y$, $LaMn_{0.1}Zn_{0.3}Cu_{0.6}O_y$, $LaMn_{0.2}Zn_{0.2}Cu_{0.6}O_y$, $LaMn_{0.3}Zn_{0.1}Cu_{0.6}O_y$, and $LaMn_{0.4}$ $Cu_{0.6}O_y$ samples were prepared, of which the subscripts were the nominal composition. The catalysts were denoted as LZC-046, LMZC-136, LMZC-226, LMZC-316, and LMC-406, respectively.

3. Results and discussion

3.1. Performance of the La–M-Mn–Cu–O (M = Mg, Y, Zn, Ce) based perovskite precursors

3.1.1. Textural and structural properties

As shown in Figure 1a [17], for the La–M-Mn–Cu–O (M = Mg, Y, Zn, Ce) based perovskite precursors, the $LaMnO_3$ (JCPDS # 75-0440) are the main phase. The diffraction peak at about 2θ=32.5° shift towards higher values when the fourth elements were doped. Small peaks at 2θ=35.6°and 38.9°assigned to CuO (JCPDS # 89-5899) appear in the doped samples but not in the P sample. In all diffraction patterns, no phase that ascribes to Mg, Y, or Zn is observed, while a new phase ascribed to CeO_2 is found for the sample of Ce–P, which demonstrates that it is difficult for all the Ce to enter the perovskite lattice, which agrees with the conclusion by Weng et al. [18].

For all reduced samples (Figure 1b), $LaMnO_3$ phase is still the main phase, which reveals that the reduction process does not destroy the perovskite structure. Meanwhile, the CuO phase disappears and the Cu phase emerges.

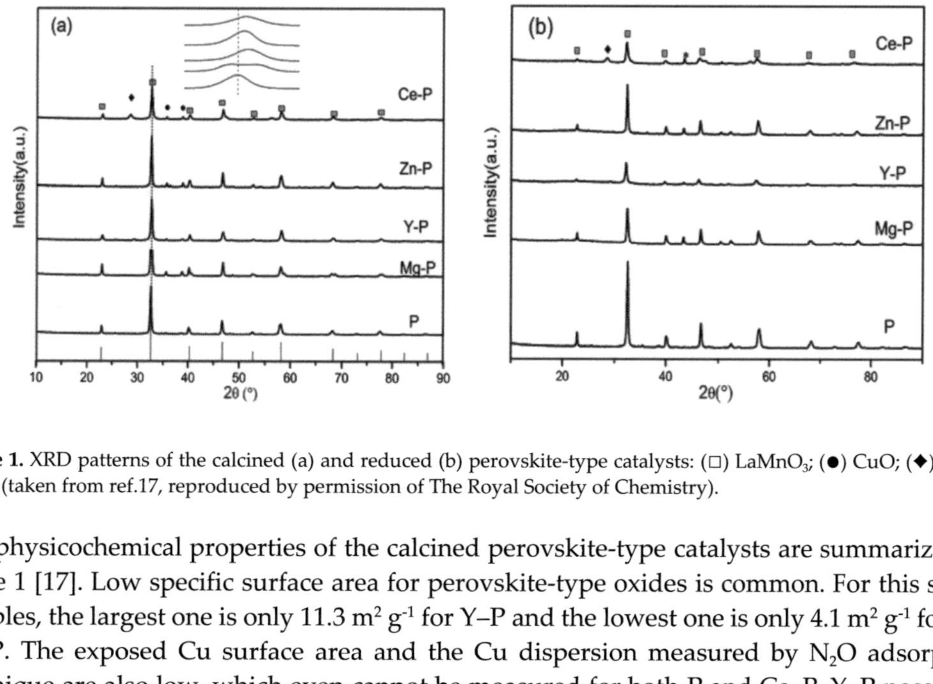

Figure 1. XRD patterns of the calcined (a) and reduced (b) perovskite-type catalysts: (□) $LaMnO_3$; (●) CuO; (◆) CeO_2; (*) Cu (taken from ref.17, reproduced by permission of The Royal Society of Chemistry).

The physicochemical properties of the calcined perovskite-type catalysts are summarized in Table 1 [17]. Low specific surface area for perovskite-type oxides is common. For this series samples, the largest one is only 11.3 m^2 g^{-1} for Y–P and the lowest one is only 4.1 m^2 g^{-1} for the Zn–P. The exposed Cu surface area and the Cu dispersion measured by N_2O adsorption technique are also low, which even cannot be measured for both P and Ce–P. Y–P possesses the largest Cu surface area and the Cu dispersion by comparison. The lower copper surface area may not be favorable for the conversion of CO_2 to methanol [19]. The ICP results show that the experimental lanthanum amount is lower than the theoretical value, and other element amounts are similar to the nominal values.

Samples	S_{BET} ($m^2 g^{-1}$)	Dispersion (%)[a]	S_{Cu} ($m^2 g^{-1}$)	Elemental composition (ICP-OES)[b]
P	6.5	-	-	$La_{0.84}Mn_{0.51}Cu_{0.50}$
Mg-P	5.4	0.9	1.2	$La_{0.67}Mg_{0.22}Mn_{0.49}Cu_{0.50}$
Y-P	11.3	3.8	4.6	$La_{0.67}Y_{0.23}Mn_{0.47}Cu_{0.50}$
Zn-P	4.1	0.7	0.9	$La_{0.67}Zn_{0.18}Mn_{0.50}Cu_{0.50}$
Ce-P	7.2	-	-	$La_{0.68}Ce_{0.19}Mn_{0.49}Cu_{0.50}$

[a] Calculated from N_2O dissociative adsorption.

[b] Subscripts came from ICP results.

Table 1. The physiochemical properties of the perovskite-type catalysts (taken from ref.17, reproduced by permission of The Royal Society of Chemistry).

3.1.2. The XPS investigations

The XPS results presented in Table 2 [17] show that lanthanum ions of the reduced perovskite-type catalysts are present in the trivalent form because the $La_{3d5/2}$ peak is close to the value of pure lanthana at 834.4 eV [20]. However, the BE of $La_{3d5/2}$ of P sample is higher than the other samples, which implies the increasing of the electron cloud density around La ions for the doped samples. It may due to the fourth elements affect the transfer of the electrons of La to O, since O has the highest electronegativity value among all elements [21]. For O_{1s}, the binding energy at around 528.9–529.1 eV is assigned to the lattice oxygen (O^{2-}) [21,22] and the binding energy at around 530.8–533.0 eV is ascribed to the adsorbed oxygen species (O_{ad}) in the surface, which contains hydroxyl (OH^-), carbonate species (CO_3^{2-}), and molecular water. The binding energy decreases after the fourth element except Ce adding, which indicates that there are more electrons around oxygen. It is likely that the fourth components transfer the electronic to the oxygen. The presence of surface adsorbed oxygen species suggests the formation of oxygen vacancies in the defected oxides [23], which is favorable for the activation of the catalyst. The O_{ad}/O^{2-} ratio is increased for the doped samples, which implies the improvement of catalysis activity. The binding energy values of $Mn_{2p3/2}$ for the perovskite-type catalysts are located at 641.3 eV–642.2 eV. The peak positons of level of MnO, Mn_2O_3, and MnO_2 are 640.6, 641.9, and 642.2 eV, respectively. The values are very similar, and the mean oxidation state of Mn ions at the surface layers is extremely difficult to detect by XPS, as reported in other studies [24,25]. However, the previous reports suggested that the BE difference between $Mn_{2p3/2}$ and O_{1s} emissions increases with about 0.6–0.7 eV for the change of the oxidation state between Mn^{3+} and Mn^{4+}. As shown in Table 2, the BE difference is in the range of 112.3–113.0 eV, i.e., increasing with 0.7 eV, which means a change of the Mn^{4+}/Mn^{3+} ratio for the perovskite-type catalysts [26,27].

Since the binding energy of the $Cu_{2p3/2}$ band in the metal (932.6 eV) and in Cu^+ (932.4 eV) is almost same, they can be distinguished by different kinetic energy of the Auger Cu LMM line position in Cu^0 (918.6 eV), Cu^+ (916.7 eV), or in Cu^{2+} (917.9 eV) [19,28]. The Auger electron spectroscopies of Cu LMM of reduced samples are shown in Figure 2 [17]. The profiles are convoluted into two peaks. It can be seen that the majority of the copper species exist as Cu^+

for all samples, which is in accordance with the report of Jia et al. [29]. The weak Cu^0 peak could be the explanation for the immeasurable of exposed Cu^0 for P and Ce–P (Table 1).

The binding energy of the fourth components shows that they exist in Mg-O binding [29], Zn-O binding [23], Y^{3+}, and Ce^{4+}, respectively.

Samples	$La_{3d5/2}$(eV)	$Mn_{2p3/2}$ (eV)	O_{1s} (eV)	O_{ad}/O^{2-}
P	835.1	642.2	529.2 (50.5%)	1.02
			531.7 (49.5%)	
Mg-P	834.0	642.0	529.1 (51.5%)	1.06
			531.5 (48.5%)	
Y-P	834.4	641.3	529.0 (54.0%)	1.18
			531.4 (46.0%)	
Zn-P	834.1	641.5	529.1 (51.2%)	1.05
			531.5 (48.8%)	
Ce-P	834.3	641.8	529.2 (51.9%)	1.08
			532.2 (48.1%)	

Table 2. The binding energy of La, Mn, O, and the ratio of different oxygen species (taken from ref. 17, reproduced by permission of The Royal Society of Chemistry).

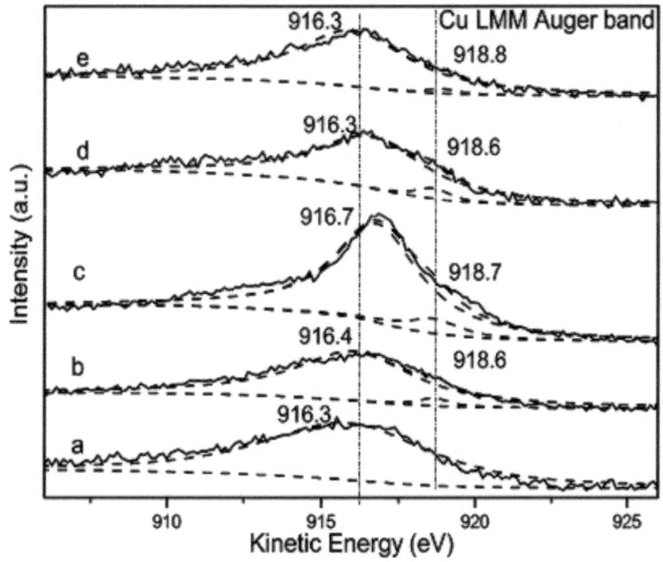

Figure 2. Cu LMM Auger electron spectroscopy of (a) P; (b) Mg–P; (c) Y–P; (d) Zn–P; (e) Ce–P samples after reduction (taken from ref. 17, reproduced by permission of The Royal Society of Chemistry).

3.1.3. H_2-TPD and CO_2-TPD analysis

The H_2 desorption over the prereduced materials on the unit surface area below 523 K (test temperature) increases apparently with the addition of the fourth element, as shown in Table 3 [17].

Two CO_2 desorption peaks are observed for all samples (Figure 3), which are denoted as peak α and peak β [17]. The peak α at around 400 K could be assigned to weak basic sites and the peak β at around 600 K could be assigned to strong basic sites. With the introduction of the fourth components, the peak α shifts to higher temperature, while the peak β shifts to lower temperature, which indicate the increase of the weak basic sites' strength but the decrease of the strong basic sites' strength. The strength for the weak basic sites of the catalysts increases in the order of: P < Ce–P < Y–P < Mg–P < Zn–P. The amount of the basic sites is also changed with the fourth element doping. The quantitative analysis for the CO_2-TPD based on the relative area of the profiles is listed in Table 3, in which the P sample is assigned as 1.00. Both the weak basic sites and the strong basic sites increase due to the alkalinity of Mg for Mg–P. For Y–P, the amount of total basic sites and strong basic sites improved remarkably with the amount of weak basic sites' decreasing. Moreover, the amount of the weak basic sites increases, but the amount of the strong basic sites and total basic sites decrease for Zn–P and Ce–P samples.

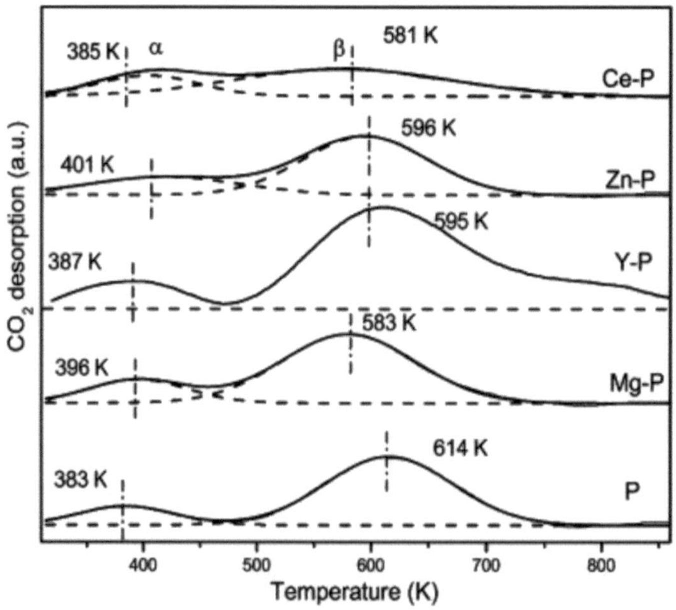

Figure 3. CO_2-TPD curves of the catalysts (taken from ref. 17, reproduced by permission of The Royal Society of Chemistry).

Samples	H_2 desorption below 523 K on per unit area (μmol g^{-1}m^{-2})	Adsoption type and distribution based in CO_2-TPD data[a]		
		Peak α	Peak β	Total
P	1.34	1.00 (383)	1.00 (614)	1.00
Mg-P	3.50	1.30 (396)	1.06 (583)	1.11
Y-P	3.56	0.87 (387)	1.67 (595)	1.53
Zn-P	5.93	1.56 (401)	0.85(596)	0.97
Ce-P	3.47	1.22 (385)	0.69(581)	0.78

[a] The amount of basicity of P is assigned as 1.00 to compare with other samples and the values in parentheses are the desorption temperature (K).

Table 3. The H_2-TPD and CO_2-TPD data (taken from ref. 17, reproduced by permission of The Royal Society of Chemistry).

3.1.4. *Catalytic performance*

The catalytic performances of the catalysts for CO_2 hydrogenation to methanol are summarized in Table 4 [17]. Both the CO_2 conversion and methanol selectivity are improved when the fourth element is added. With the introduction of Zn, the activity increased greatly, which might be due to the fact that the active site is $Cu^{\delta+}$–O–Zn^{2+} [31,32]. However, the activity improvement is slight for Ce-P. The relationship between the CO_2 conversion and the amount of H_2 desorption on unit surface area below 523 K (Table 3) is shown in Figure 4 [17]. It can be seen that the more the H_2 that is desorbed on the unit area, the more the CO_2 that is converted. Lower CO_2 conversion may result from lower copper content as well as lower surface area of copper in the system. The results of Figure 5 [17] show that the trend of the weak basic sites' strength and methanol selectivity is similar, which indicates their dependency.

Samples	CO_2 conversion (%)	Selectivity (C-mol%)		
		CH_3OH	CO	CH_4
P	1.8	0.7	93.4	5.9
Mg-P	2.8	23.7	68.1	6.5
Y-P	4.6	14.5	82.6	2.9
Zn-P	6.1	51.0	46.4	2.7
Ce-P	2.0	5.0	85.9	9.2

Reaction conditions: $n(H_2)/n(CO_2)$=3:1, T=523 K, P=5.0 MPa, GHSV=4000 h^{-1}.

Table 4. The performance for methanol synthesis from CO_2 hydrogenation over the reduced catalysts (taken from ref. 17, reproduced by permission of The Royal Society of Chemistry).

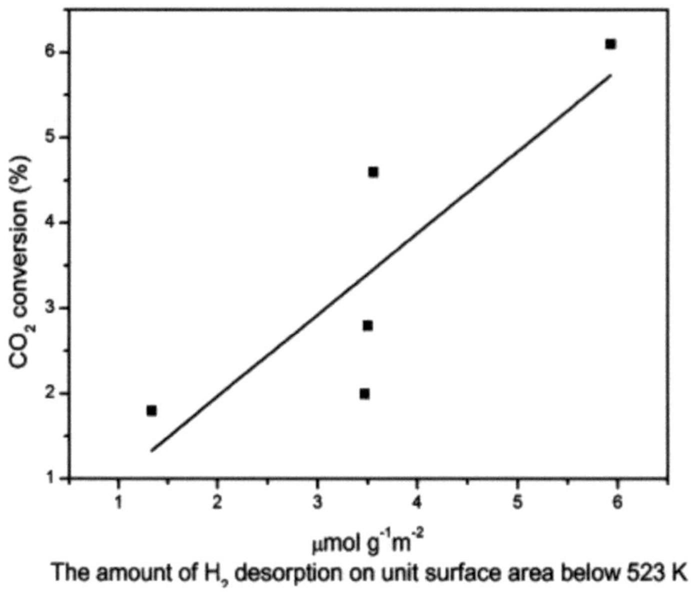

Figure 4. Relationship between the CO_2 conversion and the amount of H_2 desorbed on unit surface area below 523 K (taken from ref. 17, reproduced by permission of The Royal Society of Chemistry).

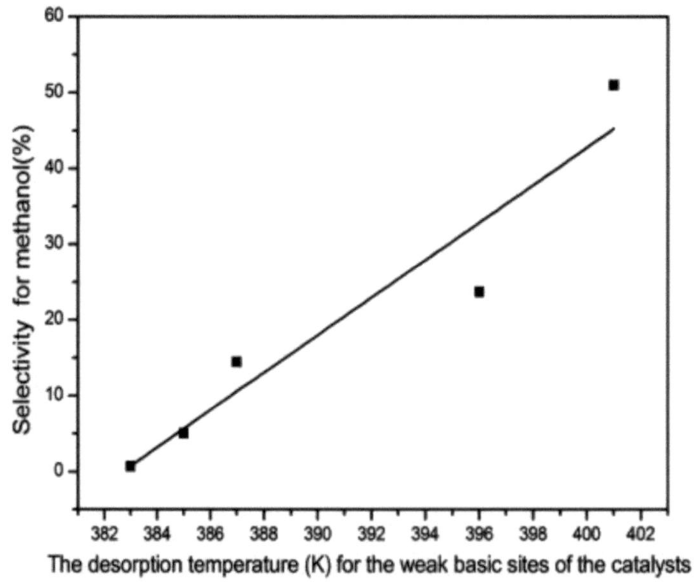

Figure 5. Relationship between the selectivity for methanol and the strength of the weak basic sites of the catalysts (taken from ref.17, reproduced by permission of The Royal Society of Chemistry).

3.2. Performance of the La-M-Cu-Zn-O (M = Ce, Mg, Zr, Y) based perovskite precursors

3.2.1. Textural and structural properties

Figure 6 [33] shows that the orthorhombic perovskite structure (A_2BO_4) with high degree of crystallinity of the La_2CuO_4 (JCPDS # 82-2142) is the main phase and two small peaks at 2θ=35.6° and 38.9° ascribed to the CuO phase (JCPDS # 89-5899) present in all samples. The weak peak at 2θ=36.3° attributed to the ZnO phase (JCPDS # 80-0075) appears in all samples except LMCZ-8273, which implies that Mg seems to have a special effect on the structure. The phase of Ce_7O_{12} and $La_2Zr_2O_7$ appear in the sample of LCCZ-8273 and LZCZ-8273, respectively, while there are no new phases containing Mg and Y appearing in the sample of LMCZ-8273 and LYCZ-8273.

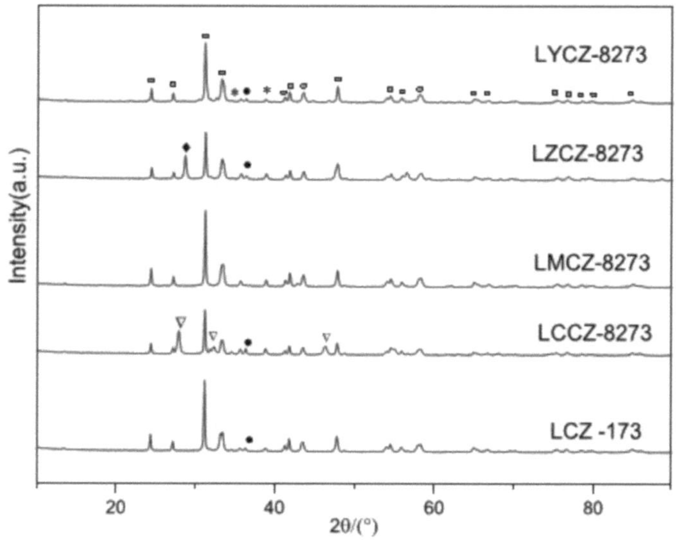

Figure 6. XRD patterns of the perovskite-type catalysts: (□)La_2CuO_4; (*)CuO; (◆)$La_2Zr_2O_7$; (●)ZnO; (∇)Ce_7O_{12} (taken from ref. 33, reproduced by permission of Elsevier B.V.)

The crystallographic parameters of the prepared materials were calculated by employing least-squares refinement, assuming an orthorhombic crystal system for the samples, and the results are listed in Table 5 [33]. A certain degree of changes of the lattice parameters occured after the fourth component was introduced. The lattice parameters a, b, and c were lower than those of LCZ-173, which can be attributed to the shrinkage of the La_2CuO_4 due to the introduction of the fourth elements. The mean grain size of La_2CuO_4 calculated by the Scherrer equation shows that the particles size of the La_2CuO_4 decreased remarkably for LCCZ-8273, LZCZ-8273, and LMCZ-8273, but slightly for LYCZ-8273. The physicochemical properties of the calcined catalysts are summarized in Table 6 [33]. The BET surface area for all calcined samples are rather low ($S_{BET} < 3\ m^2\ g^{-1}$), which is common for perovskite-type of materials [4]. It can be seen that the highest specific surface area is just only 2.3 $m^2\ g^{-1}$ for LZCZ-8273. Moreover, the tendency of the exposed Cu surface area and the Cu dispersion measured by N_2O adsorption is the same for the materials. The LZCZ-8273 shows the highest Cu surface area (S_{Cu}) and the best dispersion of copper (D_{Cu}). The physicochemical properties of LYCZ-8273 are similar to that of the LCZ-173, which indicates that the influence of the Y doping is negligible.

Samples	Lattice parameters (Å)			Volume (Å^3)	Size of La_2CuO_4 crystallites (nm)
	a	*b*	*c*		
LCZ-173	5.365	5.409	13.170	383.8	95.6
LCCZ-8273	5.350	5.392	13.137	381.8	31.9
LMCZ-8273	5.352	5.400	13.157	380.6	22.5
LZCZ-8273	5.356	5.404	13.149	379.0	24.0
LYCZ-8273	5.359	5.363	13.127	377.3	84.5

Table 5. The lattice parameters of the perovskite-type catalysts (taken from ref. 33, reproduced by permission of Elsevier B.V.).

Samples	S_{BET} (m^2g^{-1})	Dispersion[a] (%)	S_{Cu} (m^2g^{-1})
LCZ-173	0.7	5.3	3.4
LCCZ-8273	1.3	8.5	5.9
LMCZ-8273	1.2	8.5	6.2
LZCZ-8273	2.3	8.6	6.5
LYCZ-8273	0.7	4.5	3.2

[a] Calculated from N_2O dissociative adsorption.

Table 6. The physiochemical properties of the perovskite-type catalysts (taken from ref. 33, reproduced by permission of Elsevier B.V.).

3.2.2. XPS investigations

The reduced perovskite-type catalysts are analyzed by XPS, and the binding energies (BE) of $La_{3d5/2}$ and $Zn_{2p3/2}$ are presented in Table 7 [33]. According to the literature, $La_{3d5/2}$ features in perovskite structure are located at 837.5 and 834.3 eV [20,34] which are close to the values of pure lanthana at 837.8 and 834.4 eV, indicating that lanthanum ions are present in the trivalent form. A slight shift in the $La_{3d5/2}$ binding energy is observed upon introduction of the fourth elements and the values are in the range of 837.86–838.01 eV and 834.06–834.36 eV, respectively. Small differences may relate to the changes in crystal structure and/or electronic structure. In addition, small changes are also observed for the binding energy of Zn at around 1021.7 eV for different samples. The Auger electron spectroscopies of Cu LMM of reduced samples are shown in Figure 7 [33]. A broad peak appears in the range of 915.0 eV–920.0 eV for all samples, and it is hard to distinguish the Cu^+, Cu^{2+}, and Cu^0 apparently. However, the peaks at around

918.6 eV attributed to Cu^0 are distinct for all samples. A new peak appears at around 911.2–914.3 eV, lower than that of Cu^+, which is defined as peak α, implying that a special Cu species $Cu^{\alpha+}$ exists in the perovskite system. The presence of Cu^+ may accelerate the reduction of CO_2 to CO (RWGS) [34],while the $Cu^{\alpha+}$ (not Cu^{2+}, Cu^+, Cu^0) plays an important role for the methanol synthesis from CO_2/H_2 [31,35].

The X-ray photoelectron spectroscopies of the fourth elements in the reduced samples suggest that both Ce^{3+} and Ce^{4+} exist in the LCCZ-8273 and the +4 oxidation state is predominant. The result agrees with the XRD analysis. The Zr in the LZCZ-8273 sample exists in the phase of $La_2Zr_2O_7$.

Samples	Binding energy (eV)	
	$La_{3d5/2}$	$Zn_{2p3/2}$
LCZ-173	837.89 834.15	1021.67
LCCZ-8273	837.92 834.06	1021.71
LMCZ-8273	837.87 834.36	1021.34
LZCZ-8273	838.01 834.26	1021.78
LYCZ-8273	837.86 834.23	1021.69

Table 7. XPS parameters of the perovskite-type catalysts (taken from ref. 33, reproduced by permission of Elsevier B.V.).

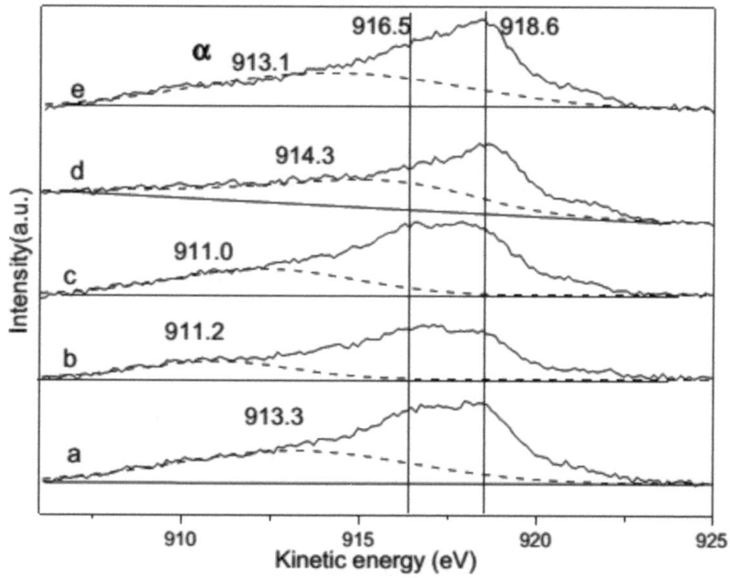

Figure 7. Cu LMM Auger electron spectroscopy of (a) LCZ-173; (b) LCCZ-8273; (c) LMCZ8273; (d) LZCZ-8273; (e) LYCZ-8273 samples after reduce (taken from ref. 33, reproduced by permission of Elsevier B.V.).

3.2.3. Catalytic performance

The catalytic performance for La-M-Cu-Zn-O (M = Ce, Mg, Zr, Y) catalysts are listed in Table 8 [33]. The LMCZ-8273 shows the highest CO_2 conversion and the maximum yield of methanol despite the lowest selectivity among all the samples. The LMCZ 8273 shows the highest methanol selectivity and the LCCZ-8273 shows moderate CO_2 conversion and methanol selectivity. The lowest CO_2 conversion and less improvement for methanol selectivity are observed for LYCZ-8273. The varying of the CO_2 conversion had the same tendency as the surface area of copper (Figure 8 [33]), indicating more surface copper existing in the catalysts may lead to higher activity, i.e., Cu^0 is the active site for CO_2 hydrogenation to methanol [10,16,19,22,35]. It is noteworthy that all catalysts show promising CH_3OH selectivity, especially for LMCZ-8273. The order of the selectivity to CH_3OH is as follows: LMCZ-8273 > LCCZ-8273 > LYCZ-8273 > LCZ173 > LZCZ-8273. The relationship between CH_3OH selectivity and the $Cu^{\alpha+}$ Auger peaks is shown in Figure 9 [33]. It can be seen that $Cu^{\alpha+}$ had a strong effect on the selectivity for methanol: the lower the binding energy of the peak α, the higher is the CH_3OH selectivity. Cu^+ is favorable for the reduction of CO_2 to CO (RWGS), so it can be derived that the farther away from 916.6 eV (the binding energy of Cu^+ in Cu LMM) for the peak α, the higher the CH_3OH selectivity that can be obtained. As discussed above, doping of Mg leads to the proper oxide state of copper, which results in the best selectivity for methanol. For LCCZ-8273 and LYCZ-8273, Ce and Y substitute La in the A-site with the same charge (+3) and similar ionic radius, which produces more defects in the perovskite structure that causes the special oxide state for copper species. With the special structure of La_2CuO_4 perovskite, the high dispersed copper species can be realized and stronger physical and electric interaction between the copper and other metal oxides can be obtained, which may lead to the formation and stabilization of the copper species with special valence [23]. However, for the LZCZ-8273, the formation of lanthanum zirconium pyrochlore has little influence on the perovskite structure but a great influence on the content of La_2CuO_4 perovskite, which may lead to the lowest selectivity of methanol. The turnover frequency (TOF), which represents the number of CO_2 molecules hydrogenated in a unit site per second (s^{-1}), is calculated from the exposed copper surface area for the perovskite-type catalysts (Table 8 [33]). The TOF values of the perovskite-type catalysts were very high compared with other catalytic systems [22,35], indicating the better efficiency for copper atoms on perovskite-type catalysts.

Samples	CO_2 conversion (%)	Selectivity (C-mol%)			CH_3OH yield (g $gcat^{-1}$ h^{-1})	TOF(Cu) × 10^3 (s^{-1})
		CH_3OH	CO	CH_x		
LCZ-173	6.4	57.9	39.5	2.5	0.05	65.1
LCCZ-8273	8.1	63.3	34.9	1.7	0.08	48.9
LMCZ-8273	9.1	65.2	33.0	1.8	0.09	52.0
LZCZ-8273	12.6	52.5	46.0	1.4	0.10	68.0
LYCZ-8273	5.0	59.6	37.0	3.5	0.04	55.7

Reaction conditions: $n(H_2)/n(CO_2)=3:1$, T=523 K, P=5.0 MPa, GHSV=3600 h^{-1}.

Table 8. Catalytic performance for methanol synthesis from CO_2 hydrogenation over the reduced catalysts (taken from ref. 33, reproduced by permission of Elsevier B.V.).

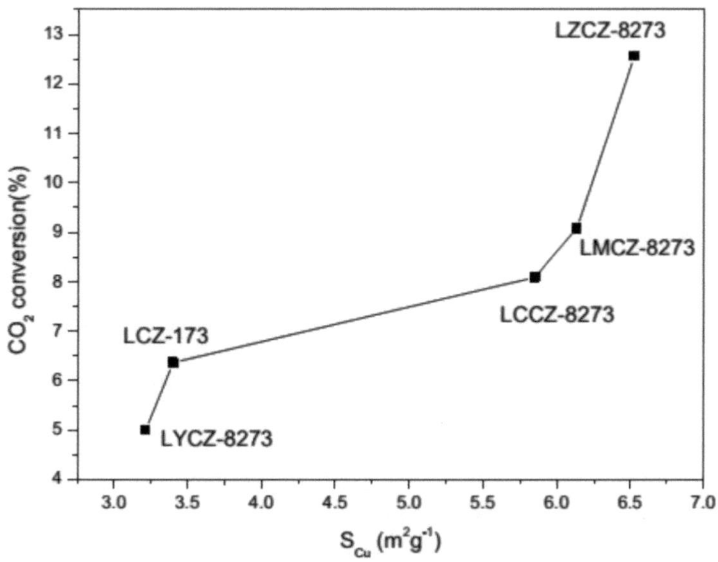

Figure 8. Relationship between copper surface area and CO_2 conversion (taken from ref. 33, reproduced by permission of Elsevier B.V.).

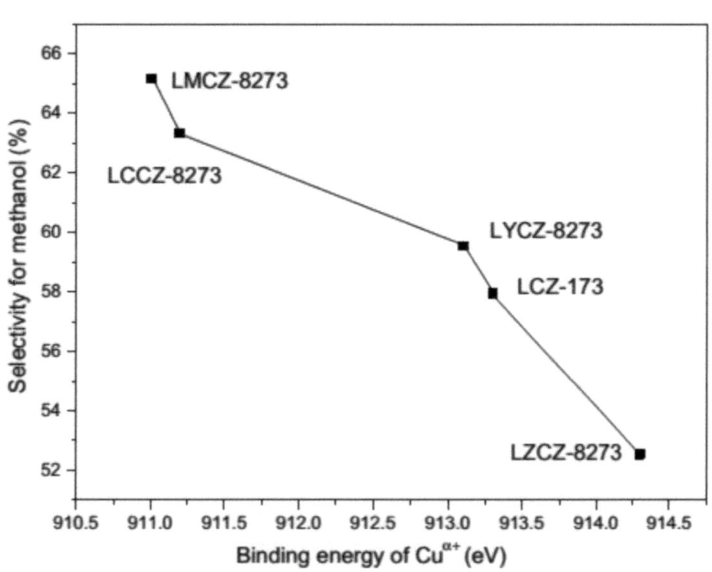

Figure 9. Relationship between methanol selectivity and the binding energy of $Cu^{\alpha+}$ (taken from ref. 33, reproduced by permission of Elsevier B.V.).

3.3. Performance of the La–Mn–Zn–Cu–O based perovskite precursors

3.3.1. Textural and structural properties

The XRD patterns of the fresh and reduced perovskites are presented in Figure 10a and b, respectively [36]. La_2CuO_4 perovskite-like structure can be observed for all fresh samples. $LaMnO_3$ phase emerges and the La_2CuO_4 phase transfers from tetragonal (JPCDS 81-2450) to orthorhombic (JPCDS 81-0872) as the manganese is introduced, which indicates that the manganese introduction distorts the structure of the La_2CuO_4. With the increasing of the manganese amount, the intensity of the La_2CuO_4 phase decreases while that of the $LaMnO_3$ phase increases, which implies that the formation of $LaMnO_3$ is easier than that of La_2CuO_4. This phenomenon reveals that the structure of $LaMnO_3$ is more stable than La_2CuO_4. Small peaks for both CuO and ZnO can also be observed except LMC-046, which indicates the perovskite structure has certain tolerance for the involved elements for this perovskite-type system. Moreover, the peak intensity of the separated CuO decreases when the value of Mn/Zn decreases, which means the formation of $LaMnO_3$ can lead to the separation of copper from the La_2CuO_4 perovskite structure. For the reduced sample (Figure 10b), the La_2CuO_4 perovskite structure disappears and the metallic copper and La_2O_3 is observed, which indicates that the "metal-on-oxide" can be attained. The appearance of $La_{0.974}Mn_{0.974}O_3$ phase reveals that the reduction progress can result in defects rather than destruction for the Mn-based perovskite.

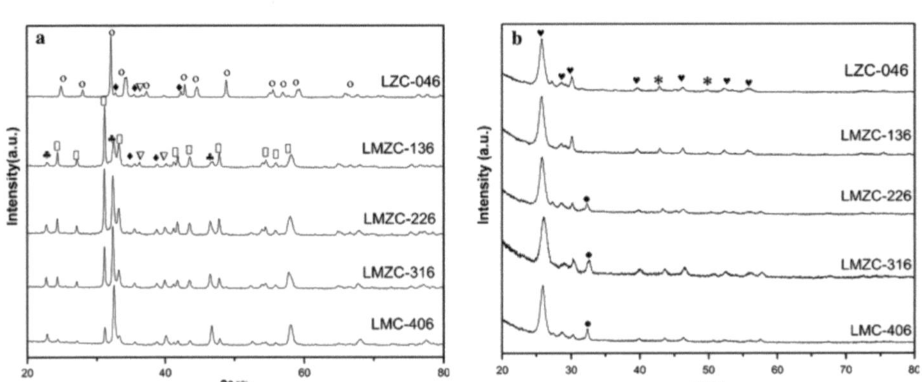

Figure 10. X-ray patterns of the fresh (a) and reduced (b) catalysts: (o) tetragonal La_2CuO_4; (□) orthorhombic La_2CuO_4; (▽) CuO; (◆) ZnO; (♣) $LaMnO_3$; (♥) La_2O_3; (•) $La_{0.974}Mn_{0.974}O_3$; (*) Cu (taken from ref. 36, reproduced by permission of Springer Science+Business Media).

The crystallographic parameters of the prepared materials were calculated by employing least-squares refinement and the results are listed in Table 9 [36]. It can be found that the axes are elongated for the four element samples, which means that doping may strut the perovskite structure. The size of La_2CuO_4 crystallites becomes smaller with the introduction of manganese. The $LaMnO_3$ phase changes from cubic to orthorhombic structure as zinc is introduced. Moreover, the LMZC-136 and LZC-046 possess the smallest $LaMnO_3$ and the largest La_2CuO_4

crystallites among all the samples, which implies that the abundant zinc species can improve the formation of small $LaMnO_3$ and large La_2CuO_4 crystallites. The change of lattice parameters of the samples implies that interaction between the involved elements might be different.

The BET specific surface area along with the exposed surface copper area and the copper dispersion measured by N_2O adsorption are summarized in Table 10 [36]. The specific surface area for all samples is low, which is common for perovskite-type materials [37]. LMC-406 possesses the largest specific surface area (S_{BET}), the exposed surface copper area (S_{Cu}) as well as the copper dispersion (D_{Cu}), while the LZC-046 sample shows the lowest S_{Cu} and D_{Cu}, which indicate that the existence of $LaMnO_3$ perovskite structure is favorable for increasing the surface copper area due to the extension of the space structure for the samples with manganese (Table 9).

Samples	Lattice parameters of $LaMnO_3$ (Å)			Size of $LaMnO_3$ crystallites (Å)	Lattice parameters La_2CuO_4 (Å)			Size of La_2CuO_4 crystallites (Å)
	a	*b*	*c*		*a*	*b*	*c*	
LZC-046	-	-	-	-	3.696	3.696	12.770	841
LMZC-136	5.472	7.754	5.513	224	5.352	5.400	13.157	820
LMZC-226	5.502	7.774	5.521	356	5.350	5.393	13.137	825
LMZC-316	5.472	7.751	5.513	475	5.350	5.393	13.157	831
LMC-406	3.88	3.88	3.88	352	5.351	5.395	13.139	658

Table 9. The lattice parameters of the perovskite-type samples (taken from ref. 36, reproduced by permission of Springer Science+Business Media).

Samples	**S_{BET} (m^2 g^{-1})**	**Dispersion[a] (%)**	**S_{Cu} (m^2 g^{-1})**
LZC-046	2.4	1.9	2.3
LMZC-136	1.1	2.3	2.7
LMZC-226	1.4	2.6	2.8
LMZC-316	1.2	2.7	3.0
LMC-406	2.5	4.2	4.5

[a] Calculated from N_2O dissociative adsorption.

Table 10. The physiochemical properties of the perovskite-type catalysts (taken from ref. 36, reproduced by permission of Springer Science+Business Media).

3.3.2. XPS investigations

The XPS results of the reduced perovskite-type catalysts are listed in Table 11 [36]. The values of $La_{3d5/2}$ binding energy (BE) are located in the range of 834.2–834.6 eV, demonstrating that La ions are in the trivalent form for all samples. Small changes of Zn (around 1021.7 eV) and Mn (around 642.0 eV) BE may relate to the small distortions in electronic structure and/or crystal structure. For the O_{1s} patterns, the peak at around 528.2–529.3 eV can be attributed to the oxygen ions in the crystal lattice (O^{2-}) and the peak at around 531.1–531.5 eV can be assigned to the adsorbed oxygen species (O_{ad}) derived from the defects or oxygen vacancies in the structure [38]. The O_{1s} BE shifts to lower value with the decreasing of the Mn/Zn ratio, which suggests the increasing of electron cloud density around O element. The value of O_{ad}/O^{2-} is max for the LZC-046, which decreases for the Mn containing samples, indicating that the $LaMnO_3$ could reduce the structural defects.

For this series catalyst, the binding energies of $Cu_{2p3/2}$ are lower than that for the copper oxide (933.0 eV) apparently, indicating that the Cu atoms are not in the simple copper oxides form. Figure 11 [36] shows the Auger electron spectroscopies of Cu LMM for the reduced samples. A broad peak can be observed, which can then be deconvoluted into three peaks. The peak at around 916.5 and 919.0 eV matched with kinetic binding energy of Cu^+ and Cu^0 within the error limit, respectively. However, a new peak at around 911.2–913.2 eV is observed which may be ascribed to the $Cu^{\alpha+}$. According to literatures and our works, $Cu^{\alpha+}$ can be appeared in perovskite-type system.

Samples	Binding energy (eV)					
	$La_{3d5/2}$	$Cu_{2p3/2}$	$Mn_{2p3/2}$	$Zn_{2p3/2}$	O_{1s}	O_{ad}/O^{2-}
LZC-046	834.6	932.2	-	1021.4	528.2	1.33
					531.1	
LMZC-136	834.3	932.4	642.0	1021.8	528.7	1.29
					531.3	
LMZC-226	834.2	932.5	641.9	1021.6	528.8	1.25
					531.3	
LMZC-316	834.3	932.7	642.0	1021.6	529.2	1.18
					531.4	
LMC-406	834.5	932.6	642.0	-	529.3	1.13
					531.5	

Table 11. XPS data of the perovskite-type catalysts (taken from ref. 36, reproduced by permission of Springer Science +Business Media).

3.3.3. *Catalytic performance*

The performance of the La–Mn–Zn–Cu–O based perovskite catalysts for methanol synthesis from CO_2 hydrogenation is shown in Table 12 [36]. The LMC-406 shows the worst performance despite the largest surface area and exposed copper surface area. The LZC- 046, which also contains three metal elements, but Zn instead of Mn, shows a moderate catalytic performance. It is well-known that the site of Cu^+-O-Zn^{2+} favors the adsorption of hydrogen that can transport to the bulk copper species via spillover [20,39]. So the lack of the site of Cu^+-O-Zn^{2+} may be the reason for the poor catalytic performance of LMC-406. Moreover, the TOF_{Cu} value increases sharply upon Zn introduction, which verifies that the copper sites are not the only active sites

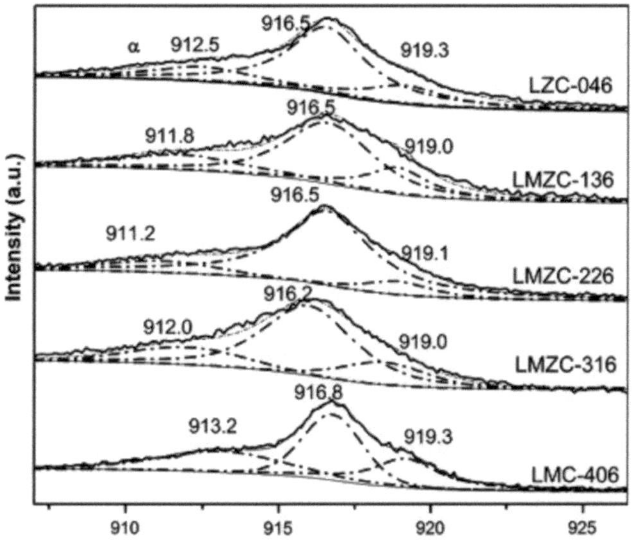

Figure 11. Cu LMM Auger electron spectroscopy of the reduced perovskite-type catalysts (taken from ref. 36, reproduced by permission of Springer Science+Business Media).

for the reaction [40]. With more reducible copper species and high TOF_{Cu} values, the four-component samples show a good catalytic performance. The ratio of both $Cu^{\alpha+}$ and Cu^0 species to the total copper species (calculated from the Auger spectroscopy (Figure 11)) shows the same change tendency with the CO2 conversion (Figure 12) [36], indicating that both $Cu^{\alpha+}$ and Cu^0 species could be the active sites for the conversion of CO_2. In addition, with the change of the ratio for Mn/Zn, the synergy between copper and the other components might vary, and then the reduction state of copper species ($Cu^{\alpha+}$ and Cu^0) changes. The four-component samples with two kinds of perovskites show better methanol selectivity, which implies that the strong synergy for different elements and different perovskite phases are significant for the improvement of the catalytic performance. In addition, it is also found that the lower the BE of the $Cu^{\alpha+}$, the higher is the CH_3OH selectivity (Figure 13) [36].

Samples	CO_2 conversion (%)	Selectivity (C-mol%)			CH_3OH yield ($g\ gcat^{-1}h^{-1}$)	$TOF_{Cu} \times 10^3$ (s^{-1})	$(Cu^{\alpha+}+Cu^0)/Cu_{Total}(\%)^a$
		CH_3OH	CO	CH_x			
LZC-046	11.0	48.2	50.4	1.3	0.07	129.4	55.31
LMZC-136	10.1	54.2	44.2	1.5	0.07	104.3	54.41
LMZC-226	10.7	55.3	43.3	1.4	0.08	161.1	54.93
LMZC-316	13.1	54.5	44.5	1.1	0.10	121.5	56.82
LMC-406	7.9	27.2	70.7	2.1	0.03	35.5	54.06

[a] The ratio value was calculated from the Auger spectroscopy (Figure 11).

Reaction conditions: $n(H_2)/n(CO_2) = 3:1$, T = 543 K, P = 5.0 MPa, GHSV = 3800 h^{-1}.

Table 12. The catalytic performance for methanol synthesis from CO_2 hydrogenation over the reduced catalysts (taken from ref. 36, reproduced by permission of Springer Science+Business Media).

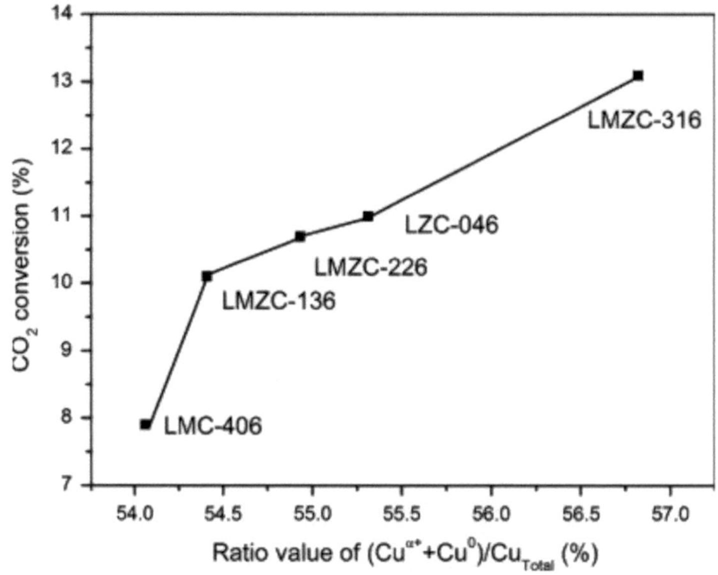

Figure 12. Relationship between CO_2 conversion and ratio value of $(Cu^{\alpha+}+Cu^0)/Cu_{Total}$ (taken from ref.36, reproduced by permission of Springer Science+Business Media).

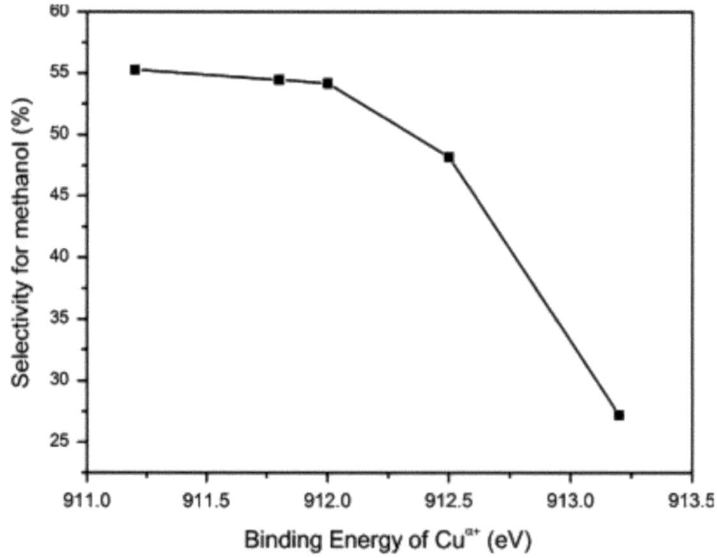

Figure 13. Relationship between methanol selectivity and binding energy of $Cu^{\alpha+}$ (taken from ref. 36, reproduced by permission of Springer Science+Business Media).

4. Conclusions

Three series of catalysts derived from perovskite-type precursors were prepared by sol-gel method, which were applied in the CO_2 hydrogenation to methanol. The conclusions are as follows:

1. The perovskite-type ABO_3, A_2BO_4, and ABO_3+A_2BO_4 can be purposive obtained in our work. Cu-based perovskite-type catalyst shows good methanol selectivity in CO_2 hydrogenation to methanol.

2. For La–M–Mn–Cu–O (M = Mg, Y, Zn, Ce) catalysts, the introduction of the fourth elements leads to the separation of copper from the $LaMnO_3$ perovskite lattice and thus produces more oxygen vacancies. Because of the increasing of defects, temperature adsorption properties are improved for the doped samples. The CO_2 conversion is related to the amount of absorbed H_2 on the unit area under 523 K, and the methanol selectivity corresponds to the strength of the weak basic sites. The catalytic performance enhanced considerably for Zn–P based on this, which also implies that Zn is important to the CO_2 hydrogenation to methanol.

3. La_2CuO_4 (A_2BO_4) perovskite structure is obtained for the La–M–Cu–Zn–O (M = Ce, Mg, Zr, Y) samples. With the addition of Ce, Mg, and Zr, good properties can be obtained: smaller particle size, higher Cu dispersion, larger amount of hydrogen desorption at low temperature, higher concentration of basic sites, and so on. The excellent methanol selectivity originates from the special copper valence that presents in the perovskite structure after reduction, and the CO_2 conversion is in correlation with the surface area of metallic copper.

4. Both La_2CuO_4 and $LaMnO_3$ perovskite structure can be observed in the $LaMn_{0.4-x}Zn_xCu_{0.6}O_y$ (x = 0, 0.1, 0.2, 0.3, 0.4) catalysts. The "metal-on-oxide" can be realized after reduction. With decreasing of Mn/Zn, more oxygen defects were formed. The perovskites exhibited better methanol selectivity due to the appearance of $Cu^{\alpha+}$ derived from the abundant defects of perovskite structure and the strong interaction between different elements. Moderate $LaMnO_3$ can balance the defects in the structure, and then lead to the perfect $Cu^{\alpha+}$, which is important for the methanol selectivity.

Acknowledgements

This work was financially supported by the Key Science and Technology Program of Shanxi Province, China (MD2014-10), the National Key Technology Research and Development Program of the Ministry of Science and Technology (2013BAC11B00), and the Natural Science Foundation of China (21343012).

Author details

Feng Li[1*], Haijuan Zhan[2], Ning Zhao[1,3*] and Fukui Xiao[1,3]

*Address all correspondence to: lifeng2729@sxicc.ac.cn; zhaoning@sxicc.ac.cn

1 State Key Laboratory of Coal Conversion, Institute of Coal Chemistry, Chinese Academy of Sciences, Taiyuan, People's Republic of China

2 School of Chemistry and Chemical Engineering, Ningxia University, Yinchuan, People's Republic of China

3 National Engineering Research Center for Coal-based Synthesis, People's Republic of China

References

[1] Tejuca L.G, Fierro J.L.G, Tascón J.M.D. Structure and Reactivity of Perovskite-Type Oxides. *Adv Catalys*. 1989; 36: 237-328. DOI: 10.1016/s0360-0564(08)60019-x

[2] Sun X.F., Komiya S., Ando Y. Anomalous damping of phonon thermal transport in lightly Y- or Eu-doped La_2CuO_4 single crystals. *Physica C*. 2003; 388-389: 355-356. DOI: 10.1016/s0921-4534(02)02500-5

[3] Jia L.s., Gao J., Fang W.p., Li Q.b. Carbon dioxide hydrogenation to methanol over the pre-reduced $LaCr_{0.5}Cu_{0.5}O_3$ catalyst. *Catalys Commun*. 2009; 10: 2000-2003. DOI: 10:2000-2003.10.1016/j.catcom.2009.07.017

[4] Pen˜a M.A., Fierro J.L.G. Chemical structures and performance of perovskite oxides. *Chem Rev*. 2001; 101: 1981-2017. DOI: 10.1021/cr980129f

[5] Zhong H., Zeng R. Structure of $LaSrMO_4$(M = Mn, Fe, Co, Ni, Cu) and their catalytic properties in the total oxidation of hexane. *J Serbian Chem Soc*. 2006; 71: 1049-1059. DOI: 10.2298/jsc0610049z

[6] Riza F., Ftikos C. Influence of A$^-$ and B$^-$site doping on the properties of the system $La_2CoO_{4\pm\delta}$. *J Eur Cera Soc*. 2007; 27: 571-573. DOI: 10.1016/j.jeurceramsoc.2006.04.069

[7] Li B., Duan Y., Luebke D., Morreale B. Advances in CO_2 capture technology: A patent review. *Appl Energy*. 2013; 102: 1439-1447. DOI: 10.1016/j.apenergy.2012.09.009

[8] Steeneveldt R., Berger B., Torp T.A. CO_2 capture and storage. *Chem Engin Res Design*. 2006; 84: 739-763. DOI: 10.1205/cherd05049

[9] Yu K.M., Curcic I., Gabriel J., Tsang S.C. Recent advances in CO_2 capture and utilization. *ChemSusChem*. 2008; 1: 893-899. DOI: 10.1002/cssc.200800169

[10] Peters M., K°ohler B., Kuckshinrichs W., Leitner W. Chemical technologies for exploiting and recycling carbon dioxide into the value chain. *ChemSusChem*. 2011; 4: 1216-1240. DOI: 10.1002/cssc.201000447

[11] Xu X.d., Moulijn J.A. Mitigation of CO_2 by chemical conversion: plausible chemical reactions and promising products. *Energy Fuels*. 1996; 10: 305-325. DOI:10.1021/ef9501511

[12] Olah G.A. Beyond oil and gas: the methanol economy. *Angewante Chemie*. 2005; 44: 2636-2639. DOI: 10.1002/anie.200462121

[13] Fujita S-i., Kanamori Y., Satriyo A.M, Takezawa N. Methanol synthesis from CO_2 over Cu-ZnO catalysts prepared from various coprecipitated precursors. *Catalysis Today*. 1998; 45: 241-244. DOI: 10.1016/S0920-5861(98)00222-3

[14] Klier K. Methanol synthesis. *Adv Catalys*. 1982; 31: 243-313. DOI: 10.1016/s0360-0564(08)60455-1

[15] Gao L.Z., Au C.T. CO_2 Hydrogenation to methanol on a $YBa_2Cu_3O_7$ catalyst. *JCatalys*. 2000; 189: 1-15. DOI: 10.1006/jcat.1999.2682

[16] Natesakhawat S., Lekse J.W, Baltrus J.P., Ohodnick P.R., Howard B.H., Deng X.Y., Matranga C. Active sites and structure–activity relationships of copper-based catalysts for carbon dioxide hydrogenation to methanol. *ACS Catalysis*. 2012; 2: 1667-1676. DOI: 10.1021/cs300008g

[17] Zhan H.j., Li F., Gao P., Zhao N., Xiao F.k., Wei W., Sun Y.h. Influence of element doping on La–Mn–Cu–O based perovskite precursors for methanol synthesis from CO_2/H_2. *RSC Adv*. 2014; 4: 48888–48896. DOI: 10.1039/c4ra07692c

[18] Weng D., Zhao H.s., Wu X.d., Xu L.h., Shen M.q. Influence of cerium on the performance of $LaMO_{3+\lambda}$ (M = Mn or Mn–Cu) perovskite-type catalyst. *Mater Sci Engin*. 2003; 361: 173-178. DOI: 10.1016/s0921-5093(03)00524-0

[19] Gao P., Li F., Zhan H.j., Ning Z., Xiao F.K, Wei W., Zhong L.S, Wang H., Sun Y.H. Influence of Zr on the performance of Cu/Zn/Al/Zr catalysts via hydrotalcite-like precursors for CO_2 hydrogenation to methanol. *J Catalys*. 2013; 298: 51-60. DOI: 10.1016/j.jcat.2012.10.030

[20] Merino N., Barbero B., Grange P., Cadus L. LaCaCoO perovskite-type oxides: preparation, characterisation, stability, and catalytic potentiality for the total oxidation of propane. *J Catalys*. 2005; 231: 232-244. DOI: 10.1016/j.jcat.2005.01.003

[21] Sutthiumporn K, Kawi S. Promotional effect of alkaline earth over Ni–La_2O_3 catalyst for CO_2 reforming of CH_4: Role of surface oxygen species on H_2 production and carbon suppression. *Int J Hydrogen Energy*. 2011; 36:14435-14446. DOI: 10.1016/j.ijhydene.2011.08.022

[22] Słoczyński J., Grabowski R., Kozłowska A., Olszewski P., Lachowska M., Skrzypek J., Stoch J. Effect of Mg and Mn oxide additions on structural and adsorptive properties

of Cu/ZnO/ZrO_2 catalysts for the methanol synthesis from CO_2. *Appl Catalys A*. 2003; 249: 129-138. DOI: 10.1016/s0926-860x(03)00191-1

[23] Li Z.q., Meng M., Zha Y.q., Dai F.f., Hu T.d., Xie Y.n., Zhang J. Highly efficient multi-functional dually-substituted perovskite catalysts $La_{1-x}K_xCo_{1-y}CuyO_{3-\delta}$ used for soot combustion, NO_x storage and simultaneous NO_x-soot removal. *Appl Catalys B*. 2012; 121-122: 65-74. DOI: 10.1016/j.apcatb.2012.03.022

[24] Rubio-Marcos F., Quesada A., García M.A., Banares M.A., Fierro J.L.G, Martin-Gonzalez M.S., Costa-Kramer J.L., Fernandez J.F. Some clues about the interphase reaction between ZnO and MnO_2 oxides. *J Solid State Chem*. 2009; 182: 1211-1216. DOI: 10.1016/j.jssc.2009.02.009

[25] Kenji Tabata Y.H., Suzuki E. XPS studies on the oxygen species of $LaMn_{1-x}CuxO_{3+\lambda}$. *Appl Catalys A*. 1998; 170: 245–254. DOI: 10.1016/j.molstruc.2014.04.065

[26] Batis N.H., Delichere P., Batis H. Physicochemical and catalytic properties in methane combustion of $La_{1-x}Ca_xMnO_{3\pm y}$ ($0 \leq x \leq 1$; $-0.04 \leq y \leq 0.24$) perovskite-type oxide. *Appl Catalys A*. 2005; 282: 173-180. DOI: 10.1016/j.apcata.2004.12.009

[27] Najjar H., Batis H. La–Mn perovskite-type oxide prepared by combustion method: Catalytic activity in ethanol oxidation. *Appl Catalys A*. 2010; 383: 192-201. DOI: 10.1016/j.apcata.2010.05.048

[28] Gao P., Li F., Xiao F.K., Zhao N., Sun N.N, Wei W., Zhong L.S, Sun Y.H. Preparation and activity of Cu/Zn/Al/Zr catalysts via hydrotalcite-containing precursors for methanol synthesis from CO_2 hydrogenation. *Catalys Sci Technol*. 2012; 2: 1447. DOI: 10.1039/c2cy00481j

[29] Jia L.s., Gao J., Fang W.p., Li Q.b. Influence of copper content on structural features and performance of pre-reduced $LaMn_{1-x}Cu_xO_3$ ($0 \leq x < 1$) catalysts for methanol synthesis from CO_2/H_2. *J Rare Earths*. 2010; 28: 747-751. DOI: 10.1016/s1002-0721(09)60193-9

[30] Aykut Y., Parsons G.N, Pourdeyhimi B., Khan S.A. Synthesis of mixed ceramic $Mg_xZn_{1-x}O$ nanofibers via Mg^{2+} doping using sol-gel electrospinning. *Langmuir*. 2013; 29: 4159-4166. DOI: 10.1021/la400281c

[31] Arena F., Italiano G., Barbera K., Bordiga S., Bonura G., Spadaro L., Frusteri F. Solid-state interactions, adsorption sites and functionality of Cu-ZnO/ZrO_2 catalysts in the CO_2 hydrogenation to CH_3OH. *Appl Catalys A*. 2008; 350: 16-23. DOI: 10.1016/j.apcata.2008.07.028

[32] Miller B.J.A., Martin-luengo M.A., Vong M.S.W., Wang Y., Self V.A., Chapmanc S.M, Sermon P.A. Junctions between CuO_x and ZnO_y in sensors for CO and catalysts for CO hydrogenation. *J Mater Chem*. 1997; 7: 2155-2160. DOI: 10.1039/a700747g

[33] Zhan H.j., Li F., Gao P., Zhao N., Xiao F.k., Wei W., Sun Y.h. Methanol synthesis from CO_2 hydrogenation over La-M-Cu-Zn-O(M=Y, Ce, Mg, Zr) catalysts derived from

perovskite-type precursors. *J Power Sources*. 2014; 251: 113-121. DOI: 10.1016/j.jpowsour.2013.11.037

[34] Maluf S.S., Nascente P.A.P., Afonso C.R.M., Assaf E.M. Study of $La_{2-x}Ca_xCuO_4$ perovskites for the low temperature water gas shift reaction. *Appl Catalys A*. 2012; 413-414: 85-93. DOI: 10.1016/j.apcata.2011.10.047

[35] Arena F., Italiano G., Barbera K., Bonura G., Spadaro L., Frusteri F. Basic evidences for methanol-synthesis catalyst design. *Catalys Today*. 2009; 143: 80-85. DOI: 10.1016/j.cattod.2008.11.022

[36] Zhan H.j., Li F., Xin Ch.l., Zhao N., Xiao F.k.,Wei W., Sun Y.h. Performance of the La–Mn–Zn–Cu–O based perovskite precursors for methanol synthesis from CO_2 hydrogenation. *Catalys Lett*. 2015; 145:1177–1185.DOI: 10.1007/s10562-015-1513-8

[37] Gao L.zh., Chua H.T., Kawi S. The direct decomposition of NO over the La_2CuO_4 nanofiber catalyst. *J Solid State Chem*. 2008; 181: 2804-2807. DOI: 10.1016/j.jssc.2008.06.051

[38] Tien-Thao N., Alamdari H., Kaliaguine S. Characterization and reactivity of nanoscale $La(Co,Cu)O_3$ perovskite catalyst precursors for CO hydrogenation. *J Solid State Chem*. 2008; 181: 006-2019. DOI: 6/j.jssc.2007.11.016

[39] Yuan Z.L., Wang L.N., Wang J.H., Xia S.X., Chen P., Hou Z.Y., Zhen X.M. Hydrogenolysis of glycerol over homogenously dispersed copper on solid base catalysts. *Appl Catalys B*. 2011; 101: 431-440. DOI: 6/j.apcatb.2010.10.013

[40] Islam Q.A, Nag S., Basu R.N. Study of electrical conductivity of Ca-substituted $La_2Zr_2O_7$. *Mater Res Bull*. 2013; 48: 103-3107. DOI: 6/j.materresbull.2013.04.081

Chapter 20

Designing Perovskite Oxides for Solid Oxide Fuel Cells

Idoia Ruiz de Larramendi, Nagore Ortiz-Vitoriano,
Isaen B. Dzul-Bautista and Teófilo Rojo

Additional information is available at the end of the chapter

http://dx.doi.org/10.5772/61304

Abstract

Perovskite-type oxides with the general formula ABO_3 have been widely studied and are utilized in a large range of applications due to their tremendous versatility. In particular, the high stability of the perovskite structure compared to other crystal arrangements and its ability, given the correct selection of A and B cations, to maintain a large oxygen vacancy concentration makes it a good candidate as electrode in solid oxide fuel cell (SOFC) applications. Utilizing this novel structure allows the engineering of advanced, effective electrolytes for such devices. This review details the development of current state-of-the-art perovskite-type oxides for solid oxide fuel cell (SOFC) applications.

Keywords: SOFC, Nanostructure

1. Introduction

The development of new energy technologies has become important with the present situation of increasing energy demand, rising energy prices, and reinforcement of countermeasures for global warming and its detrimental climatological, ecological, and sociological effects [1]. The supplies of fossil fuels are constantly decreasing and some believe that we have reached the peak for oil production. A reappearing problem in the energy field is the conversion from available to usable form. This concerns the conversion of chemical energy in the form of fossil resources and derivatives, such as hydrogen and alcohols, into electrical energy. It has become increasingly important to reduce the losses associated with the applied conversion techniques partly due to industrialization and technological progress. Traditional conversion to electrical energy is by gas turbine, steam turbine, or reciprocating engine driving a generator, where the

Carnot cycle sets a limit to the efficiency [2]. A fuel cell provides an alternative, whereby electrical energy can be made available with small losses, and no Carnot limit [3]. The effluents are in principle water, heat and CO_2 if the fuel is clean.

The energy situation has caused a push for sustainable energy technologies. Here, fuel cells play an important role in the renewable energy sector because of their highly efficient energy conversion and their especially high reliability. During the last decades, fuel cells have attracted much attention because of their potential for producing electricity more efficiently than conventional power generation like heat engines, which normally are limited by the Carnot cycle. The discovery of the fuel cell is ascribed to Sir William Grove (1839), demonstrating the reversibility of electrolytic water separation [4]. In 1899, Nernst contributed by demonstrating that certain oxides attained remarkably high electrical conductivity by doping with other oxides [5]. In the late twentieth century, the development accelerated and the interest in fuel cell technology increased. Today, there are five different types of fuel cells, all named after their electrolyte materials:

- Alkaline fuel cells (AFC)
- Polymer exchange membrane fuel cells (PEMFC)
- Phosphoric acid fuel cells (PAFC)
- Molten carbonate fuel cells (MCFC)
- Solid oxide fuel cells (SOFC)

Depending on the electrolyte, each fuel cell has certain characteristics such as operation temperature, electric efficiency, and demands for fuel composition (Table 1).

Fuel Cell	**AFC**	**PEMFC**	**PAFC**	**MCFC**	**SOFC**
Electrolyte	KOH	Polymer	H_3PO_4	Li_2CO_3+K_2 CO_3	ZrO_2+Y_2O_3
Charge carrier	OH^-	H^+	H^+	$CO_3{}^{2-}$	O^{2-}
Fuel	Pure H_2	Pure H_2	Co-free H_2	H_2, CO, CH_4	H_2, CO, CH_4, NH_3
$T_{oper\cdot}$(°C)	65%–220	60%–130	150%–220	650	700%–1000
Electrical efficiency	45%–60%	40%–60%	35%–40%	45%–60%	45%–60%

Table 1. Most important characteristics for different types of fuel cells

Low-temperature fuel cells (AFC, SPFC, and PAFC) have potential for the propulsion of cars, where a short heating time is needed and the efficiency has to be compared with about 20% for a combustion engine, whereas high-temperature fuel cells (MCFC and SOFC) are suitable for continuous power and heat production, where the cell temperature can be maintained. The high-temperature fuel cells have higher efficiency and are more tolerant to the choice of fuel compared to the low-temperature fuel cells [3].

2. What is a solid oxide fuel cell?

Solid oxide fuel cells (SOFC) are electrochemical devices which convert chemical energy directly into electrical energy with high efficiency and low environmental impact and are expected to become the emerging technology for stationary power production [6].

SOFCs are considered by many researchers to be the most suitable for generating electricity from hydrocarbons because of their simplicity, efficiency, and ability to tolerate some degree of impurities. There are many advantages of SOFC with respect to other fuel cells. Some of the main advantages are [7]:

- The highest efficiencies of all fuel cells (50–60%)
- A potential long-life expectancy of more than 40,000–80,000 h
- Constructed from readily obtainable ceramic materials
- Few problems with electrolyte management
- High-grade waste heat is produced, for combined heat and power (CHP) applications increasing overall efficiencies to over 80%, and
- Internal reforming of hydrocarbon fuels is possible

However, there are some drawbacks related to the choice of materials in relation to costs. There are roughly two design types, tubular and flat plate. For the tubular cell, the material problems are less, but fabrication costs are high, whereas for the flat plate design, fabrication costs are less but more materials problems arise [8].

The operating principle of the solid oxide fuel cells is illustrated in Figure 1.

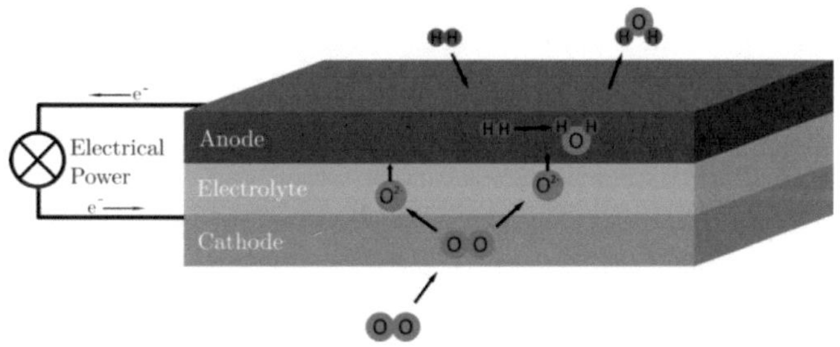

Figure 1. Sketch of solid oxide fuel cell operated on hydrogen and air.

SOFCs consist of three layers of functional materials: the anode, the electrolyte and the cathode. The anode is where the oxidation reaction takes place, and the cathode is where the reduction takes place. The cathode is fed with air or oxygen while the anode is fed with fuel gas. At the cathode, oxygen molecules are reduced to oxygen ions ($\frac{1}{2} O_2 (g) + 2 e^- \rightarrow O^{2-}$). The oxygen ions are conducted through the electrolyte to the anode. At the anode, oxygen ions oxidize the fuel gas and forms water and carbon dioxide, while the resulting free electrons are transported via the external circuit back to the cathode ($H_2 (g) + O^{2-} \rightarrow H_2O (g) + 2 e^-$).

To limit cathode overpotential, the traditional SOFC with yttria-stabilized zirconia (YSZ) electrolyte and strontium-doped lanthanum manganite (LSM) cathodes operate at relatively high temperatures (800°C–1000°C). These high temperatures encourage cell degradation—increasing cell, stack, and system maintenance. In addition, even higher temperatures are required for fabrication, encouraging electrode–electrolyte reactions, often forming undesirable, insulating secondary phases and consequently increasing cell resistance. Lowering the operating and fabrication temperature to 400°C–800°C could reduce both cell degradation and manufacturing costs. Cells which operate in this temperature range are known as intermediate temperature SOFCs (IT-SOFC). The advancement of this technology, limited by the high temperatures required, can be overcome by reducing operating temperatures, thus increasing theoretical efficiency. In order to operate efficiently at these reduced temperatures and to develop the next generation of hydrogen-related energy devices, new materials are required and the utilized processing routes must be optimized [6,9].

Among the new generation of materials, those with ABO_3-type perovskite structures stand out due to their great versatility. Through the correct choice of A and B site cations and the introduction of dopants, it is possible to obtain a large variety of materials with a wide range of properties and applications. This chapter will cover a selection of key materials developed for use in SOFC devices, their advantages and disadvantages and the optimization strategies published so far.

3. Evolution of perovskite-type structured materials as electrodes in SOFCs

The general formula unit of the perovskite is ABO_3, where A is a larger cation with a coordination number of 12 and B is a smaller cation with a coordination number of 6 (Figure 2). The large A cations can be rare earth, alkaline earth, alkali, and others [10] whereas the smaller B site accommodates many transition metals. These elements are typically the source of electronic conductivity. Cations of a rather wide range of ionic radii and valence are able to enter one or another site in the perovskite structure, exhibiting versatile physical and chemical properties and, thus, high concentrations of oxide vacancies and high ionic conductivity may be achieved [11]. This ability to hold a large content of oxygen vacancies makes them good candidates as electrodes in SOFC applications.

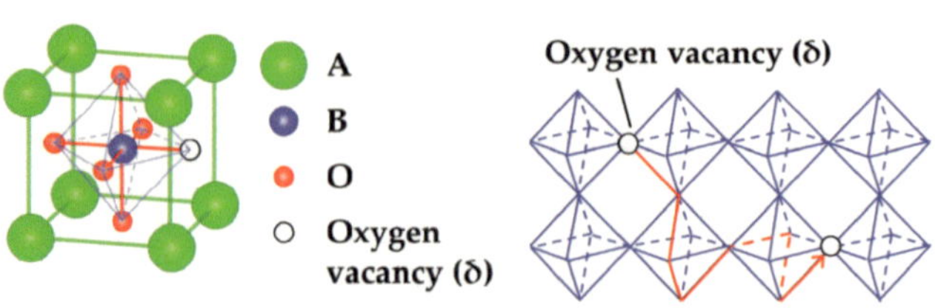

Figure 2. Atomic structure and oxygen transport in mixed conducting perovskites. On the left, the positions of the ions in a cubic structure. On the right, corner-sharing BO_6 octahedra and oxygen vacancy migration path.

The B cations are located in the corners of the cube and the A cation occupies the center. The oxygen is placed at the centers of the twelve cube edges, giving corner-shared strings of BO_6 octahedra. The A cation has the same size as the oxide ion, while the B cation is smaller. Depending on the type of cations, it is possible to create oxygen vacancies in the structure. These oxygen vacancies move along the structure giving rise to ionic conduction (Figure 2).

The composition and the microstructure of electrode materials greatly influence the performance of SOFCs. The high temperature and the reducing or oxidizing atmospheres limit the choice of these materials.

3.1. Cathode materials

One of the major research efforts in SOFC technology is directed towards decreasing the cell operating temperature. The best performance of the electrolyte, anode, and cathode materials typically is seen at an operating temperature above 800°C. This makes the use of expensive alloys as interconnectors and current collectors necessary. Decreasing the operating temperature would make it possible to use soft steel alloys with a remarkable decrease of the initial cost of the device. It is necessary, however, to develop new materials which exhibit good performance at lower temperatures in terms of conductivity, and chemical and mechanical compatibility. The widely used materials also need to be optimized in order to minimize technical problems at elevated temperatures. These problems are often associated with a mismatch between the thermal expansion coefficients of the electrolyte and cathode or as a consequence of oxygen loss from the cathode material.

The role of the cathode as the active site for the electrochemical reduction of oxygen is an important parameter to take into account in the material's design. SOFCs operate at high temperatures for long periods of time and, therefore, some requirements must be fulfilled [12]: (i) high electronic conductivity ($\sigma > 100$ S cm^{-1}); (ii) a thermal expansion coefficient (TEC) match with other SOFC components; (iii) chemical compatibility with the electrolyte and interconnect materials; (iv) adequate porosity to allow mass transport of oxygen; (v) high thermal stability; (vi) high catalytic activity for the oxygen reduction reaction (ORR); and (vii) low cost.

The cathodic reaction is quite complex and compromises a number of single steps such as diffusion, adsorption, dissociation, ionization, and finally, incorporation of oxygen into the electrolyte [13]. Oxygen molecules adsorb on the surface where they form partially reduced ionic/atomic species. These electroactive species must be transported along surfaces, interfaces, or inside the bulk of the electrode material to the electrolyte, where they are fully incorporated as O^{2-}.

The cathode materials can be classified into two groups: mainly, electronic conductors (with poor ionic conduction) and mixed ionic-electronic conductors (MIECs). Depending on the type of material, the conduction mechanism in the cathode will be different (Figure 3). In a poor ionic conductor, adsorption, dissociation, and diffusion of oxygen between the gas phase, electrode, and electrolyte [14] occurs through the triple phase boundary (TPB). In the MIECs, the conduction mechanism also occurs through the double phase boundary (DPB) [15,16]. The mixed conducting nature of the oxide ensures that electrochemical reactions occur at the MIEC/gas double phase boundary (DPB).

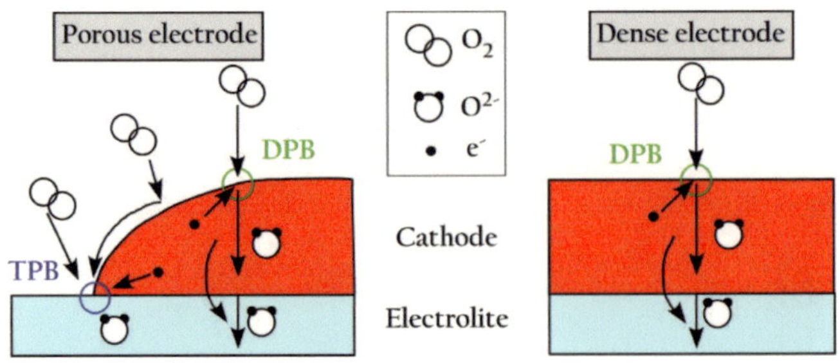

Figure 3. Schematic representation of conduction mechanisms (DPB: double phase boundary; TPB: triple phase boundary).

The deposition technique is also an important parameter which influences the electrochemical behaviour. Conduction in porous electrodes occurs through the three-phase boundary (TPB), whereas in MIECs, this also occurs through the DPB. In the case of dense electrodes, there is no direct contact between the cathode, electrolyte, and gas (hence, no TPB conduction exists). In this case, the oxygen reduction reaction occurs anywhere on the cathode surface, forming oxide anions which diffuse into the bulk of the electrode material towards the electrolyte. The material must present mixed ionic and electronic conduction [17].

$La_{1-x}Sr_xMnO_{3-\delta}$ (LSM) and $La_{1-x}Sr_xCoO_{3-\delta}$ (LSC) have been widely used as cathodes in SOFC. LSM has high electronic conductivity due to the introduction of Sr^{2+} on the La^{3+} site which will be charge-compensated by the Mn (Mn^{3+}–Mn^{4+}), making it a good cathode at high temperatures (1000°C). At intermediate temperatures (500°C–700°C), however, a large increase in the cell overpotential (from <1 to 2000 ohm cm^2) is observed, which hinders its use as a cathode [18]. One of the main limitations of LSM is its low ionic conductivity (~10^{-7} S/cm at 900°C), which is related to the Mn charge compensation where oxygen vacancies are not generated. The ease with which oxygen vacancies form is strongly influenced by the Sr content [19]. The influence of the cation vacancy on the transport properties of these materials has led to a study of the defects in these materials, such as LMO, and the effect of substitutions on the lanthanide site [20]. DeSouza et al. found that in both rhombic and rhombohedral structures, nonstoichiometry leads to the formation of vacancies on both cation positions, presenting a tendency toward the formation of vacancies [21]. These details are essential in understanding the behaviour of the materials and the degradation processes which can lead to the formation of insulating phases affecting the cathode performance. To obtain better performance from dense or porous LSM on YSZ electrolytes, a two-layer cathode is necessary—a 0.3 mm porous layer and in between a dense film of YSZ (~1 μm) [22]. This structure combines the best properties of both types of cathode in a single process and with more favorable properties than when using only the porous cathode.

Several authors have investigated other rare earths in the lanthanide position such as Pr, Sm, or Nd in order to improve the performance of manganites and avoid problems associated with the formation of the pyrochlore-insulating phases at the LSM/YSZ interface [23]. By substituting La with Sm, new materials with lower energetic barriers for adsorption and diffusion of oxygen species have been obtained [24]. In addition, they present excellent compatibility with YSZ and electronic conductivity well above that required in a SOFC (100 S/cm) [25]. The most promising electrode materials are $Pr_{0.7}Sr_{0.3}MnO_{3-\delta}$ and $Nd_{0.7}Sr_{0.3}MnO_3$ in terms of thermal expansion coefficient, reactivity, and conductivity [26]. The study of the compatibility of (Ln, Sr)$MnO_{3-\delta}$ with electrolytes has been limited to YSZ due to its widespread use. The performance of $Nd_{1-x}Sr_xMnO_{3-\delta}$ with gadolinia-doped ceria electrolyte (most suitable for use in IT-SOFC), however, revealed no reaction between the components at the interface after prolonged high-temperature treatments [27]. The same authors indicated that $Pr_{0.5}Sr_{0.5}MnO_{3-\delta}$ shows very promising properties with an electrical conductivity of 226 S/cm at 500°C [28]. Most studies on manganites have focused on doping the lanthanide site with Sr. In terms of chemical compatibility and electrical conductivity, however, the use of calcium as a dopant can offer very promising results, as in the case of $Pr_{0.7}Ca_{0.3}MnO_{3-\delta}$, which does not react with the electrolyte and has a maximum conductivity of 266 S/cm [29].

While looking for alternative materials to manganites, lanthanum cobaltites doped with bivalent metals such as Ca or Sr emerged [30]. These materials exhibit high conductivity values when 30% or 40% of Ca or Sr are added, respectively. In addition, the substitution of La by other lanthanides such as Pr and Sm has also lead to promising results. Despite the good results that these cobaltites exhibit, other drawbacks such as their high TECs and the appearance of secondary phases, limit their application.

In order to use the beneficial properties of these cobaltites, these materials are doped with Fe. Although, the addition of Fe promotes the decrease of the conductivity of the orthoferrites, lower TEC values are also achieved. Among these cobalto-ferrites, the $Ba_{0.5}Sr_{0.5}Co_{0.8}Fe_{0.2}O_3$ material stands out for SOFC cathode application. Shao et al. reported excellent performance at intermediate temperatures, proving that the orthoferrites are an interesting alternative for IT-SOFCs [30]. At the same time, the use of $La_{1-x}Sr_xFe_{1-y}Co_yO_3$ has been widely investigated [31–33]. These cobalt-based cathodes exhibit higher electrocatalytic performances than those of the conventional LSM cathodes. Unfortunately, they have high TEC, high cost of cobalt, and easy evaporation and reduction of cobalt [34,35].

Rare earth orthoferrites with perovskite-type structure are interesting materials for many electronic applications because of their mixed conductivity [36,37]. A site–doped rare earth orthoferrite compounds have been studied as candidates to replace manganites and cobaltite cathodes due to its high catalytic activity and mixed ionic and electronic conductivities at reduced temperature (σ total = 100 S/cm at 600°C–800°C) [35,38]. Lanthanum is a common element and key component of most SOFC, where several of its oxides provide the necessary electronic conductivity and high catalytic activity for oxygen reduction and some, in addition, combine ionic and electronic conductivity [39]. The choice of using Pr as the rare earth element is due to its peculiarity to show 3+ and 4+ oxidation states, which might induce interesting electrical properties [35]. Some of the Pr-doped orthoferrites display good performance as cathode materials in high-temperature solid oxide fuel cells because of their mixed, electronic and ionic, conductivity [40]. In particular, $Pr_{0.8}Sr_{0.2}Fe_{0.8}Ni_{0.2}O_{3-\delta}$ presents low electrical resistivity and good oxygen ion conductivity [41]. Fe-based perovskite oxides have also attracted much attention as possible alternatives to cobaltites due to their interesting transport properties. When the Fe fraction is higher than 0.5, the materials exhibit its highest electronic conductivity as observed for the $LaNi_{0.2}Fe_{0.8}O_3$ phase in which a conductivity of 135 S/cm, at 800°C, was obtained [42,43]. Calcium is another effective dopant for the A site, with low cost. For example, $Pr_{0.4}Ca_{0.6}Fe_{0.8}Ni_{0.2}O_3$ material exhibits an ASR value of 0.09 ohm cm^2 at 850°C [35]. Taking into account the low ASR value and the electrical conductivity of this cathode (at temperatures above 600°C, which is over 100 S/cm), this material is a promising cathode for IT-SOFC applications. $Pr_{1-x}M_xMnO_3$ (M = Ca, Sr) have also been studied where Ca-doped $PrMnO_3$ phases exhibit better performance than those doped with Sr. Ca-doped materials show higher electrical conductivity, lower cathodic overpotential and more similar thermal expansion coefficient [44]. Moreover, calcium could be a good candidate because of the similarity of its ionic radius with La^{3+} giving rise to higher stability than that of strontium-substituted phases [45].

While a large number of studies concerning cathode materials report the presence of secondary phases, they are often left unidentified [46] and are typically considered undesirable and associated with insulating phases that give rise to low conductivity values and poor performance [47]. Conversely, actual research in SOFC cathodes is focused on the development of composite materials, as the inherent requirements are so wide-ranging that no single material is capable of fulfilling every aspect [48]. The creation of composite cathodes is a good way to enhance the cathodic performance. They are composed of a solid electrolyte and an electronic conducting electrocatalytic material as in the $La_{0.6}Ca_{0.4}Fe_{0.8}Ni_{0.2}O_{3-\delta}$/samarium-doped ceria (SDC) symmetrical cells reported in a previous work [49]. The investigated composite cathodes include $Sm_{0.5}Sr_{0.5}CoO_3$ (SSC)-$Ce_{0.8}Sm_{0.2}O_{1.9}$ (SDC), $La_{0.6}Sr_{0.4}Co_{0.2}Fe_{0.8}O_3$ (LSCF)-SDC, LSCF-$Ce_{0.8}Gd_{0.2}O_2$ (CGO), and $Ba_{0.5}Sr_{0.5}Co_{0.2}Fe_{0.8}O_3$ (BSCF)-SDC, all of which exhibit low polarization resistance at 600°C [50,51]. The electrochemical performance of these cathodes is generally governed by triple phase boundary (TPB) kinetics, mass transport, and ohmic drop. The primary disadvantages of utilizing composite materials arise from the need for two independent materials; this increases both production complexity and the possibility of undesirable phases forming through inter-reaction of the component phases. Recently, the $La_{1-x}Ca_xFe_{0.8}Ni_{0.2}O_{3-\delta}$ (LCFN) system has been proposed and applied as a cathode featuring highly competitive performance [48]. Contrary to expectations, the formation of perovskite brownmillerite pseudo-composites resulted in a clear enhancement of the electrochemical performance and excellent thermomechanical compatibility with conventional electrolytes. Furthermore, new insights are gained into material surface properties controlling oxygen reduction processes.

The development of double perovskite materials such as $GdBaCo_2O_{5+x}$ [52], $PrBaCo_2O_{5+x}$ [53], $SmBaCo_2O_{5+\delta}$ [54], or $NdBaCo_2O_{5+\delta}$ [55] have also been investigated for IT-SOFCs. Their ORR activity is higher than many single perovskites [56]. They present high thermal expansion coefficients, however, which make them incompatible with the electrolytes developed so far. Table 2 summarizes the most widely study cathode materials for SOFC.

Composition	TEC (K^{-1})	Resistivity		Conductivity		Diffusion			Ref.
		ASR (Ω cm^2)	*T* (°C)	*σ* (*S/cm*)	*T* (°C)	*k** (*cm/s*)	*D** (*cm²/s*)	*T* (°C)	
$La_{1-x}Sr_xMnO_3$	$11.6\times\times10^{-6}$	0.18	750	180	800	—	3×10^{-9}	800	[57,58]
$La_{1-x}Sr_xFeO_{3-\delta}$	12.1×10^{-6}	0.1	800	125	800	1×10^{-5}	6×10^{-7}	1000	[59–61]
$La_{1-x}Sr_xFe_{0.8}Co_{0.2}O_{3-\delta}$	15.4×10^{-6}	0.34	750	404	550	6×10^{-6}	3×10^{-5}	626	[62-65]
$La_{1-x}Sr_xFe_{0.8}Ni_{0.2}O_3$	15.6×10^{-6}	0.16	700	435	800	—	—	—	[66,67]
$La_{0.6}Ca_{0.4}Fe_{0.8}Ni_{0.2}O_3$	11×10^{-6}	0.10	800	260	600	—	—	—	[6]
$La_{0.6}Sr_{0.4}Fe_{0.8}Cu_{0.2}O_3$	14.6×10^{-6}	0.138	750	135	275	---	---	---	[68]
$Pr_{0.6}Sr_{0.4}Co_{0.8}Fe_{0.2}O_3$	19.6×10^{-6}	0.046	800	1040	300	---	9.4×10^{-5}	600	[69]
$LaNi_{0.6}Fe_{0.4}O_3$	11.4×10^{-6}	0.018	800	600	600	---	2×10^{-7}	950	[70,71]
$Pr_{1-x}Sr_xCo_{0.2}Fe_{0.8}O_3$	14×10^{-6}	0.454	600	2190	600	---	---	---	[72,73]
$Sm_{0.5}Sr_{0.5}CoO_{3-\delta}$	22.8×10^{-6}	0.20	600	1000	850	6×10^{-5}	8.6×10^{-7}	890	[74,75]
$Ba_{0.5}Sr_{0.5}Zn_{0.2}Fe_{0.8}O_{3-\delta}$	16.5×10^{-6}	0.48	650	9.4	590	---	---	---	[76]
$Ba_{0.5}Sr_{0.5}Co_{0.8}Fe_{0.2}O_3$	19.9×10^{-6}	0.15	600	30	600	---	7.3×10^{-5}	---	[77,78]
$PrBaCo_2O_{5+\delta}$	20×10^{-6}	0.11	600	2000	150	1×10^{-3}	1×10^{-5}	350	[56,79,80]
$PrBaCoFeO_{5+\delta}$	21×10^{-6}	0.049	800	321	350	---	---	---	[81]
$NdBaCoFeO_{5+\delta}$	19.5×10^{-6}	0.062	800	172	350	---	---	---	[81]
$GdBaCo_2O_{5+\delta}$	20.1×10^{-6}	0.15	750	512	500	2×10^{-6}	3×10^{-7}	350	[82,83]
$SmBaCo_2O_{5+\delta}$	16×10^{-6}	0.098	750	815	500	---	---	---	[84]
$SmBa_{0.5}Sr_{0.5}Cu_2O_{5+\delta}$	14.1×10^{-6}	0.25	650	277.7	485	---	---	---	[85]
$YBaCo_2O_{5+\delta}$	16.3×10^{-6}	0.11	700	315	325	---	---	---	[86]

Table 2. The main properties of SOFC cathodes with both single and double perovskite structures

In recent years, advancements in cathode material properties have been made where a variety of perovskite oxides with a wide range of properties have been investigated. There are still, however, many chemistries to investigate which may yield superior performance in the future.

3.2. Anode materials

The requirements for a SOFC anode are quite similar to those for cathodes (changing the oxidizing atmosphere by a reducing environment), including: (i) proper electronic conductivity; (ii) thermomechanical and chemical compatibility with the electrolyte and interconnect materials; (iii) adequate porosity to allow gas transport to the reaction sites; and (iv) low cost. High ionic and electronic conductivity over a wide pO_2 range and high surface oxygen exchange kinetics and good catalytic properties for the anode reactions are also desirable properties for a good anode material. Nickel-YSZ cermet has been widely used in the SOFC anode. This material fulfills most requirements: Ni exhibits good catalytic activity for the dissociation of hydrogen while remaining stable at operating conditions and YSZ provides structural support for the Ni particles and inhibits their coarsening while matching the thermal expansion properties of the rest of the cell components [87]. Areas where this anode composition is lacking, however, include poor redox stability, low sulphur and carbon deposition tolerance when hydrocarbon fuels are used, and the tendency for nickel to agglomerate after prolonged operation [88]. In the search for alternative anode materials that are capable of withstanding sulphur contamination and carbon deposition, oxides with perovskite structure have drawn considerable attention [87]. Several authors have studied anode materials based on the perovskite structure as very promising alternatives [89,90]. Among the numerous materials with perovskite structure, $SrTiO_3$-type titanates have received a great deal of attention due to their high stability under reducing atmospheres and high temperatures [91]. This material, however, suffers from poor electronic conductivity, but this can be overcome through the introduction of appropriate dopants in the structure [92]. Marina et al. studied the effect of La doping in $SrTiO_3$ with a clear effect seen from the substitution of the Sr and the oxygen partial pressure on the total electrical conductivity with values as high as 500 S/cm at 500°C for the $La_{0.3}Sr_{0.7}TiO_3$ phase [93]. Furthermore, doping with Sr in the A site and Nb in the B site, it was possible to increase the conductivity to 10 S/cm; however, these materials exhibited a poor ionic conductivity [94]. Ti has also been substituted by other cations such as Al, Fe, Ga, Mg, Mn, or Sc, affecting significant changes on the redox properties of the material and conductivities [95]. By adding multivalent cations such as Mn or Fe, reduction or oxidation occurs in these cations in preference to or together with Ti, leading to a general decrease in the presence of Ti (III), resulting in a decrease in conductivity [96]. The effect of Co has also been analyzed, observing that after reduction with H_2, the segregation of Co nanoparticles occurs on the anode surface, which favors the oxidation of the fuel, thus reducing the resistance under anodic polarization [97]. To improve the behaviour of titanates, various dopants have been introduced on the A site (Y, La, Ce) [98,99]. The introduction of Ce^{3+} leads to decomposition into a variety of phases due to its ready oxidation in air. Some Ce-rich phases migrate to the grain boundaries, which result in an increase in the catalytic properties of up to an order of magnitude [100]. It is particularly interesting to note the results obtained by Morales-Ruiz et al., where $(La,Sr)(Ti,M)O_3$ (M = Ga, Mn) anodes exhibited comparable performance to those of conventional Ni-YSZ anodes [101]. Recently, $AMoO_3$ perovskite types (A = Ca, Sr, and Ba), containing Mo^{4+} ions have received interest for their potential use as anodes [102]. While these materials have high electrical conductivities (10^4 S/cm) they also present problems associated with the diffusion of oxide ions [102]. This limitation can be overcome by doping with other metals such as Fe or Cr on the Mo site [103,104]. Co is another suitable dopant which also creates oxygen vacancies, supplying sufficient ionic transport making a material with excellent catalytic properties for the oxidation of hydrogen [105]. Another alternative is to create composites with materials which have high ionic conductivity [106]. In this regard, the most commonly used materials are $Y_{0.08}Zr_{0.92}O_2$ (YSZ), $La_{0.8}Sr_{0.2}Ga_{0.8}Mg_{0.2}O_3$ (LSGM), and $Gd_{0.1}Ce_{0.9}O_2$ (GDC), which have been traditionally used as electrolytes, facilitating greater thermal, mechanical, and chemical compatibility with the anode composite. From these systems, a new material with the general formula $Sr_2Fe_{1.5}Mo_{0.5}O_6$ (SFMO) and perovskite-type structure has been developed, which features suitable mixed ionic-electronic conductivity and interesting catalytic activity [107]. It exhibits great stability when H_2, fuel or natural gas is used and high tolerance to sulfides [108]. Recently, Sutirakun et al. developed a theoretical model for analyzing the electrochemical oxidation of H_2 on the surface of this material [109]. One strategy for optimizing the material is to increase either the Mo content or add a small amount of active transition metal such as Ni in order to reduce the energy required for the formation of vacancies on the surface of SFMO. The other family of materials which has been widely studied are the chromites ($LaCrO_3$). These materials initially developed as interconnectors for SOFCs, which will be discussed later, because of their high stability in both oxidizing and reducing atmospheres and low tendency to accumulate carbon [110]. Their low catalytic activity, however, makes their use with some fuels such as methane impossible. To solve this problem, the effect of different dopants on the properties of this material (Ca and Sr on the A site and Mg, Mn, Fe, Co, and Ni in the B site) has been studied [111]. (La, Sr) (Cr, Fe)O_3 is the best candidate due to its high redox stability, high conductivity, and electrochemical activity [111]. In addition, this material works well as a catalyst for methane [112] and has a thermal expansion coefficient very similar to YSZ and LSGM [113]. Fowler et al. have also studied these materials, finding that the $La_{0.6}Sr_{0.4}Cr_{0.4}Fe_{0.6}O_{3-\delta}$ composition presents the best performance and highest stability [114]. In recent years, a new generation of anode materials with the double perovskite structure, such as Sr_2MgMoO_6 (SMMO) [115], $PrBaM_2O_5$ (M = Co, Mn) [116,117], and $La_4SrTi_5O_{17}$ [118] have been developed. These materials have very interesting properties which permit the use of multiple fuels.

4. Perovskite oxides as electrolytes and interconnectors for SOFC

Electrolyte ceramic materials must meet the following requirements: high ionic conductivity and low electronic conductivity; the charge carrier must either be from the oxidizer (O_2) or fuel (H_2), i.e. O^{2-} or protons [119]. To date, SOFC electrolytes are, in most commercial cases, 8 mol % yttria stabilized zirconia (8-YSZ) [120]. This material belongs to the group of fluorite-type solid solutions and up to now is the O^{2-} ion conductor most used as electrolyte material for SOFC due to its high ionic conductivity and thermomechanical stability. Doping ZrO_2 with Y_2O_3 has two main functions. First, to stabilize the cubic, fluorite phase, otherwise only stable at elevated temperatures, and second, to compensate the insertion of the trivalent Y^{3+} ions by oxygen vacancies in the zirconia lattice, giving rise to an enhancement of the oxygen ion conductivity. Generally, yttria-doped zirconia with 8 mol% Y_2O_3 exhibit an ionic conductivity higher than 0.1 S/cm at 1000°C and an electrical conductivity lower than 10^{-4} S/cm [121].

An important aspect to be considered is the chemical stability of the candidate materials at cell operation conditions, presenting high ionic conductivity (>0.1 S/cm) at intermediate temperatures and thermal expansion coefficients similar to the components of the cell [119]. The decrease in operating temperature can lead to an increase in chemical stability and cell lifetime, as well as a reduction in manufacturing costs; however, in the case of YSZ, cell performance decreases at lower temperatures due to the thermally activated ionic conductivity [122].

In order to obtain materials with improved properties at temperatures below 800°C, two families with perovskite type structure: gallates and cerates have been developed [123]. Most studies focus on Sr-doped $LaGaO_3$, but other dopants such as Ba have also been investigated [124]. Phase segregation occurs when Ba is used due to its large size compared to Sr which forms a mixture of $LaBaGa_3O_7$ and $LaGaO_3$ resulting in poor conductivities [125]. The ionic conductivity of this material can be increased by introducing Mg in the perovskite B site due to the introduction of compensating defects in the structure [126,127]. It has also been shown that the presence of a stoichiometry deficient in Ga may result in an increase in the concentration of oxygen vacancies [128]. In fact, this material is stable in CO-rich atmospheres, which allows its use as an electrolyte in direct coal solid oxide fuel cells (DC-SOFCs) [129]. Another strategy to increase the ionic conductivity of the material is through the production of composites with doped ceria electrolytes [130]. Through this method, it has been possible to optimize conduction through the grain boundaries, resulting in up to 10 times higher conductivity than ceria electrolytes at 500°C. In summary, the cells employing $(La,Sr)(Ga,Mg)O_{3-\delta}$ electrolytes have high power densities at 800°C, which makes them excellent candidates for IT-SOFC electrolytes [131,132].

The other family of materials with perovskite structures is based on $BaZrO_3$ and $BaCeO_3$, which traditionally have been developed as proton-conducting oxides, especially the cerates [133]. The production method, the temperatures employed in the process and the type and concentration of the dopant will affect proton conduction in the material. The cerates allow the introduction of precious metals such as Pd, which facilitates oxygen mobility, although the best results were obtained by using metal nanoparticles and by decorating the surface of the perovskite which acts as active catalysts [134]. Other dopants, both divalent and trivalent (Y, Yb, Gd, Sm, Nd, and La) have been employed [135–137]. This makes it possible to design materials resistant to reducing atmospheres with high conductivity and thermal and chemical stability in IT-SOFC operating conditions, even at temperatures as low as 500°C.

SOFC interconnect material requirements are as follows: (i) electronic conductivity > 100 S/cm; (ii) ionic transport number < 0.01 to avoid chemical shortcut permeation; (iii) gas tight; (iv) tolerate both reducing (H_2) and oxidizing (air/O_2) atmospheres; (v) be compatible with anode and cathode electrode materials (TEC and chemistry); and (vi) mechanical strength. $LaCrO_3$ chromites are the most widely used SOFC interconnects and, doped with other elements, their properties are shown to be improved [138]. The dopants most widely employed in the A site of the perovskite are Sr, Mg, and Ca, with dissolution limits of 50% [139], 15% [140], and 50% [141], respectively. Furthermore, the B site is also doped with Co or Fe, in order to limit the Cr content as much as possible due to its volatility [142]. Doped lanthanum chromites seem ideal materials for use as interconnects as they are highly stable in both oxidizing and reducing atmospheres and, at SOFC operating temperatures, do not react with the other cell components. Regarding thermal compatibility, through appropriate doping, it is possible to tailor the thermal expansion coefficient to that of the other components of the cell. $LaCrO_3$ is a p-type conductor that, upon divalent cation substitution on the La site, is seen to charge compensate by a valence change in the Cr (Cr^{3+}–Cr^{4+}), accompanied by an increase in the electronic conductivity of the material. The main problem associated with these compounds is their sinterability, as at high temperatures and under high oxygen pressure, volatilization of chromium oxide can occur. To avoid this, various strategies such as more reactive synthesis methods at lower temperatures, the introduction of cationic vacancies, and new fabrication techniques (including microwave sintering, freeze drying with EDTA, and more) have been analyzed [143].

5. Nanostructured perovskites for improving solid oxide fuel cells

Nowadays, the research in solid oxide fuel cells (SOFCs) is focused on lowering the operating temperature below 800°C in order to overcome problems such as the ageing of the materials. Concurrently, lowering the operating temperature has the detrimental consequences of decreasing the rate of surface reactions and bulk diffusion in the cathode, giving rise to a worsening of the cell performance. Surface exchange and ionic conduction must therefore be improved to maximize the yield of the cathode reaction. This improvement can be carried out not only by new material selection but also by the detailed control of the microstructure [144]. As stated before, a common strategy for improving the electrochemical performance of the electrodes is the fabrication of composite materials, combining ionic and mixed ionic electronic conducting materials, where the ionic conductor is homogeneously distributed [49]. Using composites, the ionic conductivity across the electrode will be enhanced and, simultaneously, a higher thermomechanical compatibility with the electrolyte can be achieved. The catalytic activity of the material is associated with its microstructure, so if the latter is improved, the active surface area will be increased, and a higher electrochemical durability will be obtained [145]. A significant number of studies have also been conducted to increase active surface area via microstructural control: the use of organic materials as pore formers, template materials such as colloidal crystals, meshes, foams or microfibers, glassy carbon microspheres, or membrane-based templates [146]. These methods look for higher catalytic activities for electrode reactions and lower annealing temperatures due to the numerous active sites and large surface areas. The same goal can be achieved using a new tool that has been developed during the last two decades: nanotechnology [147].

Recently, nanotechnology has been shown to overturn many established theories in a wide range of scientific fields, often with highly desirable properties [148]. This has naturally resulted in a great deal of interest from both scientific and industrial communities in the properties of nanostructured materials. This interest arises from new and unexpected behaviour when compared with bulk materials. Enhancements have been reported in electrical and ionic conductivity, chemical reactivity, and other properties. The shift from bulk to nanomaterials is therefore a promising approach in the development of new advanced technologies capable of achieving higher performance and low environmental impact.

Nanomaterials have typically been considered for low-temperature devices, as high temperatures could promote coarsening and therefore loss of the desired properties. This has generally kept nanomaterials from application in solid oxide fuel cells (SOFCs), but constant material advances have led to the decrease in their operational temperature. These so-called intermediate temperature SOFCs (IT-SOFCs), operating at 500°C–850°C, allow new possibilities for the use of nanomaterials.

The high surface area to volume ratio inherent in nanomaterials provides a large active area for SOFC electrodes. Nanomaterials are defined as a range of materials where at least one dimension is below 100 nm, resulting in nanostructures in zero- to three-dimensions. Nanoparticles, 0-D in nature, are well suited for use in composite electrodes, 1-D nanostructures, including nanotubes and nanowires, are thought to operate as promising electrodes, and thin films, 2-D in nature are of interest for application in micro-SOFCs (μ-SOFCs) and interlayers. Although the electrical properties of these nanostructures have been analyzed as independent 0- or 1-dimensional structures, the 2- or 3-dimesional behavior is more frequently reported, as it provides reliable results.

When working with SOFCs, there are two main approaches to develop nanomaterials with improved properties: optimizing existing composites by transforming them to nanocomposites and exploring novel nanostructured materials with high mixed ionic and electronic conductivity. In order to fabricate nanocomposites, wet impregnation/infiltration has been gaining increasing attention in recent years [149]. The infiltration process consists of placing a drop of a metal salt solution, with the chemistry required to give rise to the MIEC electrode material after decomposition, on top of a porous material (typically the ionic conductor) [150].

In parallel, MIEC nanostructured materials have also been developed in order to find further optimization of current state-of-the-art electrode materials. This kind of material features one dimension below 100 nm, giving rise to different structures such as nanoparticles, nanotubes, nanofibers, etc. There are several techniques for obtaining these unique phase nanomaterials with enhanced active areas and lower particle sizes with one of the most commonly used being via templating. Only some examples utilizing this procedure will be shown, although a large number of articles have been published on this topic. However, the excellent review by Ruiz-Morales et al. is recommended for those interested in a more detailed description of this method [146]. There are just a few materials able to act as templates because they must fulfill requirements such as removability, compatibility with the process conditions, wettability with the network forming precursor solution and a narrow particle size distribution to achieve optimal packing. Some of the most interesting templates are the organic polymer spheres of polycarbonate (PC), polystyrene (PS), and polymethyl methacrylate (PMMA) [151]. It has been demonstrated that templated porosity is maintained and highly influences electrochemical behavior, presenting an effective means of enhancing the triple phase boundary (TPB), and thus improving cell performance [152]. Another interesting method to obtain economical nanostructures is based on the use of carbon nanotubes (CNTs) as particle growth controller templates [153,154]. The growth controller material must fulfill the following two characteristics for the desired application: thermal stability and either an ease of removal or sufficient electrical properties to form a composite material. This facile and economical route allows to synthesize perovskite nanoparticles with grain sizes as small as 16 nm and surface areas of 151 m^2/g, improving electrochemical performance of the electrode by approximately one order of magnitude [153].

Of all possible nanostructures, nanotube-shaped materials seem to exhibit the most interesting improvements to electrode performance [155]. Although inorganic nanowires have been synthesized by several methods such as hydrothermal reaction [156], vapor transport [157], and electrospinning [158], the complexity of perovskite nanotube synthesis has resulted in the template-assisted synthesises becoming the most employed method. There are currently several membranes with different properties which can be used as templates, with anodized alumina (AAO) membranes being the most common [159]. The most important advantage of these membranes is their thermal stability, which allows the control of the morphology at high temperatures. Their high cost and the problems associated with their anodization do, however, limit their use. These membranes are classified in the literature as hard membranes while there is another group of templates known as soft membranes, which are primarily polymeric in nature. While not costly, they decompose between 200°C and 300°C, which means that the morphology cannot be completely controlled at higher synthesis temperatures. There are several types of polymeric membranes, with polycarbonates being the most commonly used [160]. The use of the pore wetting technique with polycarbonate membranes as templates and subsequent freeze-drying allows the fabrication of highly ordered three-dimensional nano-structures [161]. The electrospinning technique has also been used for the production of MIEC lanthanum strontium cobalt orthoferrite nanofibers [162]. This nanofiber-based cathode architecture is highly stable at intermediate temperatures (600°C–800°C) and provides continuous pathways for charge transport throughout the cathode.

Finally, there is a third generation of materials consisting of nanostructured nanocomposites. An improvement in oxygen reduction reaction (ORR) activity was reported when LSM nanoparticles were loaded on a porous YSZ framework [163–166]. Also, electrocatalytic nanoparticles could be produced in oxide anodes for solid oxide fuel cell (SOFC) by an ex-solution method, i.e., by incorporating metals into a perovskite oxide phase in air followed by the reduction of the perovskite oxide [167]. The improvement in the performance of the cell by using these nanocomposites lies in the extension of the TPB to the newly generated surfaces [168]. Templates such as polycarbonate membranes have also been used for the production of composite nanotubes with 20 nm wall thicknesses [169]. With this strategy, a clear decrease in polarization resistance of the electrode is observed, giving rise to higher efficiencies at temperatures as low as 700°C. Recently, an electrolyte-supported SOFC was fabricated with all-nanocomposite components and operated below 600°C [170]. The highly active nanocom-posite electrodes and easily sintered nanocomposite electrolyte allow an in situ low-temper-ature sintering while preserving the microstructure and electrochemical performance stability upon thermal cycling.

The performance of intermediate temperature SOFCs can be improved by engineering the electrode architecture on the nanoscale. Lowering the temperature facilitates the use of nanotechnology in synthesizing new nanostructured materials in which parameters such as porosity, the distribution of generated pores, and surface area can be closely controlled. These parameters have a significant influence on the performance of the materials used for energy conversion and storage, which means that these methods are an important starting point for the design and optimization of these types of energy devices.

6. Conclusions

The great versatility of the perovskite structure allows for different doping to obtain a variety of properties, which is a key feature in the development of materials for solid oxide fuel cells (SOFCs). This chapter not only briefly introduces the foundation and operation of SOFCs but also shows the evolution of perovskite materials for device components. Among the electrolytes, $BaCeO_3$-based cerates exhibit the highest ionic conduction at intermediate temperatures while chromites doped with Sr find potential as interconnector materials. By introducing Fe in the Cr site, these chromites become interesting alternatives to the traditional NiO anode, in addition to the new trend of double perovskite-type materials such as Sr_2MgMoO_6, which present very promising results for anode application. It is in the area of SOFC cathodes in which the most progress has been made. Initially, in the 1960s, doped manganese perovskites ($La_{1-x}Sr_xMnO_3$—LSM) were used as cathodes, but the low ionic conduction forced the development of new materials based on Co and/or Fe-containing perovskites. In addition, they have mixed ionic-electronic conductivity, allowing the active electrode area to extend across the surface. This leaves $Ba_{0.5}Sr_{0.5}Co_{0.8}Fe_{0.2}O_3$ perovskite as the most promising cathode material, although as in the case of anodes, double perovskites are also alternative materials with interesting properties.

Finally, it is worth mentioning the importance of not only the material composition but also its structure, morphology and porosity. It has been found that a higher catalytic electrode area results in a significant improvement in the electrochemical system efficiency. Reducing operating temperatures of the SOFC has allowed nanotechnology to become a useful tool for the development of future generations of materials for IT-SOFCs.

Acknowledgements

This work has been partially financed by the Ministerio de Educación y Ciencia under project MAT2013-41128-R and by the Eusko Jaurlaritza/Gobierno Vasco under project IT-570-13. N. Ortiz-Vitoriano acknowledges a Marie Curie International Outgoing Fellowship within the EU Seventh Framework Programme for Research and Technological Development (2007–2013).

Author details

Idoia Ruiz de Larramendi[1], Nagore Ortiz-Vitoriano[2], Isaen B. Dzul-Bautista[1,3] and Teófilo Rojo[1,2]

*Address all correspondence to: idoia.ruizdelarramendi@ehu.es

1 Departamento de Química Inorgánica, Universidad del País Vasco UPV/EHU, Bilbao, Spain

2 CIC energiGUNE, Parque Tecnológico de Álava, Miñano, Spain

3 Centro de Innovación, Investigación y Desarrollo en Ingeniería y Tecnología, PIIT Monterrey, CP, Apodaca, Nuevo León, Mexico

References

[1] Goto S, Oguma M, Chollacoop N, editors. EAS-ERIA Biodiesel Fuel Trade Handbook: 2010, Jakarta: ERIA, 2010. p. 6–15.

[2] Ruiz de Larramendi, I. $Ln_{1-x}Sr_xFe_{0.8}M_{0.2}O_{3-\delta}$ (Ln = Pr, Gd; M = Co, Ni, Cr, Ga; x = 0.1, 0.2, 0.3) Perovskite Type Oxides as Cathode Materials for Solid Oxide Fuel Cells [thesis]. University of the Basque Country (UPV/EHU), Bilbao; 2007.

[3] Primdahl, S. Nickel/Yttria-Stabilized Zirconia Cermet Anodes for Solid Oxide Fuel Cells [thesis]. University of Twente, Faculty of Chemical Technology, The Netherlands and Risø National Laboratory, Materials Research Department, Denmark; 1999.

[4] Grove, W.R. Philos. Mag. 1839; Ser. 3–14: 127–130.

[5] Nernst, W. Material for electric-lamp glowers. USA, US patent No.: U 00685730, 1901.

[6] Ortiz-Vitoriano N, Bernuy-López C, Ruiz de Larramendi I, Knibbe R, Thydén K, Hauch A, Holtappels P, Rojo T. Optimizing solid oxide fuel cell cathode processing route for intermediate temperature operation. Appl. Energy. 2013;104:984–991.

[7] Tu H, Stimming U. Advances, aging mechanisms and lifetime in solid-oxide fuel cells. J. Power Sources. 2004;127:284–293.

[8] Blenow, P. Strontium-Titanate–Based Anodes for Solid Oxide Fuel Cells [thesis]. Lund University, Sweden. 2007.

[9] Zhao Y, Xia C, Jia L, Wang Z, Li H, Yu J, Li Y. Recent progress on solid oxide fuel cell: Lowering temperature and utilizing non-hydrogen fuels. Int. J. Hydrogen Energ. 2013;38:16498–16517.

[10] West AR. Inorganic functional materials: optimization of properties by structural and compositional control. The Chem. Rec. 2006;6:206–216.

[11] Mogensen M, Lybye D, Bonanos N, Hendriksen PV, Poulsen FW. Factors controlling the oxide ion conductivity of fluorite and perovskite structured oxides. Solid State Ionics. 2004;174:279–286.

[12] Sun C, Hui R, Roller J. Cathode materials for solid oxide fuel cells: a review. J Solid State Electrochem. 2009;14:1125–1144.

[13] Zhou W, Ran R, Shao Z. Progress in understanding and development of $Ba_{0.5}Sr_{0.5}Co_{0.8}Fe_{0.2}O_{3-\delta}$-based cathodes for intermediate-temperature solid-oxide fuel cells: A review. J. Power Sources. 2009;192:231–246.

[14] Janardhanan VM, Heuveline V, Deutschmann O. Three-phase boundary length in solid-oxide fuel cells: A mathematical model. J. Power Sources. 2008;178:368–372

[15] Ciucci F, Chueh WC, Goodwin DG, Haile SM. Surface reaction and transport in mixed conductors with electrochemically-active surfaces: a 2-D numerical study of ceria. Phys. Chem. Chem. Phys., 2011;13:2121–2135.

[16] Muñoz-García AB, Ritzmann AM, Pavone M, Keith JA, Carter EA. Oxygen transport in perovskite-type solid oxide fuel cell materials: insights from quantum mechanics. Acc. Chem. Res. 2014;47:3340–3348.

[17] Ruiz de Larramendi I, Vivès S, Ortiz-Vitoriano N, Ruiz de Larramendi JI, Arriortua MI, Rojo T. $La_{0.6}Sr_{0.2}Ca_{0.2}Fe_{0.8}Ni_{0.2}O_3$ thin films obtained by pulsed laser ablation: effect of the substrate on the electrochemical behaviour. Solid State Ionics. 2011;192:584–590.

[18] Jacobson AJ. Materials for solid oxide fuel cells. Chem. Mater. 2010;22:660–674.

[19] Pavone M, Muñoz-García AB, Ritzmann AM, Carter EA. First-principles study of lanthanum strontium manganite: insights into electronic structure and oxygen vacancy formation. J. Phys. Chem. C. 2014;118:13346–13356.

[20] Takeda, Y.; Sakaki, Y.; Ichikawa, T.; Imanishi, N.; Yamamoto, O.; Mori, M.; Mori, N. y Abe, T. Stability of $La_{1-x}A_xMnO_{3-z}$ (A=Ca, Sr) as cathode materials for solid oxide fuel cells. Solid State Ionics. 1994;72:257–264.

[21] DeSouza RA, Islam MS, Ivers-Tiffee EJ. Formation and migration of cation defects in the perovskite oxide $LaMnO_3$. Mater. Chem. 1999;9:1621–1627.

[22] Herle JV, McEvoy AJ, Thampi KR. A study on the $La_{1-x}Sr_xMnO_3$ oxygen cathode. Electrochim. Acta. 1996;41:1447–1454.

[23] Skinner SJ. Recent advances in perovskite-type materials for solid oxide fuel cell Cathodes. Int. J. Inorg. Mater. 2001;3:113–121.

[24] Choi Y, Choi MC, Liu M. Rational design of novel cathode materials in solid oxide fuel cells using first-principles simulations. J. Power Sources. 2010;195:1441–1445.

[25] Wen TL, Tu H, Xu Z, Yamamoto O. A study of (Pr,Nd,Sm)SrMnO cathode materials for solid oxide fuel cell. Solid State Ionics. 1999;121:25–30.

[26] Sakaki Y, Takeda Y, Kato A, Imanishi N, Yamamoto O, Hattori M, Iio M, Esaki Y. $Ln_{1-x}Sr_xMnO_3$ (Ln = Pr, Nd, Sm and Gd) as the cathode material for solid oxide fuel cells. Solid State Ionics. 1999;118:187–194.

[27] Kostogloudis GC, Ftikos C. Characterization of $Nd_{1-x}Sr_xMnO_{3\pm\delta}$ SOFC cathode materials. J. Eur. Ceram. Soc. 1999;19:497–505.

[28] Kostogloudis GC, Vasihkos N, Ftikos C. Preparation and characterization of $Pr_{1-x}Sr_xMnO_{3\pm\delta}$ (x = 0, 0.15, 0.3, 0.4, 0.5) as a potential SOFC cathode material operating at intermediate temperatures (500–700°C). J. Eur. Ceram. Soc. 1997;17:1513–1521.

[29] Rim HR, Jeung SK, Jung E, Lee JS. Characteristics of $Pr_{1-x}M_xMnO_3$ (M = Ca, Sr) as cathode material in solid oxide fuel cells. Mater. Chem. Phys. 1998;52:54–59.

[30] Shao Z, Halle SM. A high-performance cathode for the next generation of solid-oxide fuel cells. Nature. 2004;431:170–173.

[31] Tai LW, Nasrallah MM, Anderson HU, Sparlin DM, Sehlin SR. Structure and electrical properties of $La_{1-x}Sr_xCo_{1-y}Fe_yO_3$. Part 2. The system $La_{1-x}Sr_xCo_{0.2}Fe_{0.8}O_3$. Solid State Ionics. 1995;76:273–283.

[32] Singhal SC, Kendal K, editors. High Temperature Solid Oxide Fuel Cells. Elsevier. 2004.

[33] Mai A, Haanappel BAC, Uhlenbruck S, Tietz F, Stöver D. Ferrite-based perovskites as cathode materials for anode-supported solid oxide fuel cells: Part I. Variation of composition. Solid State Ionics. 2005;176:1341–1350.

[34] Wang H, Tablet C, Feldhoff A, Caro J. A cobalt-free oxygen-permeable membrane based on the perovskite-type oxide $Ba_{0.5}Sr_{0.5}Zn_{0.2}Fe_{0.8}O_{3-\delta}$. Adv. Mater. 2005;17:1785–1788.

[35] Ortiz-Vitoriano N, Ruiz de Larramendi I, Gil de Muro I, Larrañaga A, Ruiz de Larramendi JI, Rojo T. A novel one step synthesized Co-free perovskite/brownmillerite nanocomposite for solid oxide fuel cells. J. Mater. Chem, 2011;21:9682–9691.

[36] Stevenson JW, Armstrong TR, Carneim RD, Pederson LR, Weber WJ. Electrochemical properties of mixed conducting perovskites $La_{1-x}M_xCo_{1-y}Fe_yO_{3-\delta}$ (M = Sr, Ba, Ca). J. Electrochem. Soc. 1996;143:2722–2729.

[37] Liu X, Cheng B, Hu J, Qin H, Jiang M. Preparation, structure, resistance and methane-gas sensing properties of nominal $La_{1-}xMgxFeO_3$. Sens. Actuator B-Chem. 2008;133:340–344.

[38] Plonczak P, Gazda M, Kusz B, Jasinski P. Fabrication of solid oxide fuel cell supported on specially performed ferrite-based perovskite cathode.J. Power Sources. 2008;181:1–7.

[39] Yoon KJ, Zink PA, Gopalan S, Pal UB, Pederson LR. Defect chemistry and electrical properties of $(La_{0.8}Ca_{0.2})_{0.95}FeO_{3-\delta}$ fuel cells and energy conversion. J. Electrochem. Soc. 2009;156:B795-B800.

[40] Hashimoto SI, Kammer K, Larsen PH, Pulsen FW, Mogensen M. A study of $Pr_{0.7}Sr_{0.3}Fe_{1-x}Ni_xO_{3-\delta}$ as a cathode material for SOFCs with intermediate operating temperature. Solid State Ionics. 2005;176:1013–1020.

[41] Ruiz de Larramendi I, López Antón R, Ruiz de Larramendi JI, Baliteau S, Mauvy F, Grenier JC, Rojo T. Structural and electrical properties of thin films of $Pr_{0.8}Sr_{0.2}Fe_{0.8}Ni_{0.2}O_{3-\delta}$. J. Power Sources. 2007;169:35–39.

[42] Chiba R, Yoshimura F, Sakurai Y. An investigation of $LaNi_{1-x}Fe_xO_3$ as a cathode material for solid oxide fuel cells. Solid State Ionics. 1999;124:281–288.

[43] Ruiz de Larramendi I, Ortiz N, López-Antón R, Ruiz de Larramendi JI, Rojo T. Structure and impedance spectroscopy of $La_{0.6}Ca_{0.4}Fe_{0.8}Ni_{0.2}O_{3-\delta}$ thin films grown by pulsed laser deposition. J. Power Sources. 2007;171:747–753.

[44] El-Himri A, Marrero-López D, Ruiz-Morales JC, Peña-Martínez J, Núñez P. Structural and electrochemical characterisation of $Pr_{0.7}Ca_{0.3}Cr_{1-y}Mn_yO_{3-\delta}$ as symmetrical solid oxide fuel cell electrodes. J. Power Sources. 2009;188:230–237.

[45] Ortiz-Vitoriano N, Ruiz de Larramendi I, Ruiz de Larramendi JI, Arriortua MI, Rojo T. Synthesis and electrochemical performance of $La_{0.6}Ca_{0.4}Fe_{1-x}Ni_xO_3$ (x = 0.1, 0.2, 0.3) material for solid oxide fuel cell cathode. J. Power Sources. 2009;192;63–69.

[46] Taguchi H, Masunaga Y, Hirota K, Yamaguchi O, Synthesis of perovskite-type $(La_{1-x}Ca_x)FeO_3$ ($0 \leq x \leq 0.2$) at low temperature. Mater Res. Bull. 2005;40:773–780.

[47] Montini T, Bevilacqua M, Fonda E, Casulla MF, Lee S, Tavagnacco C, Gorte RJ, Fornasiero P. Relationship between electrical behavior and structural characteristics in Sr-doped $LaNi_{0.6}Fe_{0.4}O_{3-\delta}$ mixed oxides. Chem. Mater. 2009;21:1768–1774.

[48] Ortiz-Vitoriano N, Ruiz de Larramendi I, Cook SN, Burriel M, Aguadero A, Kilner JA, Rojo T. The formation of performance enhancing pseudo-composites in the highly active $La_{1-x}Ca_xFe_{0.8}Ni_{0.2}O_3$ system for IT-SOFC Application. Adv. Funct. Mater. 2013;23:5131–5139.

[49] Ortiz-Vitoriano N, Ruiz de Larramendi I, Ruiz de Larramendi JI, Arriortua MI, Rojo T. Optimization of $La_{0.6}Ca_{0.4}Fe_{0.8}Ni_{0.2}O_3$–$Ce_{0.8}Sm_{0.2}O_2$ composite cathodes for intermediate-temperature solid oxide fuel cells. J. Power Sources. 2011;196:4332–4336.

[50] Dusastre V, Kilner JA. Optimisation of composite cathodes for intermediate temperature SOFC applications. Solid State Ionics. 1999;126:163–174.

[51] Wang K, Ran R, Zhou W, Gu H, Shao Z, Ahn J. Properties and performance of $Ba_{0.5}Sr_{0.5}Co_{0.8}Fe_{0.2}O_{3-\delta}$ + $Sm_{0.2}Ce_{0.8}O_{1.9}$ composite cathode. J. Power Sources. 2008;179:60–68.

[52] Tarancón A, Morata A, Dezanneau G, Skinner SJ, Kilner JA, Estradé S, Hernández-Ramírez F, Peiró F, Morante JR. $GdBaCo_2O_{5+x}$ layered perovskite as an intermediate temperature solid oxide fuel cell cathode. J. Power Sources. 2007;174:255–263.

[53] Frontera C, García-Muñoz JL, Castaño O, Ritter C, Caneiro A. The effect of oxygen disorder on magnetic properties of $PrBaCo_2O_{5.50}$layered cobaltite. J. Phys.: Condens. Matter. 2008;20:104228(8)

[54] Kim JH, Kim YM, Connor PA, Irvine JTS, Bae J, Zhou WZCS. Structural, thermal and electrochemical properties of layered perovskite $SmBaCo_2O_{5+d}$, a potential cathode

material for intermediate-temperature solid oxide fuel cells. J. Power Sources. 2009;194:704–711

[55] Zhao L, He BB, Xun ZQ, Wang H, Peng RR, Meng GY, Liu XQ. Characterization and evaluation of $NdBaCo_2O_{5+\delta}$ cathode for proton-conducting solid oxide fuel cells. Int. J. Hydrogen Energ. 2010;35:753–756.

[56] Kim G, Wang S,. Jacobson AJ, Reimus L, Brodersen P, Mims CA. Rapid oxygen ion diffusion and surface exchange kinetics in $PrBaCo_2O_{5+x}$with a perovskite related structure and ordered A cations. J. Mater. Chem. 2007;17:2500–2505.

[57] Bak T, Nowotny J, Rekas M, Sorell CC, Vance ER. A manometric method for the determination of chemical diffusion in non-stoichiometric oxides: Example of (La,Sr)MnO3. Solid State Sci. 2000;135:557–561.

[58] Marrero-López D, dos Santos-Gómez L, Canales-Vázquez J, Martín F, Ramos-Barrado JR. Stability and performance of nanostructured $La_{0.8}Sr_{0.2}MnO_3$ cathodes deposited by spray-pyrolysis. Electrochim. Acta. 2014;134:159–166.

[59] Bouwmeester HJM, Kruidhof H, Burggraaf AJ. Importance of the surface exchange kinetics as rate limiting step in oxygen permeation through mixed-conducting oxides. Solid State Ionics. 1994;72:185–194.

[60] Søgaard M, Vang Hendriksen P, Mogensen M. Oxygen nonstoichiometry and transport properties of strontium substituted lanthanum ferrite. J. Solid State Chem. 2007;180:1489–1503.

[61] Ralph JM, Rossignol Cc, Kumar R. Cathode materials for reduced-temperature SOFCs. J. Electrochem. Soc. 2003;150:A1518-A1522.

[62] Zomorrodian A, Salamati H, Lu Z, Chen X, Wu N, Ignatiev A. Electrical conductivity of epitaxial $La_{0.6}Sr_{0.4}Co_{0.2}Fe_{0.8}O_{3-\delta}$ thin films grown by pulsed laser deposition. Int. J. Hydrogen Energ. 2010;35:12443–12448.

[63] Cox-Galhotra RA, McIntosh S. Unreliability of simultaneously determining k_{chem} and D_{chem} via conductivity relaxation for surface-modified $La_{0.6}Sr_{0.4}Co_{0.2}Fe_{0.8}O_{3-\delta}$. Solid State Ionics. 2010;181:1429–1436.

[64] Lu H, Zhu L, Kim JP, Son SH, Park JH. Perovskite $La_{0.6}Sr_{0.4}B_{0.2}Fe_{0.8}O_{3-\delta}$ (B = Ti, Cr, Co) oxides: Structural, reduction-tolerant, sintering, and electrical properties. Solid State Ionics. 2012;209–210:24–29.

[65] Garcia LMP, Macedo DA, Souza GL, Motta FV, Paskocimas CA, Nascimento RM. Citrate–hydrothermal synthesis, structure and electrochemical performance of $La_{0.6}Sr_{0.4}Co_{0.2}Fe_{0.8}O_{3-\delta}$ cathodes for IT-SOFCs. Ceram. Int. 2013;39:8385–8392.

[66] Zhu G, Fang X, Xia C, Liu X. Preparation and electrical properties of $La_{0.4}Sr_{0.6}Ni_{0.2}Fe_{0.8}O_3$ using a glycine nitrate process. Ceram. Int. 2005;31:115–119.

[67] Chiba R, Yoshimura F, Sakurai Y. Properties of La1-ySryNi1-xFexO3 as a cathode material for a low-temperature operating SOFC. Solid State Ionics. 2002;152–153:575–582.

[68] Zhou Q, Xu L, Guo Y, Jia D, Li Y, Wei WCJ. $La_{0.6}Sr_{0.4}Fe_{0.8}Cu_{0.2}O_{3-\delta}$ perovskite oxide as cathode for IT-SOFC. Int. J. Hydrogen Energ. 2012;37:11963–11968.

[69] Meng X, Lü S, Ji Y, Wei T, Zhang Y. Characterization of $Pr_{1-x}Sr_xCo_{0.8}Fe_{0.2}O_{3-\delta}$ ($0.2 \leq x \leq 0.6$) cathode materials for intermediate-temperature solid oxide fuel cells. J. Power Sources. 2008;183:581–585.

[70] Hashimoto S, Kammer K, Poulsen FW, Mogensen M. Conductivity and electrochemical characterization of $PrFe_{1-x}Ni_xO_{3-\delta}$ at high temperature. J. Alloys Compd. 2007;428:256–261.

[71] Rebello J, Vashook V, Trots D, Guth U. Thermal stability, oxygen non-stoichiometry, electrical conductivity and diffusion characteristics of $PrNi_{0.4}Fe_{0.6}O_{3-\delta}$, a potential cathode material for IT-SOFCs. J. Power Sources. 2011;196:3705–3712.

[72] Guo YQ, Yin YM, Tong Z, Yin JW, Xiong MW, Ma ZF. Impact of synthesis technique on the structure and electrochemical characteristics of $Pr_{0.6}Sr_{0.4}Co_{0.2}Fe_{0.8}O_{3-\delta}$ (PSCF) cathode material. Solid State Ionics. 2011;193:18–22.

[73] Kostogloudis GC, Ftikos C. Crystal structure, thermal expansion and electrical conductivity of $Pr_{1-x}Sr_xCo_{0.2}Fe_{0.8}O_{3-\delta}$ ($0 \leq x \leq 0.5$). Solid State Ionics. 2000;135:537–541.

[74] Chang CL, Hsu CS, Hwang BH. Unique porous thick $Sm_{0.5}Sr_{0.5}CoO_3$ solid oxide fuel cell cathode films prepared by spray pyrolysis. J. Power Sources. 2008;179:734–738.

[75] Yang S, He T, He Q. Sm0.5Sr0.5CoO3 cathode material from glycine-nitrate process: Formation, characterization, and application in $LaGaO_3$-based solid oxide fuel cells. J. Alloys Compd. 2008;450:400–404.

[76] Wei B, Lü Z, Huang X, Liu M, Li N, Su W. Synthesis, electrical and electrochemical properties of $Ba_{0.5}Sr_{0.5}Zn_{0.2}Fe_{0.8}O_{3-\delta}$ perovskite oxide for IT-SOFC cathode. J. Power Sources. 2008;176:1–8.

[77] Lee S, Lim Y, Lee EA, Hwang HJ, Moon J-W. $Ba_{0.5}Sr_{0.5}Co_{0.8}Fe_{0.2}O_{3-\delta}$ (BSCF) and $La_{0.6}Ba_{0.4}Co_{0.2}Fe_{0.8}O_{3-\delta}$ (LBCF) cathodes prepared by combined citrate-EDTA method for IT-SOFCs. J. Power Sources. 2006;157:848–854.

[78] Wei B, Lü Z, Huang X, Miao J, Sha X, Xin X, et al. Crystal structure, thermal expansion and electrical conductivity of perovskite oxides $Ba_xSr_{1-x}Co_{0.8}Fe_{0.2}O_{3-\delta}$ ($0.3 \leq x \leq 0.7$). J. Eur. Ceram. Soc. 2006;26:2827–2832.

[79] Chen D, Ran R, Zhang K, Wang J, Shao Z. Intermediate-temperature electrochemical performance of a polycrystalline $PrBaCo_2O_{5+\delta}$ cathode on samarium-doped ceria electrolyte. J. Power Sources. 2009;188:96–105.

[80] Meng F, Xia T, Wang J, Shi Z, Zhao H. Praseodymium-deficiency $Pr_{0.94}BaCo_2O_{6-\delta}$ double perovskite: A promising high performance cathode material for intermediate-temperature solid oxide fuel cells. J. Power Sources. 2015;293:741–750.

[81] Jin F, Xu H, Long W, Shen Y, He T. Characterization and evaluation of double perovskites $LnBaCoFeO_{5+\delta}$ (Ln = Pr and Nd) as intermediate-temperature solid oxide fuel cell cathodes. J. Power Sources. 2013;243:10–18.

[82] Li N, Lü Z, Wei B, Huang X, Chen K, Zhang Y, et al. Characterization of $GdBaCo_2O_{5+\delta}$ cathode for IT-SOFCs. J. Alloys Compd. 2008;454:274–279.

[83] Taskin AA, Lavrov AN, Ando Y. Achieving fast oxygen diffusion in perovskites by cation ordering. Appl. Phys. Lett. 2005;86:091910.

[84] Zhou Q, He T, Ji Y. $SmBaCo_2O_{5+x}$ double-perovskite structure cathode material for intermediate-temperature solid-oxide fuel cells. J. Power Sources. 2008;185:754–758.

[85] Ding X, Kong X, Wu H, Zhu Y, Tang J, Zhong Y. $SmBa_{0.5}Sr_{0.5}Cu_2O_{5+\delta}$ and $SmBa_{0.5}Sr_{0.5}CuFeO_{5+\delta}$ layered perovskite oxides as cathodes for IT-SOFCs. Int. J. Hydrogen Energ. 2012;37:2546–2551.

[86] Xue J, Shen Y, He T. Double-perovskites $YBaCo_{2-x}Fe_xO_{5+\delta}$ cathodes for intermediate-temperature solid oxide fuel cells. J. Power Sources. 2011;196:3729–3735.

[87] Zhu WZ, Deevi SC. A review on the status of anode materials for solid oxide fuel cells. Mater. Sci. Eng. A 2003;362:228–239.

[88] Sun C, Stimming U. Recent anode advances in solid oxide fuel cells. J. Power Sources. 2007;171:247–260.

[89] Atkinson A, Barnett S, Gorte RJ, Irvine JTS, McEvoy AJ, Mogensen M, Singhal SC, Vohs J. Advanced anodes for high-temperature fuel cells. Nat. Mater. 2004;3:17–27.

[90] Boukamp BA. Fuel cells: The amazing perovskite anode. Nat. Mater. 2003;2:294–296.

[91] Verbraeken MC, Ramos T, Agersted K, Ma Q, Savaniu CD, Sudireddy BR, Irvine JTS, Holtappels P, Tietz F. Modified strontium titanates: from defect chemistry to SOFC anodes. RSC Adv. 2015;5:1168–1180.

[92] Yurkiv V, Constantin G, Hornes A, Gondolini A, Mercadelli E, Sanson A, Dessemond L, Costa R. Towards understanding surface chemistry and electrochemistry of $La_{0.1}Sr_{0.9}TiO_{3-\alpha}$ based solid oxide fuel cell anodes. J. Power Sources. 2015;287:58–67.

[93] Marina OA, Canfield NL, Stevenson JW. Thermal, electrical, and electrocatalytical properties of lanthanum-doped strontium titanate. Solid State Ionics. 2002;149:21–28.

[94] Hashimoto S, Kindermann L, Poulsen FW, Mogensen M. A study on the structural and electrical properties of lanthanum-doped strontium titanate prepared in air. J. Alloys Compd. 2005;397:245–249.

[95] Canales-Vázquez J, Ruiz-Morales JC, Irvine JTS, Zhou W. Sc-substituted oxygen excess titanates as fuel electrodes for SOFCs. J. Electrochem. Soc. 2005;152:A1458–A1465.

[96] Miller DN, Irvine JTS. B site doping of lanthanum strontium titanate for solid oxide fuel cell anodes. J. Power Sources. 2011;196:7323–7327.

[97] Cui SH, Li JH, Zhou XW, Wang GY, Luo JL, Chuang KT, Bai Y, Qiao LJ. Cobalt doped $LaSrTiO_{3-\delta}$ as an anode catalyst: effect of Co nanoparticle precipitation on SOFCs operating on H_2S-containing hydrogen. J. Mater. Chem. A. 2013;1:9689–9696.

[98] Ma Q, Tietz F. Comparison of Y and La-substituted $SrTiO_3$ as the anode materials for SOFCs. Solid State Ionics. 2012;225:108–112.

[99] Park BH, Choi GM. Ex-solution of Ni nanoparticles in a $La_{0.2}Sr_{0.8}Ti_{1-x}Ni_xO_{3-\delta}$ alternative anode for solid oxide fuel cell. Solid State Ionics. 2014;262:345–348.

[100] Périllat-Merceroz C, Gauthier G, Roussel P, Huvé M, Gélin P, Vannier RN. Synthesis and study of a Ce-doped La/Sr titanate for solid oxide fuel cell anode operating directly on methane. Chem. Mater. 2011;23:1539–1550.

[101] Ruiz-Morales JC, Canales-Vázquez J, Savaniu C, Marrero-López D, Zhou W, Irvine JTS. Disruption of extended defects in solid oxide fuel cell anodes for methane oxidation. Nature 2006;439:568–571.

[102] Chamberland BL, Danielson PS. Alkaline-earth vanadium (IV) oxides having the AVO_3 composition. J. Solid State Chem. 1971;3:243–247.

[103] Martínez-Coronado R, Alonso JA, Aguadero A, Fernández-Díaz MT. Optimized energy conversion efficiency in solid-oxide fuel cells implementing $SrMo_{1-x}Fe_xO_{3-\delta}$ perovskites as anodes. J. Power Sources. 2012;208:153–158.

[104] Martínez-Coronado R, Alonso JA, Aguadero A, Fernández-Díaz MT. New $SrMo_{1-x}Cr_xO_{3-\delta}$ perovskites as anodes in solid-oxide fuel cells. Int. J. Hydrogen Energ. 2014;39:4067–4073.

[105] Martínez-Coronado R, Alonso JA, Fernández-Díaz MT. $SrMo_{0.9}Co_{0.1}O_{3-\delta}$: A potential anode for intermediate-temperature solid-oxide fuel cells (IT-SOFC). J. Power Sources. 2014;258:76–82.

[106] Du Z, Zhao H, Yang C, Shen Y, Yan C, Zhang Y. Optimization of strontium molybdate based composite anode for solid oxide fuel cells. J. Power Sources. 2015;274:568–574.

[107] Zheng K, Świerczek K, Bratek J, Klimkowicz A. Cation-ordered perovskite-type anode and cathode materials for solid oxide fuel cells. Solid State Ionics. 2014;262:354–358.

[108] Liu Q, Dong X, Xiao G, Zhao F, Chen F. A novel electrode material for symmetrical SOFCs. Adv. Mater. 2010;22:5478–5482.

[109] Suthirakun S, Ammal SC, Muñoz-García AB, Xiao G, Chen F, zur Loye HC, Carter EA, Heyden A. Theoretical investigation of H_2 oxidation on the $Sr_2Fe_{1.5}Mo_{0.5}O_6$ (001) perovskite surface under anodic solid oxide fuel cell conditions. J. Am. Chem. Soc. 2014;136:8374–8386.

[110] Sfeir J. $LaCrO_3$-based anodes: stability considerations. J. Power Sources. 2003;118:276–285.

[111] Lü MF, Tsipis EV, Waerenborgh JC, Yaremchenko AA, Kolotygin VA, Bredikhin S, Kharton VV. Thermomechanical, transport and anodic properties of perovskite-type $(La_{0.75}Sr_{0.25})_{0.95}Cr_{1-x}Fe_xO_{3-\delta}$. J. Power Sources. 2012;206:59–69.

[112] Tao SW, Irvine JTS. Catalytic properties of the perovskite oxide $La_{0.75}Sr_{0.25}Cr_{0.5}Fe_{0.5}O_{3-\delta}$ in relation to its potential as a solid oxide fuel cell anode material. Chem. Mater. 2004;16:4116–4121.

[113] Wei T, Liu XJ, Yuan C,. Gao QY, Xin XS, Wang SR. A modified liquid-phase-assisted sintering mechanism for $La_{0.8}Sr_{0.2}Cr_{1-x}Fe_xO_{3-\delta}$—A high density, redox-stable perovskite interconnect for solid oxide fuel cells. J. Power Sources. 2014;250:152–159.

[114] Fowler DE, Haag JM, Boland C, Bierschenk DM, Barnett SA, Poeppelmeier KR. Stable, low polarization resistance solid oxide fuel cell anodes: $La_{1-x}Sr_xCr_{1-x}Fe_xO_{3-\delta}$ (x = 0.2–0.67). Chem. Mater. 2014;26:3113–3120.

[115] Huang YH, Dass RI, Xing ZL, Goodenough JB. Double perovskites as anode materials for solid-oxide fuel cells. Science 2006;312:254–257.

[116] Choi S, Yoo S, Kim J, Park S, Jun A, Sengodan S, Kim J, Shin J, Jeong HY, Choi Y, Kim G, Liu M. Highly efficient and robust cathode materials for low-temperature solid oxide fuel cells: $PrBa_{0.5}Sr_{0.5}Co_{2-x}Fe_xO_{5+\delta}$. Sci. Rep. 2013;3:2426-6.

[117] Sengodan S, Choi S, Jun A, Shin TH, Ju YW, Jeong HY, Shin J, Irvine JTS, Kim G. Layered oxygen-deficient double perovskite as an efficient and stable anode for direct hydrocarbon solid oxide fuel cells. Nat. Mater. 2015;14:205–209.

[118] Périllat-Merceroz C, Roussel P, Huvé M, Capoen E, Rosini S, Gélin P, Vannier RN, Gauthier GH. Pure and Mn-doped $La_4SrTi_5O_{17}$ layered perovskite as potential solid oxide fuel cell material: Structure and anodic performance J. Power Sources. 2015;274:806–815.

[119] Ruiz de Larramendi I, Ortiz-Vitoriano N, Acebedo B, Jimenez de Aberasturi D, Gil de Muro I, Arango A, Rodríguez-Castellón E, Ruiz de Larramendi JI, Rojo T. Pr-doped ceria nanoparticles as intermediate temperature ionic conductors. Int. J. Hydrogen Energ. 2011;36:10981–10990.

[120] McEvoy AJ. Thin SOFC electrolytes and their interfaces—A near-term research strategy. Solid State Ionics. 2000;132:159–165.

[121] Chen XJ, Khor KA, Chan SH, Yu LG. Influence of microstructure on the ionic conductivity of yttria-stabilized zirconia electrolyte. Mater. Sci. Eng. A 2002;A335:246–252

[122] Vlasov AN, Perfiliev MV. Ageing of ZrO_2-based solid electrolytes. Solid State Ionics 1987;25:245–253.

[123] Slater PR, Irvine JTS, Ishihara T, Takita Y. The structure of the oxide ion conductor $La_{0.9}Sr_{0.1}Ga_{0.8}Mg_{0.2}O_{2.85}$ by powder neutron diffraction. Solid State Ionics. 1998;107:319–323.

[124] Kim S, Chun MC, Lee KT, Lee HL. Oxygen-ion conductivity of BaO- and MgO-doped $LaGaO_3$ electrolytes. J. Power Sources. 2001;93:279–284.

[125] Sood K, Singh K, Pandey OP. Structural and electrical behavior of Ba-doped $LaGaO_3$ composite electrolyte. J. Renew. Sust. Energy. 2014;6:063112.

[126] Kurumada M, Hara H, Munakata F, Iguchi E. Electric conductions in $La_{0.9}Sr_{0.1}GaO_{3-\delta}$ and $La_{0.9}Sr_{0.1}Ga_{0.9}Mg_{0.1}O_{3-\delta}$. Solid State Ionics. 2005;176:245–251.

[127] Mineshige A, Izutsu J, Nakamura M, Nigaki K, Abe J, Kobune M, Fujii S, Yazawa T. Introduction of *A* site deficiency into $La_{0.6}Sr_{0.4}Co_{0.2}Fe_{0.8}O_{3-\delta}$ and its effect on structure and conductivity. Solid State Ionics. 2005;176:1145–1149.

[128] Ahamd-Khanlou A, Tietz F, Stöver D. Material properties of $La_{0.8}Sr_{0.2}Ga_{0.9+x}Mg_{0.1}O_{3-\delta}$ as a function of Ga content. Solid State Ionics. 2000;135:543–547.

[129] Zhang L, Xiao J, Xie Y, Tang Y, Liu J, Liu M. Behavior of strontium- and magnesium-doped gallate electrolyte in direct carbon solid oxide fuel cells. J. Alloys Compd. 2014;608:272–277.

[130] Wang XP, Zhou DF, Yang GC, Sun SC, Li ZH, Fu H, Meng J. Nonstoichiometric $(La_{0.95}Sr_{0.05})_xGa_{0.9}Mg_{0.1}O_{3-\delta}$ electrolytes and $Ce_{0.8}Nd_{0.2}O_{1.9}$–$(La_{0.95}Sr_{0.05})_xGa_{0.9}Mg_{0.1}O_{3-\delta}$ composite electrolytes for solid oxide fuel cells. Int. J. Hydrogen Energ. 2014;39:1005–1013

[131] Feng M, Goodenough JB, Huang K, Milliken C. Fuel cells with doped lanthanum gallate electrolyte. J. Power Sources. 1996;63:47–51.

[132] Lo Faro M, Aricò AS. Electrochemical behaviour of an all-perovskite-based intermediate temperature solid oxide fuel cell. Int. J. Hydrogen Energ. 2013;38:14773–14778

[133] Yamazaki Y, Blanc F, Okuyama Y, Buannic L, Lucio-Vega JC, Grey CP, Haile SM. Proton trapping in yttrium-doped barium zirconate. Nat. Mater. 2013;12:647–651.

[134] Liu Y, Ran R, Li S, Jiao Y, Tade MO, Shao Z. Significant performance enhancement of yttrium-doped barium cerate proton conductor as electrolyte for solid oxide fuel cells through a Pd ingress–egress approach. J. Power Sources. 2014;257:308–318.

[135] Stevenson DA, Jiang N, Buchanan RM, Henn FEG. Characterization of Gd, Yb and Nd doped barium cerates as proton conductors. Solid State Ionics. 1993;62:279–285.

[136] Oishi M, Yashiro K, Sato K, Mizusaki J, Kitamura N, Amezawa K, Kawada T, Uchimoto Y. Oxygen nonstoichiometry of the perovskite-type oxides $BaCe_{0.9}M_{0.1}O_{3-\delta}$ (M = Y, Yb, Sm, Tb, and Nd). Solid State Ionics. 2008;179:529–535.

[137] Nguyen NTQ, Yoon HH. Preparation and evaluation of $BaZr_{0.1}Ce_{0.7}Y_{0.1}Yb_{0.1}O_{3-\delta}$ (BZCYYb) electrolyte and BZCYYb-based solid oxide fuel cells. J. Power Sources. 2013;231:213–218.

[138] Fergus JW. Lanthanum chromite-based materials for solid oxide fuel cell interconnects. Solid State Ionics. 2004;171:1–15.

[139] Devi PS, Rao MS. Preparation, structure, and properties of strontium-doped lanthanum chromites: $La_{1-x}Sr_xCrO_3$. J. Solid State Chem. 1992;98:237–244.

[140] Flandermeyer BK, Nasrallah MM, Agarwal AK, Anderson HU. Defect structure of Mg-doped $LaCrO_3$ model and thermogravimetric measurements. J. Am. Ceram. Soc. 1984;67:195–198.

[141] Coutures JP, Badie JM, Berjoan R, Coutures J, Flamand R, Rouanet A. Stability and thermodynamic properties of rare earth perovskites. High Temp. Sci. 1979;13: 331–336.

[142] Lee HC, Kang BK, Lee JH, Heo YW, Kim JY, Kim JJ. Adhesion behavior between yttrium-stabilized zirconia added $La_{0.8}Ca_{0.2}Cr_{0.9}Co_{0.1}O_{3-\delta}$ interconnector and yttrium-stabilized zirconia-based substrate. Ceram. Int. 2013;39:8737–8741.

[143] Ruiz-Morales JC, Canales-Vázquez J, Marrero-López D, Peña-Martínez J, Pérez-Coll D, Núñez P, Rodríguez-Placeres JC, Ballesteros-Pérez B, Dorta-Martín VI, Savaniu C. Pilas de Combustible de Óxidos Sólidos (SOFC) Ed. Centro de la Cultura Popular Canaria. ISBN: 978-84-7926-567-0, 2008.

[144] Choi J, Kim B, Shin D. Fibrous mixed conducting cathode with embedded ionic conducting particles for solid oxide fuel cells. Int. J. Hydrogen Energ. 2014;39:14460–14465.

[145] Hueso L, Mathur N. Dreams of a hollow future. Nature. 2004;427:301–304.

[146] Ruiz-Morales JC, Marrero-López D, Gálvez-Sánchez M, Canales-Vázquez J, Savaniu C, Savvina SN. Engineering of materials for solid oxide fuel cells and other energy and environmental applications. Energy Environ. Sci. 2010;3:1670–1681.

[147] Pinedo R, Ruiz de Larramendi I, Ortiz-Vitoriano N, Jimenez de Aberasturi D, Rojo T. Nanotechnology for improving solid oxide fuel cells. Materials and processes for energy: communicating current research and technological developments (A. Méndez-Vilas, Ed.). 512–522. 2013.

[148] Jimenez de Aberasturi D, Serrano-Montes AB, Liz-Marzán LM. Modern applications of plasmonic nanoparticles: from energy to health. Adv. Optical Mater. 2015;3:602–617.

[149] Jiang SP. A review of wet impregnation—An alternative method for the fabrication of high performance and nano-structured electrodes of solid oxide fuel cells. Mater. Sci. Eng. A. 2006;418:199–210.

[150] Liu Z, Liu B, Ding D, Liu M, Chen F, Xia C. Fabrication and modification of solid oxide fuel cell anodes via wet impregnation/infiltration technique. J. Power Sources. 2013;237:243–259.

[151] Marrero-López D, Ruiz-Morales JC, Peña-Martínez J, Canales-Vázquez J, Núñez P. Preparation of thin layer materials with macroporous microstructure for SOFC applications. J. Solid State Chem. 2008;181:685–692.

[152] Pinedo R, Ruiz de Larramendi I, Gil de Muro I, Insausti M, Ruiz de Larramendi JI, Arriortua MI, Rojo T. Influence of colloidal templates on the impedance spectroscopic behaviour of $Pr_{0.7}Sr_{0.3}Fe_{0.8}Ni_{0.}$ or solid oxide fuel cell applications. Solid State Ionics. 2011;192:235–240.

[153] Pinedo R, Ruiz de Larramendi I, Jimenez de Aberasturi D, Gil de Muro I, Aguayo AT, Ruiz de Larramendi JI, Rojo T. A straightforward synthesis of carbon nanotube–perovskite composites for solid oxide fuel cells. J. Mater. Chem. 2011;21:10273–10276.

[154] Pinedo R, Ruiz de Larramendi I, Khavrus VO, Jimenez de Aberasturi D, Ruiz de Larramendi JI, Ritscheld M, Leonhardt A, Rojo T. Microstructural improvements of the gradient composite material $Pr_{0.6}Sr_{0.4}Fe_{0.8}Co_{0.2}O_3/Ce_{0.8}Sm_{0.2}O_{1.9}$ by employing vertically aligned carbon nanotubes. Int. J. Hydrogen Energ. 2014;39:4074–4080.

[155] Bellino MG, Sacanell JG, Lamas DG, Leyva AG, Walsöe de Reca NE. High-performance solid-oxide fuel cell cathodes based on cobaltite nanotubes. J. Am. Chem. Soc. 2007;129:3066–3067.

[156] Liu J, Wang X, Peng Q, Li Y. Vanadium pentoxide nanobelts: highly selective and stable ethanol sensor materials. Adv. Mater. 2005;17:764–767.

[157] Kuo TJ, Huang MH. Gold-catalyzed low-temperature growth of cadmium oxide nanowires by vapor transport. J. Phys. Chem. B 2006;110:13717–13721.

[158] Ostermann R, Li D, Yin Y, McCann JT, Xia Y. V_2O_5 Nanorods on TiO_2 Nanofibers: A new class of hierarchical nanostructures enabled by electrospinning and calcination. Nano Lett. 2006;6:1297–1302.

[159] Kwon CW, Son JW, Lee JH, Kim HM, Lee HW, Kim KB. High-performance micro-solid oxide fuel cells fabricated on nanoporous anodic aluminum oxide templates. Adv. Funct. Mater. 2011;21:1154–1159.

[160] Boehme M, Ionescu E, Fu G, Ensinger W. Room temperature synthesis of indium tin oxide nanotubes with high precision wall thickness by electroless deposition. J. Nanotech. 2011;2:119–126.

[161] Pinedo R, Ruiz de Larramendi I, Jimenez de Aberasturi D, Gil de Muro I, Ruiz de Larramendi JI, Arriortua MI, Rojo T. Synthesis of highly ordered three-dimensional nanostructures and the influence of the temperature on their application as solid oxide fuel cells cathodes. J. Power Sources. 2011;196:4174–4180.

[162] Zhi M, Lee S, Miller N, Menzler NH, Wu N. An intermediate-temperature solid oxide fuel cell with electrospun nanofiber cathode. Energy Environ. Sci. 2012;5:7066–7071.

[163] Huang YY, Vohs JM, Gorte RJ. Characterization of LSM-YSZ composites prepared by impregnation methods. J. Electrochem. Soc. 2005;152:A1347−A1353.

[164] Zhi MJ, Mariani N, Gemmen R, Gerdes K, Wu NQ. Nanofiber scaffold for cathode of solid oxide fuel cell. Energy Environ. Sci. 2011;4:417−420.

[165] Zhang N, Li J, Li W, Ni D, Sun K. High performance three-dimensionally ordered macroporous composite cathodes for intermediate temperature solid oxide fuel cells. RSC Adv. 2012;2:802–804.

[166] Zhang X, Liu L, Zhao Z, Tu B, Ou D, Cui D, Wei X, Chen X, Cheng M. Enhanced oxygen reduction activity and solid oxide fuel cell performance with a nanoparticles-loaded cathode. NanoLett. 2015;15:1703−1709.

[167] Park BH, Choi GM. Ex-solution of Ni nanoparticles in a $La_{0.2}Sr_{0.8}Ti_{1-x}Ni_xO_{3-\delta}$ alternative anode for solid oxide fuel cell. Solid State Ionics. 2014;262:345–348.

[168] Vohs JM, Gorte RJ. High-performance SOFC cathodes prepared by infiltration. Adv. Mater. 2009;21:943−956.

[169] Pinedo R, Ruiz de Larramendi I, Ortiz-Vitoriano N, Gil de Muro I, Rojo T. Novel $Pr_{0.6}Sr_{0.4}Fe_{0.8}Co_{0.2}O_3$:$Ce_{0.8}Sm_{0.2}O_2$ composite nanotubes for energy conversion and storage. J. Power Sources. 2012;201:332–339.

[170] Fan L, Wang C, Zhu B. Low temperature ceramic fuel cells using all nanocomposite materials. NanoEnergy. 2012;1:631–639.

Chapter 21

Perovskites Used in Fuel Cells

Diego Pereira Tarragó, Berta Moreno, Eva Chinarro and
Vânia Caldas de Sousa

Additional information is available at the end of the chapter

http://dx.doi.org/10.5772/61465

Abstract

Fuel cells are devices for energy generation with very high theoretical efficiency. Many researches were been carried out in the last few decades in order to develop reliable fuel cells. Solid oxide fuel cells (SOFC) and polymeric exchange membrane fuel cells (PEMFC) are those with more potential for commercial use. Specially for SOFC cathodes, many perovskites have been proposed as potential materials for this application. Nevertheless, other components of SOFC, such as the electrolytes, anodes and interconnects, have also been targeted with potential perovskites. More recently, the use of perovskites in PEMFC has also been proposed and studied. As many perovskite compositions can be used in SOFC components, some of the most important are discussed in this chapter and some recent works in perovskites for PEMFC are also referred. As a whole, in this chapter, the reader will find the relationship between the properties of perovskites with their compositions and the main effects of dopant agents regarding the utilization of these materials in different components of SOFC and in electrodes of PEMFC.

Keywords: SOFC, IT-SOFC, PEMFC, Nonstoichiometric compounds

1. Introduction

Fuel cells are devices that convert the chemical energy of a fuel directly into electrical energy and heat. The most common fuel is H_2, but other hydrocarbon compounds such as methanol, methane, natural gas, ethanol or others can also be used. A single cell is composed of three main components: anode, cathode and electrolyte. For the effective use of fuel cells, single cells must be interconnected to increase the power production, which requires the use of two more components: interconnects, for the serial electrical connection, and sealants, for the hermetic sealing of the set. The electrodes are permeated by the gases, fuel in the anode and oxygen (air) in the cathode, and they catalyse electrochemical reactions through electron capitation or conduction from or to the reactive sites; the electrolyte, an electrical insulator, promotes ionic conductivity. Figure 1 shows a general schematic drawing of a fuel cell operation. The residual water (if pure H_2 is used) can be produced in the anode or in the cathode and it will depend of the nature of the electrolyte. If the electrolyte is a cationic conductor, the water will form in the cathode and, conversely, if it is an anionic conductor, the water will be formed in the anode.

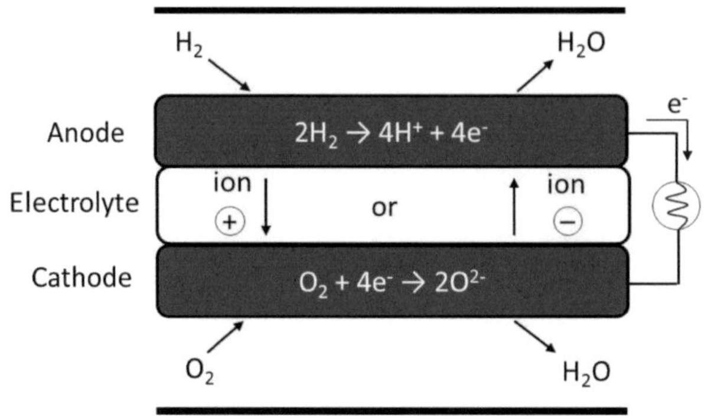

Figure 1. General operation scheme of a fuel cell running with H_2.

In general, a fuel cell works similarly to a battery; however, its energy is not stored in its electrodes, so there is no need for recharging because there is a continuous supply of fuel in the anode and oxidants (air) in the cathode. The electrical work provided by the electrochemical reactions does not consume the cell's components, which keeps on converting the chemical energy into electricity and heat while its electrodes are supplied with the gases.

There are different types of fuel cells and they are named according to the material used in the electrolyte. Besides the materials used in their components, the fuel cells also differ in other important aspects such as efficiency and operation temperature. In Table 1, the main characteristics of fuel cells are summarized.

Fuel cell type	**Alkaline (AFC)**	**Phosphoric Acid (PAFC)**	**Polymeric Exchange Membrane (PEMFC)**	**Molten Carbonate (MCFC)**	**Solid Oxide (SOFC)**
Electrolyte	Potassium hydroxide	Phosphoric acid	Polymer	Molten carbonates	Dense ceramics
Operation Temperature	50-120°C	180-210°C	60-100°C	550-650°C	550-1000°C
Transported ion	OH^-	H^+	H^+	CO_3^{2-}	O^{2-}
Expected efficiency	35-55%	35-45%	35-45%	45-55%	40-60%

Table 1. Fuel cell types and their characteristics

Among all the fuel cells, the SOFC is the one where materials with a perovskite structure are the most applied and studied. Except for the sealant, all other components of the SOFC can potentially consist of perovskite ceramics. The most common materials found in the single SOFC configuration are yttria-stabilized zirconia (YSZ) in the electrolyte, Ni/YSZ cermet in the anode and strontium-doped lanthanum manganite (LSM) in the cathode. As can be seen, only the cathode is composed of a perovskite material. However, several scientific researchers have attempted to substitute the materials of the electrolyte and the anode with perovskites. In the case of the electrolyte, their main goal is to increase the ionic conductivity at lower temperatures and, for the anode, the aim is to include hydrocarbons as potential fuels, since Ni is often poisoned by fuels containing carbon. For the interconnects, the perovskite materials used in this component were gradually substituted with stainless steel, a much cheaper material that is capable of supporting the cell; recently, perovskite-coated stainless steel has arisen as a potential interconnect material. Also, a new group of perovskites has received attention for proton-conducting SOFC, where the electrolyte works with cationic conduction, although these cells have not yet achieved a performance similar to the conventional oxygen-conducting cells.

As can be seen, the materials used in SOFC must fit many requirements in terms of electrochemical properties. Besides, chemical and physical compatibility must exist between the materials for each component. As SOFC runs in high temperature, a chemical decomposition or a chemical reaction cannot occur during its operation and the chosen materials must have a similar thermal expansion coefficient in order to avoid the formation of cracks during thermal cycling. The high operation temperature of SOFC brings some advantages: increasing the activity of the electrodes and the conductivity of the components and favours ing the kinetics of the electrochemical reactions and gaseous exchange. However, reducing their operation temperature also has its advantages: decreased densification and thermal stress and, more importantly, diversifying the materials used in their components, since many perovskites are unstable at high temperatures. SOFCs that work below 800°C can be referred to as intermediate temperature SOFC (IT-SOFC).

Recent researches indicate that besides SOFC components, perovskite materials also have potential for application in PEMFC electrodes. However, this study still is very incipient. In the coming sessions, perovskites used in the SOFC components have been addressed separately and according to the components where it can be applied. Then, in one session, perovskites for use in PEMFC electrodes were discussed.

2. Cathodes for SOFC

The cathode, in an SOFC, is the interface between the electrolyte and the oxygen, its main functions are conduct electrons to the reactive sites and catalyze the reduction of the O_2 molecules. Therefore, the material applied in this component must have specific properties in the operation temperature such as electronic conductivity, catalytic activity, chemical and physical compatibility with the electrolyte and interconnect, as well as a porous and stable microstructure. The choice of material depends mainly on the chemical composition of the electrolyte and the operating temperature.

As said, LSM is the most common material used in SOFC cathode due to its properties and compatibility with YSZ at high temperatures. Pure lanthanum manganite ($LaMnO_3$), with $A^{3+}B^{3+}O_3$ stoichiometry, is an intrinsic p-type semiconductor where the presence of cationic vacancies, primarily in the A sites, induces an oxygen nonstoichiometry. These vacancies are occupied by oxygen anions which lead to an oxygen excess that, in turn, causes the appearance of Mn^{4+} species in order to maintain the overall neutrality. The Mn^{4+}/Mn^{3+} ratio can be increased by doping both A and B sites, adding electronic holes in the structure and, hence, increasing the electronic conductivity. The most common is to dope the A site with bivalent cations such as Sr^{2+}, establishing the general formula $La_{1-x}Sr_xMnO_{3-\delta}$, according to the molar amount of strontium occupying lanthanum sites.

It is important to note that by doping the perovskites, different properties are influenced at the same time and, for application in SOFC, some of them can be disadvantageous. According to Shuk et al., doping A sites of lanthanum manganite improves its electronic conductivity and reaches its maximum at the composition $La_{0.45}Sr_{0.55}MnO_{3-\delta}$, Sr^{2+} is at 55% mol or $x = 55$, although this property also depends on the oxygen partial pressure. However, at this amount, the thermal expansion coefficient is too high, as observed by Florio et al., causing a physical mismatch with YSZ electrolytes. In amounts until 15% of strontium, there is a stabilization of the perovskite against YSZ, as reported by Yokokawa et al. but, despite this, the most common composition found in the literature is $La_{0.7}Sr_{0.3}MnO_{3-\delta}$. Doping the B sites increases the electronic conductivity and also the ionic conductivity, which is positive for the cathode performance; however, Tai et al. observed negative effects on the thermal expansion coefficient, derailing its use with YSZ electrolytes.

In the case of IT-SOFC, the temperature decrease impairs the performance of LSM, mostly because of the loss in catalytic activity. One alternative is to mix LSM with a fraction of YSZ, forming a composite cathode which can lead to better cathode/electrolyte adhesion and to an increase in the reactive sites. Mogensen et al. tested LSM/YSZ composite cathodes and showed an improvement in the catalytic activity. However, yet more promising is the use of mixed ionic electronic conductor (MIEC) perovskites. The use of this kind of material increases the amount of reactive sites, compensating the kinetic losses at temperatures below 800°C.

Doped lanthanum ferrites MIEC perovskites are a common example where the kinetic losses in the cathode, due to the low temperature, are compensated by an increase in the active reduction area. In the LSM perovskite, the B site doping with Co ions significantly increases the oxygen diffusion, and the substitution of Mn by Fe can enhance surface exchange processes. Based on this improvement, strontium- and cobalt-doped lanthanum ferrites (LSCF), with a general formula of $La_{1-x}Sr_xCo_{1-y}Fe_yO_{3-\delta}$, have excellent electrochemical properties for use in IT-SOFC. However, the thermal expansion coefficient of LSCF is not compatible with YSZ electrolytes, hence these perovskites can only be used with compatibility layers or with other compounds as electrolytes. In an attempt to overcome the thermal expansion incompatibility with YSZ electrolytes, Ko et al. used a composite cathode consisting of LSCF and GDC (gadolinea-doped ceria), another common electrolyte material, and achieved good overall performance, with a decrease of cathode polarization, for many hours of cell operation. However, other authors who tested cells with LSCF cathodes showed voltage loss with time, probably due to the thermal decomposition of this perovskite. Different amounts of dopants can be found in the literature and they report the high electronic and ionic conductivity of these compounds; they also show high activity for the O_2 reduction reaction.

Other compounds such as layered or double perovskites are also MIECs and potential materials for SOFC cathodes. These perovskites can have a structure coordinated by 4 or 5 oxygen atoms, and are typically $A_2BO_{4+\delta}$ or $AA'B_2O_{5\pm\delta}$. Some typical compositions of layered perovskites are $La_2NiO_{4+\delta}$ and $La_2CoO_{4+\delta}$, or with three cations $GdBaCo_2O_{5\pm\delta}$ and $PrBaCo_2O_{5\pm\delta}$. An interesting aspect of these materials is that they do not need to be heavily substituted to promote high oxygen diffusion. This characteristic is owed to its structure, where oxygen vacancies are concentrated in a plan between layers formed by cations, leading to anisotropic ion conduction due to a decrease in the activation energy for vacancy migration. Tarancón et al. compiled some values and showed that the oxygen self-diffusion coefficient at 500°C of $La_2NiO_{4+\delta}$ is 3.3×10^{-9} cm²/s, which is one order of magnitude higher than a typical LSCF/GDC composite cathode. The cation-ordered structure also plays a role in the oxygen surface exchange, which becomes a very important factor in cathodes with good oxygen diffusion, such as these.

Studies have also been carried out in order to determine the doping in layered perovskites; however, an undesirable increase in the thermal expansion coefficient was also observed. According to Kim et al., partial substitution with strontium to form $NdBa_{1-x}Sr_xCo_2O_{5+\delta}$ increases the specific area resistance of the cathodes and, relatively with pure layered perovskite, the doped layered perovskites showed a higher cathodic polarization due to the oxygen disorder introduced by dopant cations.

Cathodes for proton-conducting SOFC are MIEC perovskites based mainly on barium and iron and a very common general composition is $Ba_{1-x}Sr_xCo_{1-y}Fe_yO_{3-\delta}$. The ideal cathode for this kind of SOFC should conduct protons simultaneously with oxygen and electrons. Cathodes with Co dopant cations exhibit excellent performance; however, it was found that these cobalt-containing perovskites usually have problems with thermal expansion, undesired reduction and evaporation of Co. Hence, other cations are proposed, such as Ni or Nb, in order to eliminate the Co from these compositions. Another approach, as proposed by Zhang et al., is to modify the proton-conducting electrolytes, such as $BaCe_{0.8}Sm_{0.2}O_{3-\delta}$, introducing transitional elements, such as Fe. However, despite of good cationic and anionic conduction, these modified cathodes present low electrical conduction and poor catalytic activity, and need to be further improved. All compositions of cathodes for proton conducting SOFC are commonly applied with electrolytes based on cerium oxide or even with other perovskites composition, that were addressed in the Electrolytes session.

Besides the composition, the final microstructure of the cathode is of great importance. It has been demonstrated that modifications in the microstructure of the cathode can have great impact on the overall fuel cell performance. The optimum cathode porous microstructure should enhance the gas flow and have a high surface area. It also must own a minimal mechanical resistance in order to prevent cracking and collapsing of the component.

3. Interconnects for SOFC

Interconnects provide an electrical connection between cathode of one individual cell to the anode of the adjacent cell in a SOFC stack and ensure a physical barrier between the reducing atmosphere (at the anode) and the oxidizing atmospheres (at the cathode). Therefore, the material to be used as interconnect has to present a series of important characteristics, which considerably reduces the candidate material for this component. High electrical conductivity, with values higher than 1 S/cm or, in terms of area-specific resistance, values below 0.1 $\Omega.cm^2$ are required with no ionic conductivity. Chemical stability at the operation temperatures and in reducing and oxidant atmospheres, considering that atomic interdiffusion can be a recurrent problem and reaction with sealant materials can also destabilize the perovskite. Thermal expansion compatible with anode and cathode. Enough mechanical strength to bear the load of the stack and support other components. Finally, in the case of the utilization of the heat generated by the SOFC in cogeneration, a thermal conductivity of at least 5 W/mK is necessary.

The most common perovskite compositions for interconnects are based on lanthanum chromite ($LaCrO_3$). In its structure, larger-sized cations, such as Ca^{2+} and Sr^{2+}, can substitute for La^{3+} while smaller cations, such as Ni^{2+}, Cu^{2+} or Al^{3+}, can replace Cr^{3+}. Mahato et al. demonstrated that the divalent cations, such as Ca^{2+} or Sr^{2+}, increase the conductivity of $LaCrO_3$. The divalent cations will act as acceptor dopants when residing at the trivalent (La^{3+} or Cr^{3+}) sites. Thus, in order to maintain charge neutrality, holes are created as a charge compensating defect, which consequently leads to p-type conductivity. Similarly, Fergus et al. attributed the addition of trivalent cations, such as Al^{3+}, to an increase in the p-type conductivity, but in this case, occurring mainly due to an increase in carrier mobility. It may be noted that the solubility of the divalent cations decreases upon increasing pO_2.

As with other perovskites, the dopant type and amount must be carefully chosen, because they often modify thermal expansion and chemical stability at the same time as electrical conductivity. For example, pure $LaCrO_3$ has a thermal expansion coefficient of 9.5e-6 K^{-1} and an electrical conductivity of 1 S/cm at 1000°C. With the addition of 20% of cobalt in the B sites, these values increase to 14.6e-6 K^{-1} and 15 S/cm, respectively. Whereas the addition of 10% of magnesium in the A sites keeps the thermal expansion stable with an increase in the conductivity of up to 3 S/cm. Another issue related to doped $LaCrO_3$ is its poor sinterability in air, which Anderson et al. attributed to the high vapor pressure of volatile chromium components. Hence, the use of CaO, for example, creates a liquid phase during sintering and enhances the sinterability of the $(La,Ca)CrO_3$ compound.

In the case of cells operating below 700°C, there is the possibility of using metallic interconnects, which brings great advantages. With metallic interconnects, the other components can be deposited as thin films layers which can diminish costs with raw material. Besides, it can lead to better mechanical strength with more efficient accommodation of thermal tensions during heating and cooling. However, even alloys with high antioxidant capacity are proven to suffer in extreme conditions where interconnects are employed, so a ceramic coating with conventional high-temperature interconnect materials is necessary. Brylewski et al. coated a ferritic stainless steel with a $La_{0.8}Sr_{0.2}CrO_3$ ceramic film and compared the oxidation rate with an uncoated sample in wet atmosphere. In this study, it was shown that oxidation rates could be decreased with the use of conducting films; however, the authors also observed that chromium oxide could segregate from the perovskite to form undesired phases. Usually, the coating materials are perovskites, but recently, some spinel compositions have also been proposed.

4. Anodes for SOFC

The efficient use of hydrocarbons as fuels for SOFC is one of the most relevant issues concerning its current development, and their utilization depends mainly on the characteristics of the anode. Conventional SOFC anodes work only when pure H_2 is used as fuel; otherwise, the deposition of carbon in the catalyst surface poisons the cell and rapidly compromises its performance. Therefore, new materials are proposed in order to promote the direct oxidation of hydrocarbons, or even an internal reforming, considering the longevity of the SOFC lifespan. Many of these new materials are perovskites.

One potential perovskite for use in anodes is based on lanthanum chromite, a typical interconnect composition. Doping this compound on A and B sites enhances its activity towards methane oxidation. The presence of chromium at the B sites also improves redox stability and tolerance to sulphur; however, it costs the compound a decrease in the total conductivity. The use of strontium and manganese dopants helps to maintain the compound's stability and a typical composition is $La_{0.75}Sr_{0.25}Cr_{0.5}Mn_{0.5}O_{3-\delta}$ (LSCM). These dopant cations also enhance ionic/electronic conductivity and catalytic activity. The conductivity of LSCM can reach 38 S/cm at 900°C in air, but in the presence of H_2, this value can decrease to 1.5 S/cm at the same temperature.

The overall performance of LSCM anodes are compatible with the conventional Ni/YSZ cermets, showing better performance when methane is used as fuel. Also, if the fuel has sulphur impurities (H_2S), LSCM is more tolerant than the conventional cermet anodes. However, Zha et al. demonstrated the formation of sulphides within a few days when LSCM anodes operate with a 10%H_2S containing H_2 fuel. The authors also demonstrate that an increase in the Cr content on the perovskite lattice increases the number of impurity phases.

The use of cerium cations in the A sites of the LSCM lattice is also studied, with an expectation that increasing the catalytic activity would also increase the open circuit voltage (OCV) in cells fuelled with methane. The substitution with Ru in the B sites was studied due to its high catalytic activity in the steam reforming reaction with very good stability, but its application is limited by its very high cost.

Another proposed perovskite anode is based on lanthanum-doped strontium titanate, which shows good electrical conductivity in reducing atmospheres and reliable dimensional and chemical stability. Besides, it has good capacity at high temperatures to operate with CH_4 fuel without the detecting of carbon deposits formation and with high OCV, this can be a key factor in the SOFC fuel flexibility. According to Ruiz-Morales et al., anodes with the substitution of

Ti by Mn and Ga cations can put strontium titanates on par with conventional Ni-YSZ cermets when the cells are operating with H_2 fuel.

In order to enhance its electrocatalytic performance, a B site doping must be carried out and the choice of dopant must influence the redox properties and the conductivity. A potential B site dopant is chromium, for example, $La_{0.3}Sr_{0.7}Ti_{0.8}Cr_{0.2}O_{3-\delta}$ (LSTC). This composition has excellent stability and electrical conductivity; however, its catalytic activity is still very low, in order to be used as an SOFC anode, it is necessary to incorporate some catalytic materials.

5. Electrolytes for SOFC

In recent years, most of the efforts in SOFC development have been focussed on intermediate temperature solid oxide fuel cells (IT-SOFCs). The majority of them use a fluorite-structured oxide, CGO and/or YSZ in the electrolyte. In 1971, Takahashi et al. discovered that perovskite-structured materials possessed oxide ion conductivity; in 1992, Goodenough et al. reported that $LaGaO_3$ exhibited important ionic conductivity that was improved with different dopants being modified with strontium and magnesium in La and Ga sites (LSGM), respectively, showed higher oxide ion conductivity. However, these kinds of perovskites can only work under certain operating conditions, as they are not stable at low oxygen pressures. In CO and CO_2 atmosphere, they form carbonates; in reducing atmospheres, there is a Ga depletion along the grain boundaries degrading the material; and at high temperatures, they show an important solubility of Al, Ni, Co oxides. Therefore, this material (LSGM) is considered a good candidate as electrolyte for intermediate temperature SOFCs (IT-SOFCs). Since the 1970s, $LnAlO_3$ perovskites have been studied mainly taking into account their lower cost and high reduction and volatilization stability with respect to ceria and lanthanum galates; however, they exhibit problems of high electronic conductivity at high oxygen partial pressures and poor sinterability why they are constrained as an additive to composite solid electrolytes.

Most recently, there have been studies about ceramic materials that present protonic conductivity, and they are being employed as electrolytes in SOFC. In this respect, the first studies were presented by Iwahara et al. about $SrCeO_3$ materials. Some years later, the studies were directed mainly towards $BaCeO_3$ and $BaZrO_3$ perovskites due to of their higher proton conductivity, which was improved by Y doping (BCY and BZY, respectively). It is known that these barium cerate electrolytes exhibit both oxide ion and protonic conduction depending on the working temperature, changing from protonic to oxide ion transport when temperature are varied from 600°C to 1,000°C, whereas this behaviour has not been observed in strontium cerates. However, against this high conductivity is the very low chemical stability of $BaCeO_3$ materials, while Y-doped BaZrO3 with a little lower proton conductivity exhibit a great chemical stability, with the problem of resistive grain boundaries. Nevertheless, Y-doped $BaZrO_3$ could be more appropriate for its application in SOFCs.

For SOFC development, relatively high ionic conductivity of solid electrolytes is not the only requirement; an enhancement of durability is also needed. To improve both aspects, many

http://dx.doi.org/10.5772/61465

works have been carried out, some propose to decrease the thickness of the electrolyte or using the named composite electrolytes (e.g., BCY and a molten salt phase).

6. Perovskites used in PEMFC

From a catalytic point of view, the use of perovskites has been principally reserved for high-temperature fuel cells due to the improved kinetics of reactions in electrodes with temperature. Nevertheless, the possible use of ABO_3 structures as catalysts in the replacement of noble metals is an idea that has been proposed in the early 1990s, based on the mixed oxide-ion/electronic conductivity and the low cost of these materials.

These structures are a potential alternative to Pt or Au in different air cathode-electrochemical cells, where the main drawbacks to the commercial viability of those devices are the low activity of oxygen-reduction and oxygen-evolution reactions (ORR and OER), respectively. In the proton exchange fuel cell field, most of the published materials with ABO_3 structures are directed towards the ORR. In alkaline solutions, they should exhibit comparable activities to those platinum-based metal catalysts, and this activity is supposed to be directly related with the electronic configuration of their surface cations. Among the different materials studied, lanthanum is the most common A cation because, combined with other transition metals, it has demonstrated ORR activity. B site substitutions with certain transition-metal elements should impact ORR rates (e.g., with Mn, Co and Ni) and should enhance the chemical and electrochemical stability of the perovskites (e.g., Cr and Fe) in alkaline solutions. Among lanthanum-based perovskites, $LaCoO_3$ has revealed itself to be interesting due to its large ORR current density and the positive shift shown in the onset potential. In addition, $LaNiO_3$-based oxides with B cation substitution, such as $LaNi_{1-x}Fe_xO_3$ (x = 0–0.2) are promising materials in alkaline media, with an improved catalytic activity related with the increase in the valence state of Ni with the B site substitution. On this basis, perovskites have been proposed to work in temperature ranging from 60°C to 200°C, such as cathodes in alkaline fuel cells. In this media, a 2-electron pathway mechanism has been described where the HO^{2-} created is further reduced. Good performance has been reported with different compositions in the $La_{1-x}A'_xBO_3$ system with A'= Ca, Sr and B = Co, Ni and Mn.

ABO_3 oxides are also an alternative to platinum in high temperature protonic exchange fuel cells (HT-PEMFC); in these devices, new membranes based on polybenzimidazole (PBI) impregnated by H_3PO_4 allows for an increase in the operation temperature to 130°C–200°C. The increase in the operation temperature gives a chance to Pt-free catalysts, where perovskites show improved chemical resistance under these temperatures. With this aim, $LaMnO_3$ and $LaSrMnO_3$ have been prepared by combustion synthesis with high electrical conductivity at 200°C and improved resistance towards H_3PO_4 corrosion.

Although perovskites have been proposed for noble metal replacement mainly on the cathode side of fuel cells, there are strong proposals that consider its use for the oxidation of alcohols in the anode of direct methanol fuel cells (DMFCs) working at 60°C–80°C. White y Sammells proposed, in a pioneer work, perovskite electrocatalysts in, among others, $SrMO_3$ (M = Ru,Pd),

$SmCoO_3$ and $SrRuMO_3$ (M = Pt, Pd) systems. They demonstrated activity towards direct methanol oxidation during cyclic voltammetry measurements that gave methanol oxidation currents up to 28 mA/cm^2 at 0.45 V vs. SCE. Following these results, $SrRuO_3$ was prepared by the combustion method as a high specific surface area catalyst with performance comparable to PtRu towards MeOH oxidation at potentials ranging from 0.25 to 0.35 V; nevertheless, the authors proposed the addition of Pt to enhance its catalytic activity. Similarly, strontium-substituted lanthanum cobaltite and copper-based perovskite nanoparticles were synthesized by the sol–gel method. Although the methanol oxidation onset potential was 0.03 V lower for LSCo than that for LSCu, an improved electrocatalytic activity for LSCu was found. This was attributed to the higher methanol adsorption capability of Cu ions and to the oxygen ion (O^{2-}) transport into the proximity of adsorbed methanol oxidation that facilitates the formation of intermediates at the reaction site.

Finally, perovskites are being reconsidered as anode catalysts in alkaline DMFC. In contrast to acid electrolytes, the use of an alkaline media improves the kinetics of the reaction while much less expensive catalysts, particularly oxides, could potentially be used. With this idea, $La_{2-x}SrxNiO_4$ ($0 \leq x \leq 1$) was prepared using the citric acid sol–gel route producing higher current densities and negligible poisoning by the methanol oxidation products than those already reported on other perovskite oxides.

7. Synthesis and processing of perovskites for fuel cells

There are several different methods to synthetize perovskites with suitable properties and morphology for use in fuel cells and the method chosen to obtain these materials can influence the final performance of the cell. The mixture of oxides through reaction-sintering was used in the beginnings of the obtaining of oxides with multiple cations, such as perovskites for application in fuel cells and. In some cases, it is investigated until nowadays. However, some disadvantages like compositional heterogeneity, grain growth and, in the case of electrodes, low specific surface area, led to the development and utilization of new methods.

Many of these novel methods were based on a chemical route, where a solution containing the desired cations is used in order to obtain the final multiple oxide. The use of the coprecipitation method can enhance the oxidation catalysis of cathode materials; however, it can be more difficult to carry out because of external influences during the precipitation reaction. Sol–gel is a very common method to synthesize perovskites and it has a relative ease of control and requires lower temperatures of crystallization, allowing the obtaining of a single-phase homogeneous microstructure.

Drip pyrolysis can produce electrode materials with a high concentration of surface reaction sites, but with a poor surface area. One of the most applied methods is the solution combustion synthesis, due to its simplicity and quickness to produce fine powders and the possibility to vary some parameters in order to modify the final morphology.

The processing of materials is one of the great challenges in the fabrication of SOFC devices and many different processing methods were applied and reported in order to fabricate SOFC components. The use of expensive methods such as RF sputtering were carried out at the end of the last century; however, more recently, researchers and industries in this field have a great tendency of choosing simpler and less expensive techniques. Methods like screen printing, tape casting and dip coating, which are based on the preparation of a suspension containing a previously synthesized powder, are common methods to fabricate electrodes and electrolytes for SOFC and allow the control of the microstructure, which is a very important factor since electrodes must be porous and electrolytes, dense.

Figure 2 shows two scanning electron micrographs of porous thin LSM films fabricated on YSZ substrates by dip coating. The LSM powders were obtained by combustion synthesis and different solvents (a and b) were used to prepare the dispersions. Even with sintering at the same temperature, it is possible to observe that slightly different microstructures were obtained, but a strong influence on the thickness of films can be seen.

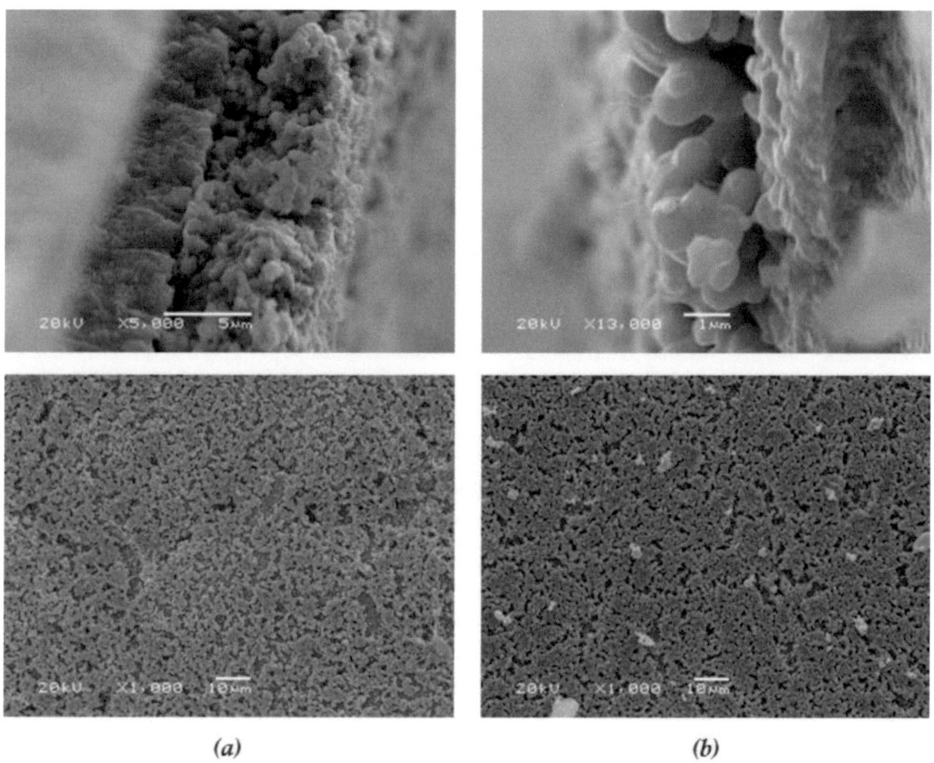

Figure 2. LSM films obtained on YSZ substrates by dip coating with different solutions.

Another approach to fabricate SOFC components is based on wet chemical techniques, where a solution is prepared and the perovskite crystallization reaction takes place on another SOFC component. The adaptation of sol–gel with dip coating or spray pyrolysis, for example, makes it possible to obtain the perovskite *in situ*. That is, already in a form of a final SOFC component instead of obtaining the powder and then coating another surface with it. These are very promising techniques because they can also reduce the processing temperature in some cases.

8. Final considerations

Ceramic perovskite-type oxides have great potential for utilization in efficient energy conversion devices such as fuel cells, especially in SOFC and PEMFC. The research and development on PEMFC is more recent and, until now, has been restricted to electrodes whereas in SOFC it has been shown that the perovskites, together with processing, will definitely play a key role towards its commercialization. The vast variety of compositions obtained with doping elements in nonstoichiometric amounts allows the modification of properties in a wide range. Very often, the use of such elements interferes in more than one property and it is common to have deleterious effects in the overall component performance.

In SOFCs, many perovskite compositions have already been tested, particularly for cathodes, and the majority are based on one or more of these cations: La, Mn, Fe, Sr Cr, Co, Ni, among others. The catalytic activity and the mixed ionic and electronic conduction are the most sought characteristics for use in the electrodes while for electrolytes only the ionic conduction is desirable, whereas for interconnects, electronic conduction is the one. Of course, besides electrochemical properties, the compositions must have chemical and physical compatibility with each other when used in a fuel cell. This means that they should not react or decompose to form undesirable phases and the thermal expansion coefficient of all components must be similar in order to avoid the formation or propagation of cracks during operation and/or thermal cycling.

Author details

Diego Pereira Tarragó[1*], Berta Moreno[2], Eva Chinarro[2] and Vânia Caldas de Sousa[1]

*Address all correspondence to: diego.tarrago@ufrgs.br

1 Universidade Federal do Rio Grande do Sul, Porto Alegre, Brazil

2 Instituto de Cerámica y Vidrio CSIC, Madrid, Spain

References

[1] EG&G Technical Services: Science and Applications International Corporation. Fuel Cell Handbook. 6th ed. United States of America: U.S. Department of Energy; 2002.

[2] Andújar J.M., Segura F.. Fuel cells: History and updating. A walk along two centuries. Renewable and Sutainable Energy Reviews. 2009;13:2309–2322. DOI: 10.1016/j.rser.2009.03.015

[3] Kirubakaram A., Jain S., Nema R.K.. A review on fuel cell technologies and power electronic interface. Renewable and Sustainable Energy Reviews. 2009;13:2430–2440. DOI: 10.1016/j.rser.2009.04.004

[4] Florio D.Z., Fonseca F.C, Mucillo E.N.S., Mucillo R.. Materiais cerâmicos para células a combustível. Cerâmica. 2004;50:275–290.

[5] Yokokawa H., Sakai N., Kawada T., Dokyia M.. Themordynamic stabilities of perovskite oxides for electrodes and other electrochemical materials. Solid State Ionics. 1992;52:43.

[6] Cortés-Gil R., Alonso J.M., Ruiz-González M.L., Gonzáles-Calbet J.M.. Topotatic migration of cationic vacancies in $La_{1-t}Mn_{1-t}O_3$. European Journal of Inorganic Chemistry. 2010;:3436–3440. DOI: 10.1002/ejic.201000086

[7] Richter J,. Holtapells P., Graule T., Nakamura T., Gauckler L.J.. Materials design for perovskite SOFC cathodes. Monatshefte für Chemie. 2009;140:985–999. DOI: 10.1007/s00706-009-0153-3

[8] Tarancón A., Burriel M., Santiso J., Skinner S.J., Kilner J.A.. Advances in layered perovskites for intermediate temperature solid oxide fuel cells. Journal of Materials Chemistry. 2010;20:3799–3813. DOI: 10.1039/b922430k

[9] Shuk P., Tichonova L., Guth U.. Materials for electrodes based on rare earth manganites. Solid State Ionics. 1994;68:177–184. DOI: 10.1016/0167-2738(94)90175-9

[10] Tai L.-W, Nasrallah M.M., Anderson H.U., Sparlin D.M., Sehlin S.R.. Structure and electrical properties of $La_{1-x}Sr_xCo_{1-y}Fe_yO_3$. Part 1. The system $La_{0.8}Sr_{0.2}Co_{1-y}Fe_yO_3$. Solid State Ionics. 1995;76(3–4):259–271. DOI: 10.1016/0167-2738(94)00244-M

[11] Mogensen M., Skaarup S.. Kinetic and geometric aspects of solid oxide fuel cell electrodes. Solid State Ionics. 1996;86–88:1151–1160. DOI: 10.1016/0167-2738(96)00280-9

[12] Beckel D., Muecke U.P., Gyger T., Florey G., Infortuna A., Gauckler L.J.. Electrochemical performance of LSCF based thin film catthodes perpared by spray pyrolysis. Solid State Ionics. 2007;178:407–415. DOI: 10.1016/j.ssi.2007.01.019

[13] Tietz F., Mai A., Stöver D.. From powder properties to fuel cell performance: a holistic approach for SOFC cathode development. Solid State Ionics. 2008;179:1509–1515. DOI: 10.1016/j.ssi.2007.11.037

[14] Torres-Garibay C., Kovar D., Manthiram A.. $Ln_{0.6}Sr_{0.4}Co_{1-y}Fe_yO_{3-\delta}$ (Ln = La and Nd; y = 0 and 0.5) cathodes with thin yttria-stabilized zirconia eletrolytes for intermediate temperature solid oxide fuel cells. Journal of power Sources. 2009;187:480–486. DOI: 10.1016/j.jpowsour.2008.11.025

[15] Siebert E., Roux C., Boréave A., Gaillard F., Vernoux P.. Oxido-reduction properties of $La_{0.7}Sr_{0.3}Co_{0.8}Fe_{0.2}O_{3-\delta}$ perovskite oxide catalyst. Solid State Ionics. 2011;183:40–47. DOI: 10.1016/j.ssi.2010.11.012

[16] Ko H.J., Myung J., Lee J., Hyun S., Chung J.S.. Synthesis and evaluation of $(La_{0.6}Sr_{0.4})(Co_{0.2}Fe_{0.8})O_3$ (LSCF) - $Y_{0.08}Zr_{0.92}O_{1.96}$ (YSZ) - $Gd_{0.1}Ce_{0.9}O_{2-\delta}$ (GDC) dual composite cathodes for high performance and durability. International Journal of Hydrogen Energy. 2012;37:17209–17216. DOI: 10.1016/j.ijhydene.2012.08.099

[17] DiGiuseppe G., Sun L.. Electrochemical performance of a solid oxide fuel cell with an LSCF cathode under different oxygen concentrations. International Journal of Hydrogen Energy. 2011;36:5076–5087. DOI: 10.1016/j.ijhydene.2011.01.017

[18] Tarancón A., Peña-Martínez J., Marrero-López D., Morata A., Ruiz-Morales J.C., Nuñez P.. Stability, chemical compatibility and electrochemical performance of $GdBaCo_2O_{5+x}$ layered perovskite as a cathode for intermediate temperature solid oxide fuel cells. Solid State Ionics. 2008;179:2372–2378. DOI: 10.1016/j.ssi.2008.09.016

[19] Kim J.H., Irvine J.T.S.. Characterization of $NdBa_{1-x}Sr_xCo_2O_{5+\delta}$ (x = 0 and 0.5) as cathode materials for IT-SOFC. International Journal of Hydrogen Energy. 2012;37:5920–5929. DOI: 10.1016/j.ijhydene.2011.12.150

[20] Huang B., Wang S.R., Liu R.Z., Ye X.F., Nie H.W., Sun X.F., Wen T.L.. Performance of $La_{0.75}Sr_{0.25}Cr_{0.5}Mn_{0.5}O_{3-\delta}$ perovskite-structure anode material at lanthanum gallate electrolyte for IT-SOFC running on ethanol fuel. Journal of Power Sources. 2007;167:39–46. DOI: 10.1016/j.jpowsour.2007.02.022

[21] Wenyi T., Qin Z., Han Y., Xiufang Z., Hongyi L.. Deactivation of anode catalyst $La_{0.75}Sr_{0.25}Cr_{0.5}Mn_{0.5}O_{3\pm\delta}$ in SOFC with fuel containing hydrogen sulfur. The role of lattice oxygen. International Journal of Hydrogen Energy. 2012;37:7398–7404. DOI: 10.1016/j.ijhydene.2012.02.008

[22] Zha S., Tsang P., Cheng Z., Liu M. Electrical properties and sulphur tolerance of $La_{0.75}Sr_{0.25}Cr_{1-x}Mn_xO_3$ under anodic conditions. Journal of Solid State Chemistry. 2005;178:1844–1850. DOI: 10.1016/j.jssc.2005.03.027

[23] Sauvet A.-L., Fouletier J., Gaillard F., Primet M.. Surface properties and physicochemical characterizations of a new type of anode material, $La_{1-x}Sr_xCr_{1-y}Ru_yO_{3-\delta}$, for a solid oxide fuel cell under methane at intermediate temperature. Journal of Catalysis. 2002;209:25–34. DOI: 10.1006/jcat.2002.3588

[24] Ruiz-Morales J.C., Canales-Vázquez J., Savaniu C., Marrero-López D., Zhou W., Irvine J.T.S.. Disruption of extended defects in solid oxide fuel cell anodes for methane. Nature. 2006;439:568–571. DOI: 10.1038/nature04438

[25] Du Z., Zhao H., Zhou X., Xie Z., Zhang C.. Electrical conductivity and cell performance of $La_{0.7}Sr_{0.7}Ti_{1-x}Cr_xO_3$- perosvkite oxides used as anode and interconnect material

for SOFCs. International Journal of Hydrogen Energy. 2013;38:1068–1073. DOI: 10.1016/j.ijhydene.2012.10.099

[26] Onuma S., Miyoshi S., Yashiro K., Kaimai A., Kawamura K., Nigara Y., et al.. Phase stability of $La_{1-x}Ca_xCrO_{3-\delta}$ in oxidizing atmosphere. Journal of Solid State Chemistry. 2003;170:68–74.

[27] Mahato N., Banerjee A., Gupta A., Omar S., Balani K.. Progress in material selection for solid oxide fuel cell technology: A review. Progress in Materials Science. 2015;72:141–337. DOI: http://dx.doi.org/10.1016/j.pmatsci.2015.01.001

[28] Fergus J.W., Hui R., Li X., Wilkinson D.P., Zhang J.. Solid oxide fuel cells: material properties and performance. Boca Raton, USA: CRC Press; 2008. 298 p. DOI: ISBN: 9781420088830

[29] Lacey R., Pramanick A., Lee J.C., Jung J., Jiang B., Edwards D.D, et al.. Evaluation of Co and perovskite Cr-blocking thin films on SOFC interconnects. Solid State Ionics. 2010;181:1294–1302. DOI: 10.1016/j.ssi.2010.07.007

[30] Sfeir J.. $LaCrO_3$-based anodes: stability considerations. Journal of Power Sources. 2003;118:276–285. DOI: 10.1016/S0378-7753(03)00099-5

[31] Anderson H.U.. Fabrication and Property Control of LaCrO3 Based Oxides. In: Hayne Palmour III, Davis R.F., Hare T.M., editors. Processing of Crystalline Ceramics. Springer US; 1978. p. 469–477. DOI: 10.1007/978-1-4684-3378-4_39

[32] Brylewski T., Dabek J., Przybylski K., Morgiel J., Rekas M.. Screen-printed (La,Sr)CrO3 coatings on ferritic stainless steel interconnects for solid oxide fuel cells using nanopowders prepared by means of ultrasonic spray pyrolysis. Journal of Power Sources. 2012;208:86–95. DOI: 10.1016/j.jpowsour.2012.02.015

[33] Moreno B., Fernández-González R., Jurado J.R., Makradi A., Nuñez P., Chinarro E.. Fabrication and characterization of ceria-based buffer layers for solid oxide fuel cells. International Journal of Hydrogen Energy. 2014;39(10):5433–5439. DOI: 10.1016/j.ijhydene.2014.01.061

[34] Takahashi T., Iwahara H.. Ionic conduction in perovskite-type oxide solid solution and its application to the solid electrolyte fuel cell. Energy Conversion. 1971;11(3): 105–111. DOI: 10.1016/0013-7480(71)90100-8

[35] Goodenough J.B., Manthiram A., Paranthaman P.. Fast oxide-ion conduction in intergrowth structures. Solid State Ionics. 1992;52(1–3):105–109. DOI: 10.1016/0167-2738(92)90096-8

[36] Ishihara T., Matsuda H., Takita Y.. Doped $LaGaO_3$ perovskite type oxide as a new oxide ionic conductor. Journal of American Ceramic Society. 1994;116(9):3801–3803. DOI: 10.1021/ja00088a016

[37] Feng M., Goodenough J.B.. A superior oxide-ion electrolyte. European Journal of Solid State and Inorganic Chemistry. 1994;31(8–9):663–672.

[38] Zhao Y., Xia C., Jia L., Wang Z., Li H., Yu J., Li Y.. Recent progress on solid oxide fuel cell: lowering temperature and utilizing non-hydrogen fuels. International Journal of Hydrogen Energy. 2013;38(36):16498–16517. DOI: 10.1016/j.ijhydene.2013.07.077

[39] Yokokawa H., Sakai N., Horita T., Yamaji K.. Recent developments in solid oxide fuel cell materials. Fuel Cells. 2001;1(2):117–131. DOI: 10.1002/1615-6854(200107)1:2<117::AID-FUCE117>3.0.CO;2-Y

[40] Thangadurai V., Weppner W.. Recent progress in solid oxide an lithium ion conducting electroytes research. Ionics. 2006;12(1):81–92. DOI: 10.1007/s11581-006-0013-7

[41] Karton V.V., Marques F.M.B., Atkinson A.. Transport properties of solid oxide electrolyte ceramics: a brief review. Solid State Ionics. 2004;174(1–4):135–149. DOI: 10.1016/j.ssi.2004.06.015

[42] Gauckler L.J., Beckel D., Buergler B.E., Jud E., Muecke U.P., Prestat M., et al.. Solid oxide fuel cells: systems and materials Chimia. 2004;58(12):837–850. DOI: 10.2533/000942904777677047

[43] Iwahara H., Uchida H., Tanaka S.. Studies on solid electrolyte gas cells with high-temperature-type proton conductor and oxide ion conductor. Solid State Ionics. 1983;11(2):109–115. DOI: 10.1016/0167-2738(83)90047-4

[44] Bi L., Traversa E.. Synthesis strategies for imporving the performance of doped-BaZrO3 materials in solid oxide fuel cell applications. Journal of Materials Research. 2014;29(1):1–15. DOI: 10.1557/jmr.2013.205

[45] Bonanos N., Knight K.S., Ellis B.. Perovskite solid electrolyte: structure, transport properties and fuel cell applications. Solid State Ionics. 1995;79:161–170. DOI: 10.1016/0167-2738(95)00056-C

[46] Fabbri E., Pergolesi D., Traversa E.. Materials challenges toward proton-conducting oxide fuel cells: a critical review. Chemical Society Reviews. 2010;39: 4355–4369. DOI: 10.1039/b902343g

[47] Santiso J., Burriel M.. Deposition and characterization of epitaxial oxide thin films for SOFCs. Journal of Solid State Electrochemistry. 2011;15(5):985–1006. DOI: 10.1007/s10008-010-1214-6

[48] Jiang S.P.. Development of lanthanum strontium manganite perovskite cathode materials of solid oxide fuel cells: a review. Journal of Materials Science. 2008;43(21): 6799–6833. DOI: 10.1007/s10853-008-2966-6

[49] Christensen J., Albertus P., Sanchez-Carrera R.S., Lohmann T., Kozinsky B., Liedtke R., et al.. A critical review of Li/air batteries. Journal of the Electrochemical Society. 2012;159(2):1–30. DOI: 10.1149/2.086202jes

[50] Suntivich J., May K.J., Gasteiger H.A, Goodenough J.B., Shao-Horn Y.. A perovskite oxide optimized for oxygen evolution catalysis from molecular orbital principles. Science. 2011;334:1383–1385. DOI: 10.1126/science.1212858

[51] Sunarso J., Torriero A.A., Zhou W.J., Howlett P.C., Forsyth M.. Oxygen reduction reaction activity of La-based perovskite oxides in alkaline medium: a thin-film rotating ring-disk electrode study. Journal of Physical Chemistry C. 2012;116(9):5827–5834. DOI: 10.1021/jp211946n

[52] Hyodo T., Hayashi M., Miura N., Yamazoe N.. Catalytic activities of rare-earth manganites for cathodic reduction of oxygen in alkaline solution. Journal of the Electrochemical Society. 1996;143:266–267. DOI: 10.1149/1.1837229

[53] Niu Y., Sunarso J., Liang F., Zhou W., Zhu Z., Shao Z.. A comparative study of oxygen reduction reaction on Bi- and La-doped $SrFeO_{3-\delta}$ perovskite cathodes. Journal of the Electrochemical Society. 2011;158:132–138. DOI: 10.1149/1.3521316

[54] Suntivich J., Gasteiger H.A., Yabuuchi N., Shao-Horn Y.. Electrocatalytic measurement methodology of oxide catalysts using a thin-film rotating disk electrode. Journal of the Electrochemical Society. 2010;157:1263–1268. DOI: 10.1149/1.3456630

[55] Jörissen L.. Bifunctional oxygen/air electrodes. Journal of Power Sources. 2006;155:23–32. DOI: 10.1016/j.jpowsour.2005.07.038

[56] Zhang D., Song Y., Du Z., Wang L., Li Y., Goodenough J.B.. Active $LaNi_{1-x}Fe_xO_3$ bifunctional catalysts for air cathodes in alkaline media. Journal of Materials Chemistry A. 2015;3:9421–9426. DOI: 10.1039/C5TA01005E

[57] Hu W.K., Kiros Y., Noreus D.. AB5-type hydrogen storage alloys as catalysts in hydrogen-diffusion electrodes for novel H_2/hydride//perovskite/O_2 alkaline fuel cells. Journal of Physical Chemistry B. 2004;108(48):18530–18534. DOI: 10.1021/jp047620x

[58] Hu W.K., Gao X.P., Kiros Y., Middelman E., Noreus D.. Zr-based AB2-type hydrogen storage alloys as dual catalysts of gas-diffusion electrodes in an alkaline fuel cell. Journal of Physical Chemistry B. 2004;108(26):8756–8758. DOI: 10.1021/jp0486548

[59] Bidault F., Brett D.J.L., Middleton P.H., Brandon N.P.. Review of gas diffusion cathodes for alkaline fuel cells. Journal of Power Sources. 2009;187:39–48. DOI: 10.1016/j.jpowsour.2008.10.106

[60] Hermann V., Dutriat D., Muller S., Comninellis C.. Mechanistic studies of oxygen reduction at $La_{0.6}Ca_{0.4}CoO_3$-activated carbon electrodes in a channel flow cell. Electrochimica Acta. 2000;46:365–372. DOI: 10.1016/S0013-4686(00)00593-4

[61] Manoharan R., Shukla A.K.. Oxides supported carbon air-electrodes for alkaline solution power devices. Electrochimica Acta. 1985;30:205–209. DOI: 10.1016/0013-4686(85)80083-9

[62] Weidenkaff A., Ebbinghaus S.G., Lippert T., Montenegro M.J., Soltmann C., Wessicken R.. Phase formation and phase transition of $La_{1-x}A_xCoO_3$ (A = Ca, Sr) applied for

bifunctional air electrodes. Crystal Engineering. 2002;5(3):449–447. DOI: 10.1016/S1463-0184(02)00056-4

[63] Villaseca L., Moreno B., Chinarro E.. Perovskites based on La(Sr)-Mn-O system as electrocatalyst in PEM fuel cell of high temperature. International Journal of Hydrogen Energy. 2012;37(8):7161–7170. DOI: 10.1016/j.ijhydene.2011.12.142

[64] White J.H., Sammells A.F.. Perovskite anode electrocatalysis for direct methanol fuel cells. Journal of the Electrochemical Society. 1993;140(8):2167–2177. DOI: 10.1149/1.2220791

[65] Deshpande K., Mukasyan A., Varma A.. High throughout evaluation of perovskite-based anode catalysts for direct methanol fuel cells. Journal of Power Sources. 2006;158:60–68. DOI: 10.1016/j.jpowsour.2005.09.025

[66] Yu H.C., Fung K.Z., Chang W.L.. Syntheses of perovskite oxides nanoparticles $La_{1-x}Sr_xMO_{3-\delta}$ (M = Co and Cu) as anode electrocatalyst for direct methanol fuel cell. Electrochimica Acta. 2004;50(2–3):811–816. DOI: 10.1073/pnas.1210315109

[67] Singh R.N., Sharma T., Singh A., Anindita, Mishra D., Tiwari S.K.. Perovskite-type $La_{2-x}Sr_xNiO_4$ ($0\leq x \leq 1$) as active anode materials for methanol oxidation in alkaline solutions. Electrochimica Acta. 2008;53:2322–2330. DOI: 10.1016/j.electacta.2007.09.047

[68] Raghuveer V., Viswanathan B.. Can $La_{2-x}Sr_xCuO_4$ be used as anodes for direct methanol fuel cells? Fuel. 2002;81:219–2197. DOI:10.1016/S0016-2361(02)00167-9

[69] Bell R.J., Millar G.J., Drennan J.. Influence of synthesis route on the catalytic properties of $La_{1-x}Sr_xMnO_3$. Solid State Ionics. 2000;131:211–220.

[70] Gaoke Z., Ying L., Xia Y., Yanping W., Shixi O., Hangxing L.. Comparison of synthesis methods, crystal structure and characterization of strontium cobaltite powders. Materials Chemistry and Physics. 2006;99:88–95. DOI: 10.1016/j.matchemphys.2005.09.078

[71] Ghosh A., Sahu A.K., Gulnar A.K., Suri A.K.. Synthesis and characterization of lanthanum strontium manganite. Scripta Materialia. 2005;52:1305–1309. DOI: 10.1016/j.scriptamat.2005.02.020

[72] Tarragó D.P., Malfatti C.F., Sousa V.C.. Influence of fuel on morphology of LSM powders obtained by solution combustion synthesis. Powder Technology. 2015;269:481–487. DOI: 10.1016/j.powtec.2014.09.037

[73] Bebelis S., Kotsionopoulus N., Mai A., Tietz F.. Electrochemical characterization of perovskite-based SOFC cathodes. Journal of Applied Electrochemistry. 2007;37:15–20. DOI: 10.1007/s10800-006-9215-y

[74] Gharbage B., Mandier F., Lauret H., Roux C., Pagnier T.. Electrical properties of $La_{0.5}Sr_{0.5}MnO_3$ thin films. Solid State Ionics. 1995;82:85–94.

[75] Arendt E., Maione A., Klisinska A., Sanz O., Montes M., Suarez S., et al.. Structuration of $LaMnO_3$ perovskite catalysts on ceramic and metallic monoliths: Physico-chemical characterisation and catalytic activity in methane combustion. Applied Catalysis A: General. 2008;339:1–14. DOI: 10.1016/j.apcata.2008.01.016

[76] Basu R.N., Das Sharma A., Dutta A., Mukhopadhyay J.. Processing of high-performance anode-supported planar solid oxide fuel cell. International Journal of Hydrogen Energy. 2008;33:5748–5754. DOI: 10.1016/j.ijhydene.2008.06.073

[77] Falcade T., Tarragó D.P., Halmenschlager C.M., Sousa V.C., Oliveira C.T., Malfatti C.F.. Elaboration of yttria-stabilized zirconia films on porous substrates. Materials Science Forum. 2010;660–661:707–711. DOI: 10.4028/www.scientific.net/MSF.660-661.707